NANONETWORK MATERIALS

Related Titles from AIP Conference Proceedings

591   Electronic Properties of Molecular Nanostructures: XV International Winterschool/Euroconference
Edited by Hans Kuzmany, Jörg Fink, Michael Mehring, and Siegmar Roth, November 2001, 0-7354-0033-4

582   Fundamental Physics of Ferroelectrics 2001: 11th Williamsburg Ferroelectrics Wkshp.
Edited by Henry Krakauer, July 2001, 0-7354-0021-0

577   Density Functional Theory and Its Application to Materials
Edited by V. Van Doren, C. Van Alsenoy, and P. Geerlings, July 2001, 0-7354-0016-4

554   Physics in Local Lattice Distortions: Fundamentals and Novel Concepts; LLD2K
Edited by Hiroyuki Oyanagi and Antonio Bianconi, February 2001, 1-56396-984-X

551   Atomic Physics 17: XVII International Conference on Atomic Physics; ICAP 2000
Edited by Ennio Arimondo, Paolo De Natale, and Massimo Inguscio, February 2001, 1-56396-982-3

550   Characterization and Metrology for ULSI Technology: 2000 International Conf.
Edited by David G. Seiler, Alain C. Diebold, Thomas J. Shaffner, Robert McDonald, W. Murray Bullis, Patrick J. Smith, and Erik M. Secula, February 2001, CD-ROM included, 1-56396-967-X

544   Electronic Properties of Novel Materials—Molecular Nanostructures: XIV International Winterschool, Euroconference
Edited by Hans Kuzmany, Jörg Fink, Michael Mehring, and Siegmar Roth, November 2000, 1-56396-973-4

535   Fundamental Physics of Ferroelectrics 2000: Aspen Center for Physics Winter Wkshp.
Edited by Ronald E. Cohen, September 2000, 1-56396-959-9

486   Electronic Properties of Novel Materials—Science and Technology of Molecular Nanostructures: XIII International Winterschool
Edited by Hans Kuzmany, Jörg Fink, Michael Mehring, and Siegmar Roth, September 1999, 1-56396-900-9

To learn more about these titles, or the AIP Conference Proceedings Series, please visit the webpage **http://proceedings.aip.org**

NANONETWORK MATERIALS

Fullerenes, Nanotubes, and Related Systems

Kamakura, Japan   15–18 January 2001

EDITORS
Susumu Saito
Tokyo Institute of Technology

Tsuneya Ando
University of Tokyo

Yoshihiro Iwasa
Tohoku University

Koichi Kikuchi
Tokyo Metropolitan University

Mototada Kobayashi
Himeji Institute of Technology

Yahachi Saito
Mie University

Melville, New York, 2001
AIP CONFERENCE PROCEEDINGS ■ VOLUME 590

Editors:

Susumu Saito
Department of Physics
Tokyo Institute of Technology
2-12-1 Oh-okayama
Meguro-ku, Tokyo 152-8551
JAPAN

E-mail: saito@stat.phys.titech.ac.jp

Tsuneya Ando
Institute for Solid State Physics
University of Tokyo
5-1-5 Kashiwanoha
Kashiwa, Chiba 277-8581
JAPAN

E-mail: ando@issp.u-tokyo.ac.jp

Yoshihiro Iwasa
Institute for Materials Research
Tohoku University
2-1-1 Katahira
Aoba-ku, Sendai 980-8577
JAPAN

E-mail: iwasa@imr.tohoku.ac.jp

Koichi Kikuchi
Department of Chemistry
Tokyo Metropolitan University
1-1 Minami-Ohsawa
Hachioji, Tokyo 192-0397
JAPAN

E-mail: kikuchi-koichi@c.metro-u.ac.jp

Mototada Kobayashi
Department of Materials Science
Himeji Institute of Technology
Kamigohri 678-1297
JAPAN

E-mail: kobayasi@sci.himeji-tech.ac.jp

Yahachi Saito
Dept of Electrical and Electronic Engineering
Mie University
1515 Kamihama-cho
Tsu 514-8507
JAPAN

E-mail: saito@elec.mie-u.ac.jp

Authorization to photocopy items for internal or personal use, beyond the free copying permitted under the 1978 U.S. Copyright Law (see statement below), is granted by the American Institute of Physics for users registered with the Copyright Clearance Center (CCC) Transactional Reporting Service, provided that the base fee of $18.00 per copy is paid directly to CCC, 222 Rosewood Drive, Danvers, MA 01923. For those organizations that have been granted a photocopy license by CCC, a separate system of payment has been arranged. The fee code for users of the Transactional Reporting Service is: 0-7354-0032-6/01/$18.00.

© 2001 American Institute of Physics

Individual readers of this volume and nonprofit libraries, acting for them, are permitted to make fair use of the material in it, such as copying an article for use in teaching or research. Permission is granted to quote from this volume in scientific work with the customary acknowledgment of the source. To reprint a figure, table, or other excerpt requires the consent of one of the original authors and notification to AIP. Republication or systematic or multiple reproduction of any material in this volume is permitted only under license from AIP. Address inquiries to Office of Rights and Permissions, Suite 1NO1, 2 Huntington Quadrangle, Melville, N.Y. 11747-4502; phone: 516-576-2268; fax: 516-576-2450; e-mail: rights@aip.org.

L.C. Catalog Card No. 2001095448
ISBN 0-7354-0032-6
ISSN 0094-243X
Printed in the United States of America

Contents

Preface xv
Symposium Organization xvii
Conference Photograph. xix

I. NANOTUBES: SYNTHESIS

SWNTs Synthesized in External Magnetic Field 3
    B. Jeyadevan, Y. Sato, K. Tohji, A. Kasuya, R. Hatakeyama, H. Ishida,
    N. Sato, K. Ueno, and T. Takagi

Production of Thick Single-Walled Carbon Nanotubes by
Arc Discharge in Hydrogen Ambience. 7
    Y. Ando, X. Zhao, K. Hirahara, and S. Iijima

Production and Measurements of Isolated Single-Wall
Carbon Nanotubes 11
    S. Arepalli, P. Nikolaev, W. Holmes, V. Hadjiev, and B. Files

In-Situ Spectroelectrochemical Study of Single-Wall
Carbon Nanotubes 15
    S. Kazaoui, N. Minami, N. Matsuda, H. Kataura, and Y. Achiba

Synthesis of Carbon Nanocoils Using Electroplated Iron Catalyst 19
    L. Pan, M. Zhang, A. Harada, Y. Takano, and Y. Nakayama

Preparation of Carbon Nanotubes by Using Mesoporous Silica 23
    S. Kawasaki, S. Komiyama, S. Ohmori, A. Yao, F. Okino, and H. Touhara

Gas Dynamic and Time Resolved Imaging Studies of Single-Wall
Carbon Nanotubes Growth in the Laser Ablation Process....................27
    R. Sen, S. Suzuki, H. Kataura, and Y. Achiba

New Simple Method of Carbon Nanotube Fabrication Using
Welding Torch. 31
    H. Takikawa, Y. Tao, Y. Hibi, R. Miyano, T. Sakakibara, Y. Ando, S. Ito,
    K. Hirahara, and S. Iijima

Temperature and Time Dependence of the Growth of Carbon
Nanotubes by Thermal Chemical Vapor Deposition 35
    H. J. Jeong, Y. M. Shin, K. S. Kim, S. Y. Jeong, Y. S. Park, Y. C. Choi,
    and Y. H. Lee

Effect of Ni-Surface Morphology on the Growth of
Carbon Nanotubes by Microwave Plasma-Enhanced Chemical
Vapor Deposition. 39
    Y. C. Choi, Y. M. Shin, H. J. Jeong, D. J. Bae, S. C. Lim, K. H. An,
    and Y. H. Lee

Growth Mechanism of Carbon Nanotubes Grown by Microwave
Plasma-Assisted Chemical Vapor Deposition 43
    T. Muneyoshi, M. Okai, T. Yaguchi, and S. Sasaki

Automation of Purification for Single-Walled Nanotubes...................47
    T. Ogawa, Y. Sato, K. Shinoda, B. Jeyadevan, K. Tohji, A. Kasuya,
    and Y. Nishina

Time and Space Evolution of Carbon Species Generated with a Laser
Furnace Technique 51
      S. Suzuki, H. Yamaguchi, R. Sen, H. Kataura, W. Krätschmer,
      and Y. Achiba

Synthesis of Vertically Aligned Carbon Nanotubes on a Large Area
Using Thermal Chemical Vapor Deposition 55
      C. J. Lee, K. H. Son, T. J. Lee, S. C. Lyu, and J. E. Yoo

Synthesis of Aligned Carbon Nanotubes by C_2H_2 Decomposition on
$Fe(CO)_5$ as a Catalyst Precursor 59
      J. H. Han, J. E. Yoo, S. C. Yoo, C. J. Lee, and K.-H. Lee

Formation of Radioactive Fullerenes by Using Nuclear Recoil 63
      T. Ohtsuki, K. Ohno, K. Shiga, Y. Kawazoe, Y. Maruyama, K. Shikano,
      and K. Masumoto

Open Nanotubes of Insulating Boron Nitride 67
      D. Golberg, Y. Bando, K. Kurashima, and T. Sato

Inter-Process Measurement of MWNT Rigidity and Fabrication of
MWNT Junctions Through Nanorobotic Manipulations 71
      L. Dong, F. Arai, and T. Fukuda

Selective Lateral Nano-Bridging of Carbon Nanowire between
Catalytic Contact Electrodes 75
      Y. H. Lee, Y. T. Jang, C. H. Choi, E. K. Kim, B. K. Ju, D. H. Kim,
      C. W. Lee, J. K. Shin, and S. T. Kim

II. NANOTUBES: ELECTRONIC AND MECHANICAL PROPERTIES

Quantum Oscillations for the Spectral Moments of Raman Spectra
from SWCNT 81
      H. Kuzmany, M. Hulman, W. Plank, A. Grueneis, C. Kramberger,
      H. Peterlik, T. Pichler, H. Kataura, and Y. Achiba

Electronic Structure Studies of Carbon Nanotubes: Aligned, Doped,
and Filled 87
      J. Fink, X. Liu, H. Peisert, T. Pichler, M. Knupfer, M. S. Golden,
      D. M. Walters, and H. Kataura

Lithium Storage in Single Wall Carbon Nanotubes 95
      B. Gao, H. Shimoda, X. P. Tang, A. Kleinhammes, L. Fleming, Y. Wu,
      and O. Zhou

Electronic Structure and Quantum Conductance of Nanotube
Structures: Defects, Crossed-Tube Junctions, and Nanopeapods 101
      S. G. Louie

Electromechanical Properties of Multiwall Carbon Nanotubes 107
      A. Zettl and J. Cumings

Field Emission from Carbon Nanotubes with Clean Surface and
Adsorbed Molecules 113
      K. Hata, A. Takakura, and Y. Saito

Hydrogen Insertion and Extraction Mechanism in Single-Walled
Carbon Nanotubes 117
      S. M. Lee, Y. H. Lee, G. Seifert, and T. Frauenheim

Optical Anisotropy of Aligned Single-Walled Carbon Nanotubes in Polymer 121
    M. Ichida, S. Mizuno, H. Kataura, Y. Achiba, and A. Nakamura

Polarized Absorption Spectra of 0.4nm-Sized Single-Wall Carbon Nanotube Arrays Formed in Channels of $AlPO_4$-5 Single Crystals 125
    Z. K. Tang, Z. M. Li, G. D. Li, N. Wang, H. J. Liu, and C. T. Chan

Structural (n, m) Determination of Isolated Single Wall Carbon Nanotubes by Resonant Raman Scattering 129
    A. Jorio, R. Saito, J. H. Hafner, C. M. Lieber, M. Hunter, T. McClure, G. Dresselhaus, and M. S. Dresselhaus

High Energy-Resolution EELS Study of the Electronic Structure of Boron Nitride Cones 133
    M. Terauchi, M. Kawana, M. Tanaka, K. Suzuki, A. Ogino, and K. Kimura

Electrical Contact with Titanium Carbide to an Individual Single-Walled Carbon Nanotube 137
    F. Nihey, T. Ichihashi, M. Yudasaka, and S. Iijima

Multiwalled Carbon Nanotubes as Single Electron Transistors 141
    M. Ahlskog, R. Tarkiainen, L. Roschier, M. Paalanen, and P. Hakonen

Conduction Mechanism in Multiwall Carbon Nanotubes 145
    S. Roche, F. Triozon, D. Mayou, and A. Rubio

Electrical Transport through Crossed Carbon Nanotube Junctions 149
    T. Nakanishi and T. Ando

Energy Barriers in Carbon Nanotube Junctions 153
    R. Tamura and M. Tsukada

Scanning Tunneling Spectroscopic Characterization of Single Walled Carbon Nanotubes in Bundles 157
    K. Suzuki, Y. Maruyama, M. Nagayama, T. Kumagai, F. Kosha, K. Tohji, H. Takahashi, A. Kasuya, and Y. Nishina

Estimation of the Mechanical Strength of Nanotube Bundle 161
    Y. Nishina, T. Maeda, A. Kasuya, K. Tohji, and Y. Sato

Optical Properties of Fullerene-Peapods 165
    H. Kataura, T. Kodama, K. Kikuchi, K. Hirahara, S. Iijima, S. Suzuki, W. Krätschmer, and Y. Achiba

Electron Transport in Carbon Nanotubes Encapsulating Fullerenes 169
    D-H. Kim, H-S. Sim, and K. J. Chang

Electronic Structure and Energetics of Carbon Nanotubes Encapsulating C_{60} 173
    S. Okada, S. Saito, A. Oshiyama, and Y. Miyamoto

Intercalation of SWNTs as Investigated by Raman and X-ray Photoemission Spectroscopy 177
    T. Ito, Y. Yatsu, H. Fudo, Y. Iwasa, T. Mitani, H. Kataura, and Y. Achiba

Adsorption and Desorption of Weakly Bonded Adsorbates from Single-Wall Carbon Nanotube Bundles 181
    T. Hertel, J. Kriebel, G. Moos, and R. Fasel

Chirality Dependent G-Band Raman Intensity of an Individual Single Wall Carbon Nanotube 185
    R. Saito, A. Jorio, J. H. Hafner, C. M. Lieber, M. Hunter, T. McClure, G. Dresselhaus, and M. S. Dresselhaus

Photoconductivity of Single-Walled Carbon Nanotubes 189
 A. Fujiwara, Y. Matsuoka, H. Suematsu, N. Ogawa, K. Miyano,
 H. Kataura, Y. Maniwa, S. Suzuki, and Y. Achiba

Electronic Structure of Multi-Walled Carbon Nanotubes Studied by
Photoemission Spectroscopy 193
 S. Suzuki, C. Bower, Y. Watanabe, S. Heun, T. Kiyokura, K. G. Nath,
 T. Ogino, W. Zhu, and O. Zhou

STM/STS on Carbon Nanotubes at Low Temperature 197
 K. Nomura, M. Osawa, K. Ichimura, H. Kataura, Y. Maniwa, S. Suzuki,
 and Y. Achiba

Dielectric Function of C_{60}-Encapsulating Nanotube 201
 N. Hamada, M. Yamaji, S. Okada, and S. Saito

Structural and Electronic Property Changes of Single-Walled
Carbon Nanotubes under Pressure 205
 L. C. Qin, J. Tang, T. Sasaki, M. Yudasaka, A. Matsushita, and S. Iijima

Preparation of Mechanically Aligned Carbon Nanotube Films and
Their Anisotropic Transport Phenomena 209
 D. J. Bae, K. S. Kim, Y. S. Park, K. H. An, J. M. Moon, S. C. Lim,
 and Y. H. Lee

Polarization Characteristics of Zeolite Single Crystals Containing
Carbon Nanotubes 213
 N. Nagasawa, I. Kudryashov, S. Tsuda, and Z. K. Tang

Electrochemical Hydrogen Storage in Single-Walled
Carbon Nanotubes 217
 W. S. Kim, K. H. An, Y. S. Park, and Y. H. Lee

Saturation of Emission Current from Carbon Nanotube Field
Emission Array 221
 S. C. Lim, H. J. Jeong, Y. M. Shin, K. S. Kim, W. S. Kim, Y. S. Park,
 Y. C. Choi, K. H. An, D. J. Bae, and Y. H. Lee

Characteristic Interactions of Gases in Solid Carbon Nanotubes 225
 C. W. Jin, K. Ichimura, K. Imaeda, and H. Inokuchi

Chemical Interaction of Rare Gases in Solid Carbon Nanotubes 229
 K. Ichimura, K. Imaeda, and H. Inokuchi

The Effect of Solvent on Electrical Transport Properties in
Single-Wall Carbon Nanotubes 233
 S. Masubuchi, H. Masubuchi, S. Kazama, H. Kataura, Y. Maniwa,
 S. Suzuki, and Y. Achiba

Analysis of C_{60} Insertion into Single Wall Carbon Nanotube by
Molecular Dynamics Simulation 237
 T. Ishii, K. Esfarjani, Y. Hashi, Y. Kawazoe, and S. Iijima

Supercapacitors Using Singlewalled Carbon Nanotube Electrodes 241
 K. H. An, W. S. Kim, Y. S. Park, H. J. Jeong, Y. C. Choi,
 J. M. Moon, D. J. Bae, S. C. Lim, and Y. H. Lee

Resonances in Deformed Carbon Nanotubes 245
 H. S. Sim, C. J. Park, and K. J. Chang

Electrochemical Lithium Insertion of Heat Treated and Chemically
Modified Multi-Wall Carbon Nanotubes.................................249
    H. Touhara, I. Mukhopadhyay, F. Okino, S. Kawasaki, T. Kyotani,
    A. Tomita, and W. K. Hsu

Spin Valve Effect in Magnetically Doped Nanotube-Based Transistors........253
    K. Esfarjani, Z. Chen, A. A. Farajian, and Y. Kawazoe

Charge Oscillation at Doped Nanotube Junctions..........................257
    A. A. Farajian, K. Esfarjani, and M. Mikami

Effective-Mass Theory of Capped Carbon Nanotubes 261
    T. Yaguchi and T. Ando

Observation of Coulomb Blockade in a Ti/Multi-Wall Carbon
Nanotube/Ti Structure..265
    A. Kanda, K. Tsukagoshi, Y. Ootuka, and Y. Aoyagi

Energy Gap Induced by Lattice Deformation in Carbon Nanotubes 269
    H. Suzuura and T. Ando

Low Temperature Magneto-Transport in Multi-Wall
Carbon Nanotubes 273
    R. Enomoto, N. Aoki, K. Ishibashi, and Y. Ochiai

Structural Transformations in Single Wall Carbon Nanotube Bundles 277
    V. Kumar, M. Sluiter, and Y. Kawazoe

In-situ Atomistic Observation of Carbon Nanotube during
Field Emission...281
    T. Kuzumaki, Y. Horiike, and T. Kizuka

Processing of Individual Carbon Nanotubes − Cutting and Joining 285
    T. Kizuka and T. Kuzumaki

III. FULLERENES AND FULLERIDES

C_{60} as Building Block for New Interesting Carbon Structures and
Species 291
    W. Krätschmer

Superconductivity in Fullerene Systems 297
    M. L. Cohen

Electronic Structure of Ba_4C_{60} and Cs_4C_{60}..................305
    K. Umemoto and S. Saito

EPR in RbC_{60} Under Pressure 309
    S. Kobayashi, H. Sakamoto, K. Mizoguchi, M. Kosaka, and K. Tanigaki

Behaviors of Metals in Production of Nanonetwork Materials
Investigated by Radiochemical Technique................................313
    K. Sueki, K. Akiyama, C. Kurata, K. Oogama, Y. L. Zhao, M. Katada,
    S. Enomoto, S. Ambe, F. Ambe, H. Nakahara, and K. Kikuchi

Ferromagnetic Transition in Europium Fullerides $Eu_xSr_{6-x}C_{60}$ 317
    K. Ishii, A. Fujiwara, H. Suematsu, and Y. Kubozono

C_{60} Molecular Configurations Leading to Ferromagnetic Exchange
Interactions in TDAE-C_{60} 321
    B. Narymbetov, A. Omerzu, V. V. Kabanov, M. Tokumoto, H. Kobayashi,
    and D. Mihailovic

Metallic Phases in Sodium Fulleride Na_xC_{70}................325
    T. Hara, M. Kobayashi, Y. Akahama, and H. Kawamura

Production of $C_{59}N$: C_{60} Solid Solution................329
    F. Fülöp, A. Rockenbauer, F. Simon, S. Pekker, L. Korecq, S. Garaj, and A. Jánossy

Dynamic Jahn-Teller Mechanism of Superconductivity in Alkali-Metal-Doped C_{60}................333
    S. Suzuki, S. Okada, and K. Nakao

Theoretical Study on the Photoemission Spectra of A_3C_{60} (A=K and Rb)................337
    T. Chida, S. Suzuki, and K. Nakao

Phase Diagrams of Alkali-Metal-Doped C_{60}: Spin- and Orbital-Polarized States................341
    J. Hirosawa, S. Suzuki, and K. Nakao

Study on the Physical Properties of Na_4C_{60}................345
    Y. Takabayashi, Y. Kubozono, S. Fujiki, S. Kashino, K. Ishii, H. Suematsu, and H. Ogata

Fabrication of C_{60}/Amorphous Carbon Superlattice Structures................349
    N. Kojima, Y. Ohshita, and M. Yamaguchi

AC Conductivity of Alkali Doped C_{60} Compounds Across the Superconductor-Insulator Transition................353
    A. Maeda, H. Kitano, R. Matsuo, K. Miwa, T. Takenobu, Y. Iwasa, and T. Mitani

Study on the Origin of Pressure-Induced Superconductivity of Cs_3C_{60}................357
    S. Fujiki, Y. Kubozono, Y. Takabayashi, S. Kashino, M. Kobayashi, K. Ishii, and H. Suematsu

Structure and Properties of $RE_{2.75}C_{60}$................361
    J. Takeuchi, K. Tanigaki, and B. Gogia

Unusual Magnetic Properties of High-Temperature Reaction Products of Cerium Metal and C_{60} Solid................365
    S. Motohashi, Y. Maruyama, K. Watanabe, K. Suzuki, S. Takagi, and H. Ogata

NMR Studies of Ammoniated Alkali Fullerides................369
    H. Tou, N. Muroga, Y. Maniwa, T. Takenobu, H. Shimoda, Y. Iwasa, and T. Mitani

Orientational and Magnetic Transitions in Ammoniated Alkali Metal Fullerides................373
    T. Takenobu, M. Miyake, T. Muro, Y. Iwasa, and T. Mitani

NMR Studies of Alkali-Doped C_{60} Superconductors with Small Lattice Constants................377
    K. Kitazume, H. Tou, Y. Maniwa, M. Kosaka, and K. Tanigaki

Magnetic Properties of $TDAE-C_{60}$ under Pressure................381
    K. Mizoguchi, M. Machino, H. Sakamoto, T. Kawamoto, M. Tokumoto, A. Omerzu, and D. Mihailovic

Synthesis and Structure of Alkaline Earth and Rare Earth Metal Doped C_{70}................385
    T. Takenobu, Y. Iwasa, T. Ito, and T. Mitani

Structural and Physical Properties of Lithium Fullerides Li_xC_{70} 389
    H. Kumada, M. Kobayashi, Y. Akahama, and H. Kawamura

The Structural and Magnetic Properties in TDAE-Fullerene System.......... 393
    K. Oshima, Y. Nogami, T. Kambe, N. Nagao, and M. Fujiwara

Synthesis and Structure of All-Carbon Bisfullerene C_{121} 397
    H. Shimotani, J. Wang, N. Dragoe, and K. Kitazawa

Photoinduced Polymerization in Crystalline C_{60} via
Multi-Photoexcitation: Lattice-Relaxation and Energy-Transfer
of Excitons 401
    M. Suzuki

The Morphology of Vapor Grown C_{60} Crystals as an Ideal Example
of the Gibbs-Wulff's Law 405
    E. Schönherr, K. Matsumoto, and K. Murakami

Synthesis and Crystal Structure of Cocrystallite with Silver
Octaethylporphyrin and C_{70} 409
    T. Ishii, R. Kanehama, N. Aizawa, M. Yamashita, H. Matsuzaka,
    H. Miyasaka, T. Kodama, K. Kikuchi, I. Ikemoto, Y. Iwasa,
    and M. Shiro

Photosensitized Oxygenation of Alkenes in the Presence of
Bisazafullerene $(C_{59}N)_2$ and Hydroazafullerene $C_{59}HN$ 413
    N. Tagmatarchis and H. Shinohara

Transition of the Heterofullerene $(C_{59}N)X$ to the Monomeric Phase
of $C_{59}N$ 417
    W. Plank, T. Pichler, S. Baes-Fischlmair, M. Krause, H. Kuzmany,
    N. Tagmatarchis, and H. Shinohara

Survey of Natural Fullerenes in Southwestern China 421
    E. Osawa, M. Ozawa, K. Chijiwa, K. Hoyanagi, K. Tanaka,
    and M. Kusunoki

Effect of Chemical Treatment on the Structure of Ultradisperse
Diamond and Onion-Like Carbon 425
    A. E. Alexenskii, M. V. Baidakova, A. T. Dideikin, V. Y. Osipov, E. Osawa,
    M. Ozawa, A. I. Shames, V. I. Siklitsky, and A. Y. Vul'

Studies of Porphyrin-Fullerene Dyads with Oligoethylene Glycols
Spacers in Solution 429
    R. Ogura, T. Toida, K. Tsunoda, H. Yajima, and T. Ishii

Magnetic-Field Induced Ferromagnetism in Bissilylated C_{60} by
Pyrolysis 433
    Y. Kajihara, K. Tanigaki, and T. Akasaka

Characterization of Actinide Metallofullerenes 437
    K. Akiyama, K. Sueki, Y. L. Zhao, H. Haba, K. Tsukada, T. Kodama,
    K. Kikuchi, T. Ohtsuki, Y. Nagame, H. Nakahara, and M. Katada

IV. FULLERENE POLYMERS

NMR Studies of Alkali-Doped C_{60} Polymers 443
    Y. Maniwa, H. Ikejiri, H. Tou, S. Masubuchi, S. Kazama, M. Yasukawa,
    and S. Yamanaka

Dimer Structure of Sm_3C_{70} 447
    H. C. Dam, X. H. Chen, T. Takenobu, T. Itou, Y. Iwasa, T. Mitani,
    E. Nishibori, M. Takata, and M. Sakata

Photopolymerization of C_{60} Crystal under High Pressure 451
    M. Sakai, M. Ichida, and A. Nakamura

Out-of-Plane and In-Plane Structures of the Cast Films of Long
Alkyl Chain-Linked C_{60} via Phenyl Ring 455
    M. Chikamatsu, K. Kikuchi, T. Kodama, H. Nishikawa, I. Ikemoto,
    N. Yoshimoto, T. Hanada, Y. Yoshida, N. Tanigaki, and K. Yase

V. ENDOHEDRALS

Structure of IPR-Violated Fullerene, $Sc_2@C_{66}$ 461
    E. Nishibori, M. Takata, M. Sakata, C. R. Wang, M. Inakuma,
    and H. Shinohara

Chemistry of Endohedral Metallofullerene Ions 465
    T. Akasaka, T. Wakahara, S. Nagase, K. Kobayashi, M. Waelchli,
    K. Yamamoto, M. Kondo, S. Shirakura, Y. Maeda, T. Kato, M. Kako,
    Y. Nakadaira, X. Gao, E. Van Caemelbecke, and K. M. Kadish

Spin Dynamics of Lanthanum Metallofullerenes 469
    S. Okubo and T. Kato

Valence Change of Tm Atom in Metallofullenes 473
    K. Kikuchi, K. Sakaguchi, N. Ozawa, T. Kodama, H. Nishikawa,
    I. Ikemoto, K. Kohdate, D. Matsumura, T. Yokoyama, and T. Ohta

Electronic Structure of $Eu@C_{60}$ 477
    S. Suzuki, M. Kushida, S. Amamiya, S. Okada, and K. Nakao

Electronic Structures of the $La@C_{82}$ Crystals by the Relativistic
LCAO Method 481
    S. Amamiya, S. Okada, S. Suzuki, and K. Nakao

Low-Energy Electron Energy Loss Spectroscopy of Monolayer and
Thick $La@C_{82}$ Films Grown on MoS_2 Substrates 485
    K. Ueno, Y. Uchino, K. Iizumi, A. Koma, K. Saiki, Y. Inada, K. Nagai,
    Y. Iwasa, and T. Mitani

Photophysical and Photochemical Properties of Higher Fullerenes and
Endohedral Metallofullerenes 489
    O. Ito, M. Fujitsuka, T. Akasaka, and K. Yamamoto

VI. CLATHRATES

Silicon and Germanium Clathrates with Magnetic Elements 495
    K. Tanigaki, T. Kawaguchi, A. Nagai, and M. Yasukawa

A New Silicon Clathrate Compound: $I_8Si_{46-x}I_x$ 499
    E. Reny, S. Yamanaka, C. Cros, and M. Pouchard

NMR Studies of Silicon Clathrate Compounds 503
    H. Sakamoto, H. Tou, Y. Maniwa, H. Ishii, E. Reny, and S. Yamanaka

A Sign of Superconductivity in Li-Doped α-Rhombohedral Boron............507
    A. Oguri, K. Kimura, A. Fujiwara, M. Terauchi, and M. Tanaka

VII. OTHER MOLECULAR MATERIALS

Laser Induced Dissociation of Linear C_6 and Reorientation of Trapping Sites in Solid Neon...513
    T. Wakabayashi, A. L. Ong, and W. Krätschmer

Spiral Carbon Nanoparticles..517
    M. Ozawa, E. Osawa, and H. Goto

Electronic Structures of Carbyne Model Compounds.......................521
    S. Hino, Y. Okada, K. Iwasaki, M. Kijima, and H. Shirakawa

Magnetic Properties of K-Absorbing Zeolite LTA.........................525
    H. Kira, H. Tou, Y. Maniwa, and Y. Murakami

Mechanism of Magnetism in Stacked Nanographite: Theoretical Study...529
    K. Harigaya, N. Kawatsu, and T. Enoki

List of Participants..533

Author Index..551

Preface

Fullerenes, nanotubes, and related covalent-bond network materials are now the subject of great interest owing to their topology-dependent novel physical and chemical properties and to their potential for a variety of applications. In this growing field of science and engineering, both theorists and experimentalists have made many significant contributions. Some of them were of essential scientific importance, others were technologically important enough to stimulate further intensive research in the field, and others were both.

In 1970, the first scientific paper on the C_{60} cage was written by E. Osawa, who theoretically discussed physical and chemical properties of this unusual all-carbon cluster having the truncated icosahedron geometry. It took as long as fifteen years for this C_{60} to be first experimentally detected in the cluster beam by H. W. Kroto, R. E. Smalley and coworkers in 1985. Thereafter, some theorists discussed one of the most important aspects of the C_{60} fullerene and its endohedral derivatives, i.e. the possibility to use them as atomlike building blocks of materials: La@C_{60} by A. Rosén and K@C_{60} by S. Saito. This dream of theorists was realized in 1990 by W. Krätschmer and coworkers, who produced the crystalline solid C_{60} where C_{60} fullerenes form a close-packed lattice. Once the existence of fullerenes became definite and they became available in macroscopic amounts, they have been studied extensively during the last decade and have proved to be actually the novel atomlike building blocks of a new class of hierachical materials: They can form van der Waals solid, molecular magnet, insulating ionic compounds, ionic but metallic compounds, covalent-bond crystalline materials (polymerized C_{60}), and even the high-transition temperature (T_c) superconductors. A recent hole doping into solid C_{60} achieved via field-effect transistor configuration by B. Batlogg and coworkers is found to give rise to the T_c of as high as 52K.

From the viewpoint of the dimensionality of materials, C_{60} and other fullerene families can be classified as zero-dimensional (0-D) network of sp^2 C atoms, being different from the two-dimensional (2-D) network graphite. Another new-dimensionality sp^2 C material, i.e. the one-dimensional carbon nanotube which fills the gap between 0-D fullerenes and 2-D graphite, was discovered by S. Iijima in 1991 without preceding work by theorists. On the other hand, immediately after the discovery, its unusual electronic transport property, i.e., the topology-dependent semiconducting or metallic electronic structure, was theoretically predicted by three groups [N. Hamada, S. Sawada, and A. Oshiyama; K. Tanaka, K. Okahara, M. Okada, and T. Yamabe; R. Saito, M. Fujita, G. Dresselhaus, and M. S. Dresselhaus]. This fascinating prediction, confirmed later experimentally, triggered further intensive research of nanotubes and they are now studied as key materials of nanoelectronics. Also their mechanical and chemical properties are of high importance in connection with various applications.

Now it is a world-wide trend to support the research on nanoscience and nanotechnology and fullerenes and nanotubes are the most important materials in the field. The Japanese Government (Ministry of Education, Culture, Sports, Science and Technology) has supported this fruitful field via a program called Grant-in-Aid for Scientific Research on Priority Area during the period of as early as 1993 to 1995, and from 1998 to present. As a part of the present project "Fullerenes and Nanotubes", the International Symposium on Nanonetwork

Materials, Fullerenes, Nanotubes, and Related Systems was held in Kamakura, Japan from January 15 to 18, 2001. This proceedings book contains most of the invited and contributed papers presented during the Symposium and consists of the following seven chapters:

I. Nanotubes: Synthesis

II. Nanotubes: Electronic and Mechanical Properties

III. Fullerenes and Fullerides

IV. Fullerene Polymers

V. Endohedrals

VI. Clathrates

VII. Other Molecular Materials

The Symposium could provide an international forum for the discussion of the current status and future prospects of the nanonetwork materials. We, the organizers, would like to express our gratitude for the support from the Ministry of Education, Culture, Sports, Science and Technology of Japan.

Susumu Saito
Tsuneya Ando
Yoshihiro Iwasa
Koichi Kikuchi
Mototada Kobayashi
Yahachi Saito

SYMPOSIUM ORGANIZATION

Organizing Committee

Susumu Saito (Tokyo Institute of Technology), Chairperson
Tsuneya Ando (University of Tokyo), Publication / Program
Yoshihiro Iwasa (JAIST), Secretary / Program
Koichi Kikuchi (Tokyo Metropolitan University), Program Chair
Mototada Kobayashi (Himeji Institute of Technology), Local / Program
Yahachi Saito (Mie University), Registration / Program

International Advisory Board

Marvin L. Cohen (University of California)
Sumio Iijima (NEC and Meijo University)
Hiroo Inokuchi (IIAS and NASDA)
Takehiko Ishiguro (Kyoto University)
Hiroshi Kamimura (Science University of Tokyo)
Yuichiro Nishina (Ishinomaki Senshu University)
Hans Kuzmany (University of Vienna)
Hiroyuki Shiba (Tokyo Institute of Technology)
Richard E. Smalley (Rice University)
Hiroyoshi Suematsu (University of Tokyo)

I. NANOTUBES: SYNTHESIS

SWNTs Synthesized in External Magnetic Field

B. Jeyadevan[1], Y.Sato[1], T. Nakano[1], K.Tohji[1], A. Kasuya[2], R. Hatekeyama[3], H. Ishida[3], N. Sato[3], K. Ueno[4] and T. Takagi[4]

[1]*Dept. Geoscience and Technology, Tohoku University, Sendai, 980-8579, Japan*
[2]*Center for Interdisciplinary Research, Tohoku University, Sendai, 980-8578, Japan*
[3]*Dept. of Electric and Electronics Tohoku University, Sendai, 980-8579, Japan*
[4]*Institute of Fluid Science, Tohoku University, Sendai, 980-8579, Japan*

Abstract This paper describes the results of the synthesis of SWNTs in weak and strong DC magnetic fields. In weak magnetic fields with the maximum intensity of 0.5 T, SWNTs were synthesized only when the applied field was parallel to the current direction. These samples consisted of tubes with a diameter as small as about 0.9 nm. Furthermore, the metal particles were smaller compared to zero field case. On the other hand, at strong magnetic fields with the maximum intensity of 2 T, the yield decreased with increasing magnetic field strengths for the arc discharge current of 70 A. This was believed due to forced convection caused by the temperature rise in the anode resulting from higher current densities due to Lorentz force acting on charged particles non-parallel to the magnetic field direction.

INTRODUCTION

Though considerable progress has been made in both production and purification of SWNTs, the morphology control is yet to be achieved. The properties of SWNTs that have great potential for nano technological applications have been found to depend on the their morphology, such as helicity[1]. Therefore, control over the diameter, length and network structure of the SWNTs is desired. Though the synthesis of the nanotubes and their characteristics have been found to depend on parameters, such as, gas atmosphere, gas pressure, type of metal catalysts, etc., the authors believe that the arc-discharge temperature and the retention time of the ions in the proximity of the arc area are also very vital in respect to the yield and morphology of the nanotubes. It is well known that the manipulation of the movement of the charged particle in the arc can be controlled by magnetic field. Depending on the magnetic field configuration, the retention time of the nucleus that forms the basis of SWNTs can be controlled. Here, we have made an attempt to study the effect of weak and strong magnetic fields on the yield and morphology of the SWNTs produced using Fe-Ni as catalysts in arc-discharge.

EXPERIMENTAL

SWNTs were synthesized by DC arc-discharge using a 6mm diameter graphite rod (99.998 %) packed with Fe and Ni powders of 99.9 % purity as anode and 20 mm diameter carbon rod (99.9%) as cathode. The total metal content in the anode was approximately 6 at. %. The arc-discharge was carried out under varying He atmosphere and discharge current of 70 A, and during the discharge, the gap between the electrode was maintained at about 1mm by advancing the consumed anode by a step motor.

The Arc-discharge were carried out under weak and strong DC magnetic fields using (a) an electromagnet that generates a uniform magnetic field of 0.0515 T (maximum) within 5 cm cubic area, and (b) a superconducting magnet that provides a maximum flux density of 5 T, respectively. An arc-discharge chamber with an internal diameter of 155 mm and 200 mm in length and a chamber with an internal diameter of 46 mm and 900 mm in length were used for the synthesis in weak and strong magnetic fields, respectively.

The SWNTs synthesized under different conditions were analyzed using Raman Spectroscopy, SEM and TEM. The Raman spectroscopy equipment used was T64000 of Dilor-Jobin Yvon-Spex with Ar ion laser (Leonix Co.). The measurements were carried out using 488.0 nm excitation wavelengths under back scattering configuration on dry samples. Hitachi S-4100 and Hitachi HF-2000 with a field emission type electron gun, were used for SEM and TEM measurements, respectively.

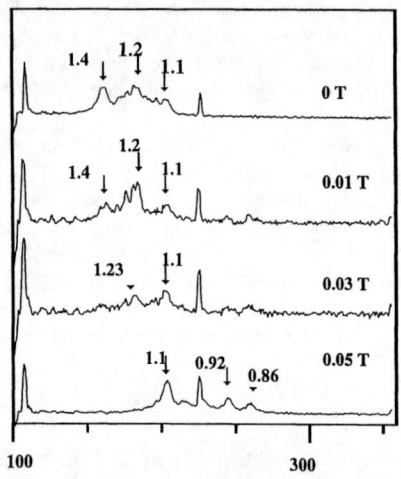

FIGURE 1. The "breathing mode" contours of nanotubes of different diameter for SWNTs grown in different magnetic field intensities

RESULTS AND DISCUSSION

Magnetic control of plasma arc[2], and energy flow density [3] has been investigated by researchers. It has been found that the plasma arc oscillates backward and forward when exposed to a transverse AC magnetic field and the oscillatory motion depends on the intensity of the magnetic field and flow rate of the plasma. Furthermore, it also has been reported that the magnetic field applied parallel and perpendicular to the plasma flow direction broadened the distribution of the plasma temperature and velocity. The broadened distribution in AC magnetic field has been attributed to the oscillation of the arc column; the oscillation was reported due to the Lorentz force induced by the external magnetic field. Similar effects could be seen in DC magnetic field, too.

Synthesis of SWNTs in Weak DC Magnetic Field

Soot samples were produced under various He pressures to determine the optimum pressure condition for the synthesis of SWNTs with no external magnetic field. Raman spectroscopy analysis suggested that the He pressure of 100 Torr was optimum. Thus, the synthesis of SWNTs soot in weak magnetic fields from 0.01 T to 0.05 T were carried out in a He pressure of 100 Torr.

Magnetic Field Direction Parallel to Arc Current

The Raman scattering characterizes a surface region of a few microns, integrating the signals from all possible nanotube fractions comprised in the material. The Raman spectroscopy of a SWNT has two main features that were different from the spectrum of planar graphite. Namely, the splitting mode at 1582 cm^{-1} and the appearance of a "breathing" radial mode in the low frequency region 150-220 cm^{-1}. Though there was no drastic difference in the splitting mode of the raw soot synthesized in different magnetic field intensities, there was considerable change in the low frequency region. We observed a shift in the "breathing" mode to higher frequencies with the increase in the magnetic field intensities applied during arc-discharge. The frequency $\omega(r)$ is almost inversely proportional to the nanotube radius r in the range of 0.3 nm < r < 0.7 nm:

$$\omega(r) = \omega_{(10,10)} (r_{(10,10)}/r)^{1.0017}$$

where, $\omega_{(10,10)} = 165$ cm^{-1}, $r_{(10,10)} = 0.6785$ nm [4]

The estimation of the tube diameters based on the above equation suggested that at 0.05 T the SWNTs of diameter as small as 0.86 nm was produced. The "breathing mode" contours and the corresponding diameter values of the low frequency profiles of the soot samples prepared at different magnetic field intensities is shown in Figure 1.

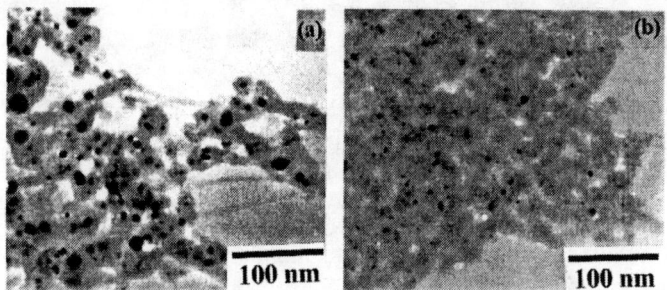

FIGURE 2. TEM photograph of the soot synthesized in 0.01 T and 0.05 T

Fig. 2 (a)&(b) shows the TEM photograph of the soot produced in 0.01 and 0.05 T. The size of the metal particles in the soot decreased with increasing magnetic field strength as shown in the figures. Furthermore, nanotubes diameters measured from the photographs decreased for any increase in the magnetic field strength and were in agreement with the results of Raman spectroscopy. It was believed that the temperature difference between the arc and its environs have increased in external magnetic field leading to the formation of fine metal particles. We also believe that the formation of fine particles have some influence on the formation of finer tube diameters.

Magnetic Field Direction Perpendicular to Arc Current

Similar experiments were carried out applying the magnetic field perpendicular to the arc current direction. Raman spectra analysis of the soot did not show the split peaks at 1582 cm^{-1} in all the samples irrespective of the magnetic field intensity. We believe that the Lorentz force acting perpendicular to both current and magnetic field produced forced convection and influenced both the arc temperature and the retention time of the ions in the vicinity of the arc area, and led to poor SWNT yield.

Synthesis of SWNTs in Strong Magnetic Field

We designed the chamber to carryout SWNTs synthesis in strong magnetic field using super conducting magnet. Since the optimum He pressure for high yield of SWNTs depends on the dimensions of the chamber, we carried out some basic experiments in different He pressure conditions. Based on the results of Raman spectroscopic analysis, the He pressure was fixed at 100 Torr. Then, experiments were carried out in external magnetic field strengths of 0, 0.05, 0.25, 0.5, 1 and 2 T. During the arc-discharge, the magnetic field was applied parallel to the arc current direction. The arc-discharge current was set at 70A while the voltage during discharge was continuously monitored. And also, the glow patterns at various magnetic field intensities were filmed. The arc configuration varied with increasing magnetic field strength as shown in Fig. 3. The glow intensity suggested the elevation of the temperature in the arc region. Furthermore, the consumption rate of the anode increased with the strength of the magnetic field.

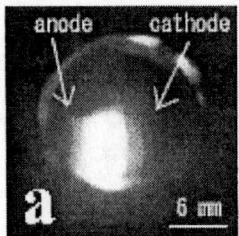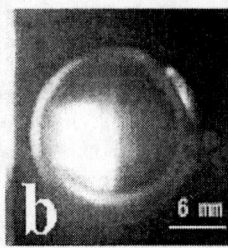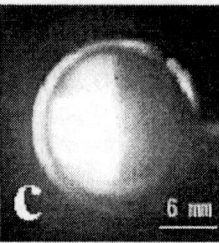

FIGURE 3. Glow pattern of the arc in (a) 0 T, (b) 0.25 T, and (c) 1 T

The Raman spectra analysis of the raw soot collected from the cathode and the wall showed the split peak at 1582 cm^{-1} only for the samples synthesized at 0.05, 0.25, 0.5 T. Furthermore, compared to zero field, the shift in "breathing mode" contours towards higher frequency with the strength of the magnetic fields up to 0.5 T was also observed. This was believed due to forced convection caused by the temperature rise in the anode resulting from higher current densities due to Lorentz force acting on charged particles non-parallel to the magnetic field direction.

CONCLUSIONS

By controlling the arc temperature and retention time of the nucleus, nanotubes with smaller diameters could be synthesized in weak magnetic fields. The yield was very poor in strong magnetic field, due to forced convection caused by the temperature rise in the anode resulting from higher current densities due to Lorentz force acting on charged particles non-parallel to the magnetic field direction. We believe that nanotubes with smaller diameter and high yield could be achieved at high magnetic field strengths by maintaining the temperature of the arc through proper control of the magnitude of the discharge current.

ACKNOWLEGEMENTS

The present study was supported by Grant-in-Aid for Priority Research (#11165202) and International Scientific Research Program (#11694119) from the Ministry of Education, Science, Culture and Sports of Japan and CREST of Japan Science and Technology Institute

REFERENCES

(1) Mintmire, J. W, Dunlap, B. I., and White, C.T., Phys. Rev. Lett. 68, 631 (1992)
(2) Takeda, K., Kouongakkaishi., 16(6), 357-367 (1990)
(3) Watanabe, T., Morimoto, Y., and Kanazawa, A., Kouongakkaishi., 18(5), 225-234 (1992)
(4) Saito, R., Takeya, T., Kimura, T., Phys. Rev. Lett. 57, 4145 (1998)

Production of Thick Single-Walled Carbon Nanotubes by Arc Discharge in Hydrogen Ambience

Yoshinori Ando[a], Xinluo Zhao[b], Kaori Hirahara[b], and Sumio Iijima[a,b]

[a]*Department of Materials Science and Engineering, Meijo University,*
Tenpaku-ku, Nagoya 468-8502, Japan
[b]*ICORP-JST, c/o Department of Materials Science and Engineering, Meijo University,*
Tenpaku-ku, Nagoya 468-8502, Japan

Abstract. High-quality single-walled carbon nanotubes (SWNTs) have been produced by DC arc discharge in hydrogen ambience. By using Fe 1% catalyst a large amount of SWNTs are produced in pure H_2 gas. Other transition metals of iron group and its combination are also used as catalyst for preparing SWNTs by adding 1% H_2S gas in H_2. Tri-metal catalyst, Fe(0.25%)-Ni(0.9%)-Co(0.9%), is found to be most effective for high yield of SWNTs. The SWNTs diameters are fairly thick ~1.4−4 nm, and some double-walled carbon nanotubes also exist with them.

INTRODUCTION

Single-walled carbon nanotubes (SWNTs) are a kind of fascinating nano-materials that can be prepared by laser ablation [1], arc discharge evaporation [2], catalytic chemical vapor deposition [3], etc. Because these SWNTs possess very small diameter d ($0.7 < d < 2$ nm), the one-dimensional quantum confine of the electrons in SWNTs can be observed [4]. For some applications, such as, hydrogen storage and synthesis of fullerene encapsulated SWNTs [5], thick SWNTs can be advantageous. Here we report the preparation of thick SWNTs with diameters of 1.4−4 nm and their analysis by scanning electron microscopy (SEM) and high-resolution transmission electron microscopy (HR-TEM). Growth mechanism of double-walled carbon nanotubes (DWNTs) coexisted with thick SWNTs is also described.

EXPERIMENTAL

The DC arc discharge evaporation apparatus, in which the metal-doped carbon anode (ϕ 6 mm) and pure carbon cathode (ϕ 10 mm) were installed horizontally, was the same as that used previously [6]. In the present set of experiments, transition metals of iron group and their combination were used as catalyst. After pure H_2 gas or H_2 including a little H_2S gas was introduced at a pressure of 60−700 torr, DC arc voltage was applied, and the arc current was changed from 30 to 70 A. The chamber soot was collected and investigated by SEM with an energy dispersive X-ray analysis system (EDX), HR-TEM and Raman spectroscopy.

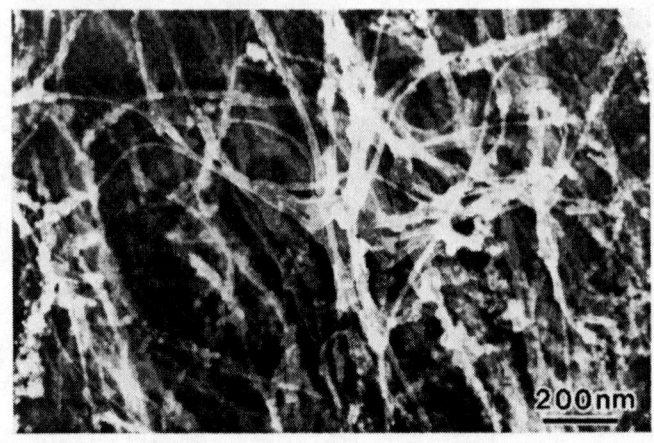

FIGURE 1. SEM image of chamber soot prepared with tri-metal catalyst, Fe-Ni-Co.

RESULTS

Using Fe (1at%) as catalyst, a large number of SWNTs could be produced in H_2 gas. The yield of SWNTs in the chamber soot became higher with increasing H_2 gas pressure from 60 to 500 torr, and got saturated thereafter. However, SWNTs could not be prepared in pure H_2 gas using the catalyst of Co, Ni and their combination, unless 1% H_2S gas was added to the ambience. The addition of H_2S in H_2 gas promoted the growth of SWNTs. Tri-metal catalyst, Fe(0.25%)-Ni(0.9%)-Co(0.9%), was the most effective to get high yield of SWNTs, as shown in a characteristic SEM image of SWNTs bundle (Fig. 1). The presence of Fe, Ni, Co and S in the soot was confirmed by EDX. The contents of Ni and Co increased from 0.9 to 3.3 at%, but Fe content did not change. The contents of S and C were 2.2 and 89.5 at%, respectively.

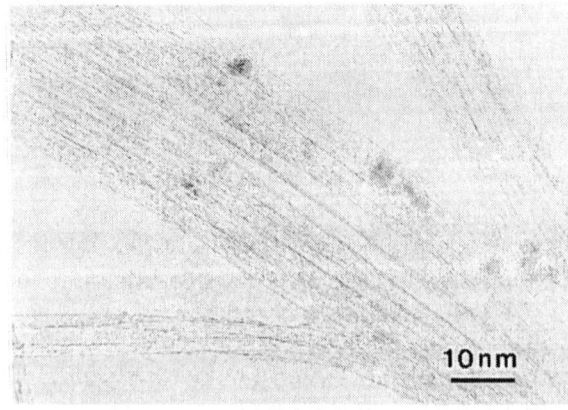

FIGURE 2. A typical HR-TEM image of chamber soot prepared with tri-metal catalyst, Fe-Ni-Co.

An example of HR-TEM image is shown in Fig. 2, in which each diameter of SWNTs is thick ~ 1.8−3 nm. Some amorphous materials are included within the thick SWNTs. From the images of SEM and HR-TEM, the quantity of SWNTs is estimated to be higher than 50%. The micro-Raman spectrum of SWNTs is shown in Fig. 3, which was obtained by using an Ar ion laser emitted at 514.5 nm. The peak at 1591 cm^{-1} with a shoulder at 1578 cm^{-1} is G mode, and the peak at 1340 cm^{-1} corresponds to D mode. The breathing modes are observed in a frequency range 100−200 cm^{-1} as inserted in left-up of Fig. 3. Observed peak values, 132, 145 and 158 cm^{-1} correspond to the diameter of SWNTs, 1.7, 1.5 and 1.4 nm, respectively. These values are clearly thicker than usual SWNTs [4].

DISCUSSION AND CONCLUSIONS

It is known that sulfur is an effective promoter for the transition metals used to prepare SWNTs [7]. In the present experiments, the SWNTs obtained by DC arc discharge evaporation of Fe-doped anode in pure H_2 gas possessed diameters of 0.84−2 nm, but the addition of H_2S gas made SWNTs thicker (1.4−4 nm). Therefore, S not only promotes the SWNTs growth but also has the effect of thickening the tube diameter.

Hydrogen arc discharge plasma can selectively etch the impurity of amorphous carbon [8], and the capping atoms of SWNTs. Because the SWNTs possess thicker diameters and no cap on their ends, some carbon materials can get into the wider ends of SWNTs easily. Furthermore, if the SWNTs including carbon materials stay in high temperature zone near the DC arc discharge plasma, the carbon materials within the SWNTs form a second layer along the outer surface. One of the evidences is given in Fig. 4, in which the second layer of DWNTs is not yet formed completely as seen in Fig.4(b), showing the growing process of DWNTs.

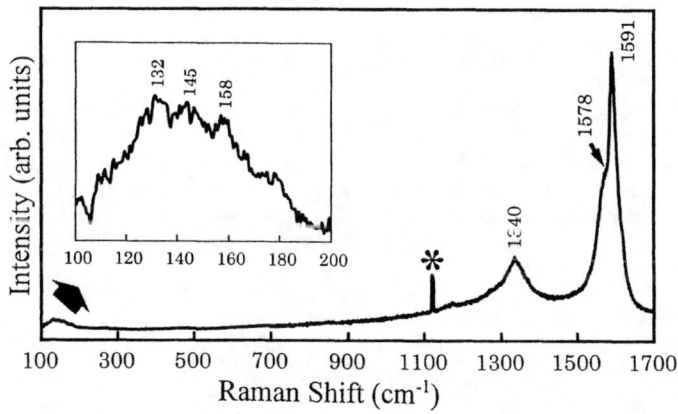

FIGURE 3. Raman spectra of SWNTs prepared with tri-metal catalyst, Fe-Ni-Co.

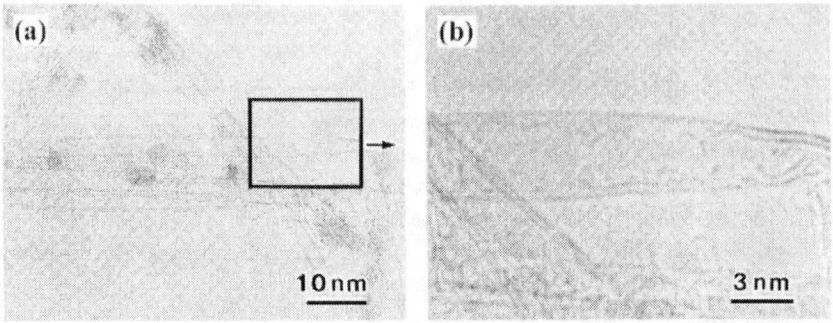

FIGURE 4. a) A HR-TEM image of DWNTs. b) Magnified image of the rectangular part of a). The second layer is not yet formed completely, showing the growing process of DWNTs.

In summary, 1.4−4 nm thick SWNTs can be produced by DC arc discharge evaporation of the anode doped with Fe(0.25%)-Ni(0.9%)-Co(0.9%) in H_2 (500 torr) including 1% H_2S, and some DWNTs also exist with these thick SWNTs. The addition of H_2S plays an important role in promoting the growth of SWNTs and increasing the diameter of SWNTs.

ACKNOWLEDGEMENTS

The authors thank Dr. M. Hiramatsu of Meijo University for allowing the use of the facilities for Raman spectroscopy. This work was partially supported by a Grant-in-Aid for Scientific Research on the Priority Area "Fullerenes and Nanotubes" by the Ministry of Education, Science, Sports and Culture of Japan.

REFERENCES

1. Guo, T., Nikolaev, P., Thess, A., Colbert, D. T., and Smalley, R. E., *Chem. Phys. Letters* **243**, 49-54 (1995).
2. Ando, Y., Zhao, X., Hirahara, K., Suenaga, K., Bandow, S., and Iijima, S., *Chem. Phys. Letters* **323**, 580-585 (2000).
3. Nikolaev, P., Bronikowski, M. J., Bradley, R. K., Rohmund, F., Colbert, D. T., Smith, K. A., and Smalley, R. E., *Chem. Phys. Letters* **313**, 91-97 (1999).
4. Rao, A. M., Richter, E., Bandow, S., Chase, B., Eklund, P. C., Williams, K. A., Fang, S., Subbaswamy, K. R., Menon, M., Thess, A., Smalley, R. E., Dresselhause, G., and Dresselhause, M. S., *Science* **275**, 187-191 (1997).
5. Hirahara, K., Suenaga, K., Bandow, S., Kato, H., Okazaki, T., Shinohara, H., and Iijima, S., *Phys. Rev. Letters* **85**, 5384-5387 (2000).
6. Wang, M., Zhao, X., Ohkohchi, M., and Ando, Y., *Fullerene Sci. Tech.* **4**, 1027-1039 (1996).
7. Kiang, C-H., Goddard III, W. A., Beyers, R., Salem, J. R., and Bethune, D. S., *J. Phys. Chem.* **98**, 6612-6618 (1994).
8. Zhao, X., Ohkohchi, M., Wang, M., Iijima, S., Ichihashi, T., and Ando, Y., *Carbon* **35**, 775-781 (1997).

Production and Measurements of Isolated Single-Wall Carbon Nanotubes

Sivaram Arepalli*, Pavel Nikolaev*, William Holmes*,
Victor Hadjiev¶, and Bradley Files$

*G. B. Tech./NASA-Johnson Space Center, 2101 NASA Road One, Houston, TX 77058, USA
¶Texas Center for Superconductivity, University of Houston, Houston, TX 77204, USA
$NASA-Johnson Space Center, 2101 NASA Road One, Houston, TX 77058, USA

Abstract. The production of isolated single wall carbon nanotubes (SWNTs) is accomplished using the laser oven process. Material is collected on quartz substrates at different locations in the laser oven for a variety of flow conditions. The lengths and diameter distributions of the nanotubes are measured directly (without additional processing steps) using AFM. Preliminary Raman data taken using 2D scans indicate the feasibility of this technique for length and diameter determination. The AFM study indicated the formation of long individual nanotubes, which then seem to coalesce into bigger bundles. The role of the inner tube of the flow-tube set up is confirmed to improve interactions between SWNTs resulting in formation of bundles. Flowing buffer gas seems to influence the dispersion of particulate material in the nanotube product.

INTRODUCTION

Carbon nanotubes are normally found as ropes/bundles of single wall or multiwall tubes. Most of the current work is on single wall nanotubes (SWNTs) because of their superior thermal, mechanical and electrical properties compared to multiwall nanotubes (MWNTs), resulting in many possible applications [1-3]. For the development of composite materials that can be reinforced mechanically by the nanotubes, issues of critical length and aspect ratio are important. Recent calculations [4] indicate that the lengths of the individual SWNTs need to be more than 10 μm in order to transfer individual tube strength to a bundle. Bundles have to be separated in order to observe and measure individual tubes. Some researchers have tried to infer the lengths of nanotubes by chromatographic methods [5], AFM and from scattering methods [6]. These results show tubes that may be too short after processing to be useful in composites. However, a recent measurement of tensile strength seems to indicate longer tubes [7]. It is therefore important to determine if the nanotubes are inherently long when they are produced and whether they get shorter during purification and processing. In order to be able to look at pristine SWNTs, "witness plates" (optical quality quartz flats and microscope slides) are placed at different locations in the laser oven. The roughness of such substrates is worse than normally used mica or silicon, but still adequate, and nanotubes bind well enough to allow AFM imaging.

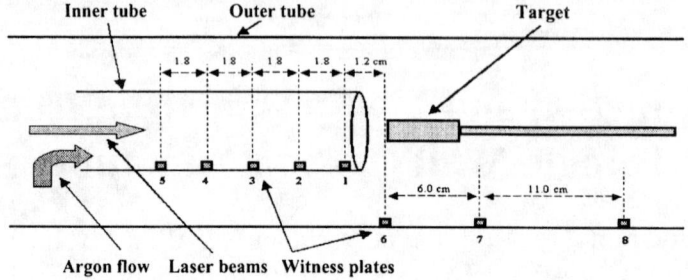

FIGURE 1. Lay out of the setup (double-pulse laser oven production process at 1473K with argon flowing at 500 Torr) for SWNT deposition directly on quartz flats that can be used as AFM substrates.

EXPERIMENTAL DETAILS

The experiments are carried out in our pulsed laser vaporization setup, which is described in an earlier communication [8]. Eight witness plates are placed as shown in Fig. 1. Once the laser furnace is running, the laser beam shutters are opened briefly (0.5 sec) to deposit nanotubes onto the plates. Similar tests are conducted under run conditions with the gas flow stopped and without the inner tube to evaluate the effect of fluid flow changes and confinement of the plume. Plates are imaged in Digital Instruments Nanoscope IIIa, operating in the tapping mode. The height resolution is about 0.1 nm and we could image individual nanotubes and ropes. Raman spectra from these witness plates are obtained on a microprobe stage (2µ laser spot of 782 nm) of a Renishaw spectrometer using 2D scans with 0.5 µ step size.

RESULTS AND DISCUSSION
AFM Data

Production runs at 1473 K have produced some nanotubes close to the target (plates

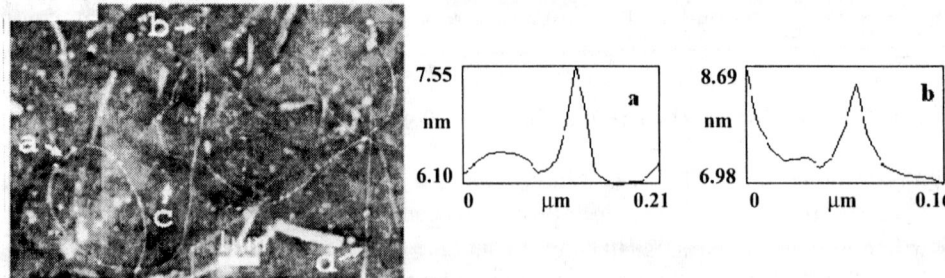

FIGURE 2. AFM image and corresponding height profiles of plate #3 exposed for 0.5 seconds, without inner tube and argon flowing at 100 sccm. a) Individual tube, 1.25 nm diameter and >22 µm long. b) Individual tube, 0.86 nm diameter, >18 µm long. c) Short tube, 0.8 nm diameter, 0.38 µm long. d) Tapered bundle.

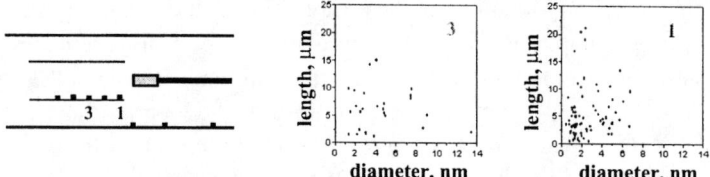

FIGURE 3. Diameter and length distributions of as prepared SWNTs on quartz substrates inside the laser oven at two different distances from the target. Notice shift towards more individual nanotubes and smaller diameter bundles closer to the target (1).

1, 2 and 3 with the inner tube and plates #3, 4 and 5 without the inner tube) and almost nothing on the rest of the witness plates (Fig. 2). Statistics on lengths and diameters obtained from AFM measurements is presented in Figure 3. It is important to emphasize that most length measurements are lower estimates, and nanotubes are definitely longer.

Far from the target we see almost exclusively bundles, while close to the target we see a significant fraction of individual tubes, and the bundles are generally much thinner (Fig. 3). This shows that nanotubes form within the ablation plume and propagate away from the target as it expands. As they fly away, they collide with each other and form bundles; hence more bundles and thicker bundles are found farther away from the target. Relative numbers of individual nanotubes vs. bundles deposited on the plates is always higher in experiments without the inner tube. In the presence of the inner tube, the volume into which the plume expands is smaller. Therefore the number density of nanotubes is higher, increasing the likelihood of tubes colliding with each other and forming bundles. Thus more and thicker bundles are produced with the inner tube.

Running at 1473 K with stopped flow produced too many particles on substrates to allow AFM imaging; nevertheless, nanotubes and bundles were seen in about the same abundance as in experiments with flow. Effective area of the nanotube greatly exceeds that of the nanoparticle of comparable weight, so they must move differently in the gas flow.

Raman Data

Figure 4 displays a 2D Raman image of a 5 μ x 5 μ area on the plate # 1 (with inner

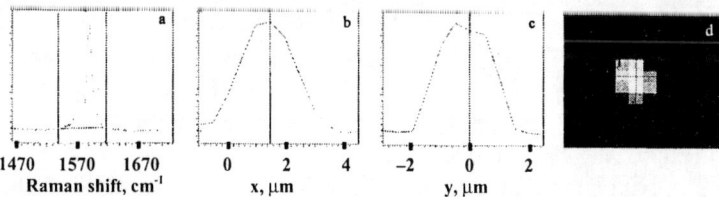

FIGURE 4. a) Tangential Raman mode of SWNT on plate #1. Indicated area under the peak was used for mapping. b, c) Raman signal profiles along X and Y axes. d) Raman intensity map of the 5 μm x 5 μm area.

tube). The image maps Raman intensity (area under the peak) of the 1590 cm^{-1} SWNT line in 0.5 µ steps. Raman intensity distribution clearly indicates an isolated nanotube less than 2 µm in size (experimental resolution). The signal is surprisingly strong for such a small sample thus conforming the importance of resonant mechanism of Raman scattering from SWNTs. We note that the Raman image does not correlate with any of the AFM scans shown here. Nevertheless it demonstrates the potential of 2D Raman scans for detection of isolated SWNTs (ropes), in particular with length larger than the laser spot size.

CONCLUSIONS

In conclusion, the current measurements indicate the formation of long individual nanotubes and their coalescence into ropes. The role of the inner tube is confirmed to control the expansion of the ablated plume and help their coalescence into bundles. The effect of argon flow seems to help distribute and dissipate bigger particulate material. Raman data indicates the need for improvement of 2D mapping by employing different excitation wavelengths and polarizations.

ACKNOWLEDGMENTS

The authors wish to acknowledge helpful support and encouragement from Bob Hauge and Rick Smalley of Rice University.

REFERENCES

1. Files, B. S. and Mayeaux, B. M., *Advanced Materials and Processes* **156**, 47-49 (1999).
2. Saito, R., Dresselhaus, G., and Dresselhaus, M. S., *Physical Properties of Carbon Nanotubes,* (Imperial College Press, London, 1998).
3. Dresselhaus, M. S., Dresselhaus, G., and Eklund, P. C., *Science of Fullerenes and Carbon Nanotubes* (Academic Press, San Diego, 1996).
4. Yakobson, B. I., Samsonidze, G. and Samsonidze, G. G., Carbon **38**, 1675-1680 (2000).
5. Duesberg, R. S., Muster, J., Krastic, V., Burghard, M. and Roth, S., *Applied Physics A* **67**, 117-120 (1998).
6. Liu, J., Rinzler, A. G., Dai, H., Hafner, J. H., Bradley, R. K., Boul, P. J., Lu, A., Iverson, T., Shelomov, K., Huffman, C. B., Rodriguez-Macias, F., Shon, Y-S., Lee, T. R., Colbert, D. L., and Smalley, R. E., *Science* **280**, 1253-1256 (1998).
7. Yu, M-F., Files, B.S., Arepalli, S. and Ruoff, R. S., *Physics Review Letters*, **84**, 5552-5555 (2000).
8. Arepalli, S., Nikolaev, P., Holmes, W., and Scott, C. D., *Applied Physics A* **70**, 125-133(2000).

In-Situ Spectroelectrochemical Study of Single-Wall Carbon Nanotubes

S. Kazaoui[*,1], N. Minami[*], N. Matsuda[*], H. Kataura[†] and Y. Achiba[†]

[*] *National Institute of Materials and Chemical Research, 1-1 Higashi, Tsukuba, Ibaraki, 305-8565,*
[†] *Faculty of Science, Tokyo Metropolitan University, 1-1 Minami-Ohsawa, Hachioji, Tokyo 192-0397, Japan*

Abstract. Electrochemical doping of single-wall carbon nanotube (SWNT) films and concomtant changes in their electronic states were investigated by in situ optical absorption and Raman spectroscopy as well as by ac resistance measurements using non-aqueous electrolytic solution. Reversible variation of these properties induced by the shift in the electrode potential demonstrated the practicability of fine-tuning of their electronic states.

INTRODUCTION

Owing to their unique one-dimensional structure and outstanding electronic properties, single-wall carbon nanotubes (SWNT) constitute a new class of materials that could contribute to the development of novel nano-scale electronic devices [1]. To fulfill such expectations, understanding their basic properties as well as establishing methods to control their valence electronic states is crucial.

In our previous studies, we elucidated the chemical doping processes in semiconducting and metallic SWNTs by monitoring the changes in their optical absorption spectra as well as of their dc resistance. In these experiments, alkali metals (electron donors) or halogens (electron acceptors) were doped into SWNT films with controlled stoichiometry. We showed that semiconducting SWNTs could be either n- or p-doped, demonstrating their amphoteric doping behavior [2].

In the present study, we carried out electrochemical doping of SWNT films while again in situ monitoring their optical absorption spectra and ac resistance. The advantage of this spectroelectrochemical approach is that the electronic structures of SWNT can be closely probed while finely tuning its electrochemical potential or in other word the Fermi level.

EXPERIMENTS

Electrochemical experiments were performed using working electrodes made of SWNT film deposited on top of quartz plates pre-coated with semi-transparent Pt film. Synthesis and deposition of SWNT film were described elsewhere [2,3]. Working

---

[1] Corresponding author's e-mail: kazaoui@nimc.go.jp

(SWNT/Pt), counter (Pt plate) and reference (Ag wire) electrodes were immersed in acetonitrile (CH_3CN) with lithium perchlorate (0.1 M $LiClO_4$) dissolved as a supporting electrolyte. We selected acetonitrile instead of water because of its stability in a large potential range and optical transparency in a wide spectral range. Nitrogen gas was flushed through the solution to purge oxygen. Measurements were carried out using a Solartron 1287 potentiostat/galvanostat apparatus. (Here after all mentioned electrode potentials are in volt versus Ag wire).

In situ optical absorption spectra of SWNT films were recorded in the transmission mode with the light beam passing through SWNT/Pt working electrodes set inside an electrochemical cell. By the use of 2-beam configuration of a Shimadzu 3100 spectrophotometer, absorption due to the SWNT film was properly extracted.

In-situ ac resistance measurements were performed on SWNT films deposited on quartz plates pre-coated with four Au electrodes. The use of ac for resistance measurements and dc for electrochemical control eliminated cross talk between the two circuits, as already described by Clay [4].

RESULTS AND DISCUSSIONS

The optical absorption spectrum of a pristine SWNT film is characterized by three main features at approximately 1800 nm (0.68 eV), 1000 nm (1.2 eV), and 700 nm (1.8 eV) superimposed on the broad absorption band centered at around 250 nm [2,3]. The features at 0.68 and 1.2 eV were attributed to electronic transitions between pairs of van Hove singularities in semiconducting SWNTs, $v_s^1 \square c_s^1$ and $v_s^2 \rightarrow c_s^2$, respectively. The feature at around 1.8 eV was assigned to the first pair of singularity $v_m^1 \rightarrow c_m^1$ in metallic SWNTs [5].

Figure 1 shows the absorption spectra of an SWNT film at constant electrode potentials. These data were collected in potentiostatic mode 500 s after the application of the potential to largely fulfill the steady state conditions for both current (generated in the electrochemical cell) and absorbance (at specific wavelength). As the electrode potential was shifted in the negative direction from -0.4 V, the absorption peaks started to diminish sequentially; first the peak at 1800, next at 1000, and then at 700 nm. When the potential was shifted in the positive direction from -0.4 V, a similar decrease of the absorption intensity was observed. It is remarkable that the optical absorption spectra of SWNT film underwent comparable changes either for anodic or cathodic polarizations.

At 1.0 V, all the absorption bands disappeared, but when the potential became even more positive (Fig. 1 inset), new broad peaks emerged: around 1070 nm (1.15 eV) at 1.4 V and 1000 nm (1.24 eV) at 1.8 V. In contrast, no new absorption features were observed at negative potentials, at least up to −1.4 V, with in our experimental conditions. At potentials larger than 1.4 V, we observed a weak but significant increase of the absorption background in the near-infrared region. The emergence of the new optical transitions in SWNT film at high electrochemical potentials is very reminiscent of doping-induced absorption observed for chemical doping from gas phase (at 1.07 and 1.3 eV for $CBr_{0.15}$ and $CCs_{0.10}$, respectively) [2]. Their origin can be rationalized since the depletion of high-lying valence states should enable new optical transitions from deep-lying valence states in both semiconducting and metallic SWNT, but their assignment to a specific electronic transition and their electrode potential dependence are still open to speculation.

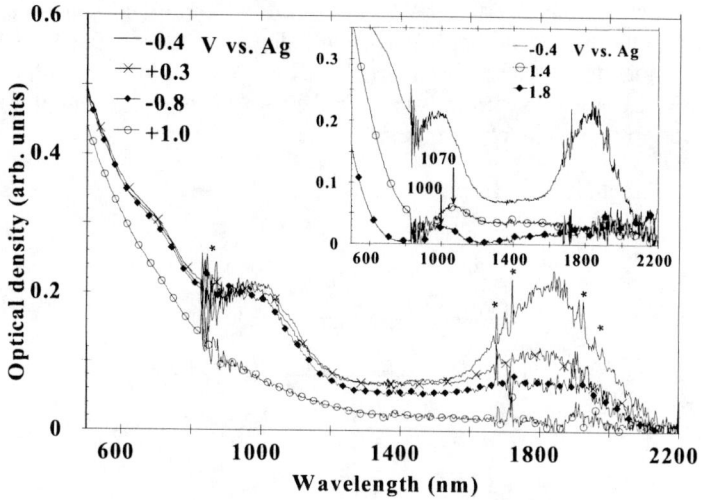

FIGURE 1. *In situ* absorption spectra of an SWNT film at several electrode potentials. * indicates features coming from the sharp absorption bands due to the solvent and also noises from the spectrometer.

Figure 2 displays the absorbance monitored at specific wavelengths versus the electrode potential. The absorbance at 0.68 eV shows a reasonably well-defined plateau (within ±5%) between approximately –0.5 and +0.1 V (0.6 eV width), and sharply decreases for both anodic and cathodic polarizations. A similar result was observed for the feature at 1.2 eV, where a larger plateau was identified between -0.9 and +0.2 V (1.1 eV width). We stress that the widths of the above-mentioned plateaus were comparable to the energies of the optical band gaps $v_s^1\text{-}c_s^1$ and $v_s^2\text{-}c_s^2$, respectively. The absorbance at 1.8 eV also showed significant potential dependence, but no clear plateaus could be resolved probably because of the larger background absorption. Figure 2 also presents the normalized ac electrical resistance (R) of an SWNT film as a function of the electrode potential. It shows a broad maximum around –0.3 V without a distinct plateau, and decreases steeply (by 70 %), on both sides.

The changes in the optical absorption and Raman spectra (not reported here) as well as ac resistance were fully reversible at least between -1.4 and 1.8 V. Beyond this range alterations were irreversible and accompanied by a degradation of the film (mainly peel off from the substrate).

Essentially the same results were obtained for different combinations of supporting electrolytes and solvents, such as sodium perchlorate ($NaClO_4$), tetraethylammonium perchlorate or tetra-n-butylammonium tetrafluoroborate in acetonitrile; and $NaClO_4$ in dimethyl sulfoxide or *N-N*-di methylformamide.

In conclusion, the sequential disappearance of the absorption features under anodic polarization (first 0.68, then 1.2 and finally 1.8 eV) reflects electron depletion of the occupied bands with the following order: first v_s^1 and then v_s^2 of the same semiconducting nanotube and finally v_m^1 in the metallic nanotube. In this context SWNT film undergoes electro-oxidation with concomitant intercalation of ClO_4^-. The decrease in R largely in parallel with the absorbance at 0.68 indicates that the conduction in the semiconducting SWNTs determined the overall electrical transport. A similar explanation can be applied for cathodic polarization, where electrons are injected into unoccupied energy levels of SWNT (electro-reduction with concomitant intercalation of Li^+). The electronic states in semiconducting and metallic SWNT are shown to undergo reversible charge transfer.

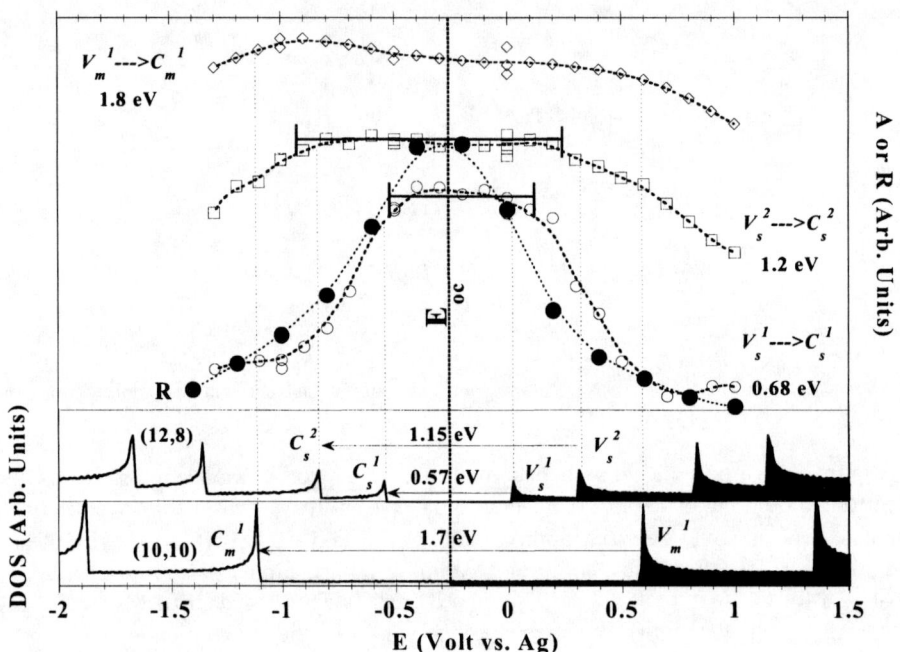

FIGURE 2. Absorbance (A) at several specific wavelengths (for convenience, curves are shifted and expanded along the vertical axis; plateaus are indicated) and normalized ac electrical resistance (R, black dotes) versus the electrode potential of an SWNT film. Electronic density of state (DOS) in a tight binding model (nearest-neighbor overlap integral γ_0=2.75 eV) for semiconducting (12,8) and metallic (10,10) nanotubes. The center of the band gap was brought into coincidence with the open-circuit potential E_{oc}=-0.26 (V vs. Ag). Lines are guides to the eye

REFERENCES

1. Dekker, C., *Phys. Today* **52(5)**, 22 (1999).
2. Kazaoui, S., Minami, N., Jacquemin, R., Kataura, H., and Achiba, Y., *Phys. Rev. B* **60**, 13339 (1999).
3. Kataura, H., Kumasawa, Y., Maniwa, Y., Umezu, I., Suzuki, S., Ohtsuka, Y., Achiba, Y., *Synth. Met.* **103**, 2555 (1999).
4. Claye, A.S., Fischer, J.E., Huffman, C.B., Rinzler, A.G., and Smalley, R.E., *J. Electrochem. Soc.* **147**, 2845 (2000).
5. Saito, R., Dresselhaus, G., and Dresselhaus, M.S., *Phys. Rev. B* **61**, 2981 (2000).

Synthesis of Carbon Nanocoils Using Electroplated Iron Catalyst

Lujun Pan[*], Mei. Zhang[*], Akio Harada[†], Yuichi Takano[†], and Yoshikazu Nakayama[*]

[*]Department of Physics and Electronics, Osaka Prefecture University,
1-1 Gakuen-cho, Sakai, Osaka 599-8531, Japan
[†]Daiken Chemical Co. LTD., 2-7-19, Nishi-hanade, Joto-ku, Osaka 536-0011, Japan

Abstract. Carbon nanocoils have been synthesized in great quantities by thermal chemical vapor deposition using the iron catalyst prepared by a convenient electroplating method on indium tin oxide substrates. The fine iron particles can be sufficiently mixed with the indium tin oxide by this method, which induces the high efficient growth of small-sized carbon nanocoils with relatively uniform diameters. The growth of carbon nanocoils is dependent on the deposition temperature, the flow rate of acetylene gas and the deposition time, which is due to the activity of the catalyst particles, the density of acetylene and/or hydrocarbon radicals in the reaction chamber and the tip growth mechanism.

INTRODUCTION

Carbon coils of nanometer-scale size or carbon nanocoils (CNC), due to their peculiar helical morphology, are expected in the fabricating of nanodevices such as a generator or detector of magnetic field, an actuator, a spring, etc. The formation of CNCs and their morphology have been reported in a number of papers [1-5]. They are synthesized by the high-temperature catalytic decomposition of hydrocarbons on a finely divided metallic catalyst such as Co, Fe, and Ni or their alloys, but the growth of CNCs was extremely accident, in very low yield and poorly reproducible.

Recently, we have successfully developed an effective method to synthesize the CNCs in high yield by catalytic thermal chemical vapor deposition (CVD) [6,7]. A CNC usually consists of two carbon nano-tubules. Excellent electromagnetic property and elasticity of CNCs can be expected. It is found that the CNCs grow in the interface of Fe and indium tin oxide (ITO) and their growth is strongly influenced by the states of Fe/ITO mixture [7]. The formation of Fe film on ITO is a key factor in the synthesis of CNCs. In this article, we report a convenient and effective electroplating method to coat Fe particles on ITO and study the growth of CNCs on such substrates. The preparation of Fe particles in this method instead of the vacuum vaporization is simple, low cost and is suitable for fabrication of the large area substrates.

Experimental

The substrates used were ITO-coated alumina plates. Fe was prepared on the substrate by the electroplating method, where the substrate was used as the cathode and a carbon rod as the anode. The electroplating liquid was a mixture of $Fe_2(SO_4)_3$, $FeCl_3$ and NH_4Cl solutions with a pH value of 4.5 to 6.0. The current density was controlled in the range of 40 to 50 mA/cm^2 and the temperature of the liquid was held on in 40 to 45°C. After Fe electroplating, the substrates were put into a tubule-shaped furnace and the thermal CVD was performed to synthesize the CNCs at atmospheric pressure [6]. We employ helium as the carrier gas and acetylene as the reaction gas. Acetylene gas with the flow rate of 30 to 60 sccm was introduced to the CVD chamber together with the 300 sccm helium gas at the reaction temperature of 650 to 800°C. The reaction time was changed from 15 to 30 min. The deposits were characterized by a scanning electron microscope (SEM) and the structures of the coils were observed by a transmission electron microscopy (TEM).

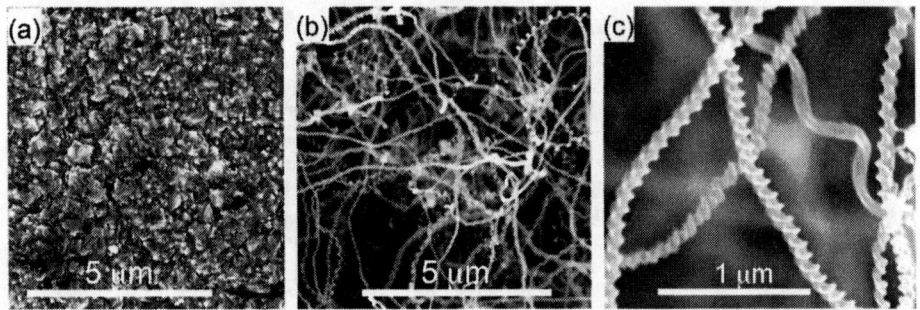

FIGURE 1. SEM micrographs of (a) a substrate after Fe electroplating, (b) the deposits on the substrate and (c) the partly enlarged image of (b).

Results and discussion

Figure 1(a) shows the SEM micrograph of a substrate after the Fe electroplating. It is clear that the ITO film is split into micro-pieces which are mixed well with the electroplated Fe particles. This state is more suitable for the growth of CNCs than that of Fe film prepared by the vacuum vaporization on ITO film, because the Fe film can not sufficiently mixed with the ITO film even at the reaction temperature[7]. This is the one of the advantages of the electroplating method. Figures 1(b) and (c) show the SEM micrographs of the coils grown by CVD at the temperature of 700°C for 30 min and its enlarged image. More than 95% deposits are carbon coils with various diameters and pitches. The external diameters of these coils are differed from each other, ranging from several tens to several hundreds of nanometers. Compared with the CNCs grown on the evaporated Fe film, the average diameter of the CNCs synthesized on the Fe electroplating substrate is smaller and their diameters are more concentrated, which are centered near 100 nm. This is due to the finely split ITO film and the relatively small and uniform diameter of the formed Fe nano-particles. This is another advantage of the electroplating method. Figure 2 reveals the TEM micrographs of some CNCs. It is clear that the coils consist of one or more coiled tubules. The spiral tubules twist well in a coil, which have the same pitch and a shift in phase. This result is similar to that of our previous study [6,7].

Figure 3 shows the SEM micrographs of the deposits grown at the different reaction

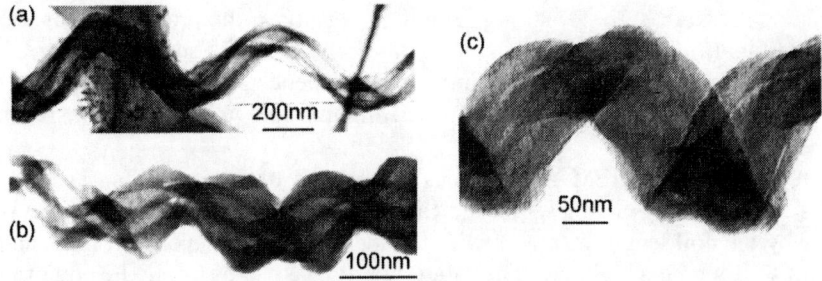

FIGURE 2. TEM micrographs of a CNC containing (a) only one spiral tubule, (b) two spiral tubules and (c) three spiral tubules, respectively.

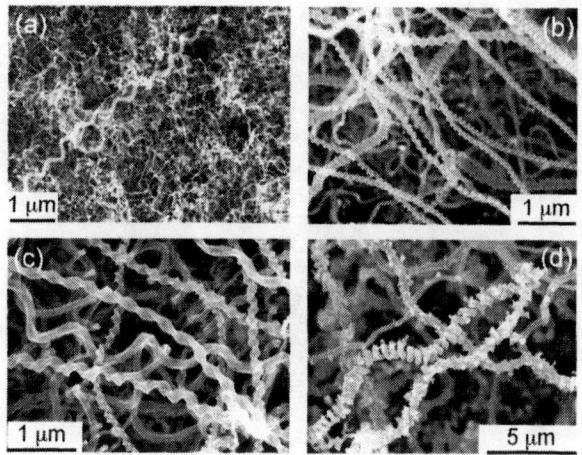

FIGURE 3. SEM micrographs of coils synthesized at the temperature of (a) 650°C, (b) 700°C, (c) 750°C and (d) 800°C, respectively.

temperatures for 30 min. In the case of 650°C (Fig. 3(a)), a large number of noodle-like nano-fibers are formed, but the yield of coils is very low. However in the cases of 700°C (Figs. 3(b) and 1(b)), a great number of CNCs grow out of the substrate maintaining their self-organization well during growth. When the reaction temperature is raised to 750°C (Fig. 3(c)), the high yield of CNCs is obtained, but the diameters of the CNCs are generally larger than those of 700°C. The coils become more flattened when the reaction temperature is raised to 800°C (Fig. 3(d)), and many amorphous carbon grains are deposited on the surfaces of the coils. The reasons for the overall results are considered to be the activity of the catalysts and the density of acetylene and/or resulted hydrocarbon radicals. At the low reaction temperature, the catalyst particles have no sufficient activity to react with acetylene and no enough resulted carbon can be precipitated to form perfect coils. At the temperature higher than 700°C, the catalyst particles become reactive to produce more carbon species, which are dissolved into and precipitated from the catalyst particles resulting in the high yield of CNCs. At the temperature of 800°C, the pyrolytic reaction of acetylene itself in atmospheric pressure becomes strong. Too high density of the produced hydrocarbon radicals results in the deposition of amorphous carbons on the surfaces of the CNCs. This growth mechanism is also confirmed by the dependence of deposits on the flow rate of acetylene gas, which shows that increasing the flow rate of acetylene results in a high yield of CNCs.

Figure 4 shows the SEM micrographs of the tips of some CNCs. The catalyst particles are at the tips of the coils indicating a tip growth mechanism. This is the reason why the coil length increases with the increase of the deposition time, and the growth of CNCs hold a high rate. The catalyst particle contains Sn, In, Fe and/or O [6]. There is a strong correlation between the constitution of catalyst particle and the structure of the resulted coil including its external diameter, pitch, number of tubules and the relative arrangement of the helix-shaped tubules. The growth mechanism of the CNCs is similar to that reported previously [7].

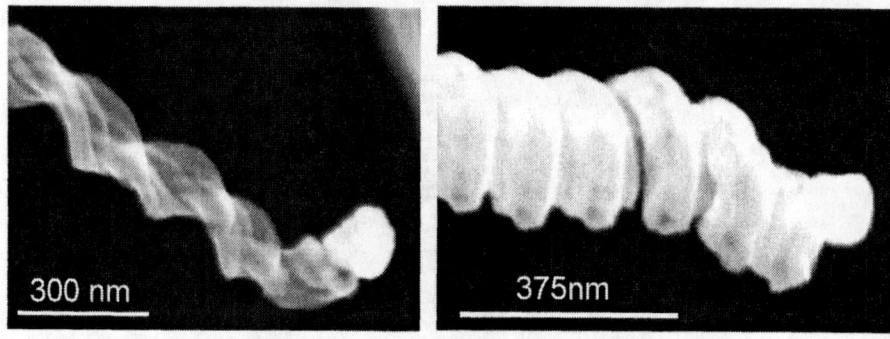

FIGURE 4. SEM micrographs of the tips of some CNCs.

Conclusions

In the process of synthesizing CNCs, the important Fe catalysts were prepared by the electroplating method instead of the vacuum vaporization. This method has the following advantages: (a) simple and convenient method without using any vacuum equipment, (b) low cost and large area substrate fabrication. (c) formation of finely divided Fe nano-particles sufficiently mixed with the ITO film, which results in the high yield of CNCs. The CNCs synthesized on such substrates have small sizes and relatively uniform diameters. This method is suitable for the industrial manufacture of CNCs.

REFERENCES

1. Davis, W. R., Slawson, R. J., and Rigby, G. R., *Nature* **171**, 756-757 (1953).
2. Baker, R. T. K., and Chludzinski, J. J., *J. Catal.* **64**, 464-478 (1980).
3. Motojima, S., Kawaguchi, M., Nozaki, K., and Iwanaga, H., *Appl. Phys. Lett.* **56**, 321-323 (1990).
4. Rodriguez, N. M., *J. Mater. Res.* **8**, 3233-3250 (1993).
5. Amelinckx, S., Zhang, X. B., Bernaerts, D., Zhang, X. F., Ivanov, V., and Nagy, J. B., *Science* **265**, 635-639 (1994).
6. Zhang, M., Nakayama, Y., and Pan, L., *Jpn. J. Appl. Phys.* **39**, L1242-L1244 (2000).
7. Pan, L., Zhang, M., and Nakayama, Y., submitted to *J. Appl. Phys.*

Preparation of Carbon Nanotubes by Using Mesoporous Silica

Shinji Kawasaki, Shingo Komiyama, Shigekazu Ohmori,
Akifumi Yao, Fujio Okino and Hidekazu Touhara

Faculty of Textile Science & Technology, Shinshu University
3-15-1 Tokida, Ueda 386-8567, Japan

Abstract. Carbon nanotubes were synthesized using several mesoporous silica with different pore sizes as the templates. TEM observation confirmed that carbon nanotubes with homogeneous diameter can be obtained by the present method. However, it was hard to find the correlation between the pore size of the templates and the diameter of the prepared carbon nanotubes.

INTRODUCTION

Carbon nanotubes have attracted much attention, not only because of their elegant structure, but also because of their promising applications demonstrated by electron field emission sources, nano-electric components etc. Although there are many methods to prepare nanotubes such as laser-ablation and arc-discharge, one of the most important problems is how to prepare nanotubes with desired form. Another significant problem is how to obtain such nanotubes in a large scale.

Recently, several techniques for making highly ordered periodic mesoporous silica with pore sizes of 2-30 nm have been developed [1,2]. If the mesoporous silica can be used as a template to prepare carbon nanotubes, it is possible to control the diameter of carbon nanotubes with the tunable mesopore size of the mesoporous silica. In the present study, we report on the preparation method of carbon nanotubes using mesoporous silica that is based on the chemical vapor deposition (CVD) procedure in order to overcome the above mentioned problems.

EXPERIMENTAL

Mesoporous silica with mesopores of 2-4 nm and 6-20 nm in diameter were synthesized in the presence of cetyltrimethylammmonium bromide (CTAB, Aldrich) surfactant and triblock poly(ethylene oxide)-poly(propylene oxide)-poly(ethylene oxide) $(P(EO)_{20}\text{-}P(PO)_{70}\text{-}P(EO)_{20})$ copolymer (Pluronic P123, BASF), respectively.

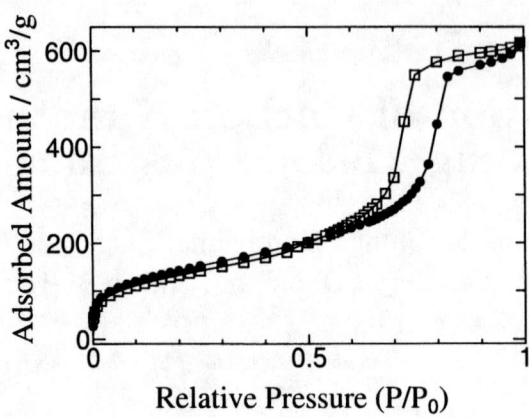

FIGURE 1. The observed N_2 adsorption-desorption isotherm for sample P-5 (Table 1).

Trimethylbenzene (TMB) was used as an organic swelling agent to enlarge the mesopore. The solution mixture was aged for 24 h at 353 ~ 383 K. Calcination was carried out by slowly raising the temperature from room temperature to 773 ~ 1073 K.

CVD experiments were performed as follows. A mesoporous silica sample was evacuated in a SiO_2 tube at 383 K for 30 min. Then the reactor was filled with nitrogen gas and the temperature of the reactor was raised to 1073 K. Propylene and nitrogen gases were flowed for 2 h at flow rate of 2.5 cm^3/min and 250 cm^3/min, respectively. After the reaction, the reactor was cooled to room temperature. The mesoporous silica with carbon deposit was washed with an excess amount of HF solution at room temperature so as to dissolve silica.

The synthesized mesoporous silica and the CVD treated samples were characterized by XRD, N_2 adsorption-desorption isotherm measurements and TEM observation carried out on Rigaku RINT2200, Shimadzu GEMINI2375 and JEOL JEM2010, respectively.

RESULTS & DISCUSSION

A typical example of the observed N_2 adsorption-desorption isotherm of the synthesized mesoporous silica at 77 K is shown in Figure 1. We determined the pore size distribution from the isotherm by Barrett-Joyner-Halenda (BJH) method. The peak values of the distributions are summarized in Table 1. It was found by XRD measurements that all the mesoporous silica prepared in the present work are hexagonal phases. $d_{(100)}$ values are also indicated in Table 1.

No carbon nanotube was observed by TEM observation in the CVD treated samples C-1, 2, 3 (Table 1) which were synthesized with CTAB. However, in the case of sample P-5 synthesized with Pluronic P123, a small amount of nanotubes was

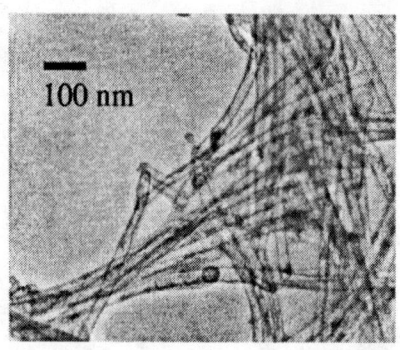

FIGURE 2. The TEM image of the prepared carbon nanotubes.

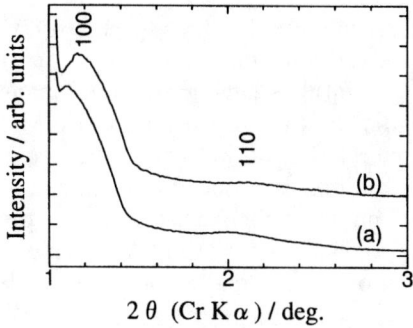

FIGURE 3. The observed XRD patterns of sample P-5 (a) before and (b) after CVD treatment.

found in the TEM image (Figure 2). EDS spectrum from the nanotubes indicated that they are made only of carbon atoms. As shown in Figure 2, the diameter and length of the nanotubes are 20 ~ 30 nm and > 1 μm, respectively. It was found by high resolution TEM observation that the wall thickness of the nanotubes is 4 ~ 5 nm and that the walls of the carbon nanotubes are not well graphitized. Figure 3 shows the XRD patterns of sample P-5 before and after CVD treatment. The ordered (hexagonal) structure of the mesoporous silica was maintained after the CVD treatment, although the diffraction peaks shifted toward the higher angle side. The shrinkage of pore-to-pore distance from 13.7 nm to 12.9 nm by the CVD treatment occurred due to the difference between the calcination temperature and the reaction temperature. However, this shrinkage is so small that the mesopores should not be closed. Therefore, it is considered that carbon deposit onto the surface of the mesopores should have occurred. However, the expected carbon nanotubes having diameter of about 10 nm which corresponds to the pore size of the template were not observed. HF treatment to remove the template may cause the destruction of the carbon nanotubes of the 10 nm diameter. Why have the carbon nanotubes whose diameters are much larger than 10 nm been produced?

TABLE 1. Physicochemical properties and synthesis conditions of the synthesized mesoporous silica

sample name[b]	TMB/surfactant weight ratio	aging temp. (K)	calcination temp. (K)	pore size[c] (nm)	$d_{(100)}$ (nm)	wall thickness[a] (nm)
C-1	0	353	773	2.0	3.4	1.9
C-2	3	353	873	2.9	4.4	2.2
C-3	5	353	873	3.5	4.9	2.2
P-4	0	353	873	6.4	9.6	4.7
P-5	0	383	873	10.0	11.9	3.7
P-6	1	373	1073	20.0		

[a] Calculated by a_0−pore size ($a_0 = 2d_{(100)}/\sqrt{3}$)
[b] Samples C-m and P-n were synthesized with CTAB and P123, respectively
[c] Calculated from adsorption branch of the N_2 isotherm by BJH method

Unfortunately, at the moment, it is hard to speculate on the mechanism of the production. We have done the same experiments for sample P-4 and P-6. In the case of P-4, some carbon nanotubes having ∼20 nm diameter were observed. On the other hand, we could not find any nanotubes in the CVD treated sample P-6. It is difficult to discuss the relation between the pore size of the template and the diameter of the prepared carbon nanotubes at present. In order to understand the mechanism of the carbon nanotube formation in the present method, not only geometrical scale such as pore size of the template but also surface structure of the pore such as concentration of OH groups should probably be taken into account.

ACKNOWLEDGEMENT

This work was supported in part by a Grant-in-Aid for COE Research (10CE2003) by the Ministry of Education, Science, Sports and Culture of Japan and in part by Tokuyama Science Foundation.

REFERENCES

1. Kresge C. T., Leonowicz M. E., Roth W. J., Vartuli J. C., and Beck J. S., *Nature* **359**, 710 (1992).
2. Zhao D., Huo Q., Feng J., Chmelka F., and Stucky G. D., *J. Am. Chem. Soc.* **120**, 6024 (1998).

Gas dynamic and time resolved imaging studies of single-wall carbon nanotubes growth in the laser ablation process

Rahul Sen[a], S. Suzuki[a], H. Kataura[b], Y. Achiba[a],*

[a]*Department of Chemistry and* [b]*Department of Physics, Tokyo Metropolitan University, 1-1 Minami-Osawa, Hachioji, Tokyo 192-0397, Japan*

Abstract Single-wall carbon nanotubes (SWNTs) were synthesized by laser ablation of Ni-Co-graphite composite targets at 1200°C under flowing argon. The effects of the temperature gradient near the target and the gas flow rate on the diameter distribution of SWNTs were studied in order to understand their growth dynamics. The diameter distribution of the SWNTs, analyzed by Raman spectroscopy, was dependent on the gas flow rate when there was a temperature gradient around the target. Time resolved scattering images from the ablated species at different flow rates indicated that velocities of backward moving species increased with increasing flow rate. These findings are used to estimate the time required for nucleation and the growth of SWNTs.

INTRODUCTION

Diameter control of single-walled carbon nanotubes (SWNTs) is important for various applications such as nanoelectronics. Laser ablation is a useful technique to grow high quality SWNTs. However, the diameters and helicities of the nanotubes vary considerably for a given set of growth conditions [1,2]. Various groups have reported that parameters such as growth temperature and type of catalyst are important in changing the diameter distribution of the SWNTs [3-5]. Despite these reports the process and mechanism of SWNT growth is not well understood. Therefore, it is important to understand the growth process of the SWNTs to obtain a better control on their structure. We have studied the effect of the temperature gradient around the target and the gas flow rate on the diameter distribution of SWNTs obtained by the laser ablation process. We also report the scattering images of the laser-ablated species recorded by time-resolved photography. From these results we can derive an estimate for the time period for the nucleation and growth SWNTs.

EXPERIMENTAL

SWNTs were grown by the laser ablation of Ni(0.6 at.%)-Co(0.6 at.%)-graphite composite targets placed in an electric furnace. The laser ablation was carried out at 500 Torr argon gas maintained at a desired flow rate and the furnace temperature was 1200°C. Second harmonics of a Nd:YAG laser (532 nm, 10 Hz, 300 mJ/pulse) focused to a 5 mm diameter spot on the target was used for the ablation. During the laser ablation the flowing argon gas carried the carbon products downstream, where a mat-like material was found to deposit on the molybdenum rod near the outlet region of the furnace. This material was collected by scraping and analyzed by Raman spectroscopy. Raman scattering was measured using a 1 m double monochromator, Jobin Yvon U-1000, and a photo-multiplier, Hamamatsu Photonics R943-02. The laser excitation line for Raman spectroscopy was 488 nm, and the spectral resolution was 4 cm^{-1}. The radial breathing mode (RBM) of the Raman spectrum was used for estimation of the diameter distribution of SWNTs. Scattering images from the laser ablated species were recorded as follows. The ablation pulse was the fundamental of Nd-YAG laser (1064 nm, ≈10 ns pulse width), and the scattering images were observed due to the second harmonic (532 nm, long pulse mode ≈200 µs pulse width) operating at a fixed time delay. The ablation and scattering lasers were co-axial, and the images were collected at a direction perpendicular to them. The images were recorded by a high-speed video camera (Kodak Ektapro HS motion analyzer 4540), and the recording rate was 9000 frames/s (≈100 µs time window for each frame).

RESULTS AND DISCUSSION

The temperature profile in the furnace was measured by a thermocouple at different positions inside it. When the target was placed at the furnace center (0 mm position), its temperature was ~1200°C and the temperature around it was uniform in a region of –40 to +40 mm. When the target was placed at –80 mm (upstream position), its temperature was ~1150°C and there was a large temperature gradient around it, with the temperature increasing towards the furnace center. When the target was placed at the furnace center, the flow rate had only a small affect on the diameter distribution. When the target was placed at –80 mm, however, the flow rate strongly affected the relative yields of SWNTs having different diameters. Fig. 1 (a), (b) and (c) shows the RBM for SWNTs grown at different flow rates by placing the target at –80 mm position. It is observed that increase in the flow rate increased the relative yields of large-diameter SWNTs and decreased the yields of small-diameter SWNTs. In the 0 mm case, the temperature gradient around the target was uniform and the diameter distribution was essentially unaffected by the flow rate. In the –80 mm case, however, the temperature gradient around the target was larger and the relative yields of SWNTs were more sensitive to the flow rate.

In order to understand this flow rate effect, we measured the scattering images of the laser-ablated species at different flow rates. Fig. 1(d) shows a series of

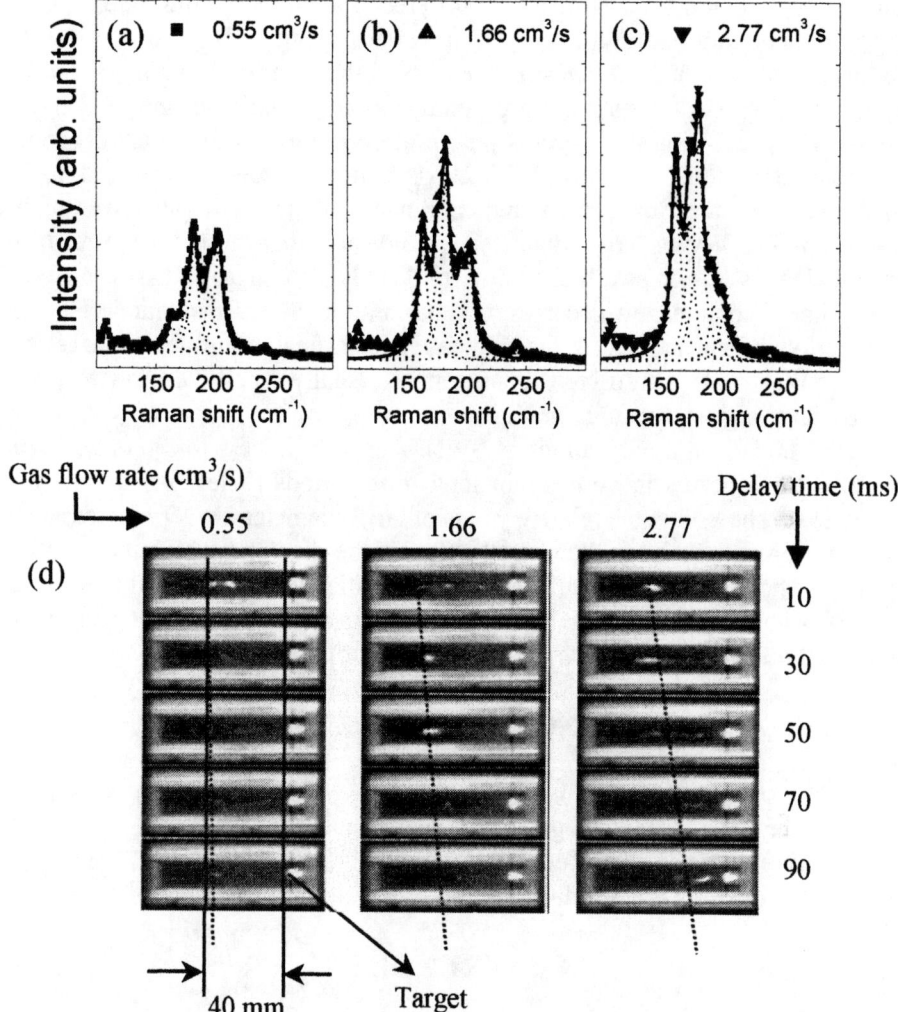

Figure 1. RBM of the Raman spectra for SWNTs grown at a target position of −80 mm at different flow rates; (a) 0.55, (b) 1.66 and (c) 2.77 cm^3/s. (d) Time-resolved scattering images from the laser-ablated products at different flow rates. The direction of the laser beam and gas flow is from left to right and the dotted line is a guide to see the movement of the species.

photographs taken by a high-speed video camera at different time intervals after the laser-ablation. Fig. 1(d) shows that 10 ms after the ablation the ablated species has moved about 40 mm from the target, and this distance is independent of the flow rate. This initial movement, caused by the ablation process, is in opposite direction to the gas flow and is termed forward movement. Images taken after 30 and 50 ms show that the particles have started moving in the direction of the gas flow, termed backward motion. From these time-resolved scattering images, the velocities of the backward moving species are estimated to be about 38, 123 and 255 mm/s for the flow rates of 0.55, 1.66 and 2.77 cm^3/s respectively. Since the velocities of the backward moving species increases with the flow rate, their positions inside the furnace after a certain time interval are different for different flow rates. This is clear from the bottom frames for different flow rates in Fig. 1(d). When the target is placed at the center, this difference in position of the backward moving species at different gas flow rates does not correspond to a large change in the temperature inside the furnace. When the target is placed at –80 mm, however, this difference means that after about 100 ms the backward moving species are at a higher temperature when the flow rate is higher. Since a higher flow rate gives higher yields of SWNTs having larger diameters, it indicates that the temperature of the growth species is higher at a higher flow rate. This also suggests that the SWNT growth process continues to occur even 100 ms after the laser ablation. Thus the growth period is long enough for the flow rate to significantly influence the relative yields of SWNTs having different diameters, especially when the growth species flow through a large temperature gradient.

Recent studies on the dynamics of SWNTs growth process [6-8] conclude that SWNTs' growth occurs in vortexes in long time periods (a few ms to a few s). Figure1 (a)-(c) shows that the relative yields of large-diameter SWNTs increase with increasing flow rate whereas the overall range of the diameter distribution remains unaffected. Therefore, we conclude here that the initial condensation and nucleation processes, which may determine the range of diameter distribution, occurs much faster (<100 ms) but the growth continues for 100 ms or more.

ACKNOWLEDGEMENTS

This work was supported by the Japan Society for the Promotion of Science, "Research for the Future Program". Partial support was obtained from the Grant-in-Aid for Scientific Research on the Priority Area "Fullerenes and nanotubes" by the Ministry of Education, Science, Sports and Culture of Japan.

REFERENCES

1. Thess, A., Lee, R., Nikolaev, P., Dai, H., Petit, P., Robert, J., Xu, C., Lee, Y.H., Kim, S.G., Rinzler, A.G., Colbert, D.T., Scuseria, G.E., Tomanek, D., Fischer J.E., and Smalley, R.E., *Science* **273,** 483 (1996).
2. Pimenta, M.A., Marucci, A., Brown, S.D.M., Matthews, M.J., Rao, A.M., Eklund, P.C.,. Smalley, R.E, Dresselhaus G., and Dresselhaus, M.S., *J. Mater. Res.* **13,** 2396 (1998).
3. Bandow, S., Asaka, S., Saito, Y., Rao, A.M., Grigorian, L., Richter E., and Eklund, P.C., *Phys. Rev. Lett.* **80,** 3779 (1998).
4. Kataura, H., Kimura, A., Ohtsuka, Y., Suzuki, S., Maniwa, Y., Hanyu T., and Achiba, Y., *Jpn. J. Appl. Phys.* **37,** L616 (1998).
5. Jost, O., Gorbunov, A.A., Pompe, W., Pichler, T., Friedlein, R., Knupfer, M., Reibold, M., Bauer, H.-D., Dunsch, L., Golden M.S., and Fink, J., *Appl. Phys. Lett.* **75,** 2217 (1999).
6. Gorbunov, A.A., Friedlein, R., Jost, O., Golden, M. S., Fink J., and Pompe, W., *Appl. Phys. A* **69,** S593 (1999).
7. Kokai, F., Takahashi, K., Yudasaka M., and Iijima, S., *J. Phys. Chem. B* **104,** 6777 (2000).
8. Puretzky, A. A., Geohegan, D. B., Fen X., and Pennycook, S. J., *Appl. Phys. A* **70,** 153 (2000).

New Simple Method of Carbon Nanotube Fabrication Using Welding Torch

Hirofumi Takikawa[*], Yoshitaka Tao[*], Yoshihiko Hibi[*], Ryuichi Miyano[*], Tateki Sakakibara[*], Yoshinori Ando[†], Shigeo Ito[‡], Kaori Hirahara[¶], Sumio Iijima[†¶]

[*]*Department of Electrical and Electronic Engineering, Toyohashi University of Technology, Toyohashi, Aichi, 441-8580, Japan*
[†]*Department of Materials Science and Engineering, Meijo University, Nagoya, 468-8502, Japan*
[‡]*Product Development Center, Futaba Corporation, Chiba, 299-4395, Japan*
[¶]*Japan Science and Technology Corporation (JST), International Cooperative Research Project (ICORP), Japan*

Abstract. A new, simple method of carbon nanotube fabrication was developed. The method employed a tungsten-electrode-inert-gas (TIG) welding arc torch, with a graphite electrode used instead of a conventional tungsten electrode in order to prevent contamination of the tungsten vapor and droplets. The substrates used as counterelectrodes for the torch arc were pure graphite and catalyst (Ni, Y)-mixed graphite. The torch arc was operated in open air with both DC and AC modes. Nanotubed surfaces were obtained on catalyst-mixed graphite with DC and AC modes, and on pure graphite with the AC mode.

INTRODUCTION

Carbon nanotubes were first discovered in the cathode deposit of low-pressure arc producing fullerene [1]. Since then, various investigations have been carried out in low-pressure arcs with a homoelectrode system of graphite (C) cathode and C anode [2-5]. The authors have examined carbon nanotube fabrication with a low-pressure heteroelectrode arc (C-molybdenum (Mo) electrode system) [6], cathodic vacuum arc with inert anode [7,8], running and jumping arcs under a magnetic field [9,10], and a catalytic heteroelectrode arc in low pressure [11]. In these experiments, the arc was discharged for 1-2 s. These experiments produced the following results. When C is used for the cathode material, the nanotubes are formed at the cathode spot, regardless of anode material, ambient gas species, and pressure. Also, the nanotubes do not readily form at the anode spot of pure C anode electrodes. However, if the C anode material contains a metal catalyst, the nanotubes can be formed at the anode spot.

These results indicate that if a catalyst-mixed C anode is used, nanotubes can be produced on the anode surface even in open air. Based on this, the present study provides a new, simple method for preparing nanotubes, using a welding arc torch operated in an open-air environment.

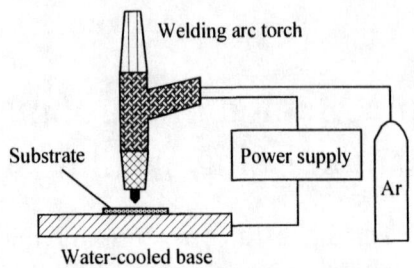

FIGURE 1. Schematic diagram of experimental setup for arc torch method. The arc torch was a conventional one for tungsten-electrode-inert-gas (TIG) arc welding. However, the tungsten (W) electrode rod of the torch was replaced with pure graphite of 3 mm in diameter in order to prevent the contamination of W vapor and droplets. When the arc was operated in DC mode, the substrate was the anode of the arc discharge.

EXPERIMENTAL

The experimental setup is depicted in Figure 1. The arc was discharged to a substrate placed on a water-cooled base electrode using a conventional arc torch. The substrates (1-2 mm in thickness), which are actually counterelectrodes of the torch arc, were pure graphite (C) and graphite containing metal catalysts (nickel (Ni), 4.2 wt% and yttrium (Y), 1.0 wt%) (C-Ni/Y). The arc was operated in open air (1 atm) in the DC or AC (60 Hz) mode at an arc current of 100 A. Shielding argon (Ar) gas of 1.8 l/m was flowed through the torch. The gap length between the torch electrode and the substrate was 1 to 3 mm. The arcing period was approximately 1-2 s.

The arced surface of the substrate was observed using a digital hi-scope (Hirox, Co. Ltd., KH2400) and high-resolution scanning electron microscope (HR-SEM; Topcon, ABT-150F), and the products on the surface were observed using a high-resolution transmission electron microscope (HR-TEM; JEOL, JEM-2010F). Hereafter, DC-C, DC-cat, AC-C, and AC-cat indicate the cases in which the C substrate was processed in the DC mode, the C-Ni/Y substrate was processed in the DC mode, the C substrate was processed in the AC mode, and the C-Ni/Y substrate was processed in the AC mode, respectively.

RESULTS

Figure 2 shows photographs and micrographs of the arc spots generated on the substrates. The diameters of the arc spots were found to be about 4 mm, 5 mm, 2 mm, and 5 mm for DC-C, DC-cat, AC-C, and AC-cat, respectively. The diameter for the AC-C was considerably smaller than that for others.

Macroscopic and microscopic views of the arced surface were noticeably different according to the current mode and substrate. With DC-C, nanotubes and fibers were rarely observed. This is consistent with the previous results that the nanotubes were

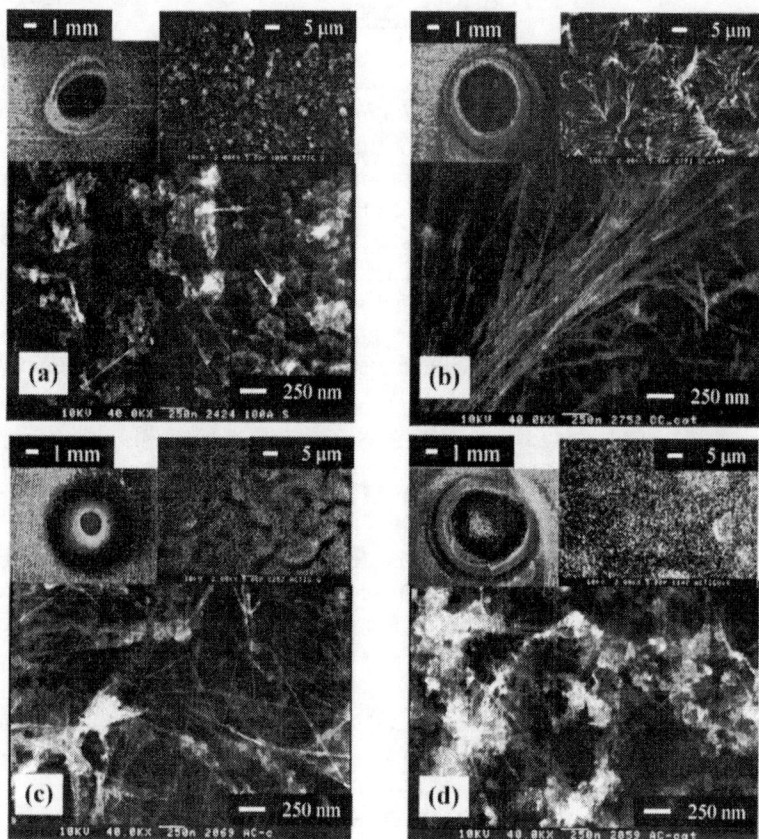

FIGURE 2. Morphology of arc spot; (a) DC-C, (b) DC-cat, (c) AC-C, and (d) AC-cat. Left upper images are hi-scope photographs of the arc spot, and right upper and lower images are low- and high-magnification HR-SEM micrographs of the arc spots, respectively, in each figure.

seldom observed on a C anode surface in a low-pressure arc [6, 11]. With DC-cat, AC-C, AC-cat, nanotubes were observed. With DC-cat, the surface consisted of abundant nanotubes and a macroscopic appearance of the surface was like sharp waves. The nanotubes were relatively fat, long, straight, and bundled, and there were few polyhedral nanoparticles. With AC-C, a macroscopic view revealed an appearance like moss and thinner nanotubes were dispersed on the surface with a considerable number of nanoparticles. With AC-cat, much thinner nanotubes were observed, though a large number of nanoparticles also existed.

All nanotubes observed in the present study were confirmed to be multi-wall carbon nanotubes (MWNTs) by HR-TEM observation. It is noted that single wall carbon nanotubes (SWNTs) are produced in a conventional low-pressure arc with C-Ni/Y anode, though SWNTs are not produced in this arc torch method even when C-Ni/Y anode is used.

CONCLUSIONS

A new method for producing carbon nanotubes on C and C-Ni/Y substrates, using a welding arc torch, was presented. The principal findings were as follows. The morphology of the arced surface on the substrate was considerably different according to the current mode of the arc and substrate species. Many nanotubes were obtained on the C-Ni/Y substrate with a DC arc, and on the C and C-Ni/Y substrates with an AC arc, but few on the C substrate with a DC arc. The greatest number of nanotubes was obtained on the C-Ni/Y substrate with a DC arc. All nanotubes obtained using this method were MWNTs, even if the catalyst was used in a carbon arc discharge. The major advantage of this method is an open-air operation. This brings a low-cost apparatus and a design flexibility of production unit, comparing with the conventional arc method.

ACKNOWLEGEMENTS

The authors would like to thank Professor Takayoshi Kubono of Shizuoka University for the use of the digital hi-scope. H. T. would like to express his appreciation to the Futaba Electronics Memorial Foundation for partial financial support.

REFERENCES

1. Iijima, S., *Nature* **354**, 56-58 (1991).
2. Takikawa, H., Coronel, A.M., Sakakibara, T., *Trans. IEE Jpn.* **119-A**, 901-902 (1999).
3. Ebbesen, T.W., "Production and purification of carbon nanotubes", in *Carbon Nanotubes: Preparation and Properties*, edited by T.W. Ebbesen, CRC Press, Inc., 1997, pp.139-162.
4. Yumura, M., "Synthesis and Purification of Multi-walled and Single Walled Carbon Nanotubes", in *The Science and Technology of Carbon Nanotubes*, edited by K. Tanaka, T. Yamabe, and K. Fukui, Elsevier, 1999, pp.2-13.
5. Harris, P.J.F., *Carbon Nanotubes and Related Structures: New Materials for the Twenty-First Century*, Cambridge University Press, Cambridge, 1999, pp.16-60.
6. Takikawa, H., Kusano, O., and Sakakibata, T., *J. Phys. D: Appl. Phys.* **32**, 2433-2437 (1999).
7. Takikawa, H., Yatsuki, M., Kusano, O., and Sakakibara, T., *Trans. IEE Jpn.* **119-A**, 1156-1157 (1999).
8. Takikawa, H., Yatsuki, M., Sakakibara, T., and Ito, S., *J. Phys. D: Appl. Phys.* **33**, 826-830 (2000).
9. Takikawa, H., Tao, Y., Miyano, R., Sakakibara, T., Ando, Y., and Ito, S., *Trans. Mater. Res. Soc. Jpn.* **25**, 873-876 (2000).
10. Takikawa, H., Tao, Y., Miyano, R., Sakakibara, Zhao, X., T., Ando, Y., and Ito, S., *Jpn. J. Appl. Phys.* (in press).
11. Takikawa, H., Tao, Y., Miyano, R., Sakakibara, T., Ando, Y., Zhao, X., Hirahara, K., Iijima, S., *J. Mater. Sci. Eng. C* (in press).

Temperature and Time Dependence of the Growth of Carbon Nanotubes by Thermal Chemical Vapor Deposition

Hee Jin Jeong[1], Young Min Shin[2], Keun Soo Kim[1], Seung Yol Jeong[1], Young Soo Park[2], Young Chul Choi[3], and Young Hee Lee[2]

[1]*Department of Semiconductor Science and Technology, Jeonbuk National University, Jeonju 561-756, Korea*
[2]*Department of Physics, Sungkyunkwan University, Suwon 440-746, Korea*
[3]*Material Tech. Lab., CRD center, Samsung SDI, Suwon 442-391, Korea*

Abstract. Multiwalled carbon nanotubes(CNTs) were synthesized on Ni-coated Si substrates by thermal chemical vapor deposition (CVD) with an acetylene gas. Ni thin film with a thickness of 120 nm was prepared at 200 °C under a pressure of 3×10^{-3} Torr by RF magnetron sputtering. The CNTs grown at a growth temperature of 650 °C were uniformly aligned with high tube density. We obtained relatively less aligned CNTs with lower tube density with increasing growth temperatures, in contrast to the general belief. This was explained by the change of the surface morphology of the metal layers, where with increasing the growth temperatures the metal grains were coalesced and thus the grain size became bigger, reducing the grain density. The length of the CNTs increased with increasing growth time and saturated at long growth time. The growth termination at large growth time is explained by the cap growth mechanism.

INTRODUCTION

Since its discovery in 1991[1], CNT has attracted great attention for its unique physical and chemical properties, and applications to nanoscale devices. There have been several methods for the synthesis of CNTs such as arc-discharge[2,3], laser ablation[4], pyrolysis[5], and thermal chemical vapor deposition(CVD)[6,7]. Especially, vertically aligned CNTs on large substrate area using thermal CVD are applied to the field emitters[8,9]. For this application, we need to grow vertically aligned CNTs on the glass substrate at low temperature. In general, vertically aligned CNTs are believed to be uniformly synthesized at high temperature rather than at low temperature, and the length of CNTs increases with increasing growth time. In this study, we synthesized relatively less aligned CNTs with lower density with increasing growth temperatures, and that the length of CNTs is saturated at high growth time, in contrast to the general belief.

EXPERIMENTAL

CNTs were synthesized by thermal CVD using C_2H_2 gas at growth temperatures of 600, 650, 700 and 750 °C. It has been commonly recognized that the growth of CNTs can be easily achieved at high temperature[10]. However, we found anomalous temperature dependence in the CNT growth. Ni thin films (with a thickness of 120 nm) deposited on TiN/Si substrates by RF magnetron sputtering were used as a catalyst metal. Ar ambient was used to avoid an oxidation of Ni during the temperature increase. No sooner than the temperature reached the growth temperature, C_2H_2 gas was injected into the chamber, resulting in the total pressure of 2.5 torr. The growth time was fixed at 20 min.

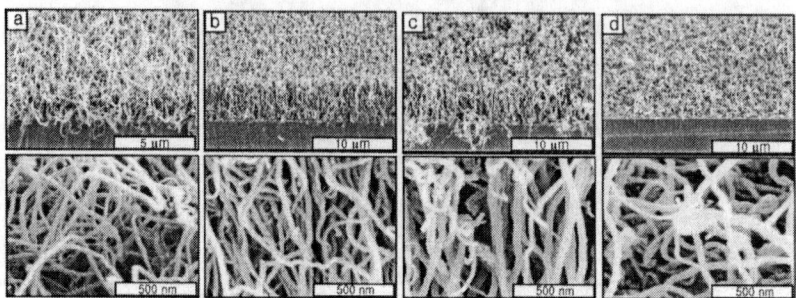

FIGURE 1. SEM images of CNTs with low and high magnifications grown at temperatures of (a) 600 °C, (b) 650 °C, (c) 700 °C and (d) 750 °C.

RESULTS AND DISCUSSION

Figure 1 shows the SEM images of CNTs grown at 600, 650, 700 and 750 °C. CNTs grown at 650 °C were more uniformly aligned with high tube density than those grown at higher temperatures. With increasing growth temperature, the density of CNTs decreases. CNTs grown at 600 °C, however, have lower tube density than those grown at 650 °C. The carbon atoms didn't diffuse enough on transition metal at 600 °C. Therefore, the growth rate and density of CNTs relatively decrease. The average diameter of CNTs grown at 600, 650, 700 and 750 °C are about 30, 30, 50 and 50 nm, as shown in the SEM images with high magnification. The observed temperature dependence of CNT growth is not in agreement with the reported results that the CNTs grown at high temperature have better quality[10]. This phenomenon can be explained by the difference in surface morphology of Ni thin films used as catalyst on Si substrate.

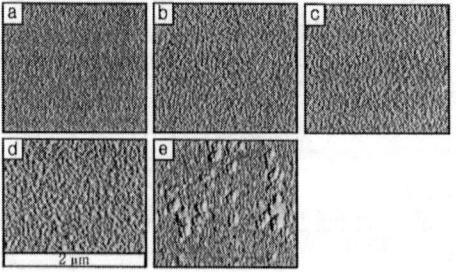

FIGURE 2. AFM images of Ni thin films after heat-treatment at various temperatures of (a) as-deposited, (b) 600 °C, (c) 650 °C, (d) 700 °C and (e) 750 °C.

Figure 2 represents the surface morphologies of the RF sputtered Ni thin films investigated by AFM. The as-deposited sample has very uniform grain size of about 20 nm [Figure 1(a)]. With increasing temperatures, the grain size and surface roughness increase, as seen in Figs. 1(b), (c) and (d). The average grain size of the heat-treated films at 600, 650, 700 and 750 °C are about 30, 30, 50 and 50 nm, respectively. The Ni films heat-treated at 600 and 650 °C still have somewhat uniform grain size and surface roughness. However, some grains are coalesced, resulting in the non-uniformly distributed large particles at 750 °C. This may be the reason that CNTs grown at 650 °C are uniformly aligned, whereas CNTs grown at 750 °C are not well aligned and sparse. It is therefore concluded that the surface morphology of catalyst metal is very important for the growth of CNTs using thermal chemical vapor deposition method, as expected from the previous reports [11,12].

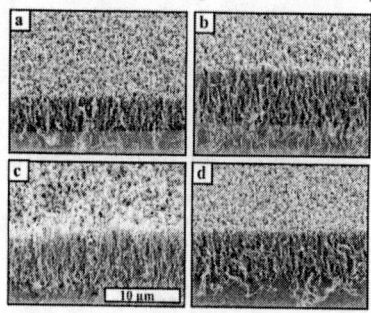
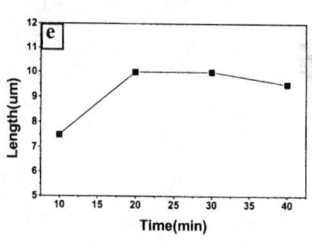

FIGURE 3. SEM images of CNTs grown at 650 °C for a growth times of (a) 10 min, (b) 20 min, (c) 30 min, (d) 40 min, and (e) the length of CNTs as a function of the growth time.

Figure 3 shows the SEM images of CNTs grown for (a) 10, (b) 20, (c) 30, and (d) 40 min at 650 °C. The lengths of the CNTs grown for 10 and 20 min are about 5.5 and 10 μm, respectively. However, the length of nanotubes did not increase further with the growth time. The length of CNTs grown for 30 min is almost the same as that grown for 20 min. Moreover, Fig. 3(d) shows that the length of CNTs is a little smaller than 10 μm. It means that the CNTs do not grow further and instead the etching process begins to occur after saturation. In order to clarify this phenomenon, the tip morphology of the nanotubes was investigated using high-resolution (HR) TEM measurements.

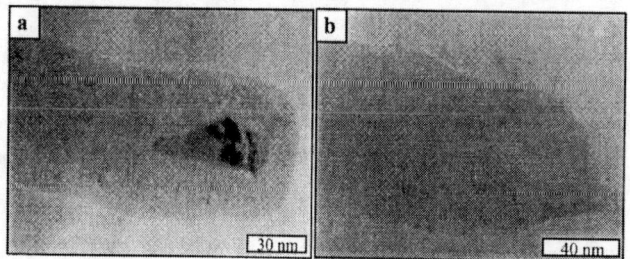

FIGURE 4. HRTEM images of CNTs grown for growth times of (a) 20 min and (b) 40 min.

Figures 4(a) and (b) show the HRTEM images of the CNT caps grown for 20 and 40 min, respectively. The Ni particles present at the top of CNTs synthesized for 20 min, whereas the CNTs grown for 40 min do not have a Ni cap. This could explain the saturation behavior of the CNT length with the growth time. When a nanotube has a transition metal cap, the CNTs keep growing the length of the tube increased with the growth time. However, when the metal cap disappears, the tube cannot be grown further, although C_2H_2 gas is continuously supplied. Amorphous phase between the Ni particle and the end of CNT is believed to be the Ni-carbon solid solution. This demonstrates the cap growth mechanism in which the carbon gases are catalytically decomposed and supplied to the edges of the CNTs through the Ni-carbon solid solution at the metal cap.

ACKNOWLEDGEMENTS

This work was supported by the Ministry of Science and Technology through National Research Lab. program and in part by the Brain Korea 21 program.

REFERENCES

1. S. Iijima, *Nature* (London) **354**, 56 (1991).
2. D. S. Bethune, C. H. Kiang, M. S. de Vries, G. Gorman, R. Savoy, J. Vazquez, R. Beyers, *Nature* (London) **363**, 603 (1993)
3. C. Journet, W. K. Maser, P. Bernier, A. Loiseau, M. Lamy de la Chapelle, S. Lefrant, P. Deniard, R. Lee, and J. E. Fischer, *Nature* (London) **388**, 756 (1997).
4. A. Thess, R. Lee, P. Nikolaev, H. Dai, P. Petit, J, Robert, C. Xu, Y. H. Lee, S. G. Kim, D. T. Colbert, G. Scuseria, D. Tománek, J. E. Fischer, and R. E. Smalley, *Science* **273**, 483 (1996).
5. M. Terrones, N, Grobert, J. Olivares, J. P. Zhang. H. Terrones, K. Kordatos, W. K. Hsu, J. P. Hare, P. D. Townsend, K. Prassides, A. K. Cheetham, H. W. Kroto, and D. R. M. Walton, *Nature* (London) **388**, 52 (1997).
6. W. Z. Li, S. Xie, L. X. Qian, B. H. Chang, B. S. Zou, W. Y. Zhou, R. A. Zhao, and G. Wang, *Science* **274**, 1701 (1996).
7. Y. C. Choi, D. J. Bae, Y. H. Lee, B. S. Lee, I. T. Han, W. B. Choi, N. S. Lee, and J. M. Kim, *Synth. Matals* **108**, 159-163 (2000).
8. S. C. Lim, Y. C. Choi, H. J. Jeong, Y. M. Shin, D. J. Bae, Y. H. Lee, N. S. Lee, and J. M. Kim, accepted to *Advanced Materials*, August, (2000).
9. W. B. Choi, D. S. Chung, J. H. Kang, H. Y. Kim, Y. W. Jin, I. T. Han, Y. H. Lee, J. E. Jung, N. S. Lee, G.-S. Park, and J. M. Kim, *Appl. Phys. Lett.* **75**, 3129-3131 (1999).
10. Y. C. Choi, D. J. Bae, T. H. Lee, B. S. Lee, G. S. Park, W. B. Choi, N. S. Lee, and J. M. Kim, *J. Vac. Sci. Technol.* A **18**, 1864 (2000).
11. Y. C. Choi, Y. M. Shin, S. C. Lim, D. J. Bae, Y. H. Lee, B. S. Lee, and D. C. Chung, *J. Appl. Phys.* **88**, 8 (2000).
12. Y. C. Choi, Y. M. Shin, Y. H. Lee, B. S. Lee, G. S. Park, W. B. Choi, N. S. Lee, and J. M. Kim, *Appl. Phys. Lett.*, **76**(17), 24 (2000).
13. Y. C. Choi, D. W. Kim, T. J. Lee, C. J. Lee, and Y. H. Lee, *Synth. Metals*, **117**(1-3), 81-86 (2001).

Effect of Ni-Surface Morphology on the Growth of Carbon Nanotubes by Microwave Plasma-Enhanced Chemical Vapor Deposition

Young Chul Choi, Young Min Shin, Hee Jin Jeong, Dong Jae Bae, Seong Chu Lim, Kay Hyeok An, and Young Hee Lee[a]

Semiconductor Physics Research Center, Jeonbuk National University, 561-756 Korea

Abstract. Vertically aligned carbon nanotubes were grown on Ni-coated Si substrates using microwave-plasma enhanced chemical vapor deposition. The surface morphology of Ni thin films was varied with the rf power density during the sputtering process. It was found that the growth of carbon nanotubes was strongly influenced by the surface morphology of Ni thin film. Pure carbon nanotubes were synthesized on the Ni thin film with uniformly distributed small grain sizes, whereas a lot of carbonaceous particles were produced in addition to the nanotubes, when the nanotubes were grown on the Ni film with widely distributed grain sizes. With decreasing Ni-grain size, the diameter and the wall-number of carbon nanotubes decreased while the length and the density of the nanotubes increased.

INTRODUCTION

Carbon nanotubes (CNTs) have been drawing a great deal of attention because of their unique and superb properties such as high field-emissivity, capability for the storage of large amount of hydrogen, high mechanical strength etc., all of which are useful in various applications [1]. However, different applications require different structures and morphologies of carbon nanotubes. Therefore, developing a way to control the structures of carbon nanotubes is prerequisite for the practical applications in various fields. In this study, carbon nanotubes were synthesized on patterned Ni substrates by microwave plasma-enhanced chemical vapor deposition (MPECVD), and their structures including diameter, length, and number of walls could be controlled by changing the surface morphology of Ni thin films. Furthermore, field emission properties of patterned carbon nanotubes that are aligned perpendicular to the substrates were investigated.

EXPERIMEMTAL PROCEDURE

Ni thin films with a thickness of 70 nm were deposited using rf magnetron sputtering on Si substrates. An array of Ni dots was prepared using a shadow mask with a substrate-size of 4 cm^2. The diameter of a Ni dot and the distance between dots

---

[a] e-mail: leeyh@sprc2.chonbuk.ac.kr

are 250 μm and 750 μm, respectively. Before the deposition of Ni thin films, the chamber was evacuated to a base pressure of 1.0×10^{-6} torr. The pressure was adjusted to 3.7 mtorr by feeding Ar gas and the substrate temperature was elevated to 350 ℃. Ni films were then deposited at various rf power densities. Carbon nanotubes were grown on Ni-coated Si substrates using MPECVD at 700 ℃ with gas mixtures of CH_4 (20 %) and H_2 (80 %). The applied microwave power and the pressure during the growth of the nanotubes were 400 W and 10 torr, respectively. The growth time of CNTs was maintained for 5 min. The surface morphologies of Ni thin films were investigated by atomic force microscopy (AFM). The growth rates and densities of nanotubes were observed by scanning electron microscopes (SEM). The field emission properties of carbon nanotubes were measured using a diode structure.

RESULTS AND DISCUSSION

Figures 1(a-c) are the AFM images in contact mode showing the surface morphologies of Ni film dots prepared at rf power densities of 0.25, 0.5, and 1.0 W/cm^2, respectively. The grain size decreased drastically with decreasing rf power density, whereas the grain density increased. Some Ni grains with larger size are also shown non-uniformly at higher rf power density, as shown in Fig. 1(c).

Figures 2(a-c) show the oblique (45°) SEM images of vertically aligned carbon nanotubes grown on Ni dots which was deposited at the previously described conditions. Vertically aligned CNTs were obtained by the MPECVD, as shown in the figures. We here note that any pretreatments for the surface of catalyst-metal thin films, i.e., the exposure to NH_3 gas [2] and the dipping in HF solution [3] were not necessary for growing CNTs in our case. In spite of the identical growth conditions of the CNTs, the growth rate increases markedly with decreasing the rf power density. The growth rate of CNTs grown on Ni film deposited at the rf power density of 0.25 W/cm^2 is as high as 8.5 μm/min. The density of CNTs is also affected by the rf power density of Ni films. The density of Ni grains increased with decreasing rf power density, as shown in Fig. 1. Hence, the density of the CNTs increases with increasing the density of Ni grains, as shown in Fig. 2. Some carbonaceous particles are placed on top of the nanotubes particularly at higher rf power density, as can be seen in Fig. 2(c). We also synthesized the nanotubes on Ni film prepared at the rf power density of 2.0 W/cm^2. CNTs were aligned vertically, but had rather larger amount of carbonaceous particles on top of nanotubes. At higher rf power density, larger size of Ni particles appears on uniformly distributed smaller particles, as shown in Fig. 1(c). The number of these larger particles increased with increasing rf power density. It seems that larger Ni particles do not act as nucleation seeds for CNTs but form carbonaceous particles, since more carbonaceous particles were observed on top of CNTs when grown on Ni films having more particles with larger size.

Using a diode structure, the field emission properties were investigated. We observed quite uniform and bright emission patterns. The current vs. voltage data are well analyzed by Fowler-Nordheim theory at low field region. However, at high field region, the saturation of the current was observed. It was found that the saturation behavior was attributed to the adsorbates.

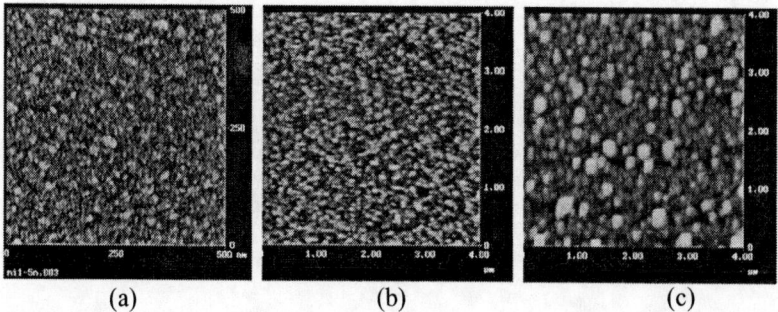

FIGURE 1. AFM images of Ni films deposited at rf power densities of (a) 0.25, (b) 0.5, and (c) 1.0 W/cm^2.

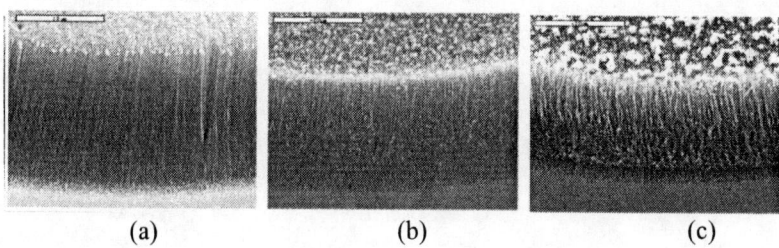

FIGURE 2. SEM images of vertically aligned CNTs on Ni films that had been deposited at rf power densities of (a) 0.25, (b) 0.5, and (c) 1.0 W/cm^2.

CONCLUSION

Vertically aligned carbon nanotubes were synthesized on patterned Ni substrates. By changing the surface morphology of Ni thin films, we could control the structure and morphology of the carbon nanotubes. Efficient field emission properties were observed from the nanotubes using diode structures.

REFERENCES

1. Tomanek, D., and Enbody, R. J., "*Science and Application of Nanotubes*", Kluwer Academic/Plenum Publishers, New York, 2000.
2. Ren, Z. F. Huang, Z. P., Xu, J. W., Wang, J. H. Bush, P., Siegal, M. P., and Provencio, P. N., *Science* **282**, 1105-1107 (1998).
3. Lee, C. J., Kim, D. W., Lee, T. J., Choi, Y. C., Park, Y. S., Lee, Y. H., Choi, W. B., Lee, N. S., and Kim, J. M., *Appl. Phys. Lett.* **312**, 461-463 (1999).

Growth Mechanism of Carbon Nanotubes Grown by Microwave Plasma-Assisted Chemical Vapor Deposition

T. Muneyoshi, M. Okai, T. Yaguchi, and S. Sasaki

Displays, Hitachi, Ltd., Japan

Abstract. To investigate the most suitable deposition conditions and growth mechanism, we grew carbon nanotubes (CNTs) by microwave plasma-assisted chemical vapor deposition under various conditions. The experimental parameters we varied were (a) the mixture ratio of methane in hydrogen, (b) the total gas pressure, and (c) the bias electric current. We found that the bias electric current was the most influential parameter in determining the shape of CNTs. We believe that the growth process of CNTs can be explained by using the solid solubility curves of metal-carbon phase diagrams. Selective growth and low-temperature growth of CNTs can also be understood from these phase diagrams.

INTRODUCTION

Carbon nanotubes (CNTs) [1,2] have attracted interest from many scientists and engineers, in terms of both theory and application. A. G. Rinzer, et al. [3] and W. A. de Heer, et al. [4] each confirmed the existence of electron field emission from CNTs. A matrix-addressable diode flat panel display has been demonstrated by using a CNT-epoxy composite as the electron emission source [5]. A well-aligned patterned array of CNTs has been fabricated by chemical vapor deposition (CVD) [6]. Many researchers have tried to utilize CNTs for field emission displays (FED) because of their excellent emission characteristics. However applying CNTs for FEDs requires shape control and selective growth of the CNTs. Our purpose is to investigate the most suitable growth conditions and growth mechanism of CNTs grown by CVD.

EXPERIMENT

A microwave plasma-assisted CVD method was used to grow CNTs. A 2.45 GHz microwave generator with 500 W of power was used to generate plasma in a quartz reaction chamber. An anode and a cathode, which served as a sample stage, were placed in the reaction chamber, and a bias electric current could be passed between the electrodes through the plasma. We used Ni, Fe, Fe-42wt% Ni-6wt% Cr (426 alloy), and Si with a 426 alloy film as substrates, and methane diluted in hydrogen as the carbon source. The experimental parameters we varied were (a) the mixture ratio of methane in hydrogen (10-50%), (b) the total gas pressure in the reaction chamber (0.15-1.3 kPa), and (c) the bias electric current (0.02-0.135 A). The total gas flow rate was fixed at 100 ccm, and CVD growth was carried out for 60 minutes. The growth temperature, measured at the sample stage, was 650-720°C. Before growing CNTs, the substrates were cleaned in the reaction chamber by hydrogen plasma for 30 minutes. The grown CNTs were investigated by high-resolution scanning electron microscopy (SEM), transmission electron microscopy (TEM), and energy dispersive X-ray (EDX) analysis.

RESULTS

Table 1 lists our evaluation of the shapes of CNTs grown under various conditions. The double and single circles indicate excellent and good fiber-like shapes, respectively, and the triangles indicate poor shapes. The X's indicate that CNTs either had very poor shapes or did not grow at all. Among the parameters, the bias electric current influenced CNT morphology the most. Figure 1 demonstrates the bias electric current (I_b) dependence of morphology, using Fe substrates as examples. When the bias electric current was high, CNTs grew on the substrate (Figure 1(a)). On the other hand, only carbon particles were generated when the bias electric current was low (Figure 1(b)). Figure 2 shows a TEM view of the top of a CNT (a) and a schematic diagram of the CNT (b). The dark area is a minute metal particle, and the gray parts are the carbon layers forming the CNT. CNTs like that in Figure 2 had knots in several places, and we recognized that the shapes of the knots were similar to the shape of the metal particle's lower part.

TABLE 1. Evaluations of CNT morphology

No.	CH_4/H_2 (ccm/ccm)	Pressure (kPa)	Current (A)	Substrate			
				Ni	Fe	426 alloy	426 on Si
1	10 / 90	0.266	0.1	×	○	○	×
2	20 / 80	0.266	0.1	×	○	○	○
3	33 / 66	0.266	0.1	×	△	△	△
4	50 / 50	0.146	0.1	×	×	△	×
5	20 / 80	0.106	0.1	×	×	×	×
6	20 / 80	1.33	0.1	○	△	○	—
7	20 / 80	0.266	0.02	×	×	×	×
8	20 / 80	0.266	0.135	△	◎	○	◎

◎ Excellent ○ Good △ Poor × Very poor or nonexistent

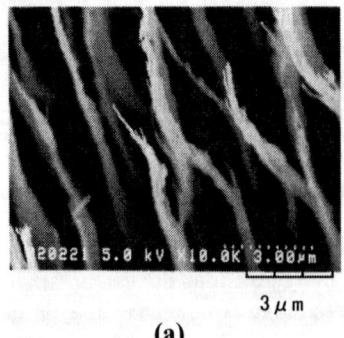
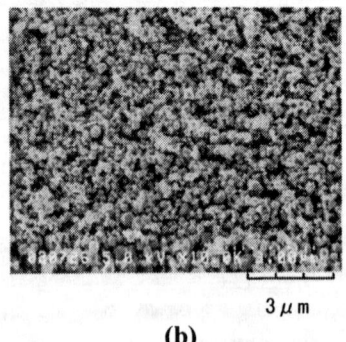

(a) 3 μm (b) 3 μm

FIGURE 1. Bias electric current dependence of morphology: (a) carbon nanotubes (I_b=0.135A), (b) carbon particles (I_b=0.02A)

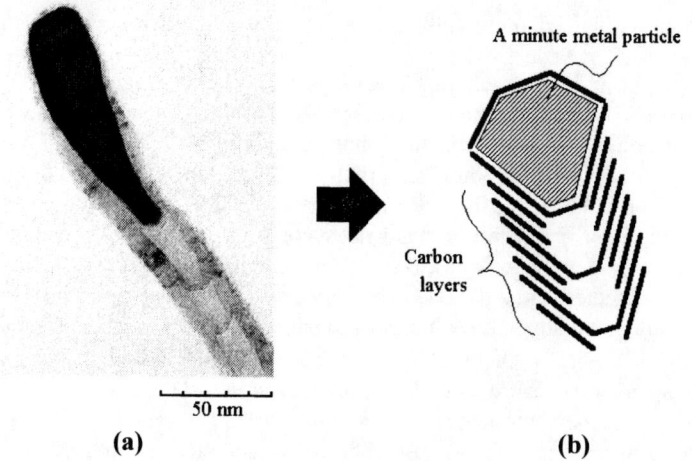

FIGURE 2. TEM view of the top of a CNT (a) and a schematic diagram of the CNT (b)

DISCUSSION

On the basis of these results, we propose a growth mechanism for CNTs. Figure 3(a) shows the CNT growth process by microwave plasma-assisted CVD. Carbon sources activated by microwaves acquired electric charges. The activated carbon sources were attracted to the top of the minute metal particle by the electric field concentration there, and they collided with the particle. The collision energy raised the top surface temperature and created an uneven temperature distribution in the particle. A higher bias electric current increased the temperature differential between the top and bottom of the particle. Carbon invaded the high-temperature top surface, then migrated along the particle, and carbon was extracted from the low-temperature bottom surface, forming a CNT. This

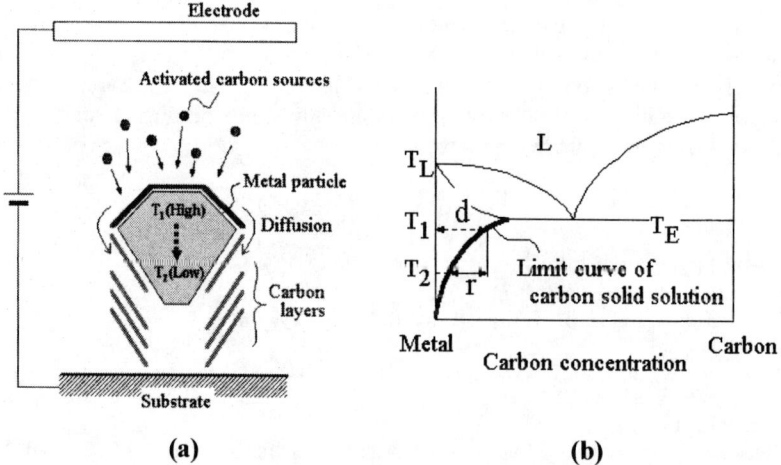

FIGURE 3. CNT growth mechanism: (a) growth process, (b) metal-carbon phase diagram

process can be explained more fully using the solid solubility curves of metal-carbon phase diagrams.

Figure 3(b) shows an example of a metal-carbon phase diagram. The "d" indicates the limit of carbon solid solution at temperature T_1, and the "r" indicates the precipitation volume of carbon at T_2. Based on this phase diagram, the requirements for a suitable catalyst are as follows: (a) the limit of carbon solid solution is high; (b) the precipitation volume of carbon is large; (c) the eutectic temperature (T_E) and melting point (T_L) of the metal are low; and (d) a stable carbide does not exist. We previously reported that a minute metal particle at the top of a CNT grown on an Fe-Ni-Cr alloy contained Fe and Ni, but Cr was not detected in the particle [7]. From the binary phase diagrams of carbon with these metals, the maximum solid solubilities of carbon in Fe, Ni, and Cr are as follows: Fe, 9.23at% (1147°C); Ni, 2.7at% (1326°C); Cr, 0.3at% (1534°C). It is thought that Cr does not work as a catalyst because its carbon solubility is low. This explains why Cr was not found in the minute metal particle at the top of the CNT [7].

The maximum solid solubilities of carbon in various other metals are as follows: Cu, 0.04at% (1085°C); Ti, 1.6at% (920°C); Pd, 5at% (1200°C); Mn, 13at% (990°C). We confirmed that CNTs grew on Pd and Mn, whose carbon solubilities are high, and that CNTs did not grow on Cu, whose carbon solubility is low. We also found that CNTs did not grow on a Ti substrate. Although the carbon solubility of Ti is not low, we believe that CNTs did not grow on Ti because a stable carbide was formed on the surface. We also confirmed that on substrates with metals of different solubilities, CNTs grew selectively on the high solubility metals.

In the case of Fe, we believe that the austenite phase has an important role as a catalyst for CNTs. The area of the austenite phase can be controlled by doping various metals. It is known that Mn, Ni, and Pt expand the area of the austenite phase, while S, Zr, and Nb narrow the area [8]. We believe that a wide austenite area is necessary for low-temperature growth of CNTs.

SUMMARY

We grew CNTs under various conditions by microwave plasma-assisted chemical vapor deposition. We found that the bias electric current has a profound effect on the shape of CNTs. We proposed a growth mechanism for CNTs by using metal-carbon binary alloy phase diagrams. Selective growth and low-temperature growth of CNTs can be understood from these phase diagrams.

REFERENCES

1. A. Oberlin, M. Endo, and T. Koyama, *J. Cryst. Growth* **32**, 335 (1976).
2. S. Iijima, *Nature* (London) **354**, 56 (1991).
3. A. G. Rinzer, J. H. Hafner, D. T. Colbert, and R. E. Smalley, *Mater. Res. Soc. Symp. Proc.* **359**, 61 (1995).
4. W. A. de Heer, A. Chatelain, and D. Ugarte, *Science* **270**, 1179 (1995).
5. Q. H. Wang, A. A. Setlur, J. M. Lauerhaas, J. Y. Dai, E. W. Seelig, and R. P. H. Chang, *Appl. Phys. Lett.* **72**, 2912 (1998).
6. H. Murakami, M. Hirakawa, C. Tanaka, and H. Yamakawa, *Appl. Phys. Lett.* **76**, 1776 (2000).
7. M Okai, T. Muneyoshi, T. Yaguchi, and S. Sasaki, *Appl. Phys. Lett.* **77**, 3468 (2000).
8. W. Hume-Rothery, *The Structure of Alloys of Iron*, Pergamon Press (1966), pp. 96-101.

Automation of purification for Single-walled Nanotubes

T.Ogawa[1], Y.Sato[1], K.Shinoda[1], B.Jeyadevan[1], K.Tohji[1], A.Kasuya[2] and Y.Nishina[3]

[1]*Dept.Geoscience and Technology, Tohoku University, Sendai, 980-8579, Japan*
[2]*Center for Interdisciplinary Research, Tohoku University, Sendai, 980-8578, Japan*
[3]*Ishinomaki Senshuu University, Ishinomaki, 986-80, Japan*

Abstract. In our earlier report, we have already proposed a novel purification method to obtain highly purified SWNTs. Considering the use of SWNTs for the industrial applications, automation of the proposed purification method is required to supply large quantities of purified SWNTs. In this study, we designed the device to automate the purification process.

INTRODUCTION

The proposed method included wet oxidation using H_2O_2 solution, concentration of SWNTs by decantation, and dissolution of metal catalysts using HCl solution[1]. Even though most of the steps involved in purification could be automated, the decantation step that depended very much on personal experience of the operator needed further investigation. During decantation, nanometer size carbon particles and micron size graphite particles were removed from the soot based on their differences in settling rates in distilled water. Since the settling rates between different types of particles are marginal, the quality of the product obtained at the end of this process was operator dependent. Furthermore, this process has to be repeated several times to remove the impurities entrapped inside aggregates of nanotube bundles. Therefore, the automation of the purification process entirely depended on the success of automating the decantation process. In this study, we introduce the features of the device designed to automate the decantation step in the purification process. And, we also discuss the performance characteristics of the same in the purification of SWNTs.

EXPERIMENT

We carried out some basic studies to obtain information necessary for the design of the device to automate the decantation process. Based on these data, we have designed a device shown in Fig.1 to automate the decantation step. In this device, the sample was dispersed using ultrasonic wave for fifteen minutes and allowed to settle for a specified period of time. Then, the diaphragm pump was used to remove nanoparticle suspension. This process was automatically repeated several times. Then, fresh

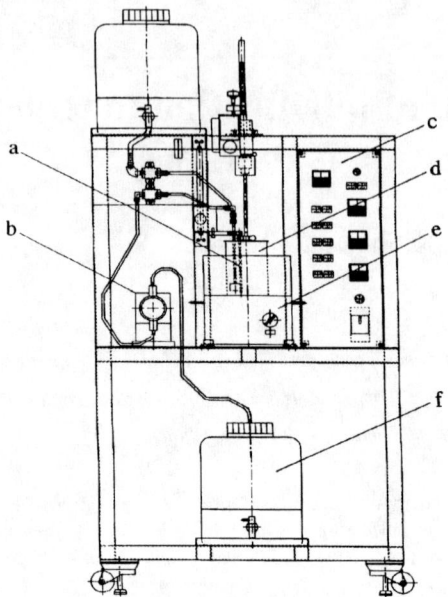

Fig.1 Automatic decantation system
a. ultrasonic sensor b. diaphragm pump
b. control panel d. sample container
e. ultrasonic bath f. tank for product storage

distilled water was introduced to the sediment that contains SWNTs and graphite particles and dispersed using ultrasonic wave. After leaving time for the graphite particles to settle, the suspension dispersing the SWNTs was removed. In the above processes, ultrasonic sensor was used to control the in and out flows of water automatically. The sequencer automated the entire process.

We attempted to concentrate SWNTs from 500mg of wet oxidized soot using this device. The volume of suspension to be drained out was set at 1 liter, and settling time was varied from 2 hours to 6 hours sequentially based on the manual decantation data. The cycle beginning from supplying the water till draining out the suspension was repeated until the solid concentration of the suspension became almost zero. It was expected that nanoparticles would be drained out at longer settling time, SWNTs at the mid settling time and the graphite particles at very short settling time. The solids in suspension as well as the ones settled at the bottom of the container were evaluated with SEM, TEM, and Raman spectra.

RESULTS AND DISCUSSION

Fig.2 shows the Raman spectra of the solids recovered from the suspension at various settling times. The intensity of the split peak of SWNTs was most remarkable for the solids recovered from the suspension allowed to settle for 4 hours, suggesting the presence of highly concentrated nanotubes. Fig.3 shows the TEM photograph of the solids recovered from the suspension that was allowed to settle for 4 hours. TEM

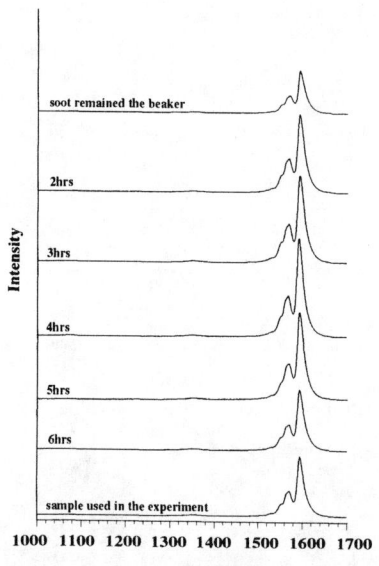

Fig.2 Raman spectra of the solids recovered from the suspension at various settling times

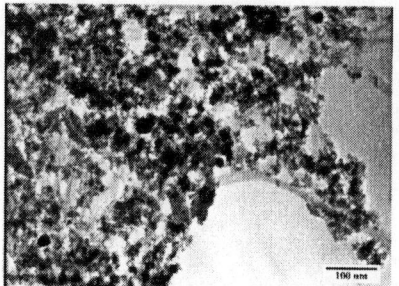

Fig.3 TEM photograph of the solids remained in the suspension after 4 hours

observations suggested that the sample contained mostly of SWNTs and metal particles with a small fractions of nanoparticles and micron size graphite particles. Results were similar for the samples withdrawn at 3 and 5 hours. Considering the data obtained with electron microscopy and Raman spectra analysis of the solids recovered from the suspension at various settling times, we conclude that the optimum settling time for removal of nanometer size particles was 6 hours, and SWNTs was between 3 and 5 hours, respectively.

Then, the solid recovered from the suspension was used for acid treatment. Fig.4 shows the Raman spectra of the sample used in this study, solids recovered from the suspension allowed settling for 4 hours and the final product obtained after acid treatment. The purity of the product increased as the purification progressed. Furthermore, the SEM and TEM photographs shown in fig.5 also suggested that the final product contained highly pure SWNTs though a little amount of nanocapsles including metal particles existed. We believe these SWNTs are pure enough to be used in many of the speculated industrial applications. We obtained ca.5mg of purified nanotubes from 500mg of wet-oxidized soot.

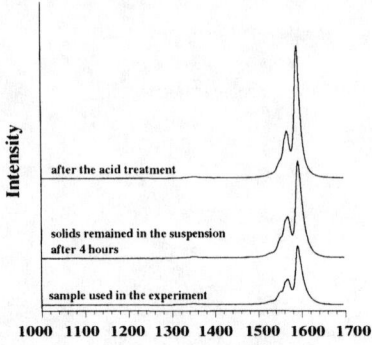

Fig.4 Raman spectra of the soot at the each purification step

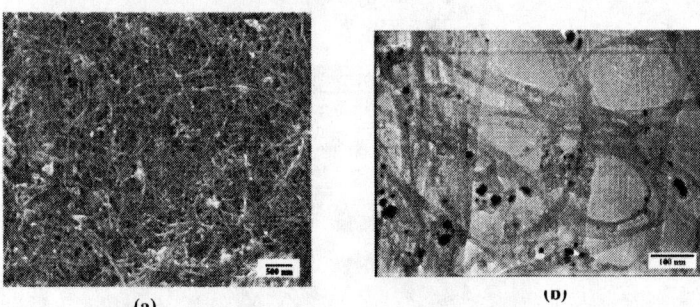

(a)   (b)

Fig.5 Photographs of SWNTs obtained after acid treatment
(a) SEM photograph (b) TEM photograph

CONCLUSION

From these results, we could conclude that we have succeeded in automating the decantation process to obtain highly concentrated SWNTs automatically the device designed in this study. This device has the merits of its simplicity and easiness to be operated, thus, anyone can purify SWNTs using the same. At present the yield of SWNTs is not high, however, larger amounts of SWNTs is believed to obtained optimizing the parameters further and also by scaling up the device.

ACKNOWLEDGMENTS

The present study was supported in part by Grant-in-Aid for Priority Research (#11165202) and International Science Research Program (#11694119) from the Ministry of Education, Science, Culture and Sports of Japan and CREST of Japan Science and Technology Institute.

REFERENCES

1. T. Ogawa et al., *Abstracts of The 16th Fullerene General Symposium*, 56-57,

Time And Space Evolution Of Carbon Species Generated With A Laser Furnace Technique

Shinzo Suzuki[a], Hirofimi Yamaguchi[a], Rahul Sen[a], Hiromichi Kataura[b], Wolfgang Krätschmer[c], and Yohji Achiba[a]

[a]*Department of Chemistry, Tokyo Metropolitan University, Tokyo 192-0397, JAPAN*
[b]*Department of Physics, Tokyo Metropolitan University, Tokyo 192-0397, JAPAN*
[c]*Max Planck Institut für Kemperphysik, Postfach 103980, D-69029, Heidelberg, GERMANY*

Abstract. Time and space evolution of carbon species generated with a laser furnace apparatus was investigated using a high-speed video camera and an ICCD system combined with a laser-induced emission technique. It was found that the blackbody emission intensity increased at $\Delta t > 400$ μsec after laser vaporization of graphite rod under high ambient temperature condition, where fullerene formation was most favored. Additionally, a tunable light from an OPO laser was introduced in order to excite C_2 ($a^3\Pi_u$) with a certain delay after laser vaporization of graphite target, and it was found that the amount of C_2 ($a^3\Pi_u$) also increased at $\Delta t > 400$ μsec under high ambient temperature condition. These findings suggest that there exists some exothermic process related to the formation of fullerene species at $\Delta t > 400$ μsec after laser vaporization of graphite target.

INTRODUCTION

After the discovery to produce C_{60} and other higher fullerenes in preparative amounts [1], the formation mechanism of them has been extensively discussed but it has not been resolved as yet. In order to attack this problem, laser furnace technique, which was first introduced by Smalley and others for the production of endohedral fullerenes and single wall carbon nanotubes [2], has been extensively applied, since it allows better control of the key parameters such as ambient temperature of the furnace, laser fluence, buffer gas pressure etc. Using this technique, one can find that the yield of C_{60} and other higher fullerenes increases if the ambient temperature of the furnace, i.e. the buffer gas temperature is increased [3, 6-8]. In particular, Wakabayashi et al. reported that the furnace temperature does not only determine the yield of C_{60} and higher fullerenes but also strongly influences the relative yields of different conformational isomers of them [3].

It is well known that carbon nanoparticles generated by laser vaporization of a graphite rod are initially rather hot. The emission spectra given by these carbon particles can be interpreted as blackbody emissions [4-7]. In this proceeding, time evolution of blackbody emission observed after laser vaporization of graphite rod with different ambient temperature condition was recorded with a high-speed video camera or an ICCD system, and carefully analyzed in correlation with the formation process of fullerene species.

EXPERIMENTAL

The experimental setup has been described elsewhere [6,7]. Briefly, the emission image from the plume of carbon nanoparticles ejected from the surface of a graphite rod (6 mm in diameter) was recorded as function of delay time (Δt) after vaporization laser was fired. We used the first or second harmonic of a Nd:YAG laser (600 mJ/cm^2, 10Hz) for vaporization and recorded the emission image using a high speed video camera (KODAK EKTAPRO H4540-S, 25 μsec time window) or an ICCD system (Princeton Instruments PI-MAX, 50 nsec gate width). Throughout the experiment, argon buffer gas (200 torr) did flow very slowly (c.a. 10ml/min) through the quartz tube surrounded by an electric furnace. In this proceeding, the blackbody emission intensity of the entire cloud of carbon nanoparticles was estimated by monitoring the image through a band pass filter at 694.3 nm ($\Delta\lambda=10$ nm). Also an OPO laser light (Quanta-Ray, MOPO-730) was introduced as a second laser in order to probe the non-emitting C_2 ($a^3\Pi_u$) species as function of delay time after laser vaporization. In this experimental setup, the emission image was recorded with a band pass filter at 514.5 nm ($\Delta\lambda=10$ nm), in order to enhance the contribution by the emission from C_2^* ($d^3\Pi_g$), which was generated by laser excitation of C_2 ($a^3\Pi_u$).

These measurements were carried out with different ambient temperature condition inside the furnace (between 300 °C and 1150 °C), because it is well known that the yield of C_{60} and other higher fullerenes drastically increases as the ambient temperature of the buffer gas increases [3,6-8].

RESULTS AND DISCUSSION

Figure 1 shows the change in the emission intensity of the entire cloud of carbon nanoparticles obtained with a band pass filter at 694.3 nm ($\Delta\lambda=10$ nm) with different ambient temperature condition. From Fig. 1, it is clearly recognized that there is an apparent increase in the emission intensity at $\Delta t > 400$ usec with high ambient temperature (1150 °C) condition. This increasing tendency was found to become less clear as the ambient temperature decreased and could hardly be seen at 300 °C.

For small (radious « wavelength) spherical particles exhibiting a $1/\lambda$ emissivity, the emission intensity at a certain wavelength (λ) is considered to be proportional to the

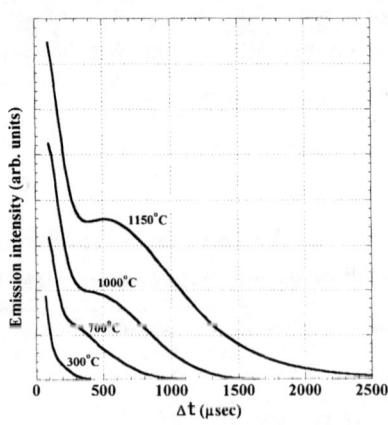

FIGURE 1. Change in the emission entensity of the entire cloud of carbon nanoparticles obtained with a band pass filter at 694.3 nm. Each set of data was obtained at different ambient temperature (1150°C, 1000°C, 700°C, and 300°C, respectively).

following expression [4], i.e.,

$$I(\lambda,T) \propto \frac{a^3}{\lambda^6(\exp(hc/\lambda k_B T)-1)} \quad (1)$$

In Eq. (1), the symbol "a" denotes the mean radius of carbon nanoparticles, which cause blackbody emission inside the furnace. This expression was successfully used for the estimate of the internal temperature of carbon nanoparticles generated after laser vaporization of graphite rod [4-7,9]. As a result, the internal temperature of them was estimated to be c.a. 2000K or below at $\Delta t > 400$ μsec after laser vaporization, showing a monotonous decreasing tendency. In consideration with the estimate performed by Mitzner et al. [5], under such low internal temperature, further fragmentation process (e.g. C_2-elimination) has only minor contribution to the cooling of these carbon nanoparticles. Our analysis also suggests that the main cause for the cooling of carbon nanoparticles is due to the collision with the surrounding rare gas [9]. Therefore, it is considered that the carbon nanoparticles giving blackbody emission cannot change the size so extensively at $\Delta t > 400$ μsec. From Eq. (1), it is suggested that the total emission intensity of carbon nanoparticles at a certain wavelength decrease monotonously as the internal temperature decreases, when the emitting species does not change its size and population. These findings indicate that new carbon species giving blackbody emission should be generated at $\Delta t > 400$μsec after laser vaporization, especially in the case of high ambient temperature condition.

Figure 2 shows a typical example of the emission image induced by different photon wavelength under high ambient temperature condition. Fig. 2 (b) was recorded with excitation by photon of 516.5 nm, with which C_2 ($a^3\Pi_u$) could be excited to C_2^*($d^3\Pi_g$). Fig. 2(c) was recorded with excitation by photon of 532.0 nm, which cannot excite C_2 ($a^3\Pi_u$). Fig. 2(a) was recorded as a reference data without a second excitation laser. It is well known that there exist spontaneous emission due to the pre-existing C_2^*($d^3\Pi_g$) and also blackbody emission caused by carbon nanoparticles, both of which were generated in the beginning of laser vaporization of graphite target [6-7,9]. However, these emission intensities rapidly and monotonously decrease as time goes by (see Fig.2 (a)). On the other hand, Fig.2 (b) indicates that the strong emission caused by excitation of C_2 ($a^3\Pi_u$) appears at around $\Delta t \approx 400$ μsec. It is interesting to note that this emission intensity strongly depends on the ambient temperature of the furnace, i.e., higher ambient temperature gives much stronger emission [10]. This behavior is similar to that of the blackbody emission at $\Delta t > 400$ μsec shown in Fig. 1.

From these experimental findings, it is most likely to consider that, under high ambient temperature condition, a certain exothermic process related to the generation of fullerene species takes place, which cause C_2-generation as well as blackbody radiation at $\Delta t > 400$ μsec after laser vaporization of graphite target.

ACKNOWLEDGMENTS

This work was supported by the fund from The Japan Society for the Promotion of the Science (JSPS) ("Future Program") and from the Ministry of Education, Sports, and Culture.

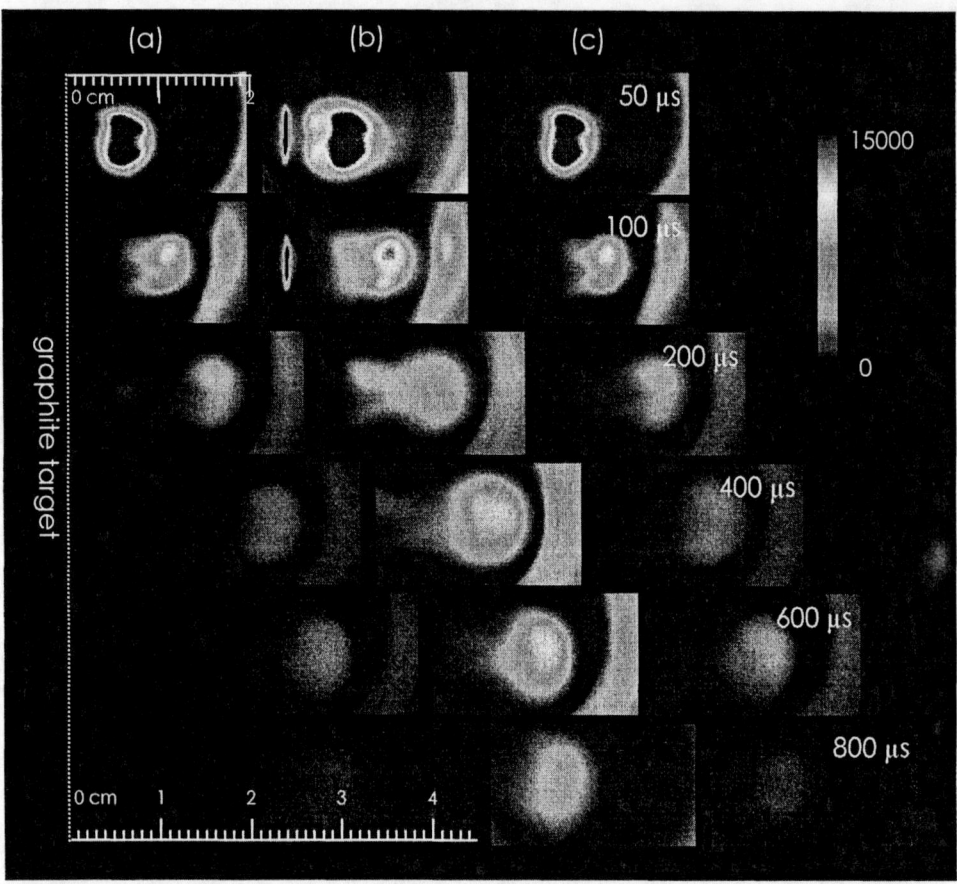

FIGURE 2. ICCD image of emission from carbon species recorded with a band pass filter of 514.5nm ($\Delta\lambda=10$ nm); (a) emission image without excitation laser (reference); (b) emission image recorded with excitation by 516.5 nm photon; (c) emission image recorded with excitation by 532.0 nm photon.

REFERENCES

1. Krätschmer, W., Lamb L.D., Fostiropoulus K., Huffmann D.R., *Nature*, **347**, 354(1990).
2. Chai Y., Guo T., Haufler R.E., Chibante L.P.F., Fure J., Wang L.H., Alford J. M., and Smalley R.E., *J. Phys. Chem.*, **95**, 7564(1991).
3. Wakabayashi T., Kasuya D., Shiromaru H., Suzuki S., Kikuchi K., and Achiba Y., *Z. Phys.* **D40**, 414(1997).
4. Rolfing E.R., *J. Chem. Phys.*, **89**, 6103(1988).
5. Mitzner R. and Campbell E.E.B., *J. Chem. Phys.*, **103**, 2445(1995).
6. Ishigaki T., Suzuki S., Kataura H., Krätschmer W., and Achiba Y., *Appl. Phys.*, **A70**, 121(2000).
7. Ishigaki T., *Ph.D. thesis*, Tokyo Metropolitan University(2000).
8. Kasuya D., Ishigaki T., Suganuma T., Ohtsuka Y., Suzuki S., and Achiba Y., *Eur. Phys. J.* **D9**, 355(1999).
9. Suzuki S., Yamaguchi H., Ishigaki T., Sen R., Kataura H., Krätschmer W., and Achiba Y., submitted
10. Yamaguchi H. et al., to be published.

Synthesis of Vertically Aligned Carbon Nanotubes on a Large area using Thermal Chemical Vapor Deposition

C. J. Lee[*], K. H. Son[*], T. J. Lee[*], S. C. Lyu[*], and J. E. Yoo[†]

[*]School of Electrical Engineering, Kunsan National University, Kunsan 573-701, Korea
[†]Iljin Nanotech Co. Ltd., Seoul 157-810, Korea

Abstract. Vertically well-aligned carbon nanotubes (CNTs) were homogeneously grown on iron deposited silicon oxide substrate by thermal chemical vapor deposition of acetylene. The CNTs have an uniform length of 100 μm and a diameter in the range from 100 to 200 nm. The CNTs reveal closed tip and very clean surface without any carbonaceous particles. The CNTs have no encapsulated iron particles at the closed tip and a bamboo structure in which the curvature of compartment layers is directed to the tip.

INTRODUCTION

Since the first observation of carbon nanotubes (CNTs) [1], extensive researches have focused on the synthesis of CNTs with high purity. Various synthetic methods such as arc discharge [2], laser vaporization [3], pyrolysis [4], and plasma-enhanced [5] or thermal chemical vapor deposition (CVD) [6] were employed. Synthesis of CNTs using CVD has attracted much attention because of the advantage that the growth of CNTs can be achieved with high purity, high yield, and vertical alignment. In this paper, vertically aligned carbon nanotubes were grown on iron deposited silicon oxide substrate by thermal CVD of acetylene. The structure and crystallinity of CNTs were investigated using a transmission electron microscope and micro-Raman.

EXPERIMENTAL

The 20 mm × 30 mm size p-type Si (100) substrate with a resistivity of 15 Ω-cm was thermally oxidized. The thickness of the silicon oxide (SiO_2) layer was estimated as approximately 300 nm. A 100 nm-thick Fe film was thermally deposited on the SiO_2 layers using a thermal evaporator under a vacuum of 10^{-6} Torr. The Fe-deposited substrates were loaded with face down direction on a quartz boat in quartz CVD reactor. Argon (Ar) was flowed into the CVD reactor to prevent the oxidation of catalytic metal while raising the temperature. In order to form the catalytic particles in nanometer size, the substrates were dipped in a diluted HF solution for 200 sec and pretreated by ammonia (NH_3) gas with a flow rate of 100 sccm for 20 min, in the temperature range 750-950 °C [7]. The CNTs were grown using C_2H_2 with a flow rate of 40-80 sccm for 10 min at the same temperature of NH_3 pretreatment. The CNTs grown on substrate were examined by a scanning electron microscope (SEM) (Hitachi S-800, 30 kV) to measure the length and the diameter. A transmission electron microscope (TEM) (Philips, CM20T, 200kV) was used to investigate the structure and crystallinity of CNTs. A Raman spectrometer (Renishaw micro-Raman 2000) was also used to identify the structure and the crystallinity of CNTs. The 632.8 nm line of a He-Ne laser was used for excitation.

RESULTS and DISCUSSION

Fig. 1. is SEM micrographs for the CNTs grown at 950 °C. Fig. 1(a) shows the vertically well-aligned CNTs with uniform length of 100 μm. A magnified top view, as shown in Fig. 1(b), reveals that the diameter is mostly in the range from 100 to 200 nm. The CNTs have closed tip and very clean surface without any carbonaceous particles. Most of tips are tilted from the vertical direction within 10 degrees.

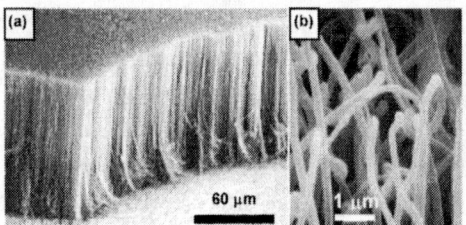

FIGURE 1. SEM images for the CNTs grown at 950 °C. (a) The vertically well-aligned CNTs with uniform length of 100 μm. (b) A magnified top view of CNTs

Fig. 2 is the TEM images of multiwalled CNTs which are consisted of hollow compartments, looking like bamboo. Fig. 2(a) reveals the closed tips with no encapsulated catalytic particles (see arrows ①), an open root separated from Fe particle (see arrow ②), and the compartment layers with a curvature directed toward the tip (see arrows ③). A CNT with an outer diameter of about 200 nm, shown in Fig. 2(b), has the compartment layers regularly at a distance of about 200 nm. The wall thickness increases due to the connection with compartment layer, but the outer diameter remains about the same for entire tube [8].

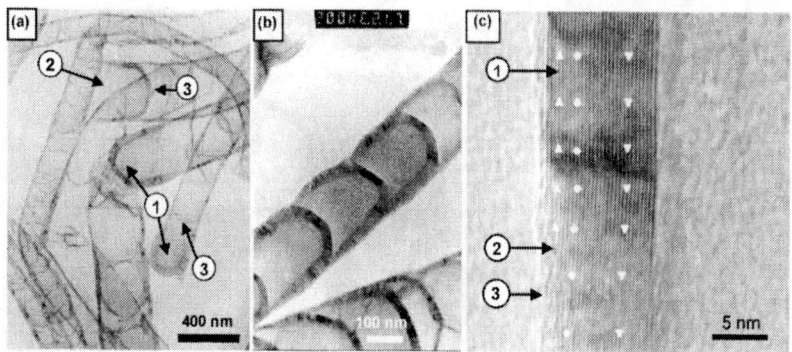

FIGURE 2. TEM images of multiwalled CNTs. (a) CNTs are consisted of hollow compartments, looking like bamboo. (b) A CNT has the compartment layers regularly at a distance of about 200 nm. (c) HRTEM image for the graphite sheets of a multiwalled CNT.

Fig. 2(c) is a HRTEM image for the wall of a CNT, revealing that the graphite sheets are aligned with a tilted angle of about 2 degrees toward the tube axis (see marks △, ◇, and ▽). The graphite sheets have a good crystallinity and a lattice distance with 0.34 nm. Especially, the graphite sheets at the outside disappear as indicated by arrow(see mark △), revealing that crystalline structure (see arrow ①), becomes defective (see arrow ②), and finally vanishes (see arrow ③).

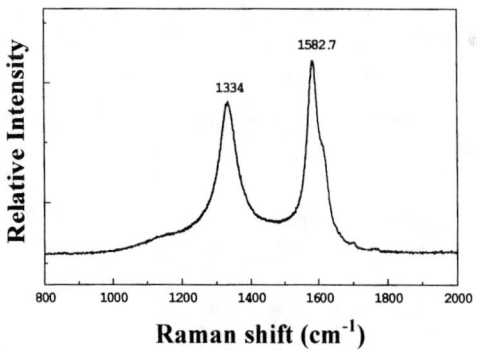

FIGURE 3. Micro-Raman spectrum for the CNTs grown on SiO$_2$ substrate at 950 °C.

Fig. 3. is the micro-Raman spectrum for the CNTs grown on SiO$_2$ substrate at 950 °C. The excitation laser is a 632.8 nm He-Ne laser. The multi-wall structure of CNTs is identified by a clear G-line at 1582.7 cm^{-1} with a small bump at 1620 cm^{-1}. There is no second order peak at ~1720 cm^{-1}. The breathing mode peak at 190.5 cm^{-1} is not appeared, confirming the multi-wall structure of CNTs. The peak at 1334 cm^{-1} could be resulted from the defective outer graphite sheets of multiwalled CNTs [9]. The Raman spectrum analysis provides definite evidence that the CNTs have multi-walls with good crystallinity and there are some defective graphite sheets on the wall surface.

In summary, we have grown the vertically well aligned multiwalled CNTs on a large area of Fe deposited SiO_2 substrates using the thermal CVD of C_2H_2 at 750-950 °C. The diameter of CNTs is 100-200 nm and the length is about 100 μm. The CNTs have closed tip and very clean surface without any carbonaceous particles. All CNTs have no encapsulated catalytic particles at the closed tip and the bamboo structure. The curvature of compartment layers is always directed to the tip. The multiwalled CNTs have a good crystalline structure.

REFERENCES

1. Iijima, S., *Nature* **354**, 56 (1991).
2. Journet, C., Maser, W. K., Bernier, P., Loiseau, A., Lamy de la Chapelle, M., Lefrant, S., Deniard, P., Lee, R., and Fischer, J. E., *Nature* **388**, 756 (1997).
3. Thess, A., Lee, R., Nikolaev, P., Dai, H., Petit, P., Robert, J., Xu, C., Lee, Y. H., Kim, S. G., Rinzler, A. G., Colbert, D. T., Scuseria, G., Tomanek, D., Fisher, J. E., and Smalley, R. E., *Science* **273**, 483 (1996).
4. Terrones, M., Grobert, N., Olivares, J., Zhang, Z. P., Terrones, H., Kordatos, K., Hsu, W. K., Hare, J. P., Townsend, P. D., Prassides, K., Cheetham, A. K., Kroto, H. W., and Walton, D. R. M., *Nature* **388**, 52 (1997).
5. Ren, Z. F., Huang, Z. P., Xu, J. W., Wang, J. H., Bush, P., Siegal, M. P., and Provencio, P. N., *Science* **282**, 1105 (1998).
6. Fan, S., Chapline, M. G., Franklin, N. R., Tombler, T. W., Cassell, A. M., and Dai, H., *Science* **283**, 512 (1999).
7. Lee, C. J., Kim, D. W., Lee, T. J., Choi, Y. C., Park, Y. S., Kim, W. S., Lee, Y. H., Choi, W. B., Lee, N. S., Kim, J. M., Choi, Y. G., and Yu, S. C., *Appl. Phys. Lett.* **75**, 1721 (1999).
8. Lee, C. J., Park, J. H., and Park, J., *Chem. Phys. Lett.* **323**, 560 (2000).
9. Kasuya, A., Sasaki, Y., Saito, Y., Kohji, K., Nishina, Y., *Phys. Rev. Lett.* **78**, 4434 (1997).

Synthesis of Aligned Carbon Nanotubes by C_2H_2 Decomposition on $Fe(CO)_5$ as a Catalyst Precursor

J. H. Han[*], J. E. Yoo[*], S. C. Yoo[†], C. J. Lee[†], and K-H Lee[¶]

[*]*Nanotechnology Center, Iljin Nanotech Co., Ltd., Seoul 157-810, Korea*
[†]*School of Electrical Engineering, Kunsan National University, Kunsan 573-701, Korea*
[¶]*Dept. of Chem. Eng., Pohang University of Science and Technology, Pohang 790-784, Korea*

Abstract. Aligned carbon nanotubes are simply synthesized in a single step by the thermal decomposition of gaseous mixture of C_2H_2 and $Fe(CO)_5$ as a catalyst precursor. Multi-walled carbon nanotubes were produced on the most of the heated zone of the furnace with high packing density. The diameter and length is 20-50nm and about 55 μm, respectively. The flow rate and temperature plays critical role in the synthesis of carbon nanotubes.

INTRODUCTION

Carbon nanotubes(CNTs) have novel electronic and structural properties, which allow for their use in a wide range of potential applications such as flat-panel display, electrochemical storage of energy and nanotube-reinforcing composite. For these practical applications, nanotube synthesis of high yield, high purity and low cost will be inevitably required on a large scale. CNTs can be produced by the catalytic methods that decompose thermally various carbon-containing gases over the transition metal particles of Fe, Co and Ni. The floating catalyst method has the advantage over the substrate method, which allows a cost-effective way to commercially synthesize the CNTs with a uniform distribution of diameter on a large scale [1,2]. This method allows the 3-dimensional dispersion of the catalyst particles in a vapor-phase by pyrolysing organometallic compounds such as $Fe(CO)_5$, $Ni(CO)_5$, $Mo(CO)_6$, $Co_2(CO)_8$, $(C_5H_5)_2Fe$, $(C_5H_5)_2Ni$ and so on. In this paper, we reported the synthesis of CNTs on the whole surface of the heated zone by the pyrolysis of mixtures of acetylene with $Fe(CO)_5$. We analyzed the structure of CNTs using the scanning electron microscopy and transmission electron microscopy.

EXPERIMENTAL

The apparatus for the synthesis of CNTs was mainly composed of gas distribution part, heating zone and gas exhaust part. Acetylene gas of purity 99.5 % was used as a carbon source and $Fe(CO)_5$ (Aldrich, 48,178-8) as a catalyst precursor. Two kinds of argon gas were used for the purpose of $Fe(CO)_5$ bubbling and carrier of gas mixture. Ar gas of purity 99.9 % was bubbled through an evaporator that contained about 5 ml of liquid $Fe(CO)_5$. The mixture of Ar and $Fe(CO)_5$ was then carried into the reactor with C_2H_2 at the same time. The flow rate of Ar gas for bubbling was fixed to 30 cc/min. The flow rate of acetylene and Ar as a carrier gas was in the range of 200-500 cc/min and 1,000-6,000 cc/min, respectively. The reactor was kept at 950 °C during the synthesis. The synthesis of CNTs was carried out for 60 min. Afterwards, the reactor was cooled down to room temperature with the flow of Ar gas to prevent the oxidation of CNTs. Scanning electron microscope (SEM) was performed using Philips XL30SFEG instrument in order to observe the morphology of as-grown samples. Transmission electron microscopy (TEM) was performed with a Philips CM20T microscope operating at 200 kV.

RESULTS AND DISCUSSION

We have produced the CNTs by flowing C_2H_2 with a small amount of $Fe(CO)_5$, which is carried by the Ar gas, through the reactor. Upon thermal decomposition at 250 °C it yield pure iron [1]. The iron atoms react to produce iron clusters through the collision in the gas phase. These clusters act as nuclei upon where nanotubes nucleate and grow. Fig.1 shows the SEM image of CNTs obtained by the pyrolysis of iron pentacarbonyl with acetylene at 950 °C. The flow rate of C_2H_2 and Ar was 30 and 1,000 sccm, respectively. The copious quantity of the carbon product was found on the most of the heated zone of the furnace depending on the synthesis conditions.

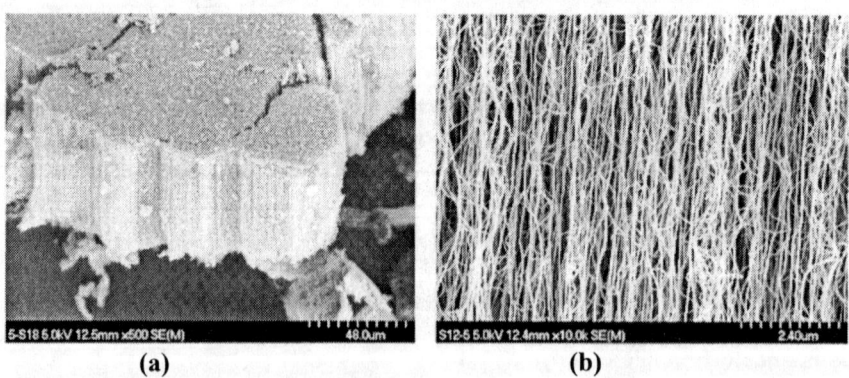

FIGURE 1. SEM images of multi-walled carbon nanotubes by the floating catalyst method.

As shown in Fig.1a), the multi-walled carbon nanotubes(MWNTs) were aligned with high packing density. The average length of the nanotubes was about 55 μm. These results were similar to those obtained by Rao et al who carried out the synthesis for ferrocene/ C_2H_2 system at 1100 °C in the two-stage furnace [3]. It is known that ferrocene begins to vaporize at about 185 °C and decompose above 400 °C. Ferrocene as a solid organometallic precursor can be less easily decomposed into metal clusters in the gas phase than iron pentacarbonyl due to its high thermal stability. Accordingly, when the ferrocene is employed as a catalyst precursor for the nanotube synthesis, it usually uses the two-stage furnace so as to effectively decompose into metal cluster. However, we obtained aligned carbon nanotubes in a single step by the thermal decomposition of gas mixture of C_2H_2 and $Fe(CO)_5$ as a catalyst precursor. Our results also showed that the flow rate and temperature plays critical role in the synthesis of carbon nanotubes.

Fig.2 shows transmission electron microscopic images of CNTs obtained under the experimental conditions of Fig.2. In the low-resolution TEM image in the left part of the figure, several MWNT are visible. From this image, a typical nanotube diameter of 30 nm can be determined. Some encapsulated Fe particles along the MWNT are detected even though these are not shown here. The high-resolution image on the right-hand side of Fig.3 reveals the tubular structure of an individual MWNT in detail. The concentric arrangement of graphitic shells is clearly visible. Unlike the typical CNTs grown by thermal CVD, as-grown nanotubes from the present method have a good crystallinity that can be observed in arc discharge method. However, the turbostratic graphitic carbon and some amorphous are observed at the exterior of the nanotube. It should be noted that the electron beam-nanotube interaction could degrade the tubular structure with loss of concentric alignment of the graphitic carbon and result in the formation of amorphous carbon.

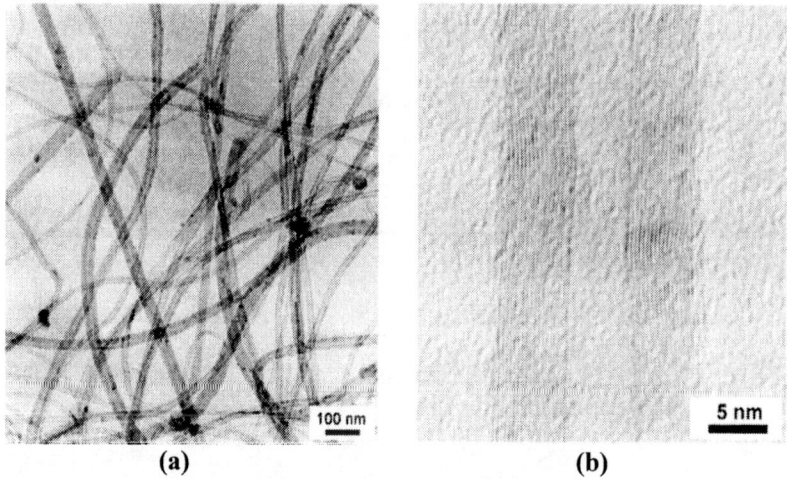

FIGURE 2. TEM images of multi-walled carbon nanotubes by the floating catalyst method.

In summary, we have synthesized aligned multi-walled CNTs on the whole surface of the heated zone in a single step by the thermal decomposition of gaseous mixture of C_2H_2 and $Fe(CO)_5$ as a catalyst precursor. The diameter and length of CNTs is 20-50 nm and about 55 μm, respectively. The multi-walled CNTs have a good crystalline structure. The gas flow rate and temperature plays important role in the structure and production of CNTs.

REFERENCES

1. Nikolaev, P., Bronikowski, M., Bradley, R.K., Rohmund, R., Colbert,D.T., Smith,K.A., and Smalley,R. E., *Chem. Phys. Letters* **313**, 91-97 (1999).
2. Cheng, H.M., Li, F., Su, G., Pan, H.Y., He, L.L.,Sun, X., and Dresselhaus, M.S., *Appl. Phys. Letters* **72**, 3282-3284 (1998).
3. Satishkumar, B.C., Govindaraj, A., and Rao, C.N.R., *Chem. Phys. Letters* **307**, 158-162 (1999).

Formation of Radioactive Fullerenes by Using Nuclear Recoil

T. Ohtsuki[1], K. Ohno[2], K. Shiga[3], Y. Kawazoe[3], Y. Maruyama[4], K. Shikano[5], K. Masumoto[6],

[1] Laboratory of Nuclear Science, Tohoku University, Taihaku, Sendai 982-0826, Japan.
[2] Department of Physics, Yokohama National University, Yokohama, 240-8501, Japan.
[3] Institute for Materials Research, Tohoku University, Aoba-ku, Sendai 980-8577, Japan.
[4] NIRIN, Kita-ku, Nagoya 462-8510, Japan.
[5] NTT Opto-Electronics Laboratories, Tokai, Ibaraki, 319-11, Japan.
[6] Radiation Science Center, KEK, Tanashi, Tokyo 188, Japan.

Abstract. The formation of As and Se atom-incorporated fullerenes has been investigated by using radionuclides produced by nuclear reactions. From the trace of radioactivities of ^{72}As, and ^{75}Se after High Pressure Liquid Chromatography (HPLC), it was found that the formation of endohedral fullerenes or heterofullerenes is possible by a recoil process following the nuclear reactions. To confirm the produced materials, *ab initio* molecular-dynamics simulations based on an all-electron mixed-basis approach were carried out.

INTRODUCTION

Chemical interaction between C_{60} and a variety of atoms is becoming a very new field of cluster research. So far, experimental studies for endohedral fullerenes with foreign atoms have been undertaken by resorting to high-pressure and radiochemical techniques[1,2,3]. In spite of the research, only partial facts for the formation process and the produced materials have been unveiled on the nature of the chemical interaction between a foreign atom and a fullerene cage. Therefore, it is intriguing to synthesize new plastic materials, such as several atom-incorporated fullerenes. In this paper, we show evidence of As and Se atom-incorporated fullerenes on the collision between a C_{60} cage and their atoms, which was generated from a recoil process following nuclear reactions. We performed *ab initio* molecular-dynamics (MD) simulations: whether the atoms can be incorporated in the fullerene with the endohedral doping or the substitutional doping.

EXPERIMENTAL PROCEDURE

In order to produce As and Se atom-incorporated fullerenes, about 10 mg of C_{60} fullerene powder was mixed homogeneously with 10 mg of GeO and As_2S_3, and used to the target material. Deuteron irradiation with beam energy of 16 MeV was performed at the Cyclotron Radio-Isotope Center (CYRIC), Tohoku University. Radioisotopes of ^{72}As, and ^{75}Se can be produced by ^{72}Ge(d,2n)^{72}As and ^{75}As(d,2n)^{75}Se reactions, respectively. The beam current was typically 5 μA and the irradiation time was about 1 hour. The sample was cooled with He-gas during irradiation. After the irradiation, the samples were left for one day to cool down the several kinds of short-lived radioactivities of byproducts. After the one-day cooling, radioactivities, such as ^{11}C or ^{13}N, e.g., ^{11}C decays to ^{11}B with $T_{1/2}$=20 min, the radioactivities of ^{72}As, and ^{75}Se could be measured with its characteristic γ-rays. The typical γ-ray and the half-life($T_{1/2}$) of ^{72}As, and ^{75}Se are 511keV, 26 hours for ^{72}As and 265keV, 120 days for ^{75}Se, respectively.

The fullerene samples were dissolved in o-dichlorobenzene after being filtrated to remove insoluble materials through a membrane filter (pore size=0.2 μm). The soluble fraction was injected into a high-pressure liquid chromatograph (HPLC) equipped with a 5PBB (silica-bonded with the pentabromobenzyl group) column of 10 mm (inner diameter)×250 mm (length), at a flow rate of 3 ml/min. The eluted solution was passed through a UV detector, the wavelength of which was adjusted to 290 nm in order to measure the amount of fullerenes and their derivatives. Downstream of the UV detector, two γ-ray detectors consisting of a bismuth germanate photomultiplier (BGO-PM) were also installed in order to count the 511 keV annihilation γ-rays emanating from ^{72}As in coincidence. The fraction was collected at 30 sec intervals, and the γ-ray activities of each fraction were measured with a Ge-detector. Therefore, the existence of ^{75}Se could be confirmed by their characteristic γ-rays.

RESULTS AND DISCUSSION

The elution curve shown by solid lines in Figure 1 indicates the absorbances monitored continuously by a UV detector for the irradiated samples of C_{60}+GeO. The horizontal axis indicates the retention time after injection into the HPLC, and vertical one indicates the absorption intensity of the UV and the γ counting rate of the ^{72}As radionuclide produced by ^{72}Ge(d,2n)^{72}As reaction. A strong absorption peak was observed at the retention time of 4.5-5.5 min in the elution curve (solid line) which was measured by the UV detector. This peak position corresponds to the retention time of C_{60} which was confirmed by the calibration run using the C_{60} sample before the irradiation. Following the first peak, two peaks at around 7.5 min and 12-14 min were consecutively observed in the UV chromatogram. This fact indicates that the second and smaller third peaks can be assigned to C_{60} dimers and C_{60} trimers, respectively, with resorting to TOFMAS measurements[4]. These materials can be produced by the interaction between C_{60}'s in coalescence reactions after ionization by incident γ-rays or produced charged particles. Four peaks appeared in the curve of the ^{72}As radioactivities in the radiochromatogram. Aside from a slight delay, the first peak (5.5 min) corresponds to the C_{60} UV

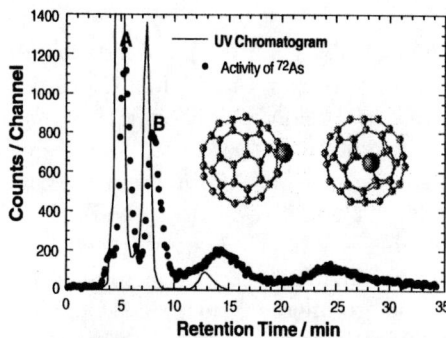 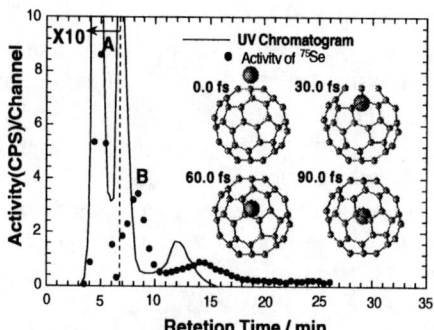

FIGURE 1. HPLC elution curves: UV chromatogram of C_{60} and radiochromatogram of ^{72}As measured with a BGO-detector. Results of MD simulation for As is also shown in the figure.

FIGURE 2. Same as Figure 1, but for the ^{75}Se measured with a Ge-detector. The difference in vertical scale between Fig.1 and Fig. 2 is due to the efficiency of each detector. Results of MD simulation for Se is also shown in the figure.

absorption peak. The second as well as the relatively broad third peaks were observed at the retention time of 7-10 min, and of 12-18 min, respectively. Though there is a delay in the elution peaks of the radioactivities against that of the UV absorption peaks, it seems that the elution behavior is similar. This result indicates that the radioactive fullerene monomers and their polymers (dimers and trimers) labeled with ^{72}As possibly exist in the final fractions. A similar trend was observed in the elution curve of ^{75}Se (see Figure 2).

In order to understand the experimental results, *ab initio* molecular-dynamics (MD) simulations were carried out. The method, which is used here, is based on the all-electron mixed-basis approach using both plane waves (PW's) and atomic orbitals (AO's) as a basis set within the framework of the local density approximation (LDA). Details of the calculation are shown in ref.[5] due to the limited space.

We show the results of the following two types of simulations; (A) insertion of As and Se atom through a six-membered ring of C_{60}, (B) insertion between one As atom and one C atom of C_{60}.

In (A), the As atom with various kinetic energies (K.E.) hits vertically the center of a six-membered ring of C_{60}. We found that an As atom can penetrate into C_{60} easily with a relatively low K.E. when its initial speed is greater than 70 eV, although, an As atom goes out again from the opposite side of the cage when its initial speed is greater than 160 eV. We also performed similar simulations for the case of a Se atom. From the simulations, we also found that a Se atom can penetrate into C_{60} easily when its initial speed is greater than 40 eV, and stop at the inside of the C_{60} cage(see Figure 2). It seems that the formation of Se@C_{60} is easier than that of As@C_{60} in this case of simulations since a Se atom can penetrate into C_{60} with a wider energy range of K.E., 40 eV~160 eV (the As case, 70 eV~160 eV). The difference can be due mainly to the nature of the covalent bonding between the C atom and the As (or Se) atom.

In (B), we shift one of the C atoms of C_{60} outward by 1.3 Å and put additionally one As atom on the same radial axis by 1.3 Å inward from the original C position of C_{60}. Then, starting the simulation with zero initial velocity. We found that there is a strong force acting on the As atom to accelerate it outward and, as a result, to repel the outward C atom. Finally, the As atom stops at 0.50 Å from the cage sphere as seen in Figure 1 (and shown in ref.[6]). Thus, it seems that the As atom put inside the cage is relatively unstable and has a strong tendency to repel the closest C atom of C_{60}, and is stabilized slightly outside the cage sphere after the removal of the closest C atom. Therefore, a heterofullerene, such as AsC_{59}, may exist stably under realistic conditions. It should be noted that similar results for the case of a Si atom have been reported by Pellarin et al[7]. But here, the possibility for endohedral doping, $As@C_{60}$, can not be completely excluded at present.

It is interesting to note that the yield ratio (the area ratio, "A/B", shown in Figs. 2), namely the monomer to the dimer, in the radiochromatogram of ^{75}Se is much larger than that (shown in Fig.1) in the radiochromatogram of ^{72}As. This fact may indicate that the shock caused by collision with higher K.E.(70 eV$\sim$) most probably induced the cage to create dimers with a higher rate.

The results of analyses of the present work have to be further supported by some other experimental data such as direct mass measurements by a TOFMAS. Finally, we briefly comment on the experimental works now in progress.

In this study, the formation of atom-incorporated fullerenes has been investigated by the trace of radioactivities of ^{72}As and ^{75}Se produced by nuclear reactions. It was found that 5B and 6B element, like As and Se, remained in the final C_{60} portion after a HPLC process. This fact suggests that the formation of ^{72}As and ^{75}Se atom-incorporated fullerenes (endohedral fullerenes or heterofullerenes) can be possible by a recoil process following nuclear reactions. Carrying out *ab initio* molecular-dynamics (MD) simulations on the basis of the all-electron mixed basis approach, we showed possibility of the formation of endohedral fullerene or heterofullerene for As and Se atom.

REFERENCES

1. M. Saunders, R.J. Cross, H.A. Jimenez-Vazquez, R. Shimshi, A. Khong, *Science* **271**, 1693(1996).
2. T. Braun, H. Rausch, *Chem. Phys. Lett.* **288**, 179(1998).
3. T. Ohtsuki, K. Ohno, K. Shiga , Y. Kawazoe, Y. Maruyama, K. Masumoto, *Phys. Rev. Lett.* **81**, 967(1998).
4. T. Ohtsuki, K. Masumoto, T. Tanaka, K. Komatsu, *Chem. Phys. Lett.* **300**, 661(1999).
5. K. Ohno, Y. Maruyama, Y. Kawazoe, *Phys. Rev.* **B56**, 1009(1997).
6. T. Ohtsuki, K. Ohno, K. Shiga, Y. Kawazoe, Y. Maruyama, K. Masumoto, Phys. Rev. **B60**, 1531(1999).
7. M. Pellarin, C. Ray, P. Mélinon, J.L. Lermé, J.L. Vialle, P. Kéghélian, A. Perez, M Broyer. *Chem. Phys. Lett.* **277**, 96(1997).

Open nanotubes of insulating boron nitride

Dmitri Golberg*, Yoshio Bando, Keiji Kurashima, and Tadao Sato

*National Institute for Materials Research,
Namiki 1-1, Tsukuba, Ibaraki 305-0044, Japan*

Abstract. Here we report on large arrays of exlusively open-ended multi-walled insulating boron nitride (BN) nanotubes synthesized through heating of carbon (C) nanotubes with metal oxides (CuO, MoO$_3$, or PbO) and boron oxide (B$_2$O$_3$) in a flow of nitrogen. The BN tubes were assembled in bundles several micrometers long. The thermal and chemical stability of the product was found to be superior to conventional C nanotubes, which opens new prospects for intra-tube chemistry at high temperatures, e.g. metal cluster encapsulation and/or entire filling of insulating BN nanotube with conductive metal.

INTRODUCTION

The nanotubes (NTs) may form in graphitic BN [1-5] as in graphite [6]. Carbon NTs have received most attention in recent years, whereas the practical uses of BN NTs have been undervalued. This is due to two facts: (i) the yields of BN NTs have usually been low [1-4]; and (ii) a BN NT is predicted to be insulating with a ~5.5 eV band gap in contrast to the semiconducting or metallic carbon NTs [7]. However, graphitic BN has a significant advantage over graphite in high-temperature applications: it is more chemically and thermally stable, although no one has managed to check this for nanotubular morphologies. We also note that C [6] and BN [1-4] NTs produced so far have normally had closed tip-ends, although C tube caps may be removed by oxidation [8].

In this paper we report on synthesis and structural analysis of BN multi-walled NTs which display exclusively open tip-ends.

EXPERIMENTAL PROCEDURE

The starting powder specimens consisted of consecutive layers of B$_2$O$_3$, metal oxide (PbO, CuO, or MoO$_3$), and C NTs preliminary synthesized by chemical vapor deposition (CVD). The specimens were placed in a crucible made of graphite, gradually heated to the synthesis temperature of ~2000 K (heating rate ~75 K/min) and annealed in a N$_2$ flow over 30 min [9]. The thermal treatment was carried out in a high frequency in-

---

*Corresponding author; GOLBERG.Dmitri@nims.go.jp

duction furnace JHF-VFX 110 QZ (JEOL). The temperature in the reaction zone was controlled by an electronic pyrometer CHINO IR-FB. After heating the reaction chamber was cooled down over 120 min in a N_2 flow.

The specimen was studied in a JEM-3000F field emission transmission electron microscope operated at 300 kV and equipped with a Noran Instruments EDX detector, and a 2D-DigiPEELS Gatan spectrometer with a CCD camera detector. During EDX and EELS analyses the electron beam was focused down to 0.5 nm in diameter. EELS spectra were taken in the diffraction mode of the microscope (camera length 12 cm) with exposure time of ≤2 s to minimize object deterioration and contamination.

Differential thermal analysis (DTA) and thermogravimetry (TG) of the starting materials and the synthesized products were carried out in air (humidity ~60 %) using a Thermo plus DTA-TG-8120 apparatus.

RESULTS AND DISCUSSION

The synthesized product consisted of BN and B-C-N graphite-like materials, e.g. hexagonal, rhombohedral and turbostratic BN, nanotubular B-C-N fibers, and long straight BN NTs. The yield of the BN NTs in the product was ~20 vol. %.

Figure 1a shows an assembly of the BN NTs produced through heating of C NTs with B_2O_3 and CuO in a N_2 flow. The BN NTs were well-graphitized and open-ended. They self-assembled into bundles (several μm long) composed of numerous tubes. A representative EELS spectrum taken from a bundle is shown in Fig. 1d. It shows that the tubes consist of B and N atoms (the peak barely visible at 284 eV is due to marginal contamination by C during analysis). The B/N ratio was calculated to be ~1.0. Similar open-ended NTs were observed if Mo or Pb oxides were substituted for Cu oxide.

The availability of open tube ends has made possible first observed placement of metal clusters, e.g. Mo, into an insulating BN tube channel (Fig. 1b). The representative EDX spectra taken from an encapsulated metal cluster and a neighboring hollow tubular area are shown in Fig. 1e. The clusters were found to freely levitate within the tube channels under electron irradiation. Continuous irradiation with ~5 A/cm^2 flux density over a few minutes forced clusters to coagulate and to form short elongated Mo fillings (~10 nm in length) inside the BN tubes. This made first natural "nanocables" made of the conductive nanowire cores and insulating tubular shields.

The preference for open BN tube-ends is thought to be due to metal atoms adsorbed at the growing BN edge and/or continuous annealing-out of non-six member BN atomic rings (which normally led to tube closure) due to highly reactive metal oxide vapors. The Cu, Mo and Pb oxides are known to be active oxidizers of C-based materials. Thus, the metal atoms at the growing edge should appear as a result of C tube interaction with metal (Me) oxides according to a chemical reaction: C + MeO → Me + CO (1). Therefore, the role of metal oxides in the present syntheses was to open a C NT template. The opening took place during the heating stage at ~1000 K. At higher T (~1700-2000 K) a chemical reaction between open C NTs and B_2O_3 vapor started. This reaction led to gradual transformation of C tubes: first, to B/N-doped C tubular fibres;

and, finally, to BN NTs *via* a substitution reaction: $B_2O_3 + C + N_2 \rightarrow BN + CO$ (2). Thus we conclude that under the oxidation tube opening, depletion of C atoms *via* outflow of CO, and crystallization of BN layers onto and within pre-existing tubular templates took place. Preliminary tube opening by metal oxide vapors *via* the reaction (1) ensured highly effective B and N substitution for C from all sides of a tubular template *via* the reaction (2) and high yield of the BN NTs.

Figure 1c shows the innermost terminated open shell of a BN NT after Pb oxide-promoted synthesis. The dark optical contrast on the edge corresponds to higher atomic density, e.g. a Pb atom or cluster. Such atoms are suggested to effectively prevent tube capping by "scootering" around the growing edge and ensuring morphological stability of tube growth. This mechanism, initially proposed for single-walled C NT synthesis [10], has until now not been verified by direct microscopic observations first presented here.

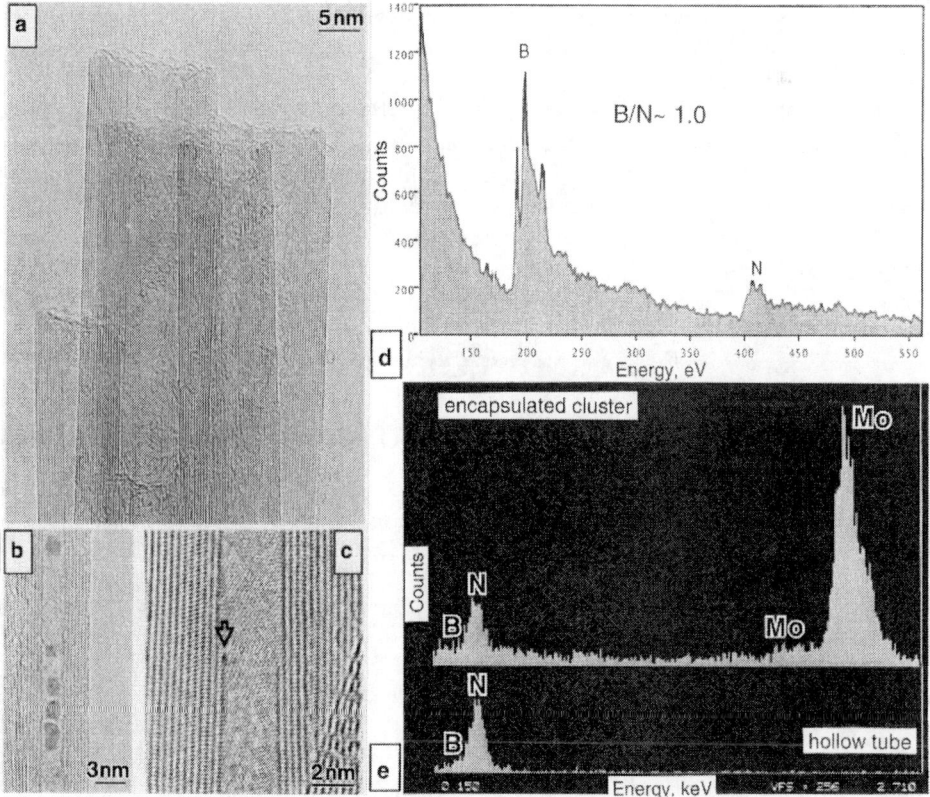

FIGURE 1. High-resolution transmission electron microscopy images of open BN nanotubes: a) A bundle produced under copper oxide-assisted synthesis; b) The outermost tube in a bundle produced under molybdenum oxide-assisted synthesis and exhibiting encapsulated molybdenum clusters levitating within the channel; c) The innermost terminated shell in a tube produced under lead oxide-assisted synthesis and exhibiting a definite dark contrast on the growing edge (marked with an arrow), assigned to a lead atom (or lead atom cluster) which may prevent tube closure. d) EELS spectrum taken from a tubular bundle; e) Comparative EDX spectra taken from an encapsulated metal cluster (i.e. Fig. 1b) and a hollow tube.

The comparative stability of BN and C NTs was explored by heating in air and measuring weight losses by thermogravimetry. We determined that the open BN tubes withstand up to ~1200-1400 K, while the C tubes start to deteriorate intensively *via* oxidation at ~800 K (chemical-vapor-deposited C tubes) or ~1000 K (arc-discharge C tubes). This suggests that BN NT use is advantageous in a high temperature environment.

The challenge is now utilization of open BN tubes, for example as extremely stable (thermally and chemically) light-weight gas accumulators, e.g. H_2, "nanofilters" for size sorting of small metal clusters, and/or natural "nanocables" made of a conductive metal core and insulating tubular shield.

CONCLUSIONS

Multi-walled nanotubes made of insulating BN and exhibiting exclusively open tip-ends were synthesized through high-temperature treatment of C nanotubes together with boron oxide and metal oxides (CuO, MoO_3, or PbO). The reason behind the existence of open tube ends was metal atoms adsorbed at the growing tube edge, which effectively prevented tube closure, and/or continuous tube oxidation by highly reactive metal oxide vapors. Thermal and chemical stability of BN nanotube product was found to be superior to conventional chemical-vapor-deposited and arc-discharge C nanotubes which may be a significant practical advantage of BN nanotubes over conventional C counterparts.

ACKNOWLEDGMENTS

This work was performed under a project of the Japan Science and Technology Corporation (JST).

REFERENCES

1. Chopra, N.G., Luyken, R.J., Cherrey, K., Crespi, V.H., Cohen, M.L., Louie, S.G., and Zettl, A., *Science* **269**, 966-967 (1995).
2. Loiseau, A., Willaime, F., Demoncy, N., Hug, G., and Pascard, H., *Phys. Rev. Lett.* **76**, 4737--4740 (1996).
3. Golberg, D., Bando, Y., Eremets, M., Takemura, K., Kurashima, K., and Yusa, H., *Appl. Phys. Lett.* **69**, 2045-2047 (1996).
4. Terrones, M., et al., *Chem. Phys. Lett.* **259**, 568-573 (1996).
5. Cumings, J., and Zettl, A., *Chem. Phys. Lett.* **316**, 211-216 (2000).
6. Iijima, S., *Nature* **354**, 56-58 (1991).
7. Blase, X., Rubio, A., Louie, S.G., and Cohen, M.L., *Europhys. Lett.* **28**, 335-340 (1994).
8. Tsang, S.C., Chen, Y.K., Harris, P.J.F., and Green, M.L.H., *Nature* **372**, 159-161 (1994).
9. Han, W., Bando, Y., Kurashima, K., and Sato, T. *Appl. Phys. Lett.* **73**, 3085-3087 (1998).
10. Thess, A., et al., *Science* **483**, 273-275 (1996).

Inter-Process Measurement of MWNT Rigidity and Fabrication of MWNT Junctions Through Nanorobotic Manipulations

Lixin Dong*, Fumihito Arai*, and Toshio Fukuda[†]

*Department of Micro System Engineering, Nagoya University
[†] Center for Cooperative Research in Advanced Sci. and Tech., Nagoya University,
Furo-cho, Chikusa-ku, Nagoya 464-8603, JAPAN

Abstract. Kinds of multi-walled carbon nanotube (MWNT) junctions including a kink junction, a cross junction, and a T-junction are fabricated by three-dimensional manipulations of MWNTs with nanorobotic manipulators operated inside a scanning electronic microscope (SEM). In order to obtain a junction with desired specification, it is required to select proper nanotubes with qualified properties, such as geometrical, mechanical and electronic ones, before assembling them into junctions. A simple inter-process measurement method is presented for determining the flexural rigidity of a single MWNT through a series of SEM image frames showing the buckling process of the nanotube, and an estimated value $EI_z = 8.641 \times 10^{-20} \mathrm{Nm}^2$ shows the high reliability of this method.

INTRODUCTION

The possibility of connecting carbon nanotubes (CNTs) with different diameters and chiralities has stimulated considerable interest recently [1,2]. This is because of the possibility of the junctions being building blocks of nanometer scale electronic devices and mechanical structures. Although junctions are randomly found in CNT samples, it is significant to find proper techniques to fabricate such basic structures. For doing so, nanomanipulation might be one of the promising ways, especially for the construction of complex junctions that will be unobtainable otherwise. Nanomanipulations of CNTs in 2-D surface with an AFM [3] and in 3-D space with a nanomanipulator [4] have been tried, and the application of such manipulations in electrical transport through nanotube junctions has been investigated [5]. Here we show two methods for the fabrication of MWNT junctions in 3-D free space with nanorobotic manipulators [6]. In order to obtain a junction with desired specification, it is required to select proper nanotubes with qualified properties, such as geometrical, mechanical and electronic ones, before assembling them into junctions. Hence, we also present a simple but accurate enough inter-process measurement method for determining the flexural rigidity of nanotubes.

MEASUREMENT OF FLEXURAL RIGIDITY OF MWNTS

Previous methods for measuring the mechanical properties of individual nanotubes include observing the thermally induced or electric field-induced vibration of "cantilevered" nanotubes inside a transmission electron microscope (TEM) [7-9], measuring the lateral bending of suspended MWNTs [10,11] or of single-walled nanotube ropes [12] with atomic force microscope (AFM), measuring axial compression with tapping-mode AFM [13], and examining the pattern of mechanically deformed MWNTs in polymer composites with a TEM [14]. However, as constructing a CNT-based structure through nanomanipulation, it is needed to test and sort proper nanotubes before assembling them into the structure. Neither an AFM nor a TEM is suitable for inter-process measurement of the mechanical properties in such cases, because it is difficult to apply AFM for 3-D manipulation, and the vacuum chamber of TEM is to narrow for holding a nanomanipulator that can perform complex operation. Presently, a SEM is the only selection. In fact, Yu et al. [4] has tried to measure the strength of MWNTs under tensile load through a "nanostressing stage" inside SEM.

By buckling a single MWNT, we evaluated the flexural rigidity of nanotubes by measuring the forces subjected to the nanotube and the deformations of the nanotube and cantilever. Fig.1(a) and (b) show two of the SEM images recording the buckling process. According to Euler's Formula and force balance conditions, the flexural rigidity of the MWNT (EI_z) can be expressed as (refer to the model shown in Fig.1(c))

$$EI_z = k\delta / [\pi^2 (h_2/l_2^3 - h_1/l_1^3)] \tag{1}$$

Substitute the values shown in Fig.1 into Eq.(1), it can be obtained that $EI_z = 8.641 \times 10^{-20} \text{Nm}^2$.

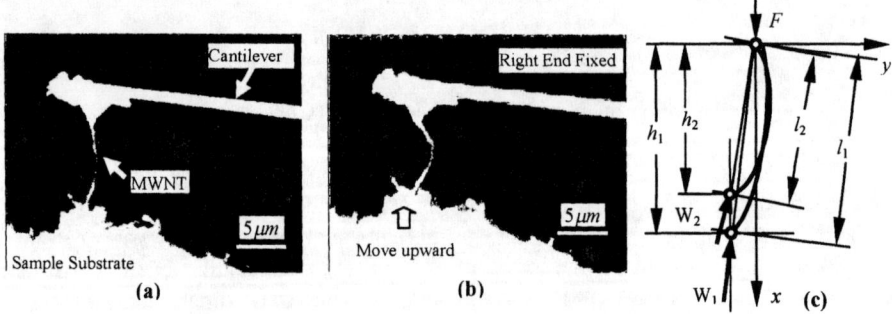

FIGURE 1. Estimate the flexural rigidity of a MWNT with the dimensions of ~φ30nm×7μm. Measured lengths: h_1=6.852 μm, h_2=5.755 μm, l_1=6.886 μm, l_2=5.796 μm, measured deformation of cantilever tip in x direction (not shown in Fig.1(c)): δ=0.244 μm, stiffness of the cantilever: k=0.03N/m.

For getting an estimation of Young's modulus further, it is reasonable to hypotheses the MWNT to be a solid cylinder, then we got E=2.17TPa. It is a little bit larger than the mean value obtained in [7], where they got an average value E=1.8TPa from a series data ranging from 0.4 to 4.15TPa.

FABRICATION OF MWNT JUNCTIONS

The types of CNT-junctions are determined by the kinds of CNTs, the configurations of CNTs and the conjunction methods:

(1) Kinds of CNTs: 1) metallic SWNTs, 2) semiconducting SWNTs, 3) semimetallic SWNTs, and 4) (metallic) MWNTs.

(2) Configurations: 1) V-junctions (kink-junctions), 2) T-junctions, 3) X-junctions (cross-junctions), 4) Y-junctions, and 5) more complex junctions (e.g. 3D).

(3) Conjunction methods: 1) van der Waals, 2) electron-beam-induced deposition (EBID), 3) chemical bonds, and 4) other methods.

Here we show a kink junction formed by bending a nanotube over its elastic limit (Fig.2), an X-junction (Fig.3(a)) and a T-junction (Fig.3(b)) constructed by joining two nanotubes with van der Waals forces.

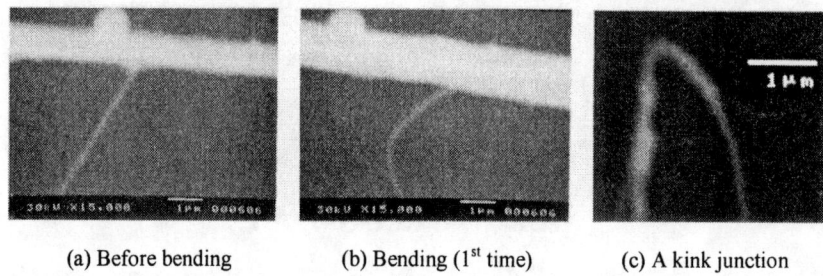

(a) Before bending   (b) Bending (1st time)   (c) A kink junction

FIGURE 2. A multi-walled carbon nanotube (~φ40nm×6μm) has been bent for 20 times between an AFM cantilever and the substrate, (a) shows the original situation, (b) shows the state as performing the first time of bending, it can be found that the contact point of the upper end of the nanotube and the cantilever slide, after then the upper end fixed, after 20 times of bending operation, an unrecoverable deformation occurred on the nanotube as shown in (c)—a kink junction formed.

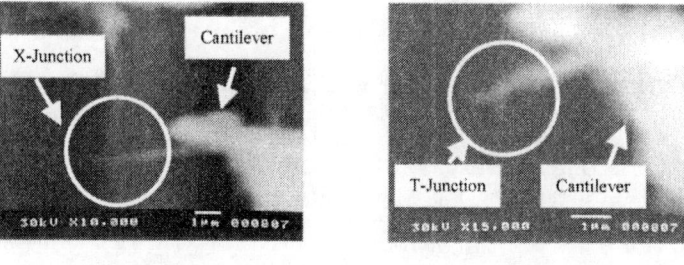

(a) X-junction   (b) T-Junction

FIGURE 3. The dimensions of the two MWNTs in the X-junction (a) are ~φ40nm×6μm (horizontal one) and ~φ50nm×7μm (vertical one), which are supported between the raw material of carbon nanotubes on the sample substrate and an AFM cantilever. Although it cannot be determined clearly that how the two MWNTs connected since the limitation of the SEM, they are most possibly linked by van der Waals forces. The T-junction (b) is formed in similar way, which is composed with two MWNTs: ~φ40nm×3μm (horizontal one) and ~φ50nm×2 μm (vertical one).

After a junction is fabricated, it can also be tested in situ. The stiffness of the X-junction is tested by pushing (as shown in Fig.4(a)) and pulling (Fig.4(b)) the vertical arm of the X-junction. The force acted on the contact point of the horizontal nanotube and the cantilever is measured as $F=F_2-F_1=314.9$nN. However, the quantitative twist stiffness of the junction cannot be obtained by this method because it is a hyperstatic problem. The application of AFM cantilevers here also provides a potential for testing the electronic properties of MWNTs and their junctions.

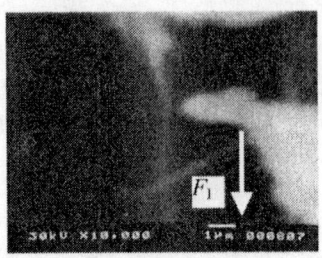
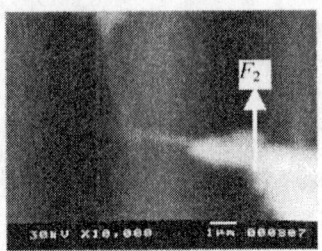

(a) Push X-junction down   (b) Pull X-junction up

FIGURE 4. The twist stiffness of the MWNT X-junction shown in Fig.3(a) is tested by pushing and pulling the vertical arm

ACKNOWLEDGMENTS

This research work was supported in part by the Scientific Research Fund of the Ministry of Education. We are grateful to Prof. Y. Saito at Mie University for providing us MWNT samples, Prof. Shinohara at Nagoya University for a helpful discussion, and the first author would like to thank Prof. R. Saito at University of Electro-Communications for presenting an instructive book on CNTs.

REFERENCES

1. Saito, R., Dresselhaus, G., and Dresselhaus, M. S., *Phys. Rev. B* **53**, 2044-2050 (1996).
2. Iijima, S., Ichihashi, T., and Ando, Y., *Nature* (London) **356**, 776-778 (1992).
3. Falvo, M. R., Taylor II, R. M., et al., *Nature* (London) **397**, 236-238 (1999).
4. Yu, M.F., Lourie, O., et al., *Science* **287**, 637-640 (2000).
5. Postma, H.W.Ch., Jonge, M. de, Yao, Z., and Dekker, C., *Phys. Rev. B* **62**, R10653-10656 (2000).
6. Dong, L.X., Arai, F., and Fukuda, T., in *Proc. of the 2000 Int'l Symp. on Micromechatronics and Human Science(IEEE)*, 151-156 (2000).
7. Treacy, M. J., Ebbesen, T. W., and Gibson, J. M., *Nature* (London) **381**, 678-680 (1996).
8. Poncharal, P., Wang, Z.L., Ugarte, D., Heer, W.A.de, *Science* **283**, 1513-1516 (1999).
9. Krishnan, A., Dujardin E., and Ebbesen, T.W., *Phys. Rev. B* **58**, 14013-14019 (1998).
10. Wong, E.W., Sheehan P.E., and Lieber, C.M., *Science* **277**, 1971-1975 (1997).
11. Salvetat, J.-P., et al., *Phys. Rev. Lett.* **82**, 944-947 (1999).
12. Walters, D.A., et al., *Appl. Phys. Lett.* **74**, 3803-3805 (1999).
13. Yu, M.F., Kowalewski, T., and Ruoff, R.S., *Phys. Rev. Lett.* **85**, 1456-1459 (2000).
14. Lourie, O., Cox D.M., and Wagner, H.D., *Phys. Rev. Lett.* **81**, 1638-1641 (1998).

Selective Lateral Nano-bridging of Carbon Nanowire between Catalytic Contact Electrodes

Yun-Hi Lee*, Yoon-Taek Jang*, Chang-Hoon Choi*, Eun-Kyu Kim*, Byeong-Kwon Ju*
Dong-Ho Kim†, Chang-Woo Lee†, Jin-Koog Shin‡, Sung-Tae Kim‡

*Korea Institute of Science & Technology, P. O. Box 131, Seoul, Korea
†Dept. of Physics, Yeungnam University, Kyungsan, Korea
‡ LG Electronics Institute of Technology, Korea

Abstract. We report selective lateral nano-bridging of carbon nanowire (CNW) between micro-sized islands using conventional photolithography technique necessary for the nanomachining and the molecular device applications compatible with Si-based process. Most distinct feature in this work is *to use a growth barrier* of Nb metal or insulating layer on the top of the catalytic Ni metal to prevent the growth of CNW from vertical direction to the substrate. As a result, CNWs of either *"straight line" or a"Y-junction"* were selectively grown between lateral sides of the catalytic metals or pre-defined electrodes without any trace of vertical growth. These results clearly indicate that this method would be one of the most feasible fabrication techniques for the nanomachines or the electronic applications with high integration level through process optimization.

INTRODUCTION

Carbon nanotubes (CNTs) as a molecular-scale device elements and nano-components for nanomachines was intensively studied. Recently, a molecular-scale device element based on a suspended, crossed nanotube geometry that leads to bistable, electrostatically switchable ON/ OFF states was introduced.[1] The device elements are naturally addressable in large arrays by the carbon nanotube molecular wires making up the devices. However, these require an ability to synthesize, isolate, manipulate and connect individual nanotubes. Dai et al. showed a novel strategy for making high-quality individual nanotubes bridging two metallic islands on silicon wafers patterned with μm - scale grown at 1000 °C and their functional applications for electronic devices.[2] Also, Papadopoulos et al. introduced Y-junction formation technique using branched nanochannel alumina templates and, however for electronic applications grown CNTs were removed from the templates and dispersed onto prepatterned electrodes.[3]

In this work, we present Selective Lateral Nano-bridging of Carbon Nanowire between Catalytic Contact Electrodes using growth barrier technology utilizing the chemical vapor deposition (CVD) method at relatively low process temperature of 650 - 750 °C].[4] Using this method CNTs bridging two parallel patterned structures with *a perfect "Y-junction" and "straight line"* were formed. The most importance of our work is to demonstrate that a good selectivity is achieved using low temperature CVD process as well as conventional photolithography, the cheapest method and easily combinable with the conventional Si - based process.

EXPERIMENTS

First, in our experiment, the layer structure was prepared by stacking SiO_2 - Ni - Nb layers on doped Si substrate, where Nb metal is introduced as a barrier layer for vertical growing, covering top of the Ni catalytic layer. A key technology of this method is to employ the thermally stable Nb layer for the CNW growth barrier as well as the electrode terminals after growth. In most cases, not only the high growth temperatures but also chemical properties of process gases limit thermal stability of barrier material laid on top of the catalytic Ni during CNW growth.

The SiO_2 layer was grown on the n - type heavily-doped Si wafer by conventional dry oxidation process followed by deposition of 10-100nm - thick catalytic film of Ni by dc sputtering method using Ar plasma. Then lift-off process was used to define patterns before CNW growth. The CVD chamber temperature was rapidly raised to the process temperature of 650 - 750 °C within 10 - 20 minutes by halogen lamps after evacuation down to 10^{-2} Torr. During this step, the substrates were first treated under hydrogen gas for nearly 10 - 20 min in order to activate the surface of the catalytic metal and to prevent the corrosion of Ni metal due to residual gases. During the growth, the total pressure of the chamber was kept constant at 10 Torr, while the total flow rate of process gases was maintained at 160 sccm. After the CNW growth, the chamber was purged continuously with a mixture of H_2 and Ar until the camber temperature reaches to room temperature.

RESULTS AND DISCUSSION

Fig 1 (a) shows a SEM image of CNWs grown laterally from one Ni pad to the adjacent pad separated by 500 nm using a mixture of C_2H_2 and N_2 as a carrier gas at 650 °C. In this case, no Nb growth barrier is used. The CNWs are severely bended through the whole length between the pads showing top-to-top growth characteristics. Furthermore both ends of CNWs, i.e., the embedded parts of CNT, showed very large bending and some of them look highly entangled one another. Bridging CNWs between two Ni pads were usually observed just after growing for 3 min, and the diameter of tubes ranges mostly from 30 to 40 nm. Figure 1(b) shows the SEM image of the laterally grown CNW bridges between two Nb passivated Ni pads after 2 min growth using a mixture of C_2H_2, H_2, and Ar at 750 °C. The CNWs grown under this condition are very straight with little bending across the whole length. The diameter of these selectively grown nanotubes is about 10 - 20 nm. CNW bridge form the side of Ni catalyst often extends through the Ni catalyst into the other Ni pad. Several interesting facts are observed : first, direction of CNW growth deviates from the opposite side, declining some degree, as indicated in Fig. 2; second, comparing the gap distance between the electrodes, the gap spacing between Ni islands may be not a deterministic factor for selectivity of the lateral growth.

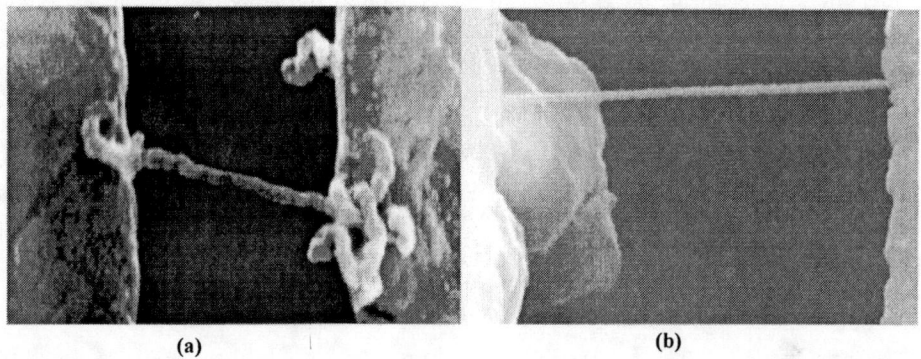

FIGURE 1. Scanning electron microscopy image of CNWs grown laterally from one Ni pad to the adjacent pad. (A) the entangled CNWs grown laterally from one Ni pad to the adjacent pad separated by 500 nm. In this case, no growth barrier is used. (B) CNW bridges two growth barrier passivated Ni pads after 2 min growth using a mixture of C_2H_2, H_2, and Ar at 750 °C.

Nearly same kinds of growth behaviors were confirmed another experiment such a replacement of Nb barrier with 50nm -thick - SiO_2 / Si_3N_4 film onto Ni pads. In Fig. 2 (b), there are three bridges forming a *"Y-junction"* between two Ni reservoirs. To our knowledge, this is the first observation yet reported on a selective site Y-junction growth of carbon wiress.

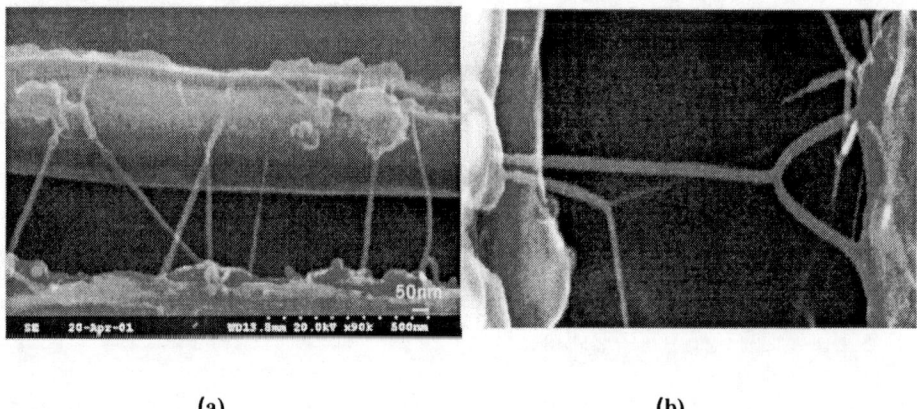

FIGURE 2. Typically observed CNW bridges between two growth barrier passivated Ni. a) direction of CNW growth deviates from the opposite side, declining some degree in most and b) SEM image for the a perfect "Y-shape" nanobridge between two catalytic Ni pads.

SUMMARY

We have successfully bridged the predefined using selective lateral growth method of CNWs. Especially, we achieved a direct nano-bridging of carbon nanowies using metallic and/or insulator growth barrier technology utilizing the chemical vapor deposition (CVD) method at relatively low process temperature of 650 - 750 °C. Using this method CNWs bridging two parallel patterned structures with *a perfect "Y-junction" or "straight line"* were formed. Our results show the promising bridging technique between electrodes for the high integration electronic and spintronic devices using growth barrier technique. These CNW bridges act as active components as well as nano-wire for the achievement of molecular - scale electronic devices with high integration density and a key block for the ultra miniaturized nanomechines.

ACKNOWLEDGEMENTS

This work was supported by the National Program for Tera-level Nanodevices of the Ministry of Science & Technology as one of the 21 century Frontier Programs (Contract No. 2N21080) and Technology for quantum dot - functional devices of the Korea Research Council of Fundamental Science & Technology(Contract No. 2N21100).

REFERENCES

[1] T. Rueckes, K. H. Kim, E. Joselevich, G. Y. Tsewng, C. L. Cheung, Charles M. Lieber, Science **289**, 94 (2000)
[2] J. Kong, H. T. Soh, A. M. Cassell, C. F. Quate, and H. Dai, Nature **395**, 878 (1998)
[3] C. Papadopoulos, A. rakitin, J. Li, A. S. Vedeneev, and J. M. Xu, Phy. Lev. Letts. **85**, 3476 (2000)
[4] Yun-Hi Lee, Yoon-Taek Jang, Chang-Hoon Choi, Eun-Kyu Kim, Byeong-Kwon Ju, Dong-Ho Kim, Chang-Woo Lee, Jin-Koog Shin, Sung-Tae Kim, Advanced Materials*(accepted)*

II. NANOTUBES: ELECTRONIC AND MECHANICAL PROPERTIES

Quantum Oscillations for the Spectral Moments of Raman spectra from SWCNT

H. Kuzmany[1], M. Hulman[1], W. Plank[1], A. Grueneis[1], Ch. Kramberger[1] H. Peterlik[1], T. Pichler[1,2], H. Kataura[3], and Y. Achiba[4]

[1] *Institut für Materialphysik der Universität Wien, Strudlhofg. 4, A-01090 Wien, Austria*
[2] *Institut für Festkörper- und Werkstofforschung Dresden, Dresden, Germany*
[3] *Department of Physics, Tokyo Metropolitan University, Tokyo, Japan*
[4] *Department of Chemistry, Tokyo Metropolitan University, Tokyo, Japan.*

Abstract.
Photoselective resonance Raman scattering is demonstrated to exhibit quantum oscillations for the spectral moments if the spectra are excited with a large number of different laser energies. The oscillations originate from the small extension of the tubes in transversal direction. An appropriate model was constructed which allows to provide a quantitative analysis of the oscillations. Assuming a Gaussian distribution of diameters values for a mean diameter and the width of the distribution were obtained. The fine structure in the Raman response is demonstrated to be due to a clustering of nanotube diameters.

INTRODUCTION

The Raman spectrum of single wall carbon nanotubes (SWCNTs) exhibits two dominating lines at 1580 cm^{-1} and around 180 cm^{-1}, respectively. The latter is unique for nanotubes and describes a radial breathing motion (radial breathing mode, RBM). This mode is of particular interest since its frequency scales as $1/d$ where d is the diameter of the nanotube. Even though this relation is generally accepted it was not possible sofar to deduce reliable average tube diameters from experiments since the respons of the mode with respect to position and shape depends dramatically on the frequency of the exciting laser. Even though a very large number of papers have reported on this behavior it remained difficult to understand. In two recent papers [1,2] we demonstrated that the mean frequency position and the peak frequency position of the mode are subjected to an oscillating behavior if the laser energy is tuned from deep red to deep blue. In the present report we analyze the oscillating behavior in more detail and investigate the origin of the oscillations as well as the way the RBM response must be analyzed in order to obtain reliable values for the tube diameter. In addition an oscillating behavior is also reported for the position of the overtone of the D-mode which is usually observed around 2600 cm^{-1}.

EXPERIMENTAL

The SWCNT's used in the present work were prepared from laser deposition in the form of bucky paper (samples BP) and by using different catalysts and different growth temperatures (samples TMU). Details of the preparation procedures were decribed previously [3,4]. The BP samples had a reported mean diameter of 1.38 nm and the TMU samples had a reported diameter between 0.9 and 1.5 nm. The as prepared material was heated in high vacuum of 10^{-7} mbar at the temperature of 1000 K for 12 hours for purification. Then the Raman spectra were recorded with up to 30 different laser lines covering the whole visible and near IR region between 413 nm and 1064 nm (1.23 - 3 eV). Measurements were performed by a triple spectrometer Dilor XY 500 with a spectral resolution 2 cm^{-1}. The Raman system was calibrated for intensities using test samples with known scattering cross sections.

RESULTS AND DISCUSSION

Figure 1 depicts a selected set of Raman spectra of BP samples as excited with laser lines between 850 nm and 455 nm.

Positions of peaks depend on the laser excitation energy and obviously exhibit an oscillating up and down-shift as the energy changes. Likewise, the width of the Raman response is very sensitive to the laser energy and does therfore not allow an immediate conclusion on the width of the distribution of nanotube diameters. Also the scattering intensities exhibit an up and down oscillation. To demonstrate these oscillations more explicitly we have plotted in Fig. 2 the peak positions and average line position (first spectral moment $\langle \nu \rangle = \int \nu I(\nu) / \int I(\nu)$) for the RBM versus the laser energy. The oscillatory behavior for both sets of data is evident. The oscillation frequency is about 0.6 eV and the oscillation amplitude is about 20 cm^{-1} in the red.

Similarly to the first moment one can also plot the second moment $\langle \nu^2 \rangle = \int \nu^2 I(\nu) / \int I(\nu)$ or equivalently the variance $\Delta(\nu)$ versus laser excitation. Such data exhibit an oscillatory behavior as well [2]. In fact we have demonstrated that these oscillations are phase shifted by $\pi/4$ to the oscillations of the first moment.

In order to explain the oscillations we make use of the fact that energy gaps between van Hove singularities (transition energies) scale as $\epsilon_i = A_i/d$ [5] where A_i characterizes different pairs of the singularities. The model is schematically explained in Fig. 3

The radial lines on the right part of the figure indicate the increase of transition energy. The hached area defines the region where nanotubes are geometrically allowed. In the left part the increase of RBM frequency is plotted versus $1/d$. The

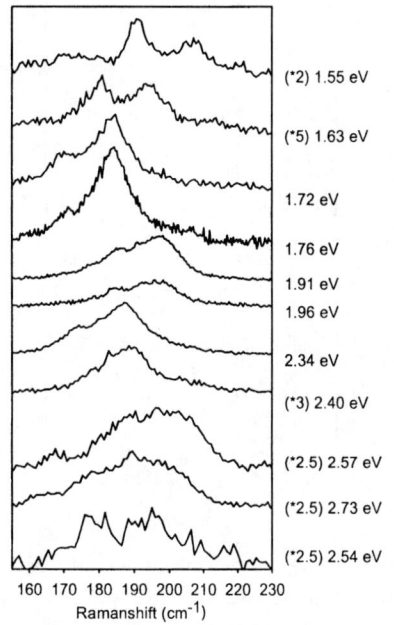

FIGURE 1. Raman line pattern for the RBM for excitation with 10 selected laser lines. Relative intensities are according to the calibration but scaled by the indicated factor.

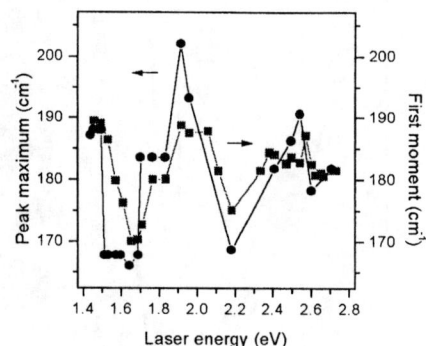

FIGURE 2. Maximum response and first moment for the RBM of bucky paper samples versus laser energy

two dash-dotted horizontal lines in the right part of the figure define the limit where standard tube material is available. The dashed arrows connect the resonance energies between the van Hove singularities on the right side of the $1/d$-axis with the frequencies of the RBM on the left side of the $1/d$-axis. When the laser energy is shifted from low to high energies tubes with decreasing diameter are tuned into resonance for a particular transition. While the laser is shifted every so often a new van Hove singularity opens up a new chanel for resonance but this resonance starts with the largest tubes available. Accordingly the first moment of the RBM response switches back to low values and the second moment shoots up.

We have recently developed two theoretical models to describe this behavior. One model uses the full density of states of all geometrically allowed tubes. The diameter distribution of the tubes was assumed to follow a Gaussian distribution. In the other model the van Hove resonances were assumed to fully dominate the cross section and the diameters were assumed quasicontinuously distributed but still weighted with a Gaussian profile. Both models reproduce the oscillations of the first and second moment well. A comparison between the two models is depicted in Fig. 4. Both models gave excellent agreement with experiments from standart

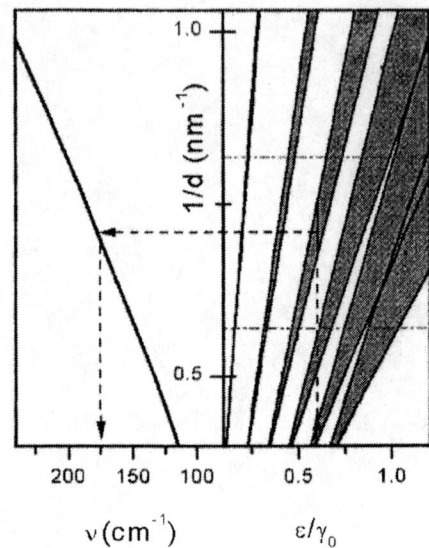

FIGURE 3.
$1/d$ versus electronic transition energy and RBM frequency. The hached area defines the range where SWCNTs are geometrically allowed.

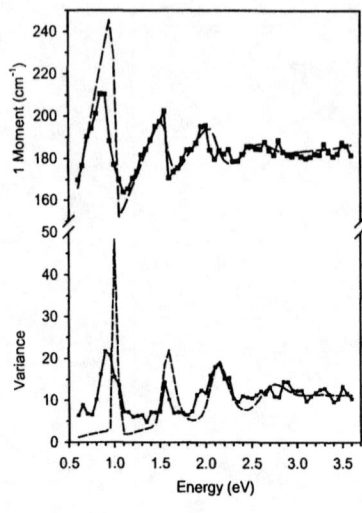

FIGURE 4. First moments and variances as calculated from two different models; dashed lines for the continuum model, dots and full drawn line for the full DOS model.

tube material with tube diameters of 1.36 nm.

According to the good results for standard tube material it was challenging to apply the model to tubes with varying diameter. We have tested tubes with diameters between 0.9 nm and 1.5 nm. The RBM was measured with only a few different laser lines and the parameters for the diameter distribution were evaluated. With these values the expected first moments and the variances were evaluated and compared to the experimental data. Results are depicted in Figure 5 for two samples.

Looking at the intensities in more detail reveals similar damped oscillations. This is demonstrated in Fig. 6 for a sample from the TMU set. Peak intensities are recorded for laser energies at 1.45 and 1.8 eV. Beyond 2 eV oscillations are damped out. The oscillations of the Raman cross section are usually less expressed if compared to the oscillations of the spectral moments.

Recently we observed an oscillatory behavior also for other Raman lines in the spectrum. Whereas the graphitic line at 1580 cm^{-1} is completely silent in that sense the defect line (d-line) between 1300 and 1360 cm^{-1} exhibits a linear up shift with increasing laser energy and very weak onsets of an oscillation. Checking the overtone of the d line a clear oscillatory behavior can be recognized on top of a rather strong scattering intensity. Experimental results are depicted in Fig. 7. The dispersion of the d-line in graphite was recently explained as a double resonance

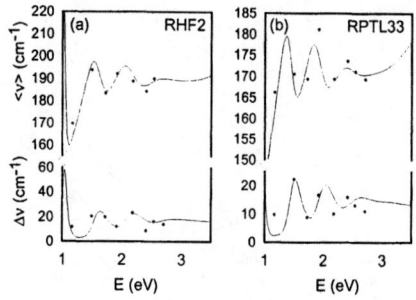

FIGURE 5. First moments and variances for two tubes with different diameter. RHF2: 1.3 nm, RPTL33: 1.47 nm. The full drawn lines are as calculated from the continuum approximation.

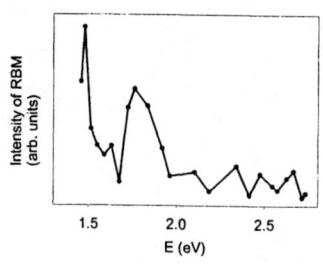

FIGURE 6. Raman cross section for sample NCKL16 from the TMU set.

process between the two bands which cross the Fermi level [6]. Most likely his explanation holds here as well.

In previous work and in the current presentation we have demonstrated that the full DOS model as well as the continuum approximation provide a good representation for the fine structure in the Raman spectra of the RBM and in the optical absorption spectra. This is still surprising due to the overlap of a rather large number of tubes with only slightly varying properties. The physical reason for the fine structure originates in fact from the nanoscopic geometry of the tubes. For such small entities the distribution of tube diameters is not really quasicontinuous but exhibits clusters. This is demonstrated in Fig. 8. The figure shows the density of diameters or more precisely the density of RBM frequencies for a certain frequency interval for two selected ranges of tube diameters. For the small tube as they are usually prepared in nanotube research clustering is evident (curve a). For large tubes (which have not been prepared so far as single wall species) the density versus frequency relation is completely smooth (curve b) If we weight curve a with a Gaussian distribution we obtain curve c which is almost a representative for the fine structure in the Raman response of the RBM.

Summarizing we have provided a consistent explanation for details of our Raman experiments. Quasi-periodic oscillations are consequences of the fact that both the transition energies and the frequency of the RBM scale as $1/d$ and that the system under consideration is subjected to severe macroscopic size quantizaiton. Fine structures in the Raman spectra are preserved even though we take into account

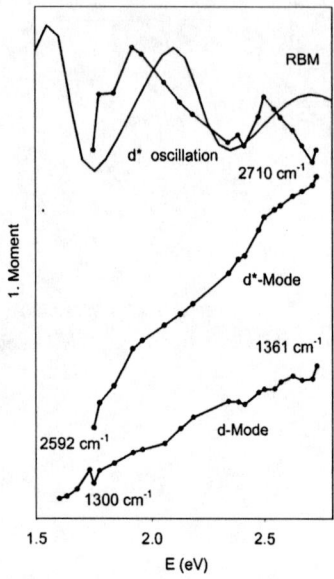

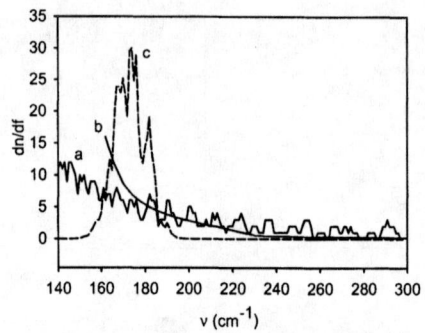

FIGURE 8. Density of RBM frequencies for standard tube material a, for very large tubes b (scaled), and after weightenig with a Gaussian distribution c.

FIGURE 7. Dispersion effect for non RBM modes; bottom: d line, center: overtone of d line, top: overtone after subtraction of a linear background. The smooth wave is as calculated from the continuum model.

all allowed tubes within a certain diameter range. They are ascribed to a clustering of tube diameters.

This work was supported by the Fonds zur Föderung der wissenschaftlichen Forschung in Austria, Project Nr. P12924, by the European network HPRN-CT 1999-00011, and by a Grant-in-Aid for Scientific Research on the Priority Area "Fullerenes and Nanotubes" by the Ministry of Education, Science and Culture of Japan.

REFERENCES

1. M. Milnera et al., Phys. Rev. Lett. **84**, 1324, (2000).
2. M. Hulman et al., Phys. Reb. B (to be published).
3. A.G. Rinzler et al., Appl. Phys. **A 67**, 29 (1998).
4. H. Kataura et al., Syntheitc Metals **103**, 2555 (1999).
5. R. Saito et al., Phys. Rev. **B 61**, 2981 (2000).
6. Ch. Thomsen, Phys. Rev Let. **85**, 5214 (2000).

Electronic Structure Studies of Carbon Nanotubes: Aligned, Doped and Filled

Jörg Fink*, Xianje Liu*, Heiko Peisert*, Thomas Pichler*[†],
Martin Knupfer*, Mark S. Golden*, Deron M. Walters[§], H. Kataura[¶]

Institute for Solid State and Materials Research Dresden, P.O. Box 270016, D-01171 Dresden, Germany

[†]*Institut für Materialphysik, Universität Wien, Strudlhofgasse 4, A-1090 Wien, Austria*

[§]*Center for Nanoscale Science and Technology, Rice University, Houston TX 77259, USA*

[¶]*Faculty of Science, Tokyo Metropolitan University, Tokyo, Japan*

Abstract. High-energy spectroscopies play an essential role in the characterization of the electronic structure of carbon nanostructures. In this contribution we report recent electronic structure studies of single-wall carbon nanotubes using electron energy-loss and photoemission spectroscopy. In particular, we have investigated the polarization dependence of interband and collective electronic excitations in single-wall carbon nanotubes. Futhermore, we have studied the character of the electron liquid in metallic undoped and doped carbon nanotubes. Finally, we have obtained information regarding the charge transfer between nanotubes and C_{60} in so-called peapods, i.e. nanotubes filled with C_{60} molecules.

INTRODUCTION

The bonding in carbon nanostructures is in most cases based on sp^2 hybrid orbitals. The electronic properties of these systems are predominantly determined by their π-orbitals which are formed by the $C2p_z$ electrons. The mechanical properties stem from the sp^2-hybrid σ-orbitals, which have three bonds within a plane. The electronic structure of carbon nanostructures also strongly depends on their dimensionality. The molecular fullerenes have a zero-dimensional (0D), while the carbon nanotubes have a quasi one-dimensional (1D) electronic structure. This should be contrasted with the quasi two-dimensional (2D) electronic structure of graphite. Although in the latter there is also a weak van der Waals interaction between the graphene sheets, the hopping integral for the π-electrons between the planes is about a factor 50 smaller than that for hopping within a single plane, thus leading to the quasi 2D electronic structure of graphite.

A similar situation also occurs when carbon nanostructures are condensed to form a solid. In the fullerides, there is only a weak interaction between the molecules. In bundles composed of carbon nanotubes, the interaction between the tubes is again weak. Therefore, the low-dimensional character of the electronic structure of carbon nanostructures is conserved after their condensation into a solid.

In this contribution, we present electronic structure studies of carbon nanostructures using electron energy-loss (EELS) and photoemission spectroscopy (PES). In particular, we have investigated the influence of the dimensionality on the electronic structure in these conjugated carbon systems. Firstly, we report studies of the polarization dependence of electronic excitations in aligned single-wall carbon nanotubes (SWNTs). Then we describe investigations of the character of the electron liquid in metallic undoped and doped SWNTs. Finally we have studied the charge transfer taking place between SWNTs and C_{60} molecules which are filled into the central hollow space within the nanotubes themselves.

ALIGNED SWNTs

Macroscopic oriented arrays of SWNTs can be produced by filtration in high magnetic fields (7-25 T) [1]. The suspended tubes were deposited under axial flow onto a filter membrane. In this way textured buckypaper films having only ~ 30 % empty volume could be prepared. From elastic electron scattering data, an angular width of the rope lattice peak of 45^0 FWHM was derived for the free-standing samples used in this study, indicating that this is the degree of texture of the SWNTs in the aligned material. Here we describe first attempts to obtain information on the anisotropic energy and momentum dependent dielectric function $\varepsilon(\omega,\mathbf{q})$. In EELS we measure the loss function, $\text{Im}[-1/\varepsilon(\omega,\mathbf{q})]$. In anisotropic systems ε is a tensor which in uniaxial systems reduces to two components $\varepsilon^{\|}$ and $\varepsilon^{\perp}$. Then the loss function is given by

$$\text{Im}\left[\frac{1}{\varepsilon^{\|}(\omega,\mathbf{q})\cos^2\Theta + \varepsilon^{\perp}(\omega,\mathbf{q})\sin^2\Theta}\right]$$

where Θ is the angle between the axis and the momentum vector $\mathbf{q}$. In Fig. 1 we show momentum dependent loss functions of textured SWNTs for various momentum transfers, parallel and perpendicular to the main orientation direction. Here we focus only on excitations of the π electron system which occur below ~ 10 eV. The spectra are dominated by the π-plasmon, which occurs for small q at ~ 6 eV. This peak in the loss function - which is observed in all conjugated carbon systems - is related to a 4 eV interband transition between a flat band region in the occupied π-band-structure and a similar region in the unoccupied π^*-bands. In addition, in the low energy region, three peaks are observed, which can be analysed in terms of transitions between van Hove singularities in the occupied π and the unoccupied π*-density of states. The first and the second transition correspond to transitions in semiconducting tubes while the third can be assigned

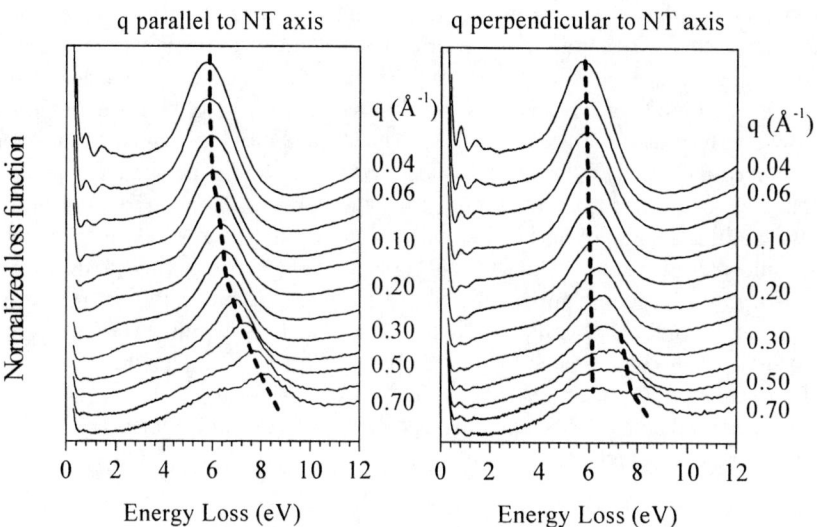

FIGURE 1. The loss functions of textured single-wall nanotubes for various momentum transfers **q**, parallel and perpendicular to the orientation axis.

to transitions across the pseudogap in metallic tubes. These transitions have been previously studied in non-oriented samples [2]. In the orientated samples, for **q** perpendicular to the orientation direction, the π-plasmon shows almost no dispersion. The dispersive feature at higher **q** probably comes from $\varepsilon^{\parallel}$-contributions, which are visible due to the not complete orientation of the sample. The fact that we see for all features no dispersion for **q** perpendicular to the orientation indicates that these excitations are related to interband transitions between narrow π and π* bands. These narrow π-bands are expected in these quasi 1D systems for wave vectors perpendicular to the tube axis, since in this direction the electronic structure is similar to a molecular fullerene. Indeed, for C_{60} no dispersion of the π-π* excitations have been observed in momentum dependent EELS measurements [3].

For **q** parallel to the tube orientation, the π plasmon shows a strong dispersion similar to that in polyacetylene (for **q** parallel to the polymer axis) [4] or graphite (for **q** parallel to the planes) [5]. This is quite reasonable, since in these systems, for the given wave-vector direction, there are wide π bands. Therefore, the 4eV-interband transitions and correspondingly the π plasmon should show a large dispersion in agreement with the experimental data.

We note that the low-energy transitions between the van-Hove singularities show no dispersion for either direction of the momentum transfer, although there is an anisotropy in their intensity. After removing contributions from the direct beam and the quasielastic peak at zero energy, our analysis shows that these transitions are twice as strong for **q** perpendicular to the orientation as compared to the case for **q** parallel to the orienation. Due to the finite orientation of the nanotubes, this would be compatible with a full polarization of these transitions for **q** perpendicular to the nanotubes. In this way it is also understandable that the low-energy interband transitions show no dispersion. Theoretical calculations [6] predict that the first three interband excitations should be seen for polarization parallel to the nanotube axis, both for nanotubes close to the zig-zag and to the armchair axis. For polarization perpendicular to the tube, only one transition should be observed, which should occur between the first and the second transition. The reason for the difference between the present experimental results and the theoretical calculations is not clear. Up to now, only vertical excitations (q=0) have been treated theoretically [6]. Possibly, inclusion of non-vertical transitions may strongly alter the transition matrix elements. From the experimental side, EELS and optical experiments should be performed on samples with a higher degree of orientation.

ALKALI METAL DOPED SWNTs

In strictly 1D metallic systems the interaction between the charge carriers is particularly strong and therefore important deviations from Fermi liquid theory or even its breakdown is expected. Theoretically this scenario has been described within the framework of Luttinger liquid models in which the spectral weight seen in PES close to the Fermi level should be suppressed [7]. Recent transport measurements indicate a Luttinger liquid behavior in metallic SWNTs [8].

X-ray induced PES (using a photoenergy hv = 1486.6 eV) of purified SWNTs shows strong O, N, and Na contaminations. After annealing at ~ 650^0C in UHV those contaminations are strongly reduced, but there remain traces of Ni (~ 0.5 at%) and Co (~ 0.5 at%) catalyst particles. For such UHV-annealed material, the valence band spectra show a clear Fermi edge, however, one can estimate from cross-section arguments that this is due to the 3d states from the catalyst particles. When using photon energies hv = 21.2 and 40.9 eV, hardly any Fermi edge is visible. In Fig. 2 we contrast such spectra of SWNTs to spectra of Pt metal where a clear Fermi edge is observed. The reason why at these photon energies no Fermi edge from the catalyst particles is observed is that for the low photon

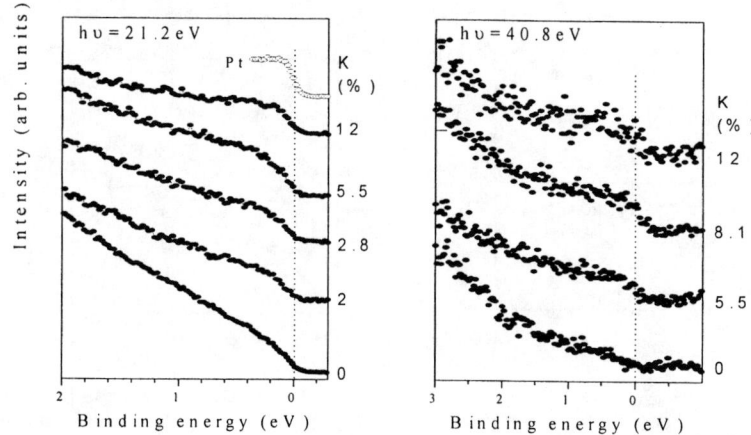

FIGURE 2. Photoemission spectra of undoped and K-doped SWNTs near the Fermi energy. For comparison we show the Fermi edge of Pt metal.

energies the photoionisation cross section for the Ni3d and Co3d states is strongly reduced compared to that of the C2p states: $\sigma(\text{Ni3d})/\sigma(\text{C2p}) = 0.65$ for $h\nu = 21.2$ eV, $= 4.5$ for $h\nu = 40.8$ eV, and $= 256$ for $h\nu = 1486.6$ eV.

In the alkali metal doped samples, due to a charge transfer from the alkali metals to the SWNTs, a filling of the unoccupied π^* states of the metallic and the semiconducting tubes is expected. It is immediately evident that in the K-intercalated samles there is a Fermi edge, even for the low photon energies(see Fig. 2). Similar results were also obtained for Na-intercalated samples (not shown). In alkali-metal doped graphite, one knows that the dimensionality is strongly reduced upon intercalation due to a reduction in the hybridization between graphene sheets. This manifests itself in the form of a strong increase of the anisoptropy of the electrical transport upon intercalation. An analogous effect can also be expected for intercalated SWNTs. Therefore, in the doped bundles, the electronic structure should have even more 1D character and, according to theoretical predictions, the spectral weight at the Fermi level should be still further suppressed. On the other hand, when the Fermi level is shifted to higher energy, the newly occupied bands do not exhibit the linear dispersion typical of 1D systems. At this stage, this would appear to be a feasible explanation for the observation of a Fermi edge in the PES of alkali metal doped SWNT bundles It would be interesting to study this transition between a pure 1D to a more 3D electronic structure by photoelectron spectroscopy in more detail.

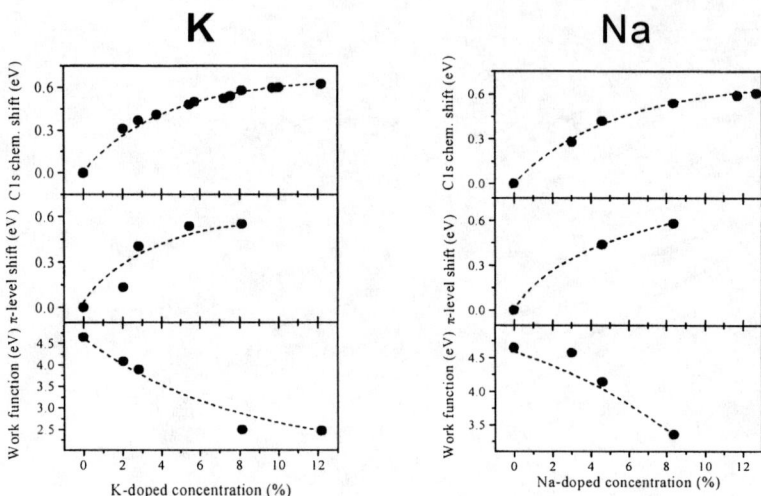

FIGURE 3. Shift of the C1s peak, the π-band related peak in the density of states and the work function for K- and Na-doped carbon nanotubes as a function of alkali metal concentration.

Similar to the case of intercalated graphite [9], the size of the shift of the Fermi level upon intercalation can also be obtained from photoelectron spectroscopy. In the spectra of the undoped SWNT samples there is a peak at 3 eV, which can be assigned to a high density of states of the π-bands due to the flat band region near the M point of the graphene Brillouin zone. Upon intercalation, this peak moves to higher binding energy because its energy is measured relative to the Fermi level and the latter moves to higher energy. This shift is shown in Fig. 3 together with a similar shift observed for the energy of the C1s level (which is also measured relative to the Fermi level). Measuring the energy difference between the photon energy and the complete width of the photoemission spectrum (from the Fermi level down to the high binding energy cut-off) the work function of the sample can be determined. For the undoped sample we arrive at a workfunction value of 4.65 ± 0.1 eV. For small alkali metal concentrations the workfunction shift is similar to the shifts observed for the C1s level and the peak due to the flat band region. The much larger workfunction shifts observed for higher dopant concentrations can be explained by the formation of alkali metal suboxides at the near surface region.

To conclude this section, from the shifts shown in Fig. 3 we derive a shift of the Fermi level into the π* states of about 0.6 eV upon doping to a level of about 10% alkali metal per C atom. This value should be compared with shifts of the Fermi level of ~ 1.2 eV in intercalated graphite for slightly larger dopant concentrations [9].

FILLED SWNTs

By opening SWNTs and heating them together with fullerenes at 650°C the nanotubes can be filled with (for example) C_{60} [10,11]. An interesting question is whether there is a charge transfer between the nanotubes and the C_{60}, the latter being generally believed to be a strong acceptor. Electron diffraction data of such samples show a peak at 0.65 Å^{-1} which corresponds to the distance between two C_{60} balls of 10 Å *within* the SWNT. The loss function (not shown) indicates the presence, besides the the SWNT-related spectral weight, of small non-dispersive features due to the interband excitations of the encapsulated C_{60} molecules. In Fig. 4 we show C1s EELS spectra of an unfilled SWNT reference sample and of filled SWNTs (peapods). These spectra give a measure, in a first approximation, of the density of unocupied states. There are measurable differences between the spectrum of the peapods and the fullerene-free reference sample. These differences can be ascribed to the density of unoccupied states of the fullerenes inside the SWNTs. From the size of the difference, we estimate a filling factor (related to the C atoms) of about 4 %. The difference spectrum (peapods minus reference) shown in the right panel of Fig. 4 shows the strong similarity between the fullerene-related peapod spectral weight to the spectrum of pure C_{60}. This

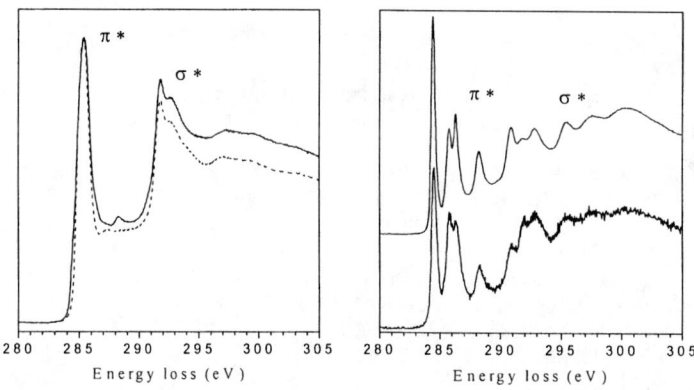

FIGURE 4 Left panel: C1s EELS spectra of C_{60} filled SWNTs (solid line) and a fullerene-free reference sample (dashed line). Right panel: the difference spectrum (peapods minus reference) (lower curve) compared to a C1s EELS spectrum of pure solid C_{60} (upper curve).

indicates that the electronic structure of the C_{60} molecules in the peapods is not very different from that of C_{60} in fullerides. In particular, the intensity of the first peak which corresponds to the unoccopied density of states of C_{60} is not strongly altered. From the comparison of the intensity of the LUMO peaks one can conclude that the charge transfer from the SWNTs to the C_{60} molecules is less than 1 electron/C_{60}. This is in line with theoretical estimates [12] which predict no charge transfer in these peapod structures.

CONCLUSIONS

At present although there are no highly crystalline samples of SWNTs with a single well-defined helicity available for bulk spectroscopic investigations of their electronic structure, there is a continuos development of our understanding of the electronic structure of these new materials by the use of high-energy spectroscopies such as electron energy-loss and photoemission spectroscopy.

ACKNOWLEDGMENTS

We acknowledge financal support from the EU under the TMR Research Network 'FULPROP' (ERBFMRXCRT-970155) and the IST Project SATURN, from the DFG (FI 439/8-1 and PO 392/10-1), and the SMWK (75 31.50-03-823-98/5). T.P. thanks the ÖAW for funding an APART grant.

REFERENCES

1. Smith, B.W., Benes, Z., Luzzi, D.E., Fischer, J.E., Walters, D.A., Casavant, M.J., Schmidt, J. and, Smalley, R.E., *Appl. Phys. Lett.* **77**, 663 (2000).
2. Pichler, T., Knupfer, M., Golden, M.S., Fink, J., Rinzler, A., and Smalley, R.E., *Phys. Rev. Lett.* **80**, 4729 (1998).
3. Sohmen, E., Fink, J., and Krätschmer, W., *Z. Phys. B* **86**, 87 (1992).
4. Fink, J., and Leising, G., *Phys. Rev. B* **34**, 5320 (1986).
5. Büchner, U., *phys. stat. sol.* **81**, 227 (1977).
6. Lin, M.F., *Phys. Rev. B* **62**, 13153 (2000).
7. Grioni, M., and Voigt, J., in "Electron spectroscopies applied to low-dimensional materials", *Kluwer Academic Publ.* (2000).
8. Bockrath, M., Cobden, D.H., Lu, J., Rinzler, A.G., Smalley, R.E., Balents, L., and McEuen, P.L., *Nature* **397**, 598 (1999).
9. Pfluger, P., and Güntherodt, H.J., in Festkörperprobleme *(Advances in Solid State Physics)(Vieweg, Braunschweig, 1981), Vol.* 21, p. 271.
10. Smith, B.W., Monthioux, M., Luzzi, D.E., *Nature* **396**, 323 (1998).
11. Kataura, H., Kodama, T., Kikuchi, K., Hirahara, K., Iijima, S., Suzuki, S., Krätschmer, W., and Achiba, Y., this volume
12. Louie, S.G., this volume

Lithium Storage in Single Wall Carbon Nanotubes

B. Gao, H. Shimoda, X.P. Tang, A. Kleinhammes, L. Fleming, Y. Wu and O. Zhou[1]

Curriculum in Applied and Materials Sciences
Department of Physics and Astronomy
University of North Carolina at Chapel Hill, Chapel Hill, NC 27599, USA

---

[1] Email: zhou@physics.unc.edu

ABSTRACT

The effects of structure and morphology on lithium storage in single wall carbon nanotubes (SWNTs) were studied by electrochemistry, x-ray diffraction and nuclear magnetic resonance (NMR) techniques. Purified SWNT bundles were chemically etched to variable lengths and were reacted with Li via the electrochemical and solid state routes. The reversible Li storage capacity increased from LiC_6 in close-end SWNTs to LiC_3 after etching. The increase is attributed to diffusion of the Li ions into the interior space of the individual SWNTs through the open ends and defects on the sidewalls.

INTRODUCTION

Materials that undergo reversible redox reaction with lithium can potentially be utilized as electrodes in lithium-based batteries. $LiMn_2O_4$ and lithium intercalated carbon (graphite or disordered carbon) are used as the two working electrodes in the commercial Li-Ion batteries. The lifetime of the battery depends on the specific Li storage capacities of the two electrodes. Nanomaterials often exhibit enhanced electrochemical properties compared to their counterparts in bulk form. For example, nanoparticles of metal oxides (i.e. SnO, NiO, CoO) are shown to have reversible Li capacities in the order of 600-800mAh/g and have excellent cycling behavior [1, 2]. The capacities are significantly higher than that of graphite (LiC_6, or 372mAh/g) and disordered carbon (~600mAh/g). The storage mechanisms are not well understood but are believed to be due to redox reaction of lithium with the surface oxides. We have recently demonstrated that nanostructured Si can electrochemically react with Li at 300K to yield a reversible capacity of over 800mAh/g [3]. The reaction temperature is about 400K lower than that for the bulk Si [4].

Carbon nanotubes can potentially accommodate guest species in both the interstitial channels (analogous to the interlayer space in graphite) and in the interior spaces, thus afford a higher storage capacity than graphite. Earlier works indeed demonstrated intercalation of alkali metals into the nanotube structures [5-9]. The saturation composition and the intercalation kinetics, however, depend on the structure and morphology of the carbon nanotubes. For example, the electrochemical properties of multi-wall carbon nanotubes (MWNTs) synthesized by the arc-discharge method differ significantly from those by the chemical vapor deposition method. The difference is related to the defect density of the nanotubes. The reactivity of close-end SWNT bundles is similar to that of graphite. Electron donors such as alkali metal Li can diffuse into the interstitial channels to a saturation composition of LiC_6 [10, 11]. The closed-shell and nearly defect-free structure of the SWNTs makes the interior space inaccessible for ion diffusion. The interior space can be exposed by post-processing such as mechanical ball-milling [12], which leads to a substantial increase in the Li/C ratio.

In this paper, we report the effects of chemical etching on lithium intercalation into SWNTs. Electrochemistry and solid state nuclear magnetic resonance measurements show unambiguously that the Li storage capacity is increased from LiC_6 to LiC_3 after etching. The increased storage capacity is attributed to Li diffusion into the inner cores of the SWNTs.

MATERIALS PREPARATION

SWNTs used in this study were synthesized by the laser ablation method, using 10wt% ^{13}C enriched targets [13]. As-grown materials were purified by a two-step process involving first refluxing SWNTs in 20vol.% of H_2O_2 in H_2O then filtration [14]. Typical purified materials contained over 90% SWNT bundles. The bundle and individual nanotube diameters were in the range of 10-50nm and 1.3-1.5nm, respectively. The purified SWNTs were chemically etched to short bundles using a previously reported procedure [15]. The nanotubes were dispersed in a beaker containing a 3:1 mixture of concentrated sulfuric and nitric acids and sonicated for 24 hours [6]. The solution was then diluted by de-ionized water and filtered. The yield was typically 20 wt.%. All the

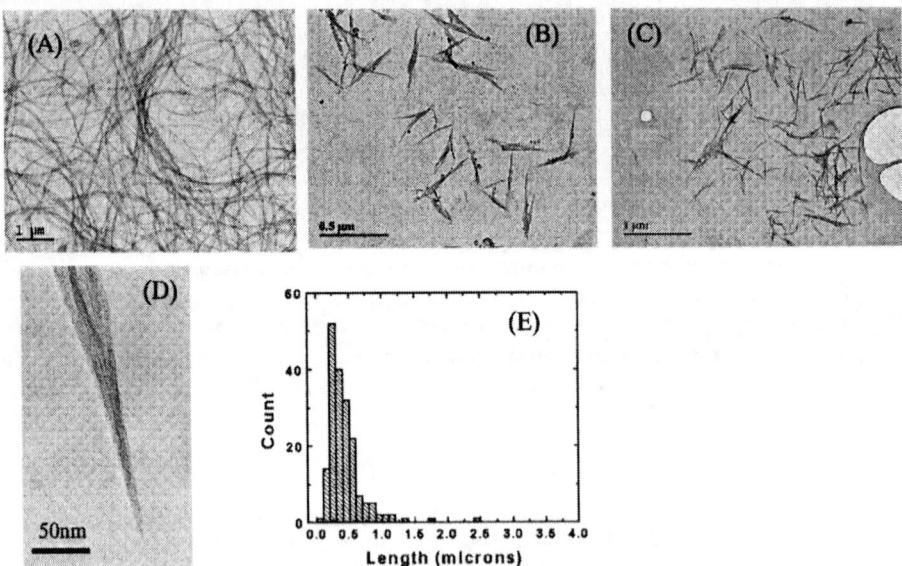

Figure 1: Transmission electron microscopy images of purified SWNTs before (A) and after oxidation (B)-(D). (E): Length distribution of the SWNTs after 24-hour oxidation.

SWNTs were vacuum dried before any measurement.

Figure 1 shows typical TEM images of purified SWNTs before and after chemical etching. In contrast to the purified materials with long (>10μm) and entangled bundles (A), etched materials (B-D) contained mostly shorter bundles with smaller bundle diameters. The bundle length is closely related to the reaction time. Figure 1 (E) shows the length distribution of a sample etched for 24 hours. The average length is about 0.5μm. Compared with materials processed by ball-milling [12], acid-treated materials are cleaner with well-separated SWNT bundles. The ends of those short bundles are typically tapered. Tubes on the outside surfaces cannot be clearly resolved. Some bundles also appear to be less compact than purified SWNTs, which could be caused by the intercalation and de-intercalation of acid molecules during process of material processing [11].

ELECTROCHEMICAL INTERCALATION

The SWNT samples were electrochemically reacted with Li using a two-electrode cell with Li foil and SWNT film deposited on a nickel plate as the two working electrodes. A polypropylene filter soaked with a 1M solution of $LiClO_4$ in 1:1 vol. ratio of ethylene carbonate and dimethyl carbonate was used as the electrolyte. Detailed experimental set up was described in a previous publication [8]. The amount of Li intercalated per unit of carbon was calculated from the time and the current used. The cell was discharged (intercalation) and charged (de-intercalation) under the galvanostatic mode at 50mA/g current between 0-3V.

Figure 2A and 2B show the second-cycle charge and discharge curves obtained from the purified SWNTs before and after etching. For purified SWNTs, a capacity of $Li_{2.5}C_6$ was obtained after the first discharge. The reversible capacity, i.e., the amount of Li that could be removed from the nanotube electrode in the first charge, is $Li_{0.89}C_6$,

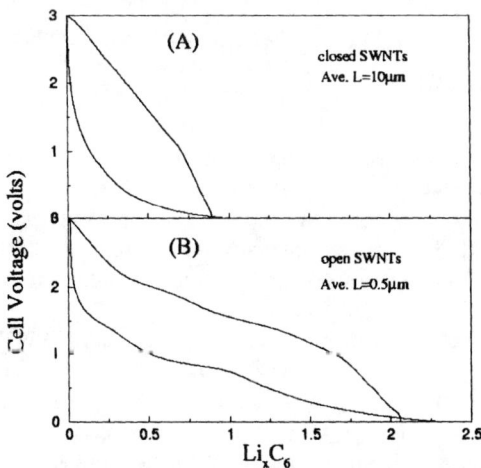

Figure 2: Second cycle cell voltage versus Li concentration data obtained from purified SWNTs (A) and samples after chemical etching (B). The data were collected at 50mA/g rate. All the capacities were calculated by normalizing the measured total capacity with the amount of carbon in the samples, not including 10 wt.% binders and carbon blacks.

similar to that of graphite. The large irreversible capacity ($Li_{1.6}C_6$) is related to the large surface area of the SWNT electrode (estimated to be 300m^2/g by a N_2 BET measurement). The reversible capacity increased to Li_2C_6 in etched SWNTs (Figure 2B). This is twice the theoretical value (LiC_6) for graphite, which is used in the current battery electrodes. A simple space-filling model indicated that the excess Li ions can only be hosted inside the individual nanotubes, as suggested by a recent theoretical calculation [Zhao, 2000 #27].

SOLID STATE REACTION AND CHARACTERIZATION

The chemical composition and properties of Li intercalated SWNTs were studied using samples prepared by the solid state rather than electrochemical method. This is to avoid contamination by the liquid electrolyte. Samples with the nominal compositions of LiC_6 and LiC_3 were synthesized by reacting SWNTs (both purified and etched) and $LiBH_4$:

$$C(SWNT) + xLiBH_4 \rightarrow Li_xC + x/2\, B_2H_6 + x/2\, H_2$$

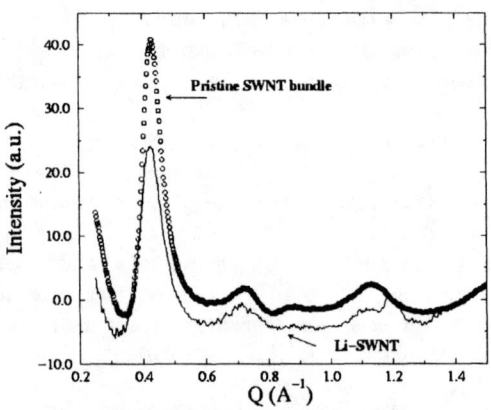

Figure 3: X-ray diffraction patterns of pristine and Li intercalated SWNT (LiC_6) after background subtraction

Proportional amount of SWNT and $LiBH_4$ were loaded into a stainless steel tube in an argon-filled glove-box, which was then sealed and heated at 500°C for a week. The sample was opened in the glove-box after the reaction and transferred to a glass tube which was sealed under 5×10^{-7} Torr pressure for x-ray and NMR measurements.

The powder x-ray diffraction pattern of a nominal LiC_6 sample (with purified SWNT with etching) is shown in Figure 3, along with the diffraction obtained from a pristine SWNT. In contrast to the heavy alkali metal intercalated SWNTs, the LiC_6 sample is crystalline. No noticeable change in the peak positions was observed, which is attributed to the intensity modulation from the nanotube scattering form factor (independent of the lattice parameter) [16]. The relative intensity of the nanotube (10) peak becomes weaker in the LiC_6 sample. Due to the weak scattering cross section of the Li ions, the diffraction pattern has not been further analysed.

The chemical compositions and electronic properties of Li intercalated SWNTs were investigated using ^{13}C and ^{7}Li NMR techniques. When as-purified SWNTs were used, the saturation composition was found to be $LiC_{5.7}$ even if excess amount of Li was used in the reaction. In the case of etched SWNTs, the saturation composition increased to $LiC_{3.2}$. Both are consistent with the results from the electrochemical studies. The SWNTs became metallic upon Li intercalation, with their electronic density of states increasing with increasing Li concentration.

SUMMARY

The results presented in this paper show that the reactivity of SWNT materials depends sensitively on the structure and morphology of individual nanotubes. By exposing the interior spaces of the nanotubes through post-processing, the specific Li storage capacity of the SWNTs is doubled to LiC_3 which is twice the value observed in Li intercalated graphite.

REFERENCES

1. P. Poizot, S. Laruelle, S. Grugeon, L. Dupont, and J.-M. Tarascon, Nature, 2000. **407**: p. 496-499.
2. Y. Idota, T. Kubota, A. Matsufuji, Y. Maekawa, and T. Miyasaka, Science, 1997. **276**: p. 1395-1397.
3. B. Gao, S. Sinha, L. Fleming, and O. Zhou, Adv. Mater. (in press), 2001.
4. R.A. Huggins, J. Power Sources, 1989. **26**: p. 81.
5. O. Zhou, R.M. Fleming, D.W. Murphy, C.T. Chen, R.C. Haddon, A.P. Ramirez, and S.H. Glarum, Science, 1994. **263**: p. 1744-1747.
6. R.S. Lee, H.J. Kim, J.E. Fischer, A. Thess, and R.E. Smalley, Nature, 1997. **388**: p. 255-257.
7. A.M. Rao and e. al., Nature, 1997 . **388**: p. 257.
8. B. Gao, A. Kelinhammes, H. Shimoda, L. Fleming, X.P. Tang, Y. Wu, and O. Zhou, Chem. Phys, Lett, 1999. **307**: p. 153.
9. A. Claye, J.E. Fischer, C.B. Huffmand, A.G. Rinzler, and R.E. Smalley, J. of Electrochem. Soc., 2000. **147**(8): p. 2845-2852.
10. S. Suzuki, C. Bower, and O. Zhou, Chem. Phys. Lett., 1998. **285**: p. 230-234.
11. C. Bower, A. Kleimhammes, Y. Wu, and O. Zhou, Chem. Phys. Lett, 1998. **288**: p. 481-486.
12. B. Gao, C. Bower, J. Lorentze, L. Fleming, A. Kleinhamme, X.P. Tang, L.E. McNeil, Y. Wu, and O. Zhou, Chem. Phys. Lett., 2000 . **327**: p. 69-75.
13. X.P. Tang, A. Kelinhammes, H. Shimoda, L. Fleming, C. Bower, S. Sinha, O. Zhou, and Y. Wu, Science, 2000. **228**: p. 492.
14. O. Zhou, B. Gao, C. Bower, L. Fleming, and H. Shimoda, Mol. Crys. and Liq. Crys., 2000 . **340**: p. 541-546.
15. J. Liu, A. Rinzler, H. Dai, J. Hafner, a.R. Bradley, P. Boul, A. Lu, T. Iverson, a.K. Shelimov, C. Huffman, F. Rodriguez-Macias, Y. Shon, R. Lee, D. Colbert, and R.E. Smalley, Science, 1998 . **280**: p. 1253-1256.
16. T. Yildirim, O. Zhou, and J.E. Fischer, in *The Physics of Fullerene-Based and Fullerene-Related Materials*, W. Andreoni, Editor. 2000, Kluwer: Dordrecht/Boston/London.

ACKNOWLEDGEMENT: This work was supported by the U.S. Office of Naval Research through a MURI program at UNC (N00014-98-1-0597).

Electronic Structure and Quantum Conductance of Nanotube Structures: Defects, Crossed-Tube Junctions, and Nanopeapods

Steven G. Louie

Department of Physics, University of California at Berkeley, Berkeley, CA 94720 USA,
and
Materials Science Division, Lawrence Berkeley National Laboratory, Berkeley, CA 94720 USA

Abstract. We present theoretical study of the electronic structure and the quantum conductance of several carbon nanotube structures, calculated using an *ab initio* approach based on the Landauer-Buttiker formalism. Systems examined include tubes with local defects, crossed nanotube junctions, and nanopeapods. Defects such as substitutional impurities and pentagon-heptagon pairs on tube walls are found to produce interesting effects, including sharp, quantized reduction in conductance at specific energies. The structural deformation and transport properties of crossed carbon nanotube junctions have been investigated as a function of applied force on the junction. Theory shows that the intertube junction conductance is very sensitive to external applied forces. Finally, some results on nanopeapods, i.e., single-walled carbon nanotube with fullerenes in the interior, are presented. We propose that, under appropriate doping conditions, these hybrid nanotube-fullerene systems may exhibit electromigration behavior and would be good superconductors.

INTRODUCTION

Since the discovery of the carbon nanotubes a decade ago [1], experimental and theoretical studies have given much information on the structural, electronic, and transport properties of nanotubes and tubes with defects and junctions [2]. Recent advances in synthesis and fabrication techniques further yielded other novel carbon nanotube structures such as the crossed-tube junctions [3] and the nanopeapods [4,5]. In this article, we present results from some recent theoretical studies on the structure and properties of several of these novel nanotube structures. The geometric and electronic structures are calculated using the *ab initio* pseudopotential potential density functional approach. The linear response conductance is evaluated with wavefunctions from the first-principles calculations, employing the Landauer formalism [6]. It is shown that defects and different structure arrangements often give rise to new electronic and transport features that are significantly different from those of the perfect nanotubes, making them valuable in applications.

In the Landauer formalism [6], the two-terminal electrical conductance of a quantum system may be obtained from the transmission coefficients of the electron waves. The conductance is given by

$$G = \frac{2e^2}{h} \text{Tr}(t^+ t) \quad (1)$$

where $G_0 = 2e^2/h$ is the quantum unit of conductance and t is the transmission matrix of the electrons at energy E. In the past, most work on the conductance of nanotubes was done within a tight-binding model. The results discussed here are based on the *ab initio* pseudopotential approach, with either a plane wave [7] or a linear combination of atomic like orbitals (LCAO) basis [8]. With the *ab initio* method, it is possible to obtain the self-consistent electronic and geometric rearrangements at the structure of interest, and obtain accurate information on the current density distribution. We present here several applications using this approach.

DEFECTS AND QUASI-BOUND STATES

Electrical transport in (n,n) metallic carbon nanotubes has been found both experimentally and theoretically to be very robust against defects and disorders. For example, backscattering by long-range disorders is suppressed in metallic tubes but not in doped or gated semiconductor tubes [7,9,10]. Moreover, away from the Fermi level E_F, calculations predicted that the maximum reduction in conductance at specific energies due to a defect is often itself quantized. This can be explained in terms of resonant backscattering by defect induced quasi-bound states [7].

Figure 1 shows the calculated conductance of a (10,10) tube with a single Stone-Wales structural defect on it [7]. For a perfect tube, the conductance is $2G_0$ since there are only 2 channels available for electrons near E_F. As seen in Fig.1, with the defect, conductance is virtually unchanged at E_F, showing the robustness of transport for carriers at the intrinsic Fermi level. On the other hand, there are two dips in the conductance, away from E_F. The depth of the dips is one G_0 and its shape is approximately Lorentzian. These features can be understood in terms of reduction in conductance due to resonant backscattering of the incoming electrons by the quasibound states derived from the Stone-Wales defect. The defect states are seen in the calculated local density of states (LDOS) as two extra peaks, corresponding to two quasibound states. Very similar results have been obtained for other defects [7] including substitutional impurities such as boron and nitrogen impurities. In particular, a boron (nitrogen) impurity acts as an acceptor (donor) with respect to the lower (upper) subbands, splitting off impurity levels into the conduction region.

The predicted phenomenon of resonant reduction in conductance has been recently observed in transport measurement using a scanned gate technique [11]. By adjusting the gate voltage on a STM tip, the Fermi level on the site of the tip was locally changed. It was found that at specific sites where there was a defect, a strong reduction in conductance at specific gate voltage was observed, corresponding to resonant scattering by quasibound defect states. The experiment further showed a measured resistance [11] of two resonant peaks, one above and one below E_F, with electron reflection coefficients of ~0.5 at resonance. These features are in good agreement with our theoretical predictions for reflection from a Stone-Wales defect.

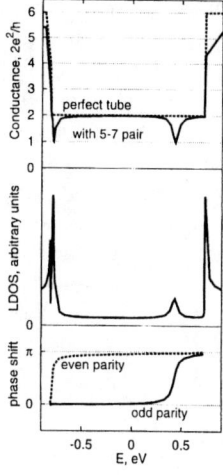

Fig. 1 Calculated conductance, local density of states, and phase shifts of a (10,10) carbon nanotube with a Stone-Wales defect [7].

CROSSED-TUBE JUNCTIONS

Since the mid-1990's, there has been much work on studying on-tube junctions, i.e., junctions formed by joining two half tubes of different chirality by pentagon-heptagon defect pairs [12,13]. These structures possess unusual properties and behave as intramolecular electronic elements. Metal-semiconductor, semiconductor-semiconductor, and metal-metal junctions have been studied theoretically [12] and verified experimentally [13]. More recently, a new class of junctions composed of crossed carbon nanotubes on a substrate has also been fabricated and studied [3,8].

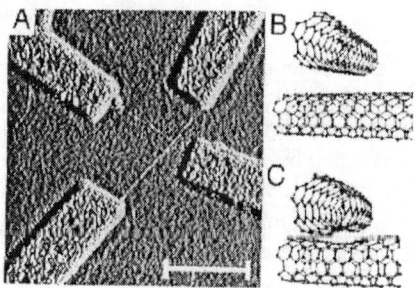

Fig. 2 AFM image of a crossed SWNT device (A). Calculated structure of a crossed (5,5) SWNT junction with a force of 0 nN (B) and 15 nN (C)
Reprinted with permission from Ref. [3]. Copyright 2000 American Association for the Advancement of Science.

Figure 2A is an AFM image of a crossed-tube junction made from two single-walled carbon nanotubes (SWNTs) of 1.4 nm diameter on a substrate [3]. Figures 2B and 2C show the calculated structure of two (5,5) SWNTs pressed against each other with zero and 15 nN force, respectively. As seen from Fig. 2C, there is considerable structural deformation when there is a contact force between the two tubes. The closest distance between atoms on different tubes is significantly closer than that of the zero force case. In panel C, the closest intertube atomic separation is only 0.25 nm [8], which is significantly smaller than the van der Waals distance of 0.34 nm. In experiment, significant contact forces (in the order of several nN's) result from the attraction between the tubes and the substrate. We find that the intertube junction conductance is very sensitive to the contact force [8]. Figure 3 shows that the calculated 4-terminal 4-probe junction conductance at E_F for two crossed (5,5) SWNTs is ~20% of G_0 at a contact force of 15 nN. The conductance, on the other hand, is orders of magnitude smaller at zero force for the same system. Figure 4 gives the calculated conductance as a function of contact force, indicating that this kind of junctions is potentially useful as nanoscale electro-mechanical devices.

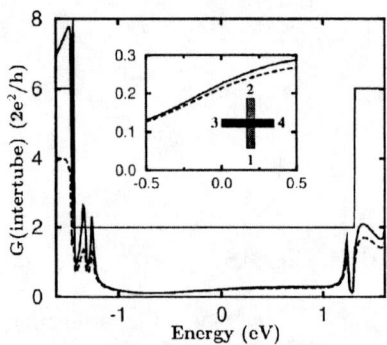

Fig. 3 Intertube conductance of a crossed (5,5) SWNT junction with contact force of 15 nN. The solid (dashed) curve is the 4-terminal 4-probe (approximate sum-of-4-transmission-coefficients) conductance. The inset shows expanded view of the conductance near $E_F = 0$ [8].

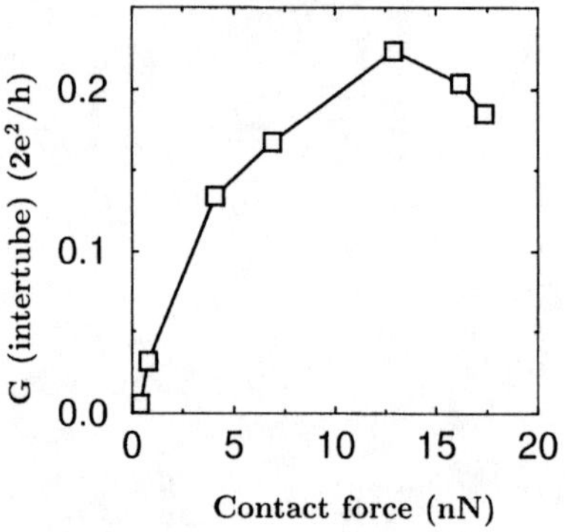

Fig. 4 Calculated 4-terminal 4-probe intertube conductance of a crossed (5,5) SWNT junction as a function of contact force [8].

Experimentally, various types of crossed-tube junctions have been measured, including metal-metal, semiconductor-semiconductor, and metal-semiconductor junctions [3]. For the metal-metal case, a conductance of 2-6% of G_0 was observed, in agreement with the theoretical results [8] for an expected substrate-induced contact force of a few nN's. Moreover, the metal-semiconductor junctions exhibit Schottky diode behavior with observed Schottky barriers in the range of 0.2-0.3 eV, consistent with a theoretical value of 0.25 eV for nanotubes with diameter of 1.4 nm.

CARBON NANOPEAPODS

Another class of structures that has generated tremendous interest lately is the nanopeapods [4], which are hybrid systems of fullerenes inside SWNTs. These systems can now be synthesized in quantity [4,5]. We carried out *ab initio* calculations on the electronic and transport properties of both individual fullerene (pure or endohedral doped) molecules and chain of molecules inside a SWNT [14]. Figure 5 depicts the calculated results of the conductance and LDOS of a single C_{60} molecule in a (10,10) carbon nanotube. The results show that the interaction between the molecule and the tube is relatively weak: the C_{60} molecular levels are intact and appear as narrow resonance peaks in the LDOS, but with their degeneracy lifted. In particular, since the Fermi level is just below the C_{60} LUMO levels, there is little charge transfer between the molecule and the tube.

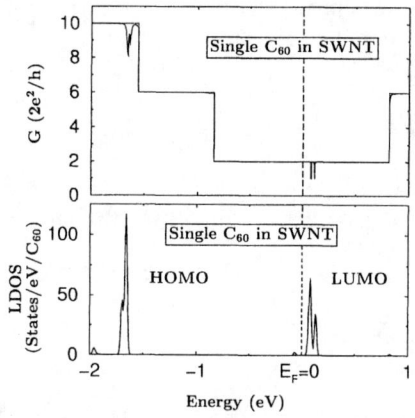

Fig. 5 Calculated conductance and local density of states of a single C_{60} molecule inside a (10,10) carbon nanotube [14].

However, the calculated conductance [14] exhibits sharp suppressions at the molecular levels, owing to strong resonant back scattering by the molecular states. This finding suggests that gating or doping E_F to the appropriate molecular levels may lead to interesting electro-migration effects, i.e., coupling of the current carriers to the motion of the molecule. Theoretical results on an endohedral K@C_{60} inside a (10,10) tube, on the other hand, reveals a transfer of electron from the K 4s level to the C_{60} molecule, resulting in a partially occupied LUMO complex that is pinned at E_F. This leads to a strong coupling of the LUMO electrons of K@C_{60} to the tube carriers at E_F, and a suppressed conductance.

For a true nanopeapod, i.e., a linear chain of C_{60} molecules in a (10,10) SWNT, we find, as expected, the C_{60} molecular orbitals form bands. Depending on the orbital character, some of the molecular bands can have widths that are about a half of an eV, but others remain quite dispersionless [14,15]. For example, within the C_{60} HOMO complex, three of the orbitals form dispersive bands but not the other two. Because of the 1D nature of the system, the density of states of the nanopeapods shows typical 1D van Hove singularities. In fact, according to our calculations [14], the density of states per molecule for the peapods is about 3 times higher than that in solid C_{60}. Thus, for a nanopeapod that is gated or doped appropriately so that E_F is at one of the molecular band complexes, one would have a quite interesting 1D narrow band metal that is embedded in the wide bands of the metallic carbon nanotube. The high DOS at E_F can then lead to interesting consequences including higher transition temperature superconductivity.

SUMMARY

In summary, we have presented first-principles calculations on the electronic structure and quantum conductance of several carbon nanotube structures. The transport properties of intrinsic (n,n) tubes are found to be quite robust against defects. Local defects in general form quasibound states, giving rise to resonant reduction in the conductance. We find that crossed-tube junctions can have sizable conductance which is sensitive to the contact force. We show that fullerenes inside nanotubes introduce resonant molecular states and may have interesting electro-migration effects. Finally, we find that nanopeapods have band dispersions of the order of 0.5 eV and may be good superconductors if appropriately doped.

ACKNOWLEDGMENTS

This work was supported by NSF Grant No. DMR-0087088 and by the U.S. Department of Energy under Contract No. DE-AC03-76SF00098. Computational resources were provided by NERSC and by the NSF.

REFERENCES

1. S. Iijima, Nature **354**, 56 (1991).
2. See, for example, articles in *Carbon Nanotubes*, Topics Appl. Phys. Vol. **80**, edited by M. S. Dresselhaus, G. Dresselhaus, and Ph. Avouris (Springer-Verlag, Berlin, 2001).
3. M. S. Fuhrer, J. Nygard, L. Shih, M. Forero, Y. G. Yoon, M. S. C. Mazzoni, H. J. Choi, J. Ihm, S. G. Louie, A. Zettl, and P. L. McEuen, Science **288**, 494 (2000).
4. B. W. Smithe, M. Monthioux, and D. E. Luzzi, Nature **396**, 323 (1998).
5. K. Hirahara, et. al., Phys. Rev. Lett. **85**, 5384 (2000), and references therein.
6. R. Landauer, Philos. Mag. **21**, 863 (1970); D. S. fisher, P. A. Lee, Phys. Rev. B**23**, 6851 (1981).
7. H. J. Choi, J. Ihm, S. G. Louie, and M. L. Cohen, Phys. Rev. Lett. **84**, 2917 (2000), and references therein.
8. Y.-G. Yoon, M. S. C. Mazzoni, H. J. Choi, J. Ihm, and S. G. Louie, Phys. Rev. Lett. **86**, 688 (2001).
9. P. L. McEuen, M. Bockrath, D. H. Cobden, Y.-G. Yoon, and S. G. Louie, Phys. Rev. Letts. **83**, 5098 (1999).
10. T. Ando, and T. Nakkanishi, J. Phys. Soc. Jpn. **67**, 1704 (1998); T. Ando, T. Nakkanishi, and R. Saito, J. Phys. Soc. Jpn. **67**, 2857 (1998).
11. M. Bockrath, W. Liang, D. Bozovic, J.H. Hafnter, C.M. Lieber, M. Tinkham, and H. Park, Science **291**, 283 (2001).
12. L. Chico, V. H. Crespi, L. X. Benedict, S. G. Louie, and M. L. Cohen, Phys. Rev. Lett. **76**, 971 (1996); P. Lambin, *et al.*, Chem. Phys. Lett. **245**, 85 (1995); R. Saito, G. Dresselhaus, and M. S. Dresselhaus, Phys. Rev. B **53**, 2044 (1996).
13. S. J. Tans, *et al.*, Nature **386**, 474 (1997); M. Bockrath, *et al.*, Science **275**, 1922 (1997); S. J. Tans, R. M. Verschueren, and C. Dekker, Nature **393**, 49 (1998); R. Martel, *et al.*, Appl. Phys. Lett. **73**, 2447 (1998); P. G. Collins, A. Zettl, H. Bando, A. Thess, and R. E. Smalley, Science **278**, 100 (1997); Z. Yao, H. W. C. Postma, L. Balents, C. Dekker, Nature **402**, 273 (1999).
14. Y.-G. Yoon, M. S. C. Mazzoni, and S. G. Louie, to be published.
15. Similar band structures for chain of C_{60} in (n,n) tubes have been obtained independently by S. Okada, S. Saito, and A. Oshiyama, Phys. Rev. Lett. **86**, 3835 (2001); see this Proceedings.

Electromechanical Properties of Multiwall Carbon Nanotubes

A. Zettl and John Cumings

*Department of Physics, University of California at Berkeley,
and Materials Sciences Division, Lawrence Berkeley National Laboratory, Berkeley, CA 94720 U.S.A.*

Abstract. We examine electrical and coupled electromechanical properties of multiwall carbon nanotubes using transport measurements performed *in-situ* inside a high resolution transmission electron microscope (TEM). In one experiment, large electrical currents are passed through the nanotubes and the failure modes for nanotube "burnout" examined . In a second set of experiments, the electrical resistance between the ends of nanotubes is measured as the tubes are either "telescoped" or partially telescoped and then severely but reversibly mechanically kinked. Our experimental results have implications for nanotube quantum charge transport mechanisms.

Multiwall carbon nanotubes (MWNTs) are comprised of concentric nanotube shells, each shell apparently "just fitting" inside the next, with an intertube spacing roughly equal to the van der Waals graphite interplane distance, 3.4Å[1]. This geometrical constraint suggests that some of the nanotube shells are individually either metals or semiconductors[2]. The composite shell structure may have a complex electrical behavior, especially if charge is transported from one concentric tube shell to the next. In addition, defect structures can affect nanotube transport. Some previous experiments have suggested that singlewalled carbon nanotubes (SWNTs) are more likely to behave as ballistic transport channels than are MWNTs[3]. On the other hand, careful "mercury dipping" experiments on MWNTs have indicated electrical conductance quantization plateaus suggestive of ballistic transport, perhaps confined to only the outer nanotube shell[4-6]. Magnetic flux quantization experiments have received similar interpretations[7].

We here report on electrical conductance measurements performed on MWNTs placed inside a high resolution transmission electron microscope (TEM) fitted with a custom-made electro-mechanical manipulation stage (with x,y,z coarse and fine mechanical motion control). The stage allows an individual MWNT to be selected, bonded to electrodes in a two-probe configuration, and mechanically manipulated while the resistance is monitored and the nanotube is viewed under high microscope magnification. The same apparatus has been previously used to "sharpen and peel" individual MWNTs[8] and to "telescope" the inner core tubes from the outer shell housing, thus forming low-friction linear bearings[9].

In the first set of experiments to be described here, a MWNT is contacted and the electrical current through it is steadily increased until the nanotube electrically and mechanically fails, i.e. "burns out". One of several different nanotube failure modes might be expected, depending on the details of the transport mechanism of the MWNT. For example, if the nanotube is indeed a ballistic conductor, then the electrical resistance is confined to the contacts. The contacts therefore are the hot spots for energy dissipation (Power = IV), and these might be expected to fail first (i.e. the contacts should "blow off"). If, on the other hand, the MWNT is a dissipative conductor, with the dissipation more or less uniformly distributed along the length of the tube (but the thermal heat sinking largely confined to the end contacts with a minimal cooling contribution from black body radiation), then the nanotube should assume a well-known temperature profile[10] with a maximum temperature realized at the half-way point along the tube. In this case tube failure would initiate exactly half-way between contacts, in a perhaps catastrophic fuse-like vaporization mode.

Experimentally, neither of these failure modes is observed! Inevitably, as the current is increased past a critical value (typically of order 200 µA) the nanotube "burns out" in a seemingly random location, at a position where even high-resolution TEM imaging (prior to failure) shows no evidence for any obvious nanotube defect structure. MWNTs are never observed to "blow off" the contacts, nor to fail exactly in the middle of the tube. Upon failure, the nanotube appears to mechanically separate over a small region (more like a cut of the tube rather than a fuse-like meltdown), and the two independent leftover pieces of the tube (still attached to the independent electrodes) appear to remain largely intact. If one assumes that the nanotube fails at the most severe defect, then one might expect that the remaining tube portions are more "defect free" than the original tube. Hence, successive burnouts of the remaining tube segments might be used to "purify" a given MWNT, in the sense that the largest remaining defects are successively cut out. We have tested this hypothesis, and indeed the remaining tube segments always have significantly higher threshold currents than the parent tube.

We now describe a second set of experiments, for which a MWNT is first sharpened and peeled[8], thus exposing core tubes. The manipulator electrode is then spot-welded to the core tubes and these are telescoped out of the housing tubes[9]. During the telescoping, the electrical resistance between the "ends" of the tube is monitored. It should be stated at the outset that the "spot welding" method of electrical contact leaves some ambiguity as to which of the concentric core tubes are actually physically contacted (similar ambiguities exist for electrical contact to the housing tubes as well). Nevertheless, at the very least we may assume that the largest diameter core tube is well contacted, and similarly is the largest diameter housing tube. A schematic for the experimental contact and mechanical manipulation configuration is shown in Fig. 1A.

One possible outcome of such an experiment is that the matrix element for charge transfer between the largest diameter core tube and the smallest diameter housing tube depends linearly on the physical "overlap area" between the tubes. In this case the resistance between the ends of the telescoping MWNT might then behave as a sliding variable resistor (a nanotrimpot!). On the other hand, one could imagine other possibilities, such as oscillations in the resistance with telescoping (as transfer matrix element resonances are encountered, depending on the chirality differences beween

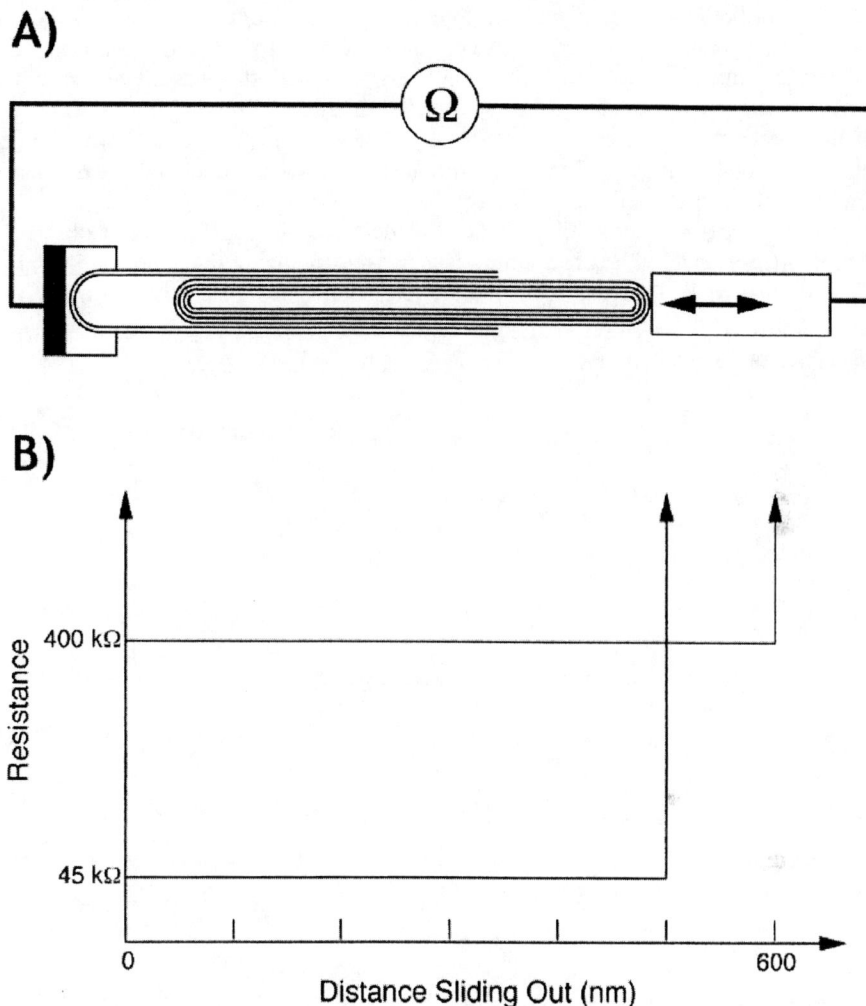

FIGURE 1. A) Schematic for in-situ resistance measurement for a "telescoping" MWNT. B) Resistance vs. core tube displacement for two independent MWNTs.

the particular tubes in sliding contact), or even a steadily *increasing* conductance with *decreasing* overlap area (due to reduced destructive interference of the electronic wavefunctions).

Fig. 1B shows the experimentally determined electrical resistance versus core sliding distance for two independent telescoping nanotubes. One nanotube had an original (fully retracted) resistance of 400kΩ while the other had an original resistance of 45kΩ. As Fig. 1B shows, the resistance of both telescoped tubes is fully *independent* of the distance the core tubes are withdrawn. Only when the core tubes are completely removed from the housing does the resistance change (it jumps to infinity, as expected). The independence of the resistance for core sliding was observed for both core withdrawal and core insertion directions for these tubes, i.e. for telescoping in and out (we were unsuccessful in attempts to reinsert the core into the original housing once the parts were completely separated; this was prevented by the inevitable thermal vibrations of the cantilevered tubes coupled with "zero clearances" for the core and housing parts). The distance-independence of the resistance data of Fib. 1B is somewhat surprising. We do not know to what extent contact resistance dominates the resistance data.

In another experiment similar to that just described in Fig. 1, we did observe a change in resistance with core sliding distance (thus realizing the nanotrimpot). For this particular MWNT the resistance before telescoping was unusually small, of order 6kΩ (almost precisely corresponding to a quantum conductance of $G=2e^2/h$), perhaps suggestive of "perfect" contacts.

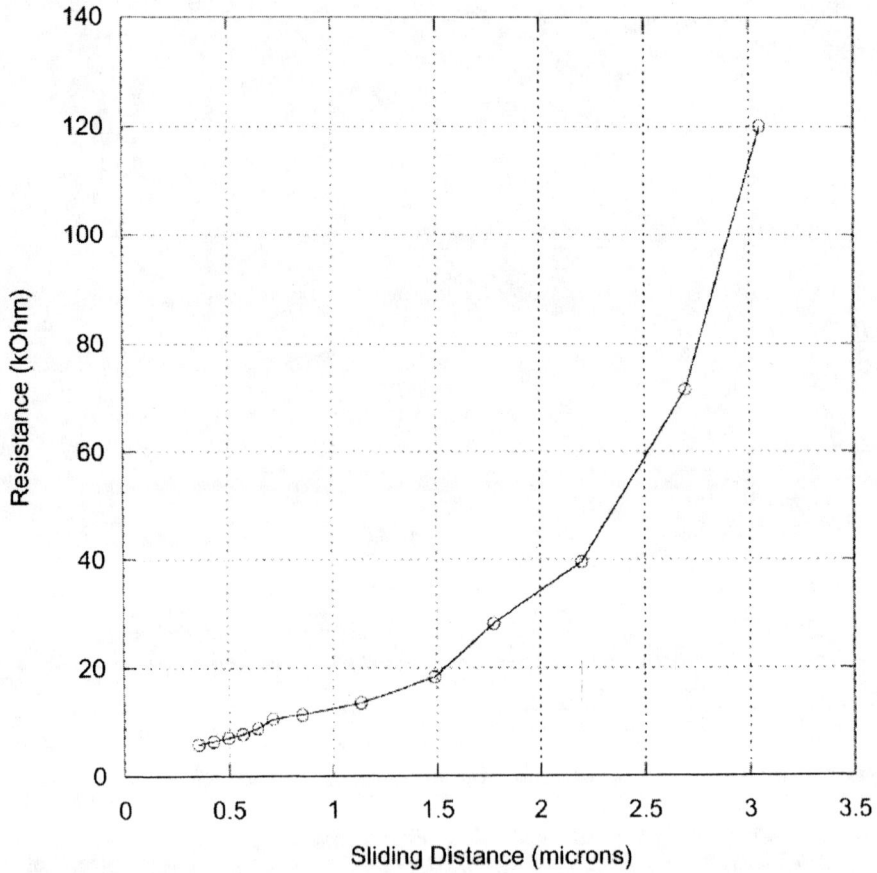

FIGURE 2. Resistance vs. core tube displacement for a MWNT that had an unusually high conductance prior to telescoping.

Fig. 2 shows the resistance between nanotube ends for this MWNT as it is telescoped out. After 3 μm of sliding distance (close to full extension), the resistance has increased by a factor of 20, to 120kΩ. The resistance increase with sliding distance for this MWNT is not linear, but exponential, reminiscent of transport in localized electronic systems.

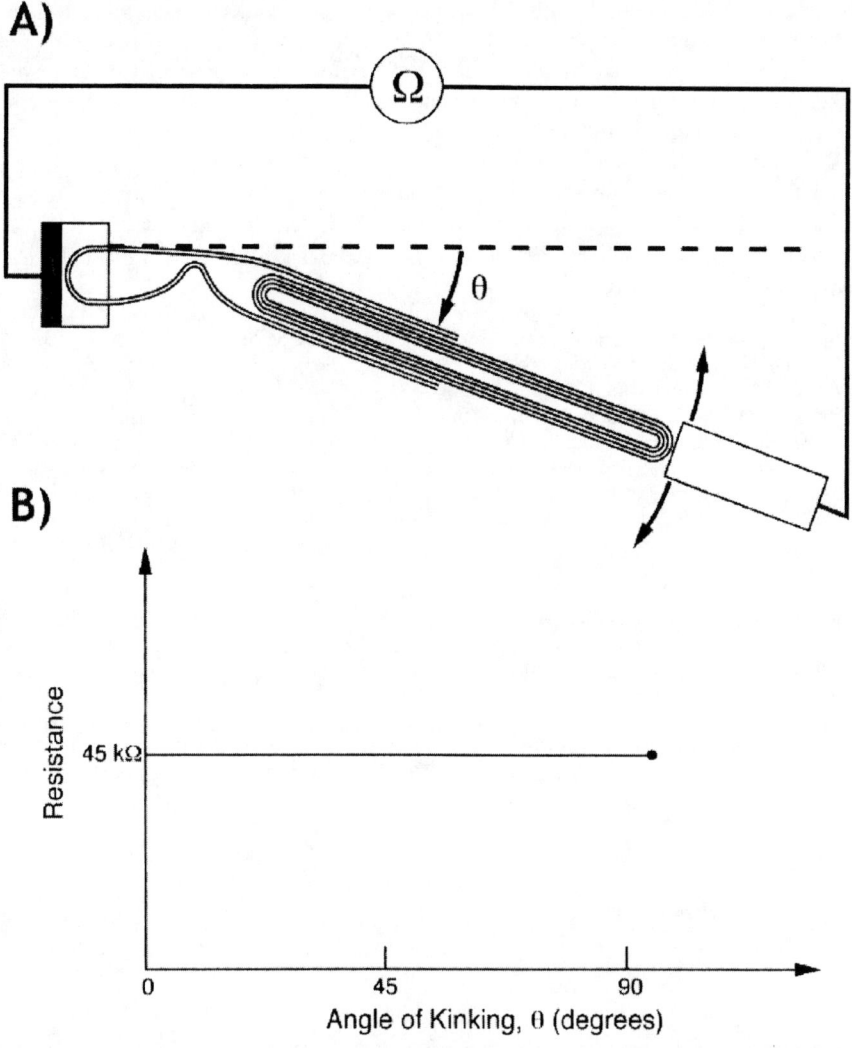

FIGURE 3. A) Schematic for in-situ resistance measurement for a telescoped and kinked MWNT. B) Resistance vs. kink angle.

Fig. 3 shows another experimental configuration we have used to investigate electromechanical response in MWNTs. As shown schematically in Fig. 3A, a MWNT is first partially telescoped out, then lateral forces are applied to the ends of the core structure to induce controlled "kinking" in nanotube fabric. The kinking occurs in the large diameter "hollowed out" region of the tube, where the housing tubes are no longer supported by the inner core tubes. The kinking position can be controlled by varying the extension of the core tubes. All the while, the resistance between the ends of the tube is monitored. Fig. 3B shows the measured electrical resistance as the tube is kinked from its unperturbed straight configuration ($\theta=0°$) to a severe right-angle kink ($\theta=90°$). Surprisingly, no resistance changes are observed even for such extreme kinking. All kinking here performed was fully reversible, with no permanent tears in the nanotube fabric. Our findings suggest that even severely mechanically deformed nanotubes are good electrical conductors (somewhat akin to flexible electromagnetic waveguides).

We thank U. Dahmen, C. Nelson, E. Stach, M.L. Cohen, and S.G. Louie for helpful interactions. This work was supported by the Director, Office of Energy Research, Office of Basic Energy Sciences, Division of Materials Sciences, of the U. S. Department of Energy under Contract No. DE-AC03-76SF00098, and by NSF Grants DMR-9801738 and DMR-9501156.

1. Iijima, S. Helical microtubules of graphitic carbon. *Nature* **354**, 56-58 (1991).
2. Hamada, N., Sawada, S. & Oshiyama, A. New one-dimensional conductors: graphitic microtubules. *Physical Review Letters* **68**, 1579-1581 (1992).
3. Bachtold, A. *et al.* Scanned probe microscopy of electronic transport in carbon nanotubes. *Physical Review Letters* **84**, 6082-6085 (2000).
4. Frank, S., Poncharal, P., Wang, Z. L. & De Heer, W. A. Carbon nanotube quantum resistors. *Science* **280**, 1744-1746 (1998).
5. Choi, H. J., Ihm, J., Yoon, Y.-G. & Louie, S. G. Possible explanation for the conductance of a single quantum unit in metallic carbon nanotubes. *Physical Review B-Condensed Matter* **60**, R14009-14011 (1999).
6. Sanvito, S., Kwon, Y. K., Tomanek, D. & Lambert, C. J. Fractional quantum conductance in carbon nanotubes. *Physical Review Letters* **84**, 1974-1977 (2000).
7. Bachtold, A. *et al.* Aharonov-Bohm oscillations in carbon nanotubes. *Nature* **397**, 673-675 (1999).
8. Cumings, J., Collins, P. G. & Zettl, A. Materials - Peeling and sharpening multiwall nanotubes. *Nature* **406**, 586 (2000).
9. Cumings, J. & Zettl, A. Low-friction nanoscale linear bearing realized from multiwall carbon nanotubes. *Science* **289**, 602-604 (2000).
10. Incropera, F. P. & DeWitt, D. P. *Introduction to heat transfer* (Wiley, New York, 1996).

Field Emission from Carbon Nanotubes with Clean Surface and Adsorbed Molecules

Koichi Hata*, Akihiro Takakura and Yahachi Saito

Department of Electrical and Electronic Engineering, Mie University, Kamihama-cho, Tsu-city 514-8507, Japan

Abstract. Adsorption and desorption on pentagons at a tip of multiwall carbon nanotube (MWNT) have been investigated by field emission microscopy (FEM) in an ultra-high vacuum and in an atmosphere of nitrogen or oxygen. MWNTs with clean surface which are obtained by heat treatment give FEM patterns consisting of six bright pentagonal rings. Adsorbates (nitrogen and oxygen) are recognized as bright spots in the FEM pattern. They reside preferentially on the pentagonal sites where the strong electric field is concentrated, and bring about stepwise increase in the emission current. Heat treatment of the MWNT emitter at about 1300 K allows nitrogen and oxygen to desorb. After the desorption of nitrogen, the original clean surface with pentagons is recovered, while the tip structure is destroyed after the desorption of oxygen.

INTRODUCTION

Recently, we observed pentagons existing on the tip of a multiwall nanotube (MWNT) by field emission microscopy (FEM) in ultra-high vacuum, and revealed that electron emission occurs preferentially through pentagons when the nanotube surfaces are clean [1]. Effect of ambient gases on the life time of MWNT field emitters has been studied [2]. Using single-wall nanotubes (SWNTs), on the other hand, Dean et al. studied the effect of adsorbed molecules on the electron emission [3]. Adsorption and desorption of gas molecules on a cathode surface affect seriously field emission properties. Here, we report the changes of FEM patterns of a MWNT exposed to nitrogen and oxygen.

EXPERIMENTAL

Carbon nanotubes used in this study were MWNTs produced by carbon arc discharge in helium or hydrogen gas [4]. A bundle of as-grown nanotubes was fixed on a hairpin-shaped tungsten filament (0.15 mm in diameter) using conducting paste. The nanotube emitter can be heated to about 1300 K by resistive heating of the tungsten filament. This heat treatment in ultra-high vacuum for a few minutes is important to clean the surfaces of carbon nanotubes. The FEM study was carried out in an ultra-high vacuum chamber with a 6×10^{-10} Torr base pressure. A phosphor screen (92 mm in diameter) for observation of emission patterns was placed at 30-40 mm from the emitter. The electrical potential applied to the nanotubes emitter relative to the screen was typically from -0.9 kV to -1.6 kV. The nanotubes emitter was exposed to nitrogen or oxygen with pressures of 1×10^{-8} Torr for 600 s during electron emission. The residual gas in the chamber and the partial pressure of introduced gases were measured by a quadrupole mass spectrometer.

RESULTS AND DISCUSSION

Figure 1 shows the changes of emission patterns with time of a MWNT exposed to nitrogen. The emission pattern before introducing nitrogen, i.e., in an ultra-high vacuum, is shown in Fig. 1 (a). A bright spot on the left of the central ring originate from an adsorbate which still remained after a heat treatment (1300 K for 60 s). When nitrogen was introduced into the FEM chamber up to the pressure of 1×10^{-8} Torr, the emission pattern changed with time as shown in Figs. 1 (b) to 1 (d). In Fig. 1 (b), a large bright spot (indicated by an arrow), showing an enhanced emission of electrons due to an adsorption of nitrogen, appeared above the central ring. In Fig. 1 (c), two large spots disappeared, and another spot (indicated by an arrow) appeared below the central ring. Furthermore, in Fig. 1 (d), the number of adsorbates increases to three.

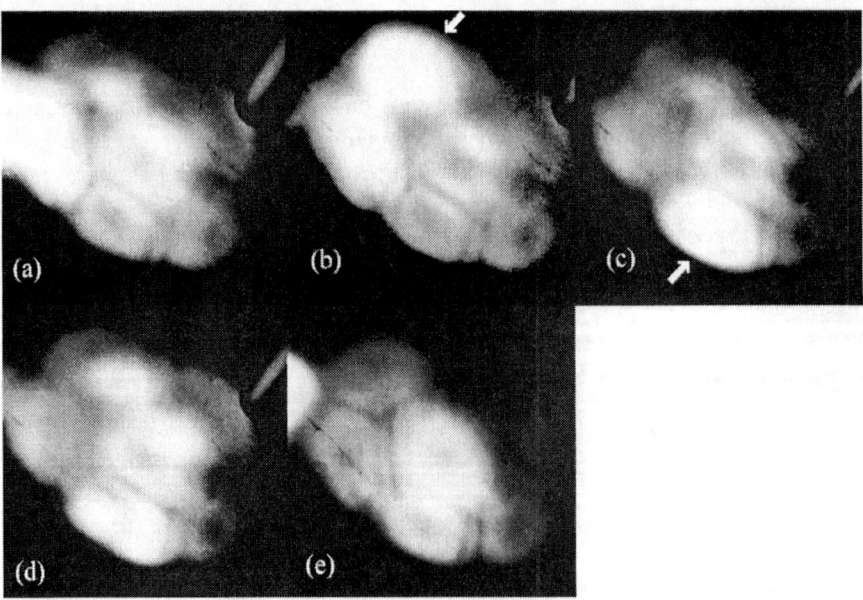

FIGURE 1. Changes of field emission patterns from a MWNT exposed to nitrogen. The patterns were taken after (a) 0 s (just after the heat treatment of 1300 K for 60 s in ultra-high vacuum) , (b) 170 s, (c) 510 s and (d) 550 s from the introduction of nitrogen up to 1×10^{-8} Torr. The emission pattern (e) was taken after the second heat treatment in ultra-high vacuum after the electron emission for 600 s.

From the random changes of the number and the position of bright spots, it is found that adsorption and desorption of nitrogen molecules occurred frequently on the pentagon sites where electric field concentrates because the pentagons are pointed like vertices of a polyhedron. The emission current also fluctuated widely due to the changes of the surface state of the nanotube. We stopped the electron emission after 600 s from the introduction of nitrogen, and evacuated the FEM chamber down to the base pressure. The emission pattern from the MWNT emitter treated by the second heating (1300 K for 60 s) in an ultra-high vacuum is shown in Fig. 1 (e). Five pentagonal rings are well observed. Though one pentagon is missing, the emission pattern indicates that the nanotube tip is free from adsorbates. It is noted that the adsorbate which was still observed after the first heat treatment (Fig. 1 (a)) was also desorbed. For common field emitters made of metals (e.g., tungsten), the emitter surfaces are damaged easily by bombardments of residual gas ions under a high pressure. For MWNTs, on the other hand, any serious damage is not observed in Fig. 1 (e). This shows us the robustness of MWNTs as field emitters.

The changes of emission patterns with time of a MWNT exposed to oxygen are shown in Fig. 2. In Fig. 2 (a) showing the emission pattern before introducing oxygen, only five pentagons arranged in a circle are observed. The cause of disappearance of one pentagon is probably due to its unfavorable position relative to the phosphor screen. After the introduction of oxygen up to 1×10^{-8} Torr, the emission pattern

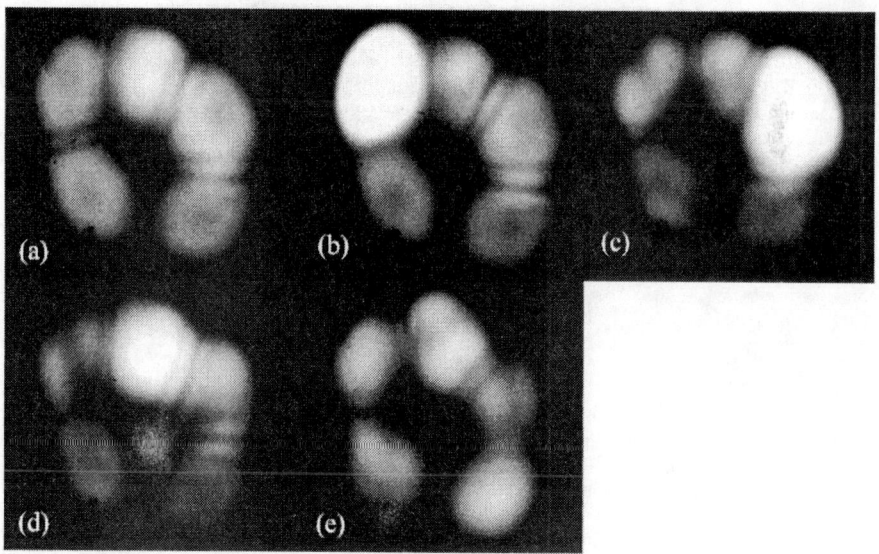

FIGURE 2. Changes of field emission patterns from a MWNT exposed to oxygen. The patterns were taken after (a) 0 s (just after the heat treatment of 1300 K for 60 s in ultra-high vacuum), (b) 210 s, (c) 375 s and (d) 420 s from the introduction of oxygen up to 1×10^{-8} Torr. The emission pattern (e) was taken after the second heat treatment in ultra-high vacuum after the electron emission for 600 s.

changed as shown in Figs. 2(b) to 2 (d). Adsorption and desorption occurred frequently at the pentagon sites, which is similar to the case of nitrogen, but the pentagonal rings were not recovered after the disappearance of bright spots. This suggests that pentagons were damaged during electron emission in an atmosphere containing oxygen. After the electron emission experiment for 600 s, the MWNT emitter was heated (1300 K for 60 s) again in an ultra-high vacuum in order to desorb oxygen molecules. Figure 2 (e) is the FEM pattern obtained after the heat treatment. All the pentagons are damaged, and the original pattern shown in Fig. 2 (a) is no longer reproduced. For the application of MWNTs as a field emitter, it is important that the partial pressure of oxygen in a vacuum chamber or a chipped-off glass tube should be maintained to be sufficiently low in order to elongate the life time of the emitter.

ACKNOWLEDGMENTS

This work was supported by the Ministry of Education, Culture, Sports, Science and Technology (Grants-in-Aids for Priority Research "Fullerenes and Nanotubes" No. 11165223, for Scientific Research (B) No. 12555006, and for University and Society Collaboration No. 11792003), and the NEDO project of "Frontier Carbon Technology".

REFERENCES

1. Y. Saito, K. Hata, T. Murata, Jpn. J. Appl. Phys. 39 (2000) L271.
2. T. Tanaka, T. Kato, Y. Saito, Presented at the Meting of Phys. Soc. Jpn. (Sept. 24-27, 1999, Iwate) 26aF-10.
3. K. A. Dean, B. R. Chalamla, Appl. Phys. Lett. 76 (2000) 375.
4. Y. Saito, R. Mizushima, S. Kondo, M. Maida, Jpn. J. Appl. Phys. 39 (2000) 1468.

Hydrogen Insertion and Extraction Mechanism in Single-walled Carbon Nanotubes

Seung Mi Lee[1,+], Young Hee Lee[1,2,*], Gotthard Seifert[3], and Thomas Frauenheim[3]

[1] Department of Semiconductor Science and Technology, and Semiconductor Physics Research Center, Jeonbuk National University, Jeonju 561-756, Korea
[2] Department of Physics, Jeonbuk National Univerisity, Jeonju 561-756, Korea
[3] Universitaet-GH Paderborn, Fachbereich Physik, Theoreticsche Physik, 33095 Paderborn, Germany

Abstract. Hydrogen insertion and extraction mechanisms in single-walled carbon nanotubes are studied using density-functional calculations. Hydrogen atoms first adsorb on the tube wall in an *arch type* and *zigzag type* upto coverage of 1.0. Hydrogen atoms can be inserted into the tube through the tube wall via *flip-in* and *kick-in* mechanism and are stored as a form of H_2 molecule at higher coverage. In the extraction process, H_2 molecule in the capillary first dissociates and adsorbs onto the inner wall, and is extracted to the outer wall by *flip-out* mechanism. Both the insertion and extraction processes require activation energy less than 2.0 eV.

INTRODUCTION

Hydrogen storage in carbon nanotubes (CNTs) has been extensively studied, for application to fuel cells, secondary batteries, or supercapacitors[1]. One of the ways of hydrogen storage is electrochemical storage method, which is more practical for an application to the secondary-hydrogen battery. The hydrogen reacts with the CNTs in form of ion in this method. However, the hydrogen storage capacity does not exceed 1 wt%[2], much less than the practical target of 6.5 wt%. A difficulty arises from the absence of theoretical model at an atomic scale. Here we theoretically describe the hydrogen adsorption and storage mechanism in CNTs in electrochemical approaches.

CALCULATIONAL METHODS

We performed a self-consistent charge density-functional-based tight-binding method (SCC-DFTB)[3] and the density functional calculations within local density approximation (LDA) and generalized gradient approximation (GGA). Structures are optimized by the SCC-DFTB and LDA schemes. The GGA calculations are done with structure optimized by LDA. The energy differences between different configurations were negligible, even the SCC-DFTB overestimate the binding energy than LDA and

---

[+] present address: Fritz-Haber-Institut der MPG, Faradayweg 4-6, D-14195 Berlin, Germany
[*] corresponding author should be addressed; E-mail: leeyh@sprc2.chonbuk.ac.kr

GGA calculations. We choose a supercell of (5,5) armchair CNT with periodic boundary conditions along the tube axis.

RESULTS AND DISCUSSION METHODS

The diameter of optimized (5,5) nanotube was similar to that of C_{60} (Fig. 1(A)). We identified several stable hydrogen adsorption geometries. Hydrogen atoms are chemisorbed in an *arch type* (Fig. 1(B)) or *zigzag type* (Fig.1(C)) until coverage 1.0. The tube diameters are expanded by the hydrogen chemisorption. Hydrogens can also exist inside the tube at higher coverage, as a form of molecule (Fig. 1(D)). It is worth noting that the *zigzag type* is more stable than the *arch type* by 0.56 eV/C-H bond, by the enhancement of sp^3 hybridization. The binding energy of a H_2 molecule (-4.57 eV) in Fig. 1(D) is weaker than that of gaseous molecule by 1.96 eV, due to the repulsive energies between H_2 molecules and that between tube wall and H_2 molecules[4].

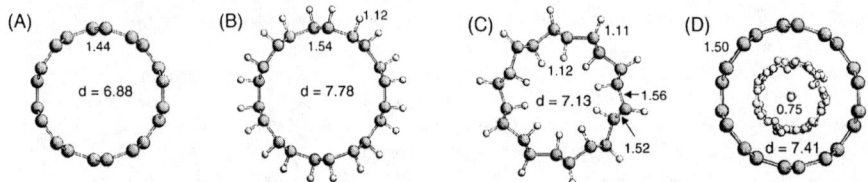

FIGURE 1. Top views of various hydrogen adsorptions in the (5,5) CNT with ball and stick forms; (A) clean, (B) arch type, (C) zigzag type, and (D) the molecular hydrogens inside the CNT with coverage θ > 1.0. The d indicates an average diameter of the CNT. Bond lengths are in units of Å. Darker bigger ball and brighter smaller ball represent the carbon and hydrogen atoms, respectively.

Although the *zigzag type* or H_2 molecules inside the CNTs are energetically more stable than the *arch type*, the insertion mechanism of hydrogen into the capillary is not clear. One may consider the capillary effect from the open ends of CNTs[5]. However, the high aspect ratio of CNTs is suggesting another insertion mechanism through the tube walls. Here, we suggest a *flip-in* mechanism and *kick-in* mechanism in this paper.

First, we consider a *flip-in* mechanism in an *arch type* (Fig. 2(I)). Hydrogen atom at the top site can be inserted inside the capillary with an activation barrier of 1.51 eV (Fig. 2(I-2)), without breaking the nanotube walls. Once one hydrogen atom is flipped-in, the adjacent hydrogen can be inserted into the nanotube more easily. The *continuous flip-in* mechanism of the first nearest neighbor hydrogen atom (Fig. 2(II)), with activation barrier of 0.74 eV (Fig. 2(II-2)), finally result in the H_2 molecules in the capillary (Fig. 2(II-5)). The *zigzag filp-in*, the continuous flip-in of second nearest neighbor hydrogen atom, with activation barrier of 0.93 eV (Fig. 2(III-2)), result in a *zigzag type* (Fig. 2(III-5)).

Additional hydrogens at *zigzag type* can be chemisorbed on the tube kicking in one hydrogen atom into the capillary, defined as *kick-in* mechanism (Fig. 2(IV)), with activation barrier of 1.95 eV (Fig. 2(IV-2)). The nanotube wall is recovered partially to the arch type (Fig. 2(IV-4)). Repeating this process induce the H_2 molecule in the capillary (Fig. 2(IV-5)).

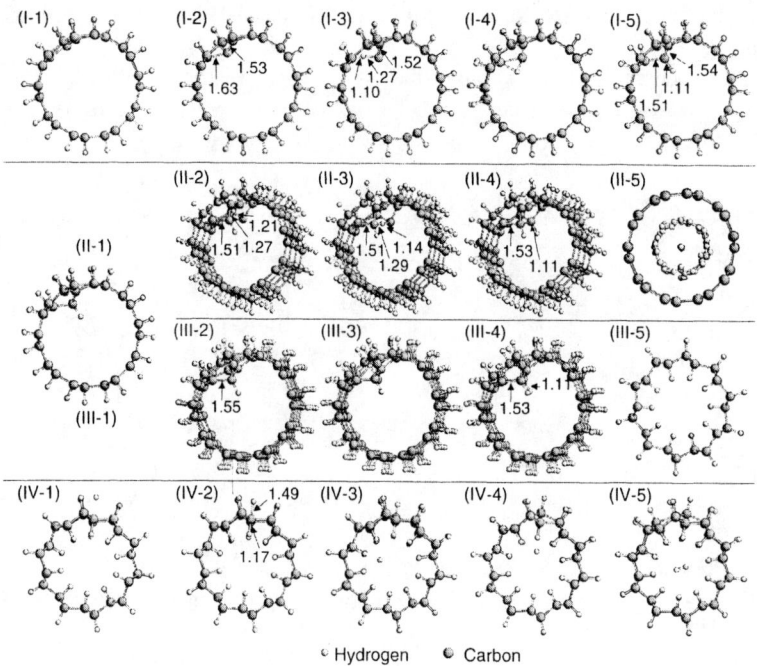

FIGURE 2. Concerted reaction pathways for hydrogen insertion. (I) *Flip-in*, (II) *continuous flip-in*, (III) *zigzag flip-in*, and (IV) *kick-in* mechanism. Bond lengths are in units of Å.

The activation barriers for reaction pathways for hydrogen insertion are shown in Fig. 3. Reaction coordinate '0' indicate the corresponding references of *arch type*, and *zigzag type* for A and C, respectively.

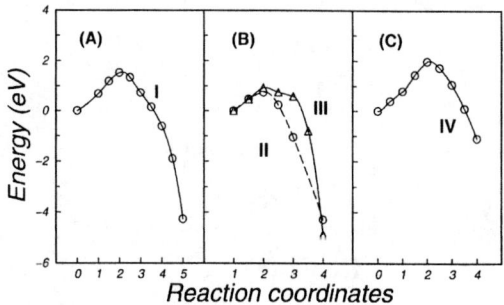

FIGURE 3. Corresponding potential energy barriers of the concerted reaction pathways of figure 2.

We also checked the extraction mechanism of hydrogen to check the reversibility of the hydrogen storage. The extraction mechanism is composed of two-step processes; H_2 dissociation and then *flip-out* processes. The H_2 dissociation process is shown in Fig. 4(I). A H_2 molecules first dissociate while approaching to the nanotube wall, with activation barrier of 1.61 eV (Fig. 4(I-3)). After step (I-3), the H_2 bond is broken and instead two C-H bonds are formed exothermally, as shown in the potential energy barrier in Fig. 4.

After adsorbed interior of the nanotube wall, hydrogens can flip-out to the exterior of the wall (Fig. 4(II)), with activation barrier of 1.96 eV (Fig. 4(II-3)). It is worth noting that the nanotube wall is not broken during this process.

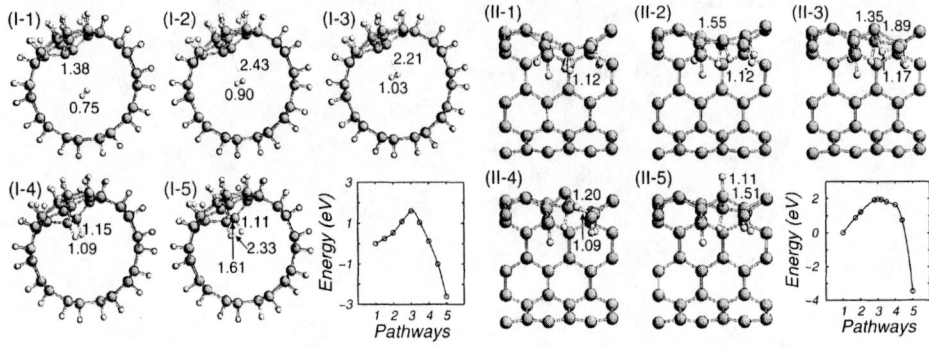

FIGURE 4. Reaction pathways and the corresponding potential barriers for (I) dissociative adsorption of molecular hydrogen in the capillary to the interior of the tube wall and (II) *flip-out* process. Bond lengths are in units of Å.

SUMMARY

We investigate the hydrogen storage mechanism in single-walled carbon nanotubes using density-functional calculations. Several hydrogen chemisorption geometries are identified. The hydrogen atoms can be inserted and extracted through the CNT wall with activation barrier less than 2.0 eV. These barrier could be overcome by the bias applied to the electrodes during electrochemical storage process. Our results describe an electrochemical storage process of hydrogen in carbon nanotubes.

ACKNOWLEDGMENTS

This work was supported by the KOSEF through the QSRC at Dongguk Univ. and in part by the BK21 program.

REFERENCES

1. Dresselhaus, M. S., Williams, K. A., and Eklund, P. C., *MRS Bulletin* **24**, 45-50 (1999), and the references therein.
2. Nutzenadel, C., Zuttel, A., Chartouni, D., and Schlapbach, L., *Electrochem. and Solid-State Letters* **2**, 30-32 (1999).
3. Elstner, M., Porezag, D., Jungnickel, G., Elsner, J., Haugk, M., Frauenheim, T., Suhai, S., and Seifert, G., *Phys. Rev. B* **58**, 7260-7268 (1998).
4. Lee, S. M., and Lee, Y. H., *Appl. Phys. Lett.*, **76**, 2877-2879 (2000).
5. Pederson, M. R., and Broughton, J. Q., *Phys. Rev. Lett.* **69**, 2689-2692 (1992).

Optical Anisotropy of Aligned Single-Walled Carbon Nanotubes in Polymer

M. Ichida[*], S. Mizuno[*], H. Kataura[†], Y. Achiba[†], and A. Nakamura[*]

[*]CIRSE and Graduate School of Engineering, Nagoya University
Furo-cho, Chikusa-ku, Nagoya 464-8603, Japan
[†]Faculty of Science, Tokyo Metropolitan University
1-1 Minami-Ohsawa, Hachioji-shi, Tokyo 192-0397, Japan

Abstract. We have investigated polarized absorption and Raman spectra of aligned single-walled carbon nanotubes in polymer. The optical transitions in nanotubes strongly depend on the angle θ between the polarization of incident light and nanotube axis: the absorbance measured at the peak energy shows a maximum at $\theta = 0°$, and they are suppressed at $\theta = 90°$. Raman spectra also show strong polarization dependence, which is mainly determined by the dependence of oscillator strength of optical transition in nanotube.

INTRODUCTION

Single-walled carbon nanotubes (SWNTs) are regarded as naturally grown quantum wires because of the high aspect ratio (length/diameter) of a fundamental structure. The high anisotropy of SWNTs results in a quasi one-dimensional electronic structure leading to optical anisotropy in absorption and Raman scattering spectra. Recently, polarized Raman spectra of SWNTs have been theoretically [1] and experimentally [2-4] studied. The anisotropic optical transitions have been theoretically predicted [5]. Hwang et al. have reported the polarization dependence of absorption spectra in the visible region [6]. Although the fundamental optical transition of semiconducting SWNTs lies in the infrared region, no measurement of polarized absorption spectra has been carried out in the infrared region. In order to investigate the anisotropy of band structure in SWNTs, we need to measure polarization dependence of absorption spectra from infrared to visible region. In this study, we report polarized absorption and Raman spectra of aligned SWNTs in polymer. Absorption and Raman spectra show the strong optical anisotropy. We analyze the spectra assuming the angle distribution of SWNTs in the sample.

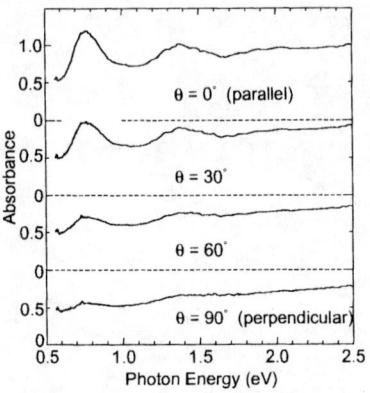

FIGURE 1. Absorption spectra of the stretched sample with different polarization angles measured at room temperature.

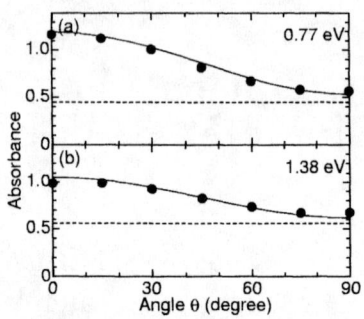

FIGURE 2. Polarization-angle dependence of absorbance measured at 0.77 eV (a) and 1.38 eV (b). Solid curves are the results fitted by using Eq. (1). Broken lines show the angle-independent component.

EXPERIMENTAL

Soot containing SWNTs were prepared by the laser vaporization method [8]. The mean diameter of SWNTs is ~ 1.2 nm. SWNTs were sonicated in toluene and polystirene mixture for 1h. The SWNT/polystirene-toluene suspension was dropped on Teflon sheet. The suspension was stretched together with the sheet. The stretching ratio was about 10. An Ar-ion laser (2.41 eV) was used as the excitation source for Raman scattering. The scattered light was detected by a liquid N_2 cooled CCD with VV and VH geometries: the scattered light polarization parallel (VV) and perpendicular (VH) to the incident laser polarization.

RESULTS AND DISCUSSION

Figure 1 shows the absorption spectrum for the stretched sample with different angles θ between the stretching axis and the polarization of incident light. In all spectra, broad absorption bands are observed at ~ 0.8 eV, ~ 1.4 eV and ~ 2.0 eV. The absorption bands at 0.8 eV and 1.4 eV are attributed to the optical transitions in semiconducting tubes, and the 2.0 eV band is due to metallic SWNTs [7,8]. The observed spectra reveal strong optical anisotropy. The absorbance at the peak decreases with increasing θ. Figure 2 shows the θ dependence of the absorbance measured at 0.77 eV and 1.38 eV. The absorbance decreases with increasing θ and shows a maximum at $\theta = 0°$ and a minimum at $\theta = 90°$. The theoretical calculation predicts suppression of the optical transition when the polarization of the incident light is perpendicular to the SWNT axis [5]. Therefore, the experimental results suggest that SWNTs in our sample are aligned to the stretched axis.

The absorption coefficient measured at the photon energy of $\hbar\omega$ in an oriented

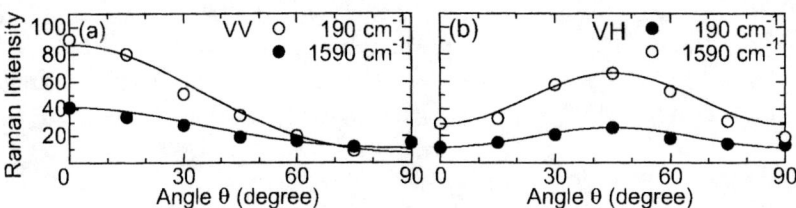

FIGURE 3. Raman intensities of breathing (open circles) and tangential (closed circles) modes as a function of polarization angle θ for VV (a) and VH (b) geometries.

SWNT is proportional to $a(\hbar\omega)\cos^2\psi$ where ψ is the angle between the SWNT axis and the incident light polarization, $a(\hbar\omega)$ is the absorption coefficient of SWNT when $\psi = 0$ [4]. In the stretched sample, there is a distribution of orientation angle ϕ between the stretching and SWNT axes. Assuming a Gaussian distribution of orientation angle, the observed absorption spectrum $\alpha(\hbar\omega,\theta)$ can be expressed by

$$\alpha(\hbar\omega,\theta) = a(\hbar\omega)\int_{-\pi/2}^{\pi/2} f(\phi)\cos^2\psi d\phi + b(\hbar\omega), \qquad (1)$$

where $f(\phi)$ is a Gaussian for the angle distribution function and $\psi = \theta - \phi$. We add an angle-independent component $b(\hbar\omega)$ which is due to nano-graphite and/or amorphous carbon remaining in the sample. We analyzed the angle dependence of absorbance measured at various photon energies using Eq. (1). The adjustable parameters are $a(\hbar\omega)$, $b(\hbar\omega)$, and the width of angle distribution function. Fitted results at 0.77 eV and 1.38 eV are shown by solid curves in Fig. 2. The calculated results well reproduce the experimental results for various photon energies using the same distribution function. The full width at half maximum of the obtained angle distribution is 46°, which means that 56% of SWNTs in the sample are aligned within the range of ± 15 to the stretching axis.

Next, we measured polarized Raman spectra of stretched sample for the VV and VH geometries. Figure 3 (a) shows the angle dependence of Raman intensity of the breathing (open circles) and tangential (closed circles) modes for the VV geometry. The Raman intensities of both modes decrease with increasing angle θ. Shown in Fig. 3 (b) is the angle dependence of Raman intensity of the breathing (open circles) and tangential (closed circles) modes for the VH geometry. Both modes show a maximum at $\theta = 45°$. The similar results have been reported by Gommans *et al.* [2-4]. The theoretical calculation predicts that the Raman intensity of tangential mode decreases with increasing ψ and shows a minimum at $\psi \sim 60°$, and the intensity of breathing mode increases with increasing ψ for the VV geometry. Both modes show a maximum at $\psi = 45°$ for the VH geometry [1]. The dependence observed for the VV geometry is in disagreement with the theoretical result which has been calculated with assumption of the non-resonant Raman scattering condition [1]. Gommans *et al.* have shown that the Raman intensity is proportional to $\cos^4\psi$ for the VV geometry and $\cos^2\psi\sin^2\psi$ for the VH geometry considering the resonant Raman effect [4]. Using the same angle distribution $f(\phi)$ which is obtained from the analysis of angle dependence of absorbance, θ dependence of Raman intensities for the VV and VH geometries can be written as

$$I_{\rm VV}(\theta) = c_{\rm VV} \int_{-\pi/2}^{\pi/2} f(\phi) \cos^4(\theta - \phi) {\rm d}\phi + d_{\rm VV} \qquad (2)$$

$$I_{\rm VH}(\theta) = c_{\rm VH} \int_{-\pi/2}^{\pi/2} f(\phi) \cos^2(\theta - \phi) \sin^2(\theta - \phi) {\rm d}\phi + d_{\rm VH}, \qquad (3)$$

respectively, where $c_{\rm VV}$ and $c_{\rm VH}$ are parameters. We add angle-independent components $d_{\rm VV}$ and $d_{\rm VH}$. We fit the experimental results by Eq. (2) and (3) with adjustable parameters c and d. Solid curves in Fig. 3 show the fitted results, and they well reproduce the experimental results, which indicates that the observed polarization dependence of Raman spectra is mainly governed by the polarization dependence of the oscillator strength of fundamental transition.

CONCLUSION

We prepared aligned SWNTs in a polymer matrix by using a mechanical stretching method. The absorption and Raman spectra showed the strong polarization dependence. The orientation angle distribution of SWNTs in polymer was obtained from the analyses of polarization angle dependence of the spectra. 56% of SWNTs in the sample are aligned within the range of $\pm$ 15° to the stretching axis. The observed polarization dependence of Raman spectra in our excitation condition is mainly determined by the resonant effect of optical transitions in SWNTs.

Acknowledgement

The authors would like to thank to Dr. Ajiki for useful comments. This work was supported by the Grant-in-Aid for General Scientific Research from the Ministry of Education, Science, Sports and Culture of Japan.

REFERENCES

1. Saito, R. et al., Phys. Rev. B **57**, 4145 (1998).
2. Sun, H. D. et al., Solid State Commun. **109**, 365 (1999).
3. Jorio, A. et al., Phys. Rev. Lett. **85**, 2617 (2000).
4. Gommans, H. H. et al., J. Appl. Phys. **88**, 2509 (2000).
5. Ajiki, H. et al., Physica B **201**, 349 (1994).
6. Hwang, J. et al., Phys. Rev. B **62**, R13310 (2000).
7. Ichida, M et al., J. Phys. Soc. Jpn. **68**, 3131 (1999).
8. Kataura, J. et al. Synth. Met. **103**, 2555 (1999).

Polarized Absorption Spectra of 0.4nm-Sized Single-Wall Carbon Nanotube Arrays formed in Channels of AlPO$_4$-5 Single Crystals

Z. K. Tang, Z. M. Li, G. D. Li, N. Wang, H. J. Liu and C. T. Chan

*Department of Physics, Hong Kong University of Science & Technology,
Clear Water Bay, Kowloon, Hong Kong*

Abstract. We report the polarized optical absorption spectra of 0.4 nm-sized single-walled carbon nanotubes which are arrayed in channels of an AlPO$_4$-5 single crystal. When the electric field (E) of light is polarized parallel to the tube direction (c), the spectra show a sharp absorption band at 1.3 eV, and two broad bands at 2.1eV and 3.1 eV, respectively. The intensities of these absorption bands gradually decrease as the light polarization is turned from $E//c$ to $E\perp c$. In the $E\perp c$ configuration, the tube shows nearly transparent in the measured energy region 0.5 – 4 eV. The measured absorption spectra could be well explained by density-of-states calculated using a local density function approximation.

INTRODUCTION

Single-walled carbon nanotubes (SWNTs) [1] have provided us an opportunity to study novel physical properties of ordered one-dimensional systems. Theoretical studies have shown that the electronic properties of SWNTs are strongly modulated by small structural variations [2]. Within a zone-folding scheme, the diameter and the chirality of a SWNT are believed to determine whether the nanotube is a metal or a semiconductor, and their energy bands have one-dimensional van Hove singularity. Indeed, the one-dimensional electronic density-of-states (DOS) of individual SWNTs have been directly observed by scanning tunneling spectroscopy [3]. Very sharp resonant feature in resonant Raman spectra showing optical transitions between spike-like DOSs seems to enhance the Raman intensity [4]. Recently, Kataura, *et al.* [5] reported optical absorption spectra in near infrared to near ultraviolet region for bundles of SWNTs. Some absorption bands related to the optical transitions between the van-Hove singularity states have been observed. More controlled experimental studies of the novel electronic properties for an individual SWNT are, however, not easy to carry out because of the technical difficulty in fabrication of *mono-sized* and well-aligned nanotubes. Very recently, we have succeeded in producing 0.4 nm-sized SWNTs which are arrayed in the channels of AlPO$_4$-5 zeolite single crystals [6]. The ability of mono-sized, well-aligned SWNTs synthesis brings the experimental situation much closer to that of the theoretical models.

In this report, we present absorption spectra for light polarized along variant directions for the 0.4nm-sized SWNTs. The nanotube contained AFI crystal behaves a good polarizer with high absorption for the light polarized parallel to the tube direction ($E//c$) and highly transparency for the light polarized perpendicular to the tube direction ($E \perp c$). The observed absorption spectrum can be well explained using the DOS of the ultra-thin nanotubes calculated using an *ab initio* process.

EXPERIMENTAL

AFI is a kind of porous aluminophosphate single crystals. Its framework consists of alternative tetrahedra of $(AlO_4)^-$ and $(PO_4)^+$ which form opened one-dimensional channels packed in the hexagonal structure. The coordinator diameter of the channel is about 0.73 nm, and the distance between two neighboring parallel channels is 1.37 nm. Fig. 1 shows schematically the framework of the AFI single crystal, viewed along (001) direction, and the carbon nanotubes stabilized inside the channels. Detailed fabrication process of the nanotubes was reported in [6,7]. The AFI single crystal itself is optically transparent from near infrared to ultraviolet region. It is an idea to study optical properties of any nanostuctures formed inside. Typical dimension of the AFI crystals used in our experiment is 100 μm in cross section diameter and 300 μm in length. In order to easily carry out the optical measurements for the tiny crystal, we first drilled a small hole on a machinable ceramic plate; the C-AFI crystal was then horizontally fixed inside the hole using epoxy resin. After the resin became hardened, both sides of the sample were polished to final thickness of 10 μm. Transmission spectra were measured at room temperature, using a tungsten-halogen incandescent lamp as a light source. The incident polarized light was focused onto the sample by a reflecting microscope objective. The transmission light was collected by another reflecting objective coupled with an optical fiber and dispersed by a 275-mm single-grating monochromator.

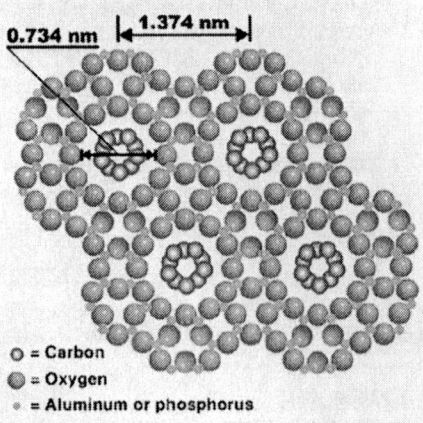

FIGURE 1. SWNTs accommodated in the channels of AFI single crystal viewed along (001) direction.

RESULTS AND DISCUSSION

Fig. 2 is a high-resolution transmission electron microscope (TEM) image (JEOL2010 electron microscope, operating at 200kV) of our nanotubes. In order to get a reasonable contrast of the nanotube image, the zeolite framework was removed using hydrochloric acid before the TEM observation. In the image, we can see many parallel doubled dark lines, which are typical images of single-walled nanotubes. All these SWNTs showed the same morphology and the same size. By measuring the separation between the doubled dark lines, we determine the diameter of the SWNTs to be 0.42 with an error ±0.02 nm. There are three possible nanotube structures with this small diameter: the zigzag (5,0) tube ($d = 0.39$ nm), the armchair (3,3) tube ($d = 0.40$ nm) and the chiral tube (4,2) ($d = 0.41$ nm), where (n,m) is the so-called chiral vector defining the tube symmetry. For such small carbon nanotubes, the large curvature effect would lead to a hybridization of σ and π orbitals, so the electronic structures can no longer be predicted by the simple band-folding picture [8]. The polarized absorption spectra (optical density) of these SWNTs are shown in Fig. 3 with different polarization angles. The top curve is the absorption spectrum for light polarized parallel to the tube axis ($E//c$). There is a sharp absorption band at 1.37 eV (labeled A) with a shoulder at 1.17 eV (labeled S). At higher photon energy region, there are two broad bands at 2.1 eV (B) and 3.1 eV (C), respectively. As the polarization angle is increased from zero-degree ($E//c$) to 90-degree ($E\perp c$) with an increment of 10 degrees, as shown from the top to the bottom curves in the figure, these absorption bands slightly shifted to higher energy, while their intensities gradually decreased and finally vanished. The nanotubes are nearly transparent in the whole measured energy region in the $E\perp c$ configuration. We could not measure the absorption spectra in the energy region higher than 4.0 eV, because of the strong absorption in ultraviolet region from the epoxy that was used to hold the sample. As a reference, the absorption of the epoxy is also shown in the same figure by the dotted curve.

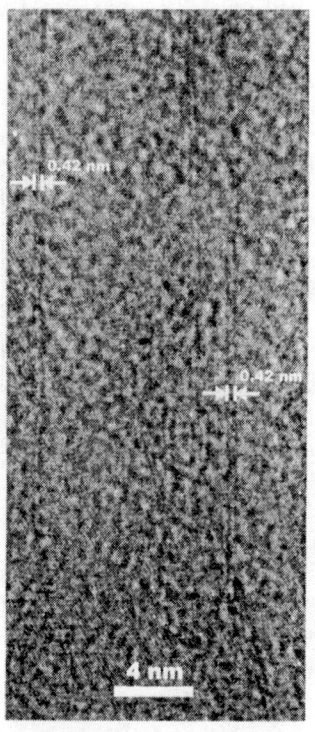

FIGURE 2. High-resolution TEM image of the SWNTs. The nanotubes were moved out from the AFI channels and nanotubes dispersed on a carbon lacey film.

It is noticed in the Fig. 3 that, in the $E//c$ polarization configuration, the nanotubes have a finite optical density in the whole photon energy region, while in the $E\perp c$ configuration the optical density is nearly zero. This finite optical density in the $E//c$ polarization implies that the ultra-small nanotubes are of metallic behavior (π-electron plasma absorption background [9]). The metallic behavior along the tube direction was also seen from the electric transport measurements [10]. To have a better understanding, we compared the measured absorption spectra with the band structural calculation using a local density function approximation (LDA). The calculation results showed that, the zigzag (5,0) tube and the armchair (3,3) tube are metallic, but the chiral (4,2) is semiconducting. As both the electric transport property [10] and the absorption spectrum ($E//c$) shown in Fig. 3 indicated the SWNT is a metal, we focus on the metallic tubes in the following discussion. The calculated DOS of zigzag (5,0) tubule is shown in the inset of Fig. 3. There are many sharp features in the DOS spectrum, which are characteristic van-Hove singularity structures of a quasi-one-dimensional system. There exists a finite density of state near the Fermi energy level (= 0), indicating the zigzag (5,0) tube is metallic. According to the DOS spectrum, the energy spacing between the pair of singularities nearest from the Fermi energy is about 1.3 eV, the next two nearest separations are 2.4 eV and 3.0 eV, respectively. These values are in good agreement with the absorption bands of *A*, *B*, and *C*, respectively. Hence, the absorption bands shown in Fig. 3 can be attributed to the electron transitions between the van Hove singularity states. More detailed theoretical studies with respect to symmetry consideration of the dipole-transition elements and polarization behavior are currently under way.

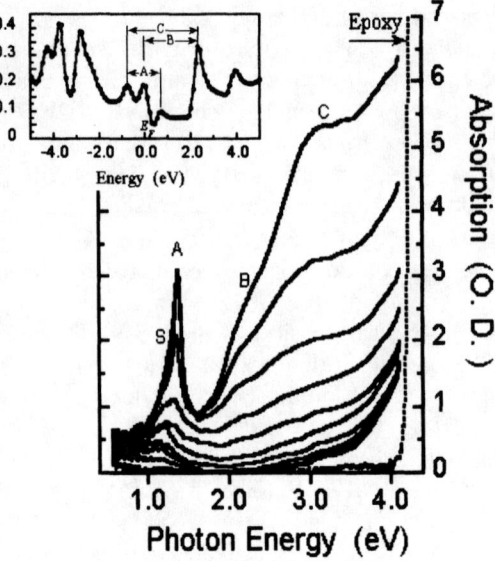

FIGURE 3. Polarized absorption spectra of the SWNTs. The top solid curve corresponds to the absorption of light polarized parallel to the tube direction (*E//c*). The bottom solid curve is for *E⊥c*. The dotted curve is the absorption of epoxy. The inset shows the DOS of the zigzag (5,0) tube.

ACKNOWLEDGEMENTS

We are grateful to Professors P. Sheng at HKUST, Prof. R. Saito at the Univ. of Electro-Communica-tions and Prof. N. Nagasawa at the Univ. of Tokyo for their useful discussions and valuable comments. This research was supported by the RGC Committee of Hong Kong, and the EHIA program from HKUST.

REFERENCE

1. S. Iijima and T. Ichihashi, *Nature*, 363, 603 (1993).
2. See, for example, M. S. Dresselhaus, G. Dresselhaus, P. C. Eklund, *Science of Fullerenes and Carbon Nanotubes* (Academic Press, New York, 1996).
3. J. W. G. Wildoer, L. C. Venema, A. G. Rinzler, R. E. Smalley, C. Dekker, *Nature* 391, 59 (1998).
4. A. M. Rao, E. Richter, S. Bandow, B. Chase, P. C. Eklund, K. A. Williams, S. Fang, K. R. Subbaswamy, M. Menon, A. Thess, R. E. Smalley, G. Dresselhaus, M. S. Dresselhaus, *Science* 257, 187 (1997).
5. H. Kataura, Y. Kumazawa, Y. Maniwa, I. Umezu, S. Suzuki, Y. Ohtsuka and Y. Achiba, *Synthetic Metals* 103, 2555 (1999).
6. N. Wang, Z. K. Tang, G. D. Li, and J. S. Chen, *Nature* 408, 50 (2000).
7. Z. K. Tang, H. D. Sun, J. Wang, J. Chen, and G. Li, *Appl. Phys. Letters* 73, 2287 (1998).
8. X. Blasé, Lorin X. Benedict, Eric L. Shirley, and Steven G. Louie, *Phys. Rev. Letters* 72, 1878 (1994).
9. S. Kazaoui, N. Minami, H. Yamawaki, K. Aoki, H. Kataura, and Y. Achiba, *Phys. Rev.* B 62, 1643 (2000).
10. Z. K. Tang, H. D. Sun, and J, N, Wang, *Physica* B 279, 200 (2000).

Structural (n,m) determination of isolated single wall carbon nanotubes by resonant Raman scattering

A. Jorio[†], R. Saito[††], J. H. Hafner[‡], C. M. Lieber[‡], M. Hunter[†], T. McClure[†], G. Dresselhaus[†], M. S. Dresselhaus[*]

[†] *Massachusetts Institute of Technology, Cambridge, MA 02139-4307*
[††] *University of Electro-Communications, Tokyo, 182-8585 Japan*
[‡] *Harvard University, Cambridge, MA 02138*
[*] *Currently on leave from Massachusetts Institute of Technology, Cambridge, MA 02139-4307*

Abstract. We performed confocal micro Raman spectroscopy of an individual single-wall nanotube. Isolated SWNTs are prepared by a chemical vapor deposition method on a Si/SiO$_2$ substrate containing iron catalyst particles of nanotube size. We show that resonant confocal micro Raman spectroscopy of an (n,m) individual single-wall nanotube makes it possible to assign its chirality uniquely by measurement of one radial breathing mode frequency ω_{RBM}.

Since there is as yet no method for selecting a specific nanotube diameter and chirality in the nanotube production process, measurement of these quantities is important because the observed properties of carbon nanotubes depend strongly on the nanotube diameter and chirality. Here we show that it is possible to obtain a Raman spectrum from one isolated SWNT by using Resonant Confocal Micro-Raman Spectroscopy (RCMRS). Such a spectrum gives the complete (n, m) atomic structural assignment for an isolated SWNT.

Isolated SWNTs were prepared by a chemical vapor deposition method on a Si/SiO$_2$ substrate containing nanometer size iron catalyst particles. We have transferred these nanotubes from the surface to AFM probe tips and have confirmed that they are individual SWNTs by transmission electron microscopy. [1] The diameter distribution of the sample ($\sim 1 < d_t <\sim 3$ nm) was characterized by AFM. Resonant Raman spectra were obtained from single isolated SWNTs on this substrate, using a Kaiser Optical Systems, Hololab 5000R: Modular Research Micro-Raman Spectrograph (1 μm laser spot) with 25 mW power, and $E_\ell = 785$ nm $= 1.58$ eV laser line excitation.

Figure 1.A shows three Raman spectra from three different locations on the Si/SiO$_2$ substrate for the low frequency region (100–350 cm^{-1}). We observe the 303 cm^{-1} peak from the Si substrate in all the three spectra, and different nanotube radial breathing modes (RBM) for each position in the sample, showing the presence

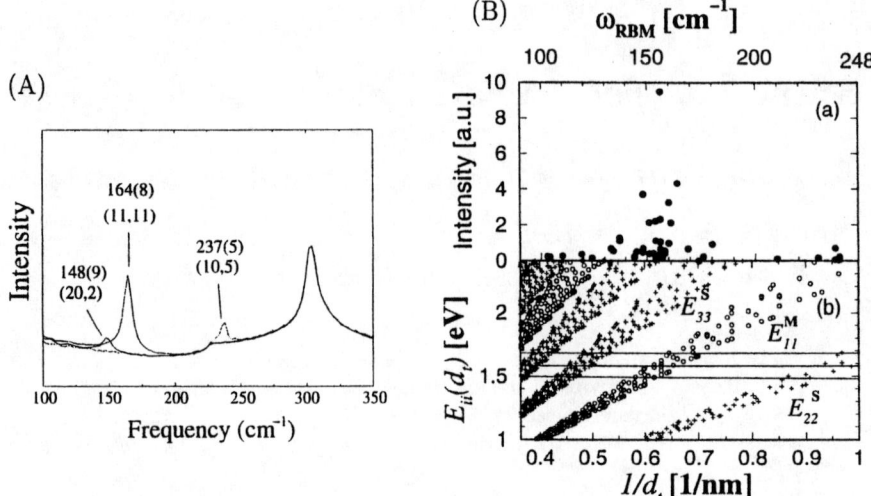

FIGURE 1. (A) Raman spectra come from three different spots on the Si substrate. The RBM frequencies (widths) and the (n,m) assignment for each resonant SWNT are displayed.
(B)(a) Measured frequency vs intensity for the RBM peaks observed at 42 different spots on the sample. (b) Calculated energy separation E_{ii} as a function of $1/d_t$, for $\gamma_0 = 2.9\,\mathrm{eV}$ and $\omega_{RBM} = 248/d_t$. Circles are for metallic SWNTs (E_{ii}^M), and crosses for semiconducting SWNTs (E_{ii}^S).

of only one resonant nanotube for each of the three spots, with $\omega_{RBM} = 148, 164,$ and $237\,\mathrm{cm}^{-1}$. The RBM peaks exhibit natural line widths of 5–10 cm^{-1} [2].

Figure 1.B(a) shows a plot of the measured ω_{RBM} vs. intensity for the RBM peak observed at 42 different spots where we found resonant SWNTs. Figure 1.B(a) has 47 points, since occasionally more than one SWNT is resonant at a single light spot. The surprising appearance of sufficient Raman intensity for a Raman signal from one isolated SWNT is due to the strong enhancement of the Raman cross section in this 1D system. The resonance condition for Raman spectroscopy is obtained when an energy separation E_{ii} between the strong, sharp van Hove singularities (vHs) [3,4] is close to the laser excitation energy $E_\ell = 1.58\,\mathrm{eV}$. Therefore, although the sample has a relatively large density of isolated SWNTs (6 ± 3 SWNTs per laser spot), the probability of finding any SWNT resonant at E_ℓ is small.

Figure 1.B(b) shows a plot of the energy separations E_{ii} between the vHs for a given (n,m) SWNT as a function of $1/d_t$. The vHs in the DOS are calculated using the tight binding calculation with the parameters $\gamma_0 = 2.9\,\mathrm{eV}$ and $a_{C-C} = 0.144\,\mathrm{nm}$, which has been shown to reproduce the resonant Raman spectra of SWNTs very well [4,5]. The SWNTs predicted to be resonant must have E_{ii} within a resonant window $E_\ell \pm 0.1\,\mathrm{eV}$. Figures 1.B(a) and 1.B(b) can be correlated since $\omega_{RBM} = \alpha/d_t$. With the value $\alpha \sim 248\,\mathrm{cm}^{-1}\mathrm{nm}$, we obtain good agreement between the observed resonant RBM frequencies and the ω_{RBM} values calculated from the E_{11}^M and E_{ii}^S

(i =2, 3, 4 for semiconducting SWNTs) interband transitions (see Fig. 1.B).

The assignment of only **one** SWNT is needed for a precise determination of α. We here show that this is possible by analyzing the observed RBM intensities. Because of the trigonal warping effect [4], each interband transition E_{ii}^M for metallic SWNTs is split into two DOS peaks, and there are two resonant conditions for each chiral ($0 < \theta < 30°$) or zigzag ($\theta = 0°$) nanotube. The separation between DOS peaks decreases with increasing θ, and is zero for armchair nanotubes (θ =30°) [4]. Thus an especially large Raman intensity is expected when: (i) an E_{ii} is close to E_ℓ, and when (ii) the DOS splitting is small. Figure 1.B(a) shows one unusually high intensity RBM peak at ω_{RBM}=156 cm^{-1}. This peak should obey the two conditions given above for large intensity. Considering theory, the observed highest intensity ω_{RBM}=156 cm^{-1} comes from a metallic SWNT. When we select metallic nanotubes and the resonant window $1.48 < E_{11}^M < 1.68$ eV, 17 different chiralities are found to be possible theoretically, corresponding to 14 different calculated ω_{RBM} frequencies that, for $\alpha = 248$ cm^{-1}nm, are in the range $\sim 144 < \omega_{RBM} < 174$ cm^{-1}, as listed in Table 1. This is consistent with the experiments where 12 different frequencies in the range $144 < \omega_{RBM} < 176$ cm^{-1} are identified within a ± 1 cm^{-1} experimental accuracy from Raman spectra taken at 42 different light spots. Table 1 shows that there are two armchair nanotubes, (11,11) and (12,12), within this resonant window, with E_{11}^M =1.63 and 1.50 eV, respectively. These values are not very close to E_ℓ and the first condition for high intensity is not satisfied. Furthermore, using $\alpha = 248$ cm^{-1}nm, their calculated ω_{RBM} are 164.0 cm^{-1} and 150.3 cm^{-1}, respectively, and these values are not close to 156 cm^{-1}. Among the 17 different chiralities in Table 1, there are only a few nanotubes with large chiral angles $\theta > 20°$. When we use $\gamma_0 = 2.9$ eV, the energies for the two E_{11}^M singularities for (13,10) are 1.58 and 1.55 eV. One of the E_{11}^M values agrees very well with E_ℓ, and the other is also close. Therefore, it is reasonable to assign the (13,10) chirality to the strongest observed intensity peak at 156 cm^{-1}. The parameter of 248 cm^{-1}nm in $\omega_{RBM} = 248/d_t$ is thus determined, so that the ω_{RBM} value for the (13, 10) nanotube becomes 156.3 cm^{-1}. Furthermore, the empirical parameter of 248 cm^{-1}nm is justified by the observation of other intense ω_{RBM} features with E_{11}^M close to E_ℓ, as shown in bold face in Table 1 [see also Fig. 1.B(a)].

A similar discussion can be made for the ω_{RBM} observed at higher frequencies (> 200 cm^{-1}). We observed 4 different RBM peaks with $\omega_{RBM} = 210, 229, 237$ and 239 cm^{-1}, and we assigned these modes, with no adjustable parameters, as coming from the (14,1), (11,4), (10,5), and (8,7) semiconducting SWNTs, respectively. Using Raman spectroscopy, unique (n, m) assignment could not be made for SWNTs with large d_t. By increasing d_t, the number of resonant SWNTs within the resonant window increases, making it difficult to make a unique (n, m) assignment.

In summary, we show that a unique (n, m) assignment for an isolated SWNT is possible by measuring the ω_{RBM} with the RCMRS technique. However, since we are discussing the structural determination of the structure and properties of a single one-dimensional molecule, it is possible that interaction with the ambient environment, or with the substrate might change the value of α. The influence of the experimental sample preparation conditions on α must be studied further.

This work utilized of MRSEC Shared Facilities supported by the National Science Foundation under award number DMR-9400334 and NSF Laser Facility grant 9708265-CHE. In addition A.J. acknowledges financial support from CNPq - Brazil, and R.S. acknowledges a Grant-in-Aid (No. 11165216) from the Ministry of Education, Japan. MIT authors acknowledge NSF Grants No. DMR 98-04734 and No. INT 98-15744 and INT 00-00408.

TABLE 1: Possible chiralities predicted for metallic nanotubes and their calculated ω_{RBM} in the resonant window $1.48 < E_{11}^M < 1.68\,\text{eV}$. In displaying the observed ω_{RBM}, the number of times each appears is between parentheses.

(n,m)	d_t [nm]	θ [°]	ω_{RBM} [cm^{-1}] (calc.)	(exp.) a)	E_{11}^M [eV] b)	
(18, 6)	1.72	13.9	144.4	144(2)	1.49	1.40
(19, 4)	1.69	9.4	146.8	-	1.53	1.42
(20, 2)	1.67	4.7	148.3	-	**1.55**	1.42
(21, 0)	1.67	0.0	148.8	**148(5)**	**1.56**	1.43
(15, 9)	1.67	21.8	148.8	-	1.51	1.46
(12,12)	1.65	30.0	150.3	**151(3)**	1.50	
(16, 7)	1.62	17.3	153.0	**154(5)**	**1.57**	1.49
(17, 5)	1.59	12.5	156.4	**156(6)**	**1.62**	1.51
(13,10)	1.59	25.7	156.4	**156(1)**	**1.58**	1.55
(18, 3)	1.56	7.6	158.8	158(1)	1.66	1.52
(19, 1)	1.55	2.5	160.0	**160(3)**	1.68	**1.54**
(14, 8)	1.53	21.1	162.0	-	1.65	**1.58**
(11,11)	1.51	30.0	164.0	**164(1)**	1.63	
(15, 6)	1.49	16.1	166.7	**165(1)**	1.72	**1.62**
(16, 4)	1.46	10.9	170.4	169(1)	1.79	1.64
(17, 2)	1.44	5.5	172.7	174(1)	1.81	1.65
(18, 0)	1.43	0.0	173.5	176(1)	1.83	1.65

a) Bold face indicates a strong intensity [See Fig. 1.B(a)].
b) Two E_{11}^M values for each *chiral* (n,m) SWNT are also given, in which bold face indicates a strong resonance.

REFERENCES

1. J. H. Hafner et al., to be published on J. Phys. Chem. B.
2. G. S. Duesberg et al., Chem. Phys. Lett. **310**, 8 (1999).
3. H. Kataura et al., Synthetic Metals **103**, 2555 (1999).
4. R. Saito et al., Phys. Rev. B **61**, 2981 (2000).
5. M. Milnera et al., Phys. Rev. Lett. **84**, 1324 (2000).

High energy-resolution EELS study of the electronic structure of boron nitride cones

M.Terauchi*, M.Kawana*, M.Tanaka*, K.Suzuki[†], A.Ogino[†], and K.Kimura[†]

*Research Institute for Scientific Measurements, Tohoku Univ., 2-1-1 katahira, Aoba-ku, Sendai 980-8577, Japan
[†]Departmentof Advanced Materials Science ,the University of Tokyo, 7-3-1 Hongo, Bunkyo-ku, Tokyo 113-0033, Japan

Abstract. Electron energy-loss spectra were obtained from a single boron-nitride cone (BN cone) with an apex angle of 20 degrees, which is made of curved BN layers. The spectra obtained from the tip region showed the π plasmon peak at 7.4eV, which is smaller than that of bulk hexagonal boron-nitride (h-BN) composed of flat BN layers. The smaller π plasmon energy indicates that the bandgap energy of the BN cone is smaller than that of h-BN. The intensity distribution of the π +σ plasmon peak is explained by the surface loss-function. The B K-shell electron excitation spectra were obtained from the bottom edge region. The spectra showed additional peak intensity compared with that of bulk h-BN.

INTRODUCTION

Graphitic cones were discovered by Ge and Sattler [1] and those with different apex angles were found by Krishnan et al. [2]. The graphitic cones were revealed to consist of a pile of mono-layer graphitic cones. BN cones were found by Bourgeois et al. [3], whose apex angles ranged from 84 to 130 degrees. They proposed a structural model, in which a conical BN layer is helically wound about the cone axis. BN cones with an apex angle of 20 degrees were produced by Kimura et al. and were reported to have a structure consisting of a pile of monolayer BN cones [4]. There is no experimental report yet about the electronic structure of those cone structure materials.

Terauchi et al. [5] measured electron energy-loss spectra of single BN nanotubes and found that BN nanotubes with smaller diameters have smaller bandgap energies than bulk hexagonal boron nitride (h-BN). BN layers of smaller diameter BN nanotubes have stronger curvatures. BN cones have different curvatures of BN layers between the tip and the bottom regions. Thus, the bandgap energy of a BN cone may be different between those regions. The BN cone has edges of BN layers. The electronic state of the edge is expected to be different from that of the h-BN layer itself.

We have investigated bandgap energies of the tip and the bottom regions of a single BN cone and compared with that of *h*-BN. The density of states of unoccupied states at the bottom region was investigated by taking B 1s electron excitation spectra.

EXPERIMENTAL

BN cones were produced by thermal annealing of a mixed powder of β-rhombohedral boron and *h*-BN at 1200 °C under lithium vapor. Almost all BN cones observed have an apex angle of about 20 degrees. Only one BN cone with an apex angle of about 85 degrees has been found. BN cones with other apex angles have not been observed. Electron energy-loss spectra were taken by a high energy-resolution EELS electron microscope (HREA80) [6]. The valence excitation spectra and B 1s excitation spectra were taken from 30nm and 90nm specimen areas, respectively. Energy resolutions of those spectra were about 0.2eV.

RESULTS AND DISCUSSION

Figure 1 shows (a) an electron microscope image of a BN cone and (b) EELS spectra obtained from the tip region (A) of about 10nm diameter and the bottom region (B) of about 40nm diameter of the BN cone. An EELS spectrum of *h*-BN is also shown for comparison.

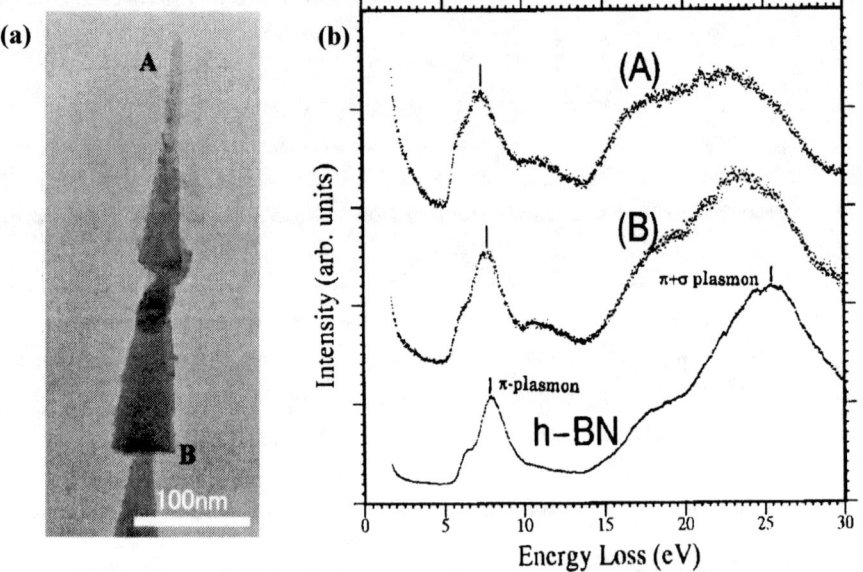

FIGURE 1. (a) an electron microscope image of a BN cone and (b) EELS spectra obtained from the tip region A and the bottom region B of the BN cone. An EELS spectrum of *h*-BN is also shown for comparison.

The spectra (A) and (B) show the π-plasmon peaks at 7.4eV and 7.6eV, respectively. Those energies are smaller than that of *h*-BN (8.2eV). The smaller energies were attributed to smaller $\pi \rightarrow \pi^*$ transition energies (bandgap), which may be due to the curving of BN layers as in the case of BN nanotubes [6]. The decrease of bandgap energy for the curved of BN layers can be explained as follows. The bandgap of a BN layer exists at point P of the Brillouin zone boundary [7]. At this point, the π and π^* bands are only constructed by N p_z and B p_z orbitals, respectively. Curving of a BN layer introduces σ–interaction between the N p_z and B p_z orbitals. This causes an increase of the widths of π and π^* bands, resulting a decrease of the minimum $\pi \rightarrow \pi^*$ transition energy (bandgap). A little smaller π-plasmon energy of (A) than that of (B) can be interpreted by the fact that the average curvature of the tip region A is larger than that of the bottom region B. The energy values and the intensity distributions of the $\pi+\sigma$ plasmon peaks of A and B are not explained by the volume loss-function Im$[-1/\varepsilon]$ but by the surface loss-function Im$[-1/(\varepsilon+1)]$ as in the case of BN nanotubes [6], where ε is a dielectric function of bulk h-BN.

Figure 2 shows a boron K-edge spectrum obtained from the bottom region B in Fig.1(a), where edges of BN layers are exposed. The spectrum of bulk h-BN is also shown for comparison. The spectrum of the BN cone shows not only 1s$\rightarrow\pi^*$ and 1s$\rightarrow\sigma^*$ peaks, which already appear in the spectrum of bulk h-BN, but also an additional peak indicated by an arrow A. The peak energy is 1.6eV smaller than that of 1s$\rightarrow\pi^*$ peak. The spectral intensity between 1s$\rightarrow\pi^*$ and 1s$\rightarrow\sigma^*$ peaks indicated by an arrow B is higher than that of bulk h-BN. The extra intensities A and B may be due to the edge states of BN sheets. The

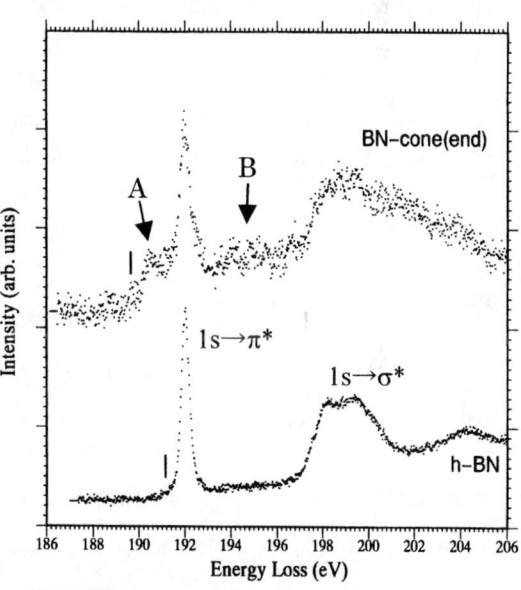

FIGURE 2. B K-edge spectra of the bottom region B of the BN cone in Fig.1(a) and bulk h-BN.

onset energy of the upper spectrum indicated by a vertical line is about 1eV smaller than that of the lower spectrum of h-BN. This smaller onset may be interpreted by a change of the binding energy of inner-shell electrons (chemical shift) because the charge distribution at the edge region should be different from that of h-BN sheets. We performed DV-Xα calculations of flat BN sheets [8] to examine whether such a chemical shift appears for the edge region. The calculated result showed that the chemical shift of B atoms at the extreme edge was about +0.7eV and that of the second nearest B atoms from the edge was about -0.7eV. The chemical shift of the third nearest B atoms from the edge was very small. Thus, the smaller onset energy of the B K-edge spectrum of the bottom region in Fig.2 may be explained by the chemical shift of the second nearest B atoms from the edge.

CONCLUDING REMARKS

The electronic structure of a single BN cone was revealed for the first time by a high energy-resolution electron energy-loss spectroscopy (EELS) microscope [6]. EELS microscopy is powerful to obtain the density of states of the conduction band as well as the dielectric properties from a specified small specimen area. However, the method to obtain the density of states of the valence band from a small specimen area has not been available yet. So, the energy-resolution of X-ray emission spectroscopy based on transmission electron microscopy should be improved for the investigation of the valence band of the fullerene materials [9].

ACKNOWLEDGEMENTS

The authors thank Mr.F.Sato of Research Institute for Scientific Measurements, Tohoku University for his skillful technical assistance. The present work was supported partly by a Grant-in-Aid for Scientific Research on the Basic Research (No.10640295) and the Priority Area "Fullerenes and Nanotubes" (No.11165204, No.11165209) from the Ministry of Education, Science, Sports and Culture of Japan.

REFERENCES

1. Ge, M., and Sattler, K., *Chem. Phys. Lett.* **220**, 192-196 (1994).
2. Krishnan, A., Dujardin, E., Treacy, M.M.J., Hugdahl, J., Lynum, S., and Ebbesen, T.W., *Nature* **388**, 451-454 (1997).
3. Bourgeois, L., Bando, Y., Shinozaki, S., Kurashima, K., and Sato, T., *Acta. Cryst.* **A55**, 168-177 (1999).
4. Terauchi, M., Tanaka, M., Suzuki, J., Ogino, A., and Kimura, K., *Chem. Phys. Lett.* **324**, 359-364 (2000).
5. Terauchi, M., Tanaka, M., Matsumoto, T., and Saito, Y., *J. Electron Microscopy* **47**, 319-324 (1998).
6. Terauchi, M., Tanaka, M., Tsuno, K., and Ishida, M., *J. Miccroscopy* **194**, 203-209 (1999).
7. Zunger, A., Katzir, A., and Halperin, A., *Phys. Rev .B* **13**, 5560-5537 (1976).
8. Terauchi, M., Kawana, M., Tanaka, M., Suzuki, K., Ogino, A., and Kimura, K., *J. Electron Microscopy,* submitted.
9. Terauchi, M., Yamamoto, H., and Tanaka, M., *J. Electron Microscopy* **50** (2001), in the press.

Electrical Contact with Titanium Carbide to an Individual Single-Walled Carbon Nanotube

F. Nihey[*], T. Ichihashi[*], M. Yudasaka[†], and S. Iijima[¶†*]

[*]NEC Laboratories, 34 Miyukigaoka, Tsukuba 305-8501, Japan
[†]Nanotubulites Project, ICORP-JST, c/o NEC Corporation, 34 Miyukigaoka, Tsukuba 305-8501, Japan
[¶]Dept. of Materials Science and Engineering, Meijo University, 1-501 Shiogamaguchi, Tempaku-ku, Nagoya 468-8502, Japan

Abstract. Titanium carbide was used as ohmic contact electrodes to an individual single-walled carbon nanotube (SWNT). The resistance of the SWNT measured by two-terminal method was drastically improved by heating due to the creation of SWNT/TiC/Ti junctions. The temperature dependence of the SWNT resistance showed one-dimensional variable range hopping behavior.

INTRODUCTION

Single-walled carbon nanotubes (SWNTs) [1] have attracted considerable attention for molecular electronic application. SWNTs are either one-dimensional metals or semiconductors, depending on their diameter and chirality [2,3], and some electronic devices based on SWNTs, such as single electron transistors and field-effect transistors, have been experimentally studied. Ohmic contacts are important ingredients for the performance of electronic devices. Especially for SWNTs, imperfect ohmic contacts sometimes make potential barriers for mobile carriers and hinder us from observing fundamental transport properties of SWNTs [4]. A method based on a controlled solid-solid reaction was reported for creating heterostructures between SWNTs and nanorods of carbides such as SiC, TiC, and NbC [5]. This method enabled us to make well-defined SWNT/carbide interfaces. Furthermore, it was reported that TiC could be low-resistance ohmic contacts to SWNTs [5]. We applied this method to make ohmic contacts to an individual SWNT. We found the drastic improvement of the resistance of SWNT by heating to create SWNT/TiC/Ti junctions. The temperature dependence of an individual SWNT resistance showed an $\exp[1/T^{1/2}]$ relation, suggesting one-dimensional variable range hopping behavior.

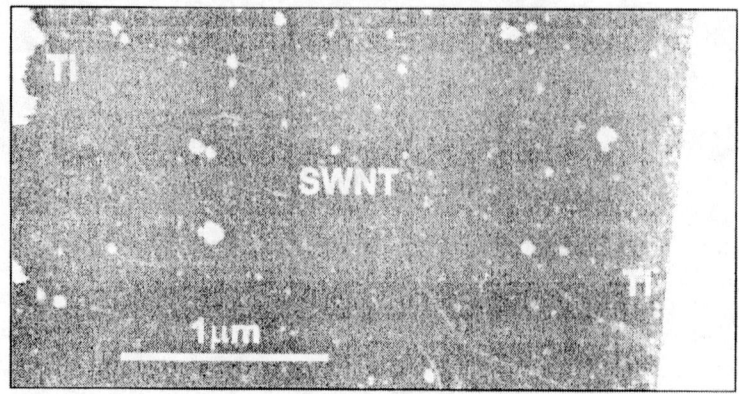

Figure 1: Atomic force microscopy image of single-walled carbon nanotubes on a sapphire substrate. One SWNT connects Ti leads located at the right and left edges of the figure.

SAMPLE PREPARATION

Laser ablation was used to produce SWNTs. The second harmonic beam of a Nd-YAG laser was focused on the surface of a target, which was made by pressing graphite powder, together with 0.6 at.% (Ni+Co) catalyst powder. The target was heated during laser ablation by an electrical tube furnace at 1200°C in a 750 Torr argon gas flow. Typical diameters of SWNTs produced by this method were 1.2 nm – 1.4 nm. We used as-grown SWNTs and omitted the purification and cutting process to avoid possible damage to the SWNTs during the process. Ti pads were defined on a sapphire substrate by photolithography process followed by e-gun deposition and a lift-off process. As-grown SWNTs were then dispersed on the substrate by tearing them with tweezers just above the sapphire substrate. We selected an SWNT with an average height of 1.2 nm and a length of 10 μm by using atomic force microscopy (AFM). We believe from the height measurement that the selected SWNT was not a bundle of SWNTs (SWNTs tend to aggregate due to the van der Waals force), but an individual SWNT. Ti leads were again defined with the same technique to connect the SWNT and the two Ti pads defined before. The position of the SWNT was identified in advance by AFM. The substrate was heated at 800°C for 30 min in a vacuum chamber ($\sim 10^{-8}$ Torr) to create SWNT/TiC/Ti junctions through the reaction of SWNT with Ti. The AFM image in Fig. 1 shows that a 3.5-μm-long SWNT connects two Ti leads located on the right and left edges of the figure.

CONDUCTIVITY OF AN INDIVIDUAL SWNT

The resistance of the individual SWNT was measured by two-terminal method. The resistance changed from 300 MΩ to 16 MΩ through the heating described before, de-

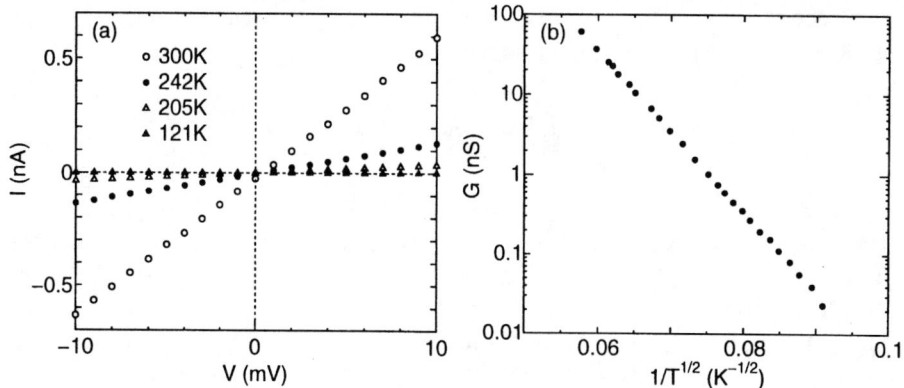

Figure 2: (a) Current-voltage characteristics at different temperatures for an individual SWNT. (b) Semi-logarithmic plot of SWNT conductance as a function of $T^{-1/2}$.

creasing to one twentieth of the initial resistance. This drastic improvement clearly demonstrated the effectiveness of the TiC electrodes for SWNTs. The resistance after heating was of the same order of magnitude as the value (30 MΩ) estimated from the in-plane resistivity of bulk graphite, implying that the SWNT resistance dominated the resistance measured by two-terminal method. Figure 2 (a) shows the current-voltage characteristics of the SWNT at temperatures from 300 K down to 121 K. No apparent nonlinearity was observed at these temperatures and in the voltage range ($|V| < 10$ mV). Tunneling events often cause nonlinearity effects in I-V characteristics. The resistance increased exponentially as the temperature decreased, reaching 50 GΩ at T=121 K. Figure 2 (b) shows the semi-logarithmic plot of the SWNT conductance as a function of $T^{-1/2}$, indicating linear dependence on the variable. The temperature dependence of the SWNT conductance revealed the relation

$$G \propto \exp\{-(T_0/T)^{1/2}\},$$

where $T_0 = 5.4 \times 10^4$ K. This temperature dependence can be explained either by (1) variable-range hopping conduction in one-dimensional systems or (2) hopping conduction in a system where the correlation effect dominates. The average distance between localized states should be quite small (0.3 nm) in order to describe the SWNT conductance with (2). We speculate that the SWNT is described by one-dimensional variable range hopping (1). This also indicates that the resistance masured by two-terminal method after heating was dominated by the individual SWNT resistance.

CONCLUSION

We have successfully used titanium carbide as ohmic contacts to an individual SWNT. Drastic improvement in the two-terminal resistance was confirmed. The temperature dependence of the SWNT resistance shows an $\exp[1/T^{1/2}]$ relation, suggesting one-dimensional variable range-hopping behavior.

REFERENCES

1. Iijima, S., and Ichihashi, T., *Nature* **363**, 603-605, 1993.
2. Hamada, N., Sawada, S., and Oshiyama, A., *Phys. Rev. Letters* **68**, 1579-1581 (1992).
3. Saito, R., Fujita, M., Dresselhaus, G., and Dresselhaus M. S., *Appl. Phys. Letters* **60**, 2204-2206 (1992).
4. Tans, S. J., Devoret, M. H., Dai, H., Thess, A., Smalley, R. E., Geerlig, L. J., and Dekker, C., *Nature* **386**, 474-477 (1997).
5. Zhang, Y., Ichihashi, T., Landree, E., Nihey, F., and Iijima, S., *Science* **285**, 1719-1722 (1999).

Multiwalled Carbon Nanotubes as Single Electron Transistors

M. Ahlskog, R. Tarkiainen, L. Roschier, M. Paalanen, and P. Hakonen

Low Temperature Laboratory, Helsinki University of Technology
Otakaari 3A, Espoo, FIN-02015 HUT, Finland

Abstract. Single electron transistors (SET) are fabricated from multiwalled carbon nanotubes (MWNT) by manipulation with an atomic force microscope. The devices consist of either a single MWNT with Au contacts at the ends or of two crossing tubes. In the latter device, the lower nanotube acted as the central island of a single electron transistor while the upper one functioned as a gate electrode. Coulomb blockade oscillations were observed on the nanotube at low temperatures. The voltage noise of the nanotube-SET was gain dependent as in conventional SETs. The charge sensitivity at 10 Hz was 6×10^{-4} e/$\sqrt{Hz}$. Furthermore, in another device where the MWNT is suspended above the substrate between the electrodes, we measure an extremely high charge sensitivity of 6×10^{-6} e/$\sqrt{Hz}$ at 45 Hz, comparable to the best of the conventional SETs.

INTRODUCTION

Carbon nanotubes are proposed as building blocks of future nanoscale electronic devices (for a review, see [1]). Single electron transistors (SET) are one possibility, since the Coulomb blockade has been observed especially in devices made from single walled nanotubes (SWNT), while in multiwalled nanotubes (MWNT) only very few results have been reported [2]. Field effect transistors based on semiconducting nanotubes have been demonstrated, made both from SWNTs and SWNT ropes.

The atomic force microscope (AFM) plays often an essential role in the fabrication of nanotube based electronic devices. Besides using the AFM for locating and imaging an individual tube, it can be used more actively for manipulating nanotubes [2]. Our group has developed a method for the manipulation of nanoscale particles using the AFM in non-contact mode. This method has the advantage that manipulation and imaging can be performed simultaneously. It is possible to move with this method a multiwalled nanotube over a distance of several micrometers. It is also possible to push a tube over an obstacle with a height of several tens of nanometers. Armed with these capabilities we have fabricated two types of nanotube-SETs: One made from two nanotubes where a MWNT has been pushed on top of another (Device1) and one where a MWNT has been pushed on top of two adjacent gold electrodes (Device2).

Results and Discussion

Figure 1(a-c) shows AFM images of how a crossing from two MWNTs separated by a few micrometers is made. The moved MWNT is ultimately positioned on top of the other nanotube. The other end of this MWNT is hanging above the substrate by 30-40 nm:s. In Figure 1(d) is shown how this "nanotube cross" is contacted with Au electrodes, resulting in a three terminal device where the upper MWNT functions as a gate electrode [3]. We call this device Device1. In the other type of a device (Device2) a MWNT has been positioned on top of two 25 nm thick Au electrodes very close to each other. The section of the nanotube between the electrodes is 275 nm:s long and is separated from the SiO_2 substrate, that is, the tube is suspended. The diameters of all nanotubes were $\cong$ 15 nm.

The lower nanotube of Device1 had a room temperature resistance $R_{300K} = 71$ kΩ, while the two-point resistance over the crossing was $\cong$ 10 MΩ, a value significantly higher than those found for the resistance between crossing metallic SWNTs. Furthermore, the zero-bias resistance increased to ~ 1 GΩ below 4 K. Thus we could utilize the upper tube for gating the current in the lower tube. Figure 2 (a) displays IV curves measured at temperatures from 300 K down to 150 mK. A Coulomb blockade develops fully only at subkelvin temperatures, with a gap of about 1 mV at 150 mK. Figure 2 (b) shows the source-drain current I as a function of the gate voltage V_g, applied from the upper tube. From the shape of the Coulomb oscillations in the I vs. V_g curves, it is concluded that $R_1 \cong R_2$, where R_1 and R_2 are the junction resistances at the opposite ends of the nanotube. We estimate [3] the corresponding capacitances as $C_1 = 0.32$ fF and $C_2 = 0.22$ fF. We get for the charging energy $E_c = \frac{1}{2}e^2/(C_1+C_2+C_{tube})$ = 0.14 meV, where we estimate the nanotube self-capacitance as $C_{tube} = 5 \times 10^{-17}$ F.

The gate modulation period was measured as $\Delta V_g = 4$ mV. We calculate the gate capacitance to the upper tube as $C_g = e/\Delta V_g = 4 \times 10^{-17}$ F. A Fourier analysis of the gate modulation curves in general (including those from a more remote side-gate [3]) revealed only one period, indicating the existence of only one island. Particularly, this implies that the lower tube is not electrically split into two parts separated by a tunneling junction at the point of crossing with the upper nanotube, where considerable mechanical forces between the tubes are known to exist [4].

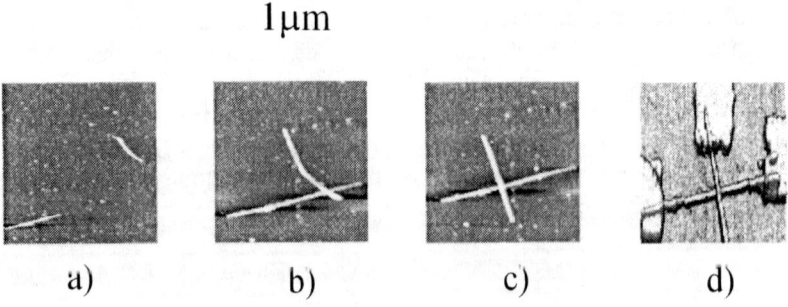

FIGURE 1. (a-c) Manipulation with AFM in non-contact mode to make a nanotube cross. (d) Gold electrodes deposited on the same nanotubes (Device1).

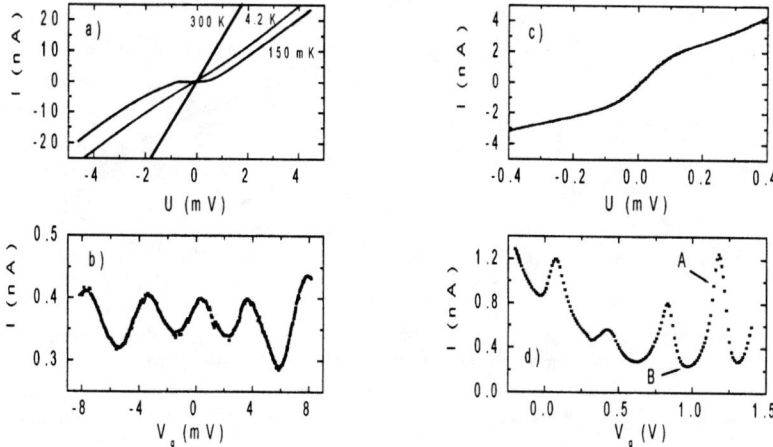

FIGURE 2. (a) IV characteristics of Device1. (b) Gate modulation from the upper tube of the current of Device1 (at U = 0.4 mV) (c) IV characteristics of Device2 at T = 150 mK (d) Gate modulation of Device2, applied from a side gate. A and B refer to the points of maximum and minimum gain where the noise data of Fig. 3(b) was measured.

The IV characteristics of Device2 is shown in Fig. 2 (c). The nanotube had a room temperature resistance of 28 kΩ. As opposed to the usual case of a Coulomb blockade at low voltages (and low temperatures), this nanotube exhibited increased conductance around zero bias, which we attribute to resonant tunnelling. Two weakly quantized steps are seen, therefore this tube can not be said to be fully ballistic. The ballisticity of freestanding samples are likely to be enhanced for several reasons; besides the absence of contact with impurity states of the surface, the plasmon speed, which is sensitive to the permittivity of the substrate, is increased. We measured a total resistance of ≅ 40 kΩ outside the Coulomb blockade regime. The junction resistance of the nanotube-Au contacts are thus less than the quantum of resistance $R_Q \cong 26$ kΩ, which means that the Coulomb blockade can not fully develop. Consequently, the Coulomb oscillations that we measure are smoothened.

We have measured the low frequency noise characteristics of these nanotube devices as charge detectors. The frequency dependence of the cross was roughly 1/f while Device2 had more of a $1/f^2$ character (f < 50 Hz). The current noise was measured for Device1 over one period of the gate modulation curve (at a bias of U_b = 0.4 mV), The 1/f noise at 10 Hz over one gate modulation period is shown in Fig. 3(a). As expected for a SET, the noise level varied with the gain of the nanotube device. The input equivalent charge noise q_n is obtained from the measured current noise i_n according to the formula $q_n = C_g i_n / (\partial I / \partial V_g)$. We obtain as the minimum charge noise at 10 Hz 6×10^{-4} e/√Hz (using i_n = 1 pA, $\partial I / \partial V_g$ = 3.5 nA/V) which corresponds to a typical value for a metallic SET device.

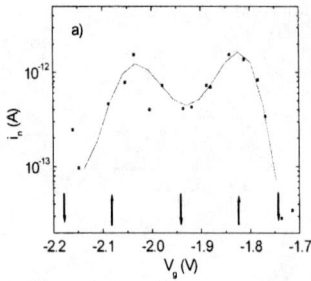
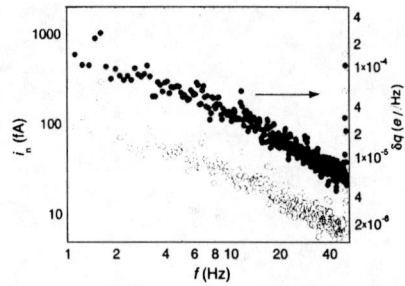

FIGURE 3. (a) Current noise (at 10 Hz) of Device1 measured over one Coulomb blockade oscillation. The up- and down-arrows refer to maximum and minimum gains, respectively. (b) Current noise of Device2 measured at minimum (lower) and maximum gain (upper trace). The right axis gives the equivalent charge noise δq for the upper trace.

Similar modulation of the noise was seen in Device2. At a frequency of 45 Hz we obtain the charge noise q_n as 6×10^{-6} e/√Hz, which is comparable to the best metallic SET devices reported to date [5]. Theoretically the minimum noise level for a SET is $\delta Q_{min} = \hbar C \Delta f R_Q / R_T$, where C is the total capacitance, Δf the frequency range and R_T is the tunneling resistance [6]. Taking $R_Q/4R_T \cong 1$ and assuming no cotunneling, we obtain the minimum noise as 1×10^{-6} e/√Hz. This implies that white noise would dominate over 1/f noise above 3 kHz.

In summary, we have demonstrated the good performance of multiwalled carbon nanotubes as building blocks of nanoscale electronic devices. Under proper conditions, such as separating the tube from the substrate, it is possible to minimize the noise level and in other aspects as well to approach the theoretical limits of performance.

ACKNOWLEDGMENTS

We thank C. Journet and P. Bernier from Universitè Montpellier II for supplying us with the nanotube material and F. Hekking, E. Sonin, and A. Zaikin for useful discussions.

REFERENCES

1. Nygård J. et al., *Appl. Phys. A* **69**, 297 (1999).
2. Roschier L. et al., *Appl. Phys. Lett.* **75**, 728 (1999).
3. Ahlskog M. et al., *Appl. Phys. Lett.* **77**, 4037 (2000).
4. Hertel T., Walkup R., and Avouris P., *Phys. Rev. B* **58**, 13870 (1998).
5. Krupenin V.A. et al., *J. Low Temp. Phys.* **118**, 287 (2000).
6. Korotkov A. N. et al. in *Single-Electron Tunneling and Mesoscopic Physics*, edited by H. Koch and H. Lubbig, Springer, Berlin, 1992, p. 45.

Conduction mechanism in multiwall carbon nanotubes

Stephan Roche[‡], François Triozon[*], Didier Mayou[*], A. Rubio[†]

[‡] *Commissariat à l'Énergie Atomique, DRFMC/SPSMS, Grenoble, France.*
[*] *LEPES-CNRS, Grenoble, France*
[†] *Departamento de Física Teórica, Universidad de Valladolid, Spain.*

Abstract. We report several exceptional consequences of intrinsic geometrical incommensurabilty of multiwalled carbon nanotubes on electronic conduction mechanisms, scaling behavior of the conductance, and temperature dependence of conductivity when dephasing mechanism is dominated by electron-electron interactions in the weak coupling limit.

INTRODUCTION

Single wall carbon nanotubes (SWNTs) [1] are expected to play a major role in the emerging field of molecular electronics devices [2]. In that perspective, their nanometer scale diameter, micron scale length, together with a helicity dependent semiconducting versus metallic status, provide unique features. In particular, metallic SWNTs behave as quasi 1D-molecular wires with a large mean-free path (~ 1 μm) [3,4]. Notwithstanding, in spite of a present fair understanding of SWNTs, the intrinsic properties of multiwalled carbon nanotube (MWNTs) are far from being fully ascertained. Different transport scenario have been reported from low resistive, ballistic, diffusive to insulating behavior [5–7]. One important feature of MWNT geometry that has been underestimated up to now, is that depending on their helicities (n,m), two consecutive shells of a MWNT are commensurate (resp. incommensurate), if the ratio between their respective unit cell lengths $T_{(n,m)}$ along the tube axis is a rational (resp. irrational) number [3] (Fig.1) Recently, such incommensurability was shown to yield anomalous corrugation properties [8], opening new perspectives for MWNTs. Unique physical feature of electronic conduction and magnetotransport have been reported [10].

QUANTUM DIFFUSION IN MWNTS

The computational method

Quantum dynamics has been investigated through the propagation of wavepackets in various long MWNTs with about 1000 unit cells ($\sim$ half million of carbon atoms). The time-dependent Schrödinger equation, solved numerically by using a polynomial expansion of the evolution operator, gives us access to the diffusion coefficients of different types of wavepacket $|\psi\rangle$. If $|\psi\rangle$ is initially localized at the center (x=0) of the nanotube, then $D_\psi(t) = L_\psi(t)^2/t$ with $L_\psi(t) = \sqrt{\langle\psi|(\hat{X}(t) - \hat{X}(0))^2|\psi\rangle}$ the spatial spreading of the WP ($\hat{X}$ is the position operator along the tube axis), and the system size is taken sufficiently long to avoid boundary effects. By averaging over many sites, energy-averaged transport properties are obtained. Alternatively, we also consider random-phase or energy-filtered initial states extented to the whole system [11], but then, periodic boundary conditions are used for the tube of length $\mathcal{L}$, and the diffusivity is approximated as $D_\psi(t) \simeq 4\pi^2/\mathcal{L}^2 I_\psi(t)/t$, where $I_\psi(t) = \langle\psi|\hat{A}^+(t)\hat{A}(t)|\psi\rangle$ and $\hat{A}(t) = \exp(2i\pi\hat{X}(t)/\mathcal{L})\exp(-2i\pi\hat{X}(0)/\mathcal{L}) - \mathbb{1}$. This estimation of $D_\psi(t)$ is accurate for diffusion lengthes smaller than $\mathcal{L}$, as checked by finite-size scaling analysis. The average spreading $L(t)$ and the average diffusion coefficient $D(t)$ are defined by $L(t) = \sqrt{\langle L_\psi^2(t)\rangle} = \sqrt{tD(t)}$, where $\langle\rangle$ denotes an average over many wavepackets. $D(\tau_\phi)$ is the average, if at τ_ϕ the electronic wavefunction looses its phase memory due to some inelastic scattering. The diffusion coefficient at τ_ϕ are connected to the conductivity through $\sigma = (e^2/h)\rho D(\tau_\phi)$, where ρ is the density of states. This approach provides a good scheme for wavepacket propagation in MWNTs constituted of conducting shells.

Anomalous conduction in incommensurate MWNTs

In commensurate systems, conduction was found ballistic along the tube-axis, i.e. $L(t) \sim vt$ similarly to the case of perfect metallic singlewall nanotube $((9,0)@(18,0)$ on Fig.1). Differently, in the incommensurate systems, a non ballistic propagation given by $L(t) \sim t^\eta$ was obtained [10]. The coefficients η were found to decrease from < 1 (non ballistic - $(9,0)@(10,10)$) to $\sim 1/2$ (diffusive like - $(6,4)@(10,10)@(17,13)$) by increasing the number of coupled incommensurate shells. Note that a more detailed presentation of the Hamiltonian describing the coupling between shells in the MWNTs, together with the energy parameters are given elsewhere [10]. Such possibility of obtaining a diffusive-like transport regime even in *a defect-free system* (with vanishingly small presence of defects such as impurities, vacancies, etc), can produce negative magnetoresistance with $\Phi_0/2$-periodic oscillation, as found in weakly disordered mesoscopic systems [10].

ELECTRON-ELECTRON INTERACTIONS AND TEMPERATURE DEPENDENCE OF INELASTIC SCATTERING TIMES AND CONDUCTIVITY

We now focus on the combined effect of anomalous propagation of wavepackets (APWP) together with a dephasing mechanism dominated by electron-electron interactions. Recently, the influence of APWP on frequency and temperature dependences of inelastic scattering times and conductivities, in systems close to a metal-insulator transition [9], or in aperiodic incommensurate and quasiperiodic structures [12,13] has been demonstrated. On the other hand, in the weak coupling limit, the perturbative effect of electron-electron interaction can be account to determine the temperature dependences of inelastic scattering times and conductivity in the low temperature regime. In such cases, the loss of phase coherence of electrons can be obtained from the evaluation of the relative phase accumulated by two interfering waves $\Psi_{i=1,2}(t) = e^{i\phi_{i=1,2}(t)} | \Psi_{i=1,2}(t) |$ which propagate coherently within the system [14]. In the limit of low temperature, the coupling with other electrons, represented by an external potential $\mathcal{V}(L_\Psi(t))$, will reduce the quantum interferences. The interference pattern will be deduced from $e^{i(\int \mathcal{V}(L_\Psi(t))dt/\hbar)}$, and the effect of e-e interaction, on the increasing of phase uncertainty between the two waves $\Psi_1(t)$ and $\Psi_2(t)$ is related to $\langle \delta\phi_{1-2}^2 \rangle$. With $\mathcal{V}(L_\Psi(t)) = -(e/c)dL_\Psi(t)/dt \cdot \mathcal{A}(L_\Psi(t),t)$, where $\mathcal{A}(L_\Psi(t),t)$ is the vector potential, the phase coherence time is shown to be given by $\langle \delta\phi_{1-2}^2 \rangle \simeq e^2 k_B T \int_0^{\tau_\phi} dt \mid L_{\Psi_1}(t) - L_{\Psi_2}(t) \mid^{2-d} \simeq 1$. In a diffusive medium,

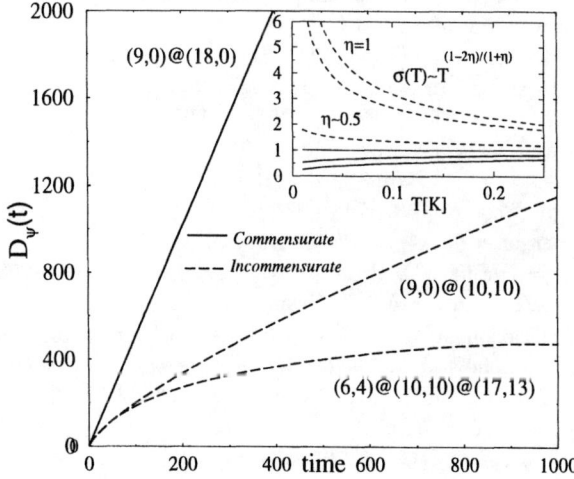

FIGURE 1. Main Frame: diffusion coefficients for typical 2-wall and 3-wall commensurate and incommensurate MWNTs as a function of propagation time (in $\hbar/\gamma_0$ units). Inset: general temperature dependence of the conductivity as a function of the anomalous exponent η.

$| L_{\Psi_1}(t) - L_{\Psi_2}(t) | \sim t^{1/2}$, and $\tau_\phi^{-1} \sim T^{2/(4-d)}$, behavior clearly identified experimentally at low temperature $(T < 4K)$ in 1D-wires.

In defect-free incommensurate MWNTs, APWP thus results in a $\tau_\phi^{-1} \sim T^{1/(1+\eta)}$ power-law for inelastic scattering times, at the origin of an anomalous temperature dependence of the conductivity given by $\sigma \sim T^{\xi(\eta)}$ with $\xi(\eta) = (1-2\eta)/(1+\eta)$ (inset of Fig.1). A temperature independent conductivity is found for the limit $\eta \to 1/2$ (diffusive-like motion). Along the same lines, one also notes that a study of the frequency dependent conductivity could be an interesting way to unveil anomalous quantum diffusion [13]. Frequency dependent study of electronic impedance has been recently demonstrated to be relevant for carbon nanotubes [15].

ANOMALOUS SCALING OF THE CONDUCTANCE

From the electronic Kubo conductivity, an anomalous length dependence of the Landauer conductance can be deduced. Since $\sigma \sim e^2 \rho \langle D \rangle \sim L^{2-1/\alpha}$, the conductance becomes $G(L) \sim \sigma/L \sim (e^2/h) L^{(\alpha-1)/\alpha}$ where one assumes that the system nanotube-electrode is quasi-1 dimensional. The above scaling should be compared with $\frac{e^2}{h} l_e/L$ *for the diffusive case-(l_e the mean free path)* and $\frac{e^2}{h}(1 - L/2l_e)$ *for the ballistic conduction* in metallic SWNTs or conducting commensurate MWNTs.

REFERENCES

‡ Author to whom correspondence should be addressed (sroche@cea.fr)
1. S. Ijima, Nature **354**, 56 (1991).
2. C. Dekker, *Physics Today* (May 1999), 22.
3. R. Saito, G. Dresselhaus, and M. S. Dresselhaus, *Physical Properties of Carbon Nanotubes* (Imperial College Press, London, 1998).
4. C.T. White and T. N. Todorov, *Nature* **393**, 240 (1998).
5. T.W. Ebbessen et al., *Nature* **382**, 54 (1996).
6. S. Frank et al., *Science* **280**, 1744 (1998).
7. C. Schönenberger et al., *Appl. Phys. A* **69**, 283 (1999).
8. A.N. Kolmogorov and V.H. Crespi, *Phys. Rev. Lett.* **85** 4727 (2000).
9. T. Brandes, L. Schweitzer and B. Kramer, *Phys. Rev. Lett.* **72**, 3582(1994).
10. S. Roche, F. Triozon, A. Rubio and D. Mayou, *submitted for publication*.
11. F. Triozon, S. Roche and D. Mayou, *RIKEN Review* **29**, 73 (2000).
12. M. Takahashi, Y. Hatsugai and M. Kohmoto, *J. Phys. Jap.* **65**, 529 (1996). S. Roche and T. Fujiwara, *Phys. Rev. B* **58**, 11338 (1998).
13. D. Mayou, *Phys. Rev. Lett.* **85** 1290 (2000).
14. A. Stern, Y. Aharonov and Y. Imry, *Phys. Rev. A* **41**, 3436 (1990).
15. Y.-P. Zhao, B. Q. Wei, P. M. Ajayan, G. Ramanath, T.-M. Lu, G.-C. Wang, A. Rubio and S. Roche, *submitted for publication*.

Electrical Transport through Crossed Carbon Nanotube Junctions

Takeshi Nakanishi* and Tsuneya Ando[†]

*Department of Applied Physics and DIMES, Delft University of Technology
Lorentzweg 1, 2628 CJ Delft, The Netherlands

[†]Institute for Solid State Physics, University of Tokyo
5-1-5 Kashiwanoha, Kashiwa, Chiba 277-8581, Japan

The conductance between two crossed nanotubes is calculated in a tight-binding model and found to depend strongly on the crossing angle with large maxima at commensurate stacking of lattices of two nanotubes. The results are in good agreement with those calculated in the lowest Born approximation in an effective-mass scheme.

1. INTRODUCTION

Carbon nanotubes (CNs) are novel quantum wires consisting of rolled graphite sheets [1]. Their electronic states are quite different from those of free electrons on a cylinder surface and give rise to various intriguing phenomena. In fact, in two-dimensional (2D) graphite two bands having an approximately linear dispersion cross each other at K and K' points of the first Brillouin zone (the Fermi level, chosen at $\varepsilon = 0$). The purpose of this paper is to study electric transport through crossed single-wall CNs and to demonstrate the importance of wave-vector as well as energy conservation [2].

Recently, experimental studies of crossed CNs with electrical leads attached to each end of both nanotubes were performed and surprisingly high conductances of $0.1 \sim 0.2(e^2/h)$ were reported [3,4]. In the tunneling region with a conductance of about $0.03(e^2/h)$ power-law behavior as a function of bias voltage and temperature was observed [4], which can be described by a Tomonaga-Luttinger liquid model for tunneling [5,6]. A tunnel conductance between a CN and a graphite substrate was also reported, which suggests the importance of momentum conservation in the tunneling process [7]. Junctions of a metallic CN and a semiconducting CN were shown to behave as Schottky diodes [3].

2. CROSSED CARBON NANOTUBE

We consider a junction consisting of two crossed CNs, CNi ($i = 1, 2$), as schematically illustrated in Fig. 1. Both CNs are metallic with a circumference L. The x_i axis is chosen to be parallel to the chiral vector $\mathbf{L}_i$ and the y_i axis is parallel to the axis of CNi, where CN2 is lying on top of CN1. According to Landauer's formula [8], conductances G_{ij} are given by the sum of transmission probabilities between the jth and ith terminals as shown in Fig. 1. The

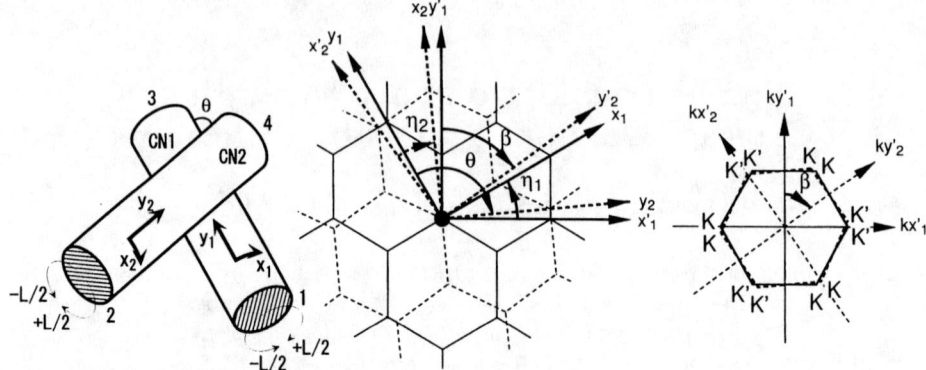

FIGURE 1. (Left) The coordinates for crossed nanotubes. θ is the angle between axes of CNs. Four terminals 1 and 3 of CN1, and 2 and 4 of CN2 are indicated.

FIGURE 2. (Middle) Lattice structure of a two-dimensional graphite sheet near the crossing of CNs. CN1 and CN2 are shown by solid and dotted lines, respectively. A view from reverse is shown for upper CN, because (x_1, y_1) and (x_2, y_2) are chosen as shown in Fig. 1.

FIGURE 3. (Right) The first Brillouin zone and K and K' points. Solid and dotted lines are corresponding to CN1 and CN2, respectively.

conductance is calculated in a single π band tight-binding model at $\varepsilon = 0$.

As a realistic model for the transfer integral between atoms in CN1 and CN2, we consider

$$t(\mathbf{R}_1, \mathbf{R}_2) = \exp\left(-\frac{|\mathbf{d}|}{\delta}\right) \left[t_\sigma \left(\frac{\mathbf{p}(\mathbf{R}_1) \cdot \mathbf{d}}{|\mathbf{d}|}\right) \left(\frac{\mathbf{p}(\mathbf{R}_2) \cdot \mathbf{d}}{|\mathbf{d}|}\right) \right. \\ \left. + t_\pi \left\{ (\mathbf{p}(\mathbf{R}_1) \cdot \mathbf{e})(\mathbf{p}(\mathbf{R}_2) \cdot \mathbf{e}) + (\mathbf{p}(\mathbf{R}_1) \cdot \mathbf{f})(\mathbf{p}(\mathbf{R}_2) \cdot \mathbf{f}) \right\} \right], \quad (2.1)$$

where $\mathbf{R}_1$ and $\mathbf{R}_2$ denote carbon sites on a 2D graphite sheet corresponding to CN1 and CN2, respectively, and $\mathbf{d} = \tilde{\mathbf{R}}_2 - \tilde{\mathbf{R}}_1$ is their distance with three-dimensional coordinates $\tilde{\mathbf{R}}_1$ of an atom on CN1 and $\tilde{\mathbf{R}}_2$ of an atom on CN2. Further, $\mathbf{p}(\mathbf{R}_1)$ and $-\mathbf{p}(\mathbf{R}_2)$ are unit vectors normal to the CN1 at $\mathbf{R}_1$ and CN2 at $\mathbf{R}_2$, respectively, and $\mathbf{d}/|\mathbf{d}|$, $\mathbf{e}$, and $\mathbf{f}$ constitute a set of three orthogonal unit vectors.

We choose the range $\delta/a = 0.325$ and the parameters $t_\sigma = 9.34\gamma_0$ and $t_\pi = -5.91\gamma_0$. They have been determined in such a way that they reproduce the structure of the π bands in bulk graphite [2]. The inter-tube distance is chosen as the inter-layer distance $D_0 = 3.35$ Å of graphite, which is larger than the distance $a/\sqrt{3} = 1.42$ Å between nearest-neighbor atoms on a graphite sheet. In this model the transfer integral decays almost exponentially with the range of about $\delta' \sim 0.8a$ as a function of the 2D distance on CN surface if the curvature is ignored.

Figure 2 shows the lattice near the crossing point. In this figure the small curvature of CNs is ignored for simplicity and η_i is the chiral angle between $\mathbf{L}_i$ and the x_i' direction fixed on each graphite plane. We introduce a parameter $\beta \equiv \theta - \eta_1 - \eta_2$ characterizing the stacking of two graphite sheets. For $\beta = \pi/3$

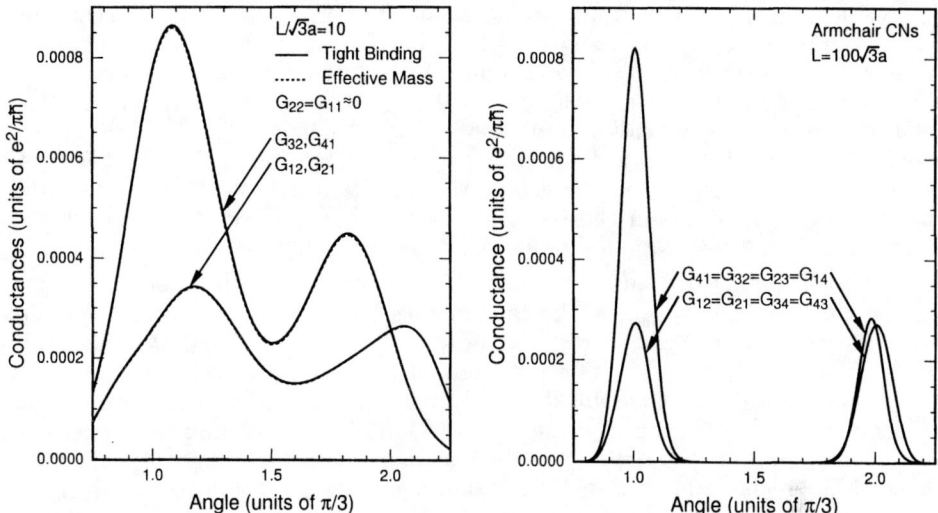

FIGURE 4. (Left) Conductances $G_{41} = G_{32}$ and $G_{21} = G_{12}$ between armchair CNs as a function of the angle θ calculated in a tight-binding model (solid lines) and in the lowest Born approximation in an effective-mass scheme (dotted lines). The reflection probabilities are vanishingly small and most of the electron wave is transmitted within each CNs, i.e., $G_{22} = G_{11} \sim 0$ and $G_{42} = G_{31} \sim 2e^2/\pi\hbar$.

FIGURE 5. (Right) Conductances $G_{41} = G_{32} = G_{23} = G_{14}$ and $G_{21} = G_{12} = G_{34} = G_{43}$ between thick CNs ($L = 100\sqrt{3}a$) calculated in an effective-mass scheme. The peak conductance is nearly independent of the circumference.

(stack I), a B site of CN2 is just above a B site of CN1 but A sites are not, just like the stacking in bulk graphite. For $\beta = 0$ (stack II), six-member rings are perfectly stacked on top of each other.

The corresponding first Brillouin zones are shown in Fig. 3. There are two Fermi points called K and K' points at the corner of the first Brillouin zone. For the stack I, the K and K' points in the (x'_2, y'_2) coordinate system are transformed to K and K' points, respectively, in the (x'_1, y'_1) coordinate system. For the stack II, the K points in the (x'_2, y'_2) coordinate system are transformed to K' points in the (x'_1, y'_1) system and vice versa. If the contact region is sufficiently large just like between graphite planes, the wave-vector conservation will determine the transmission.

3. NUMERICAL RESULTS

In explicit numerical calculations, we consider armchair nanotubes CN1 and CN2 ($\eta_1 = \eta_2 = \pi/6$), fix a B site of each CN to lie on top of each other, and rotate CNs around the B site. Figure 4 shows conductances for thin CNs with $L/\sqrt{3}a = 10$. The conductance has a broad peak at a position slightly away from $\theta = \pi/3$ and $2\pi/3$. The broadened peaks overlap and the conductance remains nonzero for all angles. The results calculated in the lowest Born approximation in an effective-mass scheme [2] are also shown in Fig. 4 by dotted lines. Both results are in excellent agreement with each other.

Numerical calculations in the realistic model become difficult for thicker nanotubes. Therefore, thicker nanotubes have been considered in a simpler model in which a transfer t_0 is possible only for atoms just on top of each other lying in a circular contact region ignoring a small curvature of CNs as in ref. [2]. The results show that tight-binding results always agree independent of the circumference with those obtained in the lowest Born approximation in the effective-mass scheme as long as $|t_0|/\gamma_0 \ll 1$. Because the inter-tube transfer integral between nearest-neighbor atoms in the realistic model is $t_0/\gamma_0 \sim -0.14$ and small, the same is expected to be applicable for the realistic model.

Figure 5 shows the conductance for thick nanotubes ($L/\sqrt{3}a=100$) calculated in the effective-mass scheme. The conductance between two nanotubes exhibits a sharp peak at an angle corresponding to the stacks I and II but vanishingly small in other cases. This shows the importance of wave-vector conservation at crossings, which allows tunneling only between states in specific directions. Although not shown explicitly, the two peaks are broadened and their positions are slightly shifted with the decrease in the diameter. The broadening and shifts become appreciable for a thin CN as shown in Fig. 4. The broadening is due to the weakening of wave-vector conservation arising from a smaller contact area and the shift of the peak position is due to a change in the optimum condition of the interlayer transfer integrals caused by curvature effects.

4. SUMMARY AND CONCLUSION

The conductance between two crossed carbon nanotubes has been calculated in a tight-binding model and found to depend strongly on the crossing angle with large maxima at commensurate stacking of lattices of two nanotubes. The results are in good agreement with those obtained in the lowest Born approximation in an effective-mass scheme.

ACKNOWLEDGMENTS

The authors wish to thank G. E. W. Bauer and Yu. N. Nazarov for helpful discussions. This work was supported in part by Grants-in-Aid for Scientific Research and for Priority Area, Fullerene Network, from Ministry of Education, Science and Culture and by NEDO. One of us (T.N.) acknowledges the support of JSPS Postdoctoral Fellowships for Research Abroad.

REFERENCES

1. Iijima, S., *Nature (London)* **354**, 56 (1991).
2. Nakanishi, T., and Ando, T., *J. Phys. Soc. Jpn.* (submitted for publication).
3. Fuhrer, M. S., Nygard, J., Shih, L., Forero, M., Yoon, Y.-G., Mazzoni, M. S. C., Choi, H. J., Ihm, J., Louie, S. G., Zettl, A., and McEuen P. L., *Science* **288**, 494 (2000).
4. Postma, H. W. Ch., de Jonge, M., Yao, Z., and Dekker, C., *Phys. Rev. B* **62**, R10653 (2000).
5. Kane, C., Balents, L., and Fisher, M. P. A., *Phys. Rev. Lett.* **79**, 5086 (1997).
6. Egger, R., and Gogolin, A. O., *Phys. Rev. Lett.* **79**, 5082 (1997).
7. Paulson, S., Helser, A., Nardelli, M. B., Taylor II, R. M., Falvo, M., Superfine, R., and Washburn, S., *Science* **290**, 1742 (2000).
8. Landauer, R., *IBM J. Res. Dev.* **1**, 223 (1957); *Philos. Mag.* **21**, 863 (1970).

Energy barriers in carbon nanotube junctions

Ryo Tamura and Masaru Tsukada

Department of Physics, University of Tokyo, Hongo 7-3-1, Bunkyo-ku, Tokyo 113-0033, Japan

Abstract. It has been studied whether the junction connecting the metallic (9,0) zigzag tube and the semiconducting (8,0) zigzag tube shows rectification based on the tight binding model, Hartree-Fock approximation and the Green's function method. The junction with considered parameters becomes a backward diode.

INTRODUCTION

A nanotube junction without a dangling bond can be formed by a pair of a pentagonal defect and a heptagonal defect (NT junction). It is likely that a NT junction connecting a metallic tube and a semiconducting tube (MS-NT junction) shows rectification due to the Schottky barrier as the recent experiment showed [1]. In this paper, we show the results of the junction between the (9,0) tube and (8,0) tube shown in Fig. 1 as a first example. Purpose of this paper is to investigate whether this MS-NT junction shows the rectification.

METHOD

Recently Odintsov discussed the rectification of the MS-NT junction theoretically by the semi-classical model with the assumption that the transmission rate through the Schottky barrier is zero [2]. In order to investigate whether this assumption is valid, the tight binding model and Hartree-Fock approximation are utilized here. The corresponding Hamiltonian H is represented by

$$H_{i,j} = t_{i,j} - \frac{1}{2}U(|\vec{r}_i - \vec{r}_j|)\rho_{i,j} + \delta_{i,j}\sum_l U(|\vec{r}_i - \vec{r}_k|)(\rho_{k,k} - 1) , \qquad (1)$$

where indexes such as j represent the positions of the carbon atoms $\vec{r}_j$. [1] The first term has a nonzero constant value $-t < 0$ when i and j are nearest neighbors

---

[1] Without a MD calculation, the positions $\{\vec{r}_j\}$ can be determined correctly enough for the discussions here, because of the strong condition that any bond has to have almost the same lengths.

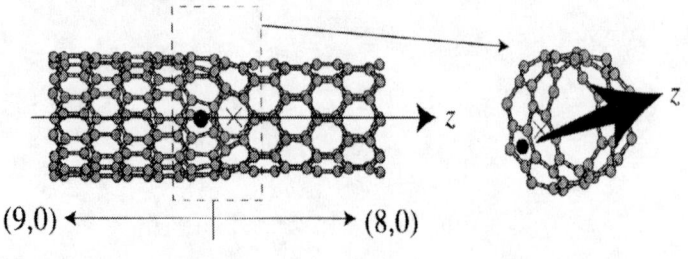

- pentagon   × heptagon

FIGURE 1. The junction connecting the (9,0) tube and the (8,0) tube.

and zero otherwise. The second term and third term are the exchange term and the Hartree term, respectively. The Coulomb interaction, $U(r)$, is assumed to be $U(r) = t/\sqrt{1 + 4(r/a)^2}$ with a being the bond length [3], and it is necessary to treat the Schottky barrier which does not appear in the MM-NT junctions [4].

We also utilize the Green's function method as follows to calculate the transmission rate [5]. The system is divided to the region near the defects (C), the (8,0) tube far from the defects (S), and the (9,0) tube far from the defects (M). Corresponding Hamiltonian H is written as $H = H_C + H_S + H_M + \tilde{t}_S + \tilde{t}_M$, where $\tilde{t}_p$ represents the elements between the C region and the p region. Effects of the two far regions $p = S, M$ are represented by the 'self energies', $\Lambda^{(p)} \equiv \tilde{t}_p(E - i\delta - H_p)^{-1}\tilde{t}_p$. When the elements of H_p and $\tilde{t}_p$ are chosen to be τ_p for those between nearest neighbors, ϵ_p for diagonal ones, and zero otherwise, $\Lambda^{(p)}/\tau_p$ becomes the non-dimensional function of a single non-dimensional parameter $(E - \epsilon_p)/\tau_p$. The values of ϵ_p and τ_p are set to be the same as the corresponding elements of H_c averaged for the atoms facing the region p. Then the Green's function $G(E) = (E - i\delta - H)^{-1}$ can be calculated as

$$G_{k,l} = (E - H_C - \Lambda_S - \Lambda_M)^{-1}_{k,l} , \qquad (2)$$

where $k, l \in C$. From eq.(2), the density matrix in the C region can be calculated as

$$\rho = \frac{1}{\pi} \int_{E_{F1}}^{E_{F2}} G\Gamma_2 G^* + \frac{1}{\pi} \int_{-\infty}^{E_{F1}} G(\Gamma_2 + \Gamma_1) G^* . \qquad (3)$$

Here $\Gamma_p \equiv \text{Im}(\Lambda_p)$, $E_{F1} \equiv \min(E_{FM}, E_{FS})$ and $E_{F2} \equiv \max(E_{FM}, E_{FS})$. The exchange terms from the far regions are neglected, but the Hartree terms from them are considered as follows. We assume that the charge in the far regions is the same as that at the corresponding edge of the near region. It is also assumed that effects of the planar gate electrode can be represented by the image charge; the nanotubes are not neutral by themselves, but the total system including the gate is neutral. Its direction and its distance from the tube axis are chosen to be the defect side and 1.06 nm , respectively. (The length 1.06 nm is three times as long as the radius of the (9,0) tube.) The potential energy caused by the assumed charge and the image charge is included in the calculation. The density matrix ρ can be calculated in this way, while Hamiltonian H is determined by the density matrix ρ by eq.(1). Therefore ρ and H can be calculated self-consistently.

After the self consistent loop becomes convergent, the current I is given by

$$I = (2e/h)\int_{E_{FM}}^{E_{FS}} T(E)dE = (2e/h)\int_{E_{FM}}^{E_{FS}} \text{Tr}[\Gamma_S G \Gamma_M G^*]dE . \qquad (4)$$

Here $T(E) \equiv \sum_{i,j} T_{i,j}(E)$ and $T_{i,j}$ is the transmission rate from i'th channel to j'th channel. Note that $T(E)$ may be larger than one, though $T_{i,j}$ cannot be larger than one.

RESULT

Figure 2 shows the transmission rate as a function of energy in the bias window for two cases. The bias defined as $V_{sd} = (E_{FM} - E_{FS})/|e|$ is negative in case (1) and positive in case (2) with common E_{FS}. The negative bias corresponds to the reverse bias for thermo-electron current. Nevertheless the resistance V_{sd}/I in case (1) is much smaller than in case (2); it is estimated to be 0.03 MΩ in case (1) and 0.88MΩ in case (2) when the hopping integral t is assumed to be 2.7 eV. It means that the junction is a backward diode as is explained below.

In Fig.3 , the circles show the electron potential at each atom in cases (1) and (2). The horizontal axis represents the position along the tube axis. Its origin is the interface between the (9,0) tube and the (8,0) tube and its range is the same as that of the near region. The local valence band edge bending along the electron potential is estimated from the band gap of the (8,0) tube, and shown by thin broken lines. The chosen Fermi levels shown by thick broken lines correspond to hole doping. Since the Schottky barrier is higher in case (1) than in case (2), the negative bias correspond to the reverse bias for the thermo-electron's flow. The tunneling current through the junction, however, cannot be neglected in case (1), while it becomes zero in most of the bias window due to the gap region in case (2), as is shown in Fig.2. This is the reason why the junction becomes a backward diode for the coherent current given by eq.(4).

DISCUSSION

In order to specify the MS-NT junctions, there are various parameters, e.g., E_{FS}, E_{FM}, the radius of the tube, the configuration of the defects, the strength of $U(r)$ and the distance from the gate electrode. The main reason why our results are different from those of Ref.2 is that the last one is smaller than in Ref.2. Smaller distance from the gate makes the width of the Schottky barrier smaller because the

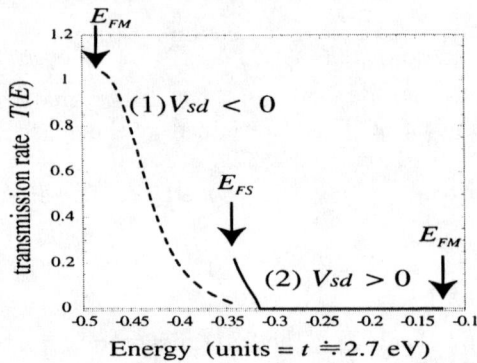

FIGURE 2. Transmission rate as a function of energy in the bias window. The broken line and the solid line correspond the negative bias (1) and the positive bias (2), respectively.

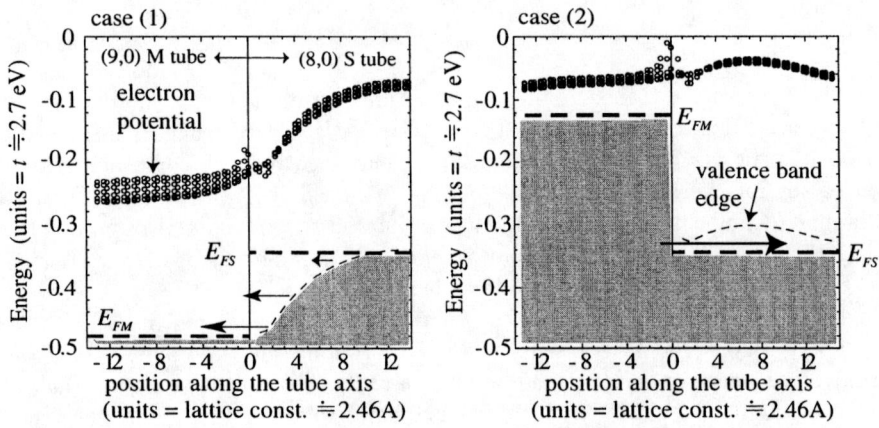

FIGURE 3. Electron potential energy at each atom in cases (1) and (2).

gate electrode screens the Coulomb repulsion in the nanotube. In our future work, calculations with more various cases are necessary to clarify the condition for the MS-NT junction to be the backword diode.

REFERENCES

1. Yao, Z., Postma, H. W. CH., Balents, L., and Dekker, C., *Nature* **402** 273 (1999).
2. Odintsov, A. A., *Phys. Rev. Lett.* **85** 150 (2000).
3. Harigaya, K., and Abe, S., *Phys. Rev. B* **49** 16746 (1994).
4. Tamura, R and Tsukada, M.,*Phys. Rev. B* **61** 8548 (2000).
5. Datta, S.,it Electronic Transport in Mesoscopic Systems, Cambridge: Cambridge University Press, 1995, ch.8, pp. 293-315.

Scanning Tunneling Spectroscopic Characterization of Single Walled Carbon Nanotubes in Bundles

K.Suzuki[1], Y.Maruyama[1], M.Nagayama[1], T.Kumagai[1], F.Kosha[1], K.Tohji[2], H.Takahashi[2], A.Kasuya[3], and Y.Nishina[4]

[1]*Department of Materials Chemistry, Hosei University, Kajinocho,Koganei,Tokyo 184-8584,Japan*
[2]*Department of Geoscience and Technology, Tohoku University, Aramaki,Aoba,Sendai 980-0845,Japan*
[3]*Center for Interdisciplinary Research, Tohoku University, Katahira,Aoba,Sendai 980-0812,Japan*
[4]*Ishinomaki Senshuu University, Minamisakai,Ishinomaki, 986-0031,Japan*

Abstract. Single walled carbon nanotubes (SWCNTs) prepared by a contact-arc-discharge followed by a hydrothermal purification are analyzed in the mat-form samples with the use of UHV scanning tunneling microscopy and spectroscopy (STM/STS). The characteristic electronic states of individual tubes in a bundle are found to be specified to the bundle.

INTRODUCTION

Electronic structures of isolated single walled carbon nanotubes (SWCNTs) have been well discussed theoretically [1] as well as experimentally[2]. However, actual SWCNTs are usually incorporated in bundles being closely packed and oriented like in a "crystal". In this case, the electronic nature of individual tube does not necessarily identical with that of isolated tube because of the inter-tube or inter-bundle interactions. Thus we have aimed to investigate the electronic structures of SWCNTs in bundles in a mat-type sample with using spatially resolved STS technique.

EXPERIMENTALS

Single walled carbon nanotubes (SWCNTs) were prepared by a contact-arc-discharge followed by a hydrothermal purification [3], which were compacted to mat-form samples. A piece of mat was fixed on a gold plate by gold paste for STM observations and STS measurements.

RESULTS AND DISCUSSION

The following findings were obtained..

1. A strong one-dimensionality in the electronic structure of SWCNTs:

Detailed observations of many individual tubes exhibit characteristic stripe lines spaced by 0.14 nm parallel to the tube axes which may correspond to the zigzag lines of armchair-type SWCNTs (Fig.1). This fact means that the π electrons are not homogeneously delocalized on the tube periphery surfaces but rather strongly localized on these lines of carbon chains. Accordingly, the tunneling spectra of this type of tubes at low temperature (25K) may tend to be insulating ones in nature.

FIGURE 1. An STM image of one-dimensional internal electronic structure in an individual SWCNT.

2. The electronic structure of individual tubes in a bundle:

Many STS analyses of individual tubes in one bundle have revealed that almost every tube in the bundle may have similar type of electronic structure (Fig.2) and the tubes in the different bundle may also have similar type in each other but different type from that in the former bundle(Fig.3). Therefore, the SWCNTs of the same type of chirality seem to be assembled in a bundle occasionally when they grew, or somewhat overall common electronic states might be dominant on all tubes in one bundle.

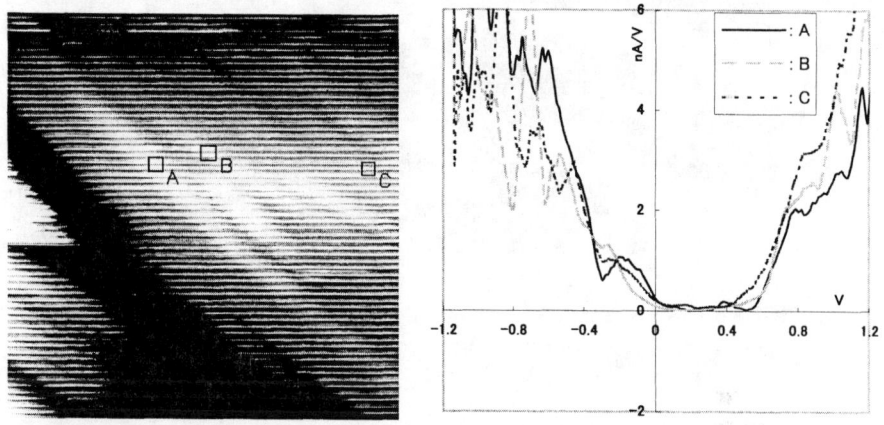

FIGURE 2. An STS image and the spatially resolved tunneling spectra of each SWCNT in one bundle.

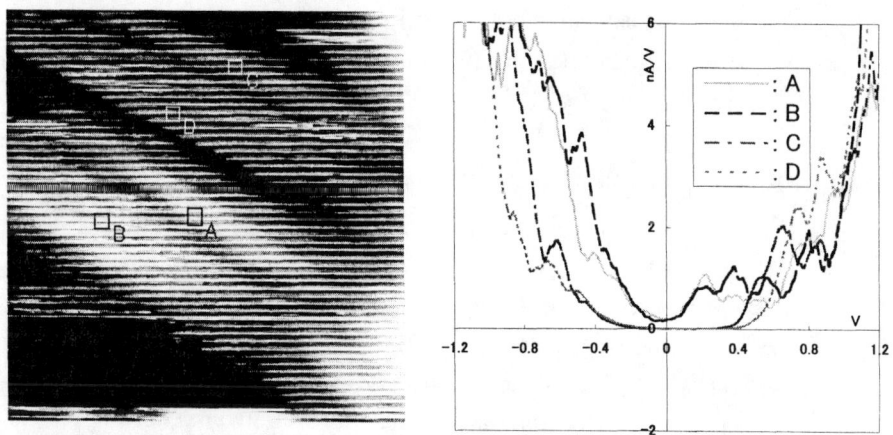

FIGURE 3. An STS image and spatially resolved tunneling spectra of each SWCNT in different two bundles.

3. STS measurement on rubidium doping to the SWCNTs mat:

0.2nm- thick Rb metal was evaporated onto the surface of the SWCNTs mat. A characteristic feature in the DOS spectra appears around the Fermi level due to doping (Fig.5). The metallic character seems to be reduced at low temperatures.

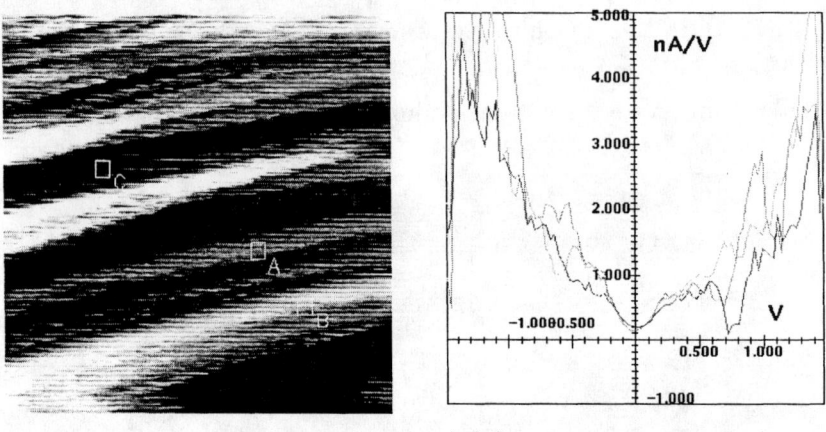

FIGURE 4. An STS image and spatially resolved tunneling spectra of Rb-doped SWCNTs at 23K.

ACKNOWLEDGEMENTS

This work was partly supported by the Grant –in-Aid for Scientific Research on the Priority Area "Fullerenes and Nanotubes Networks" by the Ministry of Education, Science and Culture of Japan (# 11165239).

REFERENCES

1. R.Saito et al., *Physical Properties of Carbon Nanotubes*, Imperial College Press, London,1998,pp.59-72.
2. T.W.Odom et al., *Nature*, 391,62(1998).
3. H.Takahashi et al., *Nature*, 383,679(1996).

Estimation of the Mechanical Strength of Nanotube Bundle

Y. Nishina[1], T. Maeda[1], A. Kasuya[2], K. Tohji[3], Y. Sato[3]

[1]*Ishinomaki Senshu University, Ishinomaki, 986-8580, Japan*
[2]*Center for Interdisciplinary Research, Tohoku University, Sendai, 980-8578, Japan*
[3]*Department of Geoscience and Technology, Tohoku University, Sendai, 980-8577, Japan*

Abstract: Mechanical strength of a nanotube bundle has been measured by using quartz glass enclosed-purified SWNTs as the specimen. These specimens were prepared by fast stretching quartz glass tube that contained purified nanotube bundles, at temperatures ranging from 900 to 1490 ℃. The maximum average critical load was 42.3 kg/mm^2 for the specimen prepared at 900 ℃. However, at temperatures higher than 900 ℃, the critical load value decreased and became almost equal to that of quartz glass at 1490 ℃. It was believed that the decrease in the critical load value at temperatures higher than 900 ℃ was due to the partial conversion of nanotubes into amorphous carbon. Since the cross-sectional area used in arriving at the critical load value was that of nanotube and the quartz glass, the cross-sectional area ratio of the quartz to that of the nanotube suggested that the mechanical strength of nanotube would be two orders of magnitude higher than the value reported here.

INTRODUCTION

The carbon nanotubes that have seamless cylindrical graphite structure are predicted to have interesting mechanical properties. The Young's modulus and stiffness of the MWNTs have been measured by researchers, and suggested for use in nanoscale fibres in strong, lightweight composite materials[1-3]. However, little has been done to measure the mechanical strength of either MWNTs or SWNTs. Particularly in the case of SWNTs, the chief obstacle for such measurements was the scarcity of the purified SWNT samples. Here, we have made an attempt to measure the mechanical strength of SWNT bundles by measuring the critical load of quartz glass clad SWNT bundles.

EXPERIMENTAL

Purified SWNTs: SWNTs were synthesized by DC arc-discharge using a 6mm diameter graphite rod (99.998 %) packed with Fe and Ni powders of 99.9 % purity as anode and 20 mm diameter carbon rod (99.9%) as cathode. The total metal content in the anode was approximately atm. 6%. The arc-discharge was carried out under varying He atmosphere and discharge current of 70 A.

The SWNTs synthesized under the above condition was purified using the hydrothermal method consisting of wet-oxidation, decantation and acid dissolution steps. First the raw soot was wet oxidized using 20% hydrogen peroxide water around 95 ℃ for 6 hours, to remove amorphous carbon from soot. Then, the wet-oxidized soot was decanted in distilled water to remove the nanometer size carbon particles and large graphite particles from nanotubes using the marginal differences in their settling rates. Then soot that was almost free of nano particles and large graphite particles was treated in 6 N HCl solution to remove the metal particles. Finally, the residue was washed to obtain the purified SWNTs.

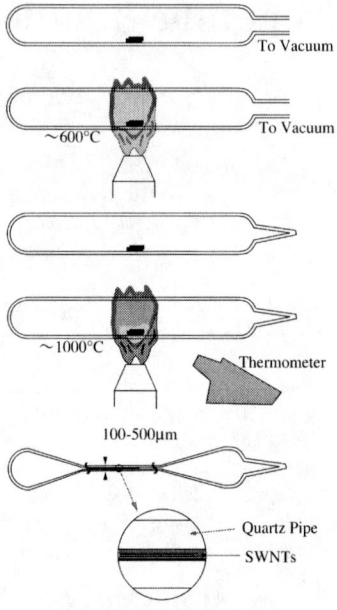

Figure 1. The schematic diagram to describe the specimen preparation for mechanical strength measurements.

Preparation of the test specimen: Five milligrams of purified SWNT was placed in an 8 mm diameter silica glass tube and was vacuumed while heating the same at 600 ℃ for one hour to remove the moisture and adsorbed gas on the SWNTs. Then, the tube was vacuum-sealed. The section of the silica glass tube with the SWNTs was heated to temperatures 900, 1000, 1350 and 1490 ℃ and stretched rapidly to form a silica glass tube of smaller diameter. Consequently, a silica tube of a few hundred microns in diameter with SWNTs enclosed inside was obtained.

Mechanical Strength Measuring System: The test specimen prepared above was used to determine the mechanical strength of the nanotube. One end of the specimen was fixed to a thin wire using a bond. Then, the other end was connected to a container that could hold water as the load. The amount of water held in the container at the time of breakage was considered the critical load of the specimen. The critical load measurements were carried out at room temperature.

RESULTS AND DISCUSSION

The study of mechanical properties of SWNTs, purified SWNTs and appropriate measuring system should be available. Fig. 2 shows the micrograph of the test specimen. SWNT sample of about few hundred nanometers in diameter is enclosed in the silica tube. Each specimen was analyzed using Raman spectroscopy to ascertain the state of the SWNTs after the heat treatment at various temperatures during rapid stretching of the quartz glass tube. The Raman spectroscopy of the SWNTs enclosed in silica glass tube treated at 900 ℃ is shown in Fig. 3. The peak split at 1582 cm^{-1} suggests the presence of SWNTs in its original form even after the heat treatment at 900 ℃. This confirmed that using the proposed method we could clad the nanotubes with quartz glass.

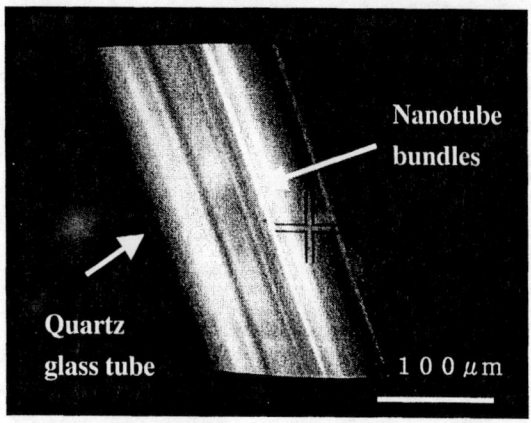

Figure 2. Micrograph of the test specimen

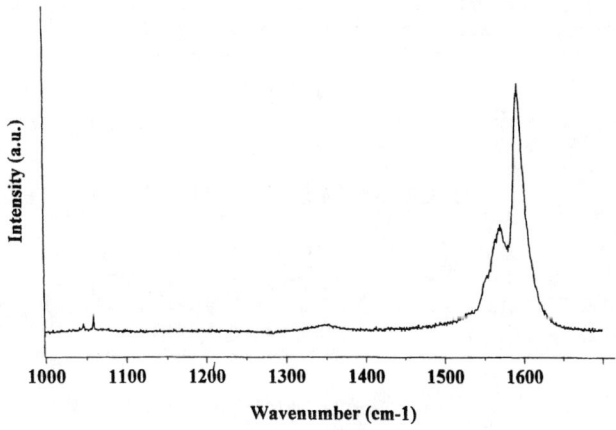

Figure 3. The Raman spectroscopy of the SWNT enclosed in silica glass tube

The results of the critical load experiment to determine the mechanical strength of the SWNT bundle is given in Table 1. The critical load took the maximum value of 42.3

kg/mm² for the sample synthesized at 900 ℃. Furthermore, the mechanical strength decreases for fabrication temperatures higher than 1000 ℃. However, the critical load values obtained in all the experiments were larger than the value for quartz glass. Analyzing the carbon form enclosed in the silica tube in detail, we find that when the silica tube is heated above 1000 ℃, the split peak at 1582 cm^{-1} becomes weaker suggesting the transformation of SWNTs into amorphous carbon. The reduction in the critical load value is due to this transformation. It should be noted that the cross-sectional area considered in the calculation includes not only the area of the bundle, but, also the area occupied by the silica glass tube. Therefore, we believe that the mechanical strength of the nanotube would be about two orders of magnitude higher than the value reported here.

Table 1

Temperature [℃]	Average Load [g]	Average Diameter [mm]	Average Critical Load [kg/mm²]
900	718	0.147	42.3
1000	557	0.217	18.4
1350	573	0.238	13.8
1490	593	0.23	14.3

*Critical Load of Quartz tube is about 11 kg/mm²

SUMMARY

We have succeeded in formulating a fabricating process for preparing the quartz glass clad SWNT bundle specimen for mechanical strength estimation. The mechanical strength of the quartz glass clad SWNT specimen was found few folds higher than that of the silica quartz. Furthermore, considering the fact that cross-sectional area used in deriving the critical load value was that of quartz glass and SWNT bundles, we believe that the mechanical strength of the nanotube bundle would be about two orders of magnitude higher than the value reported here.

ACKNOWLEGEMENTS

The present study was supported by Grant-in-Aid for Priority Research (#11165202) and International Scientific Research Program (#11694119) from the Ministry of Education, Science, Culture and Sports of Japan and CREST of Japan Science and Technology Institute. The samples were fabricated by Mr. T. Nihei of the Inst. for Materials Research, Tohoku University.

REFERENCES

(1) Falvo, M. R., Clary, G. J., Taylor II, R. M., Chi, V., Brooks Jr, F. P., Washburn, S., and Superfine, R., Nature. 389, 582-584 (1997).
(2) Treacy, M. M. J., Ebbesen, and Gibson, J. M., Nature. 381, 678-680 (1996).
(3) Wong, E. W., Sheehan, P. E, and Lieber, C. M., Science, 277, 1971-1974 (1997).

Optical Properties of Fullerene-Peapods

H. Kataura[1], T. Kodama[2], K. Kikuchi[2], K. Hirahara[3], S. Iijima[3,4], S. Suzuki[2], W. Krätschmer[5], and Y. Achiba[2]

[1] Department of Physics, Tokyo Metropolitan University, Japan
[2] Department of Chemistry, Tokyo Metropolitan University, Japan
[3] Japan Science and Technology Corporation, ICORP, Japan
[4] Department of Physics, Meijo University, Japan and NEC corporation, Japan
[5] Max-Planck-Institut für Kernphysik, Germany

Abstract. Single-wall carbon nanotubes (SWNTs) encapsulating fullerenes, so-called fullerene-peapods, were synthesized in high yield. High-resolution transmission electron microscopy revealed that almost all nanotubes are filled with high-density fullerene chains. We measured Raman spectra of C_{60}- and C_{70}-peapods. In the case of C_{60}-peapods, C_{60} Raman active mode intensity decreased rapidly by laser irradiation at room temperature. The final spectrum is similar to that of orthorhombic polymer phase, which indicates one-dimensional photopolymerization. At liquid helium temperature, no photopolymerization was observed, and the Raman spectra obtained indicates a feature of C_{60}-dimer phase. Furthermore, C_{60}-C_{60} distance estimated from electron diffraction pattern measured at room temperature is consistent with C_{60}-dimer phase. The spontaneous dimerization should be explained in part by internal high-pressure effect.

INTRODUCTION

Single-wall carbon nanotubes (SWNTs) encapsulating fullerenes, what is called fullerene-peapods, are considered to be a new solid phase of carbon constructed by zero- and one-dimensional sp^2 network systems (1,2). Since, in general, a physical property of solid is strongly depending on its network dimensionality, fullerene-peapods should have interesting properties due to the mixed dimensionality. Saito and Okada (3) have calculated the band structure of (10,10) nanotube encapsulating C_{60}, and found an anti-crossing between a flat conduction band of C_{60} and a linear band of nanotube, which indicates a considerable interaction between conduction electrons with different network dimensionality. In this work, we successfully synthesized high-yield peapod samples and measured Raman spectra of C_{60}- and C_{70}-peapods.

EXPERIMENTAL

By considering a diameter of C_{60} and van der Waals spacing, the best diameter of nanotube for fullerene encapsulation is that of (10,10). To get higher yield of encapsulations, it is desired to control diameter distribution of SWNTs to be thicker than (10,10). Laser furnace technique realizes a diameter control by changing catalysts and a furnace temperature (4). We chose NiCo (5) as a catalyst and we got thick enough SWNT samples by setting the furnace temperature to 1250 °C. Typical

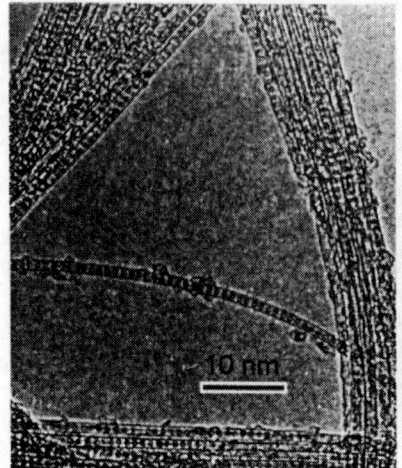
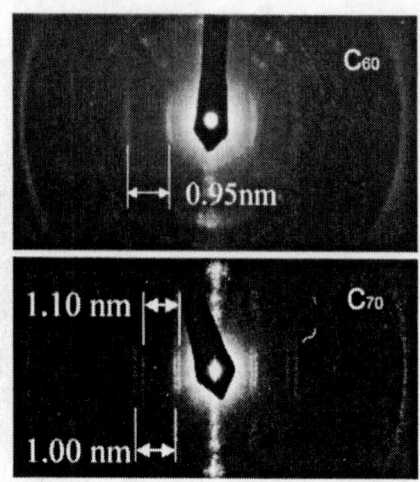

FIGURE 1. Typical HRTEM image of C_{60}-peapods (left photograph) and electron diffraction patterns of C_{60}- (upper right) and C_{70}-peapods (lower right). Inter fullerene distances were estimated from streaks originating from fullerenes one-dimensionally aligned inside nanotubes.

diameter of the sample is estimated to be 1.36 nm by X-ray diffraction. To get a high-purity peapod sample, unknown fullerenes co-produced with nanotubes are the most undesirable impurities. Fullerene-free SWNTs were obtained by in-situ heating soot at 1250 °C in vacuum. The soot was refluxed in 15 % H_2O_2 water solution at 100 °C to remove amorphous carbon and then washed in HCl solution to remove metal particles (6,7). The purity of resulting SWNT material was estimated to be higher than 90 %. Since purification processes remove caps of SWNTs (8), the purified SWNTs are ready to encapsulation of fullerenes. A sheet of SWNT paper was put in a quartz ampoule with fullerene powder and the ampoule was evacuated to 1×10^{-6} Torr. After drying process, the sheet and the fullerenes were sealed in the ampoule and were heated in a furnace up to 650 °C. After keeping the temperature for two hours, the ampoule was cooled down to room temperature. The SWNTs paper was sonicated in toluene for 1 hour to remove fullerenes coated on SWNTs surface. In this work, we used C_{60} and C_{70} in 99 % purity as fullerene sources.

RESULTS AND DISCUSSION

High purity peapod samples were successfully synthesized in both cases, C_{60} and C_{70}. Figure 1 shows typical high-resolution transmission electron microscope (HRTEM) images and electron diffraction patterns of C_{60}- and C_{70}-peapods. Although HRTEM photographs obtained are similar to those shown in the previous work (3), a yield of fullerene encapsulation of our sample is much higher. It is difficult to find empty SWNTs in HRTEM photographs. Furthermore, a cross section image evidently indicates that every nanotube is encapsulating C_{60} molecules even in case of thick bundles. We have estimated a filling rate of C_{60}- and C_{70}-peas to be higher than 70 % by HRTEM observations.

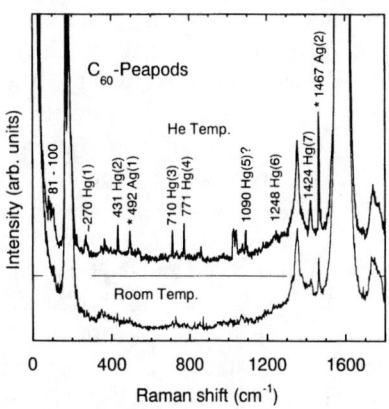 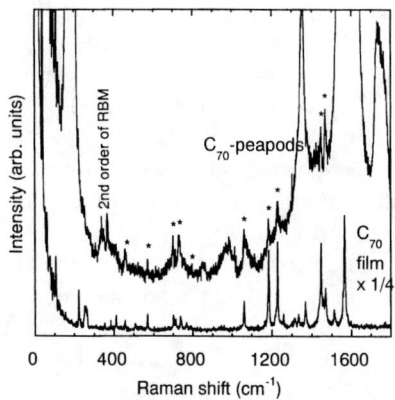

FIGURE 2. Raman spectra of C_{60}- and C_{70}-peapods. The left figure shows Raman spectra of C_{60}-peapods at liquid-helium (upper) and room temperature (lower). The right figure shows the Raman spectra of C_{70}-peapods and C_{70} thin film measured under the same condition at room temperature. Laser wavelength used is 488 nm.

New streak lines inside the first arc reflection were observed in electron diffraction patterns for peapod-bundles. These streaks mean one-dimensional fullerene crystals in nanotubes. C_{60}-C_{60} distance was estimated to be 0.95 nm, which is larger than a polymer's value, 0.92 nm, and smaller than a monomer's value, 1.00 nm (9). The intermediate C_{60}-C_{60} distance suggests a dimer phase as a possible structure of the one-dimensional C_{60}-crystal in nanotube. In case of C_{70}-peapods, we observed very sharp double streaks. The narrower streaks indicate C_{70}-C_{70} distance to be 1.10 nm and the wider streaks 1.00 nm. The double streaks mean two different crystal structures inside nanotubes, standing and lying alignment (9). Interestingly, there is no intermediate state. This result indicates C_{70} prefers standing alignment as high-density packing, if the diameter of nanotube is sufficiently large. This may cause larger interaction between C_{70}-peas and nanotubes than C_{60}-peapods.

Figure 2 shows typical Raman spectra of C_{60}-, C_{70}- peapods and a C_{70} film. In the case of C_{60}-peapod, $A_g(2)$ Raman modes of C_{60} molecule decreased very quickly by laser irradiation at room temperature. Since the final spectrum is similar to that of orthorhombic polymer (10), formations of one-dimensional photopolymers were suggested. An unpolymerized Raman spectrum was measured at liquid helium temperature. Where, all the Raman active modes of C_{60} molecule were observed. Further, additional peaks were observed around 90 cm^{-1} that is close to an external vibrational mode of C_{60} dimer at 96cm^{-1} (11). Although the frequency is slightly lower than the intrinsic dimer mode, the peak should be originating from dimers inside nanotubes because of the following reasons. The most distinctive feature of the spectrum is very weak A_g modes. Peak frequencies of A_g modes are lower than the intrinsic ones and are the same value to the orthorhombic polymer phase. It is well known that $H_g(1)$ mode is sensitive to a formation of dimer (11). A broadening of $H_g(1)$ mode observed is consistent with the dimerization. The significant reduction of A_g modes intensity and a appearing of low frequency modes strongly suggest a formation of dimers.

In the case of C_{70}-peapods, all the Raman active modes of C_{70} were observed even at room temperature and no photoreaction was observed. At sight, however, it is found that there are some anomalies in relative Raman intensity and peak widths. Raman mode intensity lying higher that 1100 cm^{-1} is strongly reduced. For example, an intensity ratio between peaks at 1229 and 702 cm^{-1} in C_{70} film is 5 times larger than that in C_{70}-peapods. Further, most of peaks show appreciable broadening. These anomalies are probably caused by anisotropic interactions between C_{70}s and SWNTs. If we use a peak at 702 cm^{-1} as a standard, however, we can estimate a filling rate of C_{70} to be higher than 50% by using mode intensities of the peapod and the thin film.

A spontaneous dimerization and high-density packing of C_{70} are probably caused in part by an internal high-pressure state. A fullerene molecule at a tip of nanotube pushes fullerenes inside driven by a van der Waals interaction. Tománek has calculated an internal pressure to be 0.1 GPa caused by this effect (12). The high-pressure state stabilizes a dimer phase (13) and standing C_{70}s at room temperature. This effect is very interesting for future applications of peapod materials as pressurized nano-cylinders.

ACKNOWLEDGEMENTS

Authors thank Prof. Susumu Saito, Prof. Okada and Prof. Tománek for fruitful discussions about peapod materials. Authors thank Mr. Misaki for electron diffraction measurements. This work was supported in part by Japan Society for Promotion of Science, Research for the Future Program and supported in part by the Grant-in-Aid for Scientific Research on the Priority Area "Fullerenes and Nanotubes" by the Ministry of Education, Science, and Culture of Japan.

REFERENCES

1 B. W. Smith, M. Monthioux, D. E. Luzzi, Nature **396** (1998) 323.
2 B. Burteaux, A. Claye, B. W. Smith, M. Monthioux, D. E. Luzzi, J. E. Fischer, Chem. Phys. Lett. **310** (1999) 21.
3 S. Saito and S. Okada, *Proc. 3rd Symposium on Atomic-Scale Surface and Interface Dynamics* (Fukuoka, 1999) p. 307.
4 H. Kataura, Y. Kumazawa, Y. Maniwa, Y. Ohtsuka, R. Sen, S. Suzuki and Y. Achiba, Carbon **38** (2000) 1691.
5 A.Thess *et al.*, Science **273** (1996) 483.
6 K. Tohji, private communication.
7 R.Rosen, W. Simendinger, C. Debbault, H. Shimoda, L. Fleming, B. Storner and O. Zhou, Appl. Phys. Lett. **76** (2000) 1668.
8 Y. Maniwa, Y. Kumazawa, Y. Saito, H. Tou, H. Kataura, H. Ishii, S. Suzuki, Y. Achiba, A. Fujiwara and H. Suematsu, Jpn. J. Appl. Phys. **38** (1999) L668.
9 M.S. Dresselhaus, G. Dresselhaus, and P.C. Eklund, *Science of Fullerenes and Carbon Nanotubes*, Academic Press, California 1996.
10 A. M. Rao, P. C. Eklund, J-L. Hodeau, L. Marques, M. N. Regueiro, Phys. Rev. B **55** (1997) 4766.
11 S. Lebedkin, A. Gromov, S. Giesa, R. Gleiter, B. Renker, H. Rietschel, W. Krätschmer, Chem. Phys. Lett. **285** (1998) 210.
12 D. Tománek, private communication.
13 V. A. Davydov *et al.*, Phys. Rev. B **61** (2000) 11936.

Electron Transport in Carbon Nanotubes Encapsulating Fullerenes

D.-H. Kim, H.-S. Sim, and K. J. Chang

Department of Physics, Korea Advanced Institute of Science and Technology, Taejon 305-701, Korea

Abstract. We study the transport properties of hybrid carbon nanotubes that contain a carbon nanotube capsule and a chain of fullerenes. We find that electron transmissions through the outer tubes are very sensitive to the alignment of inner shells. For a (5,5) capsule in the (10,10) tube, mirror symmetries in cross sections determine the lineshape of antiresonances. For a chain of C_{60} molecules inside the (10,10) tube, we find resonance peaks and barriers in transmission, depending on the existence of mirror and rotational symmetries. This feature is attributed to the fact that the coupling of incident channels with the bound states of inner shells depends on the symmetries.

INTRODUCTION

In recent experiments [1,2], single-wall carbon nanotubes (SWNTs) encapsulating fullerene-based structures have been observed. The diameters of outer tubes are in the range of 1.3 - 1.4 nm, which is equivalent to that of the (10,10) tube, while inner structures have the diameter of 0.7 nm, similar to those for the (5,5) tube and C_{60} molecule. In addition, inner structures are separated from the outer tube wall by the graphitic Van der Waals spacing of 0.34 nm, thus, weak interwall interactions may affect the electronic and transport properties of the outer tubes. Since the coupling of incident channels with the bound states of inner shells is mainly determined by the symmetries of hybrid tubes, the tranport behavior of the outer tubes is influenced by the alignment of inner shells inside the (10,10) tube.

In this work, we investigate the effects of structural symmetries on the transport properties of hybrid (10,10) tubes that contain a finite-sized capped (5,5) tube and a chain of C_{60} molecules. We calculate electron transmissions through the outer tubes using the Green's function approach [3] based on a tight-binding Hamiltonian [4]. For the (10,10) tube containing a (5,5) capsule, we find antiresonance peaks in transmission with symmetric lineshapes in the presence of mirror symmetries, while breaking mirror symmetries gives asymmetric lineshapes due to the mixing of the π and π^* states. For a chain of C_{60} molecules inside the (10,10) tube, we find resonance-like peaks when both mirror and rotational symmetries exist. If mirror symmetries are broken with preserving rotational symmetries, a transmission gap appears just above the Fermi level, while breaking all the symmetries gives resonant peaks in the transmission gap.

SWNT ENCAPSULATING A CAPSULE

Armchair single-wall nanotubes exhibit metallic conduction, with two channels characterized by π and π^* near the Fermi level, which are even and odd under mirror symmetry operations. Similarly, a (5,5) capsule has two bound states characterized by π and π^*. For the (10,10) tube containing a (5,5) capsule, the number of channels increases to 4 in the hybridized region, while the outer tube has two incident channels. Thus, we expect antiresonance peaks in transmission, regardless of the existence of mirror symmetries in the hybridized region. When the hybrid tube maintains mirror symmetries in cross sections, the π and π^* states are not coupled. Thus, the eigenchannels [5] are purely characterized by π and π^*, and periodic antiresonances appear for each channel due to the linear bands of the outer tube, as shown in Fig. 1(a). In this case, since incident channels are negligibly reflected, there is no background reflection, and antiresonance dips have the Breit-Wigner-type symmetric lineshapes [6].

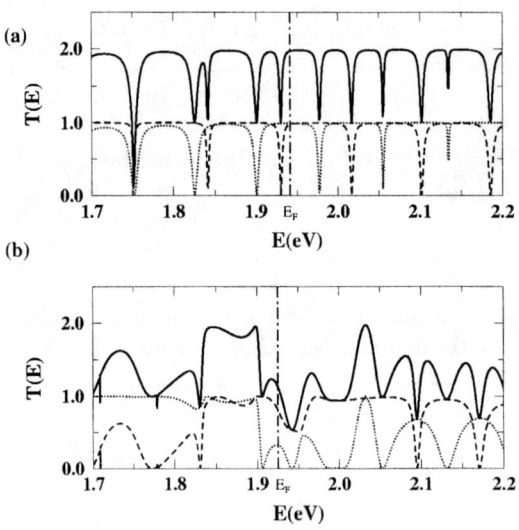

FIGURE 1. Calculated total transmission (solid) and two eigenchannel transmissions (dotted and dashed) through the (10,10) tube containing a (5,5) capsule with 60 unit cells in length, with (a) and without (b) mirror symmetries.

On the other hand, when mirror symmetries are broken, a mixing of π and π^* occurs in the hybridized region [6], giving rise to a barrier in transmission. Then, because of the background reflection, transmissions exhibit the Fano resonances with asymmetric lineshapes, as shown in Fig. 1(b). Since the mixing of π and π^* varies with the energy of incident channels, the linewidths of the resonance peaks, which is determined by the coupling of incident channels with the bound states of the capsule, also depend on energy.

SWNT ENCAPSULATING A C_{60} CHAIN

Fig. 2 shows the calculated total transmissions for the (10,10) tubes that contain a long chain of C_{60} molecules. In an isolated C_{60} molecule, the lowest unoccupied states are triply degenerate, which are characterized by $m = -1, 0,$ and $+1$, where m denotes the angular momentum quantum number [7,8]. When a C_{60} molecule is contained in the (10,10) tube, these degenerate states lie just above the Fermi level of the (10,10) tube. In this case, if both mirror and rotational symmetries are maintained, only the $m = 0$ state of C_{60} is coupled to the π state of the outer tube, because this state has even parity under mirror symmetry operations. Thus, a single antiresonance occurs in transmission.

For a chain of C_{60} molecules aligned inside the (10,10) tube with both mirror

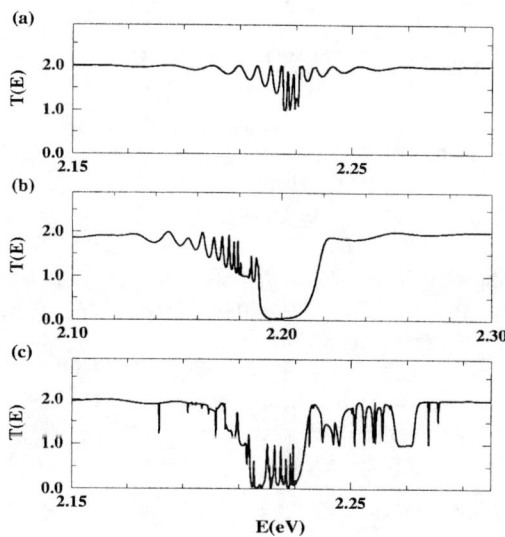

FIGURE 2. Total transmission through the (10,10) tube containing a chain of 22 C_{60} molecules; (a) both mirror and rotational symmetries are maintained in cross sections, (b) only mirror symmetries are broken, and (c) all the symmetries are broken.

and rotational symmetries, the $m = 0$ state is also found to be coupled to the π state, while the $m = -1$ and $+1$ states are not affected. From the band structure [6], we find that the incident π channel is significantly mixed with the $m = 0$ state in a very narrow range of energies near the $m = 0$ level, resulting in two channels in the hybridized region. In this case, the incident π channel is strongly reflected, thus, resonant peaks appear in transmission, as shown in Fig. 2(a).

If only mirror symmetries are broken in the hybrid tube, the $m = 0$ state is coupled to both the π and π^* states, creating a pseudogap. This pseudogap results from the mixing of π and π^* and gives rise to a transmission barrier [see Fig. 2(b)]. Although the $m = -1$ and $+1$ states are in the gap region, they do not affect the transmissions. However, when rotational symmetries are further broken, the $m = -1$ and $+1$ states interact weakly with the incident π and π^* channels, and resonance peaks occur in the barrier region [see Fig. 2(c)].

In conclusion, we find that in hybrid nanotubes encapsulating a nanotube capsule or a chain of fullerene molecules, the alignment of inner structures plays an important role in the transport behavior through the outer tubes. Depending on the symmetry of hybrid tubes, we find rich structures in transmission, such as antiresonances, Fano resonances, and transmission barriers.

ACKNOWLEDGMENTS

This work was supported by the QSRC in Dongkuk University.

REFERENCES

1. Smith, B. W., Monthioux, M., and Luzzi, D. E., *Nature(London)* **396**, 323 (1998); *Chem. Phys. Lett.* **315**, 31 (1999).
2. Sloan, J., Dunin-Borkowski, R. E., Hutchison, J. L., Coleman, K. S., Williams, V. C., Claridge, J. B., York, A. P. E., Xu, C., Bailey, S. R., Brown, G., Friedrichs, S., and Green, M. L. H, *Chem. Phys. Lett.* **316**, 91 (2000).
3. Meir, Y. and Wingreen, N. S., *Phys. Rev. Lett.* **68**, 2512 (1992).
4. Tang, M. S., Wang, C. Z., Chan, C. T., and Ho, K. M., *Phys. Rev. B* **53**, 979 (1996).
5. Büttiker, M., *IBM J. Res. Dev.* **32**, 63 (1988).
6. Kim, D.-H., Sim, H.-S., and Chang, K. J., to be published (2001).
7. Dresselhaus, M. S., Dresselhaus, G., and Eklund, P. C., *Science of Fullerenes and Carbon Nanotubes*, Academic Press, San Diego, 1996.
8. Manousakis, E., *Phys. Rev. B* **44**, R10991 (1991).

Electronic Structure and Energetics of Carbon Nanotubes Encapsulating C_{60}

Susumu Okada[1], Susumu Saito[2], Atsushi Oshiyama[3], and Yoshiyuki Miyamoto[4]

[1] *Institute of Material Science, University of Tsukuba, Tennodai, Tsukuba 305-8573, Japan*
[2] *Department of Physics, Tokyo Institute of Technology, 2-12-1 Oh-okayama, Meguro-ku, Tokyo 152-8551, Japan*
[3] *Institute of Physics, University of Tsukuba, Tennodai, Tsukuba 305-8573, Japan*
[4] *Fundamental Research Laboratory, NEC Corporation, 34 Miyukigaoka, Tsukuba 305-8501, Japan*

Abstract. We study the electronic structure and energetics of the nanotubes, (8,8), (9,9), and (10,10), encapsulating a linear chain of C_{60} by using the local density approximation in the density functional theory. We find that the encapsulating process is exothermic for the (10,10) nanotube, whereas the processes are endothermic for the (8,8) and (9,9) nanotubes. We also find that the stable C_{60}@(10,10) is a metal with multi-carriers each of which distributes either along the nanotube or on the C_{60} chain. This unusual feature is due to the nearly free electron state that is inherent to hierarchy solids with sufficient space inside.

INTRODUCTION

Discoveries of fullerenes [1] and carbon nanotubes [2] have triggered a great expansion of both theoretical and experimental studies on these new forms of carbon. Novelty of the materials is characterized by a global bond-network of carbon atoms. For instance, the fullerene has a closed network with zero dimension, whereas the nanotube has a one-dimensional tubule network. This network topology decisively affects electronic properties of the materials [3–6]. In the solids, fullerenes [1] or nanotubes [7] are regarded as constituent units or building blocks which are weakly interacting each other. Therefore, there is an infinite number of the systems to be constructed from these building blocks, which become a unique class of solids with structural hierarchy. Recently, an interesting class of the solid carbon consisting of both fullerenes and nanotubes has been synthesized: Transmission electron microscope clearly shows that fullerenes (C_{60}, C_{70} and C_{80}) are aligned in a chain and encapsulated in carbon nanotubes (peapods) [8–11]. The materials possess new hierarchy. Since, in general, the network topology of constituent units in the

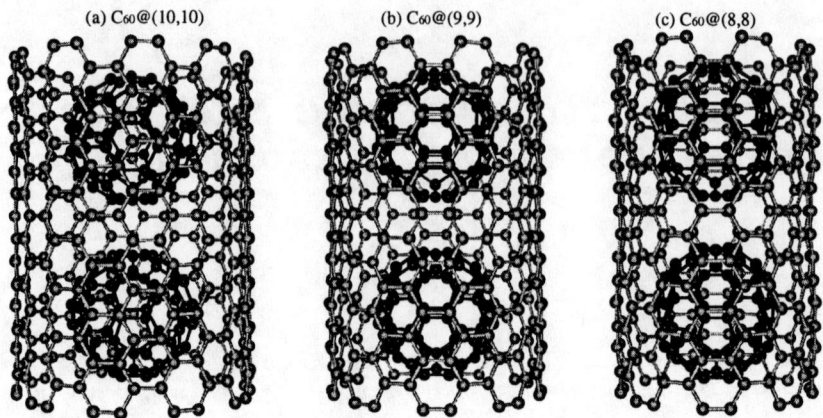

FIGURE 1. Total-energy optimized geometries of C_{60} encapsulated in (a) the (10,10), (b) the (9,9), and (c) the (8,8) nanotubes. Distances between C_{60}s are 3.14 Å, 3.15 Å and 2.45 Å, for (10,10), (9,9), and (8,8) peapods, respectively.

hierarchical solids is of importance in the physical properties of the systems, the new hierarchical systems are highly interesting carbon materials to be studied in detail.

CALCULATION METHODS

All calculations have been performed in the local-density approximation (LDA) in the density-functional theory [12,13]. To express the exchange-correlation potential of electrons, we use a functional form fitted to the Ceperley-Alder result [15,14]. Norm-conserving pseudopotentials generated by using the Troullier-Martins scheme are adopted to describe the electron-ion interaction [16,17]. In constructing the pseudopotentials, core radii adopted for C $2s$ and $2p$ states are both 1.5 Bohrs. The valence wave functions are expanded by the plane-wave basis set with a cutoff energy of 50 Ry which is known to give enough convergence of total energy to discuss the relative stability of various carbon phases [16]. We adopt supercell calculation in which each adjacent wall of nanotubes is separated by 6.5 Å to simulate the isolated nanotube-encapsulating fullerenes. In the structural optimization, we keep the lattice parameter value $c = 9.824$ Å along the tube direction which corresponds to the quadruple periodicity of the armchair-nanotube axis.

RESULTS AND DISCUSSIONS

Figure 1 shows optimized geometries of C_{60} encapsulated in the (8,8), (9,9), and (10,10) nanotubes. In the (10,10) peapods, the C_{60} and nanotubes almost keep their shapes before encapsulation. The calculated distance between wall of nanotube and

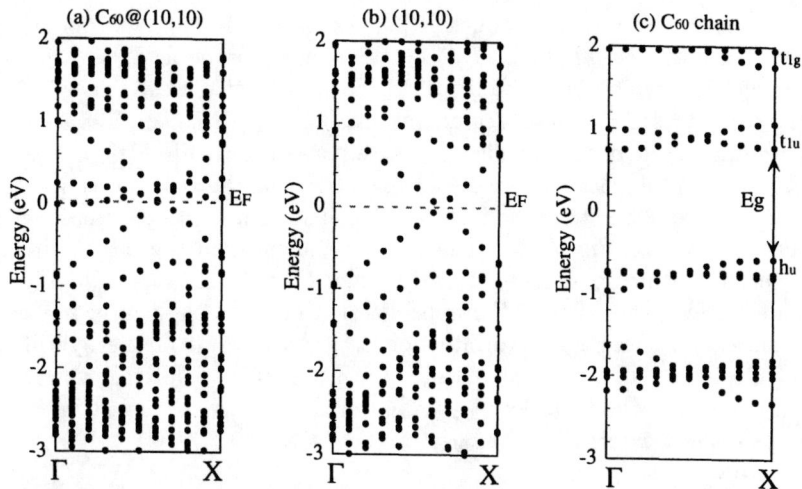

FIGURE 2. Energy band structures of (a) the C_{60}@(10,10), (b) the isolated (10,10) nanotube, and (c) the C_{60} chain.

the nearest C atom on the C_{60} is 3.31 Å which is close to the interlayer distance of graphite (3.34 Å). On the other hand, in the (8,8) and (9,9) peapods, both the C_{60} and the nanotubes substantially distorted because the tubes are too thin to accommodate C_{60}: The C_{60} are slightly elongated along tube direction and the nanotubes possess an undulating silhouette [Figs. 1 (b) and (c)]. The distances between the wall of nanotubes and the nearest atoms in the C_{60} are 2.75 Å and 2.45 Å for the (9,9) and (8,8) peapods, respectively. The optimized geometries are reflected in energetics of the peapods. The reaction energies (ΔE) on the encapsulation, (n,n)tube $+ C_{60} \rightarrow C_{60}$@$(n,n) - \Delta E$, are -0.51 eV, 0.27 eV, and 15.19 eV for (10,10), (9,9), and (8,8) peapods, respectively. Thus, the reaction on the (10,10) nanotubes is exothermic so that the encapsulated structure is stable in energy. Further, once C_{60} is trapped inside, the escaping process is unlikely to take place even if the tubes possess open end or holes on sidewall since the escaping barrier is at least more than a half eV. On the other hand, in the (8,8) and (9,9) peapods, the processes are endothermic because the encapsulation induces structural deformation not only on the C_{60} but also on the tubes. Therefore, the encapsulation is unlikely to take place on these thinner tubes even if the tubes have defect on the sidewall or open ends.

The electronic energy band (10,10) peapod is shown in Fig. 2 (a). It is well known that an isolated C_{60} has five-fold degenerated highest occupied h_u state which is completely occupied by ten electrons, and the three-fold degenerated lowest unoccupied t_{1u} state. In the solid C_{60}, both h_u and t_{1u} states show significant dispersion to form valence and conduction bands, and their gap remains finite and the material is therefore semiconducting [3]. As for the (10,10) nanotube, there are two bands near the Fermi energy E_F and the bands cross with E_F at $k = 2\pi/3$ in BZ. Therefore it may be expected in (10,10) peapods that the two bands originated from the tube cross E_F and are located in the gap of C_{60}. However, Fig. 2 (a) clearly exhibits a different feature which is unexpected in the above discussion. Four bands cross E_F: Two of them have linear dispersion possessing the π character of nanotube [Fig. 2 (b)] whereas the other two have less dispersion similar to the character of the t_{1u} band of solid C_{60} [Fig. 2 (c)]. Thus, the (10,10) peapod is a metal with multi-carriers each of which distributes mainly either on the tube or on the C_{60} chain. This unusual feature is due to the hybridization between the nearly free electron (NFE) state and the π and σ orbitals of C_{60}. As a result of the hybridization, most of electron states originated from p orbitals of C_{60} shifts downward upon encapsulation, and the shifts of C_{60} states renders the peapod a metallic system with different characters of multi-carriers.

ACKNOWLEDGMENTS

We thank Dr. H. Kataura for discussions. Computations were done at Institute for Solid State Physics, University of Tokyo and at Research Center for Computational Science, Okazaki National Institute. This work was supported in part by JSPS under contract No. RFTF96P00203 and by Grant-in-Aid for Scientific Research No. 11740219 and "Fullerenes and Nanotubes".

REFERENCES

1. W. Krätschmer et al., Nature **347**, 354 (1990).
2. S. Iijima, Nature, **354**, 56 (1991).
3. S. Saito and A. Oshiyama, Phys. Rev. Lett. **66**, 2637 (1991).
4. M. B. Jost et al., Phys. Rev. B, **44**, 1966 (1991).
5. N. Hamada, S. Sawada, and A. Oshiyama, Phys. Rev. Lett. **68**, 1579 (1992).
6. R. Saito et al., Appl. Phys. Lett. **60**, 2204 (1992).
7. A. Thess et al., Science, **273**, 483 (1996).
8. B. W. Smith, M. Monthioux, and D. E. Luzzi, Nature, **396**, 323 (1998).
9. B. Burteaux et al., Chem. Phys. Lett., **310**, 21 (1999).
10. J. Sloan et al., Chem. Phys. Lett., **316**, 191 (2000).
11. H. Kataura et al., Synthetic Metals, in press.
12. P. Hohenberg and W. Kohn, Phys. Rev. **136**, B864 (1964).
13. W. Kohn and L. J. Sham, Phys. Rev. **140**, A1133 (1965).
14. J. P. Perdew and A. Zunger, Phys. Rev. B **23**, 5048 (1981).
15. D. M. Ceperley and B. J. Alder, Phys. Rev. Lett. **45**, 566 (1980).
16. N. Troullier and J. L. Martins, Phys. Rev. B **43**, 1993 (1991).
17. L. Kleinman and D. M. Bylander, Phys. Rev. Lett. **48**, 1425 (1982).

Intercalation of SWNTs as investigated by Raman and x-ray photoemission spectroscopy

T. Ito, Y. Yatsu, H. Fudo, Y. Iwasa, T. Mitani,
H. Kataura*, and Y. Achiba*

Japan Advanced Institute of Science and Technology, Tatsunokuchi, Ishikawa 923-1292, Japan
*Tokyo Metropolitan University, Hachi-oji, Tokyo 192-0397, Japan

Abstract. Raman spectroscopy provides a unique opportunity to investigate the intercalation process of alkali metals into single walled carbon nanotubes (SWNT). Particularly, E2g mode, which is characteristic of the graphene sheet, is highly informative on the intercalation processes. X-ray photoemission spectroscopy, on theother hand, is useful for determining the atomic ratio of K and C. The depth profile of K/C ratio determined by an Ar^+ sputtering technique yields a reliable bulk stoichiometry of the sample. Combing the above information, a structural model is presented for a doped SWNT ropes.

INTRODUCTION

The intercalation properties of SWNT ropes have attracted much attention because of its fundamental interest and possible applications to Li ion secondary batteries. The smallest density of SWNT ropes among the known carbon-sp^2 allotropes, such as graphite and fullerenes, indicates that there exists intrinsically large space available for insertion of Li. From the view point of fundamental aspect, on the other hand, the intercalation process and the phase stability are interesting issues, because the SWNT is essentially a one dimensional (1D) host material, while graphite and fullerenes are 2D and 0D, respectively. However, the understanding of the intercalation process of SWNT is still quite poor except for pioneering studies of conductivity [1] and Raman spectra [2], primarily because poor crystallinity hinders a detailed structural study on the intercalated states. However, recent quasi-*in-situ* experiments on the K or Rb intercalation processes of SWNT revealed that the Raman spectra is a quite useful probe for the intercalation state [3]. In particular, the empirical correlation between the evolution of Raman spectra and the crystal structure established in the study of graphite and fullerenes enables one to speculate the structural model of the intercalated states of SWNT. For establishing the structural model, the confirmation of the K/C stoichiometry is crucial. In this paper, we show that the x-ray photoemission spectroscopy (XPS) combined with the Ar^+ sputtering technique provides reliable K/C ratio including the depth profile. Based on the K/C ratio determined by the XPS, a naive structure model of the intercalated state of SWNT is presented.

EXPERIMENTAL

Sample synthesis and film fabrication were made according to the literature methods [4]. SWNT samples were synthesized by arc-discharge using graphite rods with 4.2/1.0 atomic % Ni/Y. Amorphous carbon and short tubes were removed by filtration of ethanol suspension. The average diameter was estimated from the Raman active radial breathing mode as 1.45nm. The SWNT films were fabricated by spraying the suspension on Si or glass substrate. These films were dried in a vacuum of 2×10^{-6} torr for 12 hours to remove solvent. A typical thickness of the film determined by a stylus profiler was 600 nm. For intercalation of alkali metals, the dried films were sealed in 20cm glass tubes together with a piece of K. Intercalation of K was made by heating the sealed ampoule at 180°C in a furnace. For probing the doping process, the tube was taken out from the furnace at five minute intervals and the Raman experiment was carried out at room temperature. The XPS was performed to estimate the chemical concentration of the intercalated states. In contrast to the quasi-*in-situ* Raman experiment, the present XPS analysis requires to move samples from the reaction ampoule to the XPS chamber. Since K-doped SWNT is very unstable in air, the sample was first loaded in a removable air-tight chamber in a glove box, and tansfered to the analysis chamber. The depth profiling of the K/C atomic ratio has been carried out by repeating surface etching and XPS analysis. The sample surface was etched by sputtering of Ar+ ion with 0.2keV in energy.

RESULT AND DISCUSSION

Figure 1 displays the evolution of Raman spectra of K-doped SWNT measured with excitation at 633nm for the doping process. Since the excitation wavelength is close to the absorption peak of both metallic and semiconducting tubes, the resonance effect makes the spectra of the pristine film a superposition of the relatively sharp peak at 1590cm^{-1}, accompanied with small subpeaks, and a broader peak at about 1550cm^{-1}. The former is attributed to semiconducting SWNT, while the latter is possibly assigned to the metallic tubes [4]. Intercalation of K was found to proceed rather quickly. After the first five minute reaction, the broad component disappeared and the sharp peak was upshifted from 1590 to 1598cm^{-1}. Another five minute reaction resulted in the complete disappearance of the peaks. After a further annealing, a new peak appeared at about 1550cm^{-1}, which is close to that first observed by Rao *et al.* [2]. This peak showed a further downshift continuously to 1530cm^{-1} by a subsequent annealing for 10-20 minutes. The radial breathing mode at 163cm^{-1}, on the other hand, immediately disappeared when doping was started, and did not reappear.

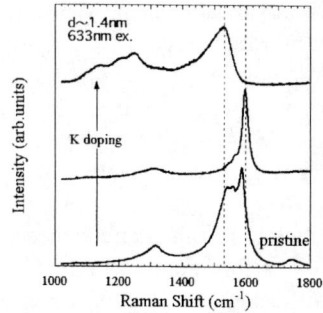

FIGURE 1. Raman spectra for the three states of K-intercalated SWNT.

The initial upshift of the E_{2g} mode followed by a downshift upon doping has been also observed in graphite intercalation compounds (GIC), displaying a sharp contrast with the simple intercalation-induced softening of the $A_g(2)$ mode in C_{60} intercalation compounds. From the latter materials, an empirical rule on the charge transfer softening of the C=C stretching region has been established. On the other hand, the upshift of the Raman peak is understood in terms of the hardening of the C=C bonds by inserting alkali metals into a narrow space between two graphene sheets. Hence, the upshift of the E_{2g} mode in the low doping state in SWNT (Fig. 1) could be evidence that K is located between two tubes rather than in a hollow site surrounded by three tubes. To establish a more detailed structural model, determination of chemical concentration of the intercalated tubes is crucial.

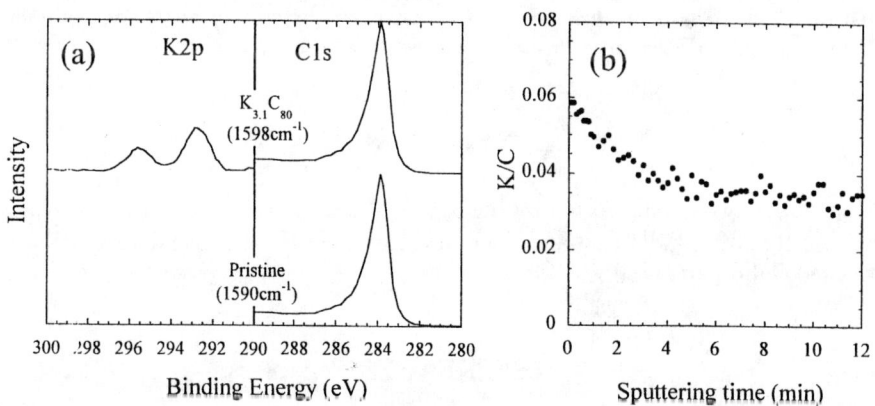

FIGURE 2. (a) XPS spectra for the pristine and low doping states of SWNT. (b) Sputtering time dependence of K/C ratio for the low doping state.

C/K concentration ratio was estimated by the relative intensity of C(1s) and K(2p) peak intensity of x-ray photoemission spectroscopy (XPS). For this measurement, the sample film was taken out from the glass tube in a glove box and transferred to an XPS apparatus using a removable air-tight chamber. The C(1s) and

K(2p) XPS spectra are shown in Fig. 2(a) for pristine and the low doping state which shows an upshifted E_{2g} mode at 1598cm^{-1}. The C(1s) spectra for the two states are almost identical. The K/C stomic ratio for the doped state can be estimated from the ratio of the integrated intensity of K(2p) and C(1s) spectra, using a normalization prefactor. However, since the XPS essentially probes only the surface the sample, the data do not necessarily correspond to the whole film. Thus, we have measured the depth profile of the K/C ratio.

Figure 2(b) shows the etching time dependence of the K/C ratio for the low doping state. The full scale of the horizontal axis approximately corresponds to the depth of 400nm, which is in the same order of the penetration depth of visible light used for the Raman experiment. Although the K/C ratio is slighly high at the surface, it decays exponentially converging at a constant value. The K/C ratio of the bulk film, determined by averaging the profile, was 0.039.

Since the average tube diameter in the present experiment is ~1.4nm, we assumed that the typical index of the tube is (10,10) which has a unit cell of C_{80}. The obtained K/C ratio of the low doping state corresponds to $K_{3.1}C_{80}$. A structural sequence of K-intercalated (10, 10) SWNT ropes was proposed by Gao et al. by means of a numerical simulations [5]. According to their calculation, up to K_2C_{80}, potassium ions occupy the hollow site surrounded by three tubes, and once the K concentration exceeds K_2C_{80}, K starts to be inserted between two pairs of tubes. (Here, the unit cell of (10, 10) tube ropes was taken as C_{80}.) At the concentration of K_3C_{80}, all K ions occupy the sites between tube pairs, and consequently, the intertube distance is dramatically expanded.

To summarize, we have shown that the XPS analysis is a useful tool for estimating the chemical concentration for the intercalated states of SWNT. Combining with the Raman data, a structural model was presented for the low doping state.

ACKNOWLEDGEMENTS

This work has been supported by the by the Grant-In-Aid for Scientific Research on the Priority Area "Fullerenes and Nanotubes" by the Ministry of Education, Sport, Science and Culture of Japan.

REFERENCES

1. Lee, R. S. *et al.*, *Nature* **388**, 255-256 (1997).
2. Rao, A. M. *et al.*, *Nature* **388**, 257-259 (1997).
3. Claye, A., Nemes, N. M., Janossy, A., and Fischer, J. E., *Phys. Rev. B* **62**, R4845-4848 (1996).
4. Kataura, H. *et al.*, *Synth. Metals* **103**, 2555-2558 (1999).
5. Gao, G., Cagin, T., and Goddard, W. A. III, *Phys. Rev. Lett.* **80**, 5556-5559 (1998).

Adsorption and Desorption of Weakly Bonded Adsorbates from Single-Wall Carbon Nanotube Bundles

Tobias Hertel[*], Jennah Kriebel[†], Gunnar Moos[*] and Roman Fasel[**]

[*]*Fritz-Haber-Institut der MPG, Faradayweg 4-6, D-14195 Berlin, Germany*
[†]*Department of Chemistry and Chemical Biology, Harvard University, 12 Oxford St, Cambridge Ma, 02138, USA*
[**]*EMPA Dübendorf, Abt. 124, Überlandstrasse 129, CH-8600 Dübendorf, Switzerland*

Abstract. We present an investigation of the kinetics of adsorption and desorption of some weakly bonded gases from single wall carbon nanotube bundles and – for comparison – from graphite. Thermal desorption spectra from CH_4, Xe and SF_6 as well as those of MeOH, EtOH and H_2O can be used to obtain adsorbate binding energies. The observed trends in the binding energies of gases with different van der Waals radii suggest that so-called groove sites on the external bundle surface are the preferred low coverage adsorption sites due to their higher binding energy. These results shed new light on the wetting properties of SWNT bundles. In addition we find that measured sticking coefficients can be related to the diffusion kinetics of adsorbates into the bulk of the nanotube samples.

INTRODUCTION

The interaction of single-wall carbon nanotubes (SWNTs) with their environment and in particular with gases or dopants adsorbed on their interior or exterior surfaces attracts increasing attention due to its anticipated influence on some of the key properties of these novel materials [1, 2]. In particular the electronic structure and consequently electronic transport properties of SWNTs should be susceptible to the presence of adsorbates. Such sensitivity was suggested to be applicable for the use of SWNTs as chemical sensors [3] but it may also prevent us from studying the intrinsic properties of SWNTs altogether because surface contamination is difficult to avoid entirely. Adsorbed gases may also have a crucial influence on the electronic properties and contact resistances of nanotube based electronic devices. Gas adsorption on SWNTs has furthermore attracted much interest due to speculations that SWNTs might be used as material for efficient gas storage.

Commonly studied nanotube samples such as bucky paper are composed of a tightly woven mat of bundles and ropes that each consist of SWNTs which are arranged in a quasi-hexagonal lattice. It is thus most appropriate to discuss adsorption on SWNT samples not in terms of adsorption on *individual* nanotubes but in terms of adsorption on the exterior or interior surfaces of such *bundles*. This raises a number of questions with respect to the preferred binding sites, the accessible sample surface area, the uptake capacity as well as the strength of the corresponding adsorbate-substrate bond. To illustrate which binding sites may be occupied by a small adsorbate like He, for

FIGURE 1. Cross-sectional view of a calculated potential energy surface for He, a small adsorbate on a SWNT bundle surface. The solid rings on the right part of the figure indicate the positions of individual nanotubes in the hexagonal close packed bundle lattice. Bright areas identify the favored high symmetry adsorption sites, *i.e.* adsorption on the external grooves, endohedral- and interstitial channel-sites.

example, we have calculated a two dimensional potential energy surface shown in Figure 1.

Here we present a comparative study on the adsorption and desorption kinetics of different weakly bonded gases from SWNT samples and highly oriented pyrolytic graphite (HOPG). The latter serves as a reference. A comparison of binding energies on both types of samples allows to identify the preferred binding sites on the nanotube bundles. The results suggest that at low coverages adsorption occurs predominantly on the exterior surface of the nanotube bundles in so-called groove sites. The corresponding binding energies are higher than on HOPG due to the higher effective coordination in these sites. The trends in the binding energies resulting from adsorption of polar molecules with stronger intermolecular forces such as methanol, ethanol and water are not yet fully understood. In particular water adsorption on SWNT samples appears to differ from that of other gases indicating that SWNT bundles are not wet by water despite the presence of higher-coordinated sites on their surface.

EXPERIMENTAL SETUP, RESULTS AND DISCUSSION

The nanotube and highly oriented pyrolytic graphite (HOPG) samples used for thermal desorption measurements were mounted back to back on a Ta disk that could be cooled to about 35 K and resistively heated to 1200 K. Nanotube samples were fabricated from commercially available nanotube suspension (tubes@rice, Houston, Texas) and where outgassed by repeated heating annealing cycles under UHV conditions. Both samples were mounted inside a UHV chamber with a base pressure of $2 \cdot 10^{-10}$ mbar. The SWNT samples used in this study have characterized in-house by optical absorption spectroscopy as well as transmission electron microscopy (TEM) to assure that samples were not excessively doped and contained only minor amounts of amorphous carbon. For thermal desorption spectroscopy the sample temperature was ramped linearly with typical heating rates of 0.5 Ks^{-1}. Sticking coefficients could be measured by a King and Wells type setup utilizing a pinhole doser. More experimental details can be found elsewhere [4].

Thermal desorption traces from both types of samples, after exposure to an equivalent of approximately 1.5 – 3 close packed HOPG-monolayers (1ML=$5.7 \cdot 10^{14}$ cm^{-2}), are shown in Figure 2a) for the inert gases CH$_4$, Xe and SF$_6$ and in Figure 2b) for the polar gases MeOH, EtOH and H$_2$O. The sample temperature during adsorption was 5-20 K below the multilayer desorption temperature for each system. The shape of the multilayer desorption features at higher coverages $\Theta \gg 1$ML is characteristic for zero order kinetics as expected for desorption from thick films. Sub-monolayer desorption of these gases from HOPG can typically also be described by zero order kinetics which is attributed to the high mobility of adsorbates between the HOPG surface and adsorbate islands. This is reflected by the exponentially increasing leading edge of the monolayer desorption features in the thermal desorption spectra of Figure 2a) and 2b) (left panels). In contrast, thermal desorption spectra from SWNT bundles are found to be quite

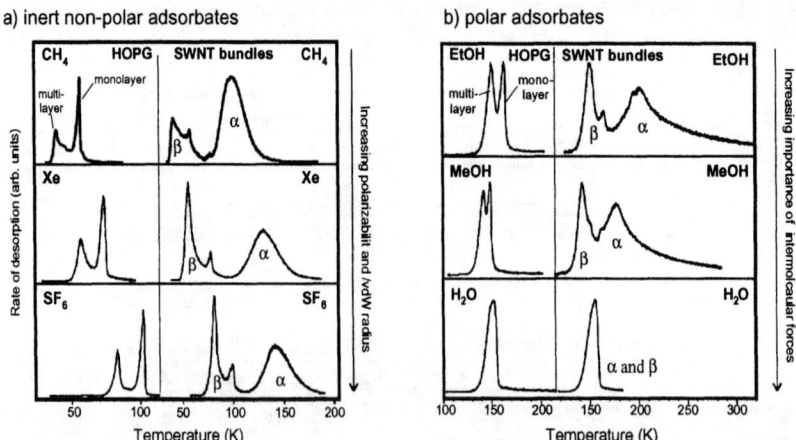

FIGURE 2. Thermal desorption spectra of different gases. a) Non-polar inert adsorbates (from top to bottom CH$_4$, Xe and SF$_6$). b) Polar adsorbates with increasingly stronger intermolecular interactions (from top to bottom EtOH, MeOH and H$_2$O). The difference in the temperatures for monolayer and multilayer desorption features indicates whether the adsorbate wets the sample surface or not. All adsorbates appear to wet these samples except for water for which no significant difference in mono- and multilayer desorption temperatures is found.

different. The low temperature b-feature is still very similar to the multilayer desorption feature observed on HOPG and corresponds to the desorption from the second and higher monolayers. The first monolayer desorption feature, however, is absent in TD-traces from the SWNT samples and is replaced by another feature at higher temperatures which is labeled a. The only exception being the case of water desorption. The small feature in SWNT spectra at the position of the HOPG monolayer desorption feature is most likely due to desorption from the back of the HOPG sample on which the tube sample is mounted.

Binding energies can be obtained from thermal desorption traces assuming Arrhenius-type desorption kinetics where the shape and position of desorption features is determined by the order of desorption n, a frequency prefactor v and the binding energy E_B [5]. The monolayer binding energies obtained for adsorption of the inert gases on HOPG are (11.5 ± 1) kJmol^{-1}, (23 ± 1) kJmol^{-1} and (28 ± 1.5) kJmol^{-1} for CH_4, Xe and SF_6 respectively - in good agreement with previous studies [6, 7]. For a comparison with binding energies on the SWNT bundles we will simply use the temperature ratio of desorption peak maxima from the HOPG monolayer- and the SWNT a-feature. This somewhat simplified analysis should be sufficient as long as we are more interested in trends rather than absolute binding energies. A more detailed analysis of the SWNT thermal desorption traces should also account for gas diffusion into and out of the SWNT sample and will be presented elsewhere [4]. We would like to point out, however, that the shape of desorption features for TDS from these porous samples cannot be used to determine the order of desorption in an analogous manner as for desorption from single crystal surfaces [8].

To determine the nature of the a-feature for desorption from the SWNT bundles we have performed simple molecular mechanics calculations using Lennard-Jones type van der Waals (vdW) pair-potentials and the vdW-parameters given in reference [9]. The calculations were performed to allow a comparison with expected inert gas binding energies on SWNT bundles and graphite (see Figure 3). We find that the observed trends for the change of experimental binding energies are best described by the calculated trends for adsorption in the groove sites on the external bundle surface which slowly approaches the HOPG binding energy as the adsorbate size is increased. In contrast to this and experimental observations, the calculated trend for adsorption in the endohedral sites would lead to an increasing effective coordination and thus continuously increasing binding energy with respect to graphite. This trend may be observed if nanotubes are opened by oxidative treatment but it is apparently not found for our samples prepared from the commercially available purified SWNT material even after vacuum annealing to 1200 K.

The thermal desorption traces of Figure 2 also shed new light onto the wetting properties of SWNT bundles. In general, the wetting properties of surfaces are determined by the competition between adhesive and cohesive forces. The difference of the desorption temperatures – and therefore binding energies – seen in our thermal desorption traces can actually be directly related to these two competing forces. The position of the multilayer desorption feature is determined by cohesive forces, i.e. sublimation energies, while the position of the monolayer or α desorption features is determined by adhesive forces, i.e. adsorbate-substrate binding energies. This means that we should expect wetting whenever monolayer or α desorption features are found at significantly higher temperature than the multilayer feature.

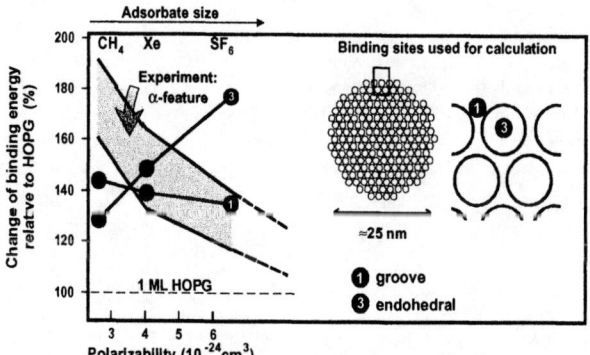

FIGURE 3. Change of binding energies on SWNT bundles if compared to adsorption on graphite. Experimental binding energies – as obtained from thermal desorption peak maxima – are compared with calculated trends for the non-polar van der Waals bonded molecules. The grey shaded area corresponds to the range of binding energies found for adsorbate coverages Θ ranging from an equivalent of less than one HOPG-monolayer to about 5 monolayers. The trends in the experimental binding energies follow those expected for adsorption in the groove sites on external bundle surface.

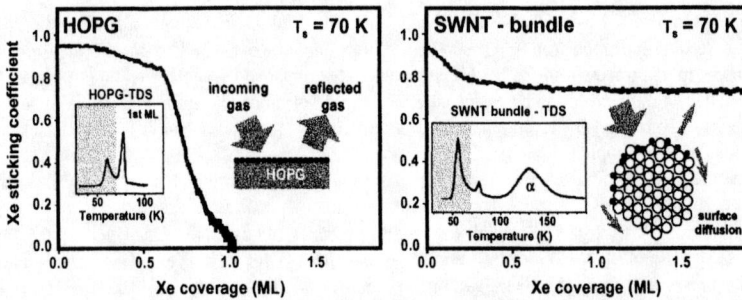

FIGURE 4. Measured sticking coefficients for adsorption on HOPG and SWNT bundles at 70 K. The grey shaded areas in the insets indicate that no multilayer adsorption is possible at this sample temperature. Sticking coefficients on the nanotube sample remain high up to many tens of monolayers exposure due to diffusion across the rope surfaces into the bulk of the SWNT mat.

As seen from Figure 2. this is the case for all adsorbates, except for water for which multi- and monolayer desorption features seem to coincide on both, HOPG and SWNT samples. Indeed, macroscopic contact angle measurements for water on HOPG give contact angles near 90± which is consistent with our observation that adhesive and cohesive forces are nearly balanced [10]. The interpretation of contact angle measurements on SWNT samples, however, is obscured due to the influence of surface roughness and porosity on the macroscopic contact angles[10]. In contrast to this thermal desorption measurements are not affected in the same way by surface roughness and reveal that SWNT bundles should not be wet by water on a microscopic scale. This is somewhat surprising since MeOH and EtOH show a pronounced increase of the desorption peak temperature associated with the low coverage desorption feature. In analogy to the case of inert gas adsorption this should be associated with the higher coordinated – and more strongly bound – groove sites that are first occupied at low adsorbate coverages. The observation that water shows no apparent increase in the low coverage binding energy is, therefore, somewhat puzzling and requires further experimental and theoretical investigations.

The kinetics of *adsorption* has been studied by measuring sticking coefficients – *i.e.* the fraction of adsorbed *vs* reflected particles – for various temperatures. In Figure 4 we show results for Xe adsorption as a function of coverage. To prevent multilayer adsorption the sample temperature was kept at 70 K – just above the onset of the Xe-multilayer desorption features. For adsorption on HOPG, the sticking coefficient is found to be close to 1 up to coverages near the completion of the first monolayer where it rapidly drops to zero. This is consistent with zero order adsorption kinetics. In contrast, the sticking coefficient on the SWNT samples was found to remain at a high level around 60-80% up to a coverage of several tens of monolayers. Such a high coverage cannot be accommodated on the first monolayer of the exposed SWNT bundles. This indicates that diffusion into the bulk keeps the coverage in the exposed surface regions well below saturation of a full monolayer and thereby allows the sticking coefficient to remain high. Further analysis of these experiments may allow to study diffusion across the external bundle surfaces in more detail.

In conclusion, we have used thermal desorption spectroscopy to study the kinetics of adsorption and desorption of some weakly bonded gases from HOPG and SWNT bundles. The results provide new insight into binding energies and sites on SWNT bundles exposed to various adsorbates. Further studies will involve more strongly, covalently bonded adsorbates where intercalation phenomena may become increasingly important.

REFERENCES

1. Collins, P. G., Bradley, K., Ishigami, M., and Zettl, A., *Science*, **287**, 1801–1803 (2000).
2. Lee, R. S., Kim, H. J., Fischer, J. E., Thess, A., and Smalley, R. E., *Nature*, **388**, 255–257 (1997).
3. Kong, J., Franklin, N. R., Zhou, C., Chapline, M. G., Peng, S., Cho, K., and Dai, H., *Science*, **284**, 622–625 (2000).
4. Kriebel, J., Hertel, T., and Moos, G., in preparation (2001).
5. King, D. A., *Surf. Sci.*, **47**, 384–402 (1975).
6. Thomy, A., and Duval, X., *J. Chim. Phys.*, **67**, 1101 (1970).
7. Ruiz-Suarez, J. C., Vargas, M. C., Goodman, F. O., and Scoles, G., *Surf. Sci.*, **243**, 219–226 (1991).
8. Kuznetsova, A., Yates, J. T., Liu, J., and Smalley, R. E., *J. of Chem. Phys.*, **112**, 9590–9598 (2000).
9. Stan, G., Bojan, M. J., Curtarolo, S., Gatica, S. M., and Cole, M. W., *Phys. Rev. B*, **62**, 2173–2180 (2000).
10. Adamson, A. W., and Gast, A. P., *Physical Chemistry of Surfaces*, Wiley-Interscience, New York, 1997.

Chirality Dependent G-band Raman Intensity of An Individual Single Wall Carbon Nanotube

R. Saito[††], A. Jorio[†], J. H. Hafner[‡], C. M. Lieber[‡], M. Hunter[†], T. McClure[†], G. Dresselhaus[†], M. S. Dresselhaus[*]

[††] *University of Electro-Communications, Tokyo, 182-8585 Japan*
[†] *Massachusetts Institute of Technology, Cambridge, MA 02139-4307*
[‡] *Harvard University, Cambridge, MA 02138*
[*] *Currently on leave from Massachusetts Institute of Technology, Cambridge, MA 02139-4307*

Abstract. The chirality-dependent G-band Raman intensity of an individual single wall carbon nanotube is presented both by a non-resonant theory for the Raman tensor and by confocal micro-Raman measurements. Theory predicts six or three intense Raman modes, respectively, for chiral or achiral nanotubes whose relative intensities depend on the chiral angle of the nanotube.

The quantum properties of single wall carbon nanotubes (SWNTs) depend on their helical geometry known as the chirality. The chirality of a SWNT is uniquely determined by the integers (n, m) [1,2], and we have been looking for physical properties which depend on the chirality. A typical example of a chirality-dependent property is that the electronic structure of a single wall carbon nanotube is either metallic ($n - m = 3q$) or semiconducting ($n - m = 3q \pm 1$), which is given by zone-folding of the Brillouin zone in the circumferential direction. Similar zone-folding for phonon dispersion relations is expected to apply to the phonon dispersion of SWNTs. However when we observe elastic and phonon properties of SWNTs, most of the properties do not depend on chirality, but only on the diameter [3]. A simple reason is that the atomic vibrations in a periodic system have a long wave length relative to the lattice constant of SWNTs, and that the mechanical properties are determined by classical elastic parameters, such as the Young's modulus, etc.

However, in the optical phonon modes, which have relatively high frequencies (between 800 and 1600 cm^{-1}), the bond stretching vibrations with short wave length give a large Raman intensity. Since the vibrating directions of the optical modes can be parallel or perpendicular to the carbon-carbon bonds, the direction of vibration is expected to depend on the chirality. The strong G-band in the Raman spectra appears at 1590 cm^{-1} and contains symmetry components A, E_1 and E_2 as given by group theory. We show here that the Raman intensity for these Raman active modes depend on the chirality in a different way. In this paper, we present theoretical results for the chirality dependent Raman intensities and explain the relationship between the direction of the phonon vibrational mode displacements and the relative intensity as a function of chiral angle. Furthermore, confocal resonant micro-Raman measurements of isolated SWNTs are presented in which the theoretical prediction of the chirality dependent G-band Raman intensity is observed for several isolated SWNTs.

Isolated SWNTs were prepared by a chemical vapor deposition method on a Si/SiO_2 substrate containing nanometer size iron catalyst particles. The details of the sample preparation are given in reference [4]. Resonant Raman spectra were obtained from a single isolated SWNT on this substrate, using a Kaiser Optical Systems, Hololab 5000R: Modular Research Micro-Raman Spectrograph (1 μm laser spot) with 25 mW power, and $E_\ell = 785$ nm $= 1.58$ eV laser line excitation. The method of calculation is for a non-resonant Raman tensor [5], and we independently consider some depolarization effects in the analysis.

In Figs. 1(a) and (b), the intense Raman-active G-band modes for (a) armchair and (b) zigzag nanotubes are shown. Here LO and TO denote one-dimensional, longitudinal or transverse optical modes, respectively, whose vibrational directions are parallel and perpendicular to the nanotube axis. In Fig. 1(c) the calculated Raman intensity as a function of chiral angle, θ, is given in which $\theta = 0$ and $30°$ correspond to zigzag and armchair nanotubes, respectively. In the region of SWNT diameters, $1.30 < d_t < 1.40$ nm, we select eleven nanotube chiralities (n, m) for our calculations: (17,0), (17,1), (16,2), (16,3), (15,3), (15,4), (14,5), (13,7), (12,7), (11,9), and (10,10), whose chiral angles [1], $\theta = \arctan\{\sqrt{3}m/(2n+m)\}$ where $0 \leq \theta \leq 30°$, are 0.0, 2.8, 5.8, 8.5, 9.0, 11.5, 14.7, 20.1, 21.4, 26.7 and 30.0°, respectively. The calculated frequencies of each G-band mode do not depend on the chiral angle [6]. However, the Raman intensity for each symmetry and LO (or TO) phonon direction does depend on the chirality. The numbers of strong Raman modes are 3 or 6, respectively, for achiral or chiral nanotubes, which is consistent with group theory [1,2].

Figure 1(d) shows a plot of the measured G-band Raman intensity observed for six different laser spots on a rough surface that is found near the broken edges of the Si/SiO_2 substrate. The Raman intensity for both the RBM and the G-band becomes strong relative to that for the smooth area in the same sample. The Raman intensities are calibrated using the Raman peak coming from the Si/SiO_2 substrate [4] observed in the range 900–1000 cm^{-1}. However, since the observed G-band Raman intensity depends on the length of a resonant SWNT in the light spot, the Raman intensities which come from different spots cannot be compared with one another directly. In our confocal micro-Raman experiment, we observed some G-band Raman spectra which exhibited only Lorentzian line shapes, as shown in the five lower spectra in Fig. 1(d), and for these traces the strong Raman peaks for each trace exhibit natural line widths from 6 to 11 cm^{-1}, which is an indication that the peaks come from only one resonant semiconducting SWNT [7]. In the top

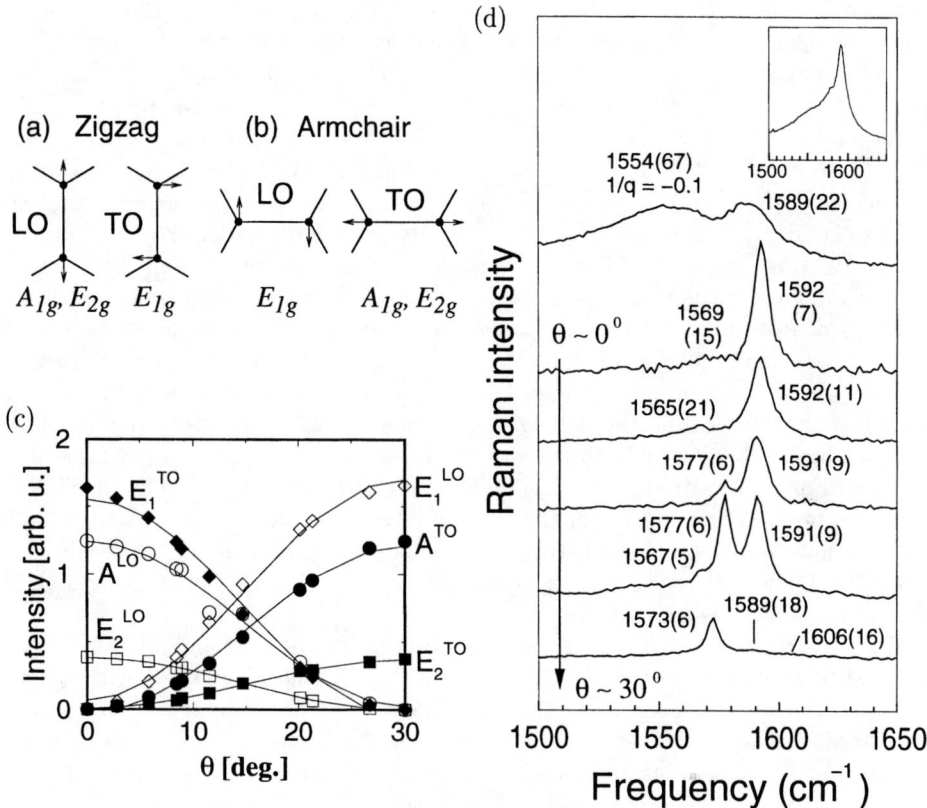

FIGURE 1. Intense Raman G-band modes for (a) armchair and (b) zigzag nanotubes. Here LO and TO denote one-dimensional, longitudinal or transverse optical modes, respectively, whose vibrational direction is parallel and perpendicular to the nanotube axis. (c) The Raman intensity of the G-band as a function of chiral angle, θ. Open and solid symbols, respectively, denote the LO and TO phonons of the nanotubes. $\theta = 0$ and $30°$ correspond to zigzag and armchair nanotubes, respectively. Solid lines are for functions fitted to the angular dependence $B + C\cos(6\theta)$. Here B and C are fitting parameters and the chiral angle θ is in units of radians. (d) G-band Raman spectra of individual SWNTs for six different light spots obtained using the confocal micro-Raman technique. The top spectrum exhibits a BWF feature and is assigned to a metallic SWNT. The lower five spectra are fit with Lorentzian lineshapes and are displayed as the nanotube chiral angle increases from $0°$ to $30°$, based on the theory of (c) [6]. The G-band frequencies (linewidths) for the two strongest G-band components are indicated and the inset shows a superposition of spectra for 25 different laser spots on the sample.

spectrum of Fig. 1(d) we also observe a broad Breit–Wigner–Fano (BWF) feature at about 1550 cm^{-1}, which we have identified as coming from a metallic SWNT [8]. With regard to the peak frequencies, based on our present calculations and recent polarization Raman studies on semiconducting SWNTs [9], we assign the peaks around 1570–1580 cm^{-1} to A^{TO} and E_1^{LO} symmetries, and the peaks around 1590–1600 cm^{-1} to A^{LO} and E_1^{TO} symmetries. Furthermore, we can see small peaks around 1550–1570 cm^{-1} and 1600–1610 cm^{-1}, which are assigned, respectively, to the E_2^{LO} and E_2^{TO} symmetries [9]. When we only look at the spectra for the semiconducting nanotubes (since we did not consider theoretically the interaction responsible for the BWF line), the Raman intensity for different symmetries shows a variety of intensity ratios for the G-bands from top to bottom, as coming from different semiconducting SWNTs, with chiral angles ranging from $\theta \sim 0°$ to $\theta \sim 30°$, respectively. The asymmetry of the G-band intensities for semiconducting SWNTs might come from the depolarization effect in which the optical absorption becomes large when the polarization of the laser light is parallel to the nanotube axis. Such an observation is possible only for measurements of an isolated SWNT. The inset shows a summation of the Raman intensities for SWNTs taken from 25 different laser spots on the substrate. The composite spectrum produced in this way reproduces well the Raman spectra observed for a bundle containing several semiconducting and metallic SWNTs.

In summary, we show that the G-band Raman intensity of an isolated SWNT is chirality dependent, which is predicted by theoretical calculations within non-resonant polarization theory. In the near future, it is desirable that the relative intensity of the G-band structure be observed for an isolated SWNT whose chirality is measured by a scanning tunneling microscope measurement.

R.S. acknowledges a Grant-in-Aid (No. 11165216) from the Ministry of Education, Japan, and A.J. acknowledges financial support from CNPq-Brazil. This work utilized of MRSEC Shared Facilities supported by the National Science Foundation under grant DMR-9400334 and NSF Laser Facility grant 9708265-CHE. MIT authors acknowledge NSF Grants DMR 98-04734, INT 98-15744 and INT 00-00408.

REFERENCES

1. R. Saito, G. Dresselhaus, and M. S. Dresselhaus, *Physical Properties of Carbon Nanotubes* (Imperial College Press, London, 1998).
2. M. S. Dresselhaus, G. Dresselhaus, and P. C. Eklund, *Science of Fullerenes and Carbon Nanotubes* (Academic Press, New York, NY, 1996).
3. M. S. Dresselhaus and P. C. Eklund, Advances in Physics **49**, 705–814 (2000).
4. A. Jorio, et al., Phys. Rev. Lett. **86**, in press (2001).
5. R. Saito, et al., Phys. Rev. B **57**, 4145–4153 (1998).
6. R. Saito, et al. (unpublished).
7. K. Kneipp, et al., Phys. Rev. Lett. **84**, 3470–3473 (2000).
8. M. A. Pimenta, et al., Phys. Rev. B**58**, R16016–R16019 (1998).
9. A. Jorio, et al., Phys. Rev. Lett. **85**, 2617–2620 (2000).

Photoconductivity of Single-Walled Carbon Nanotubes

Akihiko Fujiwara*, Yasuyuki Matsuoka*, Hiroyoshi Suematsu*,
Naoki Ogawa[†], Kenjiro Miyano[†], Hiromichi Kataura[‡],
Yutaka Maniwa[‡], Shinzou Suzuki[§], Yohji Achiba[§]

*Department of Physics, School of Science, University of Tokyo,
7-3-1 Hongo, Bunkyo-ku, Tokyo 113-0033, Japan
[†] Department of Applied Physics, School of Engineering, University of Tokyo,
7-3-1 Hongo, Bunkyo-ku, Tokyo 113-8656, Japan
[‡] Department of Physics, School of Science, Tokyo Metropolitan University,
1-1 Minami-osawa, Hachi-oji, Tokyo 192-0397, Japan
[§] Department of Chemistry, School of Science, Tokyo Metropolitan University,
1-1 Minami-osawa, Hachi-oji, Tokyo 192-0397, Japan

Abstract. We have investigated photoconducting properties of mat samples of single-walled carbon nanotubes (SWNTs) in order to clarify the mechanism of photoconductivity. Two peaks are observed in photoconductivity spectra around 0.7 and 1.2 eV for both samples of SWNTs synthesized by arc discharge and laser ablation methods, which can be interpreted as the photocurrent in the semiconductor phase of nanotubes. No threshold in applied voltage is observed for the occurrence of photoconductivity. Results show that the photoconductivity is an intrinsic feature of SWNTs and that junction areas do not play an important role on photoconductivity.

INTRODUCTION

Since the discovery of carbon nanotubes [1], they have attracted great attention as a very interesting electronic material because of the one-dimensional tubular network structure in the nanometer scale [2,3]. Actually, the findings of many functions in single-walled carbon nanotubes (SWNTs) opened up a route towards nanoscale electronic devices. Very recently, Fujiwara et al. [4] have succeeded to observe photoconducting response for mat samples of SWNTs, for the first time. However, mechanism of photoconductivity has not yet been clarified. In this paper, we present the results of study on photoconducting properties of SWNTs synthesized by arc discharge (AD) and laser ablation (LA) methods in order to solve this subject.

EXPERIMENTAL

The AD-SWNT samples were prepared by evaporation of a graphite rod with nickel (Ni) and yttrium (Y) catalysts in helium atmosphere by arc discharge [5,6]. The LA-SWNT samples were synthesized by ablating a graphite target containing Ni and cobalt (Co) catalysts at 1250 °C in argon atmosphere by using a pulsed Nd:YAG laser [7]. The diameters of SWNTs synthesized by both methods are almost the same and determined as about 1.4 ± 0.2 nm by Raman frequency of a breathing mode. On the other hand, the length of SWNTs depends on the synthesis method: The length estimated by a scanning electron microscope (SEM) observation is less than one μm and several μm for AD- and LA-SWNTs, respectively.

For preparing mat samples for photoconductivity measurements, the suspension of SWNTs in methyl alcohol was dropped on a glass substrate. Typical mat sample size is around $10^4 \mu m^2$. A pair of gold electrodes separated by a 10 μm gap were vacuum-evaporated on to the surface of the mat samples and connected to a dc regulating power supply (applied voltage: 50 - 1000 mV). This narrow separation-gap of 10 μm is used for the reduction of the number of junctions between SWNTs thorough the current pass, because resistance of junctions govern the total resistance of mat samples and conceal the intrinsic transport properties of SWNTs. The resistance of samples without optical excitation measured by this method is around 100 Ω. As light source, an optical parametric oscillator (OPO) excited by a pulsed Nd:YAG laser was used for optical excitation. The photon energy was in the rage from 0.5 to 3.0 eV and the pulse duration was 5 ns. The temporal profile of the laser pulse and the photocurrent was monitored with a digitized oscilloscope. All measurements in this work were performed at room temperature.

RESULTS AND DISCUSSION

Figure 1(a) shows the photoconductivity spectrum for an AD-SWNT sample at room temperature. Two clear peaks in the photoconductivity spectrum are observed around 0.7 and 1.2 eV. These energies are very close to the first and second interband gap of semiconductor phase of SWNTs with diameter of 1.4 nm. In addition, this spectrum is very similar to the optical adsorption spectra for the same sample of SWNTs [6,8]. From this result, we can conclude that the photoconductivity originating from the semiconductor phase of SWNTs are observed.

In this measurement, however, even in the use of the small gap of electrodes (10 μm), typical nanotube length ($\lesssim$ 1 μm) is much smaller than this gap, and resistance is still mainly governed by high resistance areas at junctions between SWNTs. Therefore, the concentration of the electric field at junctions may dissociate excitons into free carriers (electrons and holes). In this case, there should be a threshold corresponding to the binding energy of exciton in applied voltage (V) for the occurrence of photoconductivity. Figure 1(b) shows V dependence of photocurrent (ΔI) for the AD-SWNT sample. Although the ΔI shows a satura-

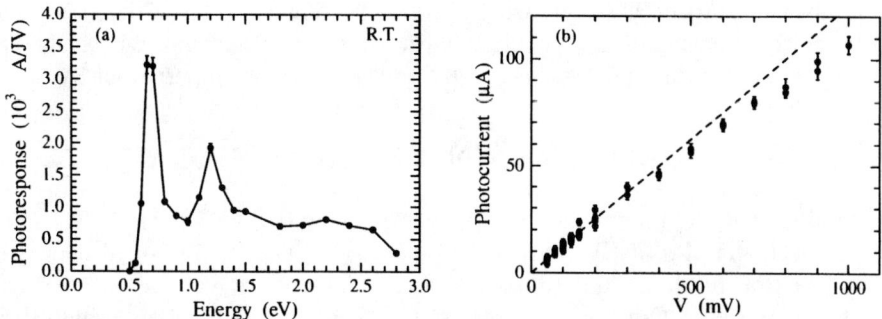

FIGURE 1. (a) Photoconductivity spectrum for a mat sample of SWNTs. (b) Applied voltage dependence of photocurrent. Dashed line shows the fitting results assuming the linear response by using data below 200 mV in V.

tion behavior at high V region as observed in the incident excitation-light-power dependence of ΔI [4], ΔI linearly decreases to zero as V decreases: the x-segment is estimated 0.0 ± 4.3 mV by linear fitting of V vs. ΔI plot by using data for 0 mV $\leq V \leq$ 200 mV. Therefore, there is no voltage threshold for photoconductivity within the experimental accuracy.

Another way to getting rid of the "junction effect" is measurements by using longer SWNT samples. Because typical nanotube length of LA-SWNTs is several μm and comparable to the distance between electrodes, the contribution of junctions to photoconductivity should be greatly reduced in this sample with comparing the AD-SWNT sample. Therefore, the strong reduction of photoconducting response in LA-SWNTs is expected, if the photoconductivity originates from the "junction effect". However, clear peaks around 0.7 and 1.2 eV are observed in photoconductivity spectrum even for the LA-SWNT sample, and it is similar to that of the AD-SWNT sample, as shown in Fig. 2.

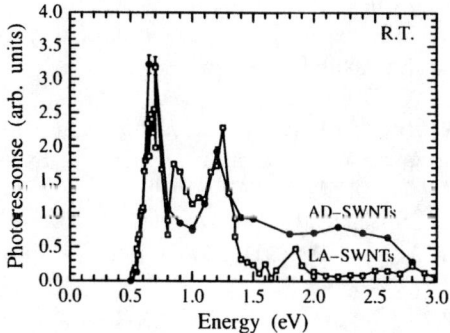

FIGURE 2. Photoconductivity spectra for mat samples of SWNTs synthesized by laser ablation methods ($\square$) with data for sample synthesized by arc discharge ($\bullet$).

From the results of two approaches to solve the "junction effect", we can conclude that the contribution of the "junction effect" to the photoconductivity is negligible and the photoconducting response is due to the intrinsic properties of SWNTs.

SUMMARY AND CONCLUSIONS

We find that the photoconductivity spectra show two peaks around 0.7 and 1.2 eV in both samples of SWNTs (d $\approx$ 1.4 nm) synthesized by arc discharge and laser ablation methods and that no threshold in applied voltage is observed for photoconductivity. Our results clearly shows that semiconductor phase of SWNTs can behave as one-dimensional semiconductor with photoconducting function and that junction areas do not play an important role for photoconductivity.

ACKNOWLEDGEMENTS

This work was supported in part by the research project, "Materials Science and Microelectronics of Nanometer-Scale Materials" (RFTF96P00104) from the Japan Society for the Promotion of Science, Japan and a Grant-in Aid for Scientific Research on the Priority Area "Fullerenes and Nanotubes" from the Ministry of Education, Science, Sports and Culture of Japan. One of authors (A.F.) was supported by a Grant-in Aid for Encouragement of Young Scientists (12740202) from The Ministry of Education, Science, Sports and Culture of Japan.

REFERENCES

1. Iijima, S., *Nature* (London) **354**, 56-58 (1991).
2. Dresselhaus, M.S., Dresselhaus, G., and Eklund, P., *Science of Fullerenes and Carbon Nanotubes* (Academic Press, New York, 1996).
3. Saito, R., Dresselhaus, G., and Dresselhaus, M.S., *Physical Properties of Carbon Nanotubes* (Imperial College Press, London, 1998).
4. Fujiwara, A., Matsuoka, Y., Suematsu, H., Ogawa, N., Miyano, K., Kataura, H., Maniwa, Y., Suzuki, S., Achiba, Y., unpublished.
5. Journet, C., Maser, W.K., Bernier, P., Loiseau, A., Lamy de la Chapelle, M., Lefrant, S., Deniard, P., Lee, R., and Fischer, J.E., *Nature* (London) **388**, 756-758 (1997).
6. Kataura, H., Kumazawa, Y., Maniwa, Y., Umezu, I., Suzuki, S., Ohtsuka, Y., and Achiba, Y., *Synth. Met.* **103**, 2555-2558 (1999).
7. Thess, A., Lee, R., Nikolaev, P., Dai, H., Petit, P., Robert, J., Xu, C., Lee, Y.H., Kim, S.G., Rinzler, A.G., Colbert, D.T., Scuseria, G.E., Tománek, D., Fischer, J.E., Smally, R.E., *Science* **273**, 483-487 (1996).
8. Ichida, M., Mizuno, S., Tani, Y., Saito, Y., and Nakamura, A., *J. Phys. Soc. Jpn.* **68**, 3131-3133 (1999).
9. Ando, T., *J. Phys. Soc. Jpn.* **66**, 1066-1073 (1997).

Electronic Structure of Multi-Walled Carbon Nanotubes Studied by Photoemission Spectroscopy

S. Suzuki, C. Bower[1,*], Y. Watanabe, S. Heun[2], T. Kiyokura[**], K. G. Nath, T. Ogino, W. Zhu[3] and O. Zhou[1,4]

NTT Basic Research Laboratories, Atsugi, Kanagawa, 243-0198 Japan
[1]*Department of Physics and Astronomy, University of North Carolina, Chapel Hill, NC 27599*
[2]*Sincrotrone Trieste, Basovizza, 34012 Trieste, Italy*
[3]*Bell Laboratories, Lucent Technologies, Murray Hill, NJ 07974*
[4]*Curriculum in Applied and Materials Science, University of North Carolina, Chapel Hill, NC 27599*

Abstract. Photoemission spectra were measured from the tips and the sidewalls of multi-walled carbon nanotubes grown on Si substrates. The results show that the Fermi level is located slightly inside the conduction band at the tips. We suggest that the Fermi level pinning is due to a large defect density at the tips of the multi-walled carbon nanotubes.

INTRODUCTION

Although the local electronic structure at carbon nanotube tips has been studied using scanning tunneling spectroscopy (STS), the energy range of the spectra is strictly limited near the Fermi level (typically ±1 eV). Therefore, information on the overall band structure at nanotube tips is still lacking. We applied photoemission spectroscopy to investigate the valence band and C 1s states and the work functions of vertically aligned and random multi-walled carbon nanotubes (MWNTs) grown on Si substrates. The spectra from the aligned MWNTs were considered to be dominated by photoemission from the tips. The results indicate that the Fermi level is located slightly inside the conduction band at the tips. We propose that defects in MWNTs play an important role in explaining the results.

EXPERIMENTAL

MWNTs aligned vertically on a Si substrate were grown using the microwave plasma enhanced chemical vapor deposition (MPE-CVD) method.[1] The diameter and length of the MWNTs were about 30 nm and 10 μm, respectively. Randomly oriented MWNTs, whose typical diameter was 20 to 50 nm, were also grown on a Si substrate using thermal CVD technique. Photoemission measurements were carried out at beamline BL-1C, Photon Factory, High Energy Accelerator Institute, Japan. Before the measurements, the samples were annealed at about 300°C in a vacuum to remove adsorbates on the surface.

RESULTS AND DISCUSSION

Because a nanotube is highly anisotropic, the photoemission from a nanotube sidewall is considered to depend on the polarization direction of the incident light. However, the valence band spectra of the aligned sample hardly depended on the polarization direction at all. This indicates that the spectra from the aligned MWNTs were dominated by photoemission from the tips that were hemi-spherically curved. On the other hand, spectra from the random MWNTs are considered to be dominated by photoemission from sidewalls because the nanotubes are much longer than the tip radii.

The valence band spectra of the aligned and the random MWNTs were quite similar. However, a slight spectral shift (~0.2 eV) to the higher binding energy side was observed in the aligned MWNTs as shown in Fig. 1. Such an energy shift was also observed in the C 1s spectra. These results indicate that the overall electronic structure at the tips is slightly shifted to the higher binding energy side.

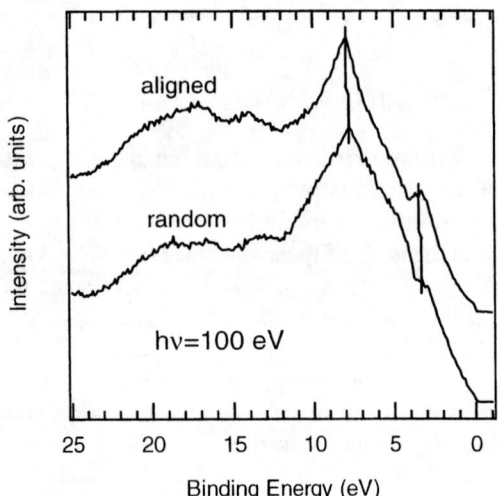

FIGURE 1. Valence band photoemission spectra of the aligned and random MWNTs.

The figure also clearly shows that the density of states at the Fermi level at the tips is considerably larger than that at the sidewalls. Photoemission spectra in the vicinity of the Fermi level (not shown) indicated that the density of states at the Fermi level decreases in order of the aligned MWNTs, random MWNTs, and random single-walled carbon nanotubes (SWNTs[2]). The rigid shift behavior and the larger density of states at the Fermi level indicate that the Fermi level is pinned slightly inside the conduction band at the tips of the aligned MWNTs.

Figure 2 shows the secondary electron threshold spectra of the aligned and random MWNTs excited by a He discharge lamp. The work function of the random MWNTs was determined to be 4.6 eV. This value is almost same as that for graphite (4.6 eV). On the other hand, the work function of the aligned MWNTs was determined to be 4.4 eV, which is slightly smaller than those of the random MWNTs and graphite. Interestingly, the work function of the SWNTs was found to be 4.8 eV in our previous study, which is slightly larger than the value for graphite.[3] The smaller work function of the aligned MWNTs is consistent with the Fermi level pinning discussed above because the shift will decrease the energy difference between the vacuum level and the Fermi level.

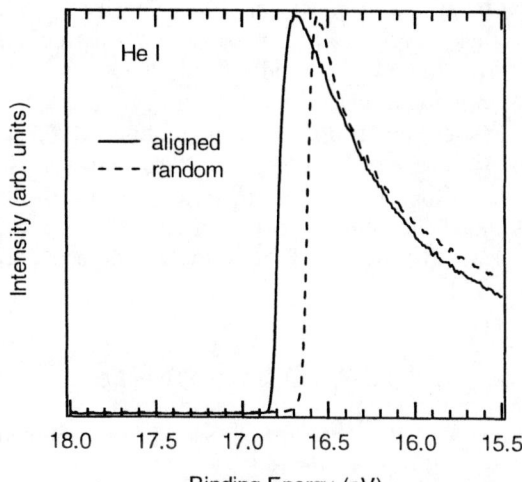

FIGURE 2. Secondary electron threshold spectra of the aligned and random MWNTs.

As discussed above, all the results can be well explained assuming the Fermi level pinning at the tips of the aligned MWNTs. The insertion of the five-member rings in the graphene network at tips and the curvature of the graphene sheets will induce an electronic structural change. However, the five-member ring effect can not explain the fact that the random MWNTs showed a larger density of states at the Fermi level than the random SWNTs, because both spectra are considered to be dominated by photoemission from the sidewalls. The curvature effect does not explain why the MWNT tips, where the electronic structure will be more affected by the curvature than that of the sidewalls, showed a smaller work function than graphite, in spite of the fact that the SWNTs showed a larger work function than graphite. Thus, it seems to be difficult to explain the results assuming that the nanotubes have ideal structures.

The Fermi level is pinned inside the conduction band if defect states are formed inside the conduction band and the defect density is sufficiently large. Our previous study revealed that most MWNTs were intercalated by alkali-metal atoms.[4] This indicates that the MWNTs had paper-mache structures that consist of small graphene sheets.[5] The MWNTs measured in this study were also considered to be defective enough to be intercalated, because a spectral shift to the higher binding energy side and a significant increase of the density of states at the Fermi level were observed.[6]

We can roughly expect that excess electrons of about 0.004 per atom (an order of 10^{20} /cm^3) are necessary at the tip region for the Fermi level shift of 0.2 eV.[2] The defect density should be at least as large as this value. From a simple geometrical consideration, the size of the graphene sheet that gives the defect density of 0.004 per atom was estimated to be about 120 nm.[2] Our previous transmission electron microscope (TEM) observation on deintercalated MWNTs revealed that most of the graphene sheets in the MWNTs had continuous length of only ~10 nm.[5] If the length corresponds to the typical size of the graphene sheet in a MWNT, the defect density will be large enough to cause the observed Fermi level shift. Thus, we think that one possible explanation for the results is that the Fermi level at the MWNT tips is pinned inside the conduction band due to the larger defect density at the tips. In contrast to MWNTs, it seems that SWNTs have more nearly perfect structures, because intercalants in an intercalated SWNT bundle reside between the tubes in the bundle rather than inside the individual tubes if the SWNTs are capped.[7,8] Thus, our finding that the density of states at the Fermi level of the random MWNTs is larger than that of the SWNTs can be similarly explained by the larger defect density in the MWNTs. That is, the Fermi level of the MWNTs is slightly shifted toward the conduction band even at sidewalls. Although the electronic properties of carbon nanotubes have been discussed assuming ideal structures, we suggest that the defects play an important role in explaining the observed results.

ACKNOWLEDGMENTS

The authors thank Dr. T. Saitoh and Prof. A. Kakizaki for their support in this work. Work done at NTT was partly supported by Special Coordination Funds of the Science and Technology Agency of the Japanese Government. Work done at UNC was supported by the Office of Naval Research through a MURI program (N00014-98-1-0597).

*Present address: Bell Laboratories, Lucent Technologies, Murray Hill, NJ 07974
**Present address: NTT Telecommunications Energy Laboratories, Atsugi, Kanagawa, 243-0198 Japan

REFERENCES

1. Bower, C. et. al., Appl. Phys. Lett. **77**, 830-832 (2000).
2. Suzuki, S. et. al., submitted to Phys. Rev. **B**.
3. Suzuki, S. et. al., Appl. Phys. Lett. **76**, 4007-4009 (2000).
4. Suzuki, S. et. al., J. Appl. Phys. **79**, 3739-3743 (1996).
5. Zhou, O. et. al., Science **263**, 1744-1747 (1994).
6. Suzuki, S. et. al., in preparation.
7. Bower, C. et. al., Appl. Phys. A **67**, 47-52 (1998).
8. Suzuki, S. et. al., Chem. Phys. Lett. **285**, 230-234 (1998).

STM/STS on Carbon Nanotubes at Low Temperature

Kazushige Nomura[*], Masato Osawa[*], Koichi Ichimura[*], Hiromichi Kataura[†], Yutaka Maniwa[†], Shinzou Suzuki[¶], and Yohji Achiba[¶]

[*]Division of Physics, Hokkaido University, Sapporo 060-0810, Japan
[†]Department of Physics, Tokyo Metropolitan University, Minami-osawa, Hachi-oji, Tokyo 192-0397, Japan
[¶]Department of Chemistry, Tokyo Metropolitan University, Minami-osawa, Hachi-oji, Tokyo 192-0397, Japan

Abstract. Single-wall carbon nanotubes (SWNT's) prepared on the graphite substrate were investigated by a low temperature STM at 290 and 77 K. At 290 K, we obtained STM images of bundle structure. SWNT's with a diameter of about 1 nm are aligned closely in the bundle. From the tunneling spectra with high energy resolution obtained at 77 K, both metallic and semiconducting SWNT's were found in the same bundle. For the metallic SWNT, the conductance curve inside the first peak is finite and quite flat. The width between the first peaks is obtained as 1300 meV. Fine structures were observed just outside the first peak. The observed gap width for the semiconducting SWNT is 600 meV. The conductance curve due to the tunneling between SWNT's was observed.

INTRODUCTION

It is well known that electronic properties of carbon nanotubes depend on their chirality. It is important to elucidate the relation between the electronic state and the chirality for understanding the conduction mechanism in carbon nanotubes. The scanning tunneling microscope (STM) is a powerful tool to probe carbon nanotubes because it can obtain the structural image with the atomic resolution and investigate the local electronic state simultaneously. The spectroscopic measurement called as scanning tunneling spectroscopy (STS) is quite useful due to much less disturbance on the electronic state at the sample surface. It is important to reduce the thermal excitation which smears the spectrum in the spectroscopic measurement. STM and STS measurements on single-wall carbon nanotubes (SWNT's) were reported by Wildoer et al. [1] and Odom et al. [2] at 4.2 and 77 K, respectively. Recently, the peak splitting for the metallic SWNT is predicted [3, 4]. Further investigation with high energy resolution is needed for STS.

We have investigated the fine structure in the tunneling spectra on SWNT's using a low temperature STM. In this paper, we present STS results with high energy resolution on carbon nanotubes in bundles at 77 K. We found both metallic and semiconducting features in tunneling spectra of SWNT's in a bundle. The fine structure in the spectra observed for the metallic SWNT is discussed.

EXPERIMENTAL

Single-wall carbon nanotubes were synthesized by the conventional arc-discharge method with catalysts of Ni-Y (Ni:Y = 4.2:1 at. %). The SWNT sample for the STM measurement was prepared by dropping SWNT's agitated ultrasonically in ethanol on the cleaved KISH graphite. The cell which contains the STM unit is filled with a low pressure of helium gas as the thermal exchange. In the STS measurement, the tunneling differential conductance was measured directly by the lock-in technique, in which 1 kHz AC modulation with the amplitude of 15 mV was superposed in the sweep bias voltage.

RESULTS AND DISCUSSION

Figure 1(a) shows topographic image of a SWNT sample on the graphite substrate at 290 K with the tunneling current I=70 pA and the bias voltage V=150 mV. A bundle of nanotubes was found as pointed by an arrow in Fig. 1(a). The image indicates that SWNT's are aligned closely in parallel in each bundle. Although the atomic resolution is not obtained, individual SWNT's are imaged. Figure 2(b) shows the scan profile between A and B in Fig. 1(a). The profile across the bundle shows the arrangement of SWNT's which are closely spaced. The diameter of SWNT's is estimated as about 1 nm from the width of the corrugation.

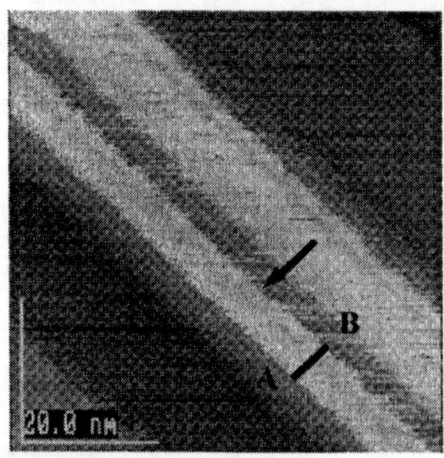
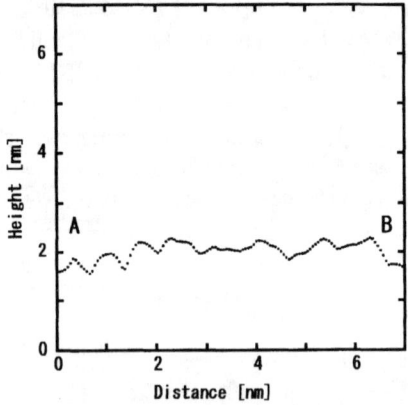

FIGURE 1(a). Topographic image of SWNT bundle. A bundle is pointed by an arrow.

FIGURE 1(b). Profile between A and B. The distance is measured from point A.

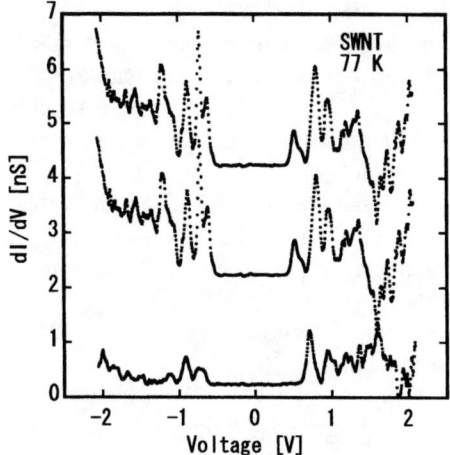
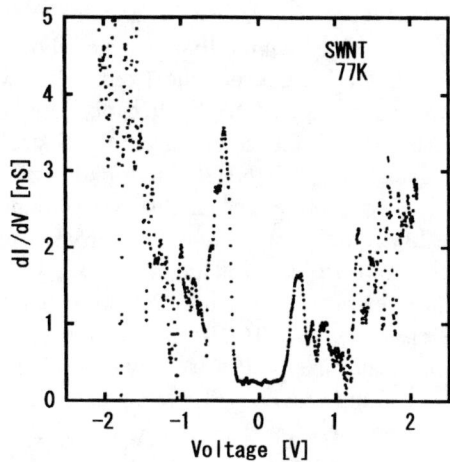

FIGURE 2. Tunneling differential conductance at 77 K. Spectra were obtained in sequence at the fixed position. The zero conductance line of each curve is shifted by two divisions for clarity.

FIGURE 3. Tunneling spectra for the semiconducting SWNT at 77 K.

We succeeded to obtain tunneling spectra at the same bundle with varying the tip position at 77 K. In most cases, we obtained spectra which have finite conductance at zero bias voltage. Figure 1 shows the tunneling differential conductance curves obtained at the fixed position in sequence. Although the spectra are a little noisy, almost the same feature is reproduced. The conductance around the Fermi energy, which corresponds to zero bias voltage, is finite and flat. This indicates that the observed spectra corresponds to the electronic density of states (DOS) for the metallic SWNT. At about V=±650 mV, the conductance shows a divergent peak like the gap edge. This peak corresponds to the van Hove singularity in the density of states for nanotubes [5]. The width between the first peak is almost consistent with that for the metallic SWNT reported by Wildoer et al. [1]. It is noteworthy that the conductance inside the first peak is much flat as compared with that in previous reports [1, 2].

Additionally, we found a few peaks located just outside the first peak at intervals of about 200 mV. Although there is a little shift in peak position, these structures are almost reproduced qualitatively as shown in Fig. 2. We point out the possibility that these fine structures are regarded as the splitting of the first peak. The first DOS peak for metallic nanotubes is predicted to split into two peaks due to the anisotropy of the Fermi surface depending on the chirality [3, 4]. Fine structures in present spectra might be related to this DOS splitting. However, the observed splitting width of about 200 meV is much larger than predicted one which is of the order of 10 meV. We need further investigation of the origin of these fine structures.

We found another type of tunneling spectra as shown in Fig. 3 at another SWNT in the same bundle. Although the conductance near the Fermi energy is finite and flat, the conductance value is smaller as compared with the metallic SWNT shown in Fig. 2. The width between the first peaks of about 600 meV is less than one half of that in Fig. 2. This width value is almost consistent with the estimation for the gap width in the semiconducting SWNT with a diameter of 1.2 nm. We consider that these spectra correspond to the semiconducting SWNT. We should note that both metallic and semiconducting SWNT's were found in the same bundle.

We point out the possibility of the tunnel junction between nanotubes. In rare cases, we found different conductance curves from above two types. The conductance is not flat near zero bias voltage. There are broad conductance peaks around $V=\pm 80$ mV. These features are partially explained by the tunneling between SWNT's which have different chirality [6]. It is likely that the tunneling junction between SWNT's is formed since nanotubes are contacted each other in a bundle.

In summary, SWNT's were studied by a low temperature STM. SWNT's aligned in a bundle were imaged at 290 K. Both metallic and semiconducting SWNT's were found in the same bundle by STS at 77 K. For the metallic SWNT, the conductance curve inside the first peak is quite flat. Fine structures just outside the first DOS peak were found in the tunneling spectra. For the semiconducting SWNT, the conductance value at zero bias voltage is much smaller than that in the metallic SWNT. The gap width of 600 meV is less than a half of the width between the first peaks in the metallic SWNT. The conductance curve, in which the peak appears at about $V=\pm 80$ mV, is partially explained by the tunneling between SWNT's in a bundle.

ACKNOWLEDGEMENT

This work was supported in part by the Grant-in-Aid for Scientific Research on the Priority Area "Fullerene and Nanotubes" by the Ministry of Education, Science, and Culture of Japan.

REFERENCES

1. Wildoer, J. W. G., Venema, L. C., Rinzler, A. G., Smalley, R. E., and Dekker, C., *Nature* **391**, 59-62 (1998).
2. Odom, T. W., Huang, J. L., Kim, P., Lieber C. M., *Nature* **391**, 62-64 (1998).
3. Saito, R., Dresselhaus, G., and Dresselhaus, M. S., *Phys. Rev. B* **61**, 2981-2990 (2000).
4. Kim, P., Odom, T. W., Huang, J. L., and Lieber, C. M., *Phys. Rev. Lett.* **82**, 1225-1228 (1999).
5. Saito, R., Fujita, M., Dresselhaus, G., and Dresselhaus, M. S., *Appl. Phys. Lett.* **60**, 2204-2206 (1992).
6. Saito, R., Dresselhaus, G., and Dresselhaus, M. S., *Physical Properties of Carbon Nanotubes*, Imperial College Press, London, 1998, pp.123-130.

Dielectric Function of C_{60}-Encapsulating Nanotube

N. Hamada[a], M. Yamaji[a], S. Okada[b] and S. Saito[c]

[a] *Science University of Tokyo, Yamazaki, Noda, Chiba 278-8510, Japan*
[b] *University of Tsukuba, Tennodai, Tsukuba, Ibaraki 305-8573, Japan*
[c] *Tokyo Institute of Technology, Oh-okayama, Meguro-ku, Tokyo 152-8551, Japan*

Abstract. The electronic band structure of C_{60}-encapsulating (10,10)-nanotube (C_{60}@(10,10) system) is calculated by using a realistic tight-binding model taking into account the s and p atomic orbitals. The imaginary part of the dielectric function is evaluated in the random phase approximation for the hexagonal close-packed crystal of the nanotubes. For the photon electric field parallel to the nanotube axis (z-polarization), three characteristic structures appear in the absorption spectrum. The low energy absorption (< 0.5eV) is due to the interband transition between the conduction bands of nanotube. The 1.9eV peak is due to the transition from the slightly occupied t_{1u} band to the h_g band of C_{60}, and the 2.1eV peak is due to the transition from the h_u band to the t_{1u} band. For x/y-polarization, a peak appears at 0.7eV, corresponding to the transition from the t_{1u} band to the t_{1g} band of C_{60}.

I INTRODUCTION

A carbon nanotube encapsulating C_{60} molecules was first observed in 1998 by Smith, Monthioux and Luzzi [1]. Several groups report that C_{60} molecules can be introduced into a carbon nanotube rather easily [2–5]. Recently, the first-principles electronic structure calculation has been done for C_{60}-encapsulating (n,n)-nanotubes with $n = 8 - 10$ [6]. The total energy calculation shows that the combined system of C_{60}'s and a (n,n) nanotube (to be called as the C_{60}@(n,n) system) is lower in energy than the separated system for $n = 10$, but not for $n = 9$ and 8. A remarkable band structure is that the C_{60} t_{1u} band is occupied by a small amount of electrons for the C_{60}@(10,10) system.

In this paper, to evaluate the optical absorption, we employ a realistic tight-binding model taking into account the s and p atomic orbitals. The s and p atomic levels are set to be -7.0eV and 0eV, respectively, for the nanotube. The atomic levels for C_{60} are shifted down by -0.75eV to reproduce the above mentioned first-principles band structure of the C_{60}@(10,10) system. We discuss characteristic peaks in the optical absorption spectrum for the C_{60}@(10,10) system

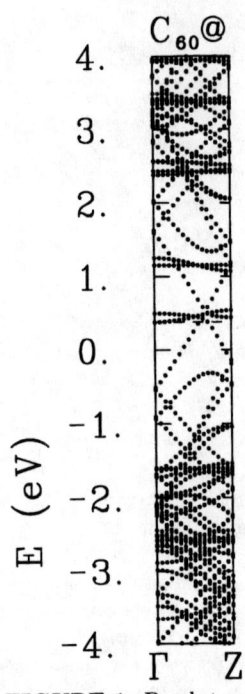

FIGURE 1. Band structure of the C_{60}@(10,10) system.

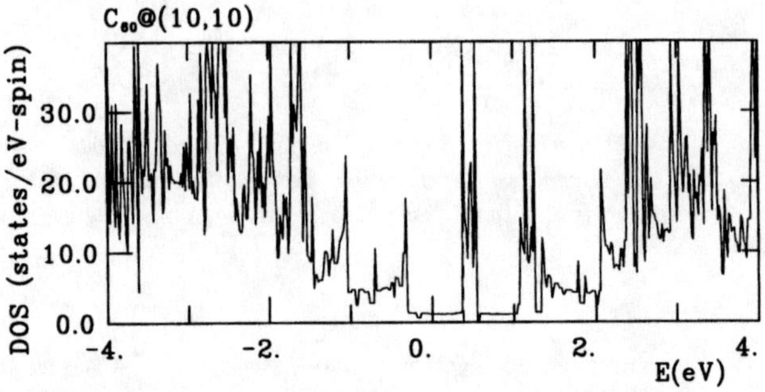

FIGURE 2. DOS of the C_{60}@(10,10) system. The Fermi level is located at 0.38eV, just above the bottom of the t_{1u} band of C_{60}.

II C_{60}@(10,10) SYSTEM

Figures 1 and 2 show the band structure and the density of states (DOS), respectively, for the C_{60}@(10,10) system. The Fermi level is located at 0.38eV. It is easy to find nondispersive bands which originate from C_{60}: The C_{60} g_g+h_g band is located at -2.7eV, h_u at -1.7eV, t_{1u} at 0.4eV, t_{1g} at 1.1eV, h_g at 2.3eV and so forth. Dispersive bands come from the (10,10) nanotube. The C_{60} t_{1u} band is occupied by 0.15 electrons.

To study the optical property, we pack nanotubes closely in a hexagonal lattice, and calculate the dielectric function in the random phase approximation (RPA). The electron transfer between nanotubes is neglected in the calculation. The Fermi surface contribution to the dielectric function is also neglected; only interband transitions are shown below.

Figure 3 shows the imaginary part of the dielectric function for the C_{60}@(10,10) system, and for the (10,10) nanotube as a comparison. The most part of the C_{60}@(10,10) spectrum is explained as a simple sum of the (10,10) nanotube spectrum and the C_{60} spectrum. We can, however, find a few characteristic peaks for the combined system.

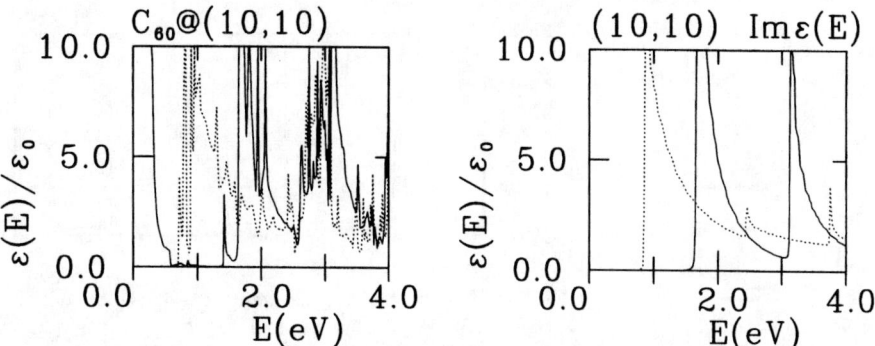

FIGURE 3. Imaginary part of the dielectric function for the $C_{60}@(10,10)$ system and the (10,10) nanotube. Solid line corresponds to the z polarization of the photon electric field, and dotted line to the x or y polarization. Nanotubes are aligned with the tube axis along the z direction, making a triangular lattice with the lattice constant of 16.9Å on the xy plane. The effect of the electron transfer between nanotubes, however, is neglected in the calculation.

The dielectric function strongly depends on the polarization of light. For the z polarized light ($\boldsymbol{E} \parallel \boldsymbol{z}$), three characteristic absorptions appear in the low-energy region, at 1.9eV and at 2.1eV. The low-energy absorption (< 0.5eV) is due to the interband transition between the nanotube conduction bands; the low energy transition is prohibited for the (10,10) nanotube only, and becomes allowed by the C_{60} encapsulation lowering the symmetry. The peak at 1.9eV is due to the transition from the slightly occupied t_{1u} band to the h_g band of C_{60}. For the x or y polarization ($\boldsymbol{E} \perp \boldsymbol{z}$), the 0.7eV peak characterizes the combined system, which corresponds to the transition from the t_{1u} band to the t_{1g} band.

III C_{60} ARRAY

C_{60} molecules make an array inside the nanotube. We take the C_{60} array out of the nanotube, and calculate the band structure and the dielectric function. In Figs. 4 and 5, we can see that each band has a considerable band width, due to the electron hopping between C_{60} molecules. Figure 6 shows the imaginary part of the dielectric function of the C_{60} array. Absorptions around 3eV are due to the transitions from the g_g+h_g band to the t_{1u} band and from the h_u band to the t_{1g} band, which are intrinsic of the C_{60} molecule. The 2.1eV peak newly appears when C_{60}'s are aligned: The absorption is due to the transition from the h_u band to the t_{1u} band, originally forbidden for a single C_{60} molecule, but becomes allowed due to the symmetry lowering by making an array.

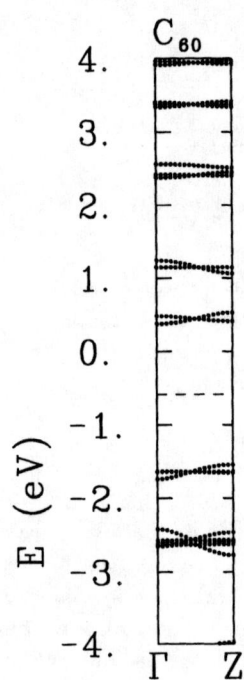

FIGURE 4. Band structure of the C_{60} array.

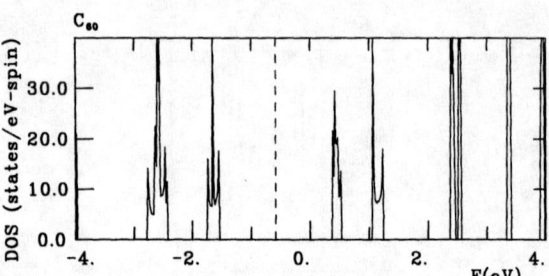

FIGURE 5. DOS of the C_{60} array. The g_g+h_g band is situated at -2.7eV, h_u at -1.7eV, t_{1u} at 0.4eV, t_{1g} at 1.1eV, h_g at 2.3eV and t_{2u} at 2.5eV. The Fermi level is located between the h_u band and the t_{1u} band.

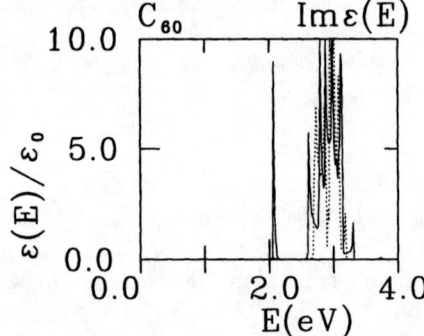

FIGURE 6. Imaginary part of the dielectric function for the C_{60} array. Solid line corresponds to the z polarization of the photon electric field, and dotted line to the x or y polarization. C_{60}'s are aligned along the z axis, and the spatial density of C_{60} is the same as that in the C_{60}@(10,10) system.

REFERENCES

1. B. W. Smith, M. Monthioux and D. E. Luzzi, Nature **396**, 323 (1998).
2. B. Burteaux *et al.*, Chem. Phys. Lett. **310**, 21 (1999).
3. J. Sloan *et al.*, Chem. Phys. Lett. **316**, 191 (2000).
4. K. Hirahara *et al.*, Phys. Rev. Lett. **85**, 5384 (2000).
5. H. Kataura *et al.*, Synthetic Metals, in press.
6. S. Okada, S. Saito, A. Oshiyama, preprint; S. Okada, S. Saito, A. Oshiyama and Y. Miyamoto, in this proceedings.

Structural and Electronic Property Changes of Single-Walled Carbon Nanotubes under Pressure

Lu-Chang Qin[1], Jie Tang[2], Taizo Sasaki[2], Masako Yudasaka[1], Akiyuki Matsushita[2] and Sumio Iijima[1,3,4]

[1] *JST-ICORP Nanotubulite Project, c/o NEC Corp., Tsukuba 305-8501, Japan*
[2] *National Research Institute for Metals, 1-2-1 Sengen, Tsukuba 305-0047, Japan*
[3] *R&D Group, NEC Corporation, Tsukuba 305-8501, Japan*
[4] *Department of Materials Science and Engineering, Meijo University, Nagoya 468-8502, Japan*

Abstract. The mechanical and electronic properties of single-walled carbon nanotubes have been studied under hydrostatic pressure up to 2 GPa. A high volume compressibility of 0.024 GPa^{-1} was obtained. The carbon nanotubes are polygonized slightly when they form bundles of hexagonal close-packed structure and the inter-tubular gap is smaller by 0.25 Å than the equilibrium spacing of graphite (002). Under high pressure, further polygonization occurs to accommodate the extra amount of volume reduction. Accompanying the polygonization, the band gap of the nanotubes increases with increasing pressure. A discontinuous change in electrical resistivity was observed at 1.5 GPa pressure, suggesting a phase transition had occurred when the polygonized nanotubes became elliptical in cross-section.

INTRODUCTION

Carbon nanotubes have been extensively studied since they were first discovered about ten years ago [1] due to their extraordinary properties and their promising potential for nanotechnology applications. While the multiwalled carbon nanotubes show greatest values of axial Young's modulus of about 2 TPa [2], approaching to the theoretical value of graphene, carbon nanotubes appear very soft in the radial directions. The van der Waals forces between neighboring carbon nanotubes can induce flattening of the otherwise circular tubules and the softness of carbon nanotubes in the radial directions has been observed experimentally [3-4]. On the other hand, single-walled carbon nanotubes are even softer than multiwalled nanotubes [5]. Molecular dynamics simulations have demonstrated that large radial compressions could be induced by a small force impact without C-C bond breakage [6]. However, there is still no quantitative measurement of the radial elasticity of carbon nanotubes reported in literature.

Hydrostatic high pressure provides an ideal condition to studying the radial elasticity of carbon nanotubes. Since the C-C bond is very strong, the axial deformation under moderate pressure is still negligible while substantial radial deformation is already induced, as indicated by the very small Poisson ratio of graphite (=0.012) [7].

The electronic structure of carbon nanotubes is very sensitive to the morphological change when their cross-sections are polygonized from the circular shape [8]. In particular, it has been suggested that the electronic band gap of a semiconducting carbon nanotube would be changed due to the facetting that should induce a σ^*-π^* hybridization and lowers the symmetry of the otherwise cylindrical tubules when the polygonization took place.

We applied hydrostatic pressure to single-walled carbon nanotubes and employed X-ray diffraction to monitor the structural changes. Four-probe measurements of the electrical resistance were conducted at various temperatures under pressure to correlate the transport properties with the structural deformation. Energetics calculations have also been carried out to establish the equilibrium inter-tubular gap and to simulate the morphological evolution under increased pressure in connection with the experimentally measured compressibility.

EXPERIMENTAL

Single-walled carbon nanotubes were produced by single-beam laser evaporation of graphite powders catalyzed by Ni/Co fine particles. The pristine carbon nanotubes often aggregated to form raft-like bundles [9]. Loosely tangled nanotube bundles were put into a gasketed diamond anvil cell and hydrostatic pressure was applied to the nanotubes via a pressure medium made of ethanol-methanol. The applied pressure was measured using the R-line emission from the ruby crystals embedded in the pressure medium. Synchrotron X-ray diffraction data were collected utilizing imaging plates with an intensity sensitivity of 10,000 scales at the Photon Factory, High Energy Accelerator Research Organization, Tsukuba, Japan using a selected wavelength of 0.8000 Å to ensure that the Bragg lattice reflections from the nanotube bundles were well preserved. Electrical resistance measurements were carried out using the four-probe method under various pressures generated by a piston apparatus. The resistivity measurements were performed at temperatures from 2 K up to the room temperature.

RESULTS AND DISCUSSION

Figure 1 shows the dependence of the lattice constant of the nanotube crystals under pressure up to 2 GPa, beyond which the characteristic Bragg reflection from the nanotube lattice becomes too weak to be identifiable. The deformation of the nanotube lattice followed a linear behavior and the volume compressibility, defined by $\kappa = -(1/V)dV/dP$, where V and P are the sample volume and the applied pressure, respectively, was obtained to be 0.024 GPa^{-1}.

The inter-tubular separation between the neighboring nanotubes in the raft-like bundles is obtained from energetics calculations [10]. It was also found that the equilibrium value of the inter-tubular gap is smaller than the graphite (002) spacing by 0.25 Å, though they reduce at almost the same rate under increasing pressure [10]. These results explain successfully the pressure dependence of the lattice constant as shown in Fig. 1.

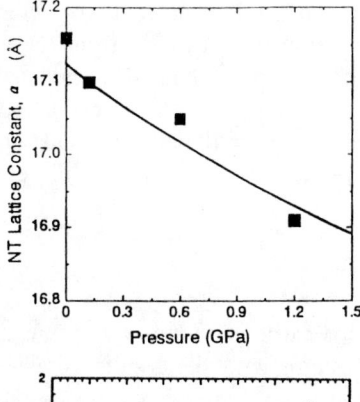

Figure 1 Linear dependence of the nanotube lattice constant on applied hydrostatic pressure. The volume compressibility of 0.024 GPa^{-1} was obtained for nanotubes of 14 Å diameter. The solid line is the calculated lattice constant as a function of pressure.

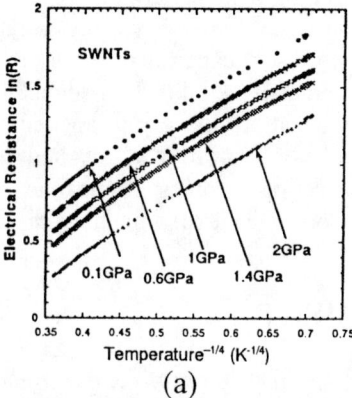
(a)

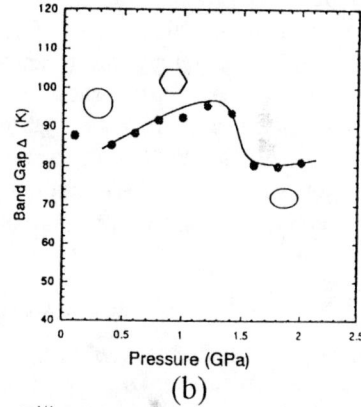
(b)

Figure 2 (a) Electrical resistance vs. temperature (plotted in $T^{-1/4}$) of single-walled carbon nanotube material under various pressures. A semiconducting behavior was observed as shown in the curves. (b) Pressure dependence of the band gap obtained from the electrical resistance measured at various temperatures above 200 K. Inserted shapes illustrate the morphological evolution of the cross-section of nanotubes with increasing pressure.

Figure 2(a) shows a set of experimental data on the electrical resistance of the single-walled carbon nanotube material, measured under various pressures up to 2 GPa, plotted as a function of $T^{-1/4}$. The relationship between the electrical resistivity and temperature of the nanotube material indicates that, under all the pressures employed in the present experiment, the nanotube material showed a semiconducting behavior [11]. The band gap deduced from the experimental data is given in Fig. 2(b). When the pressure is below 1.5 GPa, the band gap exhibited a monotonic increase with increasing pressure. This phenomenon is attributed to the fact that, as the pressure increased, the band gap of the semiconducting nanotubes increases as a result of polygonization. However, at 1.5 GPa pressure, the band gap showed a sudden drop in value and this phenomenon suggests that a structural phase transition should have occurred, in which the polygonized nanotubes become elliptical in cross-section. This feature is consistent with a similar model proposed basing on Raman spectroscopic data [12]. The morphological evolution of the nanotubes with increasing pressure is schematically illustrated with the inserted geometric shapes, from circular to hexagonal to elliptical, in Fig. 2(b). On the other hand, it should be noted that the ellipticity is still very small at this pressure range to be sufficient to accommodate the resulted volume reduction and that, as has been discussed theoretically [13], the band gap of the semiconducting nanotubes would increase only slightly as the ellipticity increased further.

CONCLUSIONS

The volume compressibility of single-walled carbon nanotubes of 14 Å diameter has been measured to be 0.024 GPa^{-1}. Combining with the results obtained from numerical computations, it was found that the nanotubes must have been polygonized under the hydrostatic pressure employed in the present experiment. Measurement of the transport properties showed that the nanotubes exhibited a semiconducting behavior with a narrow band gap. The polygonization due to the applied external pressure enlarged the band gap and a phase transition from trigonal to a new structure of lower symmetry, where the polygonal cross-section becomes elliptical, is proposed to account for the abrupt drop of band gap observed experimentally at 1.5 GPa pressure.

ACKNOWLEDGMENT

The authors thank Dr. T. Kikegawa for his assistance in the X-ray experiment.

REFERENCES

1. S. Iijima, *Nature* (London) **354**, 56-58 (1991); S. Iijima and T. Ichihashi, *ibid* **363**, 603-605 (1993).
2. M.M.J. Treacy, T.W. Ebbesen, and J.M. Gibson, *Nature* (London) **381**, 678-680 (1996).
3. R.S. Ruoff, J. Tersoff, D.C. Lorents, S. Subramoney, and B. Chan, *Nature* (London) **364**, 514-516 (1993).
4. N.G. Chopra, L.X. Benedict, V.H. Crespi, M.L. Cohen, S.G. Louie, and A. Zettl, *Nature* (London) **377**, 135-138 (1995).
5. S. Iijima, C. Brabec, A. Maiti, and J. Bernholc, *J. Chem. Phys.* **104**, 2089-2092 (1996).
6. V. Lordi and N. Yao, *J. Chem. Phys.* **109**, 2509-2512 (1998).
7. O.L. Blaksee et al., *J. Appl. Phys.* **41**, 3373-3382 (1970); E.J. Seldin and C.W. Nezbeda, *J. Appl. Phys.* **41**, 3389-3400 (1970).
8. J.-C. Charlier, Ph. Lambin, and T.W. Ebbesen, *Phys. Rev. B* **54**, R8377-8380 (1996).
9. L.-C. Qin and S. Iijima, *Chem. Phys. Lett.* **269**, 65-71 (1997); M. Yudasaka, R. Yamada, N. Sensui, T. Ichihashi, and S. Iijima, *J. Phys. Chem. B* **103**, 6224-6229 (1999).
10. J. Tang, L.-C. Qin, T. Sasaki, M. Yudasaka, A. Matsushita, and S. Iijima, *Phys. Rev. Lett.* **85**, 1887-1889 (2000).
11. J. Tang, L.-C. Qin, T. Sasaki, M. Yudasaka, A. Matsushita, and S. Iijima, *Synthetic Metals* (2001) in press.
12. M. Peters, L.E. McNeil, J.P. Lu, and D. Kahn, *Phys. Rev. B* **61**, 5939-5944 (2000).
13. C.-J. Park, Y.-H. Kim, and K.J. Chang, *Phys. Rev. B* **60**, 10656-10659 (1999).

Preparation of Mechanically Aligned Carbon Nanotube Films and Their Anisotropic Transport Phenomena

Dong Jae Bae[1], Keun Soo Kim[2], Young Soo Park[3], Kay Hyeok An[4], Jeong-Mi Moon[4], Seong Chu Lim[1], Young Hee Lee[4]

Semiconductor Physics Research Center[1], Department of Physics[2], and Department of Semiconductor Science and Technology[3], Jeonbuk National University, Jeonju 561-756, (Republic of Korea)
Department of Physics[4], Sungkyunkwan University, Suwon 440-746, (Republic of Korea)

Abstract. Thin films of aligned carbon nanotubes (CNTs) were prepared by a simple mechanical rubbing from single-walled carbon nanotube (SWNT) slurry, which was synthesized by the catalytic arc discharge. The measured electrical resistivity shows high anisotropy ρ_N/ρ_P ranging from 5 to 15. The annealed samples show a monotonic decrease in the resistivity with increasing temperature. CNTs in the mat act as strong Luttinger liquids with g values ranging from 0.18 to 0.26, similar to an isolated nanotube. We propose that the transport is dominantly governed by the formation of metal-metal crossed junctions of nanotubes in the mat.

INTRODUCTION

Transport phenomena in the CNTs' mat exhibit neither normal semiconducting nor metallic behaviors [1]. Resistivity of the mat shows typically an upturn near the room temperature from the pristine samples and this upturn disappears after degassing by annealing [2,3]. The upturn at high temperature (>400 K) remains unchanged in some cases even after degassing, indicating the evidence of metallic contribution [4]. No analysis for 1D liquid behavior has yet been reported from the CNT films. Recently magnetically aligned CNT film in a high magnetic field (25 T) has been prepared, showing the resistivity anisotropy ρ_N / ρ_P of about six [2,3]. However, the approach with high magnetic field is not easily accessible and thus some other practical approaches for aligned CNT films are always demanding.

We first provide a simple way of preparing the aligned CNT thin films by a mechanical rubbing. The anisotropy of the aligned CNT films ranges from 5 to 15, depending on the film thickness and the pristine CNT conditions. From nonlinear temperature dependence of the resistivity, we supposed that the CNTs in the mat are composed of metals revealing Luttinger liquids behavior [5] and normal metals with Fermi liquids particularly at higher temperature.

EXPERIMENTAL

SWNTs were prepared using catalytic arc discharge at low pressure [6]. The pristine samples collected from the collar part contained mostly carbon nanotubes with 20wt% of transition metals. In order to dissolve the transition metals, this sample was further refluxed for 6 h in 2.8 M HNO_3 acid. Sonication in de-ionized (DI) water for 10 min and filtering were repeated until the pH of the suspension was equal to that of the DI water. The remaining transition metals were less than 1 wt%. The slurry left on the top of the filter was coated on glass and mechanically squeezed by a bar coater with a pattern width of 50 μm in order to align CNTs of thin film to a preferred plane direction. The samples were dried in a vacuum oven at 80 °C for a day. The thickness of the aligned CNT films was about 7 μm. The resistivity was measured by using the four-probe method in a vacuum chamber at a pressure of 6×10^{-6} torr, where the contact was formed by Ag epoxy.

RESULTS AND DISCUSSION

Figure 1 shows ρ_p of the pristine and degassed samples, where the degassed sample was first annealed in air at 350 °C for an hour [7] and degassed in vacuum at 420 K for 12 h. The pristine sample shows a resistivity minimum near the room temperature, while nonmetallic behavior is shown at low temperature. However, the degassed sample shows a monotonic decrease with increasing temperature and thus nonmetallic behavior over the whole temperature range [3].

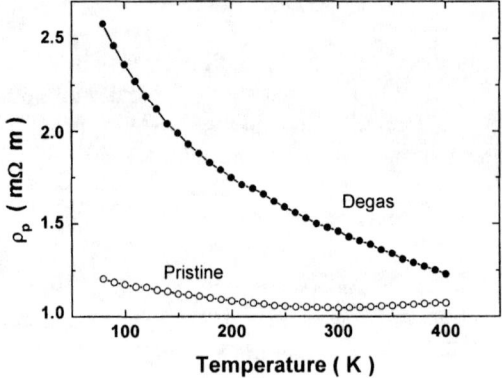

FIGURE 1. Resistivity ρ_p vs. temperature T. Resistivities in a pristine (circles) and degassed SWNTs film (dots), were measured in the parallel direction to the tube axis.

Figure 2(a) clearly shows the anisotropic resistivities in the aligned mat sample after degassing. The anisotropy is almost constant at six over the entire temperature range. The inset shows a degree of alignment of the pristine mat surface of mechanically rubbed sample. Although the resistivities of both parallel and perpendicular directions show different values and are expected to have different temperature dependence, the scaled resistivities of both directions are essentially identical to each other, as shown in Fig. 2(b). This strongly suggests that the transport mechanism of both directions is equivalent, independent of the aligned directions.

We now address transport mechanism of the aligned CNT films. The distance

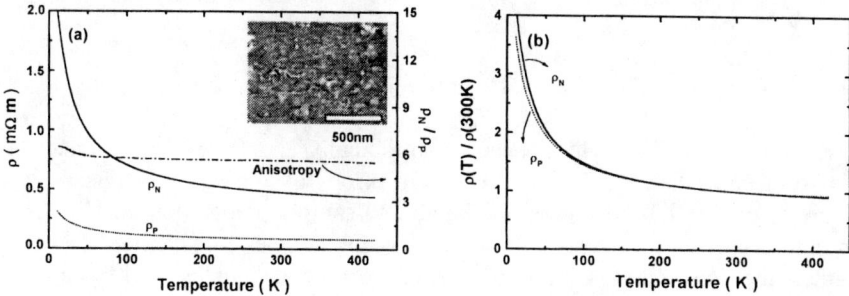

FIGURE 2. (a) ρ_p vs. T (left axis) of both parallel (dot) and perpendicular (solid) directions to the tube axis, and anisotropy (dash-dot, right axis). (b) Scaled resistivity data $\rho(T)/\rho(300K)$ of both directions.

between the measuring electrodes is far away for Luttinger liquids and screening may be neglected such that most of the CNTs cannot feel the contact. However, effective contacts between CNTs are formed randomly through the CNT-CNT crossed junctions. Since the tube-tube interactions that deviate from one-dimensional system and represent additional three-dimensional effects, may play as non-interacting Fermi liquids, we separate the linear term from Luttinger liquids. In this approach, the resistivity is expressed by adding of both linear and non-linear power laws, $\rho(T) = a\,T^{-\alpha} + b\,T$, where $\alpha \neq -1$. This prediction gives an excellent agreement with the measured value, as shown in Fig. 3. The power index $\alpha = 0.446$ gives the Luttinger parameter, g=0.186 with bulk contact [8], which indicates stronger Luttinger liquids. The inclusion of the linear term, although the coefficient b ($=4.60\times10^{-4}$) is smaller than the coefficient a ($=11.0$), particularly gives better fitting at higher temperatures.

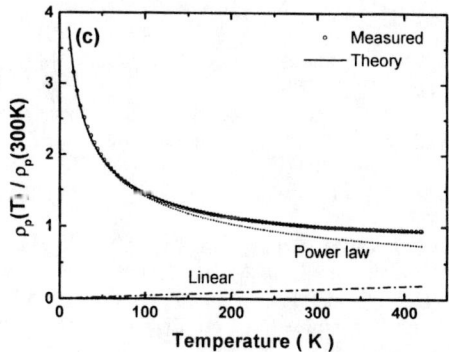

FIGURE 3. $\rho_P(T)/\rho_P(300K)$ (circle) is well reproduced by our trial equation (solid), $\rho = a\,T^{-\alpha} + b\,T$. The contribution from the power law term of T (dot) and the linear term (dash-dot) respectively.

This further implies an upturn at high temperature, where our fitting provides an upturn at 600 K, which is beyond our temperature range of the measurements and not shown in Fig. 3. Although the transport in the mat of the thin films is governed by the Luttinger liquids, the effect from the non-interacting Fermi liquids cannot be excluded particularly at temperature >100 K.

SUMMARY

In summary, the SWNTs have been anisotropically aligned by simple mechanical rubbing on film in the form of a mat. High anisotropy of 5 to 15 in the resistivity was obtained depending on the sample thickness and sample preparation conditions. The monotonic decrease of the resistivity with increasing temperature cannot be explained by either the long-range Coulomb interactions or the inclusion of short-range interactions, which is different from an isolated single walled carbon nanotube. The measured data were well fitted by including a small contribution of the linear power term. The mat of the CNT film with a mean field approximation can be modeled as a network of the effective resistors, in which each resistor is represented by a series resistance of Luttinger liquids with $g=0.186$ and Fermi liquids.

ACKNOWLEDGMENTS

This project was supported by the Ministry of Science and Technology through National Research Laboratory program and in part by Brain Korea 21 program. One of us acknowledges the support through a postdoctoral fellowship program at Sungkyunkwan University.

REFERENCES

1. J. E. Fischer, H. Dai, A. Thess, R. Lee, N. M. Hanjani, D. L. Dehaas, and R. E. Smalley, Phys. Rev. B **55**, R4921 (1997).
2. B. W. Smith, Z. Benes, D. E. Luzzi, J. E. Fischer, D. A. Walters, M. J. Casavant, J. Schmidt, and R. E. Smalley, Appl. Phys. Lett. **77**, 663 (2000).
3. J. Hone, M. C. Llaguno, N. M. Nemes, A. T. Johnson, J. E. Fischer, D. A. Walters, M. J. Casavant, J. Schmidt, and R. E. Smalley, Appl. Phys. Lett. **77**, 666 (2000).
4. G. U. Sumanasekera, C. K. W. Adu, S. Fang, and P. C. Eklund, Phys. Rev. Lett. **85**, 1096 (2000).
5. M. Bockrath, D. H. Cobden, J. Lu, A. G. Rinzler, R. E. Smalley, L. Balents, and P. L. McEuen, Nature **397**, 598 (1999).
6. Y. S. Park, K. S. Kim, H. J. Jeong, W. S. Kim, J. M. Moon, K. H. An, D. J. Bae, Y. S. Lee, G.S.Park, and Y. H. Lee, submitted to Synth. Metals, 2000.
7. X. Y. Zhu, S. M. Lee, Y. H. Lee, and T. Frauenheim, Phys. Rev. Lett. **85**, 2757 (2000).
8. Since the CNT mat are mostly composed of CNT-CNT junctions (few end contacts formed by the electrodes and CNTs), this assumption is valid.

Polarization Characteristics of Zeolite Single Crystals containing Carbon Nanotubes

N. Nagasawa, I. Kudryashov*, S. Tsuda, and Z. K. Tang[†]

Department of Physics, University of Tokyo, Hongo, Tokyo, Japan[1]
* Tokyo Instruments Inc., Nishikasai, Tokyo, Japan
[†] *Department of Physics, Hong Kong University of Science and Technology, Kowloon, Hong Kong*

Abstract. Optical polarization anisotropy of single crystals of $AlPO_4$-5 (AFI) containing carbon nanotubes is studied to develop the optical characterization method of the samples by laser micro-polarimetry at 488nm. The crystals are opaque in the configuration, $K \perp C$ with $E \| C$, but are almost transparent in the configuration, $K \perp C$ with $E \perp C$, where K and E are the wave vector and the electric field of the incident light, respectively. The degree of polarization reaches almost unity at the maximum. When $K \| C$, they are also transparent for any directions of E except for some locations including the surface. The strong luminescence bands are observed at 540nm and 570nm with the tail toward long wavelength region. The 3D imaging of the luminescence is performed by *Nanofinder* to demonstrate their extrinsic origin.

INTRODUCTION

The single-walled carbon nanotube (CN) manifests unique electronic structures depending on its geometrical structure. The electronic band structure depends on the chiral vector and the diameter of the tube. The interband transition is expected to show strong polarization anisotropy but its measurement has been difficult because of the non-aligned nature of CNs until recently. Tang et al. have succeeded in forming mono-sized, aligned single-walled CNs in channels of AFI crystals [1] and demonstrated strong polarization anisotropy in optical absorption spectra [2]. According to the direct observation of the CNs dissolved from the AFI framework, the diameter was estimated to be about 0.4nm [3]. Taking the inner diameter of a AFI channel of about 0.7nm into account, they considered that the relevant types of the CN are expected to be the zigzag (5,0) (diameter, d=0.39nm), the armchair (3,3) (d=0.41nm) and the chiral (4,2) (d=0.41nm), where (n,m) specifies the

---

[1]) N.N thanks to Mr.S.Suruga (The President of Tokyo Instruments Inc.) for his kind support to this collaboration research. He also thanks to Prof.s R. Saito and H. Suematsu for fruitful discussions. This work is partially supported by The Mitsubishi Foundation for Scientific Researches.

chiral vector. These diameters are similar to that of the innermost nanotubes in multi-walled CNs [4].

In view of optical spectroscopy, the remarkable advantage of the present CNs is not only in the aligned and mono-sized nature but also in their small diameter. Due to the strong confinement of the valence electrons, the band gap energy shifts to the visible spectral range. This situation makes the optical measurements easy and, therefore, may give us an opportunity to find new secondary optical processes associated with the relevant optical transitions. In addition, the equi-spaced and regularly aligned structure of CNs embedded in an optically transparent AFI crystal arouses our interests in their application to the micro-optical devices. However, it has been difficult to form the CNs in the AFI crystals homogeneously so far. Since the samples are prepared by pyrolysing tripropylamine (TPA) molecules in AFI, the CNs are not necessarily grown in all the channels in AFI crystals depending on the thermal treatments. This means that some intermediates toward the CNs remain in the samples as imperfections. On the other hand, most samples usually have structural defects of AFI. These situations prevent precise optical measurements. The aim of the present work is to apply micro-polarization spectroscopy to the characterization of the samples.

EXPERIMENTS AND DISCUSSIONS

Fig.1a and b show the transmission images of two samples put on a slide glass in the air at 488nm. One division of the scale corresponds to 100μm. The white lines at the top are eye-guides indicating the cracked part of the crystal. The configurations of **K** and **E** against the c-axis (**C**) of the crystal are shown in the figures. The angle of **E** against the c-axis, θ was adjusted by turning a linear polarizer. Circularly polarized light is introduced to the polarizer to make the spatial distribution of the incident light as constant as possible. Fig.1c shows the

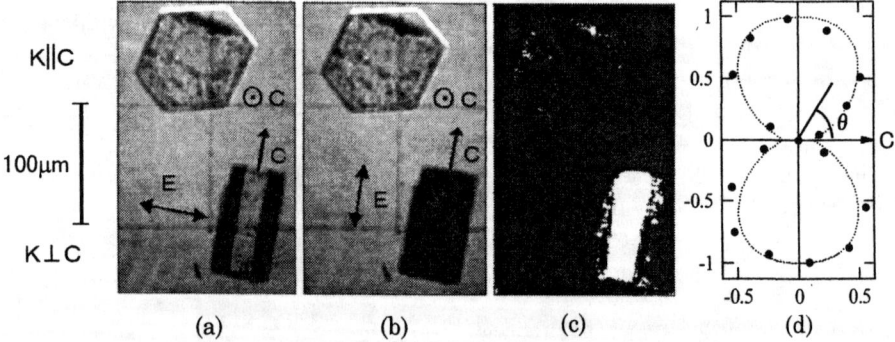

FIGURE 1. Polarization Characteristics of samples.

spatial distribution of the degree of polarization calculated from a and b. From this figure, it is clear that the polarization anisotropy is dominantly observed in the configuration, $\mathbf{K}\perp\mathbf{C}$ with $\mathbf{E}\|\mathbf{C}$ as shown by the white regions in the figure. This fact is consistent with the theoretically predicted polarization characteristics of the optical response [5]. The slight anisotropy, however, is also observed in the configuration of $\mathbf{K}\|\mathbf{C}$. The spatial distribution is not homogeneous in the sample.

Fig.1d shows polar plot of the transmitted light intensity in a region of $\sim 3 \times 10^{-8} cm^2$ as a function of θ at the maximum degree of polarization. The θ-dependence was reproduced after calculating the intensity of the transmitted light by using relevant Jones Matrixes with the optical density of ~ 2 at $\theta = 0$. The $\cos\theta$-dependent dipole matrix element was taken into account. The corresponding absorption coefficient is estimated to be $1 \times 10^3 cm^{-1}$.

Referring to the band calculation in the LDA [6] and the symmetry consideration of dipole transition matrix elements, the optical transition that gives the present optical anisotropy is not specified.

As a by-product of the polarization measurements, we observed remarkable emission peaked at 540nm and 570nm under Ar^+ laser light excitation in many samples. The emission spectra showed slight θ-dependence but had no clear correlation with the optical anisotropy mentioned above. We compared the spectra with the solar spectrum observed by the same system under the similar conditions referring to the situation as has been studied in the fullerene system [7]. The present spectral distribution was well narrower than the solar spectrum of similar peak wavelength. The emission was very stable for long-term laser excitation in the air. These facts suggest that the emission is not due to the blackbody radiation from the heated regions by the laser irradiation. Fig.2a shows an example of the 3D mapping of the

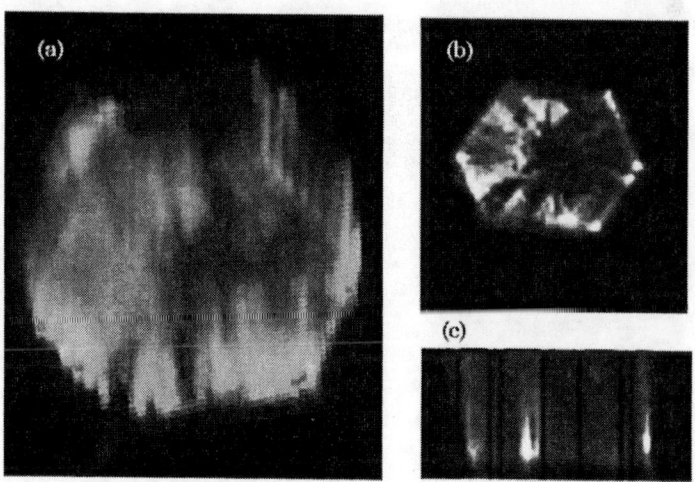

FIGURE 2. Luminescence diagnostics by *Nanofinder*.

emission at ~540nm by using a confocal micro-spectro-imaging system, *Nanofinder*, developed by *Tokyo Instruments Inc.*. The images of the horizontal and vertical cross sections of the distribution are also given in Fig.2b and c, respectively. The bright sections of b and c are computer-colored so as to reproduce the real image. The spatial resolution of the present image was $\delta x, \delta y \sim 1\mu m$, and δz (along the c-axis) $\sim 5\mu m$, respectively. The laser light was irradiated along the c-axis of the AFI crystal through the hexagonal plane by a microscope objective lens. The emission was collected by the same objective lens through the same plane. The 2D-intensity distribution of the emission at respective z was selectively accumulated to form the 3D-images. The decrease of the intensity along the c-axis is due to the extinction of the incident laser light and the re-absorption of the emission. The spatial distribution has some correlations with that of the extrinsic polarization anisotropy observed in the surface region of the samples. From these data, it is concluded that the present emission are not due to the nanotubes but to some carbon materials localized in the region where the crystal symmetry is broken.

To conclude, it was found that the 3D polarization and luminescence diagnostics of the samples are promising to monitor the transition processes of the formation of CNs in AFI crystals and to characterize the samples.

REFERENCES

1. Z. K. Tang, *Appl. Phys. Lett.* **73**, 2287-2289 (1998).
2. G. Li, Z. K. Tang, N. Wang, J. Wang, and J. Chen, in *Proceedings of ICPS, 2000* (Osaka), to be published.
3. N. Wang, Z. K. Tang, G. D. Li, and J. S. Chen, *Nature* **408**, 50-51 (2000).
4. L. C. Qin, X. Zhao, K. Hirahara, Y. Miyamoto, Y. Ando, and S. Iijima, *Nature* **408**, 50 (2000).
5. H. Ajiki and T. Ando, *Jpn. J. Appl. Phys.* **34**, 107-109 (1995).
6. Z. K. Tang, *Private Commun.*
7. C. Wen, S. Ohnishi, and N. Minami, *J. Phys. Chem.* **B102**, 2333-2338 (1998).

Electrochemical Hydrogen Storage in Singlewalled Carbon Nanotubes

Won Seok Kim, Kay Hyeok An, Young Soo Park, and Young Hee Lee

Department of Physics, Sungkyunkwan University, Suwon 440-746, Korea

Abstract. The singlewalled carbon nanotubes with a different amount of transition metals of 20 wt% and 5 wt% are synthesized by arc discharge. These samples are pre-annealed at 300 °C in air for five hours for the purpose of removing amorphous carbons, which exist in the samples, and again heated at 900 °C as many hours in Ar gas in order to improve their crystallinities. The discharge storage capacity of 210 mAh/g (0.74 Hwt%) is obtained at cutoff voltage of 0.8 V. We observed that the plateau region in the voltage scan is maintained during the constant-current discharge mode. It is very important for practical applications. The CV measurement also shows redox reaction peak near 1 V during discharge, in good agreement with the charge-discharge experiment.

INTRODUCTION

A new storage material with high capacity, light mass, and long cycle life has been strongly demanded for portable electronics, fuel cells and so forth. Large empty space inside the single-walled carbon nanotubes (SWCNTs) and large surface area shine a possibility for hydrogen storage applications. Until now, it has been reported that hydrogen can be stored in SWCNTs from high-pressure approach [1-4] that has limitation in battery application. However, electrochemical approach has structural advantage for the application [5-8].

In this report, we have investigated key factors for electrochemical hydrogen storage in SWCNTs to determine storage capability by controlling the amount of transition metals that are catalyst for SWNTs production. The influence of morphology and crystallinity of SWCNT-electrodes on storage capacity are also studied.

Results and Discussion

SWCNTs are synthesized by DC arc-discharge under a helium pressure of 100 torr. Inside the arc-charge system, a graphite rod (diameter: 6 mm), an anode, with a concentric hole (diameter: 4mm) is filled with graphite powder housing 20 or 5 wt% of Ni, Co, and FeS (1:1:1) respectively. The graphite rod is discharged at 22 V and 55 A[9]. Another graphite rod is used as a counter-electrode. We limit the weight percent of the transition metals to 5 wt% (lower limit) and 20 wt% (upper limit) during the discharge in order to investigate the effect of transition metals on the hydrogen storage capacity. Although the yield might be lowered with large amount of transition metals, uniform distribution of the transition metals in the sample may enhance the conductivity of the electrode and eventually increase the storage capacity. The samples here are collected from two different parts. One is web-like SWCNTs (called as web from now on) obtained on the wall of the chamber, and the other is taken from the cathod part (called as collar). The CNT bundles are shown in Fig. 1(a) with a diameter of a few tenths nm and with lengths of a few hundreds μm, containing the individual CNTs of a diameter of 1.4 nm. The web shows better quality of SWCNTs than the collar. However, we chose the collar for the electrochemical storage, since SWCNTs from this part usually contain more transition metals compared to the web.

It is noticeable that the collar with 5 wt % of transition metals in Fig. 1(b) has higher yield than that of 20 wt % in Fig. 1(c). This means that a larger amount of conducting materials (transition metals) are uniformly distributed in the latter.

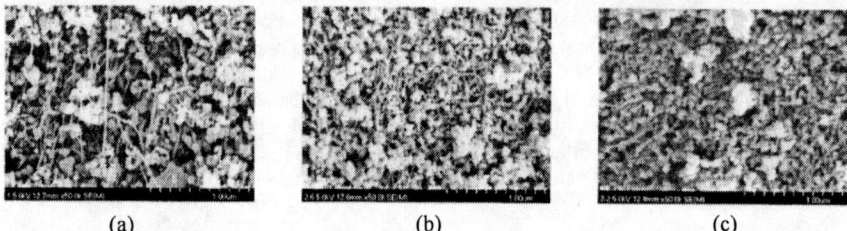

(a) (b) (c)

FIGURE 1. SEM images of as-prepared SWCNTs : (a) SWCNTs (web) with 5 wt% transition metals, (b) SWCNTs (collar) with 5 wt% transition metals, and (c) SWCNTs (collar) with 20 wt% transition metals.

To make the electrode in a form of pellet, two 5 wt% of web and collar and a 20 wt % of collar are selected first, respectively. A binder, PVA (poly vinyl alcohol), and SWCNTs in the ratio of 1:1 are mixed and casted in a pellet with a pressure of 1000 psi. The pellet is then dried at 100 °C for 12 hours to get rid of remaining moisture. The pellet is heat-treated at around 600 °C in Ar for an hour to evaporate the binder. As the last step for preparing the cathode of a cell, it is immersed in 6 M KOH for 12 hours. A unit cell for the electrochemical storage is fabricated with SWCNTs electrode as a cathode and nickel hydroxide as an anode which are in 6 M KOH aqueous solution as an electrolyte and separated by a thin polymer.

SWCNTs electrode prepared above, was tested electrochemically under static current flow of 50~100 mA/g and the cut-off voltage of 0.4 V. It is confirmed that the pellet with a large amount of transition metals, 20 wt% of collar shown in Fig. 1(c), can reserve most hydrogen up to around 140 mAh/g in Fig. 2(c), where the mass introduced here implies the total weight from SWCNTs, amorphous carbons and transition metals. The maximum capacity down at zero voltage reaches 450 mAh/g. It is expected that the maximum capacity would increase up to some levels in growing of the transition metals included the sample and saturate.

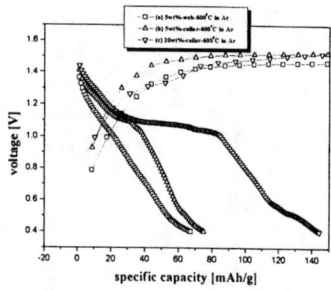

FIGURE 2. The specific Charge-Discharge tests of as-prepared sample of (a) SWCNTs (web) with 5 wt% transition metals, (b) SWCNTs (collar) with 5 wt% transition metals, and (c) SWCNTs (collar) with 20 wt% transition metal.

Existence of the amorphous carbon surrounding the transition metals is interfering the hydrogen storage because the exposed transition metals catalytically contributed to deposit the hydrogen in the above result. It is also proven that the amorphous carbons did not absorb hydrogen in previous literature [5]. We propose that the higher conductivity of the sample could flow electrons more efficiently through the circuit. It gives better possibility for electrons to react with hydrogen escaping from the anodic electrode. So, the collars containing 20 wt% transition metals is, first, pre-heated at 300 °C in open air for different time to remove amorphous carbons and then annealed at 900 °C in Ar for five hours for removing defects, which prevent SWCNTs from bending during the growth process, and improving the crystallinity of the sample resulting in higher conductivity of the electrode. It is revealed that the collar in Fig. 3(b) annealed at 300 °C in air for five hours and at 900 °C in Ar for five hours, has less defects than that of the law sample in Fig. 3(a). It is further demonstrated from Raman spectroscopies in Fig. 3(c).

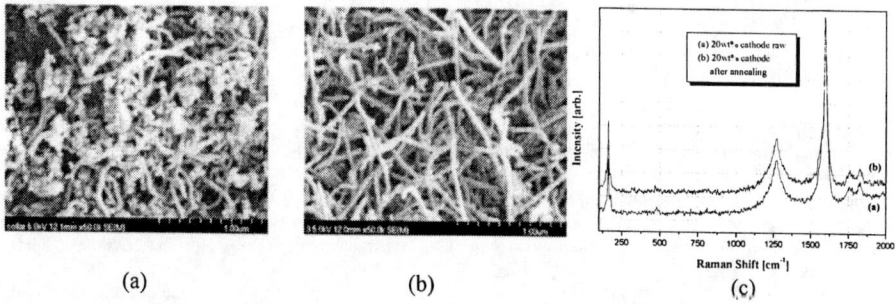

FIGURE 3. SEM figure of SWCNTs (collar) with 20 wt% transition metals of (a) the raw sample, (b) after pre-heat treatment at 300 °C in open air for five hours followed by heat treatment at 900 °C in Ar for 5 hours, and (c) RAMAN with respect to Fig. (a) and Fig. (b).

The longer period of pre-heat treatment in air leads to remarkable growing of specific capacity of hydrogen. The storage capacity of 210 mAh/g is achieved at the cutoff voltage of 0.8 V in Fig. 4(a). Cyclic voltammogram (CV) test in Fig. 4(b) confirms that the desorption and adsorption of hydrogen take place at 1.024 V and 1.346 V corresponding to two plateaus shown in Fig. 4(a).

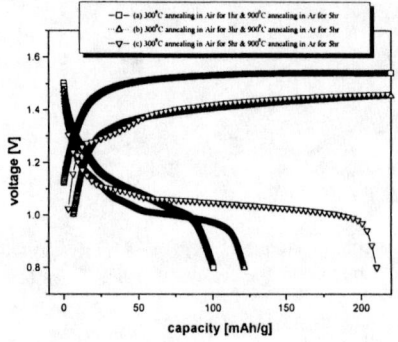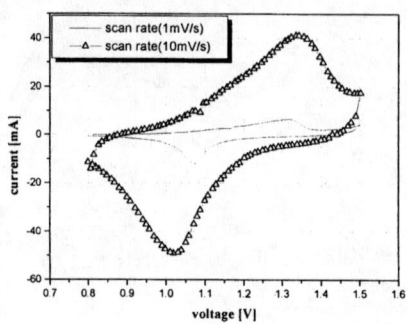

FIGURE 4. (a) Charge-Discharge tests of collar with 20 wt% transition metal with different pre-heat treatment time in air and heat treatment at 900 °C in Ar for five hours. (b) Cyclic voltammogram test of the sample pre-heat treatment at 300 °C in air for five hours followed by heat treatment at 900 °C in Ar for five hours.

ACKNOWLEDGMENTS

This project was supported by the MOST through NRL program and in part by BK21 program. One of us (KHA) acknowledges the support through a post doctoral fellowship program at Jeonbuk National University. The authors are grateful to Kwang-Ju branch of Korea Basic Science Institute for the assistance of FE-SEM.

REFERENCES

1. S. Iijima, Nature (London) **354**, 56 (1996).
2. A. C. Dillon, K. M. jones, T. A. Bekkedahl, C. H. Kiang, D. S. Bethuune, and M. J. Heben, Nature, **386**, 377 (1997)
3. C. Liu, Y. Y. Fan, M. Liu, H. T. Cong, H. M. Cheng, and M. S. Dresselhaus, Science, **286**, 1127 (1999)
4. Y. Ye, C.C. Ahn, C. Witham, B. Fultz, J. Liu, A.G. Rinzler, D.T. Colbert, K.A. Smith, R.E. Smalley, Appl. Phys. Lett. **74** (1999) 2307
5. C. Nutzenabel, A. Zuttel, D. Chartouni, L. Schlapbach, Electrochem. Solid State Lett. **2** (1999) 30.
6. Seung Mi Lee, Ki Soo Park, Young Chul Choi Young Soo Park, Jin Moon Bok, Dong Jae Bae, Kee Suk Nahm, Young Gak Choi, Soo Chang Yu, Nam-gyun Kim, Thomas Frauenheim, Young Hee Lee, Synthetic Metals, 7100 (1999)
7. X. Qin, X. P. Gao, H. Liu, H. T. Yuan, D. Y. Yan, W. L. Gong, and D. Y. Song, Electrochem Solid State Lett. **3** (2000) 532
8. N. Rajalakshmi, K.S. Dhathathreyan, A. Govindaraj, B.D. Satishkumar, Electrochimica Acta **45** (2000) 4511
9. Y. S. Park, Y. C. Choi, D. J. Bae, K. H. An, Y. H. Lee, unpublished

Saturation of Emission Current from Carbon Nanotube Field Emission Array

Seong Chu Lim[1], Hee Jin Jeong[1], Young Min Shin[3], Keun Soo Kim[2], Won Seok Kim[3], Young Soo Park[1], Young Chul Choi[4], Kay Hyeok An[3], Dong Jae Bae[1], Young Hee Lee[3*]

Semiconductor Physics Research Center[1] and Department of Physics[2], Jeonbuk National University, Jeonju 561-756, Republic of Korea

[3]*Department of Physics, Sungkyunkwan University, Suwon, 440-476, Republic of Korea*

[4]*Material Tech. Lab., CRD center, Samsung SDI, Suwon, 442-391, Republic of Korea*

Abstract. Field emission properties of carbon nanotube field emission arrays (CNT-FEA) that are exposed to various gases were studied. The existence of the hysteresis loop during the rise and fall bias sweep stems from gas adsorbates. We observed that gas adsorbates played an important role in the formation of current saturation at high field region Oxygen and nitrogen gases show different trends in the variation of the emission properties from hydrogen gases. The changes of the slopes, turn-on voltages, and current saturation are strongly correlated to the electronegativity of the individual species and the nature of adsorption. Oxygen gas dominates the filed emission properties upon adsorption and even degrades the surface morphologies with possible oxidative etching process, whereas hydrogen gases give the least effect to the field emission properties and additionally enhance the emission current by cleaning the surface of the carbon nanotube array.

INTRODUCTION

Field emission that is normally carried out under a pressure of around 10^{-6} torr or below ionizes the residual gases. The ionized residual gases inside an electric field are accelerated toward the cathode and impinge on the surface of cathode. Ion bombardment on cathode causes a modification of tip morphologies and induces an adsorbate-enhanced tunneling state. These electronic and geometric variations affect the field emission properties of a CNT-FEA and these changes in emission properties are general reflected on Fowler-Nordheim (F-N) plots. The F-N plot of CNT usually shows two characteristic regions: one is where current-voltage (I-V) curve fits well F-N equation, the other is where I-V curve deviates from the linear F-N plot due to the current suppression [1-3]. However, the origin of the current saturation from the CNTs at high field region has long been argued. Therefore, in order to achieve the emission stability and longevity of field emission array, understanding a current saturation mechanism is both technologically and fundamentally an important issue.

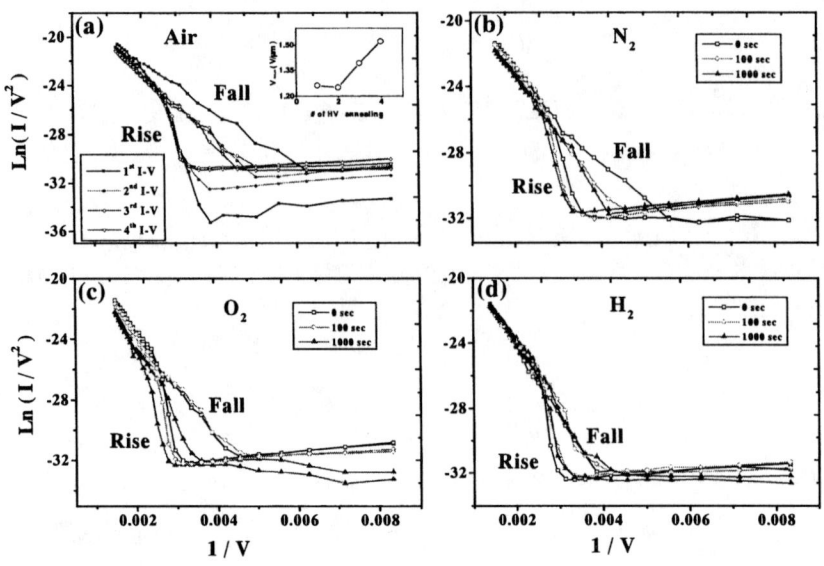

FIGURE 1. (a) Characteristic F-N curve of the air-exposed CNT-FEA measured during rise and fall sweeps. This measurement was repeated for four times. The similar I-V curves for oxygen, nitrogen, and hydrogen, are showed in (b)-(d), respectively for different gas-exposure time.

Hysteresis in Fowler-Nordheim plots and Gas Exposure

Vertically aligned CNTs were grown using thermal chemical vapor deposition (CVD) and the emission of a CNT-FEA was characterized inside a vacuum system. The emission currents are measured with increasing applied voltage, called "rise", and then with decreasing voltage, called "fall." The F-N plot obtained right after loading the FEA into our vacuum system marked by the empty squares in Fig. 1(a) shows serous fluctuations, i.e., the emission currents after a complete cycle of I-V curve are different from the starting values. However, the fluctuation in the emission currents is significantly reduced after third I-V annealing, as shown in Fig. 1(a). We note that the enhancement in field emission is attributed to the removal of extremely sharp CNT emitters by wearing out through resistive heating from high emission currents. This would improve the homogeneous distribution of emitters and result in change of turn-on voltage. The inset shows that the turn-on voltage increases with repeating of high voltage annealing.

The systematic high voltage annealing diminishes fluctuations. However, the hysteresis persistently exists and becomes more reproducible through the iterative annealings. It implies that the hysteresis is not influenced by the disappearance of sharp CNT tips. The hysteresis that is formed during the rise and fall bias sweep stems from change of the slopes. The first slope, called S_1, follows the linear F-N plot, whereas the second slope, called S_2, results from the suppression of the emission currents at high field.

The existence of the hysteresis simply negates the possibility of defect states to be the origin of the current saturation. The change of the slope at high field could be induced by the space charges screening the electric field near emitters [4]. However, this is not the case here since the emission currents consistently increase by two orders of magnitude even after the saturation and the onset voltage of the saturation is much lower than that of typical metal tips [5]. There is a possibility for the tip-tip interaction that may suppress the emission due to the screening effects [6]. Yet, typical CNT arrays do not have such high emitter density [7]. Furthermore, our samples did not show the transition metals on the tip surface [8], suggesting that the transition metals are not responsible for the saturation currents. This suggests that the current saturation is triggered by different mechanism. Since the CNT-FEA was exposed to air, gas adsorbates may be a good candidate for the origin of the current suppression. In order to investigate the effect of gas adsorbates on field emission properties of the CNT-FEA, we next performed the gas exposure experiments.

Air is composed of various gases such as nitrogen, oxygen, water vapor, hydrogen, and so on. Therefore, our observations are the integrated results from the above mentioned gases. In order to distinguish a different gas effect from air, we introduced a single gas species into a vacuum system at a time. Nitrogen, oxygen, and hydrogen gases are used for the experiments. After high voltage annealing, the gases were introduced into the chamber till the pressure increases to 2×10^{-5} torr. During gas exposures, high bias voltage of 670 V with currents of about 0.2-0.3 mA is applied for 10, 100, 500, and 1000 sec, respectively. After each exposure, the bias was turned off and the system was again pumped down to the base pressure. The I-V measurements were then done for every gas exposure times.

Discussion

We propose here that the fundamental mechanism responsible to this current saturation is the adsorption and desorption of gas adsorbates. At high electric field of ≥ 1.8 V/μm, large amount of emission current ignites desorption of adsorbates by the Joule heating at the apex of the emitter. In addition, high local electric field at the tip is expected to accelerate the reactive desorption of gas adsorbates. Figure 1(b-d) present the change of the F-N plots after three different gas exposures on the CNT-FEA. At short gas-exposure time, the emission behaviors are similar to those of air-exposed FEA for all cases, i.e., the current saturation occurs in rise sweep, whereas a single slope is observed in fall sweep. However, with large gas exposure times, oxygen and nitrogen gases will still remain at least on the tip of the CNTs in the linear F-N region, giving rise to again the current suppression that is indicated by solid triangles in Figs. 1(b,c). This effect is less significant in case of hydrogen gases, showing almost no appreciable changes with among different exposure times. The effect of gas adsorption and desorption is iterative due to the residual gases in the chamber.

Oxygen gases compared to the other gases seem to be more effective in modifying field emission properties, which is evidenced by the significant changes of slopes, turn-on voltages, and currents suppression. These behaviors are strongly correlated to the electronegativity of the materials, i.e., the electronegativity of oxygen and nitrogen atoms is stronger than that of carbon atom. It is expected that the adsorbed atoms with strong electronegativity present at the tip edge depress the electron emission, and the larger difference in electronegativity from the carbon atom changes more significantly the work function. However, in case of hydrogen gases, the emission currents do not vary appreciably with repetition of gas exposures, as shown in Fig. 1(d). This, together with the fact that the long gas-exposure time of hydrogen gases in Fig. 1(d) does not give much changes in the emission currents, indicates that hydrogen gases merely clean up the field emitters and stabilize all other physical parameters. From the cleaning, the turn-on voltage and the slope before the saturation become lower slightly after exposing to hydrogen.

ACKNOWLEDGMENTS

This work is supported by the Ministry of Science and Technology through National Research Laboratory program and in part by the Brain Korea 21 program.

REFERENCES

* To whom correspondence should be addressed: leeyoung@yurim.skku.ac.kr
1. Choi, W. B., Chung, D. S., Kang, J. H., Kim, H. Y., Jin, Y .W., Han, I. T., Lee, Y. H., Jung, J. E., Lee, N. S., Park, G. S., Kim, J. M., *Appl. Phys. Lett.* **75**, 3129 (1999).
2. Collins, P. G., Zettle, A., *Phys. Rev. B* **55**, 9391 (1997).
3. Lim, S. C., Choi, Y. C., Jeong, H. J., Shin, Y. M., Bae, D. J., Lee, Y. H., Lee, N. S., Kim, J. M., accepted to *Adv. Mater.* Nov (2000).
4. Schwoebel, P. R., Brodie, I., *J. Vac. Sci. Technol. B* **13**(4), 1391(1995).
5. Bonard, J. M., Salvetat, J. P., Stockli, T., de Heer, W. A., Forro, L., Chatelain, A., *Appl. Phys. Lett.* **73**, 918 (1998).
6. Collins, P. G., Zettle,A., *Phys. Rev. B*, **55**, 9391(1997).
7. Choi, W. B., Lee, Y. H., Lee, N. S., Kang, J. H., Park, S. H., Kim, H. Y., Chung, D. S., Lee, S. M.,Chung, S. Y., Kim, J. M., *Jpn. J. Appl. Phys.* **39**, 2560 (2000).
8. Choi, Y. C., Shin, Y. M., Lee, Y. H., Lee, B. S., Park, G. –S., Choi, W. B., Lee, N. S., Kim, J. M., *Appl. Phys. Lett.* **76**, 2367 (2000).

Characteristic Interactions of Gases in Solid Carbon Nanotubes

Chang-Wan Jin[1], Kenji Ichimura[1]*, Kenichi Imaeda[2] and Hiroo Inokuchi[3]

[1]Graduate School of Natural Science and Technology, Kumamoto University, Japan
[2]Department of Electrical and Electronic Engineering, Toyohashi University of Technology, Japan
[3]Institute for Molecular Science, Japan

Abstract. The hydrogen desorption peak for NT appears at 820K, which is lower than C_{60}. The desorption peak of nitrogen appears at around 500K for C_{60}, while NT does not have significant desorption above 300K. Carbon monoxide shows the several desorption processes both for C_{60} and NT below 400K. The weak desorption peaks appear at around 500K and 750K for C_{60}, and above 700K for NT. Van der Waals and chemical interactions depend on the carbon network and the lattice/site.

INTRODUCTION

In addition to the valence states or bonding states of rare gases such as He, Ne, Ar, Kr, and Xe in solids C_{60} and Na-C_{60}-H ternary systems under the conditions of ambient temperature and pressure, which is investigated by means of the mass-analyzed thermal desorption and X-ray photoelectron spectroscopy [1,2], we have reported the characteristic interactions of gases such as H_2, N_2 and CO in the solid endcaps-opened multi-wall carbon nanotube (NT) [3].

In this paper, the thermal desorption characteristics of diatomic molecule gases are investigated for the solids C_{60} and carbon nanotubes.

EXPERIMENTAL

C_{60} (Hoechst, 99.98% purity), endcaps and endcaps-opened multi wall carbon nanotubes (CMNT and OMNT, Bucky USA, 3-10 multi-layer with 2-10nm diameter and 3-30μm length) were used without further purification. Samples in the Ar-glove box are transferred in-situ to an ultra-high vacuum system. After vacuum heating at 653K or 1073K, samples were exposed to rare gases (Nippon Sanso, H_2, N_2 and CO, >99.9999% purity) of 1 to 1.4 atm, at 473K for 1 to 10 days. After the sample was cooled to liquid nitrogen temperature, the sample tube was evacuated to ultra-high vacuum. Desorbed gases were analyzed by using two mass-spectrometers when the sample was heated with the temperature-rise rate of 5 K/min.[1-3].

RESULTS AND DISCUSSION

Figure 1 shows the mass-analyzed thermal desorption spectra of diatomic molecule gases such as hydrogen, nitrogen and carbon monoxide from the solids C_{60}, endcaps-opened multi-wall carbon nanotube (OMNT) and endcaps multi-wall carbon nanotube (CMNT). The spectra are reproducible in the repeating use of the same batch C_{60}-samples and the same NT sample.

For C_{60} and carbon nanotubes, the desorption of hydrogen was observed below 300K. The temperature region of desorption below 300K suggests the interaction by van der Waals and/or weak chemical bonding. The further desorption peaks were observed at around 820K for C_{60} and 650K for carbon nanotubes. The temperature region of desorption above 300K suggests the interaction by strong chemical bonding. In the $KC_8H_{0.6}$ ternary system, the hydrogen desorption peak appears at 512K[2]. The desorption peaks of hydrogen in Na-H-C_{60} appear at around 650K and 900K, in which hydrogen species at around 650K has a strong correlation with super-conductivity. In these systems, hydrogen exists as H or H^{δ}. The hydrogen desorption peaks for carbon nanotubes are lower than C_{60}, indicating that the interaction of carbon nanotubes with hydrogen is weaker than that of C_{60}. The temperature region of desorption suggests that the charge transfer occurs hydrogen from C_{60} and carbon nanotubes.

For OMNT and CMNT, the absorbed amount of hydrogen for CMNT is larger than that for OMNT. As for the both, the basis structure is same and the difference of the both is only the presence of end caps. Therefore, this result that CMNT with end caps shows a larger amount of absorption indicates that sites which are composed of end caps are more active for the adsorption and absorption of hydrogen on and in the used OMNT and CMNT in the temperature region above 77K.

The desorption peak of nitrogen appears at around 500K for C_{60}, while NT does not have significant desorption above 300K. In thermal desorption spectra of nitrogen, the spectra in the temperature region from 600 to 1100K is the desorption of carbon monoxide. Although the

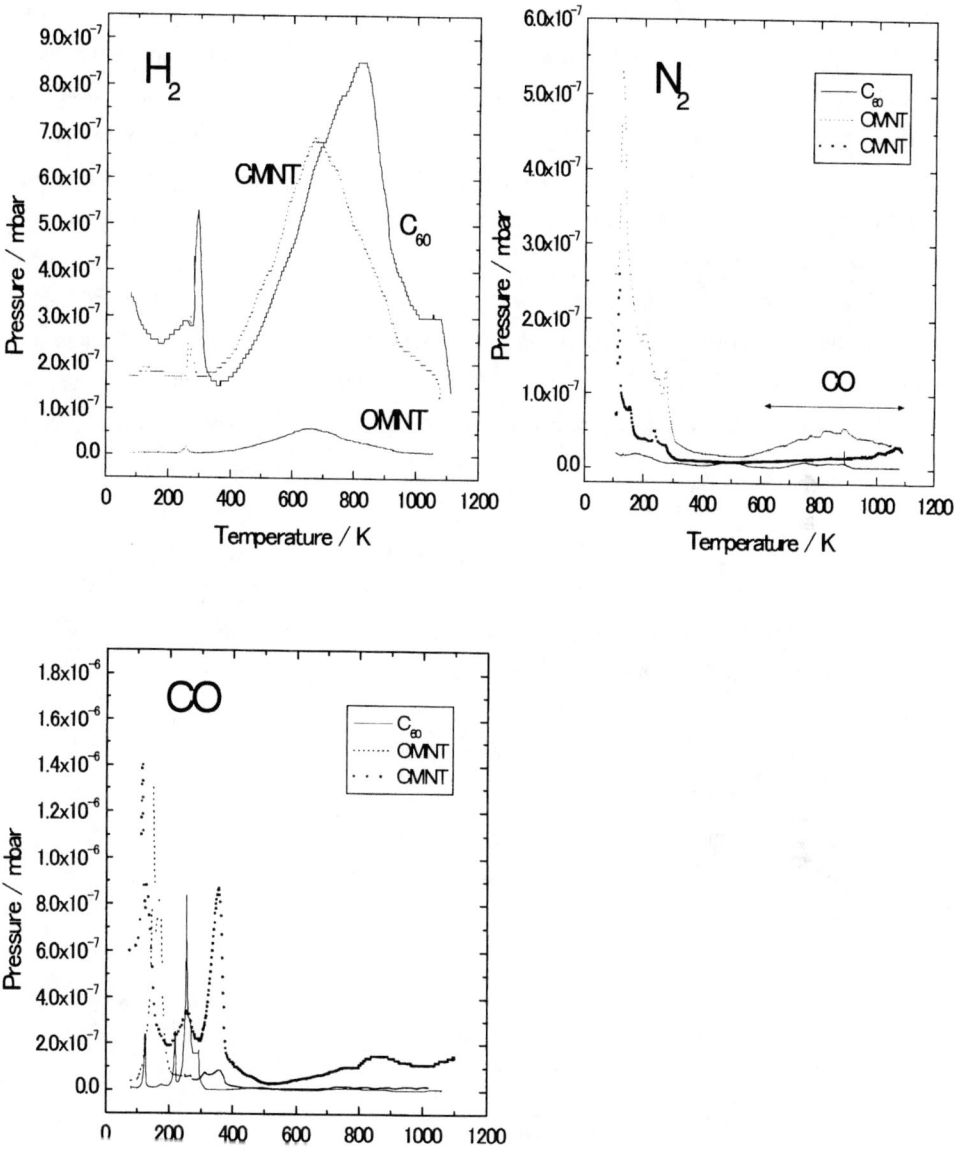

FIGURE 1. Mass-analyzed thermal desorption spectra of diatomic molecule gases such as hydrogen, nitrogen and carbon monoxide from solids C_{60}, opened multi-wall carbon nanotube (OMNT) and endcaps multi-wall carbon nanotube (CMNT).

content is small, the relatively strong interaction exists in the solid C_{60}. In the temperature region below 300K, C_{60} shows single desorption peak at around 190K, while carbon nanotubes show the several desorption peaks. Therefore, carbon nanotubes have sites by van der Waals and weak chemical interaction. For OMNT and CMNT, the absorbed amount of hydrogen for CMNT is larger than that for OCMT, while the relation is reverse for nitrogen. This result indicates that inner wall space is effective for the adsorption and absorption of nitrogen in the temperature region above 77K.

Carbon monoxide shows the several desorption processes both for C_{60} and carbon nanotubes, in which the peak profile is different from each other. The weak desorption peaks appear at around 500K and 750K for C_{60}, and above 700K for carbon nanotubes. In the case of carbon monoxide, van der Waals and chemical interactions depend on the carbon network and the lattice/site.

ACKNOWLEDGMENT

This work is partially supported by a Grant-in-Aid for Scientific Research from the Ministry of Education, Science, Sports, and Culture of Japan.

REFERENCES

1. Ichimura, K., Imaeda, K., and Inokuchi, H., *Chem. Lett.*, 196-197 (2000).
2. Ichimura, K., Imaeda, K., and Inokuchi, H., *Mol. Cryst. Liq. Cryst.*, **340**, 649-654 (2000).
3. Jin, C.-W., Ichimura, K., Imaeda, K., and Inokuchi, H., *Synthetic Metals*, in press.

Chemical Interaction of Rare Gases in Solid Carbon Nanotubes

Kenji Ichimura[1]*, Kenichi Imaeda[2] and Hiroo Inokuchi[3]

[1]Graduate School of Natural Science and Technology, Kumamoto University, Japan
[2]Department of Electrical and Electronic Engineering, Toyohashi University of Technology, Japan
[3]Institute for Molecular Science, Japan

Abstract. The mass-analyzed thermal desorption reveals that rare gases such as He, Ne and Ar in solids C_{60}, Na-C_{60}-H ternary systems and carbon nanotubes are desorbed above 400K and their desorption amounts indicate the production of non-stoichiometric compounds. X-ray photoelectron spectra in the C1s and valence band regions show different peak profiles, depending on the kinds of rare gases and materials. These results indicate that He, Ne and Ar atoms are in the bonding state. In endcaps and endcaps-opened multi-wall carbon nanotube systems, rare gases are also desorpbed above 400K and its desorption amounts in solids endcaps multi-wall carbon nanotube is larger than those in endcaps-opened multi-wall. This result indicates that sites that are composed of endcaps are more active for the chemical interaction of rare gases with the solid carbon nanotubes.

INTRODUCTION

We have reported the valence states or bonding states of rare gases such as He, Ne, and Ar in solids C_{60} and Na-C_{60}-H ternary systems under the conditions of ambient temperature and pressure [1,2]. The mass-analyzed thermal desorption reveals that the rare gases are desorbed above 400K and their desorption amounts indicate the production of non-stoichiometric compounds. X-ray photoelectron spectra in the C1s and valence band regions show different peak profiles, depending on the kinds of rare gases and materials. Ar2p and Ar3s X-ray photoelectron spectra show large chemical shifts. These results indicate that the rare gas atoms are in the bonding state under the conditions of ambient temperature and pressure.

In this paper, we report the thermal desorption characteristics of rare gases in the solids C_{60},

Na-H-C$_{60}$ ternary system and carbon nanotubes.

Experimental

Endcaps and endcaps-opened multi wall carbon nanotubes (CMNT and OMNT, Bucky USA, 3-10 multi-layer with 2-10nm diameter and 3-30μm length) were used without further purification. Samples transferred in-situ to an ultra-high vacuum system and heated at 653K or 1073K were exposed to rare gases (Nippon Sanso, He, Ne, Ar, Kr, Xe, >99.9999% purity) of 1 to 1.4 atm, at 473K for 1 to 10 days. In-situ measurements of the mass-analyzed thermal desorption with the temperature-rise rate of 5 K/min were carried out as the same manner as described in papers [1,2].

Results and Discussion

Figure 1 shows the mass-analyzed thermal desorption spectra of rare gases, He, Ne, Ar, Kr, and Xe from the solids C$_{60}$, Na-H-C$_{60}$ ternary system, endcaps and endcaps-opened multi-wall carbon nanotube systems. In all the samples studied, He, Ne and Ar are desorbed above 400K, although the desorption peak profiles strongly depend on the sorts of the gas and material. This indicates that the interaction is different among the solids C$_{60}$, Na-H-C$_{60}$ ternary system, endcaps and endcaps-opened multi-wall carbon nanotube systems. Although Teizer et al. examined the absorption/desorption of He below 30K and reported the desorption peaks of helium at around 9.9-23K from single wall carbon nanotube [3], we report characteristic desorption peaks of He, Ne and Ar above room temperature. On the other hand, the desorption peaks of krypton and xenon are observed below 300K, being considered to be due to the rate-determining step as migration. However, as shown in the spectrum of Kr from Na-C$_{60}$-H-K system, a weak desorption peak is observed at around 600K. As to carbon nanotubes, the desorption characteristics of xenon is similar to that reported by Kuznetsova et al. [4].

The desorption amounts in endcaps multi-wall carbon nanotube are larger than those in endcaps-opened multi-wall carbon nanotube. As for the both, the basis structures are same and the difference of the both is only the presence of endcaps. Therefore, this result indicates that sites that are composed of endcaps in the solid endcaps multi-wall carbon nanotube are more active for the adsorption and absorption of rare gases than the inner wall and terminal sites in the solid endcaps-opened multi-wall carbon nanotube in the temperature region above 77K. The sites that are composed of end caps show the chemical interaction with rare gases.

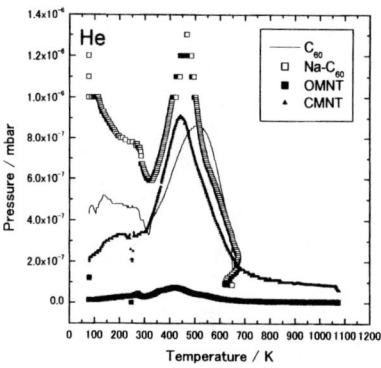
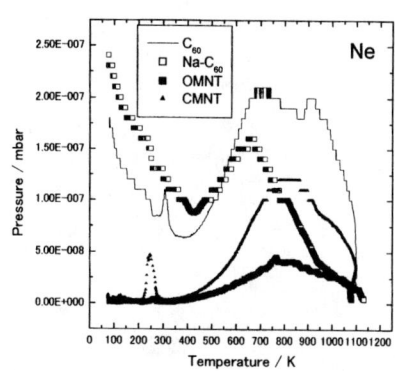
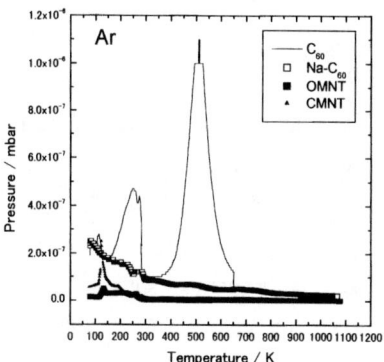
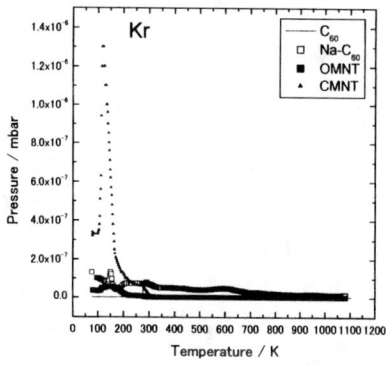
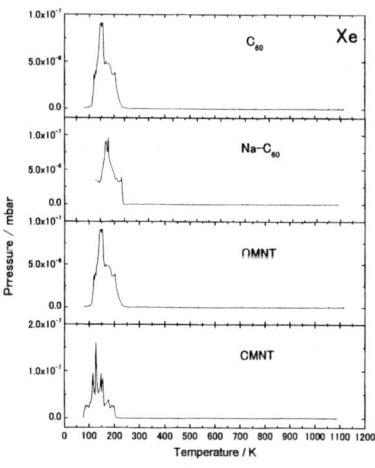

FIGURE 1. Mass-analyzed thermal desorption spectra of helium, neon, argon, kripton, and xenon from solids C_{60}, Na-C_{60}, opened multi-wall carbon nanotube (OMNT), and closed multi-wall carbon nanotube (CMNT).

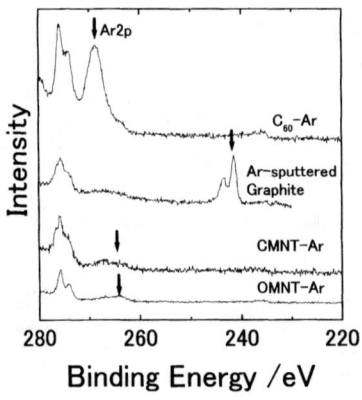

FIGURE 2. X-ray photoelectron spectra of Ar2p.

X-ray photoelectron spectra of C1s and valence band show the different peak profiles among the solids C_{60}, Na-C_{60}, opened multi-wall carbon nanotube (OMNT), and closed multi-wall carbon nanotube (CMNT) exposed to rare gases such as He, Ne and Ar. Figure 2 shows the x-ray photoelectron spectra of Ar2p. The Ar2p peak for C_{60}-Ar is located at ca. 26eV higher than that for Ar-sputtered graphite. The Ar2p peaks for carbon nanotubes show the different chemical shift. The changes in chemical shift and spin-orbit interaction splitting width indicate the changes in the valence state. The electronic states of rare gases have been calculated: the charge transfer is confirmed to occur for Ar, Kr and Xe – C_{60} systems.

ACKNOWLEDGMENT

This work is partially supported by a Grant-in-Aid for Scientific Research from the Ministry of Education, Science, Sports, and Culture of Japan.

REFERENCES

1. Ichimura, K., Imaeda, K., and Inokuchi, H., *Chem. Lett.*, 196-197 (2000).
2. Ichimura, K., Imaeda, K., and Inokuchi, H., *Mol. Cryst. Liq. Cryst.*, **340**, 649-654 (2000).
3. Teizer, W., Hallock, R.B., Dujardin, E., and Ebbesen, T.W., *Phys. Rev. Letters*, **84**, 1844-1845 (2000).
4. Kuznetsova, A., Yates, Jr., J.T., Liu, J., and Smalley, R.E., *J. Chem. Phys*, **112**, 9590-9598 (2000).

The Effect of Solvent on Electrical Transport Properties in Single-wall Carbon Nanotubes

Shin-ichi Masubuchi[*], Hisako Masubuchi[*], Shigeo Kazama[*], Hiromichi Kataura[†], Yutaka Maniwa[†], Shinzo Suzuki[¶] and Yohji Achiba[¶]

[*]*Department of Physics, Chuo University, Bynkyo-ku, Tokyo 112-8551, Japan*
[†]*Department of Physics, Tokyo Metropolitan University, Hachioji-shi, Tokyo 192-0397, Japan*
[¶]*Department of Chemistry, Tokyo Metropolitan University, Hachioji-shi, Tokyo 192-0397, Japan*

Abstract. We have measured the temperature dependence of thermoelectric power, $S(T)$, for Single-wall carbon nanotubes (SWNT) mat. The $S(T)$ for high-temperature heat-treated SWNT samples shows a strong curved T-dependence. The results indicate the interaction between conduction electrons and transition metals; i.e. Kondo effect. For dried SWNT which contained solvent such as ethanol or DMF, on the contrary, the $S(T)$ is almost proportional to T with intercept of nearly zero, indicating typical metallic behavior. The result indicates a metallic property in SWNT. We conclude that the Kondo effect is weakened when SWNT contain those solvents.

INTRODUCTION

Single-wall carbon nanotubes (SWNT) have attracted a considerable attention because of a theoretical prediction based on peculiar 1D structure that the electronic structure can be either metallic or semiconducting depending on its diameter and chirality[1]. Transport properties such as electrical resistivity and thermoelectric power in SWNT show different behaviors in temperature dependence depending on the catalyst used in synthesizing the SWNT[2,3]. The behaviors in transport properties indicate the interaction between conduction electrons and transition metals; i.e. Kondo effect. In this paper, we will report on an experimental result that the Kondo effect is weakened when SWNT sample contains remaining solvent.

EXPERIMENTAL

We used two types of SWNTs with differing synthesizing method in this work. The one was prepared by the arc discharge method using NiY as a catalyst. The other purchased from Rice University was made by the laser vaporization method using NiCo as a catalyst and was afterward chemically purified. To form an SWNT mat, the SWNTs were dispersed in ethanol (EtOH) or N,N-dimethylformamide (DMF) under sonication. The dispersed solution was filtered by suction. SWNT mat on a filter paper was peeled off and dried at 80°C for 12 hours. Further heat treatment was performed at 190°C or 350°C for 1 hour in air. Electrical resistivity and absolute thermoelectric power were measured between 4.2 and 300K. The detailed experimental procedure was previously reported[4].

RESULTS AND DISCUSSION

Electrical resistivity at 280K for the SWNT mat heat-treated at 350° is summarized in Table 1. The absolute value of resistivity for SWNT-NiY is more than 10 times larger than that for SWNT-NiCo. Since the SWNT-NiCo obtained from Rice University has been purified chemically, this difference in resistivity may depend on whether SWNT sample are chemically purified or not.

Figure 1 shows the temperature dependence of the normalized resistivity at 280K, $R(T)/R(280K)$, for SWNT with different catalyst. The $R(T)/R(280K)$ for both samples behaves as a semiconductor throughout the whole temperatures; i.e. the resistance increases with lowering temperature. We were not able to confirm the metallic temperature dependence at high temperatures (>200K) as reported by Fischer[5]. The resistance ratio between 4.2K and 280K, $R(4.2K)/R(280K)$, for SWNT-NiCo and SWNT-NiY is 15 and 48, respectively. In spite of the diverse resistivity value at 280K differing by an order of magnitude, the $R(4.2K)/R(280K)$ in SWNT-NiY is only three times larger than that in SWT-NiCo. This gives support to the idea that a factor influencing the temperature dependence of resistivity in these samples is independent of a kind of a metal catalyst used in synthesizing process and whether purification was made or not. The temperature dependence of the resistivity in our SWNTs is most likely dominated by the transport process between aggregated SWNTs (bundle or rope) through hopping or tunneling, resulting in a semiconducting dependence.

TABLE 1. Electrical Resistivity at 280K.

Sample	ρ (Ω·cm)
SWNT-NiY	0.16
SWNT-NiCo	0.021

Thermoelectric power, $S(T)$, can be measured without external current flow through a sample and the transport process between aggregated SWNTs gives much less contribution than in resistivity measurment. Thus $S(T)$ is expected to selectively observe a metallic property in SWNT.

Figure 2 shows the temperature depepndence of $S(T)$ for SWNT prepared by the arc discharge method using NiY as a catalyst. The data EtOH-NiY and DMF is for the dried SWNTs. The data HTT350-NiY and HTT190 denotes the heat-treated SWNTs at 350°C and 190°C, respectively. All the $S(T)$ data are positive throughout the whole temperatures, indicating the dominant carriers are holes.

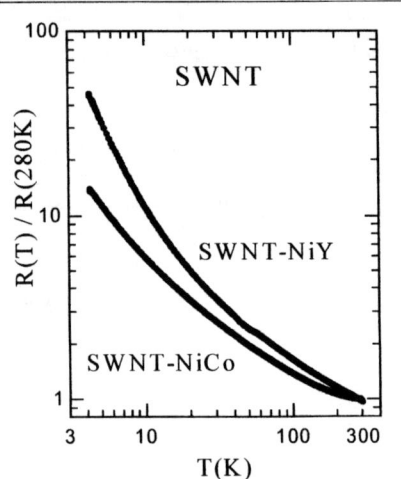

FIGURE 1. Temperature dependence of normalized resistivity for two types of SWNTs.

For HTT350-NiY and HTT190, strongly curved T-dependence appears as reported by other groups[2,3]. For DMF and EtOH-NiY, on the contrary, $S(T)$ is almost proportional to T with an intercept of nearly zero. We will discuss later paragraph the origin of these behaviors together with the results of SWNT prepared with different catalyst.

Figure 3 shows the temperature dependence of $S(T)$ for SWNTs by the laser vaporization method using NiCo as a catalyst. The data EtOH-NiCo and HTT350-NiCo are for the dried SWNT and heat-treated one at 350°C, respectively. Both data are positive throughout the entire temperatures. The relationship between the dried SWNT and the heat-treated one is qualitatively same as those of SWNT-NiY.

For the heat-treated SWNTs (HTT350-NiY, HTT190 and HTT350-NiCo), characteristic common feature is that $S(T)$ decreases rapidly below 100K and approaches zero as $T \rightarrow 0$. The degree of curvature is larger when higher heat treatment temperature was applied. We cannot interpret these three data as metallic because of no linear dependence in T. To understand such unexpected behavior in $S(T)$, Grigorian et al. have proposed a model[3] as follows: The observed $S(T)$ is sum of contributions from conduction electrons and the contribution from conduction electrons and transition metals which remained in SWNT; i.e. Kondo effect.

For dried SWNTs (EtOH-NiY, DMF and EtOH-NiCo), on the contrary, $S(T)$ is almost proportional to T with an intercept of nearly zero. It is important to note that EtOH-NiY and DMF are in fair agreement with each other throughout whole temperatures in spite of different kinds of solvent used. A further point to notice is that the slope of EtOH-NiCo is smaller than those of EtOH-NiY and DMF. Since the temperature dependence in proportion to T appears independent of synthesizing method and kinds of solvent, the results of $S(T)$ for dried SWNTs exhibit the intrinsic metallic properties in SWNT.

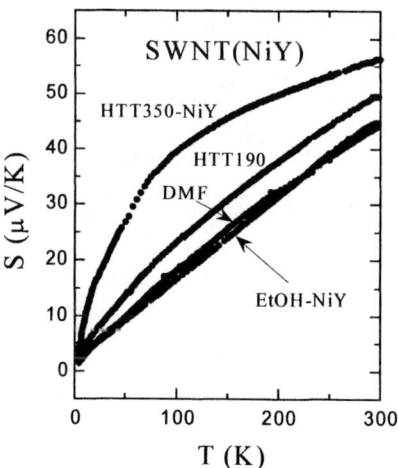

FIGURE 2. Temperature dependence of thermoelectric power for three samples: dried (EtOH-NiY, DMF), heat-treated at 190°C (HTT190) and 350°C (HTT350-NiY). NiY catalyst was used in common.

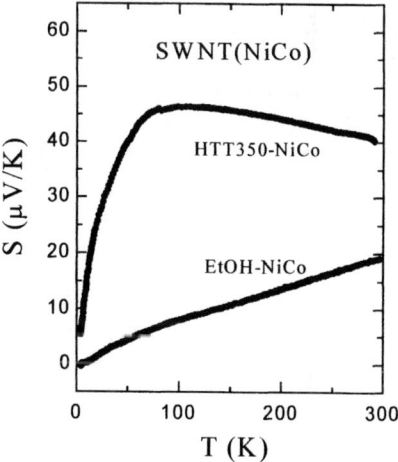

FIGURE 3. Temperature dependence of thermoelectric power for dried (EtOH-NiCo) and heat-treated at 350°C (HTT350-NiCo) SWNT using NiCo.

Next we shall consider why the temperature dependence in proportion to T appears specially for the dried SWNTs. In the heat-treated SWNT, there hardly remain solvents such as EtOH and DMF because of vaporization. In this case the $S(T)$ is sum of contributions from conduction electrons and the Kondo effect because there exists magnetic impurities such as Ni, Y and Co, as pointed out by Gurigolian et.al[3]. In dried SWNT, on the contrary, there exists the solvent to some extent. The remnant solvent molecular will make a complex with magnetic impurities and the Kondo effect is weakened. Thus $S(T)$ reflects the contribution only from conduction electrons and shows a T-linear dependence.

Lastly we shall discuss why the slopes of EtOH-NiY and EtOH-NiCo are differed. It is known that the diameter distribution of SWNT is controlled by a composition of catalyst at synthesis. A catalyst of NiY and NiCo gives SWNT tubes with a diameter of 14Å and 12Å, respectively. So the difference in the slope would reflect the difference in carrier numbers in metallic SWNT with a different diameter.

ACKNOWLEDGMENTS

Neutron activation analysis was made at Kyoto University's Research Reactor Institute (Project study No. 12p4-6). We are greatful for collaboration with Prof. T. Matsuyama, Prof. M. Seto and Dr. J. Takada.

REFERENCES

1. M. S. Dresselhaus, R. A. Jishi, G. Dresselhaus, D. Inomata, K. Nakao and R. Saito, *Molecular Materials,* **4**, 27 (1994).
2. J. Hone, I. Ellwood, M. Muno, A. Mizel, M. L. Cohen, A. Zetti, A. G. Rinzler and R. E. Smalley, *Phys. Rev. Lett.,* **80**, 1042 (1998).
3. L. Grogorian, G.U. Sumanasekera, A. L. Loper, S. L. Fang, J. L. Allen and P. C. Eklund, *Phys. Rev.,* **B60**, R11309 (1999).
4. S. Masubuchi and S. Kazama, *Synth. Met.*, **74**, 151 (1995).
5. J. E. Fischer, H. Dai, A. Thess, R. Lee, N. M. Hanjani, D. L. Dehaas and R. E. Smally, *Phys. Rev.,* **B55**, R4921 (1997).

Analysis of C_{60} Insertion into Single Wall Carbon Nanotube by Molecular Dynamics Simulation

Takafumi Ishii[1], Keivan Esfarjani[1], Yuichi Hashi[2], Yoshiyuki Kawazoe[1] and Sumio Iijima[3,4,5]

[1]*Institute for Materials Research, Tohoku University, 2-1-1 Katahira Aoba-ku, Sendai, Miyagi, Japan*
Fax: 81-022-215-2052, e-mail: tishii@imr.edu
[2]*Hitachi Tohoku Software, Ltd., 2-16-10 Honmachi Aoba-ku, Sendai, Miyagi, Japan*
Fax: 81-022-227-1411, email: hashi@hitachi-to.co.jp
[3]*JST-ICORP Nanotubelite Project, c/o NEC corporation, 34 Miyukigaoka, Tsukuba 305-8501, Japan*
[4]*R & D Group, NEC Corporation, 34 Miyukigaoka, Tsukuba 305-8501, Japan*
Fax: 81-0298-56-6136, email: s-iijima@bp.jp.nec.con
[5]*Department of Materials Science and Engineering, Meijo University, Tenpaku-ku, Nagoya 468-8502, Japan*
Fax: 81-052-834-4001, email: iijima@meijo-u.ac.jp

Abstract Classical molecular dynamics is used to simulate the insertion of C_{60} into a (10,10) single wall nano tube (SWNT). We propose that the insertion process occurs through the open end of the tube. After insertion, the energy exchange between the C_{60} and the nanotube causes the kinetic energy of the former to decrease as it moves inside the tube. This kinetic energy loss is due to a "friction force" which we calculated for several insertion conditions. The binding energy of the C_{60} with the SWNT is due to Van der Waals interaction, and is found to be about 3.5 eV.

INTRODUCTION

Encapsulation of C_{60} inside bundles of SWNT was first discovered in 1998 by Smith et. al[1] by the pulsed laser evaporation method where the laser was shone on a catalyst-containing graphite target. Also, Zhang et. al[2] and Sloan et al[3] have observed fullerenes inside SWNT grown by the arc discharge method. Later, the insertion was also achieved after reaction of C_{60} vapor under 0.1 atmospheric partial pressure with heat- treated and oxygen-treated SWNT at about 700K[2]. It was found that heat and oxygen treatment improved drastically the insertion yield. K. Hirahara et. al proposed that oxygen treating of SWNT removes the cap of it, and that C_{60} enters through the open side of SWNT[5]. On the other hand, Burteaux[4] has proposed that fullerenes enter into SWNT from the side defect of open side of the SWNT since the walls were seen to have many defects. The mechanism of encapsulation is not precisely known however. Presumably, it could change depending on the type of experiment and the condition under which it takes place. The simplest scenario would be of course to assume that C_{60} or other fullerenes enter through the end of the tubes. This is also consistent with the heat treatment at 693K and with the presence of oxygen which would react and open the tube end. A recent experiment in which Gd@C82 was inserted in a SWNT also supports this assumption[5]. On the other hand, full insertion would be less likely in ropes with even many defect walls, as only the outer tubes should be filled in this case. At a temperature of 700 K and partial pressure of C_{60} of 0.01 Pa, 153 C_{60} can hit the open ends of a SWNT per second according to the formula: insertion rate=nA<v>/4 (n=fullerene density; A=hole area; <v>=thermal speed of fullerene). This is assuming an ideal gas without any interactions. But in reality, there is Van der Waals interaction between the tube and C_{60}, and it could strongly enhance this rate. We simulated the dynamics of C_{60} insertion for the following two reasons: to observe the dynamics and the time scales of the fullerene insertion; and to compute its kinetic energy loss resulting in a "friction force".

Calculation Method

We consider a (10,10) armchair nanotube and a C_{60} of diameters 1.4 nm and 0.7 nm respectively. This allows a distance of 0.35 nm between the fullerene and the tube which is very comparable to the distance between graphene sheets in graphite. One therefore expects an attractive interaction of Van der Waals (VDW) type between the two. The SWNT is 10 nm long and open ended. The covalent interaction between the carbon atoms is modeled by the Brenner potential[7], and that between atoms of C_{60} and those of the nanotube is of Lennard-Jones (LJ) form[6]. The Brenner potential is fitted to describe the covalent bonding between carbon atoms. The LJ potential, which is successfully used to explain the thermodynamic properties of C_{60} FCC crystal by Girifalco[8], is determined to reproduce the correct interlayer distance and binding energy in graphite. Both the fullerene and the nanotube are first relaxed with the Brenner potential. Then C_{60} is placed outside the tube and given an initial kinetic energy varying between 2.0 and 10.0 eV. Its velocity is toward the tube axis. The angle between the velocity of the fullerene and the tube axis is also a parameter of the problem, and it is varied from 3.0 to 9.0 degrees. The molecular Dynamics (MD) time step is taken to be 1fs. As for the MD simulation of 26 C_{60} and 1 SWNT, we consider a (10,10) armchair nanotube with length 3.0 nm placed in a box of 4.5x4.5x6 nm in order to decrease computational time, at a temperature of 700K.

Results and Discussion

Due to the attractive VdW interaction, C_{60} is more stable inside the SWNT than separate from it by 3.5 eV. Even though the potential energy of each pair of atoms is small in our case, the large number of pairs causes the binding energy to be large. The VdW interaction between C_{60} and the end of the tube would cause an acceleration of the former towards the center of the tube if their distance becomes small enough. The 3.5 eV gain corresponds to a velocity of 0.97 km/s or 0.97nm/ps. The kinetic energy increase is, however, always less than this amount since there are collisions with the tube walls as can also be seen in Fig. 1.

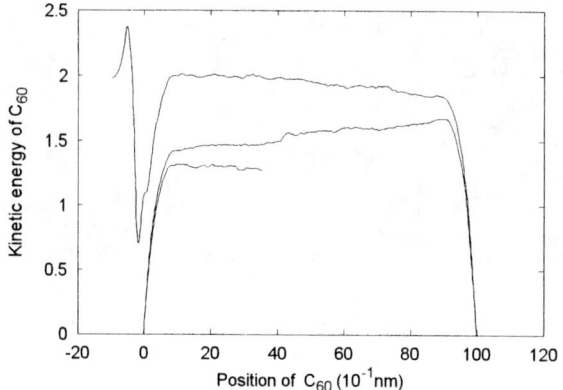

FIGURE 1. C_{60} kinetic energy variation. Insertion angle is 6.0, initial kinetic energy is 2.0 eV.

The collisions cause energy exchange between the tube and the C_{60}, so that the velocity is usually reduced and the energy is transformed to vibrations of the tube or, in other words, heat. Macroscopically, this is called friction. But after a few collisions, the motion is more smooth and the KE decreases uniformly as a function of the fullerene coordinate. The friction force is defined as the average slope of KE versus fullerene coordinate along the tube axis. A typical MD run of insertion with initial kinetic energy of 2.0 eV and incidence angle of 6.0 degree is shown in Fig.1. The energy exchange and loss can clearly be seen in this figure. Having a low kinetic energy, C_{60} is reflected at the ends of the tube before stopping.

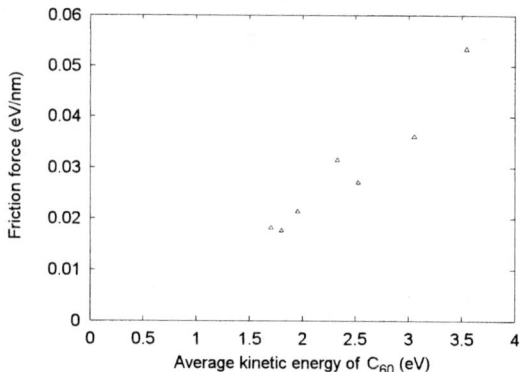

FIGURE 2. The relation between friction force and average kinetic energy of C_{60}.

Fig.2 shows the relation between the friction force and the average KE of C_{60} long after the insertion, for incidence angles varying from 20 to 40 degrees. In this case, there is no dependence on the angle since the memory of initial angle has been erased during this period. If we simply assume that the friction force linearly increases with the kinetic energy, the slope is found to be 0.0125 nm^{-1}. One should keep in mind that this result was obtained with a fully relaxed tube, i.e. at zero temperature and naturally, it always absorbs energy. So friction is usually stronger in colder tubes, and the above number could be considered as an upper bound for it.

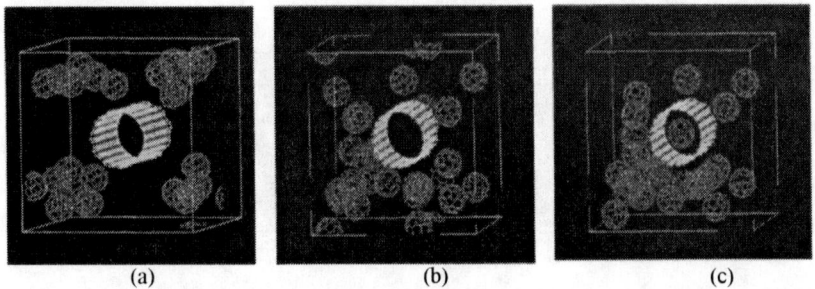

FIGURE 3. Results of M.D. simulation. Temperature is 700 K. (a) Initial condition, (b) after 20 psec, (c) after 50 psec.

Fig.3 shows the result of the MD simulation. After 25 psec, 1 C_{60} was absorbed into SWNT. After 50 psec simulation, some fullerenes surround the SWNT similar to images observed by HRTEM. In our simulation, C_{60} entered the open end from the gas phase, however, at the same time, many C_{60} are adsorbed and diffuse on the tube surface. From this simulation, we arrive at the picture that the released C_{60} gas has the possibility of wetting the outer walls of the ropes due to strong VdW forces. Encapsulation can take place either directly from the gas phase with a rate of $nA<v>/4$ per holes in the rope, or from the wetting layer covering the ropes and diffusing into the holes as they see one on their path. Penetration mechanism through the side wall defects is possible if they are large enough (> 1.4 nm) to allow encapsulation, although that could only explain filling of the outermost tubes of the rope, but not the inner ones.

Conclusion

We propose the possibility of C_{60} insertion through open end of SWNT either through gas phase or through the wetting layer of C_{60} around the ropes. In our MD simulation, one C_{60} was absorbed into SWNT after 25ps under 2.6 MPa partial pressure. The VdW interaction between C_{60} and SWNT plays a very important role in the insertion process. The friction force seems proportional to the average kinetic energy of C_{60} and its coefficient is 0.0125 nm^{-1}.

REFERENCES

[1] B.W. Smith, M. Monthioux and D. E. Luzzi, Nature **396**, 323(1998).
[2] Y. Zhang, S. Ijiima, Z. Shi and Z. Gu, Philos. Mag. Lett. **79**, 473 (1999).
[3] J. Sloan et al. Chem. Phys. Lett., **316**, 191 (2000).
[4] B. Burteaux, A. Claye, B. W. Smith, M. Monthioux, D. E. Luzzi and J. E. Fischer, Chem. Phys. Lett., **310**, 21(1999).
[5] K. Hirahara, K. Suenaga, S. Bandow, H. Kato, T. Okazaki, H. Shinohara and S. Iijima, Phys. Rev. Lett., **85**, 5384 (2000)
[6] L. A. Girifalco and R. A. Lad, J. Chem. Phys., **23**, 693 (1956)
[7] D. W. Brenner, Phys. Rev. B, **42**, 9458(1990).
[8] L. A. Girifalco, J. Phys. Chem., **96**, 858(1992).

Supercapacitors Using Singlewalled Carbon Nanotube Electrodes

Kay Hyeok An[1], Won Seok Kim[1], Young Soo Park[3], Hee Jin Jeong[3], Young Chul Choi[3], Jeong-Mi Moon[1], Dong Jae Bae[2], Seong Chu Lim[2], Young Hee Lee[1]

[1]*Department of physics, Sungkyunkwan University, Suwon 440-746, (Republic of Korea)*
[2]*Semiconductor Physics Research Center,* [3]*Department of Semiconductor Science and Technology, Jeonbuk National University, Jeonju 561-756, (Republic of Korea)*

Abstract. We have investigated the key factors determining the performance of supercapacitors using singlewalled carbon nanotube (SWNT) electrodes. Several parameters of compositions of the binder, annealing temperature, type of current collectors, charging time and discharging current density are optimized for the best performance of the energy density and power denisty. We find a maximum specific capacitance of 180 F/g and a measured power density of 20 kW/kg at the energy density of 7 Wh/kg in a solution of 7.5 N KOH. The specific surface area and the specific capacitance increase with increasing annealing temperatures of the sample. It was found that most of the BET surface area of the SWNT electrode contributed to the theoretically estimated specific capacitance. Minimization of the contact resistance is independent of the specific capacitance but directly related to the maximization of the power density.

INTRODUCTION

Recently, there have been considerable attempts to use carbon nanotubes (CNTs) for electrodes of electrochemical energy storage systems, such as Li-ion secondary battery [1-3], hydrogen storage for fuel cell and secondary batteries [4-5], and supercapacitors [6-9]. The CNTs are attractive materials for electrodes of electrochemical energy storage devices due to their superb characteristics of chemical stability, low mass density, low resistivity, and large surface area. Recent developments in massive synthesis of carbon nanotubes have accelerated a new application of these materials to the area of electrochemical energy storage systems. In the present report, the singlewalled CNTs (SWNTs) were introduced to study the relation of the preparation conditions to the equivalent series resistance (ESR), specific capacitance, energy density, and power density.

Experimental

SWNTs were synthesized by dc arc discharge under a helium pressure of 100 tor. Poly-vinylidene chloride (PVdC) were used as a binder. A pellet of diameter of 15 mm was formed by pressing the mixture of as-grown SWNT and binder with a cylindrical steel molder under 1000 psi. The pellet was heat-treated at 500~1000 °C for 30 min in argon-gas ambient. The thickness of the electrode was about 150 μm. The measured apparent mass density of the electrode was 0.75 g/cm^3.

A plane Ni-foil of a thickness of 75 μm was used as a current collector. To minimize the contact resistance between the CNT electrode and the current collector, mechanically polished Ni foil was also used. The hybrid electrode of the CNT-Ni foam was prepared by pressing the CNT sample on Ni-foam with 120 pores per inch under 1000 psi.

A unit cell for the capacitor was fabricated with two CNT electrodes separated by a thin polymer (Celgard) in 7.5N KOH aqueous solution as the electrolyte. The cell was charged at a constant voltage of 0.9 V for 30 sec ~ 6 hrs, and then discharged at a constant current density of 1 ~ 50 mA/cm^2. The discharging capacitance of the test cell was then calculated by $I_{dc} \Delta t / \Delta V$, where I_{dc} is the constant discharging current, and Δt is discharging time. The energy density was measured as a function of constant-power discharge in the range of 2 W/kg ~ 20 kW/kg.

Results and Discussion

Figure 1 shows the specific capacitances of the heat-treated electrodes at various temperatures as a function of the charging time. Capacitances increase abruptly and reach about 80 % of the maximum capacitance during the initial 10 min, regardless of the heat-treatment temperatures. The capacitances increase gradually further and saturate to the maximum values at long charging time. The saturated capacitance increases with increasing heat-treatment temperatures and saturates to 180 F/g at 1000 °C. This value is larger than the previously reported value of 113 F/g from multiwalled CNTs [6]. In the electric double layer capacitor with plane electrodes, the charge densities of about 20 to 50 μF/cm^2 are commonly realizable. In our case, the specific surface area is 357 m^2/g at 1000 °C. We can calculate a rough estimate of the theoretical capacitance to be 71~178 F/g, in good agreement with the observed values in the upper bound. This is in good contrast with the activated carbons, where the observed specific capacitance is about one-fourth the theoretical capacitance in spite of high specific surface area (2000-3000 m^2/g).

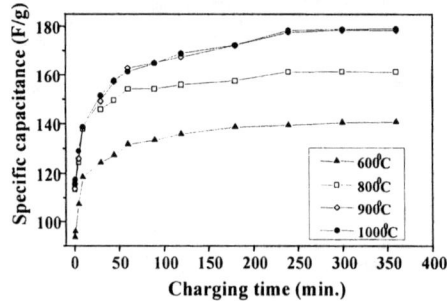

FIGURE 1. The specific capacitances of the heat-treated electrodes at various temperatures as a function of the charging time at a charging voltage of 0.9 V.

The magnitude of the ESR can be clearly shown in the complex-plane impedance plots, as shown in Fig. 2 (a). The electrolyte resistance, R_s, is constant and varies with electrolyte. Sum of the resistance of the electrode itself and the contact resistance between the electrode and the current collector is represented by the R_f. The electrolyte resistance and the contact resistance are identical in all samples. Therefore, the decrease of the R_f indicates the decrease of the CNT-electrode resistance. The CNT-electrode resistance decreases very rapidly at high temperatures of 800 and 1000 °C. The R_f is closely related to the power density.

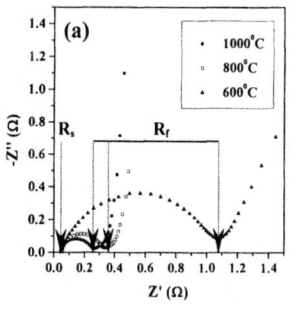 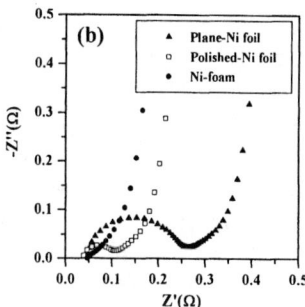

FIGURE 2. The complex-plane impedance plots for (a) SWNT electrodes heat-treated at various temperatures and (b) the fabrication method of electrodes.

The contact resistance between the CNT-electrode and the current collector is also an important factor in determining the performance of a supercapacitor. Figures 2 (b) and 3 clearly show the relations between the ESRs, particularly the contact resistance and power density. In order to change the contact resistance, we introduced a polished Ni-foil in addition to the plane Ni-foil and hybrid form of the CNT-Ni-foam. All the samples were heat-treated at 1000 °C for 30 min. The plane Ni-foil gives the largest ESR, indicating poor contact formation between the CNT and the Ni-foil. The contact resistance was reduced significantly by polishing the surface of the Ni-foil, which is attributed to increase of the contact surface area. It is interesting to see that the semi-circle almost disappears in the complex-plane impedance plot for the hybrid electrode due to the extremely small contact resistance.

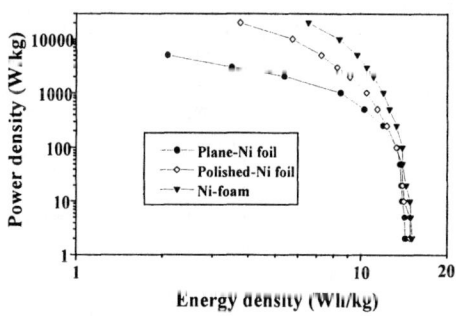

FIGURE 3. Ragone plots for various fabrication methods of electrodes

Figure 3 clearly presents how the small ESR directly affects the power density of a supercapacitor. At low power density ranging of 2 W/kg ~ 100 W/kg, the energy density in all cases of the contact preparation conditions do not alter. However, the energy density drops very rapidly with increasing power density particularly for the samples with large ESR. On the contrary, the energy density does not change appreciably with increasing the power density for the hybrid electrode, which has the smallest ESR.

Summary

We obtained a maximum specific capacitance of 180 F/g with large power density of 20 kW/kg at an energy density of 6.5 Wh/kg. The heat-treatment at high temperature was necessary to increase the capacitance and reduce the CNT-electrode resistance. The increased capacitance was well explained by the enhancement of the specific surface area and the abundant pore distributions at lower pore sizes of 30 to 50 Å estimated from the BET (N_2) measurements. The ESR should be minimized in order to obtain a high power supercapacitor. Our current approach demonstrates a possibility of the CNT application to the supercapacitor. There are still plenty of rooms to improve the performance of the supercapacitor using the CNT electrodes.

ACKNOWLEDGMENTS

This project was supported by the MOST through NRL program and in part by BK21 program. One of us (KHA) acknowledges the support through a post doctoral fellowship program at Jeonbuk National University.

REFERENCES

1. Leroux, F., Metenier, K., Gautier, S., FrackoWiak, E., Bonnamy, S., and F. Beguin, *J. Power Sources* **81-82**, 317 (1999).
2. Wu, G. T., Wang, C. S., Zhang, X. B., Yang, H. S., Qi, Z. F., He, P. M., and Li, W.Z., *J. Electrochem. Soc.* **146**, 1696 (1999).
3. Frackowiak, E., Gautier, S., Gaucher, H., Bonnamy, S., and Beguin, F., *Carbon* **37**, 61 (1999).
4. Nutzenadel, C., Zuttel, A., Chartouni, D., and Schlapbach, L., *Electrochem. Soild-State Lett.* **2**, 30 (1999).
5. Lee, S. M., Park, K. S., Choi, Y. C., Park, Y. S., Bok, J. M., Bae, D. J., Nahm, K. S., Choi, Y. G., Yu, S.C., Kim, N. g., Frauenheim, T., and Lee, Y. H., *Synthetic Metals* **113**, 209 (2000).
6. Niu, C., Sichel, E. K., Hoch, R., Moy, D., and Tennent, H., *Appl. Phys. Lett.* **70**, 1480 (1997).
7. Ma, R. Z., Liang, J., Wei, B. Q., Zhang, B., Xu, C. L., and Wu, D. H., *J. Power Sources* **84**, 126 (1999).
8. Diederich, L., Barborini, E., Piseri, P., Podesta, A., Milani, P., Schneuwly, and A., Gallay, R., *Appl. Phys. Lett.* **75**, 2662 (1999).
9. An, K. H., Kim, W. S., Lee, S. M., Park, K. S., Choi, Y. C., Park, Bae, D. J., and Lee, Y. H., *Advanced Materials* in press (2001).

Resonances in Deformed Carbon Nanotubes

H.-S. Sim, C.-J. Park, and K. J. Chang

Department of Physics, Korea Advanced Institute of Science and Technology, Taejon 305-701, Korea

Abstract. We investigate the electron coherent transport in single-wall carbon nanotubes under local flattening deformations that break mirror symmetries. Such a local deformation is found to cause transmission barrier, resonances, and anti-resonances in electron transmission. We also find that a finite perfect tube which is sandwiched between two deformed tubes behaves as a quantum dot. This nanotube device exhibits periodic and nearly degenerate resonant peak pairs in transmission, as a gating voltage applied to the dot varies. This feature of resonances is attributed to the mixing of two channels $|\pi\rangle$ and $|\pi^*\rangle$, which is derived by the breaking of mirror symmetries.

INTRODUCTION

Single-wall armchair carbon nanotubes are one-dimensional metals [1]. At the Fermi energy, they have two degenerate current-carrying states characterized by π and π^*, which have even and odd parities, respectively, under mirror reflections in the cross section of tube, resulting in a low bias conductance of $4e^2/h$. These mirror symmetries can be easily broken if tubes are flattened [2] or twisted [3]. Such symmetry breaking deformations can open the band gap by an order of 0.1 eV in armchair carbon nanotubes, so that they can strongly affect the electron transport. Thus, it is important to study the electron transport behavior in armchair nanotubes which are locally perturbed by mirror-symmetry breaking deformations.

In this work, we investigate the electron transport in (5,5) carbon nanotubes with local mirror-symmetry breaking deformations such as flattening, using a tight-binding method and the Green's function approach. We find a transmission dip below the Fermi level as well as several transmission resonances and anti-resonances. The former results from the band gap opening in the flattened region, while the latter is attributed to the formation of quasi-bound states in the flattened region. We also consider a (5,5) armchair tube which is sandwiched between two deformed regions. When the energy of incident channels matches to the energy range of the transmission dip in both the deformed regions, the undeformed tube between the two deformed regions can be considered as a quantum dot. As a gating voltage applied to the dot varies, periodic and nearly degenerate resonant peak pairs appear in transmission. This feature of resonances is attributed to the mixing of two channels $|\pi\rangle$ and $|\pi^*\rangle$ due to the breaking of mirror symmetries.

NANOTUBES WITH A LOCAL MIRROR-SYMMETRY-BREAKING DEFORMATION

In Fig. 1, the geometry and cross sections of a locally flattened (5,5) tube are drawn. The flattened region consists of N_f fully flattened unit cells between two sets of N_{buffer} (=8.5) buffering unit cells, where the cross sections of flattened cells change gradually to the undeformed ones. Employing the empirical Tersoff's potential, the deformed geometry in Fig. 1 is fully relaxed until the distance between the two facing carbon atoms in each of N_f unit cells is equal to $d = 2.71$ Å, then, the mirror symmetries are completely broken in the cross sections. Note that for the flattened tube with $d = 2.71$ Å, the band gap is about 0.095 eV and the magnitude of the band gap depends on d [2].

To calculate the electron conductance, we consider that incident channels from the left semi-infinite tube region propagate to the right semi-infinite region, passing through the flattened region. From the Landauer-Büttiker formula [4], the conductance $\mathcal{G}$ is directly obtained from the transmission probability $T(E)$ as $\mathcal{G}(E) = 2e^2 T(E)/h$, where E is the energy of the incident channels. Using the nonequilibrium Green's function technique [5] and a tight-binding method [6], $T(E)$ can be calculated from $T(E) = tr(\Gamma_L G \Gamma_R G^\dagger)$, where $\Gamma_{L(R)}$ is a coupling matrix

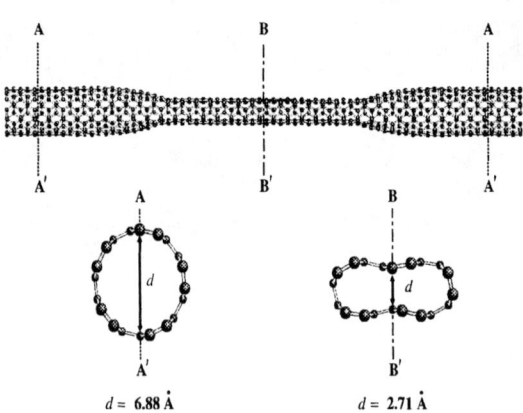

FIGURE 1. The geometry and cross sections of the (5,5) tube with one locally flattened region (N_f=15).

between the left (right) semi-infinite region and the flattened region and G is a retarded Green's function of the flattened region.

In Fig. 2(a), the calculated $T(E)$ is plotted for the flattened region with $N_f = 65$, and it exhibits a dip near $E = -0.1$ eV where the band gap opens in the flattened tube with $d = 2.71$ Å [2]. We set the Fermi energy of the (5,5) tube to be 0 eV. Because there is no allowed state in the flattened region due to the band gap opening, two incident channels are strongly reflected in that region and only transmitted through evanescent modes. Thus, the flattened region behaves as a transmission barrier. The minimum value of the dip decreases roughly exponentially as N_f increases. Other mirror-symmetry-breaking deformations also result in the same transmission barrier.

Except for $E \approx -0.1$ eV, $T(E)$ is almost 2, indicating that all incident channels well pass through the flattened region. In the ranges of $E[eV] \in [-0.6, -0.2]$ and $[0.0, 0.3]$, where there are two channels in the flattened tube with $d = 2.71$ Å, $T(E)$ shows weak and broad oscillations. The maxima of the oscillations result from resonances. At the resonances, two channels in the flattened region are multiply reflected at the boundaries of the flattened region, though the reflection probabilities are small, and each reflected channels are superposed constructively. When energies are close to 0.4 eV and 0.6 eV, $T(E)$ oscillates very rapidly between 1 and 2. In these energy ranges, there are more than two channels in the flattened tube with $d = 2.71$ Å due to the π^*-σ^* hybridization [2]. When a bound state is formed

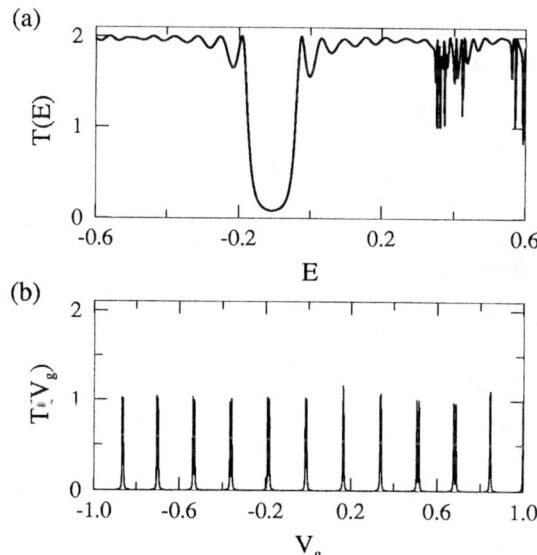

FIGURE 2. (a) The transmission probability $T(E)$ through a (5,5) tube with one flattened region ($N_f = 65$) and (b) $T(V_g)$ through a nanotube dot with $(N_d, N_f) = (30, 05)$ are drawn.

due to the additional channels, the anti-resonance [7] or the destructive interference between the directly transmitted channel and the bound state gives rise to $T(E) = 1$. The anti-resonances are observable in deformed nanotubes, because the electron transport is coherent and ballistic in single-wall nanotubes.

NANOTUBE QUANTUM DOT

Next we study a (5,5) nanotube with a finite undeformed region sandwiched between two separated flattened regions. The undeformed region and each flattened one consist of N_d and N_f unit cells, respectively. Since symmetry-breaking deformations produce transmission barriers, the nanotube device with two flattened regions is expected to behave as a quantum dot. For a dot with $(N_d, N_f)=(30, 65)$, we calculate $T(V_g)$ when a gating voltage V_g is applied to the dot. In this case, only the on-site energies of the tight-binding Hamiltonian in the undeformed N_d unit cells are shifted by V_g without breaking any symmetry, and the energy of incident channels from the lead is fixed at $E=-0.106$ eV.

In this device, periodic and nearly degenerate peak pairs appear in transmission, as shown in Fig. 2(b), resulting from resonant bound states of the dot. Using a scattering matrix approach, we derive the period of resonant peak pairs [8] as $(1/|\hbar v_{g,\pi}| + 1/|\hbar v_{g,\pi^*}|)\Delta V_g = 2\pi/L_d$, where L_d ($= 2.494 N_d$ Å) is the dot length. Using the calculated group velocities (v_g) for the $|\pi\rangle$ and $|\pi^*\rangle$ states, the period of $\Delta V_g = 5.28/N_d$ eV is obtained, in good agreement with the result in Fig. 2(b). Finally, we emphasize that these resonant peak pairs are general features in armchair nanotubes with local symmetry-breaking deformations that mix completely $|\pi\rangle$ and $|\pi^*\rangle$, independent of the detailed shape of cross sections in the flattened region [8].

REFERENCES

1. Dresselhaus, M. S., Dresselhaus, G., and Eklund, P. C., *Science of Fullerenes and Carbon Nanotubes* (Academic, San Diego, CA, 1996).
2. Park, C.-J., Kim, Y.-H., and Chang, K. J., *Phys. Rev. B* **60**, 10656 (1999).
3. Kane, C. L. and Mele, E. J., *Phys. Rev. Lett.* **78**, 1932 (1997); Rochefort, A., Avouris, Ph., Lesage, F., and Salahub, D. R., *Phys. Rev. B* **60**, 13824 (1999).
4. Büttiker, M., Imry, Y., Landauer, R., and Pinhas, S., *Phys. Rev. B* **31**, 6207 (1985).
5. Meir, Y. and Wingreen, N. S., *Phys. Rev. Lett.* **68**, 2512 (1992).
6. Tang, M. S., Wang, C. Z., Chan, C. T., and Ho, K. M., *Phys. Rev. B* **53**, 979 (1996).
7. Shao, Z., Porod, W., and Lent, C. S., *Phys. Rev. B* **49**, 7453 (1992).
8. Sim, H.-S., Park, C.-J., Chang, K. J., *Phys. Rev. B*, in press (2001).

Electrochemical Lithium Insertion of Heat Treated and Chemically Modified Multi-wall Carbon Nanotubes

H. Touhara[*], I. Mukhopadhyay[*], F. Okino[*], S. Kawasaki[*], T. Kyotani[†], A. Tomita[†] and W. K. Hsu[¶]

[*] *Faculty of Textile Science and Technology, Shinshu University, Ueda-386-8567, Japan.*
[†] *Institute of Chemical Reaction Science, Tohoku University, Sendai-980-8577, Japan*
[¶] *School of Chemistry, Physics and Environmental Science, University of Sussex, UK.*

Abstract. Highly aligned multi-wall carbon nanotubes (MWNTs), in the form of membrane (CNM) and powder (MWNTP) were examined for Li ion insertion host in aprotic medium. Effects of heat treatment and surface modification by elemental fluorine on the structure and electrochemical properties of the nanotubes were investigated. The surface structure and properties were characterized by scanning electron microscopy (SEM), transmission electron microscopy (TEM) and N_2 adsorption isotherm. The BET surface area of the nanotubes was very large compared to graphitic carbon. Heat treatment at 3000°C induced better crystallinity with a decrease in surface area. Fluorination at 50°C removed the disorder carbon from the surface of pristine sample. The reversible capacity in the first cycle was 516 mAh/g with a very high irreversible capacity. The reversible capacity increased from 22 % to 80 % after the second cycle.

INTRODUCTION

Carbon nanotubes have been extensively studied in the recent past for their usefulness as insertion material in Li ion secondary battery [1,2]. Li insertion may occur in MWNTs in the pseudo graphitic layers, in the central canal or in the lateral defect sites at the grain boundaries. However it has been observed that the extent of reversible lithium insertion depends on the preparation methods of MWNTs. Microtexture of nanotube surface and crystal structure have mainly been attributed to the observed difference. In the present investigation, we report on the studies of electrochemical Li insertion in highly aligned MWNTs, prepared by template carbonization technique [3] in non-aqueous medium. Effect of high temperature treatment and surface modification by elemental fluorine on the structure and electrochemical properties of the nanotubes was also investigated.

Experimental

MWNTs were synthesized by CVD of propylene gas in nanopore alumina template at 800°C [3]. MWNTs were studied in two different forms: (a) CNM, where the nanotubes are embedded in alumina template and (b) MWNTP, where the template alumina film was removed by dissolving in 46 % HF. Heat treated (CNM-1000) and fluorinated (F-CNM) CNM were made by heating at 1000°C for 15 min and fluorinating at 50°C by F_2 for 5 days respectively. MWNTP was treated to 1000°C (MWNTP-1000) and 3000°C (MWNTP-3000) for 15 min in Ar atmosphere. The structural characteristics and surface micro-texture were studied by SEM and TEM. Absorptive properties were investigated by N_2 adsorption isotherm at 77K. Electrochemical Li insertion was investigated by cyclicvoltammetry (CV) and galvanostatic charge-discharge (CD) measurements in the usual procedure. The working electrode (WE) out of CNM was made by connecting a Cu wire to it with conducting Ag paste and insulating the contact point by teflon paint, which was inert to electrolyte. WE for MWNTP was made by coating a slurry of 90 wt.% nanotube powder and 10 wt.% polyvinylidene difluoride (PVDF) binder in N-methyl 2-pyrrolidinone on one side of a Cu foil. The Cu foil then dried in air at 100°C. Prior to expose to electrolyte, the WE was dried under vacuum at 120°C for over night. Li foil attached to Cu wire was used as counter (CE) and reference (RE) electrode. 1 M $LiClO_4$ dissolved in 1:1 (v/v) ethylene carbonate (EC) and diethyl carbonate (DEC) was used as electrolyte for all electrochemical experiments. CV experiments were done at a sweep rate of 0.1 mV/s. CD studies were performed at 50 mA/g-C rate and a cut-off voltage of 2.5 V.

Results and Discussion

The SEM image of MWNTP is shown in Fig.1. The nanotubes are held by surface carbon coating and are opened at both ends. The interstitial space due to formation of bundle is clearly observed. The length of the nanotubes is about 70 μm. It can be seen from Fig.2 that the individual nanotube is formed by graphite-like layers (ca.10 to 12 layers) with a coating of disordered carbon on the rear surface. The outer diameter of the nanotube is 22 nm with a d spacing of 0.34 nm between the adjacent graphene layers. Shown in Fig.3 is the TEM image of MWNTP-3000 and F-CNM. Heat treatment induces better crystallinity and alignment of the graphene layers with complete removal of disordered carbon from the rear surface. Moreover it is noticed

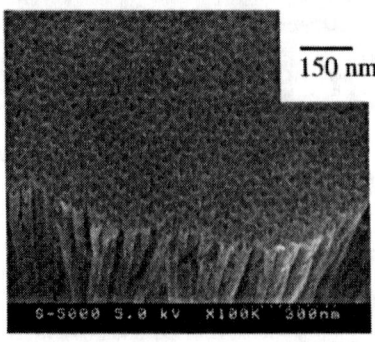

FIGURE 1. SEM of MWNTP

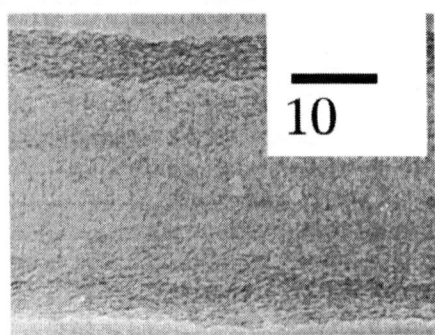

FIGURE 2. TEM of MWNTP

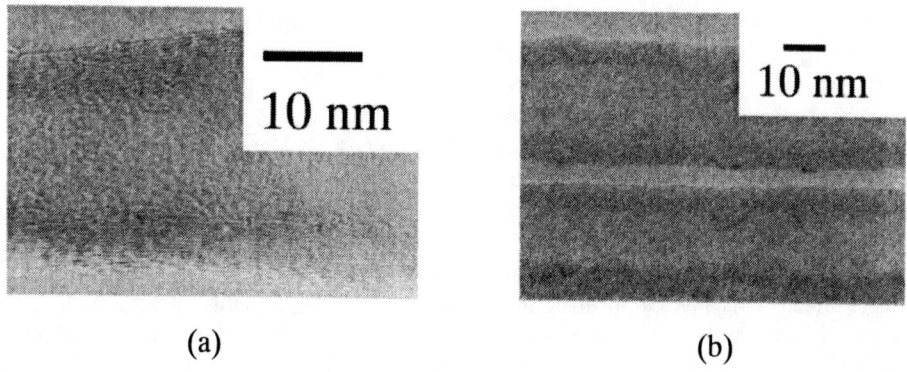

FIGURE 3. TEM of (a) MWNTP-3000 and (b) F-CNM

that the tube ends of MWNTP-3000 are closed after heat-treatment. TEM of F-CNM shows that the disorder carbon coatings removed even at 50°C. That the ends of the nanotubes are opened is confirmed from the hysteresis in the adsorption isotherm. The BET surface area of CNM and MWNTP are 28 and 224 m^2/g respectively. This high surface area may be attributed to the additional interstitial space as observed from SEM. BET surface area of both CNM-1000 and MWNTP-1000 are higher than their pristine counter part indicating only a kind of degassing on heating at 1000°C. However heat treatment at 3000°C diminishes surface area to large extent indicating better crystallinity. F-CNM also exhibits lower BET surface area than its pristine counter part indicating possibly removal of disordered carbon part from the rear surface.

The results of CV are shown in Fig.4. It can be seen from Fig.4a that CNM shows three distinguish peaks, A, B and C with sharp current due to Li intercalation into the pseudo graphitic layers at near 0 V. Peak A and B can be correlated to reduction of surface oxygenated species and the anion part of active electrolyte, $LiClO_4$. Peak C is attributed to the reduction of EC. Continuous current declination from 1 V may be due to the formation of solid electrolyte interphase (SEI). The unwanted reduction

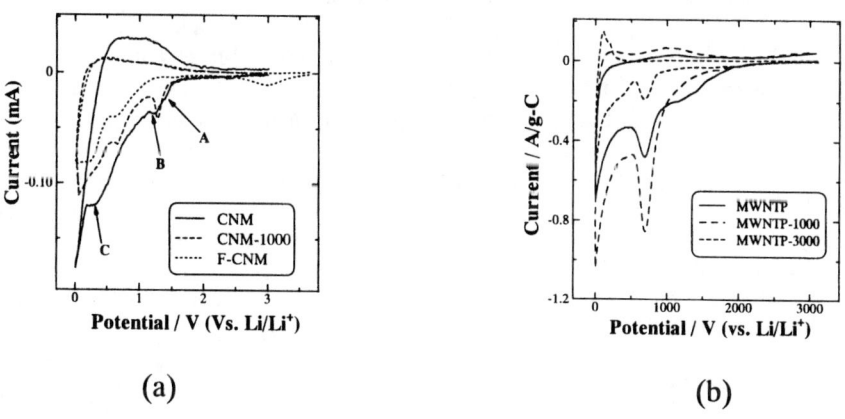

FIGURE 4. Cyclicvoltammogram of the first cycle of (a) CNM and (b) MWNTP.

251

processes associated with peak A, B and C along with the formation of SEI consumes lot of charge which finally results in very high irreversible capacity, ca.1800 mAh/g in the first cycle against reversible capacity of ca. 500 mAh/g. Peak A and B merge to a single sharp peak for CNM-1000 indicating an enhancement of the corresponding redox process. However, F-CNM does not show any peak near A and B indicating replacement of surface oxygenated species by C-F bonds which is supported by the occurrence of peak at 3 V. In the case of MWNTP, a sharp reduction peak at 0.7 V, due to reduction of EC is noticed. The irreversible consumption of charge for MWNTP-1000 and MWNTP-3000 varies consistently in accordance to their BET surface area. It is interesting to note that the irreversible capacity is very high even after avoiding the use of poly carbonate based solvent, known as the suitable media for exfoliation which is a major cause of giving irreversible capacity. Since the irreversible capacity for MWNTP-3000 is only 400 mAh/g compare to ca. 1100 mAh/g of MWNTP, at the moment it can only be speculated that EC based solvent may also favors exfoliation to occur in the carbon coating or in the interstitial sites between the individual nanotubes.

Conclusions

MWNTs, prepared by carbonization technique are highly aligned and exist as free standing bundles in the powder form. The BET surface area is very large. Heat treatment at 1000°C induces only degassing without affecting the crystal structure. However, heat treatment at 3000°C enhances crystallinity and order of the graphitic layers along c-axis. Surface chemical modification removes the amorphous carbon from the surface at low temperature. Large BET area of the nanotubes causes large irreversible capacity for Li insertion process in the first cycle. The irreversible capacity decreases for MWNTP-3000 due to better crystallinity and lower defect density than the pristine counterpart. High reversible capacity demands further investigations for making any final conclusion on its application in Li battery.

ACKNOWLEDGMENTS

One of the authors, I. M. wish to thanks JSPS for providing financial support to carry out this research. This work was supported by grant-in Aid for Research in Future Program, "Nano-carbon", for JSPS and for COE research (10CE2003)

REFERENCES

1. Gao, B., Bower, C., Lorentzen, Z. E., Fleming, L., Kleinhammes, A., Tang, X. P., McNeil, L. E., Wu, y., and Zhou, O., *Chem. Phys. Lett.* **327**, 69-75 (2000).
2. Che, G., Lakshmi, b. b., Martin, C. R., and Fisher, E. R., *Langmuir* **15**, 750-758 (1999).
3. Kyotani, T., Tsai, L., and Tomita, A., *Chem. Mater.* **10**, 1427-1428 (1995).

Spin valve effect in magnetically doped nanotube-based transistors

Keivan Esfarjani*, Z. Chen*, A. A. Farajian† and Y. Kawazoe*

IMR, Tohoku University, Sendai 980-8577 Japan
† NIMC, Tsukuba, Ibaraki 305-8565 Japan

Abstract. Transport properties of doped nanotubes double junctions forming a nanotransistor are investigated within the tight binding formalism. The effect of magnetic dopants on the electronic structure is studied by using ab initio methods. The effects of doping, gate length and gate-source hopping have been considered. It is found that in addition to the importance of rotational symmetry in determining transport properties, large gains can be achieved for semiconducting doped tubes.

I INTRODUCTION

Recent advances on GMR and TMR devices made with magnetic multilayers have prompted us in considering the effect of magnetic dopants on the electronic and transport properties of nanotubes. In this paper, we investigate the conduction properties of such tubes used as transistors in order to investigate the possibility of designing a spin valve device. First, we consider a metallic or semiconducting nanotube and dope it with a transition metal element such as cobalt to see how the cobalt "chain" inside the tube affects its electronic and magnetic properties. Next, we subject this tube to a potential difference, and at the same time, apply a voltage to its central region through a "gate". This could be done with a (nonmagnetic) STM tip. The effect of such a tip would be to raise or lower the electrostatic potential of the gate region, thus forming two junctions. In electronics, this device is called a (nano-)transistor since the gate voltage will affect the "source-drain" current produced by the applied bias. In our previous work, we have studied the electronic and transport propetries [1,2] of single junctions within the self-consistent tight binding formalism and also that of a double junction (nano-transistor) in a paramagnetic configuration [3]. Our purpose here is to study qualitatively the effects of the magnetic doping on the conductance as a function of the gate voltage and see if by tuning the latter, the device can function as a spin valve transistor.

The nanotube consists of 3 parts: the left and the right semi-infinite parts attached to a reservoir and the central part attached to a gate of potential V_G. The thickness of the gate region is variable. For simplicity, we will assume the left and right parts to be identically doped, thus having the same chemical potentials; but their magnetizations can be oriented parallel or antiparallel to each other. We will be interested in the conductance of this device as a function of the gate voltage for small applied bias (linear response).

II METHOD

For simplicity and ease of calculations, we have considered a (4,4) armchair and a (7,0) zigzag nanotube. First, we use the plane wave (ultrasoft) pseudo potential method (VASP [4]) with the spin-polarized GGA exchange-correlation potential. The cutoff energy is taken to be 21 Ryd., and it is found that the total energy converged within 5 meV with 20 k-points taken along the axis of the tube. From this calculation, we deduced the band structure (BS) and density of states (DOS) of the doped and undoped tubes.

Next, the doped tube was modeled with a tight-binding Hamiltonian with only one π orbital per atom. Since the gate is undoped, the conduction takes place only through the states of the tube and not the cobalt chain. It is assumed that the cobalt doping causes the bands of the undoped tube to be shifted.

$$\mathcal{H} = \sum_i \epsilon_{i\sigma} c_{i\sigma}^\dagger c_{i\sigma} + t \sum_{<ij>} c_{i\sigma}^\dagger c_{j\sigma} \qquad (1)$$

The on-site energy $\epsilon_{i\sigma}$ is equal to zero for the undoped tube but will be set to sign$(\sigma)\,\epsilon$ for the band of spin σ. In the gated region we have $\epsilon_{i\sigma} = V_G$. Unless explicitly mentioned, the unit for voltages or energies is the hopping integral t. In this work, the effect of self consistency has been dropped for simplicity as it does not affect qualitatively the transport phenomena in nanotubes. Our calculations have shown [5] that the screening is short-ranged with small amplitude oscillations for large steps in the junction potential.

Since we neglect any spin flipping process such as spin-orbit coupling in this model, the two up spin and down spin channels are considered as parallel, and thus the conductance will be the sum of the conductances of each channel. The retarded and advanced Green's functions (GF), for each channel, projected onto the gate region, are defined as:

$$G^{r/a}(E) = [E - \mathcal{H}_{gate} - \sum_\alpha \Sigma_\alpha^{r/a}(E) \pm i\eta]^{-1} \qquad (2)$$

where η is a small positive number, and + (resp. -) corresponds to the retarded (resp. advanced) GF. This form of the GF can be derived using the partitioning technique [6]. The self energy matrix Σ_α, representing the effect of the lead α, is:

$$[\Sigma_\alpha^{r/a}(E)]_{ij} = \sum_{kl} H_{ik}\,[g_\alpha^{r/a}(E)]_{kl}\,H_{lj} \qquad (3)$$

the index α being any of the contacts to the gate (Left or Right), i and j label two sites of the gate, and k and l belong to the lead α. g is the GF of the isolated

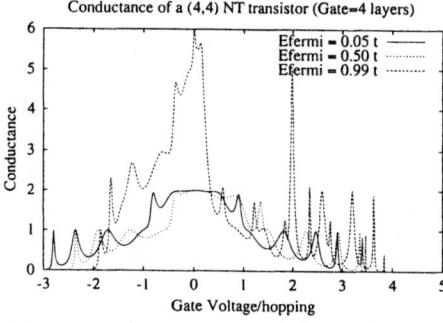
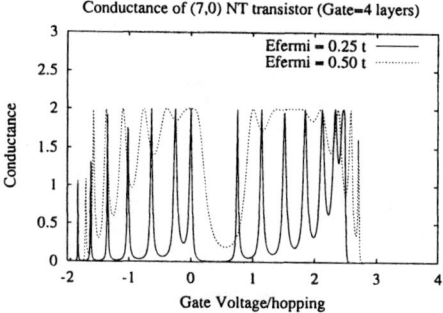

FIGURE 1. Left: Conductance of a 4 layer gated region in $2e^2/h$ versus the gate voltage in a (4,4) armchair nanotube for several Fermi energies. Right: same for a (7,0) zigzag tube.

semi-infinite left or right lead. It can be computed separately by iterative methods [7]. The conductance of the whole system can then be computed from the Landauer formula derived for the interacting systems by Meir and Wingreen [8]:

$$\mathcal{G}(E) = \mathcal{G}_\uparrow(E) + \mathcal{G}_\downarrow(E); \quad \mathcal{G}_\sigma(E) = \frac{e^2}{h} \text{Tr}(G^r_\sigma(E) \, \Gamma_{L\sigma}(E) \, G^a_\sigma(E) \, \Gamma_{R\sigma}(E)) \quad (4)$$

where the transition matrix Γ is -2 Im Σ.

III RESULTS AND DISCUSSIONS

Ab initio calculations for Co-doped infinite tubes reveal that new narrow peaks due to the d levels of Co appear near the Fermi energy. Additionally, there is a shift of the Fermi energy by 0.8 eV with respect to the undoped case, meaning that there is charge transfer from the Co atoms to the tube in both (4,4) and (7,0) tubes. The DOS of the down states at the Fermi level is almost 3 times that of the up spin states for the (4,4) and this ratio is 2 for the (7,0) tube. The magnetization per cobalt atom is about 2.4 μ_B.

We then focus our attention on the tight-binding Hamiltonian of the doped case, i.e. consider a non-zero chemical potential, and perform the conductance calculation. This study is also performed for zigzag nanotubes since, due to doping, there would be a non-zero DOS at the Fermi level in the two leads and thus a nonzero conductance. It is found that for a shift in the Fermi level of $0.25t \approx 0.67$ eV ($t = 2.7$ eV), the conductance of the semiconducting tube can vary from almost 0 to 1 (see Fig. 1 right), which is litterally an infinite gain. This occurs because for some appropriate gate voltage the local DOS at the Fermi level of the gate region becomes zero, and therefore no current can flow. This gain is about 500 in our case since the gate is of small width (4 layers). This effect is less pronounced for armchair tubes since their DOS never becomes zero (see Fig. 1).

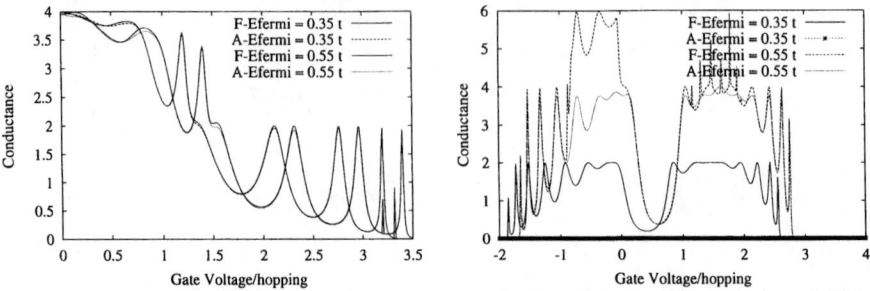

FIGURE 2. Conductance of magnetically doped (4,4) and (7,0) tubes in units of e^2/h; spin splitting $=0.3\,t$. Labels F and A refer to "F"erro- and "A"ntiferromagnetically-coupled leads. The armchair tube does not show any effect whereas the conductance ratio at $E_F = 0.35\,t$ of the zigzag is infinite!

In the magnetically-doped case, where we assumed a shift of the DOS between up and down states, this effect persists for semiconducting tubes because there are energy regions where the up-spin DOS is zero and the down-spin DOS is nonzero. In a simple model, for leads of parallel magnetizations, $G_{\text{Ferro}} \propto \text{DOS}_\uparrow(E_F)^2 + \text{DOS}_\downarrow(E_F)^2$ whereas if the two leads have antiparallel magnetizations, $G_{\text{Antiferro}} \propto 2\,\text{DOS}_\uparrow(E_F)\,\text{DOS}_\downarrow(E_F)$. If one of them is zero, then $G_{\text{Antiferro}} = 0$. Results for magnetically doped (4,4) and (7,0) tubes are displayed in Fig. 2.

IV CONCLUSIONS

Semiconducting tubes are good candidates for making nanotransistors out of doped nanotubes where we have observed that gains of more than a few hundred can be achieved for wide enough gates. Doping of the semiconducting tubes by magnetic elements can also be used to make spin valve devices. For appropriately chosen Fermi energies, the antiferromagnetic conductance becomes identically zero!

REFERENCES

1. K. Esfarjani, A. Farajian, Y. Hashi and Y. Kawazoe, Appl. Phys. Lett. **74**, 79 (1999).
2. A. A. Farajian, K. Esfarjani and Y. Kawazoe, Phys. Rev. Lett. **82**, 5084 (1999).
3. K. Esfarjani, A. A. Farajian, F. Ebrahimi and Y. Kawazoe, submitted to ISSPIC10 proceedings, to appear in Euro. Phys. Jour. D (2000).
4. The VASP package was developed at the Technical University of Vienna.
5. A. A. Farajian, K. Esfarjani and M. Mikami, to appear in the same volume.
6. D.S. Fisher and P.A. Lee, Phys. Rev. B **23**, 6851 (1981).
7. M. P. Lopez Sancho et al., J. Phys. F **15**, 851 (1985).
8. Y. Meir and N. S. Wingreen, Phys. Rev. Lett. **68**, 2512 (1992).

Charge Oscillation at Doped Nanotube Junctions

Amir A. Farajian*, Keivan Esfarjani†, and Masuhiro Mikami*

*National Institute of Materials and Chemical Research
Tsukuba, Ibaraki 305-8565, Japan
†Institute for Materials Research, Tohoku University
Sendai 980-8577, Japan

Abstract. Using a self-consistent tight-binding model, we calculate charge transfer at doped carbon nanotube junctions. This is specially interesting when one considers the effect of applying external potential difference to the junction, and there is a current flowing through it. The transferred charge is seen to be rapidly decaying at distances far from the junction, and into the bulk material at each side of the interface. Screening length is estimated to be of the order of a few link separations, even for externally-applied potential differences as large as 4 V. Calculating the variation of the total charge of the junction as a function of external bias, we conclude that the junction's relaxation time is of the order of 0.1 fs. Doped nanotube junctions are hence good candidates for very fast swiching.

INTRODUCTION

Doped carbon nanotube junctions are formed between two sides of a carbon nanotube, when there are different dopant atoms introduced in different sides of the tube, or when the substrate doping is different for the two sides. This latter case might arise from, e.g., polycrystal substrates or substrate roughness. The existence of junctions in a nanotube gives rise to the question that how the asymmetry of charge concentrations at different sides would be screened by the nanotube. A question of potential importance considering the application of external bias to and the electronic transport through the junction [1,2].

METHOD

Here, we study the screening effects and charge transfer at doped nanotube junctions using a self-consistent tight-binding formalism [3] to model the system. Within this model, the system is devided into three parts; the left and right bulk crystals, and a junction region in between which joins those semi-infinite parts together. The left and right parts are assumed to be differently doped so that the position of the Fermi level with respect to the density of states (DOS) might in general be different for the two sides. Moreover, the left and right sides could be maintained at different potentials by applying an external potential difference. The potential profile at the junction region, as well as the local density of states (LDOS) there, are calculated self-consistently. This is done in a way to include Coulomb interaction resulting from the transferred charge, so as to take Coulomb screening into account.

Using the above-mentioned model, first the surface Green's functions of the nanotube on the left and right of the junction are calculated. These surface Green's functions are then joined to the Green's function of the junction region. This way, the total Green's function of the system projected onto the junction region can be obtained [4,5]. Finally, the charge distribution on both sides of the junction are obtained through the calculation of LDOS at the junction region.

RESULTS AND DISCUSSIONS

We first consider the case of a metallic armchair tube (6,6) at bias ±1 V. The results of the charge calculation are depicted in Fig. 1. The self-consistent charge has been calculated using junctions with two different lengths; one including two

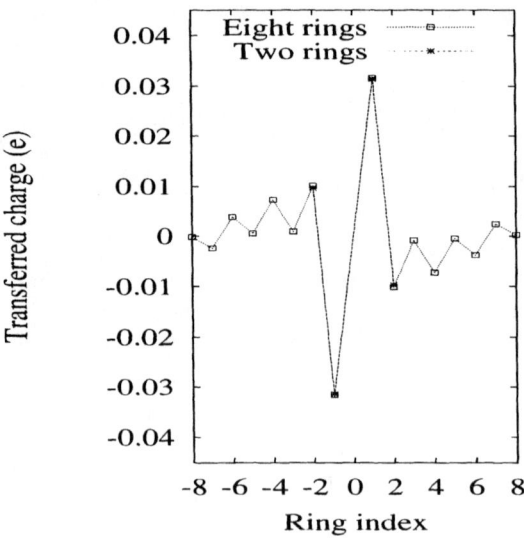

FIGURE 1. Self-consistent transferred charge per atom at the junction of a metallic armchair tube (6,6) at bias ±1 V.

carbon rings on each side, and the other including eight carbon rings on each side. First, we notice that the results of the two-ring calculation are in very good agreement with those of the eight-ring calculation. Second, it is observed that the screening length is rather small; i.e., the transferred charge reaches roughly 5% of its value at the junction – located at ring index 0 – after less than ten carbon rings away from the junction.

Next, the same nanotube is considered at bias ±2 V. The results resemble those of the Fig. 1, except for minor differences. Particularly, it is observed that the sign of the transferred charge at the second ring is the same as that at the first ring; in contrary to the case of bias ±1 V.

As an example of semiconductor tubes, we next calculate the transferred charge at the junction of a zigzag (10,0) tube. The results are shown in Fig. 2 for junctions including two and six carbon rings on each side of the interface, at bias ±1 V. It is observed that the agreement of the two-ring calculation with the six-ring one is not as good as it is for the metallic tube.

Finally, the variation of the total charge of (one side of) the junction is calculated for the metallic tube (3,3), at different externally applied potential differences. The results are depicted in Fig. 3. If one combines the resulting capacitance with the resistance obtained in a previous calculation [1], one gets a very short relaxation time of the order of 0.1 fs. The details of this calculation will be published elsewhere [6].

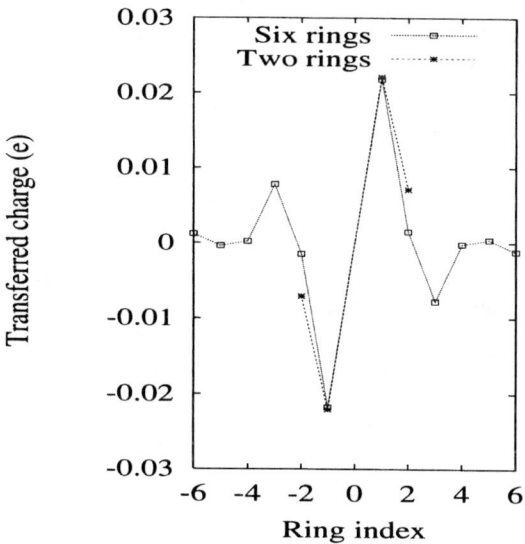

FIGURE 2. Self-consistent transferred charge per atom at the junction of a semiconductor zigzag tube (10,0) at bias ±1 V.

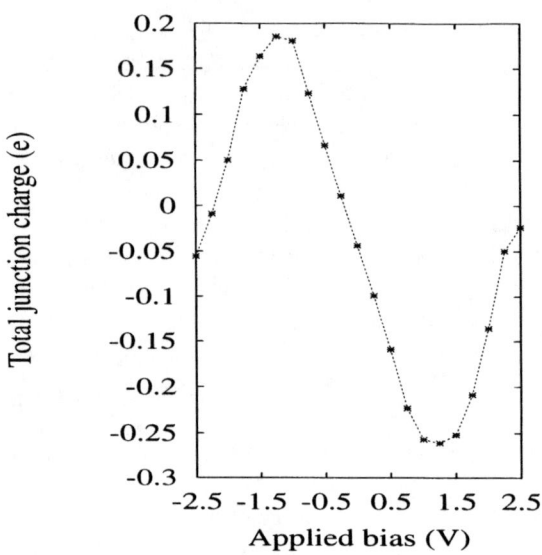

FIGURE 3. Total charge of the junction for a metallic (3,3) tube at different biases.

CONCLUSIONS

Screening length at a doped nanotube junction is observed to be of the order of a few ring separations. This being true for both semiconductor and metallic nanotubes despite minor differences. The relaxation time of a typical nanotube junction is estimated to be of the order of 0.1 fs, which indicates the very-fast-switching feature of doped nanotube junctions.

REFERENCES

1. Farajian A.A., Esfarjani K., and Kawazoe Y., *Phys. Rev. Lett.* **82**, 5084 (1999).
2. Esfarjani K., Farajian A.A., Hashi Y., and Kawazoe Y., *Appl. Phys. Lett.* **74**, 79 (1999).
3. Esfarjani K., and Kawazoe Y., *J. Phys. Condens. Matter* **10**, 8257 (1998).
4. Munoz M.C., Velasco V.R., and Garcia-Moliner F., *Prog. Surf. Sci.* **26**, 117 (1987).
5. Garcia-Moliner F., and Velasco V.R., *Theory of Single and Multiple Interfaces*, Singapore: World Scientific, 1992.
6. Esfarjani K., Farajian A.A., Chui S.T., and Kawazoe Y., in submission.

Effective-Mass Theory of Capped Carbon Nanotubes

Tatsuya Yaguchi and Tsuneya Ando

Institute for Solid State Physics, University of Tokyo
5-1-5 Kashiwanoha, Kashiwa, Chiba 277-8581, Japan

Abstract. Scattering of an electron wave by a cap is studied for several caps attached to an metallic armchair nanotube in the effective-mass approximation. The scattering phase shifts are expressed in terms of the shift in the boundary position and the effective phase shift at the boundary. The effective boundary shift is given by two ninths of the distance between the cap boundary and the cap tip, and the phase shift to be 0 and π for waves with parity $-$ and $+$, respectively.

1. INTRODUCTION

A carbon nanotube (CN) consists of coaxially rolled two-dimensional (2D) graphite sheets. Some CNs are closed by a cap at the end, sometimes consisting of a half fullerene [1]. According to Euler's theorem there are six pentagonal-membered rings in a cap if it contains no other topological defects. The local density of states of caps of a CN was calculated in a tight-binding model and the presence of states localized in a cap was predicted [2]. Scattering of the electron wave at various caps was studied in a tight-binding model and effects of a cap on electronic properties of CN's with a finite length were clarified [3]. In this paper, we study analytically the cap scattering in an effective-mass method [4].

2. EFFECTIVE BOUNDARY AND PHASE SHIFTS

In the vicinity of K and K' points of the Brillouin zone of 2D graphite, the electronic states are well described by the effective-mass equation and the envelope function $\mathbf{F}(\mathbf{r})$. The envelope function in the effective-mass equation has four components KA, KB, $K'A$, and $K'B$, where A and B denote kinds of two sublattice points in a unit cell of 2D graphite.

$$\gamma \begin{pmatrix} 0 & \hat{k}_x - i\hat{k}_y & 0 & 0 \\ \hat{k}_x + i\hat{k}_y & 0 & 0 & 0 \\ 0 & 0 & 0 & \hat{k}_x + i\hat{k}_y \\ 0 & 0 & \hat{k}_x - i\hat{k}_y & 0 \end{pmatrix} \begin{pmatrix} F_A^K(\mathbf{r}) \\ F_B^K(\mathbf{r}) \\ F_A^{K'}(\mathbf{r}) \\ F_B^{K'}(\mathbf{r}) \end{pmatrix} = \varepsilon \begin{pmatrix} F_A^K(\mathbf{r}) \\ F_B^K(\mathbf{r}) \\ F_A^{K'}(\mathbf{r}) \\ F_B^{K'}(\mathbf{r}) \end{pmatrix},$$

(2.1)

where γ is the band parameter and $\hat{k}_x$ and $\hat{k}_y$ are the wave number operator in the direction of the circumference and the axis of CN, respectively.

In the energy range $-2\pi\gamma/L < \varepsilon < 2\pi\gamma/L$ with L being the circumference of CN, the energy dispersion is $\varepsilon = \pm\gamma|k|$, where k is the wave number in the y direction chosen in the CN axis measured from K or K' points. Further, there are four traveling modes $K\pm$ and $K'\pm$ where the signs $+$ and $-$ describe incoming and outgoing mode, respectively.

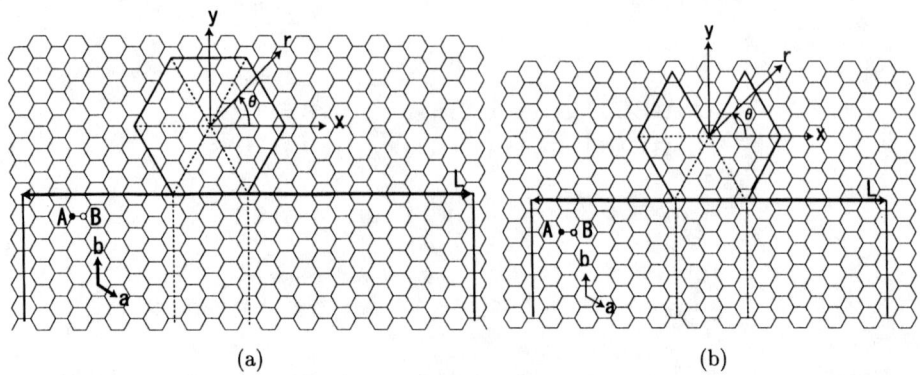

FIGURE 1. The cap structure of (a) *pencil* and (b) *bowl* caps. A pecil cap consists of 6 graphene sheets of regular triangles and a bowl cap 5 sheets. The coordinate systems (x,y) and (r,θ) are also shown.

We consider the situation that an electron incident from the left-hand side and is reflected back by the presence of a cap in an armchair CN. This process is described by the following scattering matrix:

$$S = \begin{matrix} K- \\ K'- \end{matrix} \begin{pmatrix} K+ & K'+ \\ r_{KK} & r_{KK'} \\ r_{K'K} & r_{K'K'} \end{pmatrix}, \qquad (2.2)$$

where r_{KK} and $r_{K'K'}$ describe intra-valley reflection and $r_{K'K}$ and $r_{KK'}$ intervalley reflection. The unitarity condition is written as $|r_{KK}|^2+|r_{KK'}|^2=1$ and $|r_{K'K}|^2+|r_{K'K'}|^2=1$. When a capped tube has a mirror symmetry around a plane containing the tube axis, the amplitude for intra-valley reflection vanishes identically [5] and $|r_{KK'}|=|r_{K'K}|=1$. Therefore, we can put $r_{KK'}=\exp(i\theta_{KK'})$ and $r_{K'K}=\exp(i\theta_{K'K})$ with phase shifts $\theta_{KK'}$ and $\theta_{K'K}$. The states with $K+$ and $K'-$ have parity $-$ and those with $K'+$ and $K-$ parity $+$. In the following, we shall consider such nanotubes with a mirror symmetry.

In terms of the phase shifts, the boundary conditions on the boundary $\mathbf{r}$ separating the CN and the cap are given by

$$F_C^{K'-}(\mathbf{r}) = e^{i\theta_{K'K}} F_C^{K+}(\mathbf{r}), \quad F_C^{K-}(\mathbf{r}) = e^{i\theta_{KK'}} F_C^{K'+}(\mathbf{r}), \qquad (2.3)$$

where C denotes A or B and $F_C^{K\pm}(\mathbf{r})$ and $F_C^{K'\pm}(\mathbf{r})$ are the wave functions for traveling modes. The phase shifts can be expanded around $k\sim 0$ as

$$\theta_{K'K} = \beta_{K'K} + 2\alpha_{K'K} k + \cdots, \quad \theta_{KK'} = \beta_{KK'} + 2\alpha_{KK'} k + \cdots. \qquad (2.4)$$

Then, the boundary conditions are rewritten as

$$F_C^{K'-}(\mathbf{r}+\alpha_{K'K}\mathbf{t}) = e^{i\beta_{K'K}} F_C^{K+}(\mathbf{r}+\alpha_{K'K}\mathbf{t}),$$
$$F_C^{K-}(\mathbf{r}+\alpha_{KK'}\mathbf{t}) = e^{i\beta_{KK'}} F_C^{K'+}(\mathbf{r}+\alpha_{KK'}\mathbf{t}), \qquad (2.5)$$

where $\mathbf{t}$ is a unit vector in the y direction. These equations show that we can regard $\alpha_{KK'}$ and $\alpha_{K'K}$ as an effective shift in the boundary position, and $\beta_{KK'}$ and $\beta_{K'K}$ as a phase shift at the boundary.

3. WAVE FUNCTIONS

In the following, we consider caps which have j five-membered rings lying on the circumference separating the cap and CN and a j-membered ring at the tip of the cap with $1 \le j \le 6$. These caps consist of j graphene sheets of regular triangles. For examples, a *pencil* ($j=6$) and *bowl* ($j=5$) caps have six and five graphene sheets, respectively, as shown in Fig. 1. The envelope functions in the cap should satisfy the boundary conditions [5]:

$$\mathbf{F}(r, \theta+\pi j/3) = T(\pi/3)^j \, \mathbf{F}(r, \theta), \tag{3.1}$$

with

$$T(\pi/3) = \begin{pmatrix} 0 & 0 & 0 & 1 \\ 0 & 0 & -e^{-i2\pi/3} & 0 \\ 0 & -1 & 0 & 0 \\ e^{i2\pi/3} & 0 & 0 & 0 \end{pmatrix}, \tag{3.2}$$

where r and θ are polar coordinates shown in Fig. 1. The cap has a j-fold rotation symmetry around the cap center and therefore the wave functions are specified by "angular momentum" σ, defined by

$$\mathbf{F}(r, \theta+\pi/3) = \exp(i2\pi\sigma/j) \, T(\pi/3) \, \mathbf{F}(r, \theta), \tag{3.3}$$

where $\sigma = 0, 1, \ldots, j-1$.

The effective-mass equation is solved under the above boundary conditions as

$$\mathbf{F}_{\mu+}(r,\theta) \propto \begin{pmatrix} J_\mu(\rho)e^{i\mu\theta} \\ \mathrm{sgn}(\varepsilon)iJ_{\mu+1}(\rho)e^{i(\mu+1)\theta} \\ (-1)^m e^{i\pi/3}\mathrm{sgn}(\varepsilon)iJ_{\mu+1}(\rho)e^{i(\mu+1)\theta} \\ (-1)^m e^{i\pi/3}J_\mu(\rho)e^{i\mu\theta} \end{pmatrix}, \quad \mu=3m+1+(6\sigma/j), \tag{3.4}$$

and

$$\mathbf{F}_{\mu-}(r,\theta) \propto \begin{pmatrix} J_{\mu+1}(\rho)e^{-i(\mu+1)\theta} \\ -\mathrm{sgn}(\varepsilon)iJ_\mu(\rho)e^{-i\mu\theta} \\ -(-1)^{m+1}e^{i\pi/3}\mathrm{sgn}(\varepsilon)iJ_\mu(\rho)e^{-i\mu\theta} \\ (-1)^{m+1}e^{i\pi/3}J_{\mu+1}(\rho)e^{-i(\mu+1)\theta} \end{pmatrix}, \quad \mu=3m+1-(6\sigma/j), \tag{3.5}$$

with $\rho = |\varepsilon|r/\gamma$, where $\mu > -1$ because the wave function is normalizable.

The wave functions in the tube region have also a j-fold symmetry around the tube axis and is therefore specified by σ as

$$\mathbf{F}(x+L/j, y) = \exp(i2\pi\sigma/j) \, \mathbf{F}(x, y). \tag{3.6}$$

It is clear that the wave functions with same σ are connected at the boundary of the cap and the tube.

4. TRAVELING-MODE APPROXIMATION

In the energy range $-2\pi\gamma/L < \varepsilon < 2\pi\gamma/L$, the traveling modes have $\sigma = 0$, because the wave functions are independent of x. It is sufficient therefore to consider wave functions with $\sigma = 0$ only, i.e., $\mu = 3m+1$ with $m \ge 0$ in Eqs. Eqs. (3.4) and (3.5). In the following, only the traveling modes are considered in the tube region and evanescent modes decaying away from the boundary are completely neglected, for simplicity. Further, only the most slowly-decaying wave functions $m=0$ are considered in the cap region. Then the matching conditions can be obtained simply by comparing the integral of the wave functions along the boundary of the cap and the tube.

The resulting reflection coefficient $r_{K'K}$ is

$$r_{K'K} = \frac{I_1 - \text{sgn}(\varepsilon)I_2}{I_1 + \text{sgn}(\varepsilon)I_2}$$

$$I_m = \int_{-r_0/2}^{r_0/2} dx J_m(\varepsilon\sqrt{x^2 + 3r_0^2/4}/\gamma) \left(\frac{x - i\sqrt{3}r_0/2}{\sqrt{x^2 + 3r_0^2/4}}\right)^m, \quad (4.1)$$

with $\rho_0 = |\varepsilon|r_0/\gamma$, where r_0 is the side of regular triangles constituting the cap. We have

$$r_{K'K}\big|_{k=0} = 1, \quad \frac{\partial r_{K'K}}{\partial k}\bigg|_{k=0} = i\frac{2\sqrt{3}}{9j}L, \quad (4.2)$$

Therefore, the effective shift α in the boundary position and the phase shift β at the boundary are obtained as

$$\beta_{K'K} = 0, \quad \alpha_{K'K} = (\sqrt{3}/9j)L. \quad (4.3)$$

In a same way, we obtain

$$\beta_{KK'} = \pi, \quad \alpha_{KK'} = (\sqrt{3}/9j)L. \quad (4.4)$$

The above shows that the effective boundary shift is 2/9 of the distance between the boundary and the tip of the cap and that the phase shift is 0 and π for parity − and +, respectively. These results are in good agreement with those obtain in a tight-binding model [3].

5. SUMMARY AND CONCLUSION

In summary, we have analytically calculated the scattering matrix of the electron wave at various caps which consist of j graphene sheets of regular triangles in the effective-mass method. Within the traveling-mode approximation, the effective boundary shift has been calculated to be two ninths of the distance between the cap boundary and the tip, and the phase shift to be 0 and π for parity − and +, respectively. Calculations with inclusion of evanescent modes are underway and will be published elsewhere.

ACKNOWLEDGMENTS

This work has been supported in part by Grants-in-Aid for Scientific Research and for Priority Area, Fullerene Network, from the Ministry of Education, Science and Culture, Japan. The authors thank Dr. H. Suzuura and Mr. H. Matsumura for helpful and fruitful discussion.

REFERENCES

1. Iijima, S., *Mater. Sci. Eng.* **B19**, 172 (1993).
2. Tamura, R., and Tsukada, M., *Phys. Rev. B* **52**, 6015 (1994).
3. Yaguchi, T., and Ando, T., *J. Phys. Soc. Jpn.* (submitted for publication).
4. Ajiki, H., and Ando, T., *J. Phys. Soc. Jpn.* **62**, 3542 (1993).
5. Matsumura, H., and Ando, T., *J. Phys. Soc. Jpn.* **67**, 3542 (1998).

Observation of Coulomb Blockade in a Ti/Multi-wall Carbon Nanotube/Ti Structure

Akinobu Kanda[*], Kazuhito Tsukagoshi[†], Youiti Ootuka[*], and Yoshinobu Aoyagi[†]

[*]*Institute of Physics, University of Tsukuba and CREST(JST), Tsukuba, Ibaraki, 305-8571, Japan*
[†]*The Institute of Physical and Chemical Research (RIKEN), Wako, Saitama, 351-0198, Japan*

Abstract. Coulomb blockade effect in a multi-wall carbon nanotube contacted by Ti electrodes in metal-on-tube configuration has been observed. The junction capacitance estimated from the I-V curves is proportional to the length of the nanotube segments overlapped by Ti electrodes, showing that the tunnel barrier does not locate inside the nanotube but between the nanotube and a Ti electrode, that is, the nanotube, as a whole, behaves as an island. The high resistance of the sample, which is crucial for the observation of the Coulomb blockade, can be attributed to mismatch between Ti and nanotube.

INTRODUCTION

Recently, application of carbon nanotubes to molecular-scale electronic devices has been intensively studied. One of the promising candidates is the single electron transistor (SET), which is based on the Coulomb blockade (CB) of electron tunneling [1]. Here, a nanotube is used as an island contacting two electrodes through highly resistive tunnel junctions. High resistance ($R \gg R_Q = h/4e^2 \approx 6.5$ kΩ) is crucial for observation of CB effect.

In the case of single-wall carbon nanotubes (SWNTs), the CB effect has been commonly observed [2-4]. The location of the tunnel barrier depends on whether the nanotube is deposited on the metal electrode ('tube-on-metal') or vice versa ('metal-on-tube'). In the former case, tunnel barrier is formed not only between the SWNT and an electrode [2] but also in many cases inside the SWNT due to its bending [3], while in the latter case, the contacts define the edges of the island due to the deformation of the nanotube [4]. On the other hand, for the multi-wall nanotubes (MWNTs), only very few results have been reported [5,6]. This is mainly due to the violation of the above CB condition since the contact area between a MWNT and an electrode is large in comparison to that for a SWNT.

In this article, we report observation of the CB effect in a metal-on-tube Ti/MWNT/Ti structure. We focus on the location of the tunnel barrier.

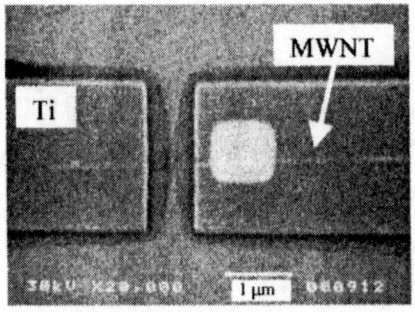

FIGURE 1. A SEM picture of a Ti/MWNT/Ti device. Ti electrodes are placed on a MWNT ('metal on tube' configuration). Bright square is a Pt/Au alignment mark.

SAMPLES

The MWNTs were synthesized using the arc-discharge method. The diameter was 10 – 20 nm. After the deposition of MWNTs on the Si/SiO$_2$ wafer with alignment markers, the position of a MWNT was determined by SEM observation. Electrodes, in a two-probe configuration, were then made by e-beam lithography and e-beam deposition of Ti. The base pressure of the evaporation chamber was 5 x 10^{-5} Pa. Figure 1 shows a SEM picture of a sample. The tube, having length of 5μm, is placed under source and drain electrodes made of Ti. The Ti electrodes with the thickness of 74 nm are separated by 0.66 μm. For comparison, we also made samples with Pt/Au electrodes (52 nm Au with sticking layer of 6 nm Pt) in the same fabrication method as for Ti electrodes.

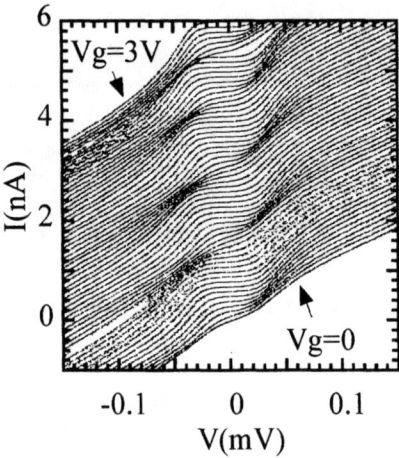

FIGURE 2. Gate voltage dependence of I-V curves at 30 mK in a sample with R=28 kΩ. Each measurement curve except for V_g=0 is offset by 0.1 nA for clarity.

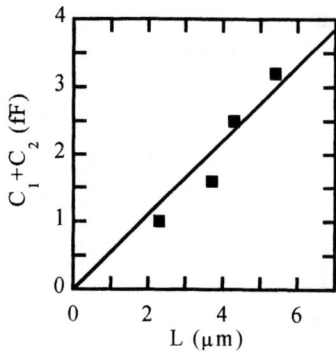

FIGURE 3. Variation of the total junction capacitance C_1+C_2 as a function of length, L, of the nanotube segments overlapped by Ti electrodes. We plotted C_1+C_2 instead of C_1 and C_2, since for some samples only C_1+C_2 was obtained from I-V curves.

We have fabricated 27 Ti samples and 13 Pt/Au samples, and measured zero-bias resistance at room temperature. Five Ti samples including one with the lowest resistance (R=28 kΩ) were cooled in a dilution refrigerator to 30 mK, and their electron transport was measured in detail. Effect of superconductivity of Ti will be reported elsewhere. Also, for cooled 4 samples we obtained length of the nanotube segments overlapped by the Ti electrodes by SEM observation.

RESULTS AND DISCUSSION

All Ti samples cooled to low temperatures showed the CB effect at T=30 mK. In Fig. 2, we show the gate voltage dependence of the I-V curves of the sample with R=28 kΩ. Gate voltage, V_g, was applied through the back gate. The features of CB, that is, zero current region near V=0 and periodic modulation of its width by V_g, are clearly seen. Following the orthodox theory of the CB effect [1], we estimate from Fig. 2 the junction capacitance C_1, C_2, and the gate capacitance C_g, to be 1.3 fF, 1.8 fF, and 0.2 aF, respectively. In the same way, we have calculated C_1, C_2, (or C_1+C_2) and C_g for cooled 5 samples.

In Fig. 3, we show C_1+C_2 vs. length, L, of the nanotube segments overlapped by Ti electrodes. The total capacitance C_1+C_2 is almost proportional to L, strongly indicating that the tunnel barrier locates between the nanotube and Ti electrodes. The proportionality constant was 0.54 fF/μm.

This result distinguishes the MWNT SET from the SWNT SET in metal-on-tube configuration. In the SWNT SET, the nanotube is strongly deformed by a metal electrode, losing the electrical conductivity. As a result, the contacts define the ends of the island [4]. On the other hand, in the MWNT SET, the nanotube does not lose its electrical conductivity and, as a whole, behaves as an island.

For observation of the CB effect, the resistance should be larger than the resistance quantum R_Q. To investigate the origin of the high resistance in Ti/MWNT/Ti structure, we compared distribution of room-temperature resistance for

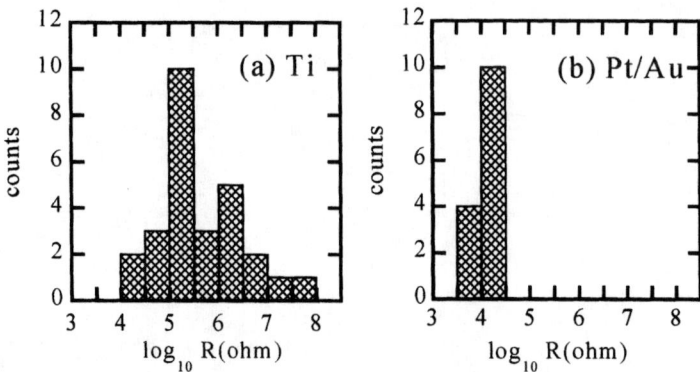

FIGURE 4. Histogram of the number of sample within a certain resistance range, plotted versus the logarithm of the resistance for (a) Ti/MWNT/Ti and (b) PtAu/MWNT/PtAu structures.

Ti/MWNT/Ti structure with that for PtAu/MWNT/PtAu. The result is shown in Fig. 4.

The resistance range of Ti/MWNT contact shifts to a high resistance in comparison with that of PtAu/MWNT contact, showing that the combination of nanotube with Ti is not suitable for good electrical contact. One possible origin is that the Ti surface contacting a nanotube is oxidized in air, resulting in the tunnel barrier.

In summary, we have observed the Coulomb blockade effect in metal-on-tube Ti/MWNT/Ti structure. The total junction capacitance is almost proportional to the length of the nanotube segments overlapped by Ti electrodes, indicating that the tunnel barrier locates between the MWNT and a Ti electrode. The high resistance is characteristic for the combination of nanotube with Ti.

ACKNOWLEDGMENT

This work was partly supported by "Research for the Future" project, by Japan Society of the Promotion of Science.

REFERENCES

1. See for instance, D. V. Averin and K. K. Likharev, in *Mesoscopic Phenomena in Solids*, edited by B. L. Altshuler, P. A. Lee and R. A. Webb (Elsevier, Amsterdam, 1991), pp.173-271.
2. Tans, S. J., Devoret, M. H., Dai, H., Thess, A., Smalley, R. E., Geerligs, L. J., and Dekker, C., *Nature* **386**, 474-477 (1997).
3. Bezryadin, A., Verschueren, A. R. M., Tans, S.J., and Dekker, C., *Phys. Rev. Letters* **80**, 4036-4039 (1998).
4. Nygard, J., Cobden, D. H., Bockrath, M., McEuen, P. L., and Lindolf, P. E., *Appl. Phys.* A69, 297-304 (1999).
5. Roschier, L., Penttila, J., Martin, M., Hakonen, P., Paalanen, M., Tapper, U., Kauppinen, E. I., Journet, C., and Bernier, P., *Appl. Phys. Letters* **75**, 728-730 (1999).
6. Ahlslog, M., Tarkiainen, R., Roschier, L., and Hakonen, P., *Appl. Phys. Letters* **77**, 4037-4039 (2000).

Energy Gap Induced by Lattice Deformation in Carbon Nanotubes

Hidekatsu SUZUURA and Tsuneya ANDO

Institute for Solid State Physics, University of Tokyo, 5-1-5 Kasiwanoha, 277-8581, Japan

Abstract. Electronic states in metallic carbon nanotubes under external force have been studied based on the effective-mass theory. Constant strains open energy gap and what type of strains can do so is dependent on the chirality of carbon nanotube. The induced gap and the total electronic energy agree well with the results by the tight-binding model and this strongly support the validity of our model.

INTRODUCTION

A carbon nanotube (CN) is a quasi-one-dimensional material made up of sp^2 carbon network [1]. It is surprising that a CN can become metallic or semiconducting dependent only on the network structure without charge doping [2]. So, CN's strongly attracts many researchers as one of the most promising candidates to support nanotechnology in the near future.

There have been several theoretical reports on energy-gap formation of carbon nanotubes under mechanical deformation [3,4]. In this paper, we study electronic states of a metallic CN under slowly-varying lattice deformation based on the effective-mass theory and show that the deformation induces the energy gap proportional to the amplitude of strain.

EFFECTIVE-MASS THEORY

First, we calculate electronic energy of metallic CN's under static and uniform strain within the effective-mass approximation. A metallic CN has two Fermi points and the Hamiltonian around Fermi energy is represented by the massless Dirac equation, or what is called Weyl equation [5]. Here, Cartesian coordinates are adopted on the nanotube surface, where x-axis is defined to be in a circumference direction with a periodic boundary condition and y-axis is in a nanotube axis. The effective Hamiltonian around one Fermi point incorporating electron-lattice interaction is derived from the tight-binding model with the modification of transfer integral due to the lattice deformation [6] and given by

$$\epsilon \begin{pmatrix} F_A(x,y) \\ F_B(x,y) \end{pmatrix} = \begin{pmatrix} h_1 & h_2 \\ h_2^\dagger & h_1 \end{pmatrix} \begin{pmatrix} F_A(x,y) \\ F_B(x,y) \end{pmatrix}, \qquad (1)$$

$$h_1 = g_1(u_{xx} + u_{yy}), \qquad (2)$$

$$h_2 = \hbar v_0(\hat{k}_x - i\hat{k}_y) + g_2 e^{i3\eta}(u_{xx} - u_{yy} + 2iu_{xy}), \qquad (3)$$

where v_0 is a Fermi velocity and $\hat{k}_x$ ($\hat{k}_y$) is defined as $-i\partial_x$ ($-i\partial_y$). Two real constants g_1 and g_2 represent coupling between electrons and strains u_{ij}. It should be noted that strains are defined as the acoustic components of lattice displacements and that off-diagonal interaction is dependent on the chiral angle η of CN's. Because the Hamiltonian around the other Fermi point has the same energy spectrum for metallic nanotubes, electronic states are discussed only around one Fermi point with the above two-component Hamiltonian.

Now that strains are static and uniform, or constants, diagonal terms give nothing but a constant energy shift to each one-particle state, and we neglect it hereafter. It is possible for off-diagonal terms to change electronic states and to generate an energy gap around Fermi energy. Actually, we analytically obtain the one-particle energy as follows:

$$\epsilon_{em}^{(\pm)} = \pm\sqrt{(\hbar v_0 k_x + U_1 \cos 3\eta - U_2 \sin 3\eta)^2 + (\hbar v_0 k_y + U_1 \sin 3\eta + U_2 \cos 3\eta)^2} \qquad (4)$$

where $U_1 = g_2(u_{xx} - u_{yy})$ and $U_2 = 2g_2 u_{xy}$. The periodic boundary condition for circumference length L makes k_x discrete as $2\pi n/L$ with n integer. The lowest conduction and the highest valence bands with $n = 0$ are in touch with each other at Fermi energy ($\epsilon = 0$). Therefore the shift to k_x opens an energy gap, while that to k_y which is a continuous variable slightly moves Fermi points with Fermi energy fixed. The energy gap induced by small strains is equal to

$$E_g = 2|U_1 \cos 3\eta - U_2 \sin 3\eta|. \qquad (5)$$

So, the energy gap grows linearly in terms of strains. What is significant here is that the induced energy gap depends on the chirality of CN's. Uni-axial strain u_{xx} or u_{yy} opens energy gap for an zigzag CN ($\eta = 0$), while torsional strain does so for an armchair CN ($\eta = \pi/6$). For a chiral CN, both types of strains induce energy gap. Electron-lattice interaction brings about the chirality dependence and this behavior of an energy gap induced in a metallic CN is closely related to its temperature dependence of phonon-limited resistivity [6].

TIGHT-BINDING MODEL

In this section, we compare the previous results with those of the tight-binding model to clarify the validity of the effective-mass theory in this problem. For simplicity, we concentrate on (N, N) armchair CN's. The one-particle energy without diagonal energy shift is given by

$$\epsilon_{tb}^{(\pm)} = \pm \left(\left[-\left(\gamma_0 + \frac{2}{3}U_1\right) \cos\frac{\pi}{N}n + \left(2\gamma_0 - \frac{2}{3}U_1\right) \cos\frac{k}{2} \right]^2 \right.$$

$$\left. + \left[-\left(\gamma_0 + \frac{2}{3}U_1\right) \sin\frac{\pi}{N}n - \frac{2\sqrt{3}}{3}U_2 \sin\frac{k}{2} \right]^2 \right)^{1/2} \quad (6)$$

for $n = 0, 1, \cdots, 2N - 1$ and $-\pi < k < \pi$. The transfer integral γ_0 is related to $\hbar v_0$ as $\hbar v_0 = 3b\gamma_0/2$ with b the equilibrium bond length. Even if $U_1 \neq 0$, or uni-axial strain is introduced, the four states with $n = 0$ remains at Fermi energy as long as $U_2 = 0$ because k can vary continuously to compensate deviations from U_1. However, torsional strain inevitably opens an energy gap proportional to its amplitude, and the energy gap is a oscillating function of strain because the bottom of higher conduction bands becomes lower as the torsional strain grows. Calculated energy gaps are plotted for $(10, 10)$ and $(20, 20)$ CN's in FIGURE 1, and both results are in excellent agreement at small strain. Deviations for large strain come from contributions of higher-order corrections in the effective-mass theory. In addition, we calculate the total electronic energy. It is necessary to make several assumptions

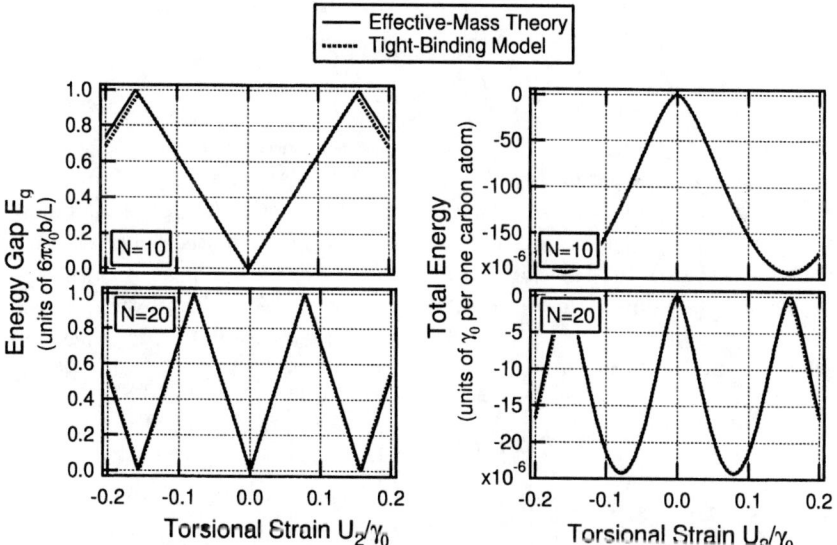

FIGURE 1. (Left) Energy gap induced by torsional strain in armchair carbon nanotubes with $N = 10$ (upper figure) and with $N = 20$ (lower figure).

FIGURE 2. (Right) Torsional-strain dependence of the total electronic energy for armchair carbon nanotubes with $N = 10$ (upper figure) and with $N = 20$ (lower figure). In this figure, the differences from the values without deformation are plotted in each data.

to compare results between the effective-mass theory and the tight-binding model. First, we should introduce an appropriate cut-off function to suppress divergence due to the n summation and k integration over infinite range [7,8]. Further, we assume that the strain dependence in the total energy is independent of what function is selected for cut-off as long as it is a smooth function decaying rapidly at infinity. Next, it is necessary to extract contributions around Fermi energy from the total energy in the tight-binding model because that effective-mass theory can describe contributions only around Fermi energy. Here, we make a conjecture that contributions away from Fermi energy are given by the total electronic energy of a graphene sheet, and we subtract the total energy of a graphene sheet under strain thorough the calculation of the tight-binding model [8]. Under these assumptions, electronic energy per one carbon atom is shown in FIGURE 2. For each data, the differences from the value without strain are plotted. Clearly, this also shows quite a good agreement, and it seems that the effective-mass theory correctly reproduces not only the band gap but also the strain dependence in the summation of electronic energy around Fermi energy. It is noted that metallic nanotubes are unstable to slowly-varying deformation because the total electronic energy decrease always overcomes the energy increase due to lattice deformation as is the case with Kekulé distortion.

DISCUSSION

Finally, we make a rough estimate of the deformation necessary for the maximum gap formation. The results in the previous section shows that the induced gap has the maximum value around $U_2 \sim 1/N$ for the $(N,N))$ armchair CN, and the ratio g_2/γ_0 is estimated to be of the order of unity. Then, u_{xy} should be of the order of $1/N$. Approximately, this strain makes an armchair CN twisted once per the length $N^2 a$ which is about 400Å for (10,10) armchair CN's. Therefore, it does not seem so difficult to open energy gap in metallic CN's by lattice deformation, and for the CN with large diameter it is promising that the gap oscillation is experimentally realized by external forces.

REFERENCES

1. Iijima, S., *Nature* **354**, 56 (1991).
2. Saito, R., Dresselhaus, G., and Dresselhaus, M.S., *Physical Properties of Carbon Nanotubes*, London: Imperial College Press, 1998.
3. Yang, L., Anantram, M.P., Han, J., and Lu, J.P., *Phys. Rev. B* **60**, 13874 (1999).
4. Yang, L., and Han, J., *Phys. Rev. Lett.* **85**, 154 (2000).
5. Ajiki, H., and Ando, T., *J. Phys. Soc. Jap.* **62** 1255 (1993).
6. Suzuura, H., and Ando, T., *Mol. Cryst. Liq. Cryst.* **340** 731 (2000).
7. Ajiki, H., and Ando, T., *J. Phys. Soc. Jap.* **62** 2470 (1993).
8. Viet, N.A., Ajiki, H., and Ando, T., *J. Phys. Soc. Jap.* **63** 3036 (1994).

Low-Temperature Magneto-Transport in Multi-Wall Carbon Nanotubes

R. Enomoto[*], N. Aoki[*], K. Ishibashi[†], and Y. Ochiai[*]

[*]*Department of Materials Science, Chiba University, 1-33 Yayoi-cho, Inage-ku, Chiba 263-8522, Japan*

[†]*The Institute for Physical and Chemical Research (RIKEN), 2-1 Hirosawa, Wako, Saitama 351-0198, Japan*

Abstract. Low temperature magnetoresistance in multi-wall carbon nanotubes prepared by d.c. arc discharge, has been studied. Those tubes show a highly distributed electrical resistance from metallic to semi-insulating conductors. No evidence for the exponential behavior is found down to 1 K, but temperature dependence of the power law is found to be clear as the conductance is reduced. We discuss these observations to try to fit Tomonaga-Luttinger theory for one-dimensional conductor.

INTRODUCTION

Recently there are great interests on a carbon nanotube for future nano-devise applications, such as a nano-scale electrical network and near-future transport devises. Carbon nanotubes are prepared by arc discharge method and multi-wall carbon nanotube (MWCNT) exists in a deposit produced on the graphite cathode [1]. Electrical transport properties of MWCNT are expected to depend on their structures such as multiplicity, chirality, imperfection and so on. Also, quantum interference behaviors are expected to appear in the low temperature magneto-resistance [2]. Theoretical studies are also reported on such novel transport properties for single wall carbon nanotubes with analyzing their band structures and transport characteristics. On the other hands, one-dimensional transports are also studied because of a fundamental attention for a recent discovery of their electron-electron interaction in quantum wire prepared by semiconductor micro-fabrication technique [3,4]. In this paper, we discuss the low temperature magneto resistance in MWCNT fabricated by d.c. glow discharge method in a dilute He gas at room temperature. Here mainly we focus on their elemental transport parameters related to weak localization of the magnetoresistance of MWCNT.

EXPERIMENTAL DETAILS

MWCNT samples were prepared by d.c. arc discharge method from graphite rod in He gas atmosphere [5]. The deposit on the cathode electrode of arc discharge seems to be a small hard material and shows a crater-like hollow at the center of the electrode. A powder material was obtained from the hard deposit by cutting and purified by heating in oxygen flow and sonicating in ethanol to remove amorphous carbon in the deposit. The graphite structure in MWCNT has been determined by the photoelectron measurements of UPS and XPS [6]. The average diameter of the MWCNT wire is scattered between 15 and 40 nm which are determined by TEM observation. We measured low-temperature magneto-resistance for single wire systems of MWCNT using two terminal measurement as shown in Fig.1a. The gap between contacts is 3 μm and the room temperature resistance is about 200 kΩ or less. Electrical contact and lead wire of the system are performed by Ti/Au thin film and attached by using a conventional photolithographic method. The sample was mounted in a conventional liq. He cryostat and resistance measurements were made using small currents to avoid electron heating.

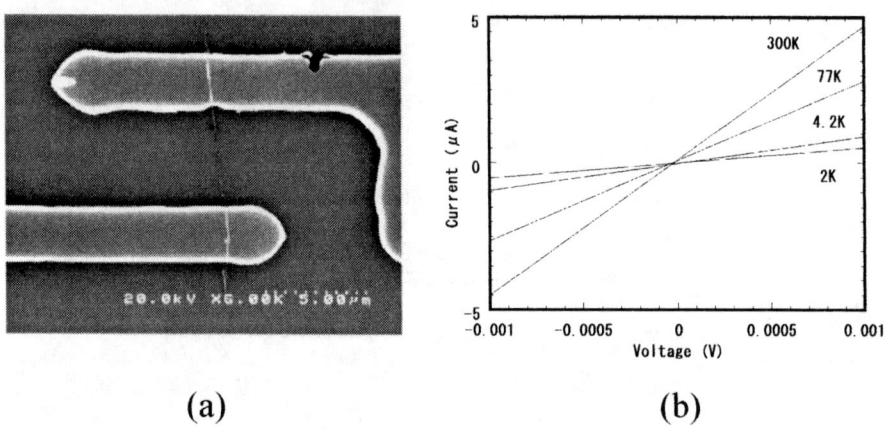

FIGURE 1. a) SEM image of the sample measured by two-terminal measurement method. b) The I-V characteristics of the sample 4.

EXPERIMENTAL RESULTS

The I-V characteristics of the MWCNT named sample 4, is shown in Fig. 1b at various temperatures. In their I-V characteristics, the all dependences have a straight line and there is no bending down to 1.7 K. Those are almost straight lines even at low temperatures so that there is no affecting on non-ohmic behavior at the contact between the metal and the sample. Also we can clearly observe universal conductance fluctuations in the magneto resistance of the sample 4 at low temperatures. The magneto resistance shows a few large peak oscillations having the period ranging from 0.75 to 2.5 T. Applying AB effect for the oscillations, we can estimate roughly the area of quantum coherent region. Two periods of the oscillations are shown in Fig. 2b. The large one is about 0.75 T and the small one is also 0.054 T. Two coherent regions are 0.55 and 7.5×10^{-14} m^2, respectively. Those oscillating behaviors are almost similar to the recent transport study of the MWCNT [2].

However, it is difficult to fit a conductance correction in two-dimensional weak localization for both the magnetic field and the temperature dependences. Especially, the power law dependence in the resistance rather fits to one-dimensional transport and shows $T^{-\alpha}$, $0<\alpha<1$. These features allow us to apply one-dimensional transport based on Tomonaga-Luttinger (T-L) model [3, 4, 7] into above MWCNT system. In

Fig. 2b, we show the conductance for two samples as a function of temperature. They show the power law behavior as shown in the figure. The slope in sample 4 is higher than that of sample 3. The room temperature resistance of samples 3 and 4 were 200 and 180 kΩ, respectively. Probably each slope value may depend on the MWCNT qualities govern by the fabrication processes.

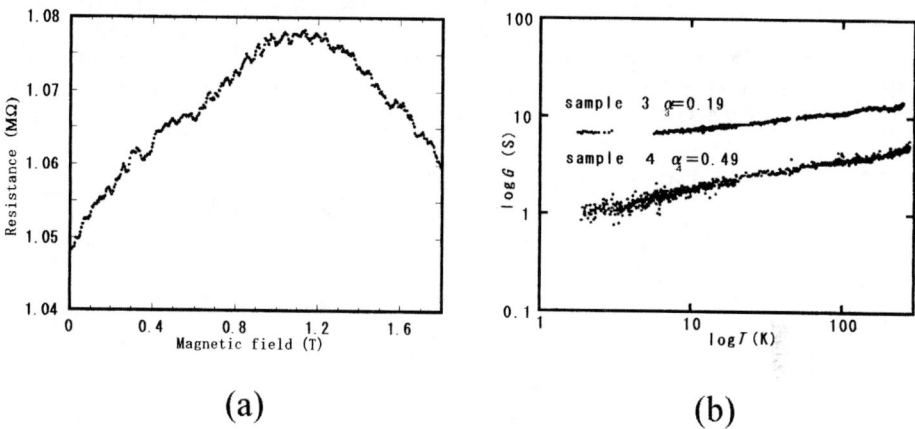

FIGURE 2. a) Low field magnetoresistance of the sample 4. b) Temperature dependence of the conductance of the two MWCNT.

According Ref.7, the Luttinger parameter, g, was 0.29 and the theoretical value was 0.28. In our results, the values are 0.56 and 0.38 from each slope of sample3 and 4, respectively. Comparing with the theoretical values, our results are slightly large. However our samples are MWNTs, it would be exist a few different properties from SWNTs. Here, we apply to another theoretical expression reported by Ogata et al. They discuss using three parameters for the fitting. Therefore, our results on the fitting to the theory are more reasonable than previous discussions Ref.7.

As mentioned above, we have applied T-L model into the result of both samples. We can determine the boundary temperature, T_{TL}, from the edge of the flat temperature dependence region in Fig 4 and T_{TL} is about 6 K. The result is indicated

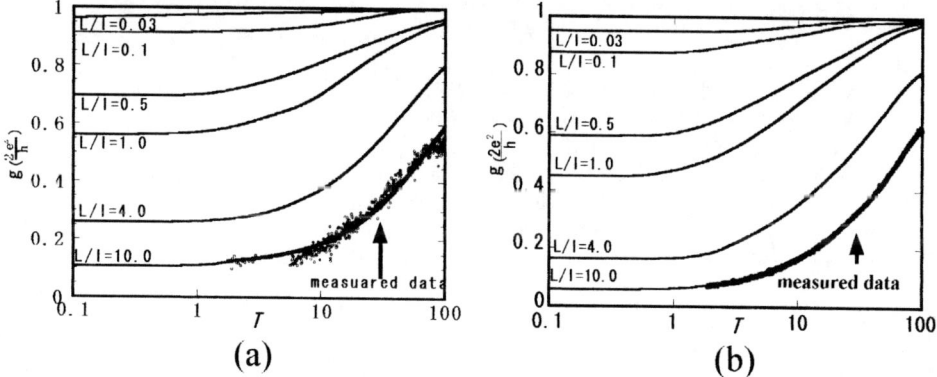

FIGURE 3. T-L fitting of a) sample 3 and b) sample 4.

in the Figs. 3a and 3b. Since the T_{TL} is defined as hv_F/Lk_B, where v_F is Fermi velocity and L is the length of MWCNT, we can estimate the v_F from L. In the case of the sample 3, v_F is about 2.4×10^5 m/s. We can also determine the mean free path, l, from fitting parameter of the theoretical calculation of T-L model [3, 4]. The l can be determined as about 0.3 μm for the sample 3. This is about 1/10 of the sample length of L. Therefore, we can estimate the transport relaxation time τ as 1.27×10^{-12} s. As for sample 4, similar behaviors are observed. The comparison between samples 3 and 4 is found in Table 1. The carrier concentration of the sample 4 can be estimated by the SdH oscillations in the magnetoresistance

TABLE 1. Transport Parameters for Comparison between Samples 3 and 4.

Sample Name	Fermi Velocity v_F (m/s)	Relaxation Time τ (s)	Carrier Density n_s (/m^2)	Fermi Velocity v_F^* (m/s)
Sample 3	2.4×10^5	1.3×10^{-12}	—	—
Sample 4	7.9×10^4	3.8×10^{-12}	1.0×10^{16}	9.8×10^5

CONCLUSIONS

The resistance of the MWCNT fabricated by arc discharge method is found to show a quasi-metallic transport. This indicates that the electron transport should be largely sensitive to the imperfections introduced during fabrication processes of the tube. As for better conductance MWCNT we can fit T-L model and consider as one-dimensional conductor. Although no evidence that for the fitting is applicable for such wires, basic transport parameters determined are almost laying on realistic value. Even at the result of two terminal measurement the observed low temperature magnetoresistance shows a weak localization behavior as well as mesoscopic quantum wires. These results lead us to suggest that the quantum transport properties of MWCNT give very important and wide possibilities for future nano-devise applications.

REFERENCES

1. Hino, S. et al., J. Phys. Chem. A **101**, 4346 (1997).
2. Fujiwara, A. et al., Phys. Rev. **60**, 13492b (1999).
3. Ogata, M. et al., Phys. Rev. Lett. **73**, 468 (1994).
4. Ogata, M. et al., Jpn. J. Appl. Phys. **34**, 3363 (1995).
5. Ago, H. et al., J. Phys. Chem. B **103**, 8116 (1999).
6. Umishita, K. et al., ICSM 2000, Austria July 19,2000.
7. Bockrath, Marc. et al., Nature. **397**, 598 (1999).

Structural Transformations in Single Wall Carbon Nanotube Bundles

Vijay Kumar[1,2], M. Sluiter[1], and Y. Kawazoe[1]

[1]*Institute for Materials Research, Tohoku University, 2-1-1 Katahira Aoba-ku, Sendai 980-8577, Japan*

[2]*Dr. Vijay Kumar Foundation, 45 Bazaar Street, K.K. Nagar (West), Chennai-600078*

Abstract. Recently there has been experimental evidence for structural transformations in single wall carbon nanotube bundles (SWCNTB) under pressure. We have performed *ab initio* electronic structure calculations on (10,10) and (17,0) SWCNTBs using the generalized gradient approximation for the exchange-correlation potential. When the cell shape is fixed to an hexagonal one, we find flattening of the nanotube walls in both the cases at low pressures. At higher pressures, the flattened walls buckle and ultimately cross-link. However, relaxation of the cell shape leads to a reversible first order transition of the (10,10) SWCNTB lattice from near hexagonal to monoclinic. These results are discussed in the light of the available experimental results.

INTRODUCTION

Currently studies of single wall carbon nanotube bundles are attracting increasing attention due to their potential technological applications such as in hydrogen storage media [1], batteries [2], etc. Their low density has added advantage. There are indications [3-6] that structural transformation occurs in SWCNTB under hydrostatic pressure. However, the nature of this transition is still not clear. Raman spectroscopy studies have provided conflicting results. Venkateswaran *et al.* [3] reported disappearance of radial breathing modes between 150 and 200 cm^{-1} in Raman spectra at around 1.5 GPa. They interpreted this to be due to a hexagonal distortion of the initially cylindrical nanotubes. Recently Tang *et al.* [4] also reported a similar observation from synchrotron X-ray diffraction and an irreversible transformation at around 5 GPa. Peters *et al.* [5] have reported a reversible structural transformation at ~ 1.7 GPa from Raman spectroscopy. Their classical molecular dynamics simulations showed a structural phase transition from near hexagonal to monoclinic phase. The coupling between the nanotubes is affected by the application of pressure due to changes in the electronic structure. Therefore, a proper understanding of the pressure effects needs first principles calculations. More recently, Venkateswaran *et al.* [6] reported Raman spectroscopy studies on SWCNTB upto 9 GPa. The frequency of the radial modes was found to upshift while the intensity decreases with an increase in pressure. This is a general result of all Raman spectroscopic studies. The value of the shift and the positions of the peaks, however, differ. This could be due to the presence of nanotubes of different diameters. In one sample, a discontinuous change has been obtained in the radial mode intensity at 2 GPa, while in another sample, the radial modes could be observed upto 7 GPa. One of the possible reasons of such behavior could also be due to the fact that in these experiments there could be penetration of the pressure transmitting media in the nanotubes that could affect the phase transitions. First principles studies could provide useful information for understanding the experimental results. We report here results of such an *ab initio* study of pressure effects on SWCNTB of (10,10) and (17,0) type. (10,10) nanotubes are reported to be abundant in such bundles and (17,0) has nearly the same diameter.

METHOD OF CALCULATION

We have used [7] the ultrasoft pseudopotentials for the electron-ion interaction and a plane wave basis for the wave functions. The cut-off energy for the plane waves is taken to be 358.4 eV. The exchange and correlation energy has been calculated within the generalized gradient approximation. The interaction between the nanotubes is weak similar to graphite or fullerenes. In bundles nanotubes arrange in a triangular lattice. However, in experimental conditions, variations in nanotube diameters are possible which could effect the structural transformations. We consider the ideal case where all the tubes are identical. Both (10,10) and (17,0) are not commensurate with hexagonal lattice. Therefore, it is likely that there would be distortions in this structure that could lead to more than one nanotubes per cell. However, in order to keep the calculations in a manageable limit, we consider here only one nanotube per cell. The cell dimensions perpendicular to the nanotube axis is about 17.5 A without application of pressure. There are 40 atoms per unit cell for (10,10) nanotubes and 68 for (17,0) nanotubes. In this large cell, **k**-space integrations were carried out using initially 1x1x7 **k**-points. Subsequently calculations with upto 29 **k**-points along the nanotube axis were carried out near the transition region for the (10,10) nanotubes. The convergence was achieved when the force on each ion became less than < 5 meV/A.

RESULTS AND DISCUSSION

We first performed calculations keeping the hexagonal symmetry for the unit cell. At low pressures facetting occurs for both (10,10) and (17,0) nanotubes,. Increasing the pressure, the nanotubes get deformed. There is no abrupt transformation in the structure, but a gradual polygonization of the nanotubes. Detailed calculations without the constraint of hexagonal symmetry show that even at 0 pressure there is a slight difference in the lengths of the *a* and *b* lattice vectors. It is due to the fact that the (10,10) and (17,0) nanotube bundles are commensurate neither with 6-fold nor with 3-fold symmetry. The nanotubes, however, are cylindrical as shown in Fig. 1. In (10,10) SWCNTs the pressure was increased in steps of 0.5 GPa. Upto a pressure of 2.5 GPa, the nanotubes retained their shapes and predominantly, there was a reduction in the inter-tube separation and a marginal increase in the difference between *a* and *b*.

When the pressure was increased further to 3 GPa, there was a sudden transformation of the nanotubes to an oval shape (Fig. 2) accompanied by a transformation to a monoclinic structure. The interatomic distances within a tube and the lattice vector along the nanotube axis remain nearly the same in the whole pressure range.

In order to locate the transition pressure, further calculations were done by decreasing the pressure in steps of 0.1 GPa on (10,10) bundles. It is found that the monoclinic structure continues at least upto 1.3 GPa for which calculations have been done. Figure 3 shows the variation of the lattice parameters perpendicular to the nanotube axis as a function of pressure. The hysteresis in the lattice constant is an indication for a first order transition. Figure 4 shows the difference of the enthalpy in the near hexagonal and monoclinic structures with 7 **k**-points. The two phases are in equilibrium at around 1.3 GPa hydrostatic pressure. This is in good agreement with the findings of structural transition around 1.5 GPa. Our calculations with 29 **k**-points at 16 GPa show the monoclinic structure to have lower enthalpy as compared to the near hexagonal lattice. Therefore, the present calculations suggest that the structural transition should occur below 16 GPa. Further calculations with large number of **k**-points at other pressures are in progress. It is possible that SWCNTBs with nanotubes having symmetries commensurate with the hexagonal lattice may show different behavior under pressure. We are currently studying such systems.

CONCLUSIONS

We have studied the behaviors of the (10,10) and (17,0) SWCNTBs under hydrostatic pressure from first principles calculations. Our results using no constraints on the cell shape suggest a transition from near hexagonal to a monoclinic structure at around 1.3 GPa for (10,10) nanotube bundles. These results are in good agreement with those obtained from Raman spectroscopy. Further work is in progress on bundles of nanotubes with other symmetries.

ACKNOWLEDGEMENTS

VK thankfully acknowledges the kind hospitality at the Institute for Materials Research, Tohoku University and the cooperation of the staff of the IMR-Tohoku University Computer Center.

REFERENCES

1) Liu, C., Fan, Y.Y., Liu, M., Cong, H.T., Cheng, H.M., and Dresselhaus, M.S Science **286**, 1127 (1999).

2) Winter, M., Besenhard, J.O., Spahr, M.E., and Novak, P., Adv. Mater. **10**, 7259 (1998).

3) Venkateswaran, U.D., Rao, A.M., Richter, E., Menon, M., Rinzler, A., Smalley, R.E., and Eklund, P.C., Phys. Rev. B**59**, 10928 (1999).

4) Tang. J., Qin, L.-C., Sasaki, T., Yudasaki, M., Matsushita, A., and Iijima, S., Phys. Rev. Lett. **85**, 1887 (2000).

5) Peters, M.J., McNeil, L.E., Lu, J.P., and Kahn, D., Phys. Rev. B**61**, 5939 (2000).

6) Venkateswaran, U.D., Brandsen, E.A., Schlecht, U., Rao, A.M., Richter, E., Loa, I., Syassen, K., and Eklund, P.C., Phys. Stat. Sol. (b) **223**, 225 (2001).

7) Kresse, G., Furthmueller, J., Comp. Mat. Sci., **6**, 15 (1996).

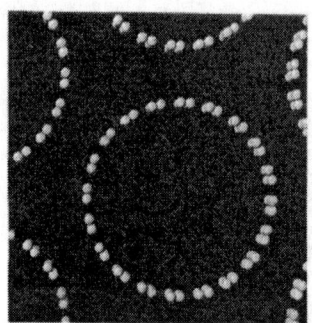
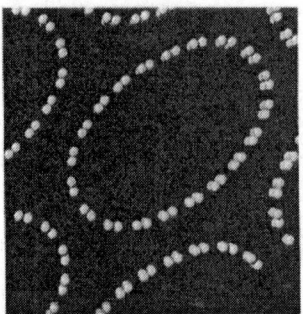

Fig. 1. SWCNTB at zero hydrostatic pressure.   Fig. 2. SWCNTB at 3.0 GPa.

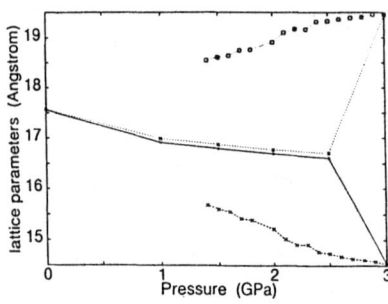
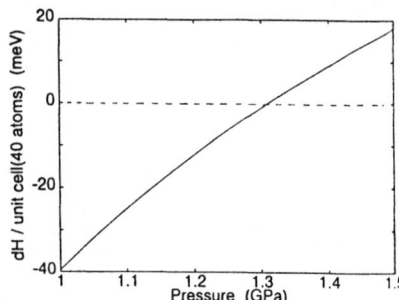

Fig. 3. Variation of the lattice parameters as a function of the hydrostatic pressure increasing from 0 to 3 GPa and then decreasing.

Fig. 4. Enthalpy of hexagonal phase minus the enthalpy of the monoclinic phase as a function of the hydrostatic pressure. The two phases are in equilibrium at ~1.3 GPa.

In-situ Atomistic Observation of Carbon Nanotubes during Field Emission

Toru Kuzumaki[*], Yasuhiro Horiike[*] and Tokushi Kizuka[†]

[*] Department of Materials Science, The University of Tokyo, Japan
[†] Department of Applied Physics, Nagoya University, Japan
[†] Research Center for Advanced Waste and Emission Management, Nagoya University, Japan
[†] Precursory Research for Embryonic Science and Technology, Japan Science and Technology Corporation, Japan

Abstract. The structural variation in the tip of carbon-nanotube during field emission was in-situ observed at an atomic scale by high-resolution transmission electron microscopy equipped with newly designed two specimen holders system. The nanotubes which were sticking out of the rod specimen were fixed in a specimen holder. In another specimen holder, a gold coated silicon tip was fixed as an opposite electrode. Dynamic in-situ observation revealed that the graphene layers of the tip protruded along the electric field direction. The protrusion then recovered during the field emission. A semi-empirical molecular orbital calculation suggested that the protrusion structure resulted from the exchanges of several hexagonal carbon rings for pentagonal-heptagonal carbon rings pairs at the tip.

INTRODUCTION

Much attention has recently been paid to carbon-nanotubes as electron sources for novel flat panel displays owing to their excellent field electron emission characteristics [1-3]. The emission characteristics are closely affected by the tip structure of the nanotubes. We have found that the protrusion formed at the tip after the field emission [4]. Structural simulations performed by semi-empirical molecular orbital calculations suggested that the deformed structure of the nanotube wall such as buckling was formed by introducing of the pentagonal and heptagonal carbon ring pair into the hexagonal network [5].

In this report, we demonstrated the in-situ observation of the nanotube tip during the field emission using high-resolution transmission electron microscope (HRTEM) equipped with newly designed two specimen holders system [6]. The atomistic deformation process of the tip was discussed.

EXPERIMENTAL PROCEDURES

Multi-walled carbon nanotubes were synthesized by the carbon DC arc-discharge method [5]. In order to use the nanotubes as an electron source, they were bundled and aligned along one direction by drawing using C_{60} crystals of ~ 10 μm in diameter. The C_{60} crystals were produced by the arc-discharge and the high performance liquid chromatography. A mixture of the C_{60} crystals with 40 mass% nanotube was packed in a silver sheath and was drawn to produce a multi-core wire [7]. The C_{60} crystals were brittle and their size reduced down to 20 nm by the drawing. The nanotubes were oriented same direction due to the flow of the refined C_{60} crystals during the drawing. The silver sheath was evaporated by heat treatment at 1243 K for 54 ks and then the nanotubes/C_{60} rod was obtained. The nanotubes were sticking out of the fractured surface of the rod tip.

The nanotubes/C_{60} rod was mounted on a specimen holder as a cathode. In another specimen holder, a gold coated silicon cantilever for atomic force microscopy was fixed as an opposite electrode. Both specimen holders were inserted in the specimen chamber of TEM with an accelerating voltage of 200 kV as shown in Fig.1 [6]. The tips of the gold electrode approached to the nanotube at the distance from 5 mm to 5 nm. Applied voltage between both tips ranged from 0 to 200 V. The structural variation during the field emission was in-situ observed at the spatial resolution of 0.1 nm using a TV rate system. The emission current was recorded simultaneously with the image observation.

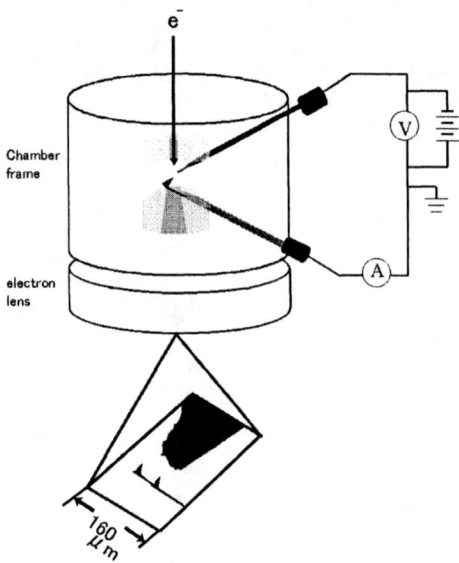

FIGURE 1. Illustration of the specimen chamber of TEM.

RESULTS AND DISCUSSION

The nanotubes were aligned along the longitudinal direction of the rod within the angular deviation of 30 deg. In the I-V measurement, the emission current was about

16 μA when the applied voltages was 200V and the distance between the electrodes was 30 nm. Figure 2 shows a time-sequence series of the HRTEM images of the tip structure of a nanotube. The images reveal that the protrusion was formed at the tip by bending the outer layers of the nanotube toward the anode direction. The protrusion then recovered. This process is repeated during the field emission. The total observation time was about 270 sec and damage was hardly observed in the tip structure.

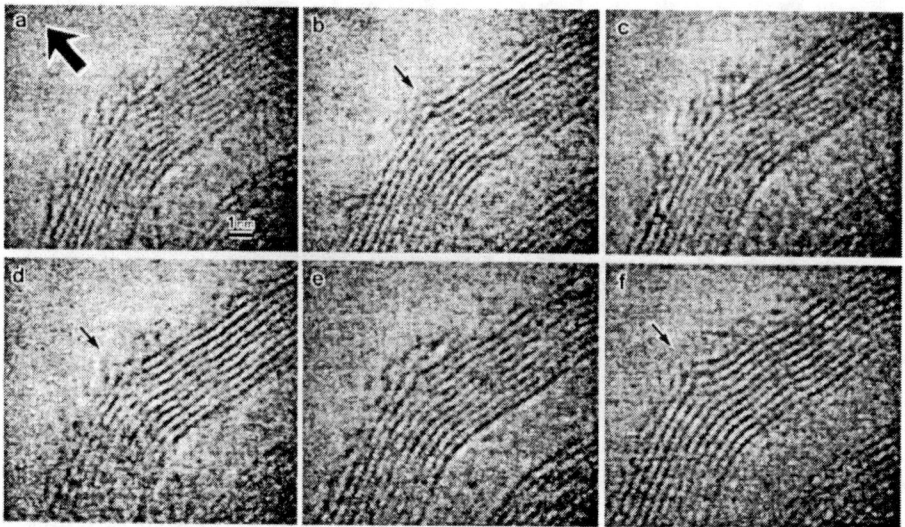

FIGURE 2. In-situ HRTEM images of the tip structure. Thick arrow shows the anode direction. The protrusion is observed at the tip in Fig.2 (b), (d) and (f). The time in each image is (a) 0/60s, (b) 382/60s, (c) 706/60s, (d) 1074/60s, (e) 1175/60s, (f) 1544/60s.

The present observation successfully revealed that the structural change occurred at the tip of the nanotube during the field emission. A kind of bending structure of the nanotube such as buckling is explained by the nucleation of pentagonal and heptagonal carbon ring pair in the hexagonal carbon network [5]. If the pentagon and heptagon carbon ring pair is introduced into the hexagonal network around a pentagonal carbon ring by heterogeneous nucleation mechanism [4], it is possible to make protrusion structure form.

To investigate the proposed mechanism, we have performed semi-empirical molecular orbital calculations (PM3) on a piece of a graphene sheet containing a pentagonal carbon ring. As a result of the calculation, the angle of the corner gradually decreased with the increase in the number of pentagonal-heptagonal carbon ring pair addition. The atomic structure model shown in Fig.3 (b) (the angle of the corner: 109°) was well in agreement with observed protrusion shown in Fig.2 (f).

The stress caused by the concentration of the electric field and thermal activation due to temperature increment by the electron emission may contribute to the formation and the motion of the pentagon-heptagon carbon ring.

The fluctuation of the distribution of electric field intensity near the tip surface occurs by the reconstruction of the atomic bonding or the absorption of hydrocarbons on the emission sites. The reversible process of the protrusion and recovery is caused if such fluctuation occurs during the electron emission.

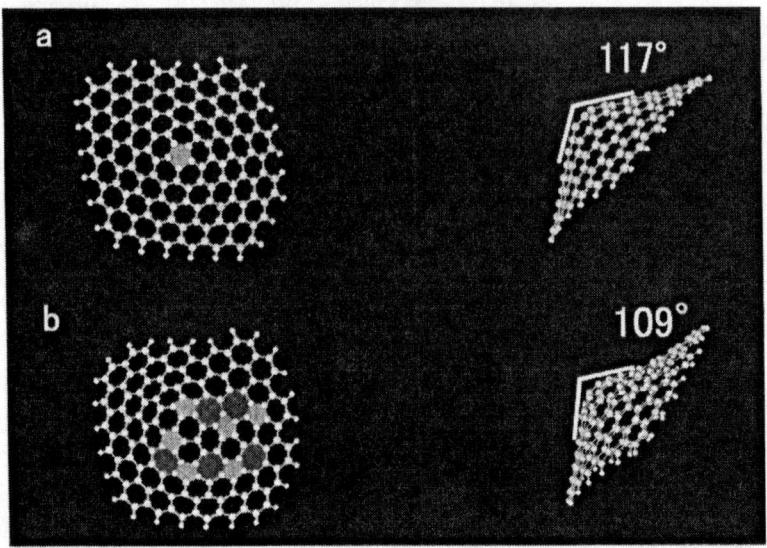

FIGURE 3. Proposed deformation model of the nanotube tip. (a) Hexagonal carbon network with a pentagonal carbon ring (a part of the closed tip), (b) Deformation of the closed tip simulated by introducing pentagonal - heptagonal carbon ring pairs into the (a).

ACKNOWLEDGMENTS

This work was partially supported by a Grant-in-Aid (No.12450289) for Scientific Research from the Ministry of Education, Science, and Culture of Japan.

REFERENCES

1. Rinzer, A. G., Hafner, H. H., Nikolaev, P., Lou, L., Kim, S. G., Tomanek, D., Nordlander, P., Colbert, D. T., and Smalley, R. E., Science, **269** 1550-1553 (1995)
2. de Heer, W. A., Chatelain, A., and Ugaarte, D., Science, **270** 1179-1180 (1995)
3. Saito, Y., Uemura, S., and Hamaguchi, K., Jpn. J. Appl. Phys. **37** L346-L348 (1998)
4. Kuzumaki, T., Takamura, Y., Ichinose, H., and Horiike, Y., Appl. Phys. Lett., to be submitted.
5. Kuzumaki, T., Hayashi, T., Ichinose, H., Miyazawa, K., Ito, K., and Ishida, Y., Phil. Mag. A**77**, 1461-1469 (1998)
6. Kizuka, T., Ohmi, H., Sumi, T., Kumazawa, K., Deguchi, S., Naruse, M, Fujisawa, S., Sasaki, S., Yabe, A., and Enomoto, Y., Jpn. J. Appl. Phys. **40** (2001) in print.
7. Kuzumaki, T., Hayashi, T., Ichinose, H., Miyazawa, K., Ito, K., and Ishida, Y., Mater. Trans. JIM, **39** 574-577 (1998)

Processing of Individual Carbon Nanotubes - Cutting and Joining

Tokushi Kizuka[*], and Toru Kuzumaki[†]

* Department of Applied Physics, Nagoya University, Japan
* Research Center for Advanced Waste and Emission Management, Nagoya University, Japan
* Precursory Research for Embryonic Science and Technology, Japan Science and Technology Corporation, Japan
† Department of Materials Science, The University of Tokyo, Japan

Abstract. Cutting and joining of individual carbon nanotubes were performed using a high-resolution transmission electron microscope equipped with the specimen chamber for atomic scale mechanics of materials. The cap of the nanotubes were burst and evaporated by the contact with nanometer-sized tips at the applied voltage more than 2V. After the burst, each carbon layer at the opened cap was terminated with the neighbor layers in the same nanotube. Two nanotubes were joined at subsequent contact at the applied voltage less than 2V.

INTRODUCTION

After the discovery of carbon nanotubes, various types of the application to structural elements or electronic devices have been proposed [1-5]. The development in the processing technology of individual nanotubes has been required to assemble them, while the synthesis and properties have been studied. The bending and fatigue of single- and multi-walled carbon nanotubes have been investigated [6-10]. However, cutting and joining are essential techniques to assemble the nanotubes, fullerenes and other nanometer-sized elements into the composite structures. For example, one nanotube was laid across another, and conductance of the joint was measured [11,12]. In this study, we manipulated individual nanotubes at an atomic scale, and the cutting and joining were performed inside a high-resolution transmission electron microscope (HRTEM).

EXPERIMENTAL PROCEDURES

Multi-walled carbon nanotubes were synthesized by the carbon DC arc-discharge method. Nanotubes were bundled and aligned along one direction by drawing using C_{60} crystals. The C_{60} crystals of ~1 μm in diameter were produced by the arc-discharge and the high performance liquid chromatography. A mixture of C_{60} crystals and the nanotube was packed in a silver sheath and was drawn to produce a multi-core rod [13]. The mixture ratio of the nanotubes was 40 mass%. The C_{60} crystals were brittle and their size reduced down to approximately 20 nm by the drawing. The nanotubes were oriented same direction due to the flow of the refined C_{60} crystals during the drawing. The silver sheath was evaporated and the composite rod was obtained by heat treatment at 1243 K for 54 ks.

In this study, we used the HRTEM equipped with the specimen chamber for atomic scale mechanics of materials [14,15]. The composite rod was mounted on a specimen holder. In another specimen holder, silicon (Si) cantilevers coated with amorphous carbon or gold (Au) for atomic force microscopy was fixed as an opposite electrode. Both specimen holders were inserted in the specimen chamber of the HRTEM. The tips of the cantilever approached to the nanotube at the distance from 5 mm to 5 nm, and then contacted. The applied voltage between both tips ranged from 0 to 10 V. The structural variation during the contact and retraction was in-situ observed at the spatial resolution of 0.1 nm using a TV rate system.

RESULTS AND DISCUSSION

Figure 1 shows a time-sequence series of the HRTEM images of the contact process between tips of the nanotube (A) and an amorphous carbon (B) at the applied voltage of 5 V. First, two tips (A, B) are separated. The cap of tip A is closed. Tip B approaches, and is contacted with tip A (Fig.1 (a-f)). The cap of the tip A is burst and disappears at the contact. The current measured during the contact is 0.4 µA.

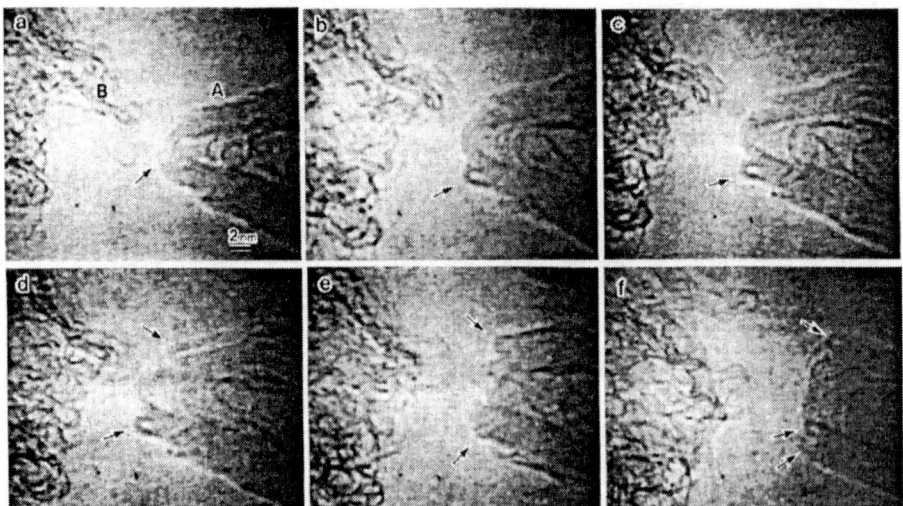

FIGURE 1. In-situ high-resolution images of the tip of the nanotube during a contact process at applied voltage of 5V. The tip (A) is burst and evaporated by contact. Arrows show the carbon layer terminated with the neighboring layer.

The closed cap of the nanotube A is opened as shown in Fig.1 (e). Each carbon layer of tip A is terminated with the neighboring layer. The length of the nanotube is reduced by about 1~10nm at one contact, showing that cutting of the nanotubes can be performed by this operation.

After several contacts of the Au coated tip and the nanotubes, the protruded tip of the nanotubes in the composite rod was fixed on the opposite tip at the applied voltage of less than 2 V. The nanotube fixed on the opposite tip was pulled out of the composite rod. We could then contact another tip of the nanotube with the other nanotubes. The nanotubes were tightly joined at the contact when the applied voltage is 2V. Figure 2 shows the high-resolution images of the joining process. The current through junction is 25 nA. The conductance is $\sim 7 \times 10^{-4}$ ($2e^2/h$) (e; electron charge, h; Plank constant). Both tips are separated by the piezo-driving, and the junction is fractured. The spacing between the layers near the junction is spread before the fracture. The joining-strength of the junction is estimated from the critical bending of the cantilever at the fracture and is ~0.6 MPa. This strength is similar to the exfoliation strength of the graphite layers, i.e. 0.4 MPa. The nanotubes was also joined with the Si, Au and amorphous carbon at the same voltage.

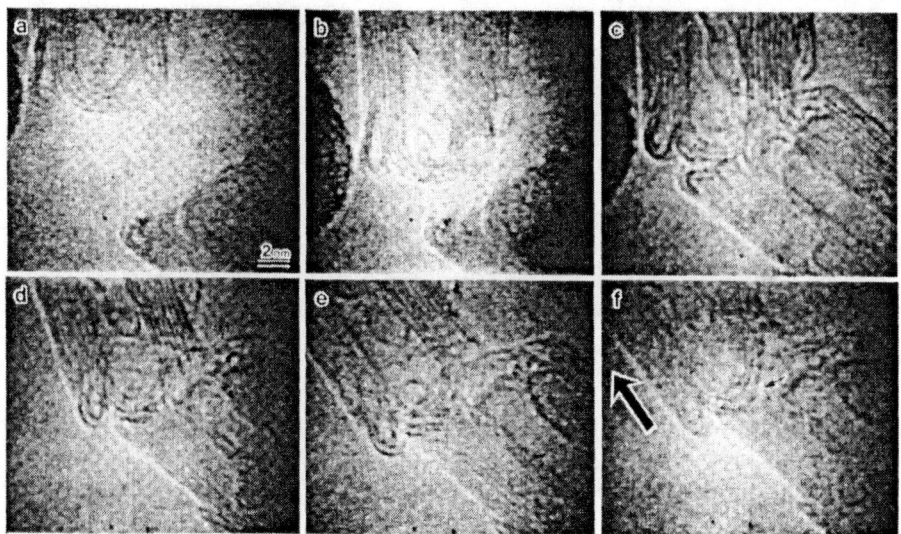

FIGURE 2. In-situ high-resolution images of the joining process between two nanotubes. The applied voltage is 2V. A larger arrow in Fig.2 (f) shows the tensile stress direction. A smaller arrow in Fig.2 (f) indicates the expansion of the layer spacing.

At the contact, the cutting and joining of the nanotubes are determined by the bias voltage. We observed the melting of Si tip after the contact at 5 V. It shows that the temperature of the junction increased at least up to the melting temperature (~1687K) by Joule heating. It is deduced that the joining of the nanotubes, i.e., the bonding of the carbon atoms of both tip surfaces, is caused by the heating. For the cutting, electromigration will contribute in addition to the evaporation due to the heating.

In summary, we demonstrated the processing of individual nanotubes. By control of the applied voltage between nanotube and the opposite tip, we could cut or join the nanotubes. The threshold voltage of the processing was 2 V. The processing will be applied to molecular electronic devices production.

ACKNOWLEDGMENTS

This research was partially supported by a Grant-in-Aid for Scientific Research from Ministry of Education, Science and Culture of Japan, The Mitsubishi Foundation, and The Fund of Toyota Physical and Chemical Research.

REFERENCES

1. Iijima, S., Nature, **354**, 56-58 (1991).
2. Sander, J. T., Alwin, R. M. V., and Dekker, C., Nature, **393**, 49-52 (1998).
3. Frank, S., Poncharal, P., Wang, Z. L., and de Heer, W. A., Science, **280**, 1744-1746 (1998).
4. Tans, S. J., Devoret, M. H., Groeneveld, R. J. A., and Dekker, C., Nature, **394**, 761-764 (1998).
5. Yao, Z., Postma, H. W. Ch., Balents, L., and Dekker, C., Nature, **402**, 273-276 (1999).
6. Iijima, S., Brabec, C. J., Maiti, A., and Bernholc, J., J. Chem. Phys., **104**, 2089-2092 (1996).
7. Tersoff, J., and Ruoff, R. S., Phys. Rev. Lett., **73**, 676-679 (1994).
8. Weldon, D. N., Blau, W. J., and Zandbergen, H. W., Chem. Phys. Lett., **241**, 365-372 (1995).
9. Kuzumaki, T., Hayashi, T., Ichinose, H., Miyazawa, K., Ito, K., and Ishida, Y., Phil. Mag. A**77**, 1461-1469 (1998).
10. Kizuka, T., Phys. Rev., **B59**, 4646-4649 (1999).
11. Lefebvre, J., Lynch, J. F., Llaguno, M. Radosavijevic, M., and Johnson, A.T., Appl. Phys. Lett., **75**, 3014-3016 (1999).
12. Fuhrer, M. S., Nygard, J., Forero, M., Yoon, Y. G., Mazzoni, M. S. C., Choi, H. J., Ihm, J., Louie S. G., Zettl, A., and McEuen, P. L., Science, **288**, 494-497 (2000).
13. Kuzumaki, T., Hayashi, T., Ichinose, H., Miyazawa, K., Ito, K., and Ishida, Y., Mater. Trans. JIM, **39**, 574-577 (1998).
14. Kuzumaki, T., Kizuka, T., " In-situ Atomistic Observation of Carbon Nanotube during Field Emission" in Proc. *ISNM 2001*.
15. Kizuka, T., Ohmi, H., Sumi, T., Kumazawa, K., Deguchi, S., Naruse, M., Fujisawa, S., Sasaki, S., Yabe, A., and Enomoto, Y., Jpn. J. Appl. Phys. **40** (2001) in print.

III. FULLERENES AND FULLERIDES

C_{60} as Building Block for New Interesting Carbon Structures and Species

Wolfgang Krätschmer

Max-Planck-Institut für Kernphysik, P.O.Box103980, D-69029-Heidelberg, Germany

Abstract. C_{60} is the most prominent fullerene and its features are well known. It is thus tempting to use C_{60} as starting compound in building up larger cage structures by chemical processes. Conversely, by controlled degradation or fragmentation of C_{60} one may obtain large linear, cyclic, or cage-like carbon molecules, which are difficult to access otherwise. The building of larger - though not perfect sp^2 cages - was accomplished by synthesizing various dimeric C_{60}-compounds, e.g. $C_{120}O$, $C_{120}O_2$. As a final product, the peanut shaped all-carbon cluster C_{119} could be prepared and its structure determined. Larger species were also obtained but could not be well characterized. Clusters like C_{119} may help to bridge the gap which so far exists between nanotubes and fullerenes and may contribute to materials research. On the other hand, C_{60} fragment molecules and their derivatives may be also of interest in organic and interstellar chemistry.

INTRODUCTION

Since 1990 when C_{60} became available in bulk amounts, the remarkable features of this molecule were studied in great detail, particularly by chemists. At the same time, the molecular crystals formed by C_{60} became research object of solid state physics and material sciences. A certain peak in the fullerene excitement was reached when the electron-doped C_{60} superconductors were discovered: C_{60} was chosen as the "molecule of the year 1991"[1]. Very recent works on hole-doped C_{60} superconductors may revive the interest in this fascinating topic [2]. A review of the major results obtained in the fullerene field until 1996 can be found in the book of Dresselhaus et al. [3].

Over the years, methods of producing endohedral fullerene compounds were developed. Unfortunately, to the present day the efficiency in preparing such species remained rather limited. In endohedral compounds, the enclosed guest atom(s) induce modifications of the electronic structure of the host-cage. Molecular solids based on such species may exhibit striking properties.

With regard to the "classical" forms of carbon, the extreme hardness of diamond has been challenged by materials which consist of fullerenes exposed to high pressure and temperature [4]. A review on the effects of pressure on fullerenes can be found in [5]. Since the sp^2 bond is potentially stronger than the sp^3 bond of diamond, carbon-based solids harder than diamond in principle should exist. The existence of a lattice of 3-D connected sp^2 carbon atoms had been already suggested by Hoffmann many years ago

[6]. In this context I may mention that in 1968 ElGoresy and Donnay in the vicinity of an ancient meteorite crater in Bavaria (Germany) discovered graphite flakes which contained tiny lamellae of a form of carbon they named chaoite [7]. This crystalline material, of which only trace amounts could be gained and which thus could not be sufficiently well characterized, may be a high pressure and/or temperature form of graphite. In passing I may also point out the possible existence of solids composed entirely of sp carbon chains, an issue which from time to time surfaces in the literature (see, e.g.[8]). It appears that in the field of carbon structures nature has still further surprises to offer.

The discovery of S. Iijima in 1991 of nanotubes grown on the graphite electrodes of a fullerene generator initiated a new development [9], which gained momentum when in the following years, with the help of metal catalysts, these tube structures could be produced in single-walled form and almost mono-dispersed in diameter. The fast development of this filed raised the problem whether carbon nanotubes can be made accessible for the analytical tools of solution based chemistry. In attacking this problem from the other end, one may try to build nanotube-like structures from soluble fullerenes, e.g. from C_{60}. Surprisingly, by using "peapods", i.e. C_{60} filled single-walled nanotubes, where the latter served as kind of reaction vessels, the thermal fusion of C_{60} cages into elongated units was reported, though only in nanoscopic scales [10]. In our approach, we in a first step produced a variety of C_{60} dimers, and tried to fuse these chemically into single cages. So far, all our attempts terminated in producing C_{119} which has a peanut structure [11]. I will report this work in the first part of my contribution.

Fullerenes smaller than C_{60} exist under collision free conditions in molecular beams. Recently, C_{20} the smallest possible fullerene could be produced and characterized in the gas phase [12]. However, such small fullerenes which cannot satisfy the celebrated "isolated pentagon rule" (IPR) seem to be too reactive and thus are difficult to prepare in quantity – disregarding the still unconfirmed report on the preparation of the fullerene C_{36} [13]. Never the less, small fullerenes may exhibit intriguing features. They may be regarded as pseudo-atoms, the valences of which may be saturated by other elements. Hydrocarbon cages may form in contact with hydrogen, or polymeric diamond-like solids may result from cage-cage linking [14]. Small fullerenes may also be formed by degradation or fragmentation of C_{60}. Recent experiments conducted by Taylor et al. indicate the formation on C_3 fragments upon thermolysis of fullerenes embedded in matrices of KBr [15]. I will report our attempts to reproduce these results.

C_{60} DIMERS AND RELATED STRUCTURES

Various strategies have been applied to obtain C_{60} dimers. Komatsu et al. used C_{60} and an electron donating salt, both subjected to mechanical milling [16], or Dragoe et al. dimerized C_{60}- fulleroid compounds [17], to mention two examples. Interestingly, both approaches lead to all-carbon C_{60} dimers. In our work we exclusively used $C_{60}O$ as starting compound. The reason is simple: this compound can be easily prepared in bulk quantities, and the epoxy-oxygen on the cage [18] serves as activation center, at which consecutive reactions can occur, leading to dimerization or, usually unwanted, also to polymerization. $C_{60}O$ is produced by a method worked out by Sergei Lebedkin [19]. At first, ozone containing oxygen is bubbled for a certain time period through a toluene C_{60} solution which during this process assumes a dirty-brownish color. When higher oxides, as e.g. $C_{60}O_2$ or $C_{60}O_3$ start to form, the bubbling should be stopped, since higher oxides in the following procedures promote polymer formation, leading to unsolvable compounds. After evaporation of the solvent, a powder is obtained consisting of a mixture of unreacted C_{60} (80-90 %) and $C_{60}O$ (10-20 %). It is important that the following reactions are carried out in solid state, i.e. the reactants are finely dispersed mixed powders. The same reactions carried out in solutions are, if technical feasible at all, much less efficient. The powder is heated under inert conditions to various temperatures ranging from 200-600°C for 1 hour, re-dissolved and analyzed by HPLC and MS. At temperatures around 200°C, the dimer $C_{120}O$ forms and can be readily separated from the solution by fractionated precipitation, in which use is made from the much lower solubility of $C_{120}O$ as compared to that of the excess C_{60}. The $C_{120}O$ yield with respect to the initial $C_{60}O$ is high, about 80-90%. Figure 1 shows the structure of $C_{120}O$ which exhibits C_{2v} symmetry, as has been confirmed by ^{13}C NMR.

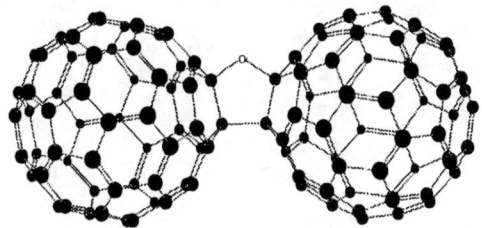

FIGURE 1. Structure the dimer $C_{120}O$ exhibiting C_{2v} symmetry. The furane-like bridge connects 6,6 bonds of either cage. The 2-fold axis is located in the plane of the paper.

The results I report in the following are based on the work by Andrei Gromov [20,21]. At temperatures up to 400 °C, $C_{120}O$ almost completely decays back into C_{60}, and the sample releases CO_2 and CO. In the soluble portion of the powder, a variety of dimeric, trimeric, and also oligomernic C_{60} oxides of the composition $(C_{60})_mO_n$ were detected. With the exception of $C_{120}O_2$, which occurs in a low (200-250 °C) and a high temperature (350-400 °C) isomer, most of the oxides have not yet been characterized in greater detail. The $C_{120}O_2$ isomer which is formed at the higher temperature with 10-20 % yield (relative to $C_{60}O$), features two adjacent furan-like bridges between the C_{60} cages, as shown in Fig. 2. Further heating to 500-600 °C produces oxides in which the carbon cages are slightly "eaten up" by further CO and CO_2 release, as testified by e.g. the compounds $C_{118}O$, $C_{178}O_3$ and others. Unfortunately, the yields of these interesting species are rather low (< 1%) precluding further characterization. Under such conditions, but having yields of a few percent, also the enigmatic cluster C_{119} can be detected, a species which was already known from early mass spectroscopy of fullerenes and fullerene-oxides [22]. We could isolate this cluster by HPLC. Using C_{60} samples enriched in ^{13}C we also could obtain a ^{13}C NMR spectrum, which suggested that C_{119} has C_2 symmetry and contains three sp^3 carbon atoms, two of which are equivalent [11]. A comparison of experimental Raman spectra with synthetic spectra

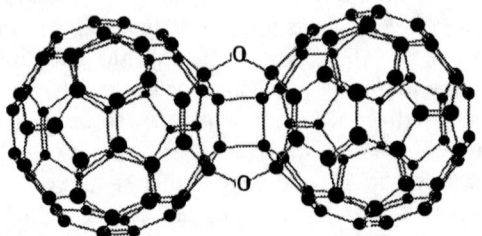

FIGURE 2. The dimer $C_{120}O_2$ featuring a pair of furan-like bridges between the C_{60} cages. The molecule as C_{2v} symmetry like $C_{120}O$. The 2-fold axis is facing perpendicular to the paper plane.

calculated by quantum-chemical methods confirmed that the thermodynamical most stable structure of Fig. 3 is indeed correct (see, e.g. [23]).

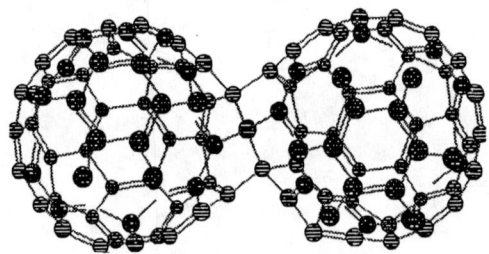

FIGURE 3. The structure of C_{119} features three bridging sp^3 carbon atoms.

As expected from the odd number of atoms, C_{119} is not a fullerene. The two cages are not opened to each other and no passage from one cage to the other exists.

To obtain a cluster which at least has an even number of carbons, the thermolysis of $C_{120}O_2$ (see Fig. 2) may be more promising. Here the release of two CO units may lead to C_{118} which, when formed, may relax directly to an elongated, tube-like fullerene. However, so far all our attempts of such a C_{118} synthesis failed, and upon heating C_{60} - $C_{120}O_2$ mixtures, we obtained nothing but an unsolvable polymer.

FULLERENE DECOMPOSITION

In our previous studies we noticed the thermal decomposition of dimeric C_{60}-oxyd samples, indicated by the development of CO_2 and CO, i.e. gasses which can be easily detected by IR spectroscopy. In our work we employed gas cells for this purpose. In a series of experiments conducted by Roger Taylor et al., these researchers embedded fullerene-oxide samples in KBr pellets, and upon heating the KBr pellet to 250 °C observed the occurrence of the sharp CO_2 (and also CO) IR lines in the sample transmission [15]. Most strikingly, however, a few samples showed a further sharp line at 2035 cm^{-1}, which in position is close to the known IR absorption of the C_3 molecule. Roger Taylor was so kind to send us such a sample for further examination. We confirmed the presence of this line and, in addition we tried to detect the also well known UV transition of C_3 at 405 nm, which in cryogenic rare gas matrices and in the gas phase is a relatively sharp line. Our attempts to detect this line failed, even when we cooled down the sample to 20 K. However, we cannot rule out that this UV line is heavily broadened by interaction of C_3 with the KBr matrix, and for that reason may have escaped our detection. In any event, the presence of an IR line so close to the C_3 band is very suggestive, and strongly indicates the decay of fullerene cages. Other fullerene fragments, like C_2, are IR silent but may be detectable in the UV. As described above, in our heating experiments we also detected dimeric oxides with damaged cages, as e.g. $C_{118}O$, however at more elevated temperatures. To gain more insight into this problem, we started similar experiments, i.e. embedded various fullerene and fullerene-oxide samples under 10 tons of pressure in KBr (i.e. applying the standard IR pellet preparation technique), and heated these samples to temperatures ranging from 200 to 500 °C. We also added Fe, Co, and Ni powder to the samples in the hope that these known catalytic metals may promote and even control fullerene decomposition. We could produce and detect CO_2 and CO in our samples, but never could see traces of a 2035 cm^{-1} line. The issue remains open, but I want to encourage researchers to work into this direction. As reward, intriguing new compounds, e.g. "shrunk" cages may be obtained. The reactivity of such non IPR-fullernens may be quenched by other elements, like e.g. hydrogen. Finally a new class of cage-hydrocarbons, and also large cyclic and linear hydrocarbons may become accessible.

CONCLUSIONS

Especially the work on large fullerenes has shown that the classical methods based on solution chemistry have serious limitations. Likewise, powerful analytical methods like e.g. ^{13}C NMR in solutions are difficult to apply in such research. Common reason for these shortcomings is the low solubility of larger fullerenes in conventional solvents. The new fullerene chemistry may better use "solid state solvents", like e.g. C_{60} itself [24]. We at least made some progress by this. For structural characterization, some methods may be adapted which are already routinely applied in nanotube research, namely the use of electron, tunnel, or force microscopy. Here the structure can be directly seen, sometimes in sufficient detail. Reference [17] contains a striking example for such kind of characterization of a fullerene dimer.

It appears that the methods which are applied in the research of carbon clusters, nanotubes, peapods, and other structures come together, and specific tools in synthesis, analysis and preparation are developing. From that a new, advanced carbon chemistry seems to emerge.

ACKNOWLEDGEMENTS

The author thanks Dr. Sabine Giesa and Ruth Alberts for sample preparation, Dr. Stefan Kalhofer for carrying out the IR and UV measurements, and Dr. William E. Hull (Deutsches Krebsforschungszentrum Heidelberg) for carrying out the rather demanding ^{13}C MNR measurements. Dr. Roger Taylor generously sent us samples. I further thank Dr. Nita Dragoe for fruitful discussions and valuable suggestions. The work was supported in part by a grant from the Deutsche Forschungsgemeinschaft.

REFERENCES

1. Issue of *Science* **254**, 1697-1860 (1991)
2. Schön J.H., Kloc Ch., Batlogg B, *Nature* **408**, 549-552 (2000)
3. Dresselhaus M.S., Dresselhaus G. and Eklund P.C., *Science of Fullerenes and Carbon Nanotubes*, Academic Press. San Diego, USA (1996)
4. Chernozatonskii L.A., Serebryanaya N.R., Mavrin B.N., *Chem. Phys. Lett.* **316**, 199-204 (2000)
5. Sundqvist B., *Advances in Physics* **48**, 1, 1-134 (1999)
6. Hoffmann R., Hughbanks T. and Kertesz M., Bird P.H., *J. Am. Chem. Soc.* **105**, 4831-4832 (1983)
7. ElGoresy A., Donnay G., *Science* **161**, 363-364 (1968)
8. Lagow R.J., Kampa J.J., Wei H.-C., Battle S.L., Genge J.W., Laude D.A., Harper C.J., Bau R., Stevens R.C., Haw J.F., Munson E., *Science* **267**, 362-367 (1995)
9. Iijima S., *Nature* **354**, 56-58 (1991)
10. Smith W.B., Luzzi D.E., *Chem. Phys. Lett.* **321**, 169-174 (2000)
11. Gromov A., Ballenweg S., Giesa S., Lebedkin S., Hull W.E., Krätschmer W., *Chem. Phys. Lett.* **267**, 460-466 (1997)
12. Prinzbach H., Weiler A., Landenberger P., Wahl J., Wörth J., Scott L.T., Gelmont M., Olevano D., Issendorff von B., *Nature* **407**, 60-63 (2000)
13. Piskoti C., Yarger J. & Zettl A., *Nature* **393**, 771-774 (1998)
14. Kroto H. W. and Walton D.R.M., *Chem. Phys. Lett.* **214**, 3,4, 353-356 (1993)
15. Taylor R., Penicaud A., Tower N J., *Chem. Phys. Lett.* **295**, 481-486 (1998)
16. Wang G.W., Komatsu K., Murata Y., Shiro M., *Nature* **387**, 583-586 (1997)
17. Dragoe N., Tanibayashi S., Nakahara K., Nakao S., Shimotani H., Xiao L., Kitazawa K., Achiba Y., Kikuchi K., and Nojima K., *Chem. Comm.* **1999**, 85-86 (1999)
18. Creegan K.M., Robbins J.L., Robbins W.K., Millar J.M., Sherwood R.D., Tindall P.J., Cox D.M., Smith III A.B., McCauley J.P. Jr., Jones D.R., Gallagher R.T., *J. Am. Chem. Soc.* **114**, 1103-1105 (1992)
19. Lebedkin S., Ballenweg S., Gross J., Taylor R., Krätschmer W., *Tetrahedron Lett.* **36**, 28, 4941-4974 (1995)
20. Gromov A., Lebedkin S., Ballenweg S., Avent A.G., Taylor R., Krätschmer W., *Chem. Comm.* 209-210 (1997)
21. Gromov A., Lebedkin S., Hull W.E., Krätschmer W., *J. Phys. Chem. A*, **102**, 26, 4997-5005 (1998)
22. McElvany S.W., Callahan J.H., Ross M.M., Lamb L.D., Huffman D.R., *Science* **260**,1632-1634 (1993)
23. Lebedkin S., Rietschel H., Adams G.B., Page J.B., Hull W., Hennrich F.H., Eisler H.-J., Kappes M.M., Krätschmer W., *J. Chem. Phys.***110**, 24, 11768-11778 (1999)
24. N. Dragoe (2000) suggested this technique.

Superconductivity in Fullerene Systems

Marvin L. Cohen

Department of Physics, University of California, Berkeley, CA 94720, USA;
Materials Sciences Division, Lawrence Berkeley National Laboratory, Berkeley, CA 94720, USA

Abstract. Using the "standard model" for calculating electronic structure and BCS theory as a basic theoretical framework, some results, predictions, discussions, and speculations are presented related to superconductivity in C_{60}, C_{36}, and $C_{24}N_{12}$ molecular solids, $B_xC_yN_z$ nanotubes, and peapods. For gated C_{36} superconductors, it is suggested that higher levels of doping will cause superconductivity in other subbands.

INTRODUCTION

The discovery [1] of superconductivity in 1911 has had an enormous influence on science. In addition to the practical applications for technology, the fields of theoretical physics and of materials synthesis and characterization owe much to investigations of superconductivity and the search for superconducting materials. One of the most active subareas of this field is the search for higher superconducting transition temperatures. Although theoretical considerations have been useful, success in this field is primarily due to experimental searches guided by insight and in some cases methodical testing. The maximum transition temperature T_c^{max} had only increased by 20K to around 25K in the 75 years since the discovery of superconductivity when the 1986 breakthrough on copper oxide materials by Bednorz and Müller [2] occurred. At this point, their work has led to an increase in the transition temperature for oxide materials from less than 1K [3] in $SrTiO_3$ to around 165K for Hg-Ba-Ca-Cu-O materials under pressure [4].

The discovery and identification [5] of C_{60} molecules and of the carbon nanotube [6] resulted in a broad spectrum of research. For superconductivity, as in the case of the oxides, T_c^{max} was raised from less than 1K for intercalated graphite to the 30–40K range [7] in chemically doped C_{60} systems and to 52K for charges introduced through gated geometries [8]. Nanotube superconductivity has been predicted [9] and there are recent experimental results [10] for small diameter carbon nanotubes.

Although there is no consensus on the underlying theory and mechanisms for superconductivity in the copper oxide superconductors, it will be argued here that the properties of fullerene superconductors such as the C_{60} based materials and carbon nanotubes are, in the main, explainable using the "Standard Model" for electronic structure [11] and BCS theory [12] with electron-phonon induced pairing. This latter statement is probably also correct for oxide superconductors without Cu such as $BaPb_{1-x}Bi_xO_3$ and the newly discovered MgB_2 [13].

If one accepts the arguments for a BCS electron-phonon induced pairing model for C_{60} and carbon nanotubes, then speculations [14] about obtaining even higher T_c's for C_{36} and $C_{36-x}N_x$ based systems or superconductivity in doped BN nanotubes [15,16] should be taken seriously. Although detailed calculations for the BN nanotubes have not been presented, similar arguments to those made for carbon nanotubes are appropriate.

The Standard Model and BCS Theory

For a broad class of solids and clusters having itinerant electrons it is possible to explain and predict their electronic and structural properties. Techniques such as the plane-wave pseudopotential approach together with density functional theory constitute a scheme or standard model [11] which is applicable when electron correlation is not too large. This approach also allows the calculation of electronic, structural, vibrational, and electron-lattice coupling properties for use with the BCS theory of superconductivity [12] to predict and explain properties of superconductors [11].

Because electron correlation is moderately large in fcc C_{60} solids and in $A_x C_{60}$ where A is an alkali atom, the question of whether the standard model and BCS theory are appropriate descriptions was called into question. However, calculations of the electronic structure [17,18] electron-phonon matrix elements [19], and transport calculations [20] give a consistent picture of these properties and the measurements of the normal and superconducting properties. It can be argued that for $A_3 C_{60}$ and for a C_{60} crystal with electrons or holes induced through external gates that the picture of C_{60} molecules emersed in a sea of fairly itinerant electrons is appropriate. Using the standard model and the view that electron pairing is achieved through intermolecular phonon excitations leads to a theoretical picture consistent with the experimental observations.

The input to the BCS theory are the phonon frequencies ω, the electron-phonon coupling parameter λ or λ^* where $\lambda^* = \dfrac{\lambda}{1+\lambda}$, and the Coulomb repulsion parameter μ or $\mu^* = \dfrac{\mu}{1+\mu \ln \dfrac{\omega_{ph}}{E_F}}$ where E_F is the Fermi energy and ω_{ph} is an appropriate phonon cutoff frequency. Using the standard model, λ can be calculated from a sum over the Fermi surface of the electron-phonon matrix element $M_{kk'}$ for scattering an electron from state k to k' with associate energies ε_k and $\varepsilon_{k'}$;

$$\lambda = \frac{\sum_{kk'} \frac{|M_{kk'}|^2}{\hbar \omega_{k-k'}} \delta(\varepsilon_k)\delta(\varepsilon k')}{\sum_k \delta(\varepsilon_k)} \qquad (1)$$

where $\omega_{k-k'}$ is the phonon frequency involved in the scattering. Equation (1) can be used to compute λ, but an alternate form is more useful for discussing λ physically

$$\lambda \sim \frac{N(E_F)\langle I^2 \rangle}{M\langle \omega^2 \rangle} \qquad (2)$$

where $\langle I^2 \rangle$ is the Fermi surface average of the square of the displacement induced electron scattering matrix element, M is the atomic mass, $\langle \omega^2 \rangle$ is an average of the square of the phonon frequencies, and $N(E_F)$ is the electronic density of states at E_F. Equation (2) can be described as the ratio of an electronic spring constant and a phonon spring constant. Hence for harmonic systems λ is a dimensionless coupling constant which is independent of M and increases when the electronic spring constant becomes stiffer and the lattice is "softer".

The modern theory of T_c for conventional superconductors is based on Eliashberg theory [21]. Approximate analytic expressions exist for moderate λ such as the McMillan Equation [22]

$$T_c = \frac{\hbar \omega_{\log}}{1.2 k_B} \exp\left[\frac{-1.04(1+\lambda)}{\lambda - \mu^* - 0.62\lambda\mu^*} \right] \qquad (3)$$

where $\omega_{\log}$ is an appropriate average of the phonon frequencies, and k_B is Boltzmann's constant. The McMillan equation for $T_c(\lambda)$ saturates for large λ whereas numerical solutions of the Eliashberg equation lead to a $\sqrt{\lambda}$ dependence for large λ. Although simple analytic expressions for $T_c(\lambda, \mu^*)$ over a wide range of λ are not available, it is possible to obtain an expression for the case where $\mu^* = 0$. The Kresin-Barbee-Cohen form [23,24]

$$T_c = 0.25 \frac{\sqrt{\omega^2}}{\sqrt{e^{\frac{2}{\lambda}} - 1}} \qquad (4)$$

is exponential for small λ and has the correct $\sqrt{\lambda}$ behavior at large λ. The quantity $\sqrt{\langle \omega^2 \rangle}$ is an average of the phonon frequencies.

Another useful parameter to use as input to the theory is the isotope effect parameter α where $T_c \propto M^{-\alpha}$ for ionic mass M. In the simplest McMillan type model

$$T_c \sim \omega_{ph} e^{-\frac{1}{\lambda^* - \mu^*}} \text{ and} \tag{5}$$

$$\alpha = \frac{1}{2}\left[1 - \left(\frac{\mu^*}{\lambda^* - \mu^*}\right)^2\right]. \tag{6}$$

The C_{60} Superconductors

Using the above as a theoretical basis, the experimental data on the normal transport properties [20] and the superconducting properties are consistent with a phonon induced pairing model for the superconductivity where the dominant phonons are intramolecular in nature. The measurements of T_c and α support this description. For example [25], using Rb_3C_{60} as a prototype, $\alpha_{Rb} \approx 0$ indicating that the Rb vibrations are not important for the phonon mechanism, while $\alpha_C \approx 0.2$ is consistent with reasonable values for the parameters λ, μ^*, and the average phonon frequencies [25]. In addition, photoemission spectra [26] yield a superconducting gap Δ to T_c ratio for Rb_3C_{60} of $\frac{2\Delta}{k_B T_c} = 3.53$ which is consistent with BCS theory. Similar results are obtained for K_3C_{60}.

Recently both electron and hole doping has been achieved using gated geometries [8]. In this study the maximum T_c found for electron doping of a C_{60} layer is 11K while hole doping yields values of T_c up to 52K. Because of the dependence of λ on the density of states $N(E_F)$ in Eq. (2), it is reasonable to investigate whether the dependence of T_c on doping is directly associated with changes in $N(E_F)$. For A_3C_{60} systems it has been demonstrated that larger lattice constants yield higher T_c's. This has been interpreted in terms of monotonic increases in $N(E_F)$ with lattice constant since $N(E_F)$ is inversely related to the bandwidth which in turn is inversely related to the lattice constant. The measured negative dependence of T_c on pressure is supportive of this idea. However it should be pointed out that peaks in the density of states do not necessarily imply large increases in T_c. This point can be illustrated using a total dielectric function model to evaluate the pairing interactions where both the attractive and repulsive contributions to the pairing are included. Then the argument can be made using a BCS model for illustration. The "NV" BCS parameter representing the density of states times the pairing potential is inversely related to the dielectric function which itself is roughly proportional to N for large N. Hence peaks in the density of states tend to cancel out.

For systems such as the C_{60} based materials the argument is not operative using the current model [27] for the superconductivity where N is determined by the band structure and carriers from alkali atoms or other doping whereas V is dominated by intramolecular vibrations. This separation is consistent with the lattice constant and pressure dependence of T_c and it results in a more straightforward dependence of T_c on $N(E_F)$ arising from the dependence of λ on $N(E_F)$.

If we examine the electronic structure of C_{60} for the fcc [18] structure, the supporting photoemission results [28] and the further theoretical studies [29] of the density of states for the Fm3, Pa3 and isotropically, merohedrally disordered structures, characteristic peak structures in the density of states are evident. In particular the H_u states for the holes (HOMO) has a two peak structure each having a width of about 0.5eV. For the electron states the T_{1u}, (LUMO) state also has a two peak structure with similar widths. It is not clear at this point how to line up the charge per C_{60} molecule axis (CPM) of figure 4 of ref. 8 with the energy scale of the density of states. However, instead of assuming [8] that the peak in T_c for the hole doping case plotted against CPM mimics the entire density of states for the H_u band, here we suggest that this peak in T_c is associated with only the first or highest energy HOMO peak. This interpretation also predicts that further doping could push the Fermi level through the valley between the peaks and then T_c could rise again. The electronic structure calculations suggest that the next peak is higher than the first, and this could result in a high T_c.

It is important to note that the coupling of the phonons to the holes in the H_u band changes as E_F moves through the band. This originates from the changes in the character of the electronic states at each $\bar{k}$-point in the band and also from their coupling to the phonons for different wavevectors $\bar{q}$. In addition, as the density of states gets lower, correlation effects play a larger role. At this time, we do not have access to data on the temperature dependent resistivity for different doping levels so these are open questions. The tendency toward a metal-insulator transition would decrease T_c for the lower density of states regions resulting in a faster rise and decline in T_c as a function of doping than one obtains with just an Eliashberg calculation of T_c assuming a robust metallic system. Also the two peaked structure may not be evident in T_c(CPM) if correlation effects are strong.

Similar considerations to those presented above for hole doping would be operative for electron doping. Here the T_{1u} band is of interest and the lower T_c arises because of the lower values for $N(E_F)$ or electron-phonon coupling in this band.

Using a Lorentzian density of states and a six election or hole maximum with λ proportional to the density of states, we have performed an Eliashberg calculation with ω_{log} = 1200K and μ^* used as a parameter to refine the fit to the data. For the electrons, the peak maximum T_c = 10K while the maximum T_c for the holes = 47K. The variation in μ^* used is small and within the accepted values for these systems. As discussed above, this calculation suggests another peak in T_c for both electron and hole doping when the doping level is increased.

Higher T_c's?

As stated in the last section, the rise in T_c from below 1K for graphite to the 50K range for C_{60} is similar to what occurred for the oxide superconductors from $SrTiO_3$ to $La_{2-x}Ba_xCuO_4$. There is, of course, considerable motivation to attempt to mimic the next rise found in the oxides to T_c's above the boiling point of liquid nitrogen and beyond. Another goal is to continue the exploration of the study of these materials systems to find more novel superconductors.

One natural extension of the current studies is to explore methods of increasing $N(E_F)$. For the gated C_{60} systems, doping into other peaks in the density of states is desirable. Also, increasing the lattice constant of the C_{60} layer may lead to larger T_c's as in the case of the A_3C_{60} systems. Besides doping into regions of higher density of states, increasing the density of states of the system by introducing different electronic character is an option. A good example is the work of Umemoto and Saito[30] on Ba_4C_{60} where the d-electron states of Ba can lead to higher $N(E_F)$ and hence higher T_c's.

Another path to higher T_c's for fullerene-like systems is to maximize the coupling of the electrons to the phonons. The T_c increase in going from graphite to the C_{60} systems is believed to result from the increase in the number of electron-phonon coupling channels. For graphite, the π orbitals with p_z character do not couple to s-like orbitals for various displacements. When curvature is introduced as in the case of C_{60} systems, this breaks the mirror symmetry and mixes the p_z and s orbitals which leads to a fivefold increase in the coupling.

A similar curvature argument can be made for evaluating the couplings for the case of nanotubes [9] and smaller fullerene molecules such as C_{36} and $C_{24}N_{12}$ [14,31,32]. An approximate expression for the electron-phonon pairing potential can be written [9] in terms of a flat component appropriate for graphite and a curved contribution which scales with the square of the radius of curvature. Many aspects of the results of the first-principles calculations can be recovered using this simple model for the decomposition of the pairing potential. For example, in the nanotube case T_c can be evaluated [9] as a function of tube radius. For large radii, the T_c curve approaches values appropriate for graphite while T_c increases rapidly as the radius is reduced. For very small radii, T_c increases dramatically. The results are consistent with the recent reports [10] of superconductivity in very small tubes ~ 4 Å.

In general, there is good agreement between several authors on the first principles calculation of the electron-phonon coupling strength for C_{60} and its fivefold increase from graphite [19]. Hence the application of this approach to the case of C_{36} should have credibility. An important caveat is the fact that crystal structure can play an important role, and at this point a solid C_{36} sample of known structure is not available. However, for the structures studied, a comparison of T_c for solid C_{36} with that of K_3C_{60} predicts a sixfold increase in T_c which would suggest superconductivity at temperatures above the boiling point of nitrogen [14]. An interesting companion compound is $C_{24}N_{12}$ which in addition to having the increased curvature of C_{36} over C_{60}, there are added positive features arising from the chemical changes such as changes in bond lengths.

The recent synthesis [33] of peapods (C_{60} molecules in carbon nanotubes) suggest interesting modifications of the C_{60} density of states and hence T_c. These effects are likely to be even more dramatic in C_{36} peapods if they are fabricated.

Hence modifications of the density of states and the electron-phonon couplings appear to be possible in fullerene systems through geometric and chemical changes. $B_xC_yN_z$ tubes and molecules with chemical or gated dopings in films or peapods or other geometric configurations may lead to higher T_c's. In any case, the search should prove to be scientifically interesting.

ACKNOWLEDGMENTS

I would like to thank Professor Vincent H. Crespi for calculating the Eliashberg solutions for T_c on C_{60} reported here. This work was supported by National Science Foundation Grant No. DMR00-87088 and by the Director, Office of Science, Office of Basic Energy Sciences of the U.S. Department of Energy under Contract DE-AC03-76F0098.

REFERENCES

1. Onnes, H.K., *Leiden Comm.*, 120b, 122b, 124c (1911).
2. Bednorz, J.G., and Müller. K.A., *Z. Phys.* **B64**, 189 (1986).
3. Schooley, J.F., Hosler, W.R., and Cohen, M.L., *Phys. Rev. Lett.* **12**, 474 (1964).
4. Gao, L., Xue, Y.Y., Chen, F., Xiong, Z., Meng, R.L., Ramirez, D., Chu, C.W., Eggert, J.H., and Mao, H.K., *Phys. Rev. B* **50**, 4260 (1994).
5. Kroto, H.W., Heath, J.R., O'Brien, S.C., Curl, R.F., and Smalley, R.E., *Nature* **318**, 162 (1985).
6. Iijima, S., *Nature* **354**, 56 (1991).
7. Hebard, A.F., Rosseinsky, M.J., Haddon, R.C., Murphy, D.W., Glarum, S.H., Palstra, T.T.M., Ramirez, A.P., and Kortan, A.R., *Nature* **350**, 600 (1991).
8. Schön, J.H., Kloc, Ch., and Batlogg, B., *Nature* **408**, 549 (2000).
9. Benedict, L.X., Crespi, V.H., Louie, S.G., and Cohen, M. L., *Phys. Rev. B* **52**, 14935 (1995).
10. Tang, Z.K., Zhang, L., Wang, N., Zhang, X.X., Wen, G.H., Li, G.D., Wang, J.N., Chan, C.T., and Sheng, P., to be published.
11. Cohen, M.L., *Physica Scripta* **T1**, 5 (1982).
12. Bardeen, J., Cooper, L.N., Schrieffer, J.R., *Phys. Rev.* **108**, 1175 (1957).
13. Nagamatsu, J., Nakagawa, N., Muranaka, T., Zenitani, Y., and Akimitsu, J., *Nature* **430**, 63 (2001).
14. Côté, M., Grossman, J.C., Cohen, M.L., and Louie, S.G., *Phys. Rev. Lett.* **81**, 697 (1998).
15. Rubio, A., Corkill, J.L., and Cohen, M.L., *Phys. Rev. B* **49**, 5081 (1994).
16. Chopra, N.G., Luyken, R.J., Cherrey, K., Crespi, V.H., Cohen, M.L., Louie, S.G., and Zettl, A., *Science* **269**, 966 (1995).
17. Saito, S. and Oshiyama, A., *Phys. Rev. Lett.* **66**, 2637 (1991).
18. Shirley, E.L., and Louie, S.G., *Phys. Rev. Lett.* **71**, 133 (1993).
19. Gunnarsson, O., *Rev. Mod. Phys.* **69**, 575 (1997).
20. Crespi, V.H., Hou, J.G., Xiang, X.-D., Cohen, M.L., and Zettl, A., *Phys. Rev. B* **46**, 12064 (1992).
21. Eliashberg, G.M., *JETP* **11**, 696 (1960).
22. McMillan, W.L., *Phys. Rev.* **167**, 331 (1968).

23. Kresin, V.Z., *Bull. Am. Phys. Soc.* **32**, 796 (1987).
24. Bourne, L.C., Zettl, A., Barbee III, T.W., and Cohen, M.L., *Phys. Rev. B* **36**, 3990 (1987).
25. Fuhrer, M.S., Cherry, K., Zettl, A., Cohen, M.L., and Crespi, V.H., *Phys. Rev. Lett.* **83**, 404 (1999).
26. Hesper, R., Tjeng, L.H., Heeres, A., and Sawatzky, G.A., *Phys. Rev. Lett.*, **85**, 1970 (2000).
27. Cohen, M.L., and Crespi, V.H.,"Theory of Electronic and Superconducting Properties of Fullerenes," in *Buckminsterfullerenes,* edited by W. E. Billups and M. A. Ciufolini, VCH Publishers, New York, 1993, p. 197.
28. Carlisle, J.A., Terminello, L.J., Hamza, A.V., Hudson, E.A., Shirley, E.L., Himpsel, F.J., Lapiano-Smith, D.A., Jia, J.J., Callcott, T.A., Perera, R.C.C., Shuh, D.K., Louie, S.G., Stöhr, J., Samant, M.G., and Ederer, D.L., *Mol. Cryst. Liq. Cryst.* **256**, 819 (1994).
29. Louie, S.G., and Shirley, E.L., *J. Phys. and Chem. Solids* **54**, 1767 (1993).
30. Umemoto, K., and Saito, S., *Proceedings ISNM 2001* (2001).
31. Grossman, J.C., Piskoti, C., Louie, S.G., Cohen, M.L., and Zettl, A. "Molecular and Solid C_{36}", in *Fullerenes: Chemistry, Physics, and Technology* , edited by K.M. Kadish and R.S. Ruoff, John Wiley & Sons, Inc., 2000, p. 887.
32. Cohen, M.L., "Possibility of High Temperature Superconductivity in C_{36} and $C_{24}N_{12}$" in *High Temperature Superconductivity*, edited by S. E. Barnes, J. Askenazi, J. L. Cohn, and F. Zuo, Am. Inst. of Phys., New York, 1999, p.359.
33. Hirahara, K., Suenaga, K., Bandow, S., Kato, H., Okazaki,T., Shinohara, H., and Iijima, S., *Phys. Rev. Lett.* **85**, 5384 (2000).

Electronic structure of Ba_4C_{60} and Cs_4C_{60}

Koichiro Umemoto and Susumu Saito

*Department of Physics, Tokyo Institute of Technology
2-12-1 Oh-okayama, Meguro-ku, Tokyo 152-8551, Japan*

Abstract. We study the electronic structure of body-centered-orthorhombic Ba_4C_{60} superconductor and Cs_4C_{60}. In both fullerides, it is found that the band structure is metallic and that the low symmetry of the lattice gives rise to the lifting of the threefold degeneracy of t_{1u} and t_{1g} bands even at the Γ point. We study also the electronic structure of Cs_4C_{60} under pressure and find it to be a promising candidate of a superconducting Cs fulleride.

INTRODUCTION

A recent research identified the true superconducting phase of Ba fullerides to be body-centered-orthorhombic (bco) Ba_4C_{60} (Fig. 1) [1]. Ba_4C_{60} is a very important meterial to understand nature of superconducting fullerides, because (1) it is the first noncubic superconducting fullerides among alkali and/or alkaline-earth fullerides, (2) the t_{1g} conduction band is stoichiometrically expected not to be half-filled, and (3) there is no disorder unlike other superconducting fcc A_3C_{60} and bcc $A_3Ba_3C_{60}$.

On the other hand, superconductivity at 40 K under pressure was reported in Cs_3C_{60} [2]. However, it has been still under discussion which phase of Cs fullerides

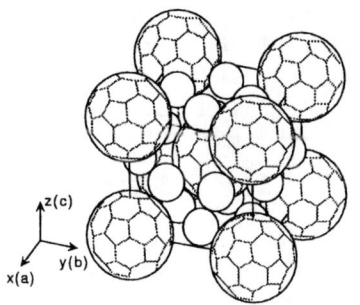

FIGURE 1. Structure of bco Ba_4C_{60} and Cs_4C_{60} studied here.

goes through the superconducting phase transition under pressure. We consider Cs_4C_{60} to be a candidate of the superconductor under pressure, because (1) Cs_4C_{60} has a bco geometry [3] similar to that of superconducting Ba_4C_{60} and the major phase of metallic Cs_3C_{60} [4], (2) the t_{1u} conduction band in Cs_4C_{60} is stoichiometrically expected to be occupied by *four* electrons, while the t_{1g} band in superconducting Ba_4C_{60} is occupied by two electrons, in other words, *four* holes.

In the present paper, we study the electronic structure of bco Ba_4C_{60} superconductor and Cs_4C_{60}, using the local-density approximation (LDA) within the framework of the density-functional theory [5]. Furthermore, we investigate the electronic structure of bco Cs_4C_{60} under the external pressure of 12 kbar which was reported to induce superconductivity at 40 K in Cs_3C_{60} [2].

COMPUTATIONAL METHOD

In the electronic-structure calculations, we adopt the Ceperley-Alder exchange-correlation potential [6] in the LDA. The norm-conserving pseudopotentials [7] with the Kleinman-Bylander separable approximation [8] are also adopted. A plane-wave basis set with a cutoff energy of 50 Ry is used. We treat $5p$ states of Cs and Ba as not core states but valence states in order to take into account a considerable spatial overlap between $5p$ states and other valence states [9,10]. As for Cs_4C_{60}, atomic coordinates are optimized until forces acting on atoms become negligible.

RESULTS AND DISCUSSION

Figure 2(a) shows the band structure and density of states (DOS) of Ba_4C_{60} in which lattice constants ($a = 11.6101 Å$, $b = 11.2349 Å$, $c = 10.8830 Å$) and atomic coordinates determined experimentally [1] are used. Fermi level lies in the t_{1g} band, which are occupied by two electrons. The calculated value of the DOS at the Fermi level, $N(E_F)$, is 5.3 states/eV spin, which is in good accord with the experimental value of 6.0 states/eV spin [1]. It should be noted that the threefold degeneracy of the t_{1g} band at the Γ point, which should appear in cubic fullerides, are completely lifted. Since the threefold degeneracy is found to be lifted even in the hypothetical pristine bco C_{60} shown in Fig. 2(b), this lifting is caused by the lower symmetry of the bco lattice. By making a comparison between two band structure, furthermore, the presence of four Ba atoms is found to widen each band considerably, which indicates the strong hybridization between Ba and C_{60} states. Due to a combination of the lifting of the degeneracy and the hybridization, the t_{1g} bandwidth is as wide as 0.88 eV, which is close to that of bcc Ba_6C_{60}.

The band structure and DOS of Cs_4C_{60}, in which atomic coordinates are optimized under lattice constants determined experimentally ($a = 12.1496 Å$, $b = 11.9051 Å$, $c = 11.4520 Å$) [3], are shown in Fig. 3(a). The Fermi level is found to cross the t_{1u} band and therefore the band structure is metallic, being in contrast to experimental reports that bct A_4C_{60} films (A=K, Rb, and Cs) were not metallic

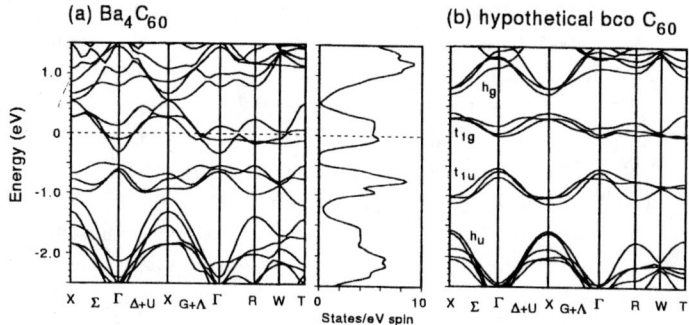

FIGURE 2. (a) The band structure (square panel) and the density of states (rectangular panel) of Ba_4C_{60} and (b) the band structure of the hypothetical bco C_{60} in which lattice constants and atomic coordinates are identical to those of Ba_4C_{60}. In (a), energy is measured from the Fermi level denoted by the horizontal broken line. The density of states is broadened by the Gaussian-distribution function with $\sigma^2 = 0.001$ eV^2.

[11]. The threefold degeneracy of the t_{1u} band at the Γ point is lifted due to the low symmetry, just like Ba_4C_{60}. The t_{1u} bandwidth is very narrow, 0.36 eV, and consequently $N(E_F)$ is rather high, 14.6 states/eV spin. The high $N(E_F)$ value should be a preferable factor for superconductivity. On the other hand, however, the residual electron correlation beyond the LDA may make Cs_4C_{60} non-metallic, because of the narrow t_{1u} bandwidth.

Next, we study the electronic structure of Cs_4C_{60} under the pressure of 12 kbar. We estimate lattice constants of Cs_4C_{60} under 12 kbar from the bulk modulus. Although each lattice constant along a, b, and c should vary independently, we assume that the ratio of lattice constants remains the same as the experimental

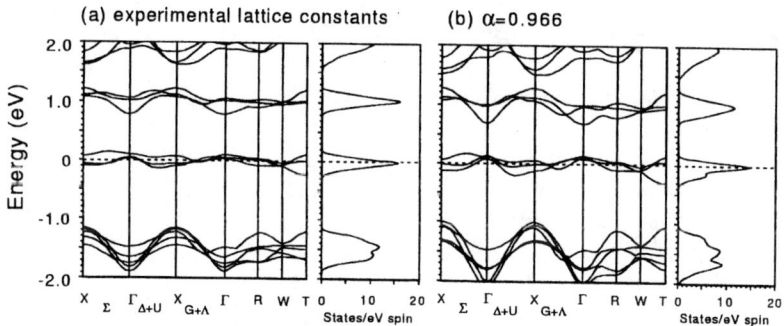

FIGURE 3. The band structure (square panel) and the density of states (rectangular panel) of Cs_4C_{60} (a) at experimental lattice constants and (b) at $\alpha = 0.966$.

value under ambient pressure, i.e., (a, b, c) is represented as $\alpha(a_{\text{exp}}, b_{\text{exp}}, c_{\text{exp}})$. From total energies calculated at several α, we estimate the bulk modulus of Cs_4C_{60} to be 3.5×10^2 kbar, which indicates that setting $\alpha = 0.966$ corresponds to applying the pressure of 12 kbar. The band structure and DOS at $\alpha = 0.966$ are shown in Fig. 3(b) and are found to be still metallic. Shorter lattice constants give rise to the wider t_{1u} bandwidth, 0.46 eV, which is comparable to that of the fcc K_3C_{60} superconductor [12]. The value of $N(E_F)$ is found to remain high, 14.2 states/eV spin, in spite of the wide t_{1u} bandwidth. Because of the metallic band structure, the wide t_{1u} bandwidth, and the high $N(E_F)$, Cs_4C_{60} under the pressure is expected to be a metal and a promising candidate of a superconducting Cs fulleride.

ACKNOWLEDGMENTS

We would like to thank Professor Y. Iwasa and co-workers, Professor Y. Kubozono, and Dr. S. Fujiki for providing their work prior to publication. We have used programs for electronic-structure calculations by Professors A. Oshiyama, T. Nakayama, and N. Hamada and Drs. M. Saito and O. Sugino. This work was supported by Grant-in-Aid for Scientific Research on the Priority Area "Fullerenes and Nanotubes" by the Ministry of Education, Science, and Culture of Japan, the Nissan Science Foundation, and the Japan Society for the Promotion of Science (Project No. 96P00203). Numerical calculations were performed on NEC SX-5 and Fujitsu VPP5000 at Reseach Center for Computational Science, Okazaki National Institute.

REFERENCES

1. C. M. Brown et al., Phys. Rev. Lett. **83**, 2258 (1999).
2. T. T. M. Palstra et al., Solid State Commun. **93**, 327 (1995).
3. P. Dahlke et al., J. Mater. Chem. **8**, 1571 (1998).
4. Y. Yoshida et al., Chem. Phys. Lett. **291**, 31 (1998).
5. P. Hohenberg and W. Kohn, Phys. Rev. **136**, B864 (1964); W. Kohn and L. J. Sham, Phys. Rev. **140**, A1133 (1965).
6. J. P. Perdew and A. Zunger, Phys. Rev. B **23**, 5048 (1981); D. M. Ceperley and B. J. Alder, Phys. Rev. Lett. **45**, 566 (1980).
7. N. Troullier and J. L. Martins, Phys. Rev. B **43**, 1993 (1990).
8. L. Kleinman and D. M. Bylander, Phys. Rev. Lett. **48**, 1425 (1982).
9. L. H. Yang et al., Phys. Rev. B **47**, 16 101 (1993).
10. T. Charpentier et al., Phys. Rev. B **54**, 1427 (1996).
11. F. Stepniak et al., Phys. Rev. B **48**, 1899 (1993); H. Suematsu et al., Mater. Sci. Eng. B **19**, 141 (1993); R. C. Haddon et al., Chem. Phys. Lett. **218**, 100 (1994); M. Knupfer and J. Fink Phys. Rev. Lett. **79**, 2714 (1997).
12. J. L. Martins and N. Troullier, Phys. Rev. B **46**, 1766 (1992).

EPR in Rb_1C_{60} under Pressure

Shigenori Kobayashi[a], Hirokazu Sakamoto[a], Kenji Mizoguchi[a], Mayumi Kosaka[b], and Katsumi Tanigaki[c]

[a]*Department of Physics, Tokyo Metropolitan University, Hachioji, Tokyo 192-0397, Japan*
[b]*Fundamental Research Laboratory, NEC Corporation, 34-Miyukigaoka, Tsukuba, Ibaraki 305-0841, Japan*
[c]*Department of Material Science, Faculty of Science, Osaka-city University, 3-3-138 Sugimoto, Sumiyoshi-ku, Osaka 558-8585, Japan and PRESTO, 4-1-8 Motomachi, Kawaguchi-city, Saitama 332-0012, Japan*

Abstract. The orthorhombic one-dimensional polymer phase of Rb_1C_{60} is proposed to be a Mott-Hubbard insulator, having three-dimensional electronic states, on the border of the metal-insulator transition at ambient pressure. That is confirmed from the transition temperature T_N vs pressure phase diagram, which was obtained on the basis of the pressure dependence of T_N by electron paramagnetic resonance (EPR), together with the electrical resistivity data under pressure reported by Khazeni et al. [*Phys. Rev.* **B56**, 6627-6630 (1997)]. The EPR linewidth can be understood from the Elliott mechanism viewpoint.

INTRODUCTION

Intensive research has been done on the orthorhombic one-dimensional polymer phase of Rb_1C_{60} stabilized by the slow cooling from face centered cubic phase above 350 K [1, 2, 3]. EPR measurement showed a rapid decrease of its intensity around 50 K [4, 5] and antiferromagnetic resonance was observed at ambient pressure [6]. Along with the data indicating relatively high electrical conductivity [4], those results were considered to show that the ground state is insulating spin density wave (SDW) state that is characteristic of quasi-one-dimensional (Q1D) electronic states. Almost experimental results including NMR and μSR were mainly regarded as the evidence for the electronic states of o-Rb_1C_{60} to be 1D. In contrast to that, band calculations suggested that this material has 3D electronic structure and would be a Mott-Hubbard insulator with strong electronic correlation [7,8]. Therefore the electronic states of o-Rb_1C_{60} is still open question. Khazeni et al. reported the pressure dependence of the electrical resistivity on a single crystal [9]. They showed that the resistivity in Rb_1C_{60} is semiconducting at ambient pressure and below 200K the transition occurred from semiconductor to metal in the range of 0.5kbar and 1.4kbar

In this study EPR intensity and linewidth were measured under several hydrostatic pressures to elucidate the electronic states of o-Rb_1C_{60}.

EXPERIMENTAL

EPR signals of Rb_1C_{60} powder sample were measured around 50 MHz under hydrostatic pressure using a clamp-type CuBe cell and Daphne 7373 oil as a pressurizing medium. The intensities were calibrated with *in-situ* ^{1}H NMR of the Daphne oil as reference. The linewidths were estimated by a least squares fit of the signals to Lorentzian functions. Pressure values at 50 K used in this report were corrected with the reported data [10] to account for the thermal contraction of the medium.

RESULTS AND DISCUSSION

The transition temperature T_N was estimated as the intensity peak temperature. The Curie-Weiss temperature Θ was also obtained from the temperature dependence of the intensity in the paramagnetic region above T_N [11]. Figure 1 shows T_N's and Θ's as functions of pressure. One can see that T_N approaches zero and Θ becomes large above 12 kbar, indicating that the pressure induced the transition from insulating to metallic state, which is consistent with the data of the electrical resistivity under pressure by Khazeni *et al.* mentioned above [9].

Combining the present and the resistivity [9] results, we constructed the temperature-pressure phase diagram as shown in Fig. 2. It should be noted that this diagram is typical of a Mott-Hubbard system. Both of interchain and intrachain transfer energy are so weak that the electronic structure is semiconducting at ambient pressure. The pressure could enhance the interchain transfer energy enough to make the interchain electonic states metallic since the interchain coupling is dominated by weak van der Waals interaction, but not covalent-bonds as within the chain. As a result, the three dimensional network is developed through C_{60} molecules in the nearest chains under pressure. This conclusion well agrees with the theoretical prediction that Rb_1C_{60} has a half filled 3D band. Contrary to that, SDW ground state with the energy gap at the Fermi energy of 1D metal is in contradiction to the paramagnetic insulator phase above T_N and the antiferromagnetic metal phase below T_N. It is noteworthy that the antiferromagnetic metal phase is characteristic of magnetically frustrated system such as the present one forming a triangle lattice [12]. Thus Rb_1C_{60} is concluded to be

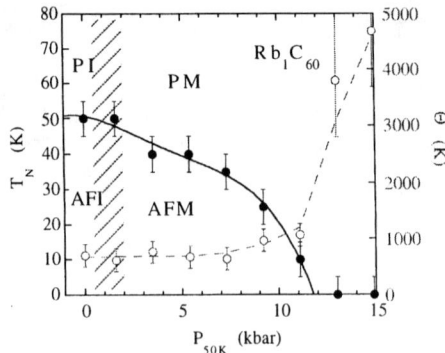

FIGURE 1. The T_N vs pressure phase diagram of Rb_1C_{60}. Closed circles: transition temperature T_N, Open circles: Curie-Weiss temperature Θ, PI: paramagnetic insulator, PM: paramagnetic metal, AFI: antiferromagnetic insulator, AFM: antiferromagnetic metal. Pressure is the estimated values at 50 K. The solid and dashed curves are guides for the eyes.

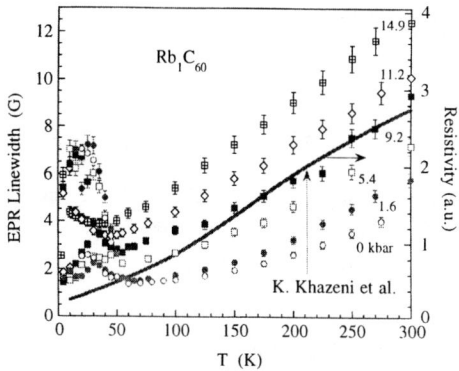

FIGURE 2. The temperature dependence of EPR linewidth. The pressures estimated at 50 K are indicated near each plot. The solid curve is the electrical resistivity (in arbitrary unit) at 10.6 kbar at 300K from Ref.9.

a Mott-Hubbard system on the border of metal-insulator transition at ambient pressure.

The temperature dependence of EPR linewidths under various pressures is shown in Fig. 2. We discuss the linewidths in the paramagnetic range above T_N, where they increase with increasing temperature. The solid curve in Fig. 2 indicates the electrical resistivity at 10.6 kbar at 300 K reproduced from Ref. 9. The agreement of the temperature variations of the linewidth and the resistivity around this pressure, where the electronic state is metallic, implies that the linewidth is governed by Elliott mechanism, applicable to the free charge carriers bearing spins with spin-phonon scattering via spin-orbit interaction [13]. Then linewidth in metallic sample is proportional to the inverse of a momentum relaxation time, i.e., the electrical resistivity. Even in the semiconducting phase, however, the linewidth could increase with temperature as observed at the low pressure data in Fig. 2, since there exist thermally excited electrons in the conduction band, as Elliott discussed originally [13]. The resistivity, however, decreases with temperature since the number of free charge carriers rises as temperature goes up.

The linewidth of ≈ 5 G at 300K at ambient pressure is about 100 times narrower than that of 3D superconducting Rb_3C_{60}, ≈ 450G. This was regarded, so far, as an evidence that the electronic structure of Rb_1C_{60} is 1D, because Elliott mechanism is suppressed in the 1D systems [13]. On the contrary, this difference in the linewidths can be understood as the difference of the g-shifts Δg in these materials, $\Delta g \approx 0.0011$ for Rb_1C_{60} and $\Delta g \approx 0.0137$ for Rb_3C_{60} [14], which dominates the Elliott linewidth proportionally to $(\Delta g)^2$. Thus it could be considered that the linewidth is dominated by Elliott mechanism even in RbC_{60}.

CONCLUSION

EPR measurements under pressure were performed to reveal the electronic structure of 1D polymer phase of o-Rb_1C_{60}. From the pressure dependences of the transition temperature obtained from the temperature variation of EPR intensity and the electrical resistivity from Ref. 9, we constructed the phase diagram, which shows that this material is a 3D Mott-Hubbard system near the metal-insulator transition at ambient pressure. The origin of the EPR linewidth is the spin-phonon scattering via spin-orbit interaction.

ACKNOWLEDGEMENTS

This work is supported by Grant-in-Aid for Scientific Research on the Priority Area "Fullerenes and Nanotubes" by the Ministry of Education, Science, Sports and Culture of Japan.

REFERENCES

1. Pekker, S., Forro, L., Mihaly, L., and Janossy, A., *Solid St.Commun.* **90**, 349-352(1994).
2. Stephens, P. W., Bortel, G., Faigel, G., Tegze, M., Jannosy, A., Pekker, S., Oszlanyi, G., and Forro, L., *Nature* **370**, 636-639(1994).
3. Launois, P., Moret, R., Hone, J., and Zettl, A., *Phys. Rev. Lett.* **81**, 4420-4423(1998).
4. Bommeli, F., Degiorgi, L., Wachter, P., Legeza, O., Jannosy, A., Oszlanyi, G., Chauvet, O., and Forro, L., *Phys. Rev.* **B51** 14794-14797(1995).
5. O. Chauvet, L. Forro, J. R. Cooper, L. Mihaly, and A. Jannosy, *Synth. Met.* **70**, 1333 -1336(1995).
6. Janossy, A., Nemes, N., Feher, T., Oszlanyi, G., Baumgartner, G., and Forro, L., *Phys. Rev. Lett.* **79**, 2718-2721(1997).
7. Erwin, S. C., Krishna, G. V., and Mele, J., *Phys. Rev.* **B51**, 7345-7348(1995).
8. Ogitsu, T., Sectional Meeting of Phys. Soc. Jpn. Sep. 1999.
9. Khazeni, K., Crespi, V. H., Hone, H., Zettl, A., and Cohen, M. L., *Phys. Rev.* **B56**, 6627-6630(1997).
10. Murata, K., Yoshino, H., Yadav, H. O., Honda, Y., and Shirakawa, N., *Rev. Sci. Instrum.* **68**, 2490-2493(1997).
11. Sakamoto, H., Kobayashi, S., Mizoguchi, K., Kosaka, M., and Tanigaki, K., *Phys. Rev.* **B62**, R7692-R7694(2000).
12. Kotliar, G., and Moeller, G., "The Mott Transition: Results from Mean-Field Theory" in *Spectroscopy of Mott Insulators and Correlated Metals*, edited by A. Fujimori and Y. Tokura, Springer Series in Solid-State Sciences Vol.119, Springer, Tokyo, 1994, pp15-27.
13. Elliott, R. J., *Phys. Rev.* **96**, 266-279(1954).
14. Mizoguchi, K., Sasano, A., Sakamoto, H., Kosaka, M., Tanigaki, K., Tanaka, T., and Atake, T., *Synth. Met.* **103**, 2395-2398(1999).

Behaviors of Metals in Production of Nanonetwork Materials Investigated by Radiochemical Technique

Keisuke Sueki*, Kazuhiko Akiyama*, Chieko Kurata*, Kazuya Oogama*, Yuliang Zhao*, Motomi Katada*, Syuichi Enomoto[†], Shizuko Ambe[†], Fumitoshi Ambe[†], Hiromichi Nakahara*, and Koichi Kikuchi*

*Department of Chemistry, Tokyo Metropolitan University, Hachiouji 192-0397 JAPAN
† RIKEN, Wako, Saitama 351-0198, JAPAN

Abstract. The iron atoms in the SWNTs soot produced from a carbon rod with Fe-Ni metal by the arc-discharge method were studied for their chemical species by Mössbauer spectroscopy. They were found present as a mixture of some Fe-Ni alloys and the α-Fe metal particles of sizes more than 10nm that were dispersed. No change was observed in the Mossbauer spectra by the treatment of the soot with 1M HCl but a drastic temperature dependence of the spectra observed for the samples burned in air indicated that more than 50 % of the iron was oxidized to Fe_2O_3 in small particles of less than 10nm. Effects of acid leaching and baking commonly used for separation of SWNTs were quantitatively studied by use of a radiotracer. Effectiveness of various kinds of metal atoms for forming nanonetwork materials were also investigated by taking advantage of the use of multi-radiotracers. A clear dependence of the yield of metal-containing fullerene species on the boiling point of the metal was indicated, and the yields were found sensitively affected by the amount of the current at the time of dc arc-discharge for the metal elements that were known to take +2 oxidation state in $M@C_{82}$.

INTRODUCTION

The property of metal atoms in relation to their capability of producing metallofullerenes and single-walled carbon nanotubes (SWNTs) remains as one of the major puzzles for basic understanding of the formation mechanism and for a large-scale production of nanonetwork materials. Macroscopic quantities of metallofullerenes has been produced in air for the metallic elements of Group II (Ca,Sr,Ba), Group III (Sc,Y), lanthanides, light actinides (Th,U), and Group IV (Hf) by use of the ordinary method of arc-discharge and/or laser ablation of carbon rods containing those metal atoms. On the other hand, SWNTs have been successfully synthesized from the carbon electrodes containing more than one kind of metallic atoms, such as Ni with Fe or Co and Pd with Rh. The reason for such selectivity of metallic atoms for formation of either metallofullerenes or SWNTs under similar production conditions is utterly unknown.

The metal atoms in the production of SWNTs are considered to act as catalyzers but they remain intricately mixed with the SWNTs and the soot in the product, and make the separation and purification of the SWNTs extremely difficult even though separation of the latter in high purity is essential for detailed characterization of the nanotubes and assessment of their potential utility. Recently, some separation and purification methods have been reported for SWNTs that mainly rely on acid-leaching and baking.

In this work, we have first investigated chemical states of the metal atoms in the webbed soot of the SWNTs by Mössbauer spectroscopy and their reactions toward acid leaching and baking by use of the radiotracer technique. Then, the selectivity of various metals in the formation of fullerene spiecies containing metallic atoms has been studied by use of the multi-radio tracers.

THE CHEMICAL STATES OF METAL ATOMS IN SWNT

Carbon rods of 6 mm diameter that contained mixed metal oxides of Ni and Fe or Ni and Co were used as the anode in dc arc-discharge. The mixing ratio of the metal and graphite powders was roughly 1: 1: 100 (Ni: Fe (or Co): C). The discharge was carried out in the He atmosphere under the pressure of 500 Torr with the current of 70 A. The webbed soot produced was peeled off from the upper part of the chamber wall and homogenized. Production of SWNTs in the webbed soot was confirmed by TEM and Raman absorptions. One portion of the webbed soot produced with the (Fe/Ni) catalyst was leached by 1M HCl solution for 90 min with sonic vibration. The other portion was baked at 623 K for 2 hours in air followed by leaching for 90 min by 1M HCl. Mössbauer absorption spectrum of ^{57}Fe was observed for the sample at each step of the separation procedure. The soot produced with the Co/Ni catalyst was irradiated in St. Paul's reactor for production of ^{60}Co. It was used for quantitative observations of the effects of acid leaching (6M HNO_3, conc. HNO_3 and 6M HCl) and baking (at 573, 623 and 673 K in an electric furnace) on the Co metal. Amounts of the metal were determined by γ-ray spectrometry using a HPGe detector.

The ^{57}Fe Mössbauer spectra of various samples observed at 77 K are shown in Figs.1a-1f. Figure 1a shows the spectrum of α-Fe used as a standard material which exhibits IS= 0 mm/s and 6 splitting peaks with the internal magnetic field of H(T)=33.1. Figure 1b shows the spectrum of the iron atoms in the initial carbon rod which tells the iron being in the α-Fe state. Figure 1c is the spectrum of the iron atoms in the initial webbed soot. Existence of α-Fe is revealed from the 6-fold splitting (H(T)=33.9) and that of Fe-Ni alloys can be deduced from the broadness of the widths of those peaks. The spectrum observed at room temperature was the same as that of Fig.1c (H(T)=31) and this invariance of the H value with temperature indicates that most of the α-Fe in the webbed soot exist with particle sizes larger than 10 nm. Figure 1d shows the spectrum of the soot treated with 1M HCl. The similarity of the spectra in Fig.1c and 1d indicates that the iron atoms remained in the webbed soot after 1M HCl acid leaching are in the α-Fe state. Figure 1e is the spectrum of the webbed soot after baking for 2h at 623 K in air. It shows the existence of two kinds of chemical states for iron atoms: metallic state of α-Fe with H(T)=32.5 and hematite(iron oxide) with H(T)=48.4. The magnetic splitting of the latter component was not observed when the temperature was raised to room temperature. This kind of disappearance of the strong magnetic splitting at elevated temperature is due to super paramagnetic phenomena and it indicates that the particle size of the hematite in the webbed soot is less than 10 nm. It is to be noted that the iron atoms in the metallic state of α-Fe do not show such phenomena, and they must be present in larger particle sizes. Figure 1f is the spectrum of the baked soot after 1M HCl acid leaching for 90 min. The iron oxide is completely dissolved although some α-Fe metal still remains.

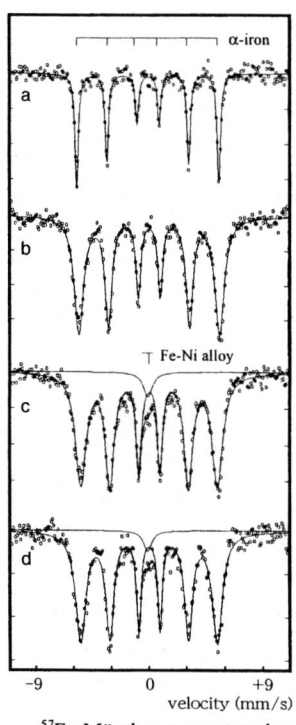
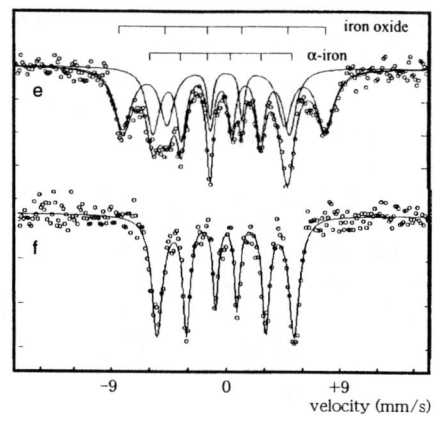

Figure 1. ^{57}Fe Mössbauer spectra observed at 77 K for the sample of (a) α-Fe metal, (b) the carbon rod containing Fe and Ni used as an anode for arc-discharge, (c) the produced webbed soot with no treatment, (d) the webbed soot leached by 1 M HCl for 90 min, (e) the webbed soot baked at 623 K for 2 h, (f) the soot of (e) leached for 90 min by 1 M HCl.

From the ^{60}Co radiotracer experiment of acid leaching on the webbed soot produced from the Co-Ni carbon rod, it was found that only less than 50 % of the Co metal was leached either by HNO$_3$ or by HCl even after 6 hours. The effect of baking on acid leaching became noticeable only at the temperature above 623 K and the percentage of the dissolved Co after the 90 min acid leaching(1M HCl) was 65-70 % for the sample baked at 623 K for 2 hrs and 75-80 % at 673 K for 2 hrs. However, observations of Raman absorptions and TEM betrayed that most of the nanotubes were also destroyed both by heating at high temperature and by the strong acid leaching of long duration.

SELECTIVITY OF VARIOUS METALS FOR PRODUCTION OF CARBON NANONETWORK MATERIAL

In previous report [1], formation of metallofullerenes is examined for 23 kinds of elements, and it is found that the yield of the metallic atoms in the solvent-extracted fraction(we call this as "crude fraction") decreases exponentially as the group number of the element in the periodic table advances from II to V (including lanthanides and actinides), and rather insensitive to the number of period. Beyond group VI, the metal elements do not make endohedral metallofullerenes and they seem to function as catalysts for formation of single-walled nanotubes.

In the present work, experiments similar to the one reported in ref.[1] were performed by use of the carbon rods with radio-multitracers that were embedded into a hole in the rod in the form of solutions and metal oxides while in the previous work, they were adsorbed to porous carbon rods by immersing them in a radiotracer solution. The aim of changing the fabrication method of carbon rods was to see the effect of the arc-discharge current, namely, the effect of the temperature of the anode on the yields of the crude products.

The results showed that yields of the crude relative to that for La were about the same for most elements although they drastically decreased to one tenth of the previous data for most elements of alkaline earth elements and Eu, Yb, namely, the elements that were known to take +2 oxidation state in the carbon cage of C_{82}. In fig. 2 are plotted the relative yields of the metallic atoms in the crude against the boiling point of the element in the metal phase where present results are depicted in open squares and the those taken from ref.[1] by solid circles. A clear correlation between the yields of the crude and the boiling points is suggested by the plot. The lowering of the yields for the elements with +2 oxidation state suggests that the formation mechanism of metallofullerenes is likely related to the vapor pressure of the metal. The present and the previous experiments both indicate that the yields of the crude for Fe and Co, namely group VIII and IX elements, are less than one-hundredth of those for Ln@C_{82}. The propensity of some elements for forming metallofullerenes while other for acting as catalysts in the SWNTs production indicates that the affinity of the metal element toward carbon atoms, such as the tendency of forming various kinds of carbides, is important for production of various kinds of nanonetwork materials. For further elucidation of such relations, another experiment on the production yields of SWNTs for various kinds of elements by use of the multi-radiotracer method is being planned.

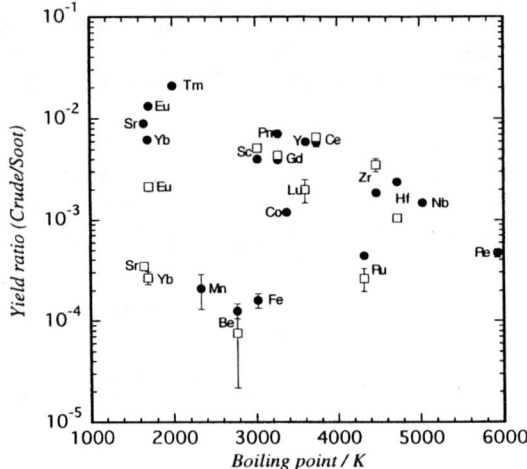

Figure 2. The yield of each element extracted into the CS_2 solvent ("crude") relative to that for La. Results of yields using two kinds of anodes are shown: ● the anode prepared by the multitracers being adsorbed into a porous carbon rod from a solution [1], and □ the anode prepared by packing mulititracers and metal oxide powders within the carbon rod.

REFERENCE

1. Sueki, K., Kikuchi, K., Akiyama, K., Sawa, T., Katada, M., Ambe, S., Ambe, F., and Nakahara, H., *Chem. Phys. Lett.* 300, 140-144 (1999).

Ferromagnetic transition in europium fullerides $Eu_xSr_{6-x}C_{60}$

K. Ishii*, A. Fujiwara*, H. Suematsu*, and Y. Kubozono[†]

*Department of Physics, The University of Tokyo, Tokyo 113-0033, Japan
[†]Department of Chemistry, Okayama University, Okayama 700-8530, Japan

Abstract. Crystal structure and magnetic properties of $Eu_xSr_{6-x}C_{60}$ were studied. Ferromagnetic transition was observed at 10-14 K, and magnetic moment is ascribed to Eu^{2+}. The divalent state of Eu is also confirmed by Eu L_{III}-edge XANES experiments. The fact that ferromagnetic transition is observed even at a dilute case ($x=1$) suggests the magnetic interaction is caused through C_{60} molecules, that is, π-f interaction, rather than direct exchange interaction between Eu atoms.

INTRODUCTION

C_{60} is known to make compounds with various atoms and molecules. As for the rare earth C_{60}, superconductivity was observed in the compounds with Yb [1] and Sm [2], in which metal atoms were non-magnetic. On the other hand, the compounds with magnetic atoms attract little attention so far, though magnetism is one of the interesting issues in rare earth compounds. Recently we have succeeded to synthesize C_{60} compounds with europium which has a magnetic moment of $7\mu_B$ ($S=7/2$) in the divalent state, and determined the crystal structure by x-ray diffraction experiments [3]. Eu_6C_{60} has a *bcc* structure, which is an isostructure of other M_6C_{60} (M is an alkali or alkaline earth metal atom), and is suitable for a prototype study of europium fullerides because of the structural simplicity.

In this paper, we report the crystal structure and magnetic properties of Eu_6C_{60} and its strontium-substituted compounds, $Eu_3Sr_3C_{60}$ and $Eu_1Sr_5C_{60}$. Sr is adopted for the study of substitution to the non-magnetic atoms, because it has a similar ionic radius to that of Eu^{2+} and the crystal structure of Sr_6C_{60} is the same as that of Eu_6C_{60}.

EXPERIMENTAL

The polycrystalline samples of $Eu_xSr_{6-x}C_{60}$ ($x=1, 3, 6$) were synthesized by heat treatment of stoichiometric mixture of metal and C_{60}. Powder x-ray diffraction

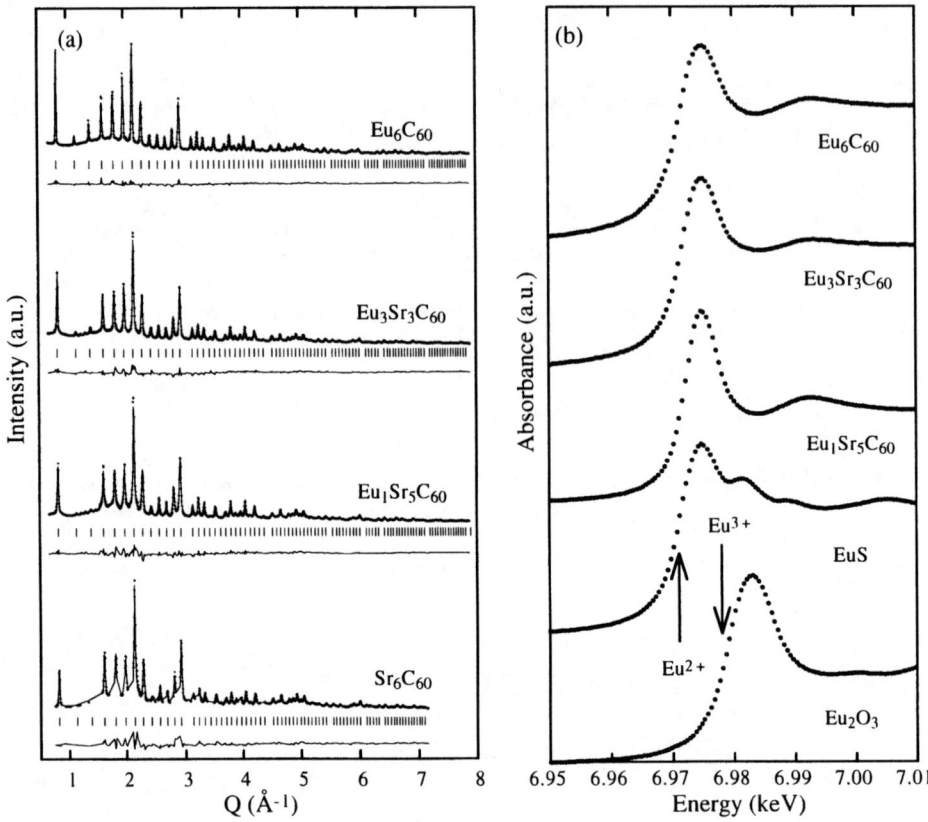

FIGURE 1. (a) : Powder x-ray diffraction spectra of $Eu_xSr_{6-x}C_{60}$. The cross marks represent observed intensity, and solid lines are the results of Rietveld refinements. The spectrum of Sr_6C_{60} was collected at the wavelength of 0.801 Å and others were at 0.851 Å. (b) : Eu L_{III}-edge XANES spectra of $Eu_xSr_{6-x}C_{60}$. The arrows indicate the absorption edges of EuS (Eu^{2+}) and Eu_2O_3 (Eu^{3+}).

experiments were carried out with use of synchrotron radiation x-rays at Photon Factory (BL-1B), KEK, Tsukuba. Eu L_{III}-edge XANES was measured in the fluorescence method at BL01B1 of SPring-8, Harima. Magnetic measurements were performed using a SQUID magnetometer.

RESULTS AND DISCUSSION

Figure 1(a) shows powder x-ray diffraction spectra of $Eu_xSr_{6-x}C_{60}$. The spectra of Eu_6C_{60} and Sr_6C_{60} are consistent with the previous works [3,4] and those of $Eu_3Sr_3C_{60}$ and $Eu_1Sr_5C_{60}$ are quite similar. The spectra can be understood by a

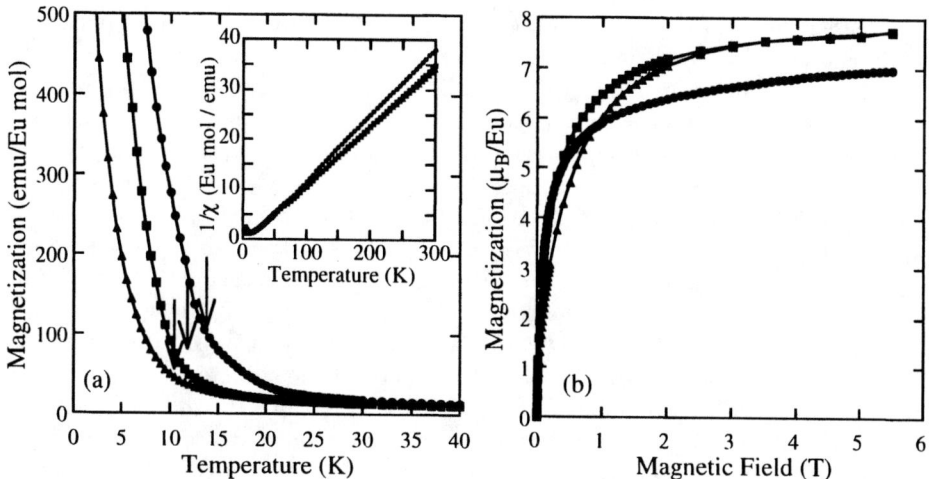

FIGURE 2. (a) : Temperature dependence of magnetization for $Eu_xSr_{6-x}C_{60}$ measured at 3 mT. The arrows indicate the ferromagnetic transition temperature. The inset shows the inverse magnetic susceptibility. (b) : Field dependence of magnetization. The circles, squares, and triangles correspond to the data of $x = 6$, 3, and 1, respectively.

bcc structure and the lattice constants obey the Vegard's law; that is, the lattice constant changes linearly with the Eu concentration x. This confirms the formation of solid solution at $x = 1$ and 3. Eu concentrations (x) obtained from the Rietveld analysis are in good agreement with the nominal ones. Eu L_{III}-edge XANES spectra are shown in Fig. 1(b). The spectra of EuS and Eu_2O_3 are also presented as a reference of divalent and trivalent Eu, respectively. The results suggest all Eu atoms are divalent in $Eu_xSr_{6-x}C_{60}$. The divalent state of Eu was also observed in Eu-C_{70} compounds [5] and metallofullerene Eu@C_{60} [6].

As shown in the inset of Fig. 2(a), the magnetic susceptibility of $Eu_xSr_{6-x}C_{60}$ obeys the Curie-Weiss law above 30 K. The effective Bohr magneton obtained from Curie constant is consistent with the theoretical value ($7.94\mu_B$) of Eu^{2+} state ($S = 7/2$, $L = 0$, and $J = 7/2$). The temperature dependence of magnetization at a weak field of 3 mT (Fig. 2(a)) shows a steep increase below 10-14 K, indicating a ferromagnetic transition. The transition temperatures (T_C) determined from the Arrott plot are 13.7 K for $x = 6$, 12.8 K for $x = 3$, and 10.4 K for $x = 1$. The field dependence of magnetization at 2 K gives the saturation moment is close to $7\mu_B$, which is also equal to the magnetic moment of Eu^{2+}. The divalent state of Eu is consistent with the Eu L_{III}-edge XANES experiments mentioned above. The present results for Eu_6C_{60} are different from those of Kasri-Habiles et al. [7], where they observed a mixed valence state of Eu (Eu^{2+} and Eu^{3+}) and successive magnetic anomalies. This disagreement might be attributed to some contamination in their sample. Our samples of Eu_6C_{60} have good quality in the sense that there are few unassigned impurity peaks in the x-ray diffraction spectrum and that Rietveld analysis gives very low R-factor ($R_{wp} = 3.81$ %).

In the *bcc* structure of Eu_6C_{60}, the number of nearest neighbor of Eu atoms (N) for an Eu site is 4. The percolation threshold (p_c) in this structure can be assumed as 0.428, the value in the site process of the diamond lattice [8], which has the same number of nearest neighbors. Though Eu ratio in $Eu_1Sr_5C_{60}$ (1/6) is smaller than both values $1/N$ and p_c, T_C is not so different from that of Eu_6C_{60}. From this result we consider that the ferromagnetic correlation between Eu atoms is caused through C_{60} (π-f interaction) rather than the direct exchange interaction between Eu atoms. On the analogy of the band calculation of Ba_6C_{60} and Sr_6C_{60} [9] which have an isostructure of Eu_6C_{60}, the hybridization between metal and C_{60} orbitals may exist. This is a quite contrast with the case of magnetic semiconductor, EuO, where the direct exchange interaction between Eu atom is important [10] and substitution of Ca for Eu significantly reduces the ferromagnetic transition temperature [11].

The π-f interaction also affects the electric resistivity. In the resistivity measurement of a compressed pellet of $Eu_xSr_{6-x}C_{60}$, we found a huge negative magnetoresistance below around T_C. The ratio of resistivity under magnetic field to that without field (ρ/ρ_0) reaches almost 10^{-3} in Eu_6C_{60}. The details will be reported elsewhere [12].

In summary, we have found a ferromagnetic transition in $Eu_xSr_{6-x}C_{60}$ ($x = 1$, 3, 6) which is attributed to the ordering of magnetic moment of $7\mu_B$ in Eu^{2+}. The study of substitution effect suggests that the magnetic correlation is caused by π-f interaction.

ACKNOWLEDGMENTS

This work was supported by "Research for the Future" of Japan Society for the Promotion of Science (JSPS), Japan.

REFERENCES

1. Özdaş, E., et al., *Nature (London)*, **375**, 126-129 (1995).
2. Chen, X. H., and Roth, G., *Phys. Rev. B*, **52**, 15534-15536 (1995).
3. Ootoshi, H., et al., *Mol. Cryst. and Liq. Cryst.*, **340**, 565-570 (2000).
4. Kortan, A. R., et al., *Chem. Phys. Lett.* **223** 501-505 (1994).
5. Takenobu, T., et al., private communication.
6. Inoue, T., et al., *Chem. Phys. Lett.* **316** 381-386 (2000).
7. Ksari-Habiles, Y., et al., *J. Phys. Chem. Solids*, **58**, 1771-1778 (1997).
8. Stauffer, D., *Introduction to Percolation Theory*, Taylor & Francis, London, 1985.
9. Saito, S., and Oshiyama, A., *Phys. Rev. Lett.*, **71**, 121-124 (1993).
10. Kasuya, T., *I.B.M. J. Res. Develop.*, **14**, 214-223 (1970).
11. Samokhvalov, A. A., et al., *Sov. Phys. Solid State*, **9**, 555-557 (1967).
12. Ishii, K., et al., to be published.

C_{60} Molecular Configurations Leading to Ferromagnetic Exchange Interactions in TDAE-C_{60}

Bakhyt Narymbetov,[1] Ales Omerzu,[2,3] Victor V. Kabanov,[2] Madoka Tokumoto,[1,3] Hayao Kobayashi,[1] and Dragan Mihailovic[2]

[1] *Institute for Molecular Science, Okazaki 444-8585, Japan,*
[2] *Institute Josef Stefan, Jamova 39, 1000 Ljubljana, Slovenia*
[3] *Electrotechnical Laboratory, 1-1-4 Umezono, Tsukuba, Ibaraki 305-8568, Japan*

Abstract. The charge-transfer salt tetrakis(demethylamino)ethylene-fullerene [C_{60}] or TDAE-C_{60} is a rare exception among pure organic crystalline systems because it shows a transition to a ferromagnetic (FM) state with fully saturated s=1/2 molecular spins at a respectable T_c=16 K. In spite of extensive experimental and theoretical work in the last ten years, the origin of the ferromagnetism in TDAE-C_{60} has remained mysterious. To resolve this problem we have performed a comparative structural study of two different magnetic forms of TDAE-C_{60} crystals, one being magnetic and the other nonmagnetic, at low temperatures, fully correlating the structural properties – and particularly the inter-molecular orientations – with the magnetic properties. We have identified the relative orientations of C_{60} molecules along the c-axis as the primary variable controlling the ferromagnetic order parameter and have shown that both FM and low-temperature spin-glass-like ordering are possible in this material, depending on the orientational state of C_{60} molecules. Thus we have resolved the apparent contradictions posed by different macroscopic measurements and have opened the way to a microscopic understanding of π-electron FM exchange interactions.

INTRODUCTION

The room temperature structure of ferromagnetic TDAE-C_{60} crystals has been determined to be monoclinic (C2/c)[1] with two formula units per unit cell and rapidly rotating C_{60} molecules.[2] As the temperature is lowered, their rotation gradually slows down and, although NMR data suggest that the rotational motion freezes out below about 150K, the onset of the FM state occurs at a significantly lower temperature, suggesting that stationary C_{60} molecules may be a necessary but not sufficient condition for the occurrence of a FM state. Fortunately, from the point of view of understanding the low-temperature magnetic interactions, TDAE-C_{60} exists in two modifications, one ferromagnetic and one paramagnetic (PM). We have identified the relative orientations of C_{60} molecules along the c-axis and have shown that both FM and low-temperature spin-glass-like ordering are possible in this material, depending on the orientational state of C_{60} molecules.

EXPERIMENTAL

For crystal growth we have used the diffusion method. A solution of C_{60} (2 mg ml^{-1}, Hoechst gold grade) and a 3:1 mixture of toluene and TDAE (Aldrich, 95% pure) were poured into two compartments of a growth cell separated by fritted glass. The cell was carefully closed and thermostated at 8 °C for six months, whereupon crystals were extracted from it. The measurements were performed on a selected sample in the PM phase without annealing, and compared with the same sample after transformation to the FM phase by annealing for 6 h at 70 °C.

Single crystal X-ray diffraction studies were carried out by using an Imaging Plate (IP) system (DIP 320S, MAC Science Co., Inc.) equipped with a liquid helium-cooling device. The standard oscillation and Weissenberg type diffraction patterns were used to control the crystal structures of both samples (unannealed PM and annealed FM) at different temperatures. Monochromatized MoK$_\alpha$ radiation (λ=0.7107 Å) was used in the X-ray experiments. The intensity data for structure analyses were collected by the Weissenberg type IP system. The crystal structure of TDAE-C$_{60}$ was solved by direct method on data obtained from the unannealed crystal at 7 K and refined by least-squares method in *C2/c* space group.[3] The averaged structure of the annealed sample at 7 K was firstly refined by using the structural model of the unannealed crystal and then refined further by taking into account the presence of two orientations of the C$_{60}$ molecules with half-occupation of the positions.

RESULTS AND DISCUSSION

On cooling below 50 K, remarkable changes are found in the X-ray diffraction patterns of both PM and FM samples with the appearance of new diffuse lines. Those lines gradually, in a period of a few hours, disappear for PM sample, while they transform into additional sharp diffraction spots for FM one, as the samples are kept at low temperature. The X-ray oscillation patterns are shown in Figs. 1(a-d), where the diffuse lines are indicated by arrows. (The data used here were obtained at 11 K and 7 K, on PM and FM samples, respectively, after a 4-hour cooling period.)

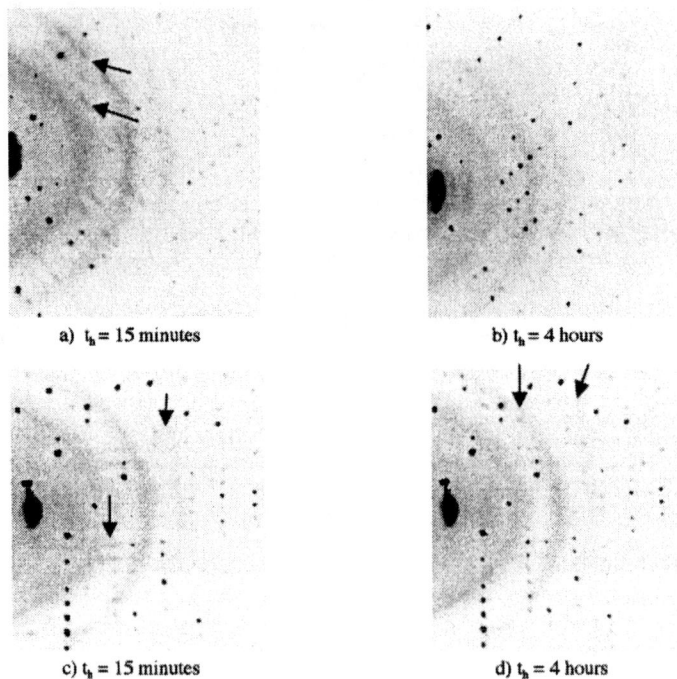

a) t_h = 15 minutes   b) t_h = 4 hours

c) t_h = 15 minutes   d) t_h = 4 hours

FIGURE 1. The fragments of X-Ray oscillation patterns of TDAE-C$_{60}$ single crystal: (a) and (b) – from the unannealed sample at 7K; (c) and (d) – from the annealed sample at 20K; t_h is a time of holding the samples at low temperatures before starting the measurements.

The X-ray structure analysis of PM sample has revealed the presence of some degree of molecular orientational disorder of the C_{60} molecules due to their rotations along the threefold molecular axis.[3] In the FM phase, the positions of the additional diffraction spots coincide with those of a primitive unit cell suggesting that the crystal transforms from the C-centered structure to a primitive one. Our attempts to solve the structure in primitive unit cell failed to be satisfactiorily and resulted in high values of R-factor (~ 0.16) and large divergence of temperature factors of individual atoms. Refinement of the structure in C-centered unit cell has also resulted in high R-factors (~0.20) but with the reasonable thermal ellipsoids of individual atoms of C_{60} which testified to the presence of a high degree of orientational disorder of C_{60} molecules (Fig. 2a)). Further analysis of the obtained C_{60} molecular structure has shown that in the FM sample the molecules are statistically distributed in two orientations related to each other by 60° rotation about their threefold axis. Using of this model, taking into account two molecules at the same positions with half occupation, in refinement procedure allowed us to essentially improve the R-factor and it's final value was 0.066. The experimentally obtained molecular structures of C_{60}s in two orientations are shown in Fig. 2b).

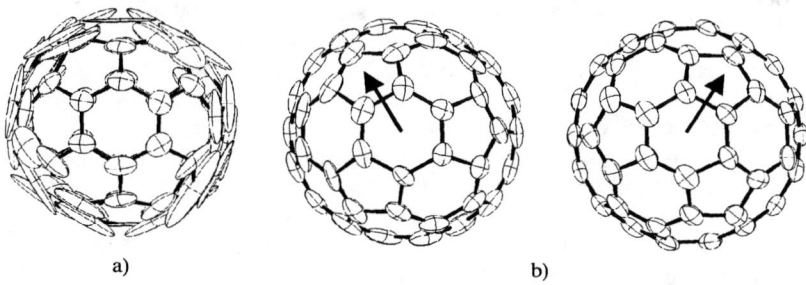

a) b)

FIGURE 2. Molecular structures of C_{60} in the annealed crystal of TDAE-C_{60} in $C2/c$ space group (ORTEPIII, 50% probability): a) averaged structure; b) obtained by refinement with taking into account two orientations of C_{60} related by ±60° rotations. The arrows denote relative orientations of C_{60}.

Thus, in the FM phase, in addition to the conventional 120° rotations found in other C_{60}-based crystal structures, we find evidence of additional positions with C_{60} rotated by ± 60° about its threefold molecular axis. This leads to a set of new inter-fullerene contact configurations not previously observed in C_{60}-based solids.

The relative C_{60} contact configurations corresponding to the new low-temperature orientations, projected along the c axis (which is also the direction of closest contact between the C_{60}s) are shown in Figs. 3a-d. In the PM sample, the relative C_{60} orientations are similar to those encountered in other C_{60} solids, namely the 6-6 double bond (nearly) faces the center of the hexagon on the neighbouring molecule (Fig. 3a), minimising the electronic overlap[4]. In the FM samples on the other hand, a new orientation appears (II), which leads to three different possible *relative* orientations of the C_{60}s as shown in Figs. 3b-d. In the first configuration, two C_{60}s of orientation I face each other, with the molecules slightly rotated about the c-axis as shown in Fig. 3b. In the second possible configuration, two C_{60}s with type II orientations are in contact as shown in Fig. 3c. This orientation essentially corresponds to a slightly displaced PM configuration (Fig. 3a), with the double bond displaced to the side. The third configuration involves two C_{60}s, with orientations I and II. In this case the double bond on one molecule approximately faces the center of

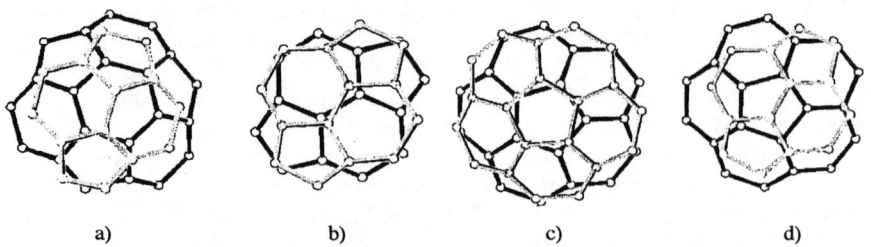

a)   b)   c)   d)

FIGURE 3. Near-neighbour molecular environments of C_{60} viewed along the [001] direction. (a) for unannealed crystal; (b) for annealed crystal, the I-I configuration of C_{60} molecules; (c) for annealed crystal, the II-II configuration; (d) for annealed crystal, mixed, I-II configuration.

the pentagon of its neighbour (Fig. 3d). The II-II configuration can be eliminated from further considerations of the possible low-temperature equilibrium positions since the occurrence of two 6-6 double bonds in close proximity is clearly energetically very unfavourable. We can conclude therefore with reasonable confidence that only one of these configurations is compatible with the near 50% occupancy of configurations I and II and primitive unit cell determined from the structural refinement (Fig. 3d). [5]

We note that the observed arrangement associated with the FM state is in excellent agreement with calculations of the angular dependence of the effective FM exchange coupling strength J_{eff}[6], which also shows a minimum corresponding to the I-II configuration in Fig. 3d, corresponding to the c-axis Euler angle $\gamma \approx 30°$.

With two equilibrium configurations for the C_{60} molecules, we can describe the magnetic behaviour of the system in terms of a Hamiltonian for a non-interacting two level system in which the coupling between the configurational and magnetic degrees of freedom appears because the exchange interaction J depends on the overlap between adjacent molecules along the c axis. [5]

SUMMARY

We have identified the relative orientations of C_{60} molecules in TDAE-C_{60} along the c-axis as the primary variable controlling the ferromagnetic order parameter and have shown that both FM and low-temperature spin-glass-like ordering are possible in this material, depending on the orientational state of C_{60} molecules.

REFERENCES

1. Golic, L., Blinc, R., Cevc, P., Arcon, D., Mihailovic, D., Omerzu, A. & Venturini, P., *Fullurenes and Fullerene Nanostructures* (Eds. H. Kuzmany, J. Fink, M. Mehring, S. Roth) 531-534 (World Scientific, Singapore, 1996).
2. Arcon, D., Dolinsek, J. & Blinc, R., *Phys. Rev. B* **53**, 9137-9142 (1996).
3. Narymbetov, B., Kobayashi, H., Tokumoto, M., Omerzu, A. & Mihailovic, D., *Chem. Commun.* (Cambridge) **1999**, 1511.
4. Dressselhaus, M. S. Dresselhaus, G. & Eklund, P. C. *Science of fullerenes and carbon nanotubes*. (Academic Press, San Diego, 1996).
5. Narymbetov, B., Omerzu, A., Kabanov, V.V., Tokumoto, M., Kobayashi, H. & Mihailovic, D., *Nature* **407**, 883 (2000).
6. Sato, T., Saito, T., Yamabe, T. & Tanaka, K., *Phys.Rev. B* **55**, 11052-55(1997).

Metallic Phases in Sodium Fulleride Na_xC_{70}

Toshifumi Hara, Mototada Kobayashi, Yuichi Akahama, and Haruki Kawamura

Department of Material Science, Himeji Institute of Technology
3-2-1 Koto, Kamigori, Ako, Hyogo 678-1297, Japan

Abstract. We synthesized sodium fulleride Na_xC_{70} and measured its electrical resistivity and magnetic susceptibility. The Na-saturated phase in Na_xC_{70} was determined to $Na_{11}C_{70}$. Following Na doping, the resistivity of Na_xC_{70} film decreased to 5.4×10^{-3} Ωcm at $x = 8$, and increased slightly with further doping. The contribution from Pauli paramagnetism was observed in the magnetic susceptibilities of Na_xC_{70} at $x = 8$ and 11 from both SQUID and ESR measurements. Na_8C_{70} and $Na_{11}C_{70}$ are considered to be possible candidates of metallic phases in Na_xC_{70}.

INTRODUCTION

Since the discovery of superconductivity in K_3C_{60} [1], various fulleride superconductors have been found in alkali, alkaline earth and lanthanide doped C_{60}. No superconductivity has been verified in doped higher fullerenes. According to the local density approximation for C_{70}, there are some bunching levels around the energy gap, which is the same as the case of C_{60} [2]. It is possible that doped C_{70} solid also form the metallic and superconducting phase with high density of state at the Felmi level.

We have been studied structure and physical property of A_xC_{70} (A = K, Rb, Cs). These A_xC_{70} have the phases with $x = 1, 2, 3$ (K-doped C_{70} only), 4, 6, 9. Here we report the synthesis, the electrical resistivity and the magnetic susceptibility for sodium fulleride Na_xC_{70} in order to search the metallic phase. Because of small ionic radius of Na, it is expected that a larger number of Na atoms can be introduced into C_{70} lattice than the other alkali metal species. As for C_{60} fulleride, the synthesis of $Na_{11}C_{60}$ was reported [3].

EXPERIMENT

Electrical resistivity measurement was carried out for Na_xC_{70} film. Pristine C_{70} film of 1000 Å thick was prepared by thermal evaporation of C_{70} powder (99+ % purity, MER Corp.) on a glass substrate with pre-deposited gold electrodes. Base pressure in the evaporation chamber was about 10^{-5} Pa. Substrate temperature (T_s) during the deposition was 423 K. Na metal (99.9 % purity, Rare Metallic Co., Ltd.) was used as dopant. We deposited Na on the film and monitored the resistance of the film by four-probe method. The amount of deposited Na was checked using a quartz oscillator thickness monitor. The composition of Na_xC_{70} film was estimated from the thickness ratio of C_{70} to Na.

Structural change of Na_xC_{70} was investigated by bulk doping. C_{70} powder and excess Na metal were put into each ends of a Pyrex glass cell with a capillary after evacuating, and the specimen was heated in a furnace. X-ray powder diffraction study was performed using a Mo rotor x-ray generator and an image plate.

To prepare the sample for magnetic susceptibility measurements, stoichiometric amount of C_{70} powder and Na metal were enclosed in a copper tube under an argon atmosphere and it was sealed into a Pyrex glass tube after evacuating. The Pyrex tube was heated at 623 K for several weeks. Magnetic susceptibility measurements were carried out for Na_xC_{70} at x = 8 and 11 by SQUID and ESR. SQUID measurement was performed using Quantum Design MPMS-5SH SQUID magnetometer under the applied field of 5 T. The specimen for SQUID was sealed into a quartz cell with a partition in the center together with helium gas for thermal exchange. ESR measurement was performed using Bluker ESP-300E ESR spectrometer (Institute for Molecular Science) at X band. The specimen for ESR was sealed into an ESR sample tube together with helium gas for thermal exchange.

RESULT AND DISCUSSION

Figure 1 shows the reaction time dependence of x-ray powder diffraction profile for Na_xC_{70} reacted at 623 K. The starting C_{70} had face-centered cubic structure with the lattice constant of a = 14.95 Å. By Na doping, the lattice of Na_xC_{70} was swelled and finally came to simple cubic structure with the lattice constant of a = 15.18 Å. The composition of Na-saturated phase in Na_xC_{70} was determined to $Na_{11}C_{70}$ from weight uptake.

Figure 2 shows Na concentration dependence of electrical reistivity for Na_xC_{70} film. The resistivity of undoped C_{70} film was more than 10^5 Ωcm. Following Na doping, resistivity decreased rapidly and marked the two minima of 7.5×10^{-2} Ωcm and 5.4×10^{-3} Ωcm at x = 2 and 8, respectively. Then slight resistivity increase was found by

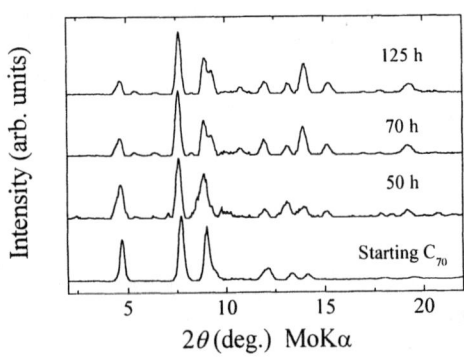

FIGURE 1. Reaction time dependence of x-ray powder diffraction profile for Na_xC_{70} reacted at 623 K.

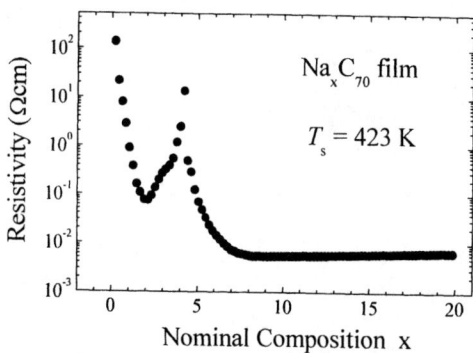

FIGURE 2. Composition dependence of the electrical resistivity for Na_xC_{70} film.

further doping. This behavior suggests the charge transfer between Na and C_{70}. The second minimum resistivity value at x = 8 is close to that of the metallic K_3C_{60} film [4]. This suggests that Na_xC_{70} film $x \geq 8$ is possibly metallic. The *in situ* measurement for the temperature dependence of the resistivity is now in progress.

Figure 3 (a) and (b) show the temperature dependence of the magnetic susceptibility for Na_8C_{70} and $Na_{11}C_{70}$ powder, respectively, by SQUID measurements. The contribution from the quartz cell has already been corrected. The data were fit by the formula

$$\chi = \frac{C}{T-\theta} + \chi_0, \quad (1)$$

where C is the Curie constant, θ is the Weiss temperature and χ_0 is the temperature independent component in the susceptibility. The temperature dependent component was attributed to the impurities and/or defects of the sample because the spin

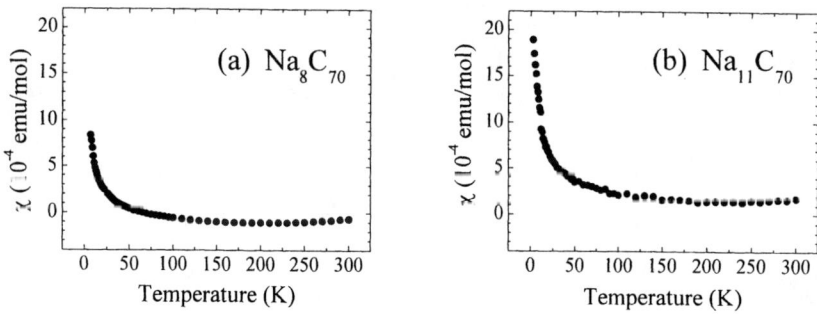

FIGURE 3. Temperature dependence of the magnetic susceptibility for (a) Na_8C_{70} and (b) $Na_{11}C_{70}$ measured using SQUID.

concentrations calculated from the C were less than 5×10^{-2} spins per C_{70} molecule for both Na_8C_{70} and $Na_{11}C_{70}$. The χ_0 contains the diamagnetic contribution from core susceptibilities of C_{70}^{6-} and Na^+ corresponding to the composition of Na_xC_{70} and other temperature independent susceptibility χ_s, that is,

$$\chi_0 = \chi(C_{70}^{6-}) + \chi(Na^+) \times x + \chi_s. \qquad (2)$$

The $\chi(C_{70}^{6-})$ was estimated to -1.61×10^{-4} emu/mol from the results of SQUID and ESR measurement for insulating K_6C_{70}. The $\chi(Na^+)$ of -6.8×10^{-6} emu/mol was used. In consequence, the χ_s of 1.51×10^{-4} and 3.09×10^{-4} emu/mol were obtained for Na_8C_{70} and $Na_{11}C_{70}$, respectively. Un-reacted Na metal in the samples wasn't detected on XRD measurements. We consider the temperature-independent paramagnetic susceptibility χ_s to the contribution from Pauli paramagnetism. It is considered that Na_8C_{70} and $Na_{11}C_{70}$ are possible candidates to have metallic character in Na_xC_{70}. From ESR results, Pauli paramagnetic contributions for Na_8C_{70} and $Na_{11}C_{70}$ were estimated to 1.48×10^{-4} and 2.28×10^{-4} emu/mol, respectively. These are consistent with the SQUID results mentioned above.

CONCLUSION

The Na-saturated phase in Na_xC_{70} was determined to $Na_{11}C_{70}$ which had simple cubic structure with the lattice constant of a = 15.18 Å. By evaporating Na onto C_{70} film, the electrical resistivity of Na_xC_{70} film was decreased to 5.4×10^{-3} Ωcm at x = 8 and little increased with further doping. This minimum resistivity is close to the resistivity for metallic K_3C_{60} film. The magnetic susceptibilities measured using SQUID and ESR for Na_8C_{70} and $Na_{11}C_{70}$ suggested the contribution from Pauli paramagnetism. We consider that Na_8C_{70} and $Na_{11}C_{70}$ are possible candidates to have metallic character in Na_xC_{70}.

ACKNOWLEDGMENTS

This work was partly supported by the Grant-in-Aid for Scientific Research on the Priority Area "Fullerene and Nanotubes" by the Ministry of Education, Science, and Culture of Japan.

REFERENCES

1. Hebard, A. F., Rosseinsky, M. J., Haddon, R. C., Murphy, D. W., Glarum, S. H., Palstra, T. T. M., Ramirez, A. P., and Kortan, A. R., *Nature* **350**, 600-601 (1991).
2. Saito, S. and Oshiyama, A., *Phys. Rev.* **B 44**, 11532-11535 (1991).
3. Yildirim, T., Zhou, O., Fischer, J. E., Bykovetz, N., Strongin, R. A., Cichy, M. A., Smith III, A. B., Lin, C. L., and Jelinek, R., *Nature* **360**, 568-571 (1992).
4. Kochanski, G. P., Hebard, A. F., Haddon, R. C., and Fiory, A. T., *Science* **255**, 184-186 (1992).

Production of $C_{59}N:C_{60}$ Solid Solution

Ferenc Fülöp[1], Antal Rockenbauer[2], Ferenc Simon[1], Sándor Pekker[3],
László Korecz[2], Slaven Garaj[4], András Jánossy[1]

[1]*Budapest University of Technology and Economics, Institute of Physics, Budapest H-1521 POB 91 Hungary, Research Group of Condensed Matter in Magnetic Fields Hungarian Academy of Sciences*
[2]*Chemical Research Center, Institute of Chemistry, Budapest H-1525 POB 17 Hungary*
[3]*Research Institute for Solid State Physics and Optics Budapest H-1525 POB 49 Hungary*
[4]*Ecole Polytechnique Federale de Lausanne, Institut de Genie Atomique, CH-1015 Lausanne, Switzerland*

Abstract. We describe a simple way to produce large quantities of solid solutions of monomer $C_{59}N$ in pure C_{60} using an electric gas discharge tube. Typical concentrations are 10^{-5} to 10^{-4} $C_{59}N$ with respect to C_{60}. The ^{14}N and several ^{13}C hyperfine constants were measured by ESR. These are a sensitive test for electronic structure calculations of the monomer. As the temperature is raised towards the *sc* to *fcc* structural transition at 261 K, the ESR spectrum motionally narrows and the activation energy for reorientation is measured. The rotational dynamics of the $C_{59}N$ monomer between 130 and 600 K parallels that of C_{60} in the bulk thus interactions between $C_{59}N$ and C_{60} are surprisingly weak.

INTRODUCTION

Azafullerene, $C_{59}N$, is one of the most interesting chemical modifications of the fullerene C_{60} as it carries magnetic and electric dipole moments and is only slightly deformed[1]. The solid is formed of $(C_{59}N)_2$ dimers and is non-magnetic[2]. Dimerization has been a major obstacle to the study of monomeric $C_{59}N$ and although several calculations of the electronic and molecular structure of $C_{59}N$ were published[3] the extent of delocalization of the charge over the cage could not be measured as the monomer was not available in sufficient amounts.

Here we describe a simple way to produce solid solutions of $C_{59}N$ in C_{60} ($C_{59}N:C_{60}$) in an electric gas discharge tube. We find that monomeric $C_{59}N$ is stable in this solid solution and its rotational dynamics parallels that of C_{60} in the bulk. The production of $C_{59}N:C_{60}$ in large quantities allowed us to measure the various ^{14}N and ^{13}C hyperfine constants that are a sensitive test of electronic structure calculations. A more detailed account of this work is published in Ref. 4.

PRODUCTION OF $C_{59}N$ IN A N_2 GAS DISCHARGE TUBE

We produced $C_{59}N$ in a N_2 gas discharge tube originally designed for the production of endohedral $N@C_{60}$ following the method of Pietzak et al.[5]. The set-up is shown in Fig. 1. Typical experimental conditions were: 1 mbar N_2 pressure, 1 mA discharge current, 10 cm electrode distance. The tube is heated to about 550 °C at one electrode (usually the anode) while the other electrode is water-cooled and is at ambient temperature. C_{60} powder placed between the electrodes at the bottom of the tube is sublimed and deposited onto the wall of the quartz tube and the cooled electrode. As the electric discharge is turned on, N ions in the plasma react with C_{60} to form $N@C_{60}$ and $C_{59}N$ simultaneously. Endohedral $N@C_{60}$ is collected from the water-cooled electrode. $C_{59}N:C_{60}$ is collected from the heated surface of the quartz tube. The highest concentration of $C_{59}N$ in C_{60} of about 100 ppm (determined from the ESR intensity) was found in narrow stripes on the quartz tube at the hottest regions where deposition still occurred i.e. where the temperature is roughly 400 °C. The yield of $C_{59}N:C_{60}$ (with concentration of 100 ppm) is about 5% of the starting C_{60} material. After deposition, the discharge tube was opened to air before sealing the samples in quartz tubes under helium. The material is stable in He atmosphere. In air, $C_{59}N$ decays at ambient temperatures in about a week. X-ray diffraction showed that the deposited powder mainly consists of crystalline C_{60}. The ESR spectrum shows a low concentration of unknown free radicals formed during the production. We have not attempted to purify the material.

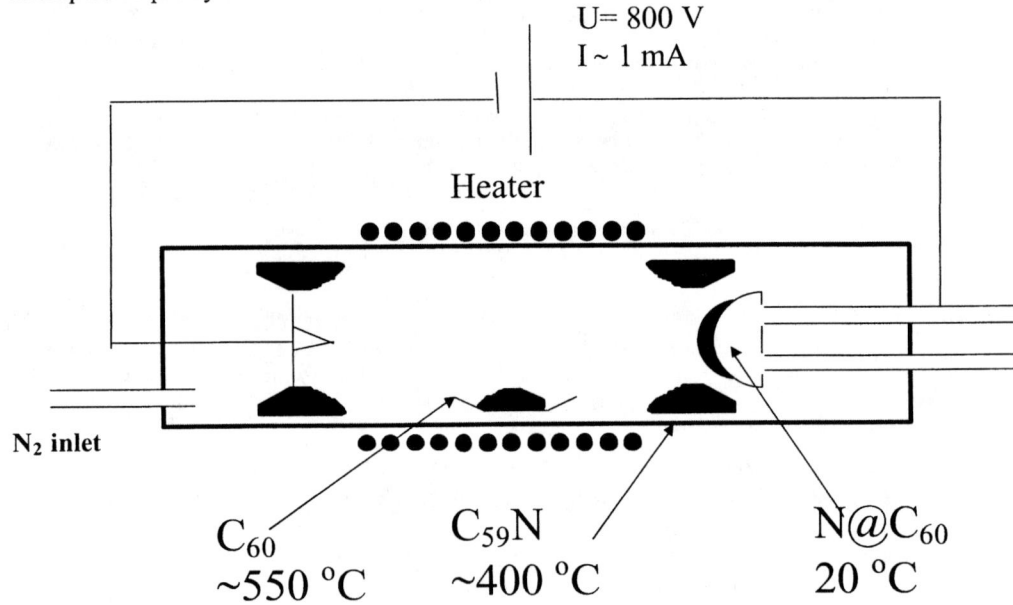

FIGURE 1. $C_{59}N:C_{60}$ production in a N_2 gas electric discharge tube.

ESR SPECTRUM AND ROTATIONAL DYNAMICS

Electron spin resonance (ESR) was performed at 9 GHz (X-band), 75 and 225 GHz. We identified the material as a solid solution of $C_{59}N$ in C_{60} from the similarity of the ESR spectrum to that of monomers produced by photo-[6] or thermolysis[7]. At 300 K

(Fig. 2) the three ^{14}N (I=1) hyperfine lines characteristic of rapidly tumbling free radicals are observed with g_{av}=2.0014(2) and an isotropic hyperfine coupling constant, A_{iso}=0.363(1) mT.

The ESR spectrum is temperature independent between 257 and 290 K. As shown in Fig. 2, the spectrum consists of narrow ^{14}N hyperfine and a series of well resolved ^{13}C lines. In this temperature range the $C_{59}N$ molecules are rapidly reorienting in all three directions together with the C_{60} molecules of the matrix. The reorientation is fast on the time scale of the ESR measurement and all anisotropic interactions are averaged to zero. The motional narrowing of the ESR spectrum of the rapidly reorienting $C_{59}N$ molecule remains effective even at high ESR frequencies where relaxation due to g-factor anisotropy is more important. At 290 K and 225 GHz the linewidth was less than the instrumental resolution of 0.05 mT.

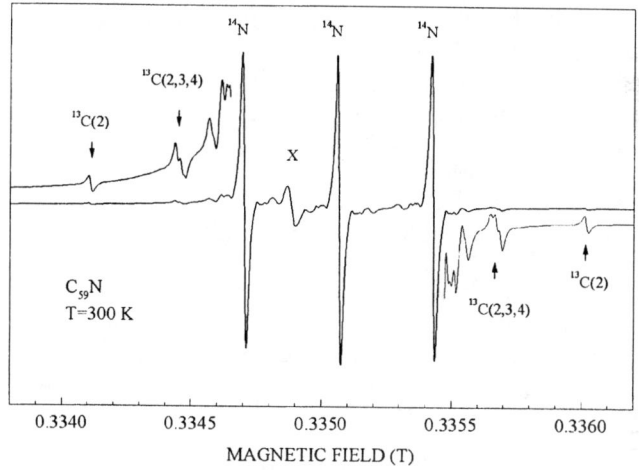

FIGURE 2. Motionally narrowed ESR spectrum of tumbling $C_{59}N$ substituted in C_{60} at 290 K and 9 GHz. Wings are 10 times magnified. The ^{14}N and ^{13}C hyperfine lines of various sites are indicated. Other lines are unassigned ^{13}C lines and an unidentified free radical line, denoted by "X".

Isotropic Fermi contact couplings were calculated by the density functional *ab initio* method, B3LYP with 6-31G* basis set on a PM3 optimized molecular geometry of $C_{59}N$. The calculations were carried out using the Gaussian 98 software[8]. The electronic structures of the monomer and the dimer forms of $C_{59}N$ were determined previously by Andreoni et al.[1] applying a similar method. The agreement between calculated and measured hyperfine constants is not particularly good. However, the calculation shows, in agreement with the experiment, that the extra spin density is not localized to the first C neighbors of N.

Pure C_{60} has a first order phase transition at 261 K from a low temperature simple cubic (*sc*) to the high temperature face centered cubic (*fcc*) structure. At low temperatures the molecular orientations are arranged preferentially in two standard configurations with respect to each other and there is a merohedral disorder of regions with differing configurations. As the temperature is raised, molecules fluctuate more and more rapidly between the possible orientations with large angle jumps. The fluctuations shorten the spin lattice relaxation time, T_1, and above about 200 K motionally narrow the anisotropic ESR spectrum. The width of the spectral components is proportional to the spin-spin relaxation rate $1/T_2$. The activation energy of monomeric $C_{59}N$ in C_{60} measured from T_1 and $1/T_2$ is about $\Delta E_{C_{59}N} / k_B = 2300$ K. This value is remarkably close to $\Delta E_{C_{60}} = 2980$ K, the activation energy of the correlation time of the rotation of C_{60} molecules measured[9] in pure C_{60}. The close agreement between $\Delta E_{C_{59}N}$ and $\Delta E_{C_{60}}$ indicates that the reorientation dynamics of $C_{59}N$ follows that of the bulk. This would not be so if interactions between $C_{59}N$ and C_{60} were strong.

SUMMARY

In summary, $C_{59}N$ is a stable substituent molecule in C_{60} crystals at all temperatures below the formation of the solid solution at 400 C. Hyperfine interactions of the free radical with ^{14}N and ^{13}C show a partial delocalization of the extra electron over the cage. Interactions with neighboring C_{60} molecules is small and thus rotational dynamics are similar.

ACKNOWLEDGMENTS

Support from the Hungarian State grants OTKA T032613, OTKA T029150, and FKFP 0352/1997 and the Swiss National Science Foundation are acknowledged. We thank G. Oszlányi for the X-ray characterization of the samples.

REFERENCES

[1] W. Andreoni, F. Gygi, M. Parrinello, *Chem. Phys. Lett.*, **190** 159 (1992).
[2] J. C. Hummelen, B. Knight, J. Pavlovich, R. Gonzalez, F. Wudl, Science **269** 1554 (1995).
[3] See the review by J. C. Hummelen, C. Bellavia-Lund, F. Wudl Heterofullerenes, in: Topics in Current Chemistry, Vol.199, Springer Verlag, Berlin, Heidelberg 1999. pp. 93-134 and references therein.
[4] F. Fülöp et al., *Chem. Phys. Lett.*, in print.
[5] B. Pietzak, M. Waiblinger, Almeida T. Murphy, A. Weidinger, M. Höhne, E. Dietel, A. Hirsch *Chem Phys. Lett.*, **279** 259 (1997).
[6] K. Hasharoni, C. Bellavia-Lund, M. Keshavarz-K., G. Srdanov, F. Wudl, *J. Am. Chem. Soc.*, **119** 11128 (1997). A. Gruss, K.-P. Dinse, A. Hirsch, B. Nuber, U. Reuther, *J. Am. Chem. Soc.*, **119** 8728 (1997).
[7] F. Simon, D. Arcon, N. Tagmatarchis, S. Garaj, L. Forró, and K. Prassides, *J. Phys. Chem. A.*, **103** 6969 (1999).
[8] Gaussian 98, Revision A.7, Gaussian, Inc., Pittsburgh PA, 1998.
[9] K. Mizoguchi, Y. Maniwa and K. Kume, *Mater. Sci. Eng.*, **B19** 146 (1993).

Dynamic Jahn-Teller Mechanism of Superconductivity in Alkali-Metal-Doped C_{60}

Shugo Suzuki, Susumu Okada, and Kenji Nakao

Institute of Materials Science, University of Tsukuba, Tsukuba 305-8573, Japan

Abstract. We study the mechanism of superconductivity in alkali-metal-doped C_{60} by using a model which takes account of both the electron-electron and electron-phonon interactions. It is shown that the dynamic Jahn-Teller effect is responsible for superconductivity in this material. The importance of the pair-transfer interaction is also pointed out. Furthermore, we show that the dynamic Jahn-Teller mechanism can be regarded as the Suhl-Kondo mechanism in the wave-number space. It is emphasized that the multiplicity of the conduction bands is essential in both mechanisms.

INTRODUCTION

Superconductivity in alkali-metal-doped C_{60}, $A_x C_{60}$ where A is alkali metal and x is the concentration of alkali metal, has been one of the topics that have attracted much attention of the researchers in condensed matter physics. The transition temperatures are unusually high although the materials are molecular solids made up of carbon and alkali metals. So far, several mechanisms of superconductivity have been proposed: the electron correlation mechanism, the alkali-metal optical phonon mechanism, the intramolecular phonon mechanism, etc. In spite of extensive studies in the last decade, the mechanism has not been clarified yet.

Of great importance for $A_x C_{60}$ are the electron-electron and electron-phonon interactions. The most striking fact is that these interactions cooperatively make $A_2 C_{60}$ and $A_4 C_{60}$ insulating [1]. This accordingly implies that both interactions play important roles in superconductivity in $A_3 C_{60}$. In fact, there exists some evidence for this expectation. One is that photoemission spectra are very broad in contradiction to the results of band calculations. Another is that optical conductivity measurements show the existence of mid-infra-red absorption whose origin is still unknown.

The purpose of the present study is to investigate the role of the electron-electron and electron-phonon interactions in superconductivity in $A_3 C_{60}$ by using a model which takes account of both interactions. We first explain our model and then show that superconductivity can be described in terms of the dynamic Jahn-Teller effect. Finally, we point out that the dynamic Jahn-Teller mechanism can be interpreted as the Suhl-Kondo mechanism in the wave-number space.

MODEL

We employ the following model Hamiltonian in this study on the basis of the threefold degeneracy of the lowest-unoccupied-molecular orbitals, t_{1u} orbitals, say x, y, and z, of the C_{60} molecule [1,2].

Firstly, the one-electron part is given by

$$H_0 = \sum_{m,n,a,b,\sigma} t_{ab}^{mn} a_{m\sigma}^\dagger b_{n\sigma} = \sum_{\alpha \mathbf{k} \sigma} \xi_{\alpha \mathbf{k}}^0 \alpha_{\mathbf{k}\sigma}^\dagger \alpha_{\mathbf{k}\sigma} , \qquad (1)$$

where t_{ab}^{mn} is the transfer integral between the t_{1u} orbital a in the mth C_{60} molecule and the t_{1u} orbital b in the nth C_{60} molecule. Also, $a_{m\sigma}^\dagger$ ($a_{m\sigma}$) is the creation (annihilation) operator of the σ-spin electron in the t_{1u} orbital a in the mth C_{60} molecule. Furthermore, $\xi_{\alpha \mathbf{k}}^0$ represent the energies of the Bloch electrons obtained by diagonalizing H_0. The corresponding creation (annihilation) operators are denoted by $\alpha_{\mathbf{k}\sigma}^\dagger$ ($\alpha_{\mathbf{k}\sigma}$), where α (α=1, 2, 3) is the band index and $\mathbf{k}$ is the wave number.

Next, the effective electron-electron interaction is introduced by taking account of both the electron-electron and electron-phonon interactions:

$$H_{\text{int}} = \frac{1}{2} \sum_{\sigma,\tau} \sum_{m,a,b,c,d} V_{abcd} a_{m\sigma}^\dagger c_{m\tau}^\dagger d_{m\tau} b_{m\sigma} . \qquad (2)$$

The interaction constants V_{abcd} in the above expression are classified into the intraorbital repulsion, V_{intra}, the interorbital repulsion, V_{inter}, the exchange interaction, J, and the pair-transfer interaction, K. The details of these interactions are given elsewhere [1,2]. In $A_x C_{60}$, the order of magnitude of the effective interactions can be estimated as $V_{\text{intra}} \simeq 0.2$ eV, $V_{\text{inter}} \simeq 0.4$ eV, $J \simeq -0.1$ eV, and $K \simeq -0.1$ eV.

SUPERCONDUCTIVITY

We now show that the attractive interaction between the t_{1u} electrons is caused by the dynamic Jahn-Teller effect resulting from the pair-transfer interaction.

We first consider two electrons on two C_{60} molecules, neglecting the electron transfer between the molecules. There are no interactions when the two electrons reside on the molecules separately. The interaction between the two electrons takes place when they occupy a single molecule. The ground state of the two electron system is given by the linear combination of the doubly occupied states:

$$\frac{1}{\sqrt{3}}(|x\uparrow x\downarrow\rangle + |y\uparrow y\downarrow\rangle + |z\uparrow z\downarrow\rangle). \tag{3}$$

This is because the pair-transfer interaction, K, is negative. We refer to this kind of state as the orbital-singlet state. We can interpret this state as the realization of the dynamic Jahn-Teller effect because the three doubly occupied states are accompanied by different Jahn-Teller distortions in the static case. The energy of the above state is given by $V_{\text{intra}} + 2K$ and thus there arises attractive interaction between the t_{1u} electrons if the quantity,

$$g \equiv -(V_{\text{intra}} + 2K), \tag{4}$$

is positive. The similar consideration gives us the same criterion for the occurrence of the attractive interaction for A_3C_{60} by examining the relative stability of $C_{60}^{2-}+C_{60}^{4-}$ to $2C_{60}^{3-}$ [2].

We now show the above criterion is actually obtained by constructing the BCS ground-state wavefunction for A_3C_{60}, $|\Psi\rangle$, as follows:

$$|\Psi\rangle = \prod_{\alpha\mathbf{k}}(u_{\alpha\mathbf{k}} + v_{\alpha\mathbf{k}}\alpha_{\mathbf{k}\uparrow}^\dagger \alpha_{-\mathbf{k}\downarrow}^\dagger)|0\rangle. \tag{5}$$

Here we assume that $|\Psi\rangle$ is of the A_g type, i.e.,

$$u_{\alpha R\mathbf{k}} = u_{\alpha\mathbf{k}} \quad v_{\alpha R\mathbf{k}} = v_{\alpha\mathbf{k}} \tag{6}$$

for any symmetry operation R of the system. We then calculate the expectation value of the Hamiltonian H in $|\Psi\rangle$. This consists of two parts; one is of the Hartree-Fock type and the other is of the BCS type. We take account of the former one by replacing $\xi_{\alpha\mathbf{k}}^0$ by $\xi_{\alpha\mathbf{k}}$ which can be obtained by the Hartree-Fock calculations. The BCS part is found to be given as follows:

$$\langle\Psi|H_{\text{int}}|\Psi\rangle_{\text{BCS}}$$
$$= \tfrac{1}{2}\sum_\sigma \sum_{\alpha\mathbf{k},\beta\mathbf{l}} V_{\alpha\beta\alpha\beta}^{\mathbf{k}\mathbf{l}-\mathbf{k}-\mathbf{l}} \langle\Psi|\alpha_{\mathbf{k}\sigma}^\dagger \alpha_{-\mathbf{k}-\sigma}^\dagger|\Psi\rangle$$
$$\langle\Psi|\beta_{-\mathbf{l}-\sigma}\beta_{\mathbf{l}\sigma}|\Psi\rangle. \tag{7}$$

Here, $V_{\alpha\beta\alpha\beta}^{\mathbf{k}\mathbf{l}-\mathbf{k}-\mathbf{l}}$ denote the interaction constants represented by the band indexes and the wave numbers. The result of the calculations shows that the ground-state energy is given by

$$\langle\Psi|H_0 + H_{\text{int}}|\Psi\rangle = 2\sum_{\alpha\mathbf{k}}\xi_{\alpha\mathbf{k}}v_{\alpha\mathbf{k}}^2 - 3NgP_{Ag}^2, \tag{8}$$

where

$$P_{Ag} = \frac{1}{3N}\sum_{\alpha\mathbf{k}} u_{\alpha\mathbf{k}}v_{\alpha\mathbf{k}}, \tag{9}$$

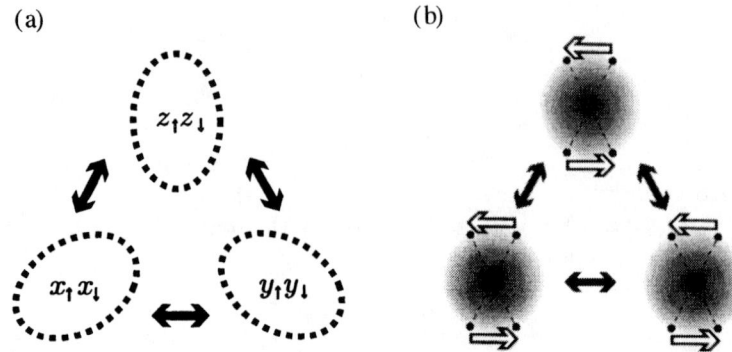

FIGURE 1. The mechanism of superconductivity in A_3C_{60}: (a) the dynamic Jahn-Teller mechanism and (b) the Suhl-Kondo mechanism. The former shows that the C_{60} molecule undergoes a tunneling motion among three different Jahn-Teller distortions. The latter shows that there exist not only usual intraband scattering process of the Cooper pairs (white arrows) but also the interband transfer process of the Cooper pairs (black arrows).

being N to be the total number of the C_{60} molecules in the system. Furthermore, $g = -(V_{\text{intra}} + 2K)$ is exactly the same as that defined by Eq. (4). Finally, by minimizing the ground-state energy, we find the gap equation. The solution exists only if g is positive and is given as follows:

$$\Delta = gP_{Ag} \simeq 2\hbar\omega_c \exp\left(-\frac{1}{gN_F}\right). \tag{10}$$

We denote the cut-off energy in the above expression by $\hbar\omega_c$ and the density of states per spin per orbital at the Fermi level by N_F.

The expression (7) shows that there exist not only the energy gain due to intraband scattering process of the Cooper pairs but also the energy gain due to the interband transfer process of the Cooper pairs. The latter is characteristic of the multiband superconductor and can stabilize the superconducting order in such systems. This mechanism is known as the Suhl-Kondo mechanism and has been proposed as the key mechanism in superconductivity in A_3C_{60} by Rice et. al. and also by Asai and Kawaguchi [3,4]. We can thus interpret the dynamic Jahn-Teller mechanism as the Suhl-Kondo mechanism in the wave-number space as shown in Fig. 1.

REFERENCES

1. Suzuki, S., and Nakao, K., *Phys. Rev. B* **52**, 14206 (1995).
2. Suzuki, S., Okada, S., and Nakao, K., *J. Phys. Soc. Jpn.* **69**, 2615 (2000).
3. Rice, M. J., Choi, H. Y., and Wang, Y. R., *Phys. Rev. B* **44**, 10414 (1991).
4. Asai, Y., and Kawaguchi, Y., *Phys. Rev. B* **46**, 1265 (1992).

Theoretical Study on the Photoemission Spectra of A_3C_{60} (A=K and Rb)

T. Chida, S. Suzuki, and K. Nakao

Institute of Materials Science, University of Tsukuba, Japan

Abstract. We calculate the spectral density for A_3C_{60} based on the dynamical mean-field theory and compare our results with the experimental photoemission spectra. We show that the charge fluctuation occurs in A_3C_{60} due to the competition between the electron-electron and electron-phonon interactions and consequently the material remains metallic although both interactions are considerably strong. The charge fluctuation in A_3C_{60} means the resonance between $C_{60}^{3-} + C_{60}^{3-}$ and $C_{60}^{2-} + C_{60}^{4-}$ because these two states have almost the same energy. We find a satellite derived from the charge fluctuation in the spectral density for A_3C_{60}. It is pointed out that this satellite corresponds to the shoulder observed only for A_3C_{60}. Also, we find that the broadening of the lowest-unoccupied-molecular-orbital-derived band is originated in the multiplet splitting. Thus, our results can successfully explain the experimental photoemisson spectra for K_3C_{60} and Rb_3C_{60}.

Many experiments have shown that the electron-electron and electron-phonon interactions play important roles in alkali-metal-doped C_{60}, A_xC_{60} where A is an alkali metal [1-4]. A_2C_{60} and A_4C_{60} are nonmagnetic insulators [5] in contradiction to the results of band calculations that these materials are metals. The theoretical studies predict that these materials favor the nonmagnetic insulator due to the cooperation of the electron-electron and electron-phonon interactions [6-9]. Furthermore, there exists experimental evidence for A_3C_{60} to be an anomalous metal although the metallic behavior is in agreement with the results of band calculations [10]. In particular, the photoemission spectra for A_3C_{60} have some anomalies [11-14]. Firstly, the lowest-unoccupied-molecular-orbital (LUMO)-derived band is much broader than predicted by band calculations. Also, a shoulder at about -1.5 eV below the Fermi level appears on the highest-occupied-molecular-orbital (HOMO)-derived band. As well as in A_2C_{60} and A_4C_{60}, it is expected that the electron-electron and electron-phonon interactions play important roles in the electronic structures of A_3C_{60}. In this paper, we show the calculated spectral density for A_3C_{60} and compare our results with the experimental photoemission spectra.

In the present study, we employ a model which takes account of both the electron-electron and electron-phonon interactions within antiadiabatic approximation [6,15]. We also consider only t_{1u} orbitals, which are three-fold degenerate. We calculate the spectral densities for A_3C_{60} based on the dynamical mean-field theory (DMFT) by using the exact diagonalization method, which is a powerful method to investigate many-body effects [16,17]. In the limit of an isolated molecule, our model can be diagonalized and the spectral density can be calculated analytically.

We first show that the charge fluctuation occurs in A_3C_{60} due to the competition between the electron-electron and electron-phonon interactions. The calculated spectral density for A_3C_{60} is shown in Fig. 1(a). The spectral density for an isolated C_{60}^{3-} molecule can be calculated analytically as shown in Fig. 1(b). One can understand the occurrence of the charge fluctuation and the metallic behavior in A_3C_{60} by considering an isolated C_{60}^{3-} molecule as follows. In Fig. 1(b), the energy gap between occupied and unoccupied states is very small. In general, the energy gap is given by $\Delta E = E_{gs}^{N+1} + E_{gs}^{N-1} - 2E_{gs}^{N}$, where E_{gs}^{M} is the ground-state energy for M-electron system. Thus, the energy gap in an isolated C_{60}^{3-} molecule represents the energy difference between the two states, $2C_{60}^{3-}$ and $C_{60}^{2-}+C_{60}^{4-}$. In the former state, $2C_{60}^{3-}$, the energy loss due to the Coulomb repulsion is minimized because of the uniform charge distribution while the energy gain due to the Jahn-Teller effect is not maximized because of the unpaired electrons. In the latter state, $C_{60}^{2-}+C_{60}^{4-}$, on the contrary, the energy loss due to the Coulomb repulsion is not minimized because of the nonuniform charge distribution while the energy gain due to the Jahn-Teller effect is maximized because of the absence of unpaired electrons. As a result, the energy difference between the two states is very small. That is, the charge fluctuation, $2C_{60}^{3-} \rightleftharpoons C_{60}^{2-}+C_{60}^{4-}$, occurs in A_3C_{60} due to the competition between the electron-electron and electron-phonon interactions.

We find that the satellites at ± 1.3 eV in Fig. 1(a) are derived from the charge fluctuation in A_3C_{60}. It is useful for understanding the calculated spectral density to consider the photoemission processes as follows. As we show in the above, the two states, $2C_{60}^{3-}$ and $C_{60}^{2-}+C_{60}^{4-}$, resonate with each other in A_3C_{60}. The photoemission process produced from the former state, $2C_{60}^{3-}$, is only one, $2C_{60}^{3-} \rightarrow C_{60}^{2-}+C_{60}^{3-}$. This process produces the continuous features ranged from -1 eV to $+1$ eV in Fig. 1(a). However, there are two processes produced by the latter state, $C_{60}^{2-}+C_{60}^{4-}$. One process, $C_{60}^{2-}+C_{60}^{4-} \rightarrow C_{60}^{2-}+C_{60}^{3-}$, enhances the continuous feature around the Fermi level in Fig. 1(a). The other process, $C_{60}^{2-}+C_{60}^{4-} \rightarrow C_{60}^{-}+C_{60}^{4-}$, produces the satellite at about -1.3 eV below the Fermi level in Fig. 1(a). Similarly, the satellite at about $+1.3$ eV in Fig. 1(a) can be explained by considering the inverse photoemission process, $C_{60}^{2-}+C_{60}^{4-} \rightarrow C_{60}^{2-}+C_{60}^{5-}$. Thus, we conclude that these satellites arise due to the charge fluctuation.

We next find that the continuous features around the Fermi level in Fig. 1(a) is originated in the multiplet splitting. This continuous features are directly derived from the peaks for the isolated C_{60}^{3-} molecule in Fig. 1(b); each peak is broadened by the band formation. We find six peaks in the spectral density for an isolated C_{60}^{3-} molecule as shown in Fig. 1(b). In this paper, we denote the six peaks by the T_{1g}, H_g, A_g, $\bar{A}_g$, $\bar{H}_g$ and $\bar{T}_{1g}$ peak as shown in Fig. 1(b). Here, we consider the photoemission processes for understanding the three peaks for the occupied states.

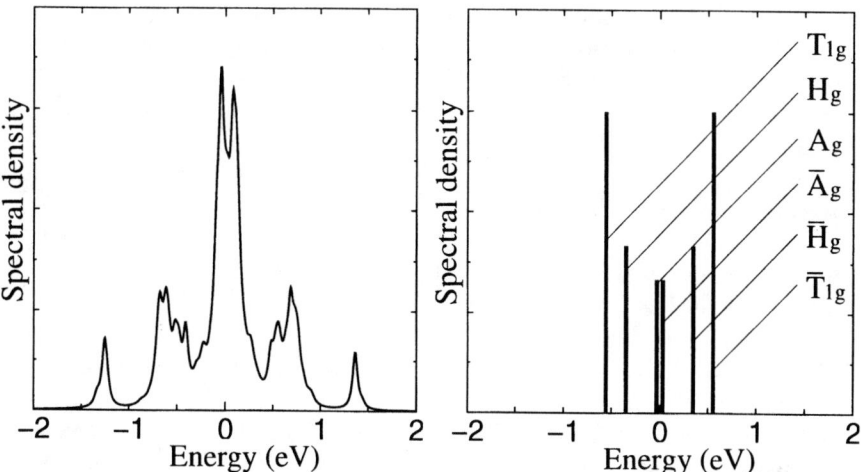

FIGURE 1. Spectral densities. (a) the DMFT result for A_3C_{60} and (b) the result for an isolated C_{60}^{3-} molecule.

The initial state in the photoemission processes is the ground state of an isolated C_{60}^{3-} molecule. On the other hand, there are three final states in the photoemission processes because the eigen states for an isolated C_{60}^{2-} molecule split into three multiplets; $t_{1u} \times t_{1u}$ is reduced to $A_g + H_g + T_{1g}$. Firstly, the ground state for an isolated C_{60}^{2-} molecule is of the A_g symmetry and spin-singlet. Thus, the A_g peak at about -0.03 eV in Fig. 1(b) is produced by the process from the initial state to this state. Secondly, the first excited state of an isolated C_{60}^{2-} molecule is of the H_g symmetry and spin-singlet. The process to this state produces the H_g peak at about -0.3 eV. Finally, the highest excited state of an isolated C_{60}^{2-} molecule is of the T_{1g} symmetry and spin-triplet. The process to this state produces the T_{1g} peak at about -0.55 eV. Similarly, we can understand three peaks for the unoccupied states by considering the inverse photoemission processes.

Our results can explain the experimental photoemisson spectra for K_3C_{60} and Rb_3C_{60}. The satellite at -1.3 eV in our result has been observed as a shoulder at about -1.5 eV on the HOMO-derived band in the photoemission spectra [11,12]. Furthermore, the LUMO-derived band in the experiments is much broader than predicted by band calculations with the width of about 1.0 eV [11–13]. Our results reproduce this LUMO derived band as the continuous features below the Fermi level ranged from -1 eV to 0 eV. In the experimental photoemission spectra, two features are observed below the Fermi edge; one feature exists at about -0.3 eV and the other feature exists at about -0.6 eV [11–13]. These features may originate in the H_g peak and the T_{1g} peak in Fig. 1(b), respectively. However, it is difficult to resolve these features in Fig. 1(a). This may be due to the finite size effect in our calculations.

Finally, we propose the interpretation of the optical conductivity spectra from our results. Iwasa et al. report that there are not only the Drude component but also the midinfrared absorption at 0.4–0.5 eV in the optical conductivity spectra of K_3C_{60} and Rb_3C_{60} [5]. Furthermore, Degiorgi et al. observe the midinfrared absorption at 0.1 eV in Rb_3C_{60} [18]. We suggest that these midinfrared absorption is produced by the intermolecular charge transfer transition; it should be noted that the intramolecular transition is forbidden due to the selection rule. In Fig. 1(b), the energy difference between one peak for unoccupied states and that for occupied states is the energy needed to produce one C_{60}^{2-} and one C_{60}^{4-} molecules from two C_{60}^{3-} molecules. In other words, this energy difference corresponds to the energy needed for the intermolecular charge transfer. We find that there are several transitions, which need the energy of about 0.4–0.5 eV. For example, the energy difference between the $\bar{T}_{1g}$ and A_g peak is about 0.5 eV. The transition associated with these peaks can produce the midinfrared absorption at 0.4–0.5 eV in the optical conductivity spectra. On the other hand, the energy difference between the $\bar{A}_g$ and A_g peak is about 0.1 eV. Thus, the transition associated with these peaks can produce the midinfrared absorption at about 0.1 eV.

REFERENCES

1. Weaver, J. H., *J. Phys. Chem. Solids* **53**, 1433 (1992).
2. Lof, R. W., van Veenendaal, M. A., Koopmans, B., Jonkman, H. T., and Sawatzky, G. A., *Phys. Rev. Lett.* **68**, 3924 (1992).
3. Kuzmany, H., Matus, M., Burger, B., and Winter, J., *Adv. Matter.* **6**, 731 (1994).
4. Pintschovius, L., *Rep. Prog. Phys.* **59**, 473 (1996).
5. Iwasa, Y., and Kaneyasu, T., *Phys. Rev. B* **51**, 3678 (1995).
6. Suzuki, S., and Nakao, K., *Phys. Rev. B* **52**, 14206 (1995).
7. Fabrizio, M., and Tosatti, E., *Phys. Rev. B* **55**, 13465 (1997).
8. Han, J. H., Koch, E., and Gunnarsson, O., *Phys. Rev. Lett.* **84**, 1276 (2000).
9. Chida, T., Suzuki, S., and Nakao, K., *J. Phys. Soc. Jpn.* **70**, (to be published).
10. Saito, S., and Oshiyama, A., *Phys. Rev. B* **44**, 11536 (1991).
11. Knupfer, M., Merkel, M., Golden, M. S., Fink, J., Gunnarsson, O., and Antropov, V. P., *Phys. Rev. B* **47**, 13944 (1993).
12. Benning, P. J., Stepniak, F., and Weaver, J. H., *Phys. Rev. B* **48**, 9086 (1993).
13. Goldni, A., Friedmann, S. L., Shen, Z. -X., and Pamigiani, F., *Phys. Rev. B* **58**, 11023 (1998).
14. Chida, T., Suzuki, S., and Nakao, K., *J. Phys. Soc. Jpn.* **69**, 1249 (2000).
15. Suzuki, S., Okada, S., and Nakao, K., *J. Phys. Soc. Jpn.* **69**, 2615 (2000).
16. Caffarel, M., and Krauth, W., *Phys. Rev. Lett.* **72**, 1545 (1994).
17. Georges, A., Kotliar, G., Krauth, W., and Rozenberg, M. J., *Rev. Mod. Phys.* **68**, 13 (1996).
18. Degiorgi, L., Grüner, G., Wachter, P., Huang, S. -M., Wiley, J., Whetten, R. L., Kaner, R. B., Holczer, K., and Diederich, F., *Phys. Rev. B* **46**, 11250 (1992).

Phase Diagrams of Alkali-Metal-Doped C_{60}: Spin- and Orbital-Polarized States

J. Hirosawa, S. Suzuki, and K. Nakao

Institute of Materials Science, University of Tsukuba, Japan

Abstract. We investigate the phase diagrams of alkali-metal-doped C_{60}, A_3C_{60} and A_4C_{60}, concerning the strength of interactions, employing a model in which both the electron-electron and electron-phonon interactions are taken into account. We find that alkali-metal-doped C_{60} has five kinds of possible phases: paramagnetic metal, superconductor, orbital-polarized paramagnetic metal, spin- and orbital-polarized insulator, and orbital-polarized insulator. Both materials have two phases of paramagnetic metal and superconductor in common. Only A_3C_{60} has the phases of orbital-polarized paramagnetic metal and spin- and orbital-polarized insulator. Also, only A_4C_{60} has the phase of orbital-polarized insulator.

INTRODUCTION

In the last decade, the electronic structures of alkali-metal-doped C_{60}, A_xC_{60} where A is an alkali metal, have been studied extensively [1]. In particular, A_3C_{60} is one of the most attractive materials in this field because this is a superconductor with the transition temperatures beyond 30 K [2,3]. Furthermore, in addition to A_3C_{60}, A_4C_{60} has been investigated as well [4-9]. It has been shown by the extensive studies that there is anomalous behavior of these materials as described below and the elucidation of its origin is desired strongly.

Since the conduction bands of C_{60} crystal consist of threefold degenerate molecular orbitals which are the lowest unoccupied molecular orbitals of C_{60} molecule, the t_{1u} orbitals, A_xC_{60} for $0 < x < 6$ are expected to be metallic. Also, according to the theoretical study employing the density functional method performed by Saito and Oshiyama, the band width is about 0.4 eV [10]. It is one of the greatest successes that the band calculations are able to explain the relation between the superconducting transition temperature and the lattice constant in A_3C_{60} [1].

However, it has been shown that the rigid band model cannot be applied to the electronic states of A_4C_{60} according to many experiments. For instance, the magnetic susceptibility measurement employing NMR and ESR has shown that A_4C_{60} are nonmagnetic insulators [4-6]. In the infrared reflection experiment, it has been observed that the spectrum of A_3C_{60} shows the Drude behavior but that of A_4C_{60} does not [7]. Furthermore, the photoemission and inverse photoemission study have shown that there exists an extraordinary change of spectrum near the Fermi level as the valence is varied [8]. This cannot be explained by the rigid band model. Also, according to the μSR experiment, it has been shown that the lowest electronic excitation in K_4C_{60} occurs at 0.33 eV [9]. As for the case of A_3C_{60}, it has been recently revealed that $(NH_3)A_3C_{60}$ is the Mott-Hubbard insulator with antiferromagnetic spin order [11]. These experiments reveal that alkali-metal-doped C_{60} has not only metallic phases but also both magnetic and nonmagnetic insulating phases.

In the present study, we investigate the phase diagrams of alkali-metal-doped C_{60}, A_3C_{60} and A_4C_{60}, concerning the strength of interactions, employing a model which takes account of both the electron-electron and electron-phonon interactions. As shown by the above experiments, the rigid band model is not applicable to the electronic state of A_4C_{60}. By considering both the electron-electron and electron-phonon interactions, we can elucidate that alkali-metal-doped C_{60} has not only metallic or superconducting phases but also magnetic or nonmagnetic insulating phases.

RESULTS AND DISCUSSION

We investigate possible phases in A_3C_{60} and A_4C_{60} as follows. We adopt the effective Hamiltonian. This Hamiltonian has the effective electron-electron interaction in which the electron-phonon interaction is included in the second order perturbation method [12,13]. Also, we employ the Hartree-Fock (HF) approximation using the transfer integrals that reproduce the band structure given in Ref. 10. Furthermore, in A_4C_{60}, we take account of the crystal field splitting of 0.2 eV. In the calculated phase diagrams, the horizontal axis is the intraorbital repulsion V_{intra} and the vertical axis is the pair transfer interaction K with changing its sign. Also, we draw a line, which is the condition for superconductivity, in the phase diagrams according to Ref. 13

In consequence, we are able to classify possible phases in A_3C_{60} and A_4C_{60}. We find that A_3C_{60} have four phases. The obtained phases are paramagnetic metal (PM), superconductor (SC), orbital-polarized paramagnetic metal (OPM), and spin- and orbital-polarized insulator (SOI). Furthermore, it is found that OPM and SOI exist only for A_3C_{60}. It is worth while to note that SOI is the Mott-Hubbard insulator. As shown in Fig. 1(a), this phase exists in the region of large V_{intra}. On the other hand, three metallic phases, PM, SC, and OPM, exist in the region of small V_{intra}. Furthermore, OPM exists in the region of large K. Many experiments shows that A_3C_{60} is a superconductor and $(NH_3)A_3C_{60}$ is a magnetic insulator. Therefore, in this phase diagram, A_3C_{60} should belong to SC and $(NH_3)A_3C_{60}$ should belong to SOI. Nevertheless, since estimated values of V_{intra} and K are about 0.2 eV and about -0.1 eV, respectively, A_3C_{60} belongs to OPM in this phase diagram in contradiction to the experimental results. Next, we show that A_4C_{60} have three phases. The phase diagram is shown in Fig. 1(b). In addition to PM and SC, there is a phase of orbital-polarized insulators (OI). This insulator is nonmagnetic and exists only for A_4C_{60}. Also, A_4C_{60} has PM and SC in common with A_3C_{60}. As in A_3C_{60}, PM and SC exists in the region of small V_{intra} and K. The experiments employing NMR, ESR, and μSR have shown that all known A_4C_{60} are nonmagnetic insulators [4–6]. Therefore, all of A_4C_{60} belong to OI. Since estimated values of V_{intra} and K are about 0.2 eV and about -0.1 eV, respectively, A_4C_{60} belongs to OI in this phase diagram in agreement with the experimental results. However, it is possible that some of them can be metallic or even superconducting, for example, under high pressures because there exist PM and SC in the region of small V_{intra} and K in our phase diagram.

Finally, we discuss the reason why our results fail to explain experimental observation for A_3C_{60} in spite of the success in explaining A_4C_{60}. A most critical deficiency in the HF approximation is the ignorance of fluctuations. In particular, A_3C_{60} possesses many low energy solutions of the HF equation with almost the same energy. Accordingly, the HF approximation is not suitable for A_3C_{60} because the fluctuations are expected to play important roles when we study beyond the HF approximation. On the other hand, the HF approximation is suitable for A_4C_{60} because it possesses only one HF ground state much lower in energy well separated from the exited states. In the future study, it is thus necessary to consider many body effects for obtaining the reliable phase diagram of A_3C_{60}.

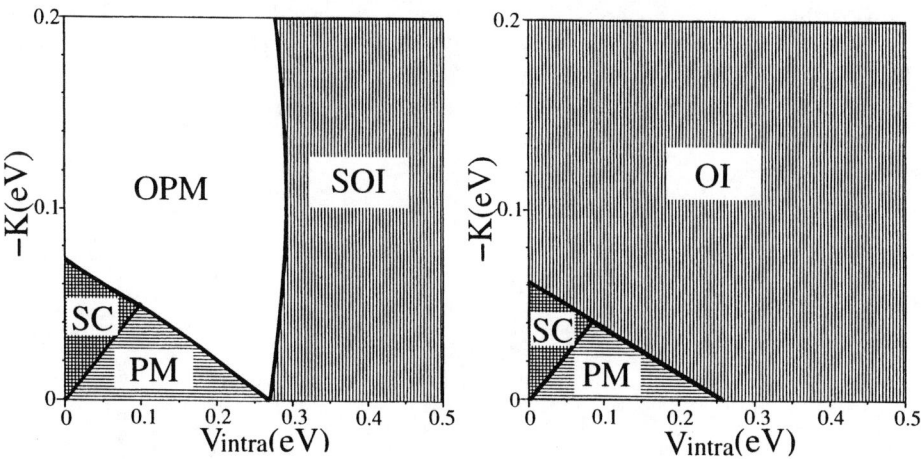

FIGURE 1. Phase diagrams of (a) A_3C_{60} and (b) A_4C_{60}. The horizontal axis is the intraorbital repulsion V_{intra} and the vertical axis is the pair transfer interaction K with changing its sign: PM denotes a paramagnetic metal. SC denotes a superconductor. OPM denotes an orbital-polarized paramagnetic metal. SOI denotes a spin- and orbital-polarized insulator. OI denotes an orbital-polarized insulator.

REFERENCES

1. Ehrenreich, H. and Spaepen(eds.) F., *Solid State Physics, Vol. 48*, New York: Academic Press, 1994 and references therein.
2. Rosseinsky, M. J., Ramirez, A. P., Glarum, S. H., Murphy, D. W., Haddon, R. C., Hebard, A. F., Palstra, T. T. M., Kortan, A. R., Zahurak, S. M., and Makhija, A. V., *Phys. Rev. Lett.* **66**, 2830 (1991).
3. Tanigaki, K., Ebbesen, T. W., Saito, S., Mizuki, J., Tsai, J. S., Kubo, Y., and Kuroshima, S., *Nature* **352**, 222 (1991).
4. Tycko, R., Dabbagh, G., Rosseinsky, M. J., Murphy, D. W., Ramirez, A. P., and Fleming, R. M., *Phys. Rev. Lett.* **68**, 1912 (1992).
5. Kosaka, M., Tanigaki, K., Hirosawa, I., Shimakawa, Y., Kuroshima, S., Ebbesen, T. W., Mizuki, J., and Kubo, Y., *Chem. Phys. Lett.* **203**, 429 (1993).
6. Lukyanchuk, I., Kirova, N., Rachdi, F., Goze, C., Molinie, P., and Mehring, M., *Phys. Rev. B* **51**, 3978 (1995).
7. Iwasa Y., and Kaneyasu, T., *Phys. Rev. B* **51**, 3678 (1995).
8. Weaver, J. H., *J. Phys. Chem. Solids* **53**, 1433 (1992).
9. Kiefl, R. F., Duty, T. L., Schneider, J. W., MacFarlane, A., Chow, K., Elzey, J. W., Mendels, P., Morris, G. D., Brewer, J. H., Ansaldo, E. J., Niedermayer, C., Noakes, D. R., Stronach, C. E., Hitti, B., and Fischer, J. E., *Phys. Rev. Lett.* **69**, 2005 (1992).
10. Saito S., and Oshiyama, A., *Phys. Rev. Lett.* **66**, 2637 (1991).
11. Tou, H., Maniwa, Y., Iwasa, Y., Shimoda, H., and Mitani, T., *Phys. Rev. B* **62**, 775 (2000).
12. Suzuki, S., and Nakao, K., *Phys. Rev. B* **52**, 14206 (1995).
13. Suzuki, S., Okada, S., and Nakao, K., *J. Phys. Soc. Jpn.* **69**, 2615 (2000).

Study on the Physical Properties of Na_4C_{60}

Yasuhiro Takabayashi,[a] Yoshihiro Kubozono,[a] Satoshi Fujiki,[a] Setsuo Kashino,[a] Kenji Ishii,[b] Hiroyoshi Suematsu,[b] Hironori Ogata[c]

[a]*Department of Chemistry, Okayama University, Okayama 700-8530, Japan*
[b]*Department of Physics, University of Tokyo, Tokyo 113-0033, Japan*
[c]*Institute of Molecular Science, Okazaki 444-8585, Japan*

Abstract. Temperature dependence of resistivity, ρ, of a two-dimensional polymer phase of Na_4C_{60} has been studied from 190 to 300 K. The ρ was 6.9 x 10^3 Ω cm at 300 K, which is higher by six orders than those of metallic fullerides. The temperature dependence of ρ shows a semiconducting behavior in this temperature region. This result is different from the previous result derived from ESR which suggests a metallic property above 100 K. The gap energy, E_g, was estimated to be 0.8 eV. This value is close to those of other A_4C_{60} (A: K, Rb and Cs).

INTRODUCTION

Study on a low-dimensional polymeric fullerides is one of the most interesting research-subject in chemistry and physics of fullerenes because low-dimensionality leads to very interesting physical properties. AC_{60} (A: K, Rb and Cs) forms a one-dimensional (1D) polymer between C_{60} molecules through a [2+2] cycloaddition [1-4]. The polymeric phase of RbC_{60} shows a metallic behavior above 50 K, and it transforms to an insulating state below 50 K [2]. The temperature dependence of ESR suggests that this transition is caused by an SDW instability of 1D metal [2]. On the other hand, a recent study on pressure dependence of ESR suggests that RbC_{60} is a 3D Mott-Hubbrad insulator [3]. The polymeric phase of KC_{60} showed no metal-insulator transition at low temperature, suggesting that KC_{60} is a 3D metal [4].

Na_4C_{60} forms a two-dimensional (2D) polymer between C_{60} molecules through single C-C bonds [5]. The crystal structure is body-centered monoclinic (bcm) [5], and the phase transforms to a monomeric phase with a body-centered tetragonal (bct) structure above 500 K [6]. It has been believed so far that both phases are metallic because of Pauli-paramagnetic behavior in spin susceptibility, χ_{spin}, determined from ESR [5,6]. The metallic behavior in the monomeric phase above 500 K, whose structure is the same as insulating A_4C_{60}, is interpreted based on the picture of Mott-Hubbard and Jahn-Teller effect [7]. Recently, we reported an existence of magnetic transition below 100 K in Na_4C_{60} on the basis of temperature dependence of ESR [8]; the ESR study also showed the Pauli-paramagnetic behavior above 100 K.

In the present study, we measured temperature dependence of resistivity, ρ, of Na_4C_{60} in order to confirm the metallic property suggested from ESR. The result showed a semiconduction bahavior with the close energy gap, E_g, to those of A_4C_{60}. The electronic property of Na_4C_{60} is discussed based on both results of ρ and ESR.

EXPERIMENTAL

The Na_4C_{60} sample was prepared by annealing stoichiometric amounts of C_{60} and Na metal at 723 K for 234 h under a vacuum of 10^{-5} Torr; a trace of benzene contained in commercially available C_{60} was removed before the annealing. The Na_4C_{60} sample was introduced into a glass capillary for X-ray powder diffraction and Raman measurements to check the sample quality. The X-ray diffraction pattern (FIGURE 1) is similar to those reported previously [5,8]. The Rietveld refinement for the X-ray diffraction pattern showed that this sample was $Na_{3.90(4)}C_{60}$ of the bcm 2D structure. The $A_g(2)$ Raman peak (FIGURE 2) was observed at center frequency, ω_0, of 1439 cm^{-1}, supporting the formation of $Na_{3.90(4)}C_{60}$ as is shown from the X-ray diffraction. The measurement of ρ was performed with a pellet by a standard four-probe technique.

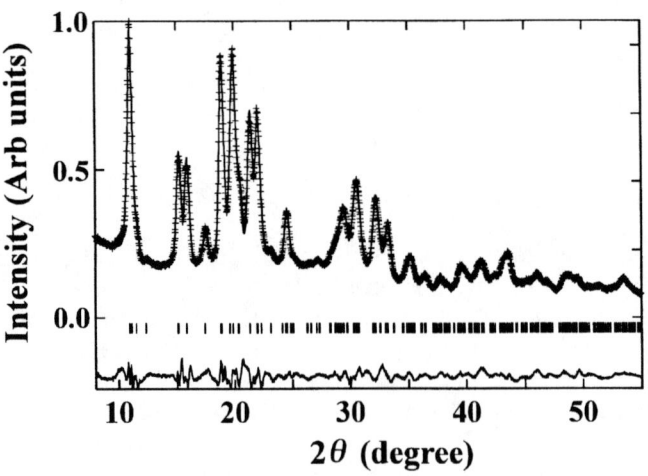

FIGURE 1. X-ray diffraction pattern at 300 K.

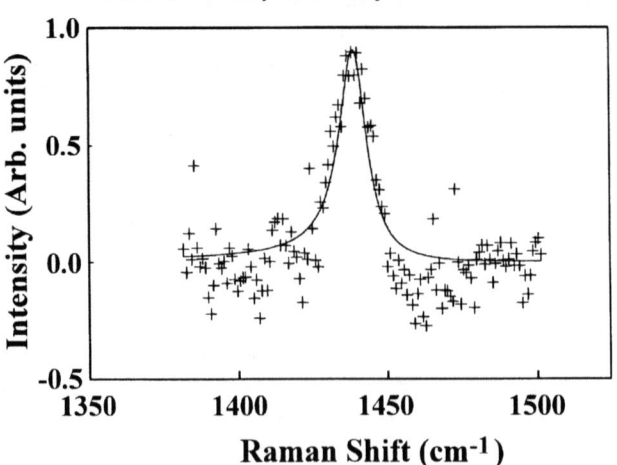

FIGURE 2. $A_g(2)$ Raman spectrum at 300 K.

RESULTS AND DISCUSSION

The ρ for the $Na_{3.90(4)}C_{60}$ sample was 6.9×10^3 Ω cm at 300 K, which is higher by six order than those of metallic fullerides (5.2×10^{-3} Ω cm for K_3C_{60} and 4.7×10^{-3} Ω cm for Rb_3C_{60} [9]. Further, the ρ of 6.9×10^3 Ω cm is higher by four order than that of Cs_3C_{60} (0.52 Ω cm) [10]; the ρ of Cs_3C_{60} is according to the granular metal theory based on hopping conductivity, $\rho = \rho_0 \exp(T_0/T^{1/2})$ [11]. The temperature dependence of ρ for the $Na_{3.90(4)}C_{60}$ sample in a temperature region from 190 to 300 K is shown in FIGURE 3. This result shows a semiconducting behavior. The ρ at 190 K is 4.5×10^6 Ω cm which is extremely large among fullerides.

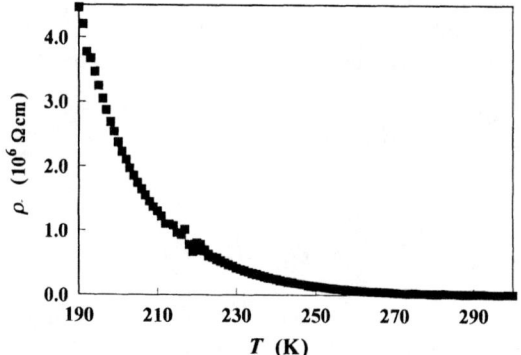

FIGURE 3. Temperature dependence of ρ.

The plots of $\ln \rho$ vs $1/T$ in the temperature region from 250 to 300 K is shown in FIGURE 4. The plots show a linear relationship in this temperature region. The E_g was estimated to be 0.8 eV from this plots according to the relationship, $\rho = \rho_0 \exp(E_g/2k_BT)$, while the E_g was estimated to be 0.5 eV from the plots in the low temperature region from 190 to 250 K. The E_g values are smaller than that of C_{60}, 2.1 eV [12], and they are close to those of A_4C_{60} (0.5 eV for K_4C_{60}, 0.6 eV for Rb_4C_{60} and 0.6 eV for Cs_4C_{60} [7]). The E_g values for the 2D Na_4C_{60} were larger than that for face-centered cubic (fcc) monomeric phase of Na_4C_{60} estimated from the optical

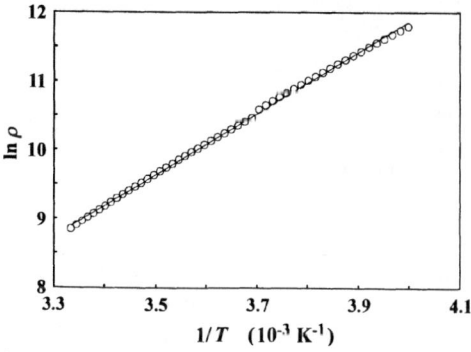

FIGURE 4. Plots of $\ln \rho$ vs $1/T$ in a temperature region from 250 to 300 K.

conductivity, 0.25 eV [7]. This result implies that electronic property of the 2D Na_4C_{60} is similar to those of the other semiconducting A_4C_{60}. Further, we tried to fit the ρ according to the granular metal theory, and the ρ could be fitted by this relationship. Consequently, it has been concluded that the 2D Na_4C_{60} is not a normal metal but a semiconducting material with a relatively small E_g or a weak localization system with electron-electron interactions.

It was expected that the short C_{60}-C_{60} distance in the 2D Na_4C_{60} results in a broad band witdth, W, which exceeds the electron-electron Coulombic repulsion U; this picture should lead to metallic Na_4C_{60} [7]. In fact, the density of state, $N(\varepsilon_F)$, on the Fermi level estimated from the χ_{spin} of 1.4×10^{-4} emu/mol at 290 K was 2 state/eV-spin-C_{60} [8], which is smaller than those of normal metallic fullerides (14 state/eV-spin-C_{60} for K_3C_{60} and 19 state/eV-spin-C_{60} for Rb_3C_{60} [13]), suggesting a large W in the 2D Na_4C_{60}. The metallic behavior suggested from the ESR results could well be explained by this picture. However, the result of ρ is different from that of ESR. The direct measurement of ρ with a thin film of the 2D Na_4C_{60} is now in progress in order to obtain the data of electric transport with high accuracy. This will give a clue for explaining the inconsistency between the results of ρ and ESR.

ACKNOWLEDGMENTS

This work is supported by a Grant-in-Aid (11165227) from the Ministry of Education, Science, Sports and Culture, Japan. A part of this work is supported by a Grant-in-Aid of WESCO Science Foundation.

REFERENCES

1. Stephens, P. W., Bortel, G., Faigel, G., Tegze, M., Janossy, A., Pekker, S., Oszlanyi, G., and Forro, L., *Nature* **370**, 636-639 (1994).
2. Chauvet, O., Oszlanyi, G., Forro, L., Stephens, P. W., Tegze, M., Faigel, G., and Janossy, A., *Phys. Rev. Lett.* **72**, 2721-2724 (1994).
3. Sakamoto, H., Kobayashi, S., Mizoguchi, K., Kosaka, M., and Tanigaki, K., *Phys. Rev.* **B62**, R7691-R7694 (2000).
4. Bommeli, F., Degiorgi, L., Wachter, P., Legeza, O., Janossy, A., Oszlanyi, G., Chauvet, O., and Forro, L., *Phys. Rev.* **B51**, 14794-14797 (1995).
5. Oszlanyi, G., Baumgartner, G., Faigel, G., and Forro, L., *Phys. Rev. Lett.* **78**, 4438-4441 (1997).
6. Oszlanyi, G., Baumgartner, G., Faigel, G., Granasy, L., and Forro, L., *Phys. Rev.* **B58,** 5-7 (1998).
7. Knupfer, M., and Fink, J., *Phys. Rev. Lett.* **79**, 2714-2717 (1997).
8. Kubozono, Y., Takabayashi, Y., Kambe, T., Fujiki, S., Kashino, S., and Emura., S., *Phys. Rev.* **B**, in press.
9. Stepniak, F., Benning, P. J., Poirier, D. M., and Weaver, J. H., *Phys. Rev.* **B48**, 1899-1906 (1993).
10. Fujiki, S., Kubozono, Y., Takabayashi, Y., Kashino, S., Kobayashi, M., Ishii, K., Suematsu, H., in Nanonetwork Materials: Fullerenes, Nanotubes, and Related Systems, edited by S. Saito, AIP Conference Proceedings, New York: American Institute of Physics, submitted.
11. Sheng, P., Abeles, B., and Arie, Y., *Phys. Rev. Lett.*, **31**, 44-47 (1973)
12. Takahashi, T., Suzuki, S., Morikawa, T., Katayama-Yoshida, H., Hasegawa, S., Inokuchi, H., Seki, K., Kikuchi, K., Suzuki, S., Ikemoto, K., and Achiba, Y., *Phys. Rev. Lett.*, **68**, 1232-1235 (1992).
13. Ramirez, A. P., Rosseinsky, M. J., Murphy, D. W., and Haddon, R. C., *Phys. Rev. Lett.*, **69**, 1687-1690 (1992).

Fabrication of C_{60} / Amorphous Carbon Superlattice Structures

Nobuaki Kojima, Yoshio Ohshita and Masafumi Yamaguchi

Toyota Technological Institute, 2-12-1 Hisakata, Tempaku, Nagoya 468-8511, JAPAN

Abstract. The nitrogen doping effects in C_{60} films by RF plasma source was investigated, and it was found that the nitrogen ion bombardment broke up C_{60} molecules and changed them into amorphous carbon. Based on these results, formation of C_{60} / amorphous carbon superlattice structure was proposed. The periodic structure of the resulted films was confirmed by XRD measurements, as the preliminary results of fabrication of the superlattice structure.

INTRODUCTION

The optical properties of C_{60} incorporated in rigid solid matrix, such as zeolites, have been investigated in order to study the confinement effect. Some researchers reported white light emission from such confinement structures. However, superlattice structure made of C_{60} has not been realized so far.

In our previous work, we have investigated [1] the nitrogen doping effects in C_{60} films by radio frequency (RF) plasma source, and found that the nitrogen ion bombardment breaks up C_{60} molecules and changes them into amorphous carbon. Based on these results, it is thought that the periodic structure of C_{60} and amorphous carbon layers can be realized by intermittent supply of nitrogen ions during C_{60} deposition. In this paper, we propose carbon-based superlattice structure and its fabrication technique.

Superlattice structure has advantages to control the optical and electrical properties of the material. The band gap energy of superlattice can be controlled by the deposition condition of amorphous carbon layers and the thickness of each layer. Furthermore, the conductivity control is enabled by impurity doping in the amorphous carbon layers. Our objective is the property control of carbon-based materials and to investigate the confinement effect by superlattice structures.

EXPERIMENTAL

An ultra-high vacuum chamber equipped with the RF plasma source (SVT Model RF 4.5) was used for the deposition of N-doped C_{60} films. The base pressure of the chamber was 2×10^{-9} Torr. The pure (99.98%) C_{60} powder was evaporated from a Knudsen cell. N_2 source gas was introduced to the plasma source through a mass flow controller. C_{60} and excited nitrogen species were supplied simultaneously onto the Si(100) or glass substrates during the deposition. The deposition rate was about 2.2 nm/min.

N_2 source gas was excited to N^+ ions, atomic N, 1st excited neutral N_2, higher excited neutral N_2, and so on by RF plasma source, and these excited nitrogen species were supplied into the deposition chamber through the (Pyrolytic Boron Nitride) PBN aperture. Nitrogen ions can be deflected from a substrate by applying the high voltage to the deflection plates placed at the tip of the RF plasma source. The quantity of nitrogen ions onto the substrate was controlled by the aperture size, applied voltage to the deflection plates and the plasma discharge condition (RF power and N_2 flow rate). The PBN aperture of $\phi1.0$ mm or $\phi2.5$ mm in diameter was used in this work. The quantity of nitrogen ions increases with the diameter of the aperture.

RESULTS AND DISCUSSION

Structural Changes in N-doped C_{60} Films

Raman scattering spectroscopy of N-doped C_{60} films was measured to confirm the film structure, as shown in Fig. 1. The pure C_{60} film has a strong Raman peak at 1469 cm^{-1} and weak Raman peaks at 1426 and 1573 cm^{-1}. These Raman peaks are assigned to A_g vibrational mode of C_{60} molecule for 1469 cm^{-1} peak and to H_g vibrational mode of C_{60} molecule for 1426 and 1573 cm^{-1} peaks. The intensity of these C_{60} Raman peaks decreases in order of (b), (c) and (d) in Fig. 1. The broad peaks at around 1570 and 1450 cm^{-1} are observed in (c) and (d). These broad peaks correspond to the G- and D-peaks of amorphous carbon. The quantity of nitrogen ions onto the substrate increases in order of (b), (c) and (d). Therefore, it can be concluded that the nitrogen ion bombardment breaks up C_{60} molecules and changes them into amorphous carbon.

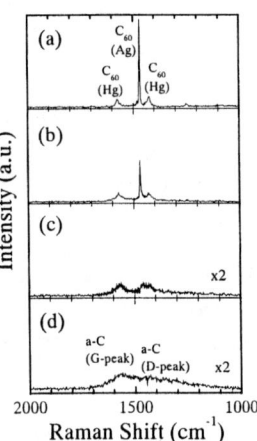

FIGURE 1. Raman spectra of pure C_{60} and N-doped C_{60} films. (a) pure C_{60} film, (b) N-doped film (RF power (P_{RF}): 350 W, N_2 flow rate (F_N): 0.4 sccm, Aperture size (H): 1.0mm, Deflection voltage (V_{DEF}): 750 V) (c) N-doped film (P_{RF}: 350 W, F_N: 0.4 sccm, H: 1.0mm, V_{DEF}: 0 V) (d) N-doped film (P_{RF}: 350 W, F_N: 2.0 sccm, H: 2.5mm, V_{DEF}: 0 V)

The chemical composition of resulted amorphous carbon was determined by X-ray photoelectron spectroscopy (XPS). Nitrogen concentration is ranged from 2.4 to 7.7 at. %.

The absorption coefficient spectra of the N-doped C_{60} films were calculated from the uv-visible reflectance / transmittance measurements, and optical band gaps were estimated by Tauc plots as shown in Fig. 2. The optical band gap decreases with the formation of amorphous carbon, and 1.30 ~ 1.52 eV are obtained for N-doped C_{60} films, while pure C_{60} has a band gap of 1.7 ~ 1.8 eV.

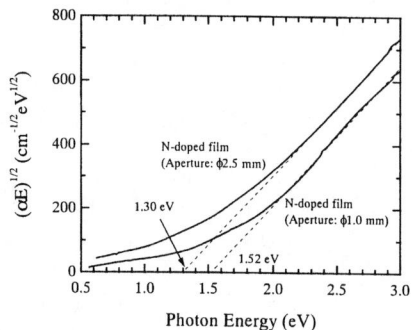

FIGURE 2. Tauc plots of optical absorption spectra of N-doped C_{60} films

C_{60} / Amorphous Carbon Superlattice Structures

Based on the above results, it is thought that the periodic structure of C_{60} and amorphous carbon layers can be realized by intermittent supply of nitrogen ions during C_{60} deposition. Figure 3 shows the shutter sequence for the fabrication of superlattice structure of C_{60} and amorphous carbon. C_{60} layers are deposited during T_1 (C_{60}: on, N: off), and amorphous carbon layers are formed during T_2 (C_{60}: on, N: on) due to the C_{60} cage breaking. The interval times T" and T' are necessary for the nitrogen gas shutting and ignition of nitrogen plasma.

Figure 4 shows the expected band diagram of C_{60} / amorphous carbon superlattice structure. The valence band discontinuity is unknown.

The periodic structure of resulted films was confirmed by X-ray diffraction (XRD) measurements. Figure 5 shows XRD pattern of low angle 2θ-ω scan of the films grown with shutter sequence of $T_1=T_2$ for (a), (b) and $T_1=6\,T_2$ for (c) and 45 cycles. Diffraction peaks from periodic layer structure were observed, corresponding to the periodic distance (C_{60} layer + a-C layer) of 7.96, 5.51 and 4.03 nm for #1, #2 and #3 sample, respectively. These values are very close to the designed value. The

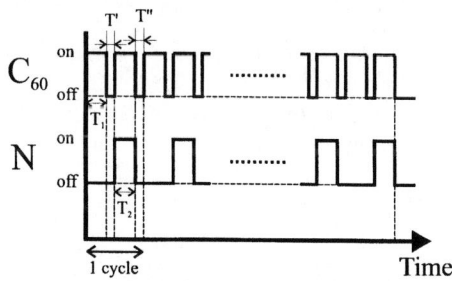

FIGURE 3. Shutter sequence for the fabrication of C_{60} / amorphous carbon superlattice structure.

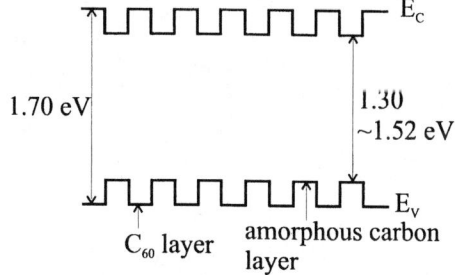

FIGURE 4. Band diagram of C_{60} / amorphous carbon superlattice structure.

each layer thickness of C_{60} and a-C layer is unknown from this measurement. Designed thickness from shutter sequence is a-C(4.0 nm)/C_{60}(4.0 nm), a-C(2.8 nm)/C_{60}(2.8 nm), and a-C(0.6 nm)/C_{60}(3.4 nm) for #1, #2 and #3 sample, respectively.

Furthermore, diffraction peaks show splitting for #1 and #2 sample. The differences between the splitting are around 0.4 ~ 0.5 nm, which is near to the C_{60}(220) or (311) spacing. In our previous work, we observed that the XRD peaks of C_{60}(220) and (311) survived in N-doped C_{60} films, even if the C_{60}(111) peak almost vanished [1]. Therefore, it is thought that the splitting of diffraction peak is related to the molecular arrangement. The detailed analysis will be discussed elsewhere. However, such fluctuation of periodic distance affects band structure of superlattice. To control thickness of each layer, further optimization is necessary. Furthermore, transmission electron micrograph (TEM) observation is now under study to clarify the detailed structure of this film.

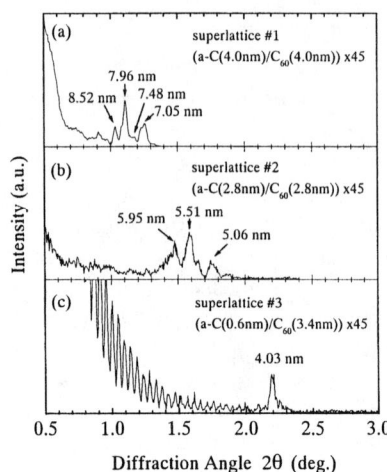

FIGURE 5. X-ray diffraction (XRD) pattern of low angle 2θ-ω scan of the C_{60} / amorphous carbon superlattice structure.

CONCLUSION

The nitrogen doping effects in C_{60} films by RF plasma source were investigated, and it was found that the nitrogen ion bombardment broke up C_{60} molecules and changed them into amorphous carbon. Based on these results, formation of C_{60} / amorphous carbon superlattice structure was proposed by intermittent supply of nitrogen ions during C_{60} deposition. The periodic structure of the resulted films was confirmed by XRD measurements, as the preliminary results of fabrication of the superlattice structure.

ACKNOWLEDGEMENTS

This work was supported in part by the Japan Society for the Promotion of Science as a program entitled Research for the Future (JSPS- RFTF97P00902: Study of New Carbon Based Materials and Solar Cells), and by the Ministry of Education as a Private University High-Tech. Research Center Program.

REFERENCES

1. Kojima, N and Yamaguchi, M, "Radical Beam Doping in C_{60} Films For Solar Cell Application" in *16th European Photovoltaic Solar Energy Conference-2000*, proceeding is to be published.

Ac conductivity of alkali doped C_{60} compounds across the superconductor-insulator transition

Atsutaka Maeda*, Haruhisa Kitano*, Rhoji Matsuo*, Kazuhiko Miwa*, Taishi Takenobu†, Yoshihiro Iwasa† and Tadaoki Mitani†

*Department of Basic Science, The University of Tokyo [1]
3-8-1, Komaba, Meguro-ku, Tokyo 153-8902, JAPAN
†Japan Advanced Institute of Science and Technology, Tatsunokuchi, Ishikawa 923-1292, JAPAN

Abstract.
We investigated the electrical conductivity of alkali fullerides K_3C_{60}, $(NH_3)K_3C_{60}$ and $(NH_3)_x NaRb_2C_{60}$ by the microwave cavity perturbation method. We found that $(NH_3)K_3C_{60}$ showed a semiconductive behavior of resistivity between 4.2 and 250 K. We did not find any anomaly at temperatures where phase transitions were suggested to take place in other techniques. On the other hand, the resistivity of $(NH_3)_x NaRb_2C_{60}$ is metallic both for x=0.8 and x=0.9. These results suggest that the disappearance of superconductivity in $(NH_3)K_3C_{60}$ is due to the distortion of the crystal structure from the cubic structure.

INTRODUCTION

Superconductivity in alkali-doped fullerides attracts much attention in terms of both high-temperature superconductivity and metal-to-insulator transition. The host material, C_{60}, is a typical insulator. By doping alkali atoms, superconductivity appears in limited conditions. A_3C_{60} (A= K), which has a cubic structure, is a superconductor with the critical temperature, T_c, of 19 K [1]. By increasing the lattice constant of the cubic structure via the change of the "A" material in the formula, T_c increases up to 40 K [1-3]. Introduction of NH_3 ion increases the lattice constant further [4]. At the same time, however, superconductivity disappears in $(NH_3)K_3C_{60}$. Instead, an antiferromagnetic order was found to occur at low temperatures [5,6]. Another series of materials $(NH_3)_x NaRb_2C_{60}$ [7] have larger lattice constant than $(NH_3)K_3C_{60}$. These materials show superconductivity. However, T_c decreases with increasing lattice constant. These results suggest important factors

---

[1] This work is partially supported by the Grant-in-Aid for Scientific Research on Priority Area (A) No. 258, "Vortex Electronics" sponsored by the Ministry of Education, Science, Sports and Culture.

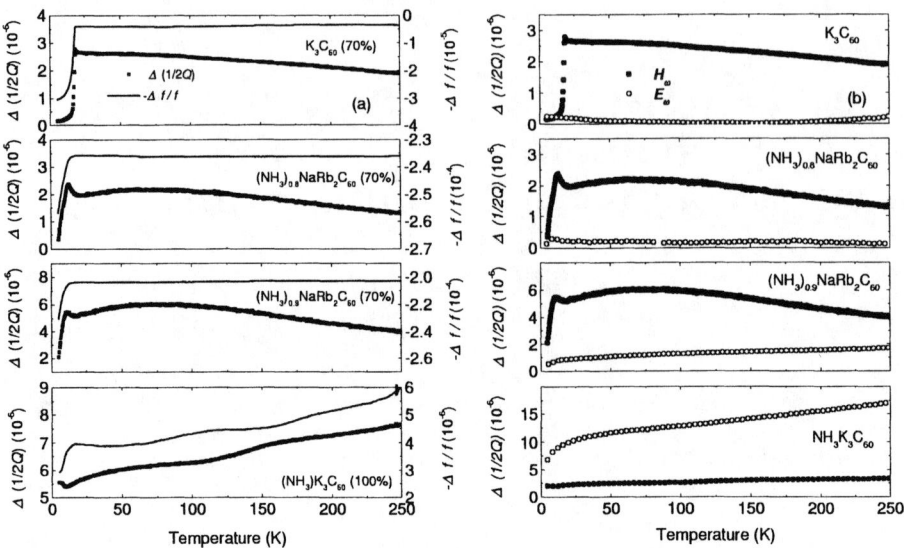

FIGURE 1. (a) $\Delta(1/2Q)$ and $\Delta f/f$ of K_3C_{60}, $(NH_3)K_3C_{60}$ and $(NH_3)_x NaRb_2C_{60}$ with $x=0.8$ and 0.9 as a function of temperature. (b) $\Delta(1/2Q)$ of K_3C_{60}, $(NH_3)K_3C_{60}$ and $(NH_3)_x NaRb_2C_{60}$ with $x=0.8$ and 0.9 measured in the E- and H- fields, respectively, as a function of temperature. Closed symbols are the H-field data, and the open symbols are the E-field data.

other than the lattice constant seriously affect superconductivity. To clarify the mechanism of the disappearance of superconductivity in $(NH_3)K_3C_{60}$ and the decrease of T_c in $(NH_3)_x NaRb_2C_{60}$ with increasing lattice constant, it is expected that the conductivity data will be obtained. However, the samples were obtained only in the form of anaerobic powders, and the electrical conduction of these materials has not yet been determined. Microwave cavity perturbation method is a potentially powerful tool for such a purpose. Since this method does not require an electrical contact, we can measure the electrical conductivity for such samples. In this paper, we applied this method to K_3C_{60}, $(NH_3)K_3C_{60}$ and $(NH_3)_x NaRb_2C_{60}$.

EXPERIMENTAL

Sample preparation was described elsewhere [7]. Sample powders were sealed in a glass tube within a He atmosphere at 300 Torr.

The microwave response was obtained by placing the sample tube at a selected position in a cylindrical cavity, and by measuring the central frequency and the Q value of the resonance of the mode [9]. We selected a sample position where only the microwave magnetic field or microwave electric field was present at the center of the sample. If we placed the sample at the point of maximum microwave electric field, a complex dielectric response was obtained (E-field measurement), whereas at the position of maximum microwave magnetic field, a complex magnetic response

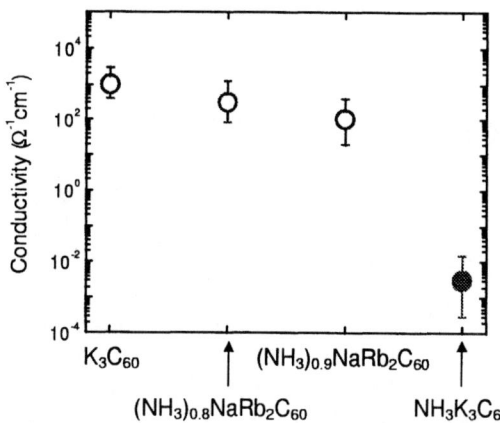

FIGURE 2. Conductivity of K_3C_{60}, $(NH_3)K_3C_{60}$ and $(NH_3)_xNaRb_2C_{60}$ with $x=0.8$ and 0.9 at 150 K.

was obtained (H-field measurement). Typically, we used the TE_{011} mode with the resonance frequency of 10.7 GHz for measurements both in microwave magnetic and electric fields. The details on the analysis were described in a separate publication. [8,10] The net microwave response of the sample was obtained by subtracting the response of an empty tube of almost the same size.

RESULTS AND DISCUSSION

Figure 1 is the change in the inverse of the Q values, $\Delta(1/2Q)$, and that of the relative change of the resonant frequency, $\Delta f/f$, of K_3C_{60}, $(NH_3)K_3C_{60}$, and $(NH_3)_xNaRb_2C_{60}$ with $x=0.8$ and 0.9 as a function of temperature. It was found that all materials other than $(NH_3)K_3C_{60}$ show larger losses in the H-field measurement than in the E-field measurement. This definitely shows that these materials are metallic. On the ohter hand, $(NH_3)K_3C_{60}$ shows larger loss in the E-field measurement, suggesting that the material is semiconducting. It is remarkable that the resistivity behavior is semiconductive between 4.2 K and 250 K, and there were any signs of anomaly at the temperatures where the occurrence of the phase transitions were reported (Structural transition at $\sim$150 K [11] and antiferromagnetic transition at $\sim$40 K [12]). Thus, the data suggests that semiconductive behavior of resistivity is not related to these phase transitions.

The resistivity value obtained from Fig. 1 is shown in Fig. 2. This figure shows that all of the three materials other than $(NH_3)K_3C_{60}$ show rather high conductivity values. It should be noted that all of these materials have cubic crystal structure, and exhibit superconductivity. This is quite contrast to $(NH_3)K_3C_{60}$, which has a non-cubic crystal structure and does not exhibit superconducting transition. Since this material is the only one that does not show superconductivity and has a non-cubic crystal structure, we speculate that the origin of the disappearance of superconductivity in $(NH_3)K_3C_{60}$ is the change of the electronic structure due to the distortion of the crystal structure from the cubic symmetry.

CONCLUSION

We investigated the electrical conductivity of alkali fullerides K_3C_{60}, $(NH_3)K_3C_{60}$ and $(NH_3)_xNaRb_2C_{60}$ with $x=0.8$ and 0.9 by the microwave cavity perturbation method. For $(NH_3)K_3C_{60}$, we found that the material shows a behavior of resistivity which is typical of semiconductor between 4.2 K and 250 K. We did not find any anomaly at the temperatures where phase transitions were suggested to take place in other techniques. On the other hand, the resistivity of $(NH_3)_xNaRb_2C_{60}$ was found to be rather metallic both for x=0.8 and x=0.9. These results suggest that the disappearance of superconductivity in $(NH_3)K_3C_{60}$ is due to the distortion of the crystal structure from the cubic symmetry.

REFERENCES

1. Fleming, R. M., Ramirez, A. P., Rosseinsky, M. J., Murphy, D. W., Haddon, R. C., Zahurak, S. M., and Makhija, A. V., Nature **352**, 787-788 (1991).
2. Tanigaki, K., , Ebbesen, T. W., Saito, S., Mizuki, J., Tsai, J. S., Kubo, Y., and Kuroshima, S., Nature **353**, 222-223 (1991).
3. Palstra, T. M., Zhou, O., Iwasa, Y., Sulewski, P. E., Fleming, R. M., and Zegraski, B. R., Solid State Commun. **93**, 327-330 (1995).
4. Rosseinsky, M. J., Murphy, D. M., Fleming, R. M., and Zhou, O., Nature **364**, 425-427 (1993).
5. Iwasa, Y., Shimoda, H., Palstra, T. T., Maniwa, Y., Zhou, O., and Mitani, T., Phys. Rev. **B53**, R8836-R8839 (1996).
6. Allen, K. M., Heyes, S. J., and Rosseinsky, M. J., J. Mater. Chem. **6**, 1445-1447 (1996).
7. Shimoda, H. et al., Phys. Rev. **B54**, R15653-R15656 (1996).
8. Maeda, A. et al., Jpn. J. Appl. Phys. **39**, 6459-6464 (2000).
9. Klein, O., Donovan, S., Dressel, M., and Grüner, G., Int. J. Infrared Millim. Waves **14**, 2423-2457 (1993).
10. Porch, A., Waldram, J. R., and Cohen, L., J. Phys. F. Metal Phys. **18**, 1547-1562 (1988).
11. Ishii, K. et al., Phys. Rev. **B59**, 3956-3960 (1999).
12. Tou, S. et al., Phys. Rev. **B62**, R775-R778 (2000).

Study on the Origin of Pressure-Induced Superconductivity of Cs_3C_{60}

S. Fujiki,[a] Y. Kubozono,[a] Y. Takabayashi,[a] S. Kashino,[a] M. Kobayashi,[b] K. Ishii,[c] H. Suematsu[c]

[a]*Department of Chemistry, Okayama University, Okayama 700-8530, Japan*
[b]*Department of Material Science, Himeji Institute of Technology, Kamigori 678-1297, Japan*
[c]*Department of Physics, University of Tokyo, Tokyo 113-0033, Japan*

Abstract. Physical properties of bco phase of $Cs_{3+\alpha}C_{60}$ ($\alpha = 0.0 - 1.0$) and A15 phase of Cs_3C_{60} are studied by X-ray diffraction, ESR, AC susceptibility, resistivity and Raman. The ESR of $Cs_{3.00(6)}C_{60}$ showed a broad peak of ~ 380 G due to conduction electron, while no broad peak was observed in the ESR of bco $Cs_{3+\alpha}C_{60}$ ($\alpha \neq 0.0$) and A15 phase. This shows that only bco phase of Cs_3C_{60} is metallic. The AC susceptibility of bco phase of $Cs_{3.2(3)}C_{60}$ and $Cs_{3.5(1)}C_{60}$ showed no superconducting transition above 1.3 K even under high pressure, and the resistivity was 0.52 Ω cm for both samples.

INTRODUCTION

It is reported by Palstra *et al.* [1] that Cs_3C_{60} is a pressure-induced superconductor of $T_c = 40$ K at 14.3 kbar. The value of T_c increases with increasing pressure, in contrary to normal superconducting fullerides [1]. The crystals of Cs_3C_{60} take both structures of body-centered orthorhombic (bco) and A15 at 1 bar [2]. The ESR of bco enriched sample of Cs_3C_{60} showed a metallic behavior above 1.9 K at 1 bar [2]. The electron-phonon coupling constant, λ, of the sample was ~ 0.18 at 1 bar [3] which was smaller than those of Rb_3C_{60} and K_3C_{60} [4], and the λ showed no linear increase when increasing pressure [5]. The A15 phase in Cs_3C_{60} disappeared above 20 kbar [1,6], suggesting that the bco phase is a superconducting phase under high pressure. However, the origin of a pressure-induced superconductivity in Cs_3C_{60} has not yet been clarified. Further, the superconducting phase under high pressure is still controversial. The crystal structure of Cs_4C_{60} is the same as that of Cs_3C_{60} except for occupancy fraction, w_F, of Cs atom at $4f$ and $4h$: $w_F = 1.0$ for Cs_4C_{60} and $w_F = 0.75$ for Cs_3C_{60} [2,7]. This suggests possibility for the formation of $Cs_{3+\alpha}C_{60}$ ($\alpha = 0.0 - 1.0$). In the present study, we investigated the physical properties of the bco phase of $Cs_{3+\alpha}C_{60}$ ($\alpha = 0.0 - 1.0$) and the A15 phase in a wide temperature region at various pressure in order to specify the pressure-induced superconducting phase.

EXPERIMENTAL

The samples of the bco phase $Cs_{3+\alpha}C_{60}$ ($\alpha = 0.0 - 1.0$) were prepared according to the method described elsewhere [1-3, 6, 7]. A trace of NH_3 in the sample was removed by a dynamical pumping at 100 ℃ under 10^{-6} Torr. The values of α in the samples were determined by the Rietveld analyses for X-ray powder diffraction patterns of the samples. The center frequency of $A_g(2)$ Raman peak in the samples supported the α value determined from the X-ray powder diffraction. The rapid removal of NH_3 from the sample led to the formation of A15 enriched sample in which the fraction of A15 is ~30 %; the composition of A15 phase is Cs_3C_{60}. On the other hand, the slow removal led to the formation of bco enriched sample in which the fraction of A15 is ~ 2 %.

RESULTS AND DISCUSSION

The ESR spectra of the samples of $Cs_{3.00(6)}C_{60}$ and $Cs_{3.4(2)}C_{60}$ are shown in figures 1(a) and (b), respectively; the fractions of A15 phase in these samples are 2% for $Cs_{3.00(6)}C_{60}$ and 20% for $Cs_{3.4(2)}C_{60}$ sample. The ESR spectra of bco $Cs_{3.00(6)}C_{60}$ is composed of three components of very narrow, narrow and broad peaks. The broad peak of line width, ΔH_{pp}, = ~380 G was assigned to that due to conduction electron based on the temperature dependence of the ESR; the ΔH_{pp} shows a linear increase with an increase in temperature. On the other hand, the ESR spectrum of the $Cs_{3.4(2)}C_{60}$ sample showed no broad peak, implying that the bco phase of $Cs_{3.4(2)}C_{60}$ is not metallic. Further, the A15 phase is also suggested to be nonmetallic because this sample contains 20 % of the A15 phase.

The plots of logarithm of resistivity, $\ln\rho$, vs $1/T^{1/2}$ for the samples of $Cs_{3.2(3)}C_{60}$ and

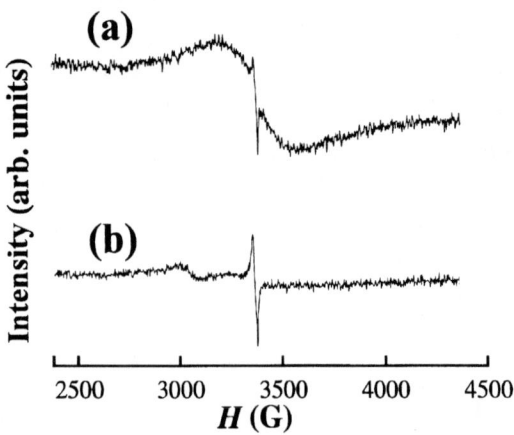

Figure 1. ESR spectra of (a) $Cs_{3.00(6)}C_{60}$ and (b) $Cs_{3.4(2)}C_{60}$ at 300 K.

$Cs_{3.5(1)}C_{60}$ are shown in figures 2(a) and (b), respectively; the fractions of A15 are 9% for $Cs_{3.2(3)}C_{60}$ and 31% for $Cs_{3.5(1)}C_{60}$. The ρ of both samples is the same value of 0.52 Ω cm which is larger by two orders than those of K_3C_{60} and Rb_3C_{60} [8]. The plots of $\ln\rho$ vs $1/T^{1/2}$ show a liner relationship in a temperature region from 40 to 300 K. The ρ of bco $Cs_{3+\alpha}C_{60}$ ($\alpha \neq 0.0$) and the A15 phases are in accord with the granular metal theory ($\rho \sim \exp(T_0/T^{1/2})$) [9]. This implies that the phases of bco $Cs_{3+\alpha}C_{60}$ ($\alpha \neq 0.0$) and A15 are not a normal metal, in consistent with the result of ESR.

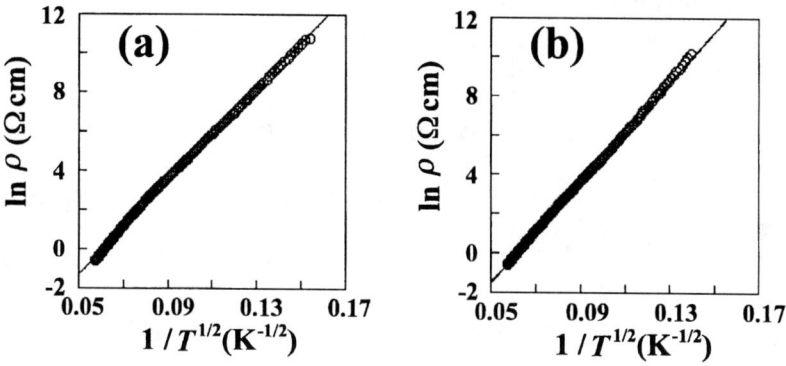

Figure 2. Plots of $\ln \rho$ vs $1/T^{1/2}$ in (a) $Cs_{3.2(3)}C_{60}$ and (b) $Cs_{3.5(1)}C_{60}$ Cs at 1 bar. Solid lines are the fitted ones.

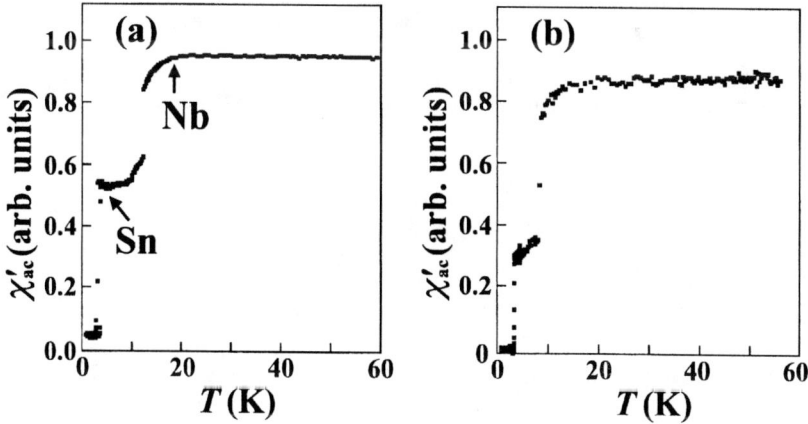

Figure 3. AC susceptibilities of (a) $Cs_{3.2(3)}C_{60}$ at 6.9 kbar and (b) $Cs_{3.5(1)}C_{60}$ at 8.9 kbar. These samples are the same as those used in resistivity measurements.

The AC susceptibility of the samples of $Cs_{3.2(3)}C_{60}$ and $Cs_{3.5(1)}C_{60}$ are shown in figures 3(a) and (b); the applied pressures were 6.9 kbar for $Cs_{3.2(3)}C_{60}$ and 8.1 kbar $Cs_{3.5(1)}C_{60}$. The superconducting transition was not observed above 1.3 K for both samples. This result shows that the bco phase of $Cs_{3+\alpha}C_{60}$ ($\alpha \neq 0.0$) is not superconducting phase under high pressure. Further, this suggests that the A15 phase is also not a superconducting phase under high pressure because these samples contain 9 – 31% of A15 phase. The fcat that superconducting transition is not observed for the bco phase of $Cs_{3+\alpha}C_{60}$ ($\alpha \neq 0.0$) and the A15 phase of Cs_3C_{60} is reasonable judging from the fact that these phases are not a normal metal above 40 K, because all superconducting fullerides are metallic above T_c. The results derived from ESR, resistivity and AC susceptibility indicates that the superconducting phase under high pressure is the bco phase of $Cs_{3+\alpha}C_{60}$ ($\alpha = 0.0$), i.e., Cs_3C_{60}. The study on AC susceptibility of the bco Cs_3C_{60} under high pressure is now in progress.

ACKNOWLEDGMENT

This work was supported by a Grant-in-Aid (12640557) from the Ministry of Education, Science, Sports and Culture, Japan.

REFERENCES

1. Palstra, T. T. M., Zhou O., Iwasa, Y., Sulewski, P. E., Fleming, R. M., and Zegarski, B. R., *Solid State Commun.* **93**, 327-330 (1995).
2. Yoshida, Y., Kubozono, Y., Kashino, S., and Murakami, Y., *Chem. Phys. Lett.* **291**, 31-36 (1998).
3. Kubozono, Y., Fujiki, S., Hiraoka, K., Urakawa, T., Takabayashi, Y., Kashino, S., Iwasa, Y., Kitagawa, H., and Mitani, Y., *Chem. Phys. Lett.* **298**, 335-340 (1998).
4. Zhou, P., Wang, K.-A., Eklund, P. C., Dresselhaus, G., and Dresselhaus, M. S., *Phys. Rev. B* **48**, 8412-8417 (1993).
5. Fujiki, S., Kubozono, Y., Emura, S., Takabayashi, Y., Kashino, S., Fujiwara, A., Ishii, K., Suematsu, H., Murakami, Y., Iwasa, Y., Mitani, T., and Ogata, H., *Phys. Rev. B* **62**, 5366-5369 (2000).
6. Kubozono, Y., Fujiki, S., Takabayashi, Y., Yoshida, Y., Kashino, S., Ishii, K., Fujiwara, A., and Suematsu, H., in *Electronic Properties of Novel Materials-Science and Technology of Molecular Nanostructures*, edited by Kuzmany, H., Fink, J., Mehring, M., and Roth, S., AIP Conference Proceedings 486, New York: American Institute of Physics, 1999, pp. 69-72.
7. Dahlke, P., Henry, P. F., and Rosseinsky, M. J., *J. Mater. Chem.* **8**, 1571-1576 (1998).
8. Stepniak, F., Benning, P. J., Poirier, D. M., and Weaver, J. H., *Phys. Rev. B* **48**, 1899-1906 (1993).
9. Sheng, P., Abeles, B., and Arie, Y., *Phys. Rev. Lett.* **31** 44-47 (1973).

Structure and Properties of $RE_{2.75}C_{60}$

Junji Takeuchi[†], Katsumi Tanigaki[†1] and Balvinder Gogia[‡]

[†]*Material Science, Graduate School of Science, Osaka City University*
3-3-138, Sugimoto, Sumiyoshi, Osaka 558-8585, Japan
[‡]*Department of Materials Science and Engineering, Rensselaer Polytechnic Institute, Troy, New York 12180-3590, U.S.A.*

Abstract.
Rare-earth doped C_{60} fullerides (RE=Yb and Sm) are re-examined. It is clearly shown that $RE_{2.75}C_{60}$ with superlattice structure, which so far believed to be the superconductor, is not superconducting. The possibility of the true superconducting phase is discussed with a comparison between Yb and Sm.

INTRODUCTION

The doped C_{60}'s with a half-filling band, identical to the trivalent state of C_{60}, are metallic and superconducting, and no superconductivity is observed in the other filling in the t_{1u} associated C_{60} fullerides. Such features have been exemplified for alkali and alkaline-earth metal doped C_{60} in the face centered cubic (fcc) crystals isostructural to that of pristine C_{60} (for example, see simplified band filling situations of C_{60} fullerides shown in Fig.1). On the contrary, rear-earth metal doped C_{60}'s are reported not to show an fcc lattice, but instead form $RE_{2.75}C_{60}$'s (RE= Yb and Sm) with superlattice structure [1,2]. This has been so far believed to be the superconducting phase. However, this assignment seems to be unreasonable since the divalent state of RE intercalants gives rise to a band filling in the vicinity of the t_{1u} upperband edge. In view of the fundamental knowledge achieved from the electronic states of alkali and alkaline-earth metal doped C_{60}'s as described above, such a situation can not be understood and is warranted to be clarified.

The present study will report the detail examinations about the electronic properties of the $RE_{2.75}C_{60}$ crystal phases, and give an answer of whether this superlattice phase is really superconducting.

[1)] To whom correspondence should be addressed.

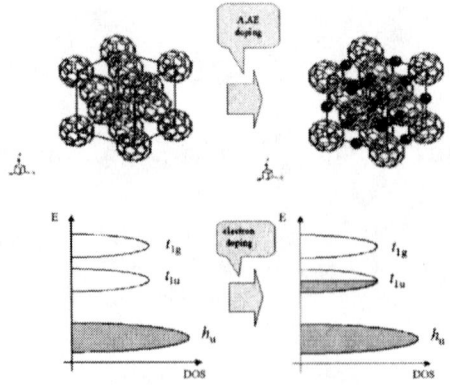

FIGURE 1. A simplied band fillling of doped C_{60} fullerides. Left shows semiconducting C_{60} with a quintetly degenerate h_u-derived valence band and right a metallic phase with a partial filling in the triply degenerate t_{1u}-derived first conduction band

EXPERIMENTAL

$RE_{2.75}C_{60}$ (RE=Yb and Sm) fullerides were prepared from the direct reaction. X-ray diffraction measurements were performed using high-energy synchrotron radiation at SPring-8. Magnetic properties were measured with a Quantum Design MPMS*XL* apparatus.

RESULTS AND DISCUSSION

A C_{60} fulleride with nominal stoichimetry $Yb_{2.75}C_{60}$ was successfully prepared from the C_{60} powder mixed with Yb. Fig.2 shows the Reitvelt analysis using the structural parameters reported previously. Almost all the observed peaks were well indexed as well as the specific low angle peaks associated with the superlattice structure. This result evidently shows that the $Yb_{2.75}C_{60}$ crystal has a superlattice structure due to the deficiency of Yb, being in good agreement with the previous literature [1]. We have also made C_{60} fullerides with Yb by changing the nominal stoichiometry ranging from 1 to 6 per C_{60}. When the concentration of Yb is less than 2.75, no other stable crystal phases were not detected, and the phase separation between pristine C_{60} and $Yb_{2.75}C_{60}$ was observed. When the Yb amount was increased larger than three, some other phases formed, which can be figured out from the appearance of other new diffraction peaks. Most likely, the crystal phases made in the concentration of x=3 to 6 will be body-centered tetragonal (bct), body centered orthorhombic (bco) and/or body-centered cubic (bcc) phases. However, more accurate study is needed for confirmation in the future.

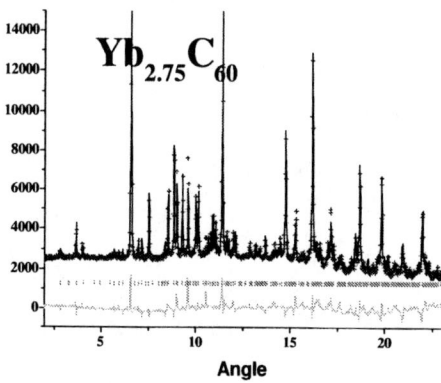

FIGURE 2. Xray diffraction pattern of $Yb_{2.75}C_{60}$ collected with synchrotron radiation. Rietvelt fittings are included in the figure

For confirmation of the superconductivity, magnetic susceptibility measurements have been carried out using SQUID under a low-magnetic field of 15 Gauss. The measurements showed that any important superconductivity does not appear, except for only a trace amount of superconducting fraction. As clearly seen in Fig.3, the superconducting fraction of $Yb_{2.75}C_{60}$ with superlattice structure is only less than 1 %. Considering that the crystal quality of this phase prepared in the present research is sufficiently high as described above, the observed superconductivity should not be assigned to the superlattice $Yb_{2.75}C_{60}$ and to be ascribed to other accompanying minor crystal phases. However, considering the experimental evidence described later that the Tc does not vary with the Yb concentrations in the feed, the superconducting phase is most probably only one.

In order to have further information, we have measured SQUID for Yb_xC_{60} fullerides with various concentration of x. When x=4 is used, superconducting diamagnetic susceptibility was markedly increased. It is also important that the magnetic susceptibility is decreased slightly again, when x=6 was used. Such situation can be seen in Fig.3. As shown in X ray diffraction studies described earlier, it is evident that some other crystal phases exit. All the data presented here clearly displays that Yb2.75, which has so far been believed to be the superconducting, is not a true superconducting phase, but that another phase of Yb_xC_{60} with x lager than three would be superconducting. From the viewpoint of SQUID measurement, Yb_4C_{60} seems to be real superconducting phase.

In the case of Sm, the $Sm_{2.75}C_{60}$ superlattice phase was also observed but no superconductivity was also detected as seen in the inset of Fig.3. It should be noted that any symptom of superconductivity is not implied even if the Sm stoichiometry is increased larger than 3, this being in strong contrast to the case of Yn doped C_{60} situation. These results indicate that the real superconducting phase of RE doped C_{60}'s is not $RE_{2.75}C_{60}$.

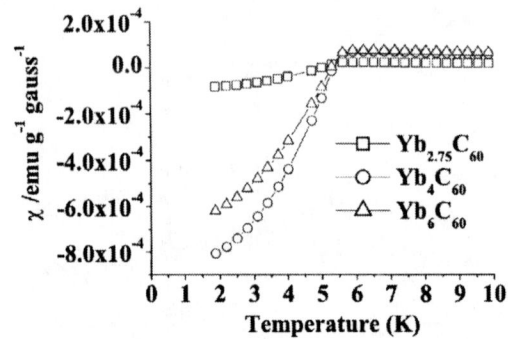

FIGURE 3. Magnetic moment measured by SQUID under a low magnetic-filed of 15 G for RE_xC_{60}

CONCLUSION

The superconductivity of $Yb_{2.75}C_{60}$ was re-examined by making a high quality sample. It was evidently shown that this superlattice phase, with a band filling in the vicinity of the upper t1u band edge, is not superconducting. At present, we do not have an accurate determination of the superconducting crystal phase, our experiments suggests that it could be Yb_4C_{60}. Further study is now being in progress.

Acknowledgements

We are grateful to the staff members at SPring8 (beamline BL02B2) and KEK (PF-BL1B) for X-ray measurements. The project was supported by The Japan Society for the Promotion of Science (MIRAIKAITAKU-Project), the Grant-in-Aid for Scientific Research on the Priority Area "Fullerenes and Nanotubes" by the Ministry of Education, Science, and Culture of Japan.

REFERENCES

1. E. Ozdas *et al.*, *Nature* **375**, 126 (1995).
2. X. H. Chen and G. Roth, *Phys. Rev.* B. **52**, 15534 (1995).

Unusual Magnetic Properties of High-Temperature Reaction Products of Cerium Metal and C_{60} Solid

S.Motohashi[1], Y.Maruyama[1], K.Watanabe[1], K.Suzuki[1], S.Takagi[1], and H.Ogata[2]

[1]Department of Materials Chemistry, Hosei University, Kajinocho, Koganei, Tokyo 184-8584, Japan

[2]Institute for Molecular Science, Myoudaiji, Okazaki 444-8585, Japan

Abstract. Relatively high temperature reaction products of cerium metal and C_{60} solid with nominal molar ratio of 3:1 have shown rather unusual magnetism which may relate to a coexistence or competition of ferro- and dia-magnetism below 13.5K. The crystal structures of the samples of various preparation conditions are analyzed with IP-XRD.

INTRODUCTION

Intercalation of C_{60} with the lanthanide metals Yb and Sm has so far been reported to be led to the superconducting phases $Yb_{2.75}C_{60}$ ($T_c = 6$ K)[1] and $Sm_{2.75}C_{60}$ ($T_c = 8$ K)[2]. In the search for new lanthanide metal intercalation compounds we have targeted cerium metal doping because of its lowest melting point, 798℃, among the lanthanide metals and also its interesting outermost electronic configuration, $4f^1 5d^1 6s^2$. The low melting temperature may lead to a rather low temperature reaction in the synthesis of Ce-fulleride, and the possible $4f^1$ electron of Ce^{3+} in the fulleride may exhibit interesting magnetic properties[3].

EXPERIMENTALS

The samples were prepared by heating the mixture of Ce metal fine powder and C_{60} powder in the sealed quartz tube. The reaction temperatures ranged from 500 to 600℃ with 50℃ interval and the reaction times range from 7 to 24 hrs. Typical nominal mol ratio of Ce to C_{60} is 3 to 1.

RESULTS AND DISCUSSION

Temperature dependence of the magnetic susceptibilities and X-ray powder diffractions of each sample were investigated. The optimum reaction condition seemed to be 600℃, 24 hrs. The zero-field-cooling susceptibility curve of this sample showed clear diamagnetic behavior below 13.5 K and on the contrary the field-cooling curve exhibited a ferromagnetic rise below 15 K as shown in Fig.1

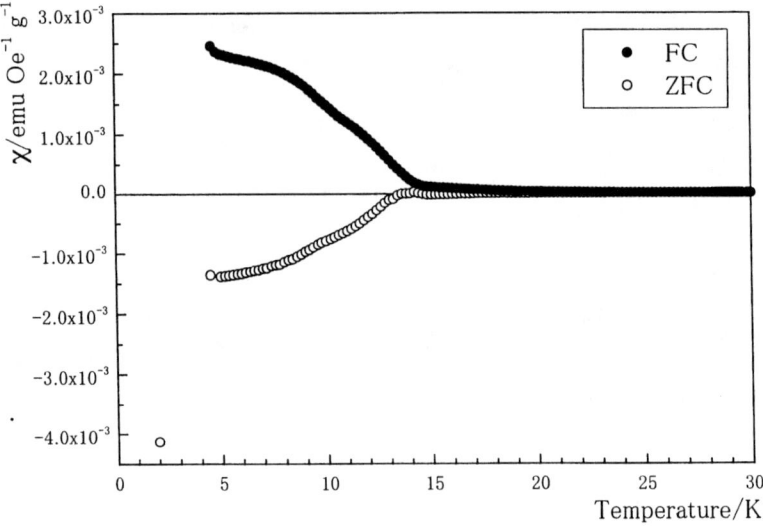

FIGURE 1. Temperature dependence of the magnetic susceptibility(χ) under 2 Oe of Ce_3C_{60} (nominal composition) treated at 600℃ for 24 hrs.

The magnetization curves in Fig.2 clearly indicated the existence of hysteresis response which is larger at 12 K than that at 4.5 K. This fact means the coexistence or competition of ferromagnetism and diamagnetism in this system.

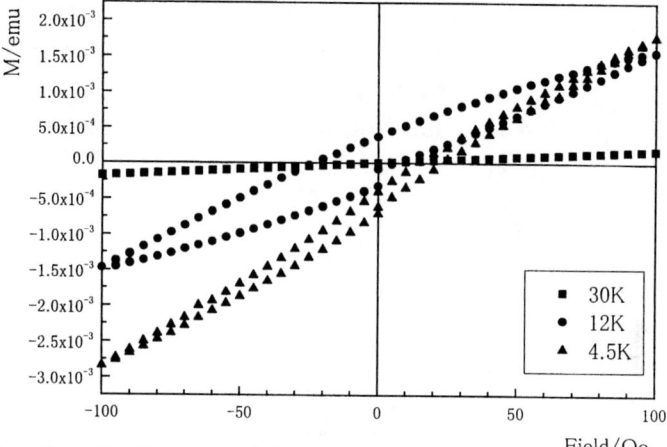

FIGURE 2. Magnetization curves of the same sample as in Fig.1

The structural analysis by X-ray diffraction indicated that the fcc structure of C_{60} crystal was conserved with a very little expanded lattice constant and a few new lines. Rather strong peaks at higher angles might originate in some Ce-compounds (possibly carbides or oxides).

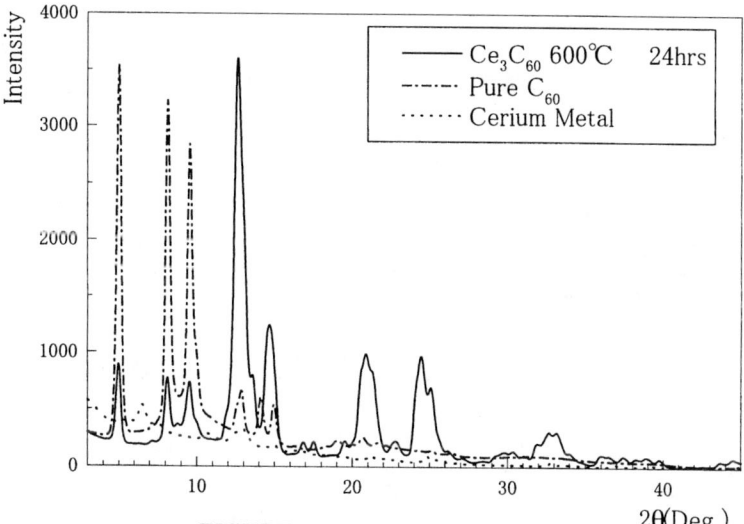

FIGURE 3. X-ray diffraction patterns

We tried a blank experiment in which active charcoal and Ce mixture was heated at the same condition. The resultant sample showed very weak ferromagnetic behavior but never diamagnetism.

In summary, we found out that the solid reaction (600°C, 24hrs) product of Ce_3C_{60} (nominal composition) seems to have significant diamagnetism and ferromagnetism in the similar low temperature region. The nature of the diamagnetism should be examined more precisely.

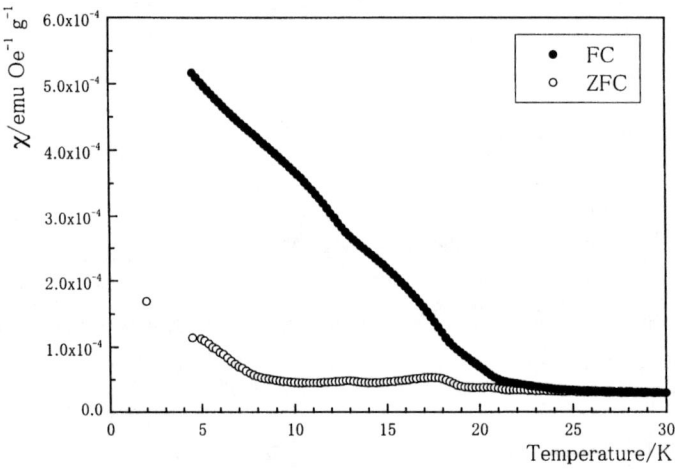

FIGURE 4. Magnetic susceptibility(χ) vs. Temperature of Ce-active charcoal

ACKNOWLEDGEMENTS

This work was partly supported by the Grant-in-Aid for Scientific Research on the Priority Area "Fullerenes and Nanotubes Networks" by the Ministry of Education, Science and Culture of Japan (# 11165239).

REFERENCES

1. E.Ozdas et al., *Nature*, 375, 126(1995).
2. X.H.Chen et al., *Phys.Rev.* B52, 15534(1995).
3. Y.Maruyama et al., *Solid State Commun.* 115, 457(2000).

NMR Studies of Ammoniated Alkali Fullerides

Hideki Tou*, Nariyasu Muroga*, Yutaka Maniwa*, Taishi Takenobu†, Hideo Shimoda¶, Yoshihiro Iwasa†, Tadaoki Mitani†

*Department of Physics, Tokyo Metropolitan University, Minami-osawa 1-1, Hachi-oji, Tokyo 192-0397, Japan[1]
¶Department of Physics and Astronomy, University of North Carolina at Chapel Hill, Chapel Hill, NC27599, U.S.A
†Japan Advanced Institute of Science and Technology, Tatsunokuchi, Ishikawa 923-1292, Japan

Abstract. ^{13}C-NMR measurements of $(NH_3)_x K_3 C_{60}$ were carried out for three samples with different NH_3 content of $x = 1.14(\#1)$, $x = 1.05(\#2)$, and $x = 0.98(\#3)$. ^{13}C-NMR spectrum for $x = 0.98(\#3)$ shows an anomalous spectral broadening, whereas a distinct anomaly is hardly observed for the off-stoichiometric sample with $x = 1.14(\#1)$. In the magnetic ordered state at 4.2 K, although unusual spectral broadening ascribed to the AF order was observed for all sample. We found, however, the linewidth depends on the sample-quality, i.e., stoichiometry of NH_3 content (∼0.8, 1.3, and 3 MHz for $x = 1.14(\#1)$, $1.05(\#2)$, and $0.98(\#3)$, respectively). Present studies reconfirmed that the lines A and B are inherent in AF ordering and the off-stoichiometric ammoniation obstruct the AF ordering.

1. INTRODUCTION

It is well known that superconductivity in fullerides is possible to be interpreted by the phonon mechanism, i.e, the SC transition temperature T_c goes up by increasing the unit-cell volume [1,2]. The relation between T_c and the cell volume was also confirmed by ammoniation of $Na_2 CsC_{60}$ where neutral NH_3 molecules are intercalated as a spacer and successfully increase the cell volume without symmetry lowering [3]. Meanwhile, ammoniation of superconducting $K_3 C_{60}$ reduces crystal symmetry and suppresses T_c [4]. In order to clarify the reason why the superconductivity is suppressed in $(NH_3)K_3 C_{60}$, extensive experimental studies have been carried out so far. Iwasa et al., reported that the ground state of $(NH_3)K_3 C_{60}$ is not superconducting, but some magnetic insulating state from ESR measurements

---

[1] This work was supported by the fund for Special Research Projects of Tokyo Metropolitan University and in part by grants from the Ministry of Education, Sport, Science and Culture in Japan. H.T. was supported by Research Aid of the Sumitomo Foundation for Science.

[5]. Allen et al. reported that the electronic state of this material falls into a spin-density wave state below 40 K [6]. Brown et al. provided unambiguous evidence for a long-range antiferromagnetic (AF) order from μSR measurements [7]. Recent our NMR studies clarified that the ground state is the three dimensional (3D) antiferromagnetic (AF) ordered state with a relatively large moment of $\sim 1\mu_B/C_{60}$. We suggested that the AF order is accomplished by coupling through orbital degree of freedom of the degenerated t_{1u} orbits, i.e. "molecular orbital order". Anyhow, these experimental efforts established that $(NH_3)K_3C_{60}$ is the 3D antiferromagnetic Mott-Hubbard insulator.

However, magnetic character strongly depends on the sample-quality. In order to check microscopically, we carried out ^{13}C-NMR for $(NH_3)_xK_3C_{60}$ with x =1.14(#1), 1.05(#2), and 0.98(#3). We report here the relationship between sample-quality and magnetic properties in this system.

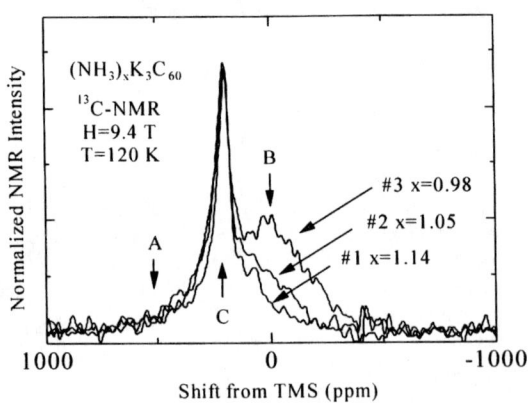

FIGURE 1. ^{13}C-NMR spectra at $T = 4.2$ K (AF ordered state) and H =9.4 T for various NH$_3$ content.

3. EXPERIMENTAL RESULTS

Figure 1 shows the ^{13}C-NMR spectra of $(NH_3)_xK_3C_{60}$ with x =1.14(#1), 1.05(#2), and 0.98(#3) at 120 K. The NMR experiments were carried out by using a conventional pulse NMR spectrometer at magnetic field of $H \sim 9.4$ T. The NMR spectra above T_N were obtained by a Fourier transform technique. The ^{13}C powder NMR spectrum for $x = 0.98$(#3) has a hump around -50 ppm in addition to the sharp line around 195 ppm. The spectrum broadens over the range from -500 to +700 ppm consisting of several lines (lines A, B, and C indicated by arrows according to Ref. [8]). This spectral broadening cannot be only attributed to the freezing of C$_{60}$ molecular rotation, which exceeds the range of -100$\sim$+400 ppm for various fullerides. Although off-stoichiometric samples (#1 and #2) show spectral broadening, there exist no clear hump around -50 ppm but smooth shoulders. We can easily see that the signal intensity around -50 ppm (line B) and 500 ppm (line A) becomes weak as the NH_3 content deviates from the stoichiometry. Thus the lines A and B are very sensitive for stoichiometry of NH_3 content. According to the previous paper, abnormally broadened spectrum is due to the interaction between C_{60} and the octahedral site potassium (K_O), associated with the *molecular orbital order* of t_{1u} orbits. Thus it is strongly suggested that off-stoichiometry of NH_3 content causes molecular orbital disorder.

Since the molecular orbital order is deeply involved with the AF order according to Ref. [8], the magnetic character is considered to be sensitive for the stoichiometry. Figure 3 shows the T dependence of the integrated intensity of ^{13}C-NMR spectra multiplied by temperature $I.I. \times T$. $I.I. \times T$ decreases below $T_N \sim 45$ K for all samples. However, the loss of the NMR intensity, which is due to a wipeout effect in the ordered state, depends on the stoichiometry: $I.I. \times T$ decreases down to ~ 30 % for $x = 0.98(\#3)$, whereas ~ 65 % for $x = 1.14(\#1)$.

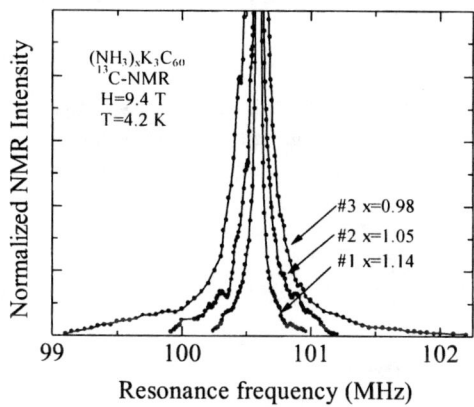

FIGURE 2. ^{13}C-NMR spectra at $H = 9.4$ T for various NH_3 content.

Figure 3 shows the NMR spectra at 4.2 K. As clearly seen in figure, the linewidth of ^{13}C-NMR spectrum for $x = 0.98$ is ~ 3 MHz, corresponding to the magnetic moment of $1\mu_B/C_{60}$, whereas the linewidths for the off-stoichiometric samples, #1 and #2 are ~ 0.8 MHz and ~ 1.3 MHz, respectively. For the off-stoichiometric samples, we examined signals in the frequency range 99~100 MHz and 101~102 MHz, however no signals could be observed within the experimental accuracy because of the poor signal to noise ratio. Anyhow, it is clear that the magnetism of $(NH_3)_x K_3 C_{60}$ is quite sensitive to the stoichiometry of NH_3 content.

SUMMARY

^{13}C-NMR measurements were performed for $(NH_3)_x K_3 C_{60}$. Present comparative study on ammonia content reconfirmed that the lines A and B are inherent in AF

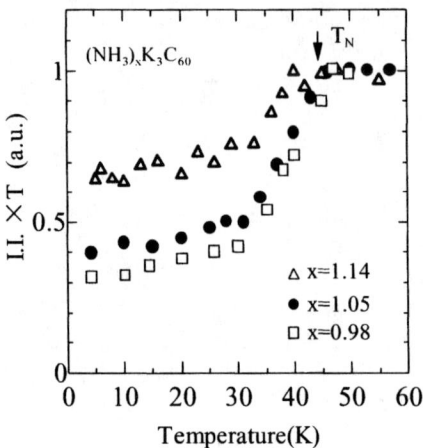

FIGURE 3. Temperature dependence of the FFT spectral intensity of ^{13}C-NMR multiplied by temperature for various NH$_3$ content.

ordering, whereas line C originates in the impurity phase. Present studies clearly show that the off-stoichiometry of NH$_3$ content causes a molecular orbital disorder. Namely, the stoichiometry is very important to stabilize the molecular orbital order as well as AF magnetic order.

REFERENCES

1. C.H. Pennington and V.A. Stenger, Rev. of Mod. Phys. **68**,855(1996).
2. Y. Maniwa et al., Phys. Rev. **B54**, R6861 (1996).
3. O.Zhou et al., Nature (London) **362**,433(1993).
4. M.J. Rosseinsky et al., Nature (London) **364**, 425 (1993).
5. Y. Iwasa et al., Phys. Rev. **B53**, R8836 (1996)
6. L.M. Allen et al., J. Matter. Chem. **6**, 1445 (1996);
7. K.Prassides et al.,J. Am. Chem. Soc. **121**,11227(1999).
8. H. Tou et al., Physica **B259-261**, 868(1999), Phys. Rev. **B62**, R775 (2000) .

Orientational and Magnetic Transitions in Ammoniated Alkali Metal Fullerides

Taishi Takenobu, Masaaki Miyake, Tsunehiro Muro, Yoshihiro Iwasa, and Tadaoki Mitani

Japan Advanced Institute of Science and Technology, Tatsunokuchi, Ishikawa 923-1292, Japan

Abstract. Monoammoniated alkali fulleride salts $(NH_3)K_{3-x}Rb_xC_{60}$ ($0 \leq x \leq 3$), forming an isostructual orthorhombic series, show a Mott-Hubbard type antiferromagnetic transition. The Néel temperature T_N was found to first increase with the inter-fullerene spacing and then decrease for $(NH_3)Rb_3C_{60}$, forming a maximum at 76K. Moreover, a structural phase transition, which is associated with the ordering of metal-NH_3 pair, was found at 150K in all samples and a correlation between the magnetic and structural transition is expected. The identical structural transition temperature T_S is a striking contrast with T_N.

INTRODUCTION

Recent progress in the synthesis of fullerene intercalation compounds has afforded a huge variety of materials. Among them, the ammoniated alkali fulleride $(NH_3)K_3C_{60}$ provides a novel opportunity to investigate the correlation between molecular rotation and electronic properties, which is one of the most unique aspects of fullerene based solids [1]. $(NH_3)K_3C_{60}$ is synthesized by intercalation of neutral ammonia molecules into fcc K_3C_{60}, which is a superconductor with a superconducting transition temperature T_C=19K [2]. The structure of $(NH_3)K_3C_{60}$ is very similar to that of K_3C_{60} except for a slight orthorhombic distortion induced by ammoniation. Interestingly, despite the similar crystal structures, the electronic ground state of $(NH_3)K_3C_{60}$ is an antiferromagnetic (AF) insulator with a Néel temperature of T_N=40K, as shown by zero-field(ZF)/longitudinal-field(LF) muon spin rotation/relaxation (μ^+SR) [3], nuclear magnetic resonance (NMR) [4], and electron paramagnetic resonance (EPR) [5,6]. Moreover, a structural phase transition was found at 150K in $(NH_3)K_3C_{60}$ [7]. While the octahedral K-NH_3 group rotates at higher temperature, the orientation of this group is ordered in an antiferroelectric fashion below a structural phase transition temperature T_S=150K, associated with the unit cell doubling.

In this paper, we present synthesis, structural and EPR experiments of a new series of the $(NH_3)K_3C_{60}$ type compounds $(NH_3)K_xRb_{3-x}C_{60}$ (x=0, 1, 2, 3), in order to uncover the nature of AF state. We found that T_S is identical for all samples, while T_N systematically changes, which has a maximum as a function of the interfullerene distance. The constant T_S indicates that the origin of structural phase transition is not alkali metals but C_{60} molecules.

EXPERIMENTAL

While $(NH_3)K_3C_{60}$ is synthesized by exposing preformed K_3C_{60} to ammonia gas [1], $(NH_3)K_{3-x}Rb_xC_{60}$ ($0 < x \leq 3$) were obtained by removing of ammonia from the ammonia rich phase [8]. After dissolving stoichiometric amount of C_{60} and alkali metals into dry liquid ammonia kept at $-65°C$, an unidentified ammonia rich phase was obtained by a slow evaporation of ammonia. To obtain purified ammonia rich sample, additional heat treatment under 700 torr ammonia gas atmosphere was carried out at 100°C for a few weeks. Ammonia was further removed by a ten-minute annealing at temperature between 80 and 100°C to obtain high-quality $(NH_3)K_{3-x}Rb_xC_{60}$ samples. The sample was sealed in a thin glass capillary. A synchrotron radiation x-ray powder diffraction experiment was made with imaging plates as detectors at the BL-1B, Photon Factory, KEK and BL02B2, SPring-8. 9GHz EPR spectra have been measured from 40K to 300K using a JEOL EPR spectrometer equipped with an APD cryostat.

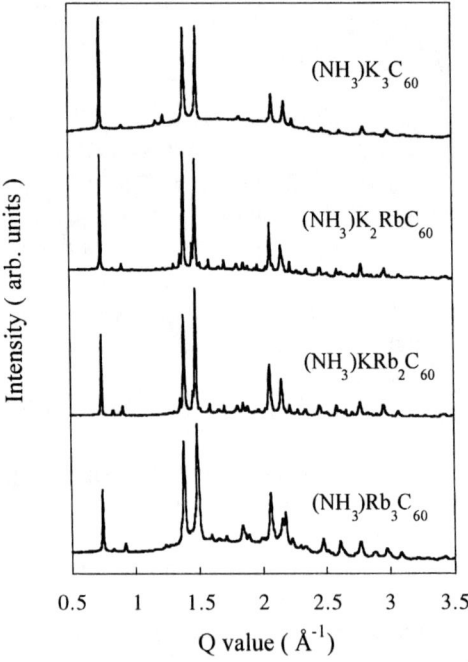

FIGURE 1. The x-ray powder diffraction patterns of $(NH_3)K_{3-x}Rb_xC_{60}$ ($0 \leq x \leq 3$) at room temperature.

RESULT AND DISCUSSION

As shown in Figure 1, the x-ray powder diffraction patterns of $(NH_3)K_{3-x}Rb_xC_{60}$ ($0 \leq x \leq 3$) at room temperature are very similar to each other, indicating that they are isostructural. The observed x-ray patterns of these samples agreed fairly well with the simulation by the Lazy-Pulverix software (K. Yvon, W. Jeitschko, E. Parthe, unpublished) on the orthorhombic $(NH_3)K_3C_{60}$ type structural model, where a single A-NH_3 pair occupies every octahedral site [1]. Since the structural phase transition at 150K has been observed in $(NH_3)K_3C_{60}$, which is attributed to the orientational order-disorder transition of the K-NH_3 pair with the cell doubling, the similar transition is expected for isostructural $(NH_3)K_xRb_{3-x}C_{60}$ ($x<3$) compounds [7]. For all samples, we observed a number of extra peaks below 150K and these extra peaks are possible to be indexed on half an integer or the forbidden reflections in the face-center lattice. The appearance of these superlattice and forbidden peaks indicates that the unit lattice vectors of the low-temperature phase should have a double size of the fundamental ones. These results are the direct evidence of the $(NH_3)K_3C_{60}$ type structural phase transition in all samples with T_S=150K.

The ESR signals of samples consist of two components. Through a Lorentzian fit, we deduced the integrated intensities and the full width at half maximum of ESR spectra linewidth $\Delta H_{1/2}$ for the two components. Since the intensity of the narrow line ($\Delta H_{1/2}$ = 3.2mT at 300K for $(NH_3)KRb_2C_{60}$) displays a Curie-type temperature dependence throughout 40-300K, the narrow line is ascribed to a paramagnetic impurity possibly due to lattice imperfections. The intrinsic broad peak shows dramatic anomalies associated with the intensity drop and rapid line broadening at 40K for $(NH_3)K_3C_{60}$, 67K for $(NH_3)K_2RbC_{60}$, 76K for $(NH_3)KRb_2C_{60}$ and 58K for $(NH_3)Rb_3C_{60}$, receptively. Importantly, the decrease of intensity at 40K for the standard compound $(NH_3)K_3C_{60}$ is caused by the AF transition, which has been confirmed by other measurements [3-6]. The drop in intensity does not necessarily mean the decrease of spin susceptibility, but that the signal becomes invisible due to the anomalous magnetic broadening, as observed in $(NH_3)K_3C_{60}$. Taking into consideration the close correlation of chemical, structural and EPR properties with those of $(NH_3)K_3C_{60}$, we concluded that the newly compounds $(NH_3)K_2RbC_{60}$, $(NH_3)KRb_2C_{60}$ and $(NH_3)Rb_3C_{60}$ are antiferromagnets with T_N = 67K, 76K, and 58K, respectively.

Finally, we should discuss the anomalous volume dependent of T_N. Figure 2 shows the correlation between $\Delta H_{1/2}$ and $(\Delta g)^2$ at room temperature in $K_{3-x}Rb_xC_{60}$ and $(NH_3)K_{3-x}Rb_xC_{60}$, where Δg is the g-value shift from the free electron value g_e=2.00232. It is useful to compare this result with the case of *fcc* $K_{3-x}Rb_xC_{60}$ compounds, where $\Delta H_{1/2}$ increases with Rb concentration accompanied with the increase of $(\Delta g)^2$. This behavior has been understood in terms of the Elliot mechanism of conduction electron EPR [9]. In $(NH_3)K_{3-x}Rb_xC_{60}$ compounds, both $\Delta H_{1/2}$ and $(\Delta g)^2$ increase with the Rb concentration between x=0 and x=2 in a similar manner to $K_{3-x}Rb_xC_{60}$. $(NH_3)Rb_3C_{60}$ (x=3), however, displays an abnormality of $(\Delta g)^2$. This is the clear evidence of somewhat difference in $(NH_3)Rb_3C_{60}$ between $(NH_3)K_{3-x}Rb_xC_{60}$, and the detailed research of $(NH_3)Rb_3C_{60}$ is necessary to clarify the causality of T_N reduction.

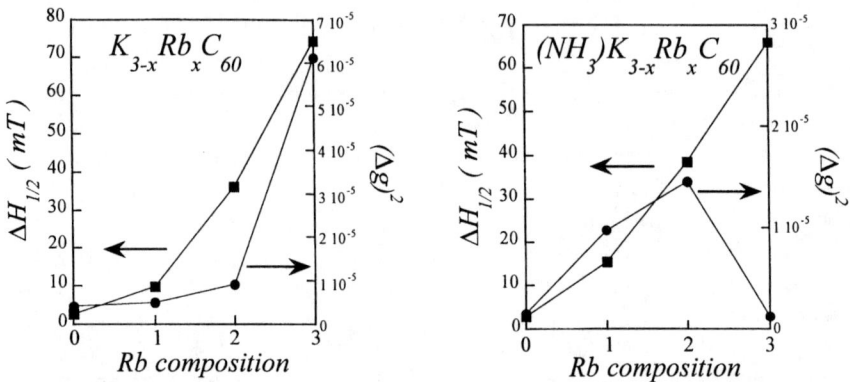

FIGURE 2. The relationships between $\Delta H_{1/2}$ and $(\Delta g)^2$ at room temperature in $K_{3-x}Rb_xC_{60}$ and $(NH_3)K_{3-x}Rb_xC_{60}$.

In summary, we have synthesized an orthorhombic series of alkali ammonia trivalent fulleride compounds, and showed that T_S is identically 150K and AF transition T_N exhibits a systematic change against the inter-fullerene spacing.

ACKNOWLEDGMENTS

We thank E. Nishibori, M. Takata, M. Sakata, Y. Murakami, H. Shimoda, T. Ito, H. C. Dam, and S. Moriyama for experimental collaborations. This work has been supported by the JSPS "Future Program", RFTF96P00104, and also by the Grant-In-Aid for Scientific Research on the Priority Area "Fullerenes and Nanotubes" by the Ministry of Education, Sport, Science and Culture of Japan.

REFERENCES

1. Rosseinsky, M. J. et. al., *Nature (London)* **364**, 425-427 (1993).
2. Prassides, K. et. al., *J. Am. Chem. Soc.* **121**, 11227-11228 (1999).
3. Tou, H. et. al., *Phys. Rev. B* **62**, R775-778 (2000).
4. Iwasa, Y. et. al., *Phys. Rev. B* **53**, R8836-8839 (1996).
5. Allen, K. M. et. al., *J. Mater. Chem.* **6**, 1445- (1996).
6. Simon, F. et. al., *Phys. Rev. B* **61**, R3826-3829 (2000).
7. Ishii, K. et. al., *Phys. Rev. B* **59**, 3956-3960 (1999).
8. Takenobu, T. et. al., *Phys. Rev. Lett.* **85**, 381-384 (2000).
9. Tanigaki, K. et. al., *Chem. Phys. Lett.* **240**, 627-632 (1995).

NMR Studies of Alkali-Doped C_{60} Superconductors with Small Lattice Constants

K. Kitazume[*], H. Tou[*], Y. Maniwa[*], M. Kosaka[¶] and K. Tanigaki[¶¶]

[*]*Department of Physics, Tokyo Metropolitan University, Minami-osawa, Hachi-oji, Tokyo 192-0397, Japan.*
[¶]*Fundamental Research Laboratories, NEC Corporation, 34 Miyukigaoka, Tsukuba 305-8501, Japan.*
[¶¶]*Dep. of Material Science, Osaka City Univ., Sugimoto, Osaka 558-8585, Japan..*

Abstract. The electronic states of the non-superconducting Li_2CsC_{60} and the superconducting Li_3CsC_{60} were studied by ^{133}Cs-, ^7Li- and ^{13}C-NMR. The results were compared with those of other alkali(A)-doped C_{60} compounds, A_3C_{60}. We could not find any reason for the suppression of superconductivity in Li_2CsC_{60} in the NMR data taken at room temperature. Alternatively, the sp^3-like carbons were suggested to exist in Li_2CsC_{60} below ~160K.

INTRODUCTION

The superconducting transition temperature, T_c, in alkali fullerides, A_3C_{60} where A is alkali metal, can be controlled through the Fermi-level density of states, $N(E_F)$, by the unit-cell parameters. However, it is known that the materials with the small lattice constant, a, have lower T_c than expected. Possible candidates for the origin have been proposed as follows; (1) the low T_c material with a simple cubic structure, Na_2AC_{60}, would have a steeper a vs $N(E_F)$ relation compared to the conventional fcc A_3C_{60}, and (2) in the case of non-superconducting Li_2CsC_{60}, the rotational disorder of C_{60} molecule or strong Li-C_{60} interaction [1] may suppress the superconductivity. For the first case, the ^{13}C-NMR T_1 measurements at low temperatures, T, actually indicated that the apparent a vs. $N(E_F)$ relation is steeper in Na_2AC_{60} than in fcc A_3C_{60}. However, it was not enough to explain the observed T_c suppression [2]. As for the second case, it was proposed that the valence of C_{60} must be trivalent for the occurrence of superconductivity. Thus, because the valence of C_{60} in Li_2CsC_{60} is smaller than -3, it becomes non-superconducting. On the other hand, that of Li_3CsC_{60} is –3, leading to a normal T_c of ~10 K [3]. It was also indicated that the Fermi level density of states, $N(E_F)$, at low temperatures are smaller than at high temperatures in all the low-T_c materials with the small lattice constant in group I material [4,5,6], as shown in Fig. 1. Combined with an observation that the localized spins are present in the group I low-T_c materials, this result strongly suggests that an electron localization starts to occur in these materials [6]. Then, we faces an question why the metallic states of the group I materials are so unstable. Although it has been pointed out that the structural imperfections play an important role in determination of the electronic states and the onset of superconductivity [2], this problem has been still unsettled. Thus, we performed a comparative NMR study on the non-superconducting Li_2CsC_{60} and the superconducting Li_3CsC_{60} in order to obtain further insight on this problem.

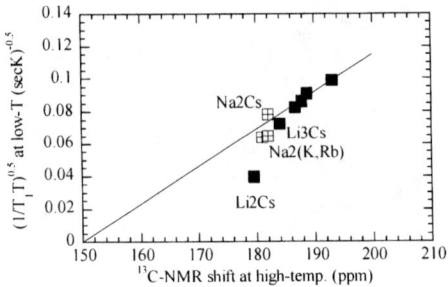

FIGURE 1. Relationship between the ^{13}C-NMR Knight shift around room temperature and the ^{13}C-$(T_1T)^{-0.5}$ at low-temperature (<100K). Both the quantities are expected to be proportional to the Fermi level density of states, $N(E_F)$, in metals, as shown by a straight line. It is found that the low-T_c materials, $Na_2(K,Rb)C_{60}$ and Li_2CsC_{60}, are substantially deviated from the straight line.

EXPERIMENTAL RESULTS

Figure 2 shows the ^{13}C-NMR spectra as a function of temperature in Li_2CsC_{60} and Li_3CsC_{60}. With decreasing temperature, the line-width steeply broadens below ~200K in both the samples. The full width at half maximum (FWHM) is shown in Fig. 3 as

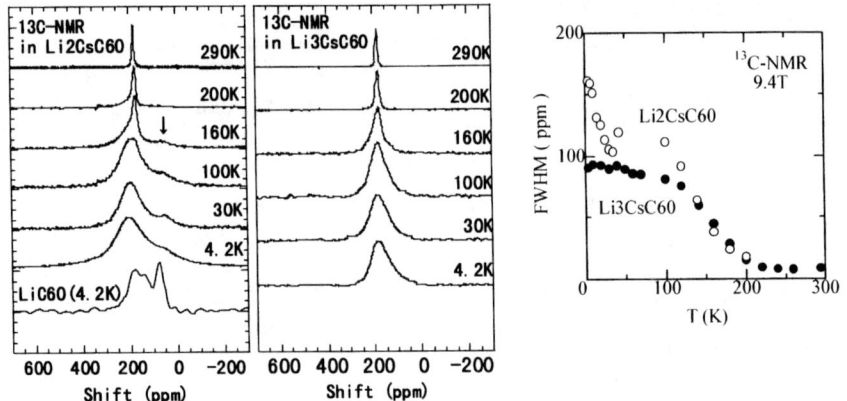

FIGURE 2 (left). ^{13}C-NMR spectra of Li_2CsC_{60}, and Li_3CsC_{60} as a function of temperature. For a comparison, a spectrum for 3R-LiC_{60} polymer is also shown. A small hump indicated by an arrow was assigned to the sp^3-like carbon. The applied magnetic field was 9.4T for Li_2CsC_{60} and Li_3CsC_{60}, and 3.9T for LiC_{60}.

FIGURE 3 (right). The full width at half maximum of the ^{13}C-NMR spectra in Li_2CsC_{60} and Li_3CsC_{60} as a function of temperature. The measurements were performed at an applied magnetic field of 9.4T. An increase below ~50K in Li_2CsC_{60} is due to the localized spins in the crystal.

a function of temperature. The broadening is naturally assigned to the freezing of C_{60} molecular rotation. This is a well-known behavior often observed in C_{60} solid and C_{60} compounds. However, from a detailed inspection at the spectra of Li_2CsC_{60}, we found a small unusual hump indicated by an arrow. The shift ~ 50ppm is in a range for sp^3 carbons; for example, ~30ppm and ~70ppm for the sp^3 carbon in diamond and C_{60} polymers, respectively [7, 8]. The intensity of the sp^3 signal was roughly estimated to be ~5%. At the bottom of Fig. 2, the spectrum of rhombohedral C_{60} polymer (3R-LiC_{60}) is also shown for a comparison. Therefore, the ^{13}C-NMR spectra of the Li_2CsC_{60} strongly suggest the existence of the sp^3 carbons in the crystal. These carbons may be assigned to those atoms bridging two C_{60} molecules as in C_{60} polymers and C_{60} dimmers, although there is no evidence from previous structural studies in this system. Another possibility for it would be carbon atoms hybridized with Li or Cs atoms.

Next, we discuss the electronic states at the alkali-sites. This is important because the charge transfer from each Li and Cs atom is believed to be smaller than unity in Li_xCsC_{60}. There must be a partial charge on the alkali-sites. Here, we show ^{133}Cs- and ^7Li-NMR spectra around room temperature (RT) in Figs. 4 and 5, respectively. From Fig. 5, it is found that the ^{133}Cs-NMR shift of group I materials (Li_xCsC_{60}, Na_2CsC_{60}) substantially deviates from that of group II materials with fcc structure (K_2CsC_{60}). Because the deviation from the K_2CsC_{60} is positive (in the direction of the positive Knight shift), this is qualitatively consistent with the charge transfer ratio smaller than unity in this group. However, the magnitude for the non-superconducting Li_2CsC_{60} is almost the same as that of superconducting Na_2CsC_{60} with T_c=12K.

On the other hand, the ^7Li-NMR shift of Li_2CsC_{60} and Li_3CsC_{60} are shown in Fig. 5. The observed shifts are very close to 0ppm (the resonance position of 1M-LiCl solution), indicating that the amplitude of the conduction electron wave functions at the Fermi level is very small at the Li sites in both the compounds. Therefore, combined with ^{13}C-NMR data reported so far, we cannot find any reason why Li_2CsC_{60} is non-superconductor in the NMR spectra taken at high temperatures.

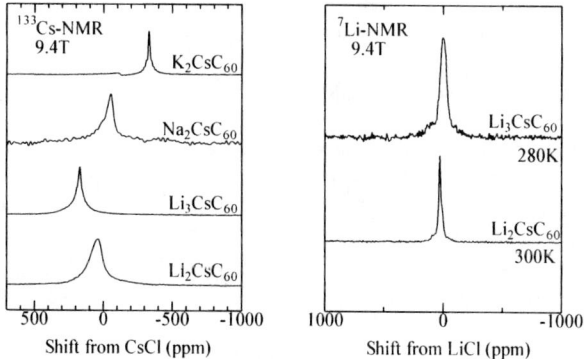

FIGURE 4 (left). ^{133}Cs-NMR spectra in Li_2CsC_{60}, Li_3CsC_{60}, Na_2CsC_{60} and K_2CsC_{60} around RT. The measurements were done at RT. The lattice constant of the sample increases from bottom to top.

FIGURE 5(right). ^7Li-NMR spectra in Li_2CsC_{60} and Li_3CsC_{60} around RT.

DISCUSSIONS

We studied the electronic states of the non-superconducting Li_2CsC_{60} and the superconducting Li_3CsC_{60}, and compared the results with those of other alkali-doped C_{60} compounds, A_3C_{60}. We could not find any anomaly in the ^{133}Cs-, ^{7}Li- and ^{13}C-NMR shift data taken around RT. Alternatively, the sp^3-like carbons were suggested to exist in Li_2CsC_{60} below ~160K. Thus, in all the low T_c materials (Na_2RbC_{60}, Na_2KC_{60} and Li_2CsC_{60}), it was shown that the hybridized sp^3-like carbon atoms can exist in the crystal, although it depends on the cooling rate. These low-T_c materials have the C_{60}-C_{60} distance shorter than the shortest distance of Li_3CsC_{60} among the conventional group II superconductors with a normal T_c. Therefore, it can be speculated that the strong C_{60}-C_{60} interaction, probably combined with local lattice distortions and imperfections, can remove the degeneracy of the t_{1u} orbital responsible for the metallic properties. As a result, the Fermi level density of states decreases, leading to the lower T_c. Very recently, it was proposed from a theoretical point of view that the degeneracy of the t_{1u} orbital is important in the occurrence of superconductivity [9]. Thus, the removal of the orbital degeneracy may lead to a further suppression of superconductivity.

ACKNOWLEDGMENTS

This work was supported in part by the Grant-in Aid for Scientific Research on the Priority Area "Fullerenes and Nanotubes" by the Ministry of Education, Science and Culture of Japan.

REFERENCES

1. Hirosawa, I., Prassides, K., Mizuki, J., Tanigaki, K., Gevaert, M., Lappas, A., Cockcroft, J.K., *Science* **264**, 1294 (1994); Margadonna, S., Brown, C.M., Prassides, K., Fitch, A.N., Knudsen, K.D., Bihan, T.L., Mezouar, M., Hirosawa, I., Tanigaki, K., *Int. J. Inorganic Materials.*, **1**, 157 (1999).
2. Maniwa, Y., Saito, T., Kume, K., Kikuchi, K., Ikemoto, I., Suzuki, S., Achiba, Y., Hirosawa, I., Kosaka, M., Tanigaki, K., *Phys. Rev.* **B52**, R7054 (1995).
3. Kosaka, M., Tanigaki, K., Prassides, K., Margadonna, M., Lappas, A., Brown, C.M., Fitch, A.N., *Phys. Rev.* **B 59**, R6628 (1999).
4. Maniwa, Y., Muroga, N., Tou, H., Kikuchi, K., Suzuki, S., Achiba, Y., M. Kosaka, K. Kosaka, M., Tanigaki, K., Shimoda, H., Iwasa, Y., Mitani, T., *Synthetic Metals* **103**, 2458 (1999).
5. Maniwa, Y., Sugiura, D., Kume, K., Kikuchi, K., Ikemoto, I., Suzuki, S., Achiba, Y., Hirosawa, I., Kosaka, M., Tanigaki, K., Shimoda, H., Iwasa, Y., *Phys. Rev.* **B54**, R6861 (1996).
6. Kitazume, K., Tou, H., Maniwa, Y., Kosaka, M., and Tanigaki, K., to be appeared in *Synthetic Metals*
7. Maniwa, Y., Ikejiri, H., Tou, H., Yasukawa, M., and Yamanaka, S., to be appeared in *Synthetic Metals*.
8. Maniwa, Y., Sato, M., Kume, K., Kozlov, M.E., Tokumoto, M., *Carbon* **34**, 1287 (1996).
9. Suzuki, S., Okada, S., and Nakao, K., *J. Phys. Soc. Jpn.*, **69**, 2615 (2000).

Magnetic Properties of TDAE-C_{60} under Pressure

Kenji Mizoguchi*[1], Masayoshi Machino*, Hirokazu Sakamoto*,
Tohru Kawamoto[†], Madoka Tokumoto[†],
Ales Omerzu[‡], and Dragan Mihailovic[‡]

*Department of Physics, Tokyo Metropolitan University, Hachioji, Tokyo 192-0397, Japan
[†]Electrotechnical Laboratory, AIST, Umezono, Tsukuba, 305-8568, Japan
[‡]Jozef Stefan Institute, Jamova 39, Ljubljana, Slovenia

Abstract. Electron Spin Resonance (ESR) of α-TDAE-C_{60} in a crystalline form was performed under pressure to investigate the mechanism of ferromagnetic phase transition, where TDAE is tetrakis-dimethylamino-ethylene. Curie temperature T_C decreased parabolically, reaching zero around 9 kbar. It is demonstrated that an antiferro-orbital ordering model of Jahn-Teller distorted C_{60}^- anion is consistent with the experimental data, not only qualitatively but also quantitatively. Enough higher pressure than 10 kbar transforms α-TDAE-C_{60} to β-type polymerized-form along c-axis with [2+2] cycloadditive mechanism at room temperature. The β-form is stable even at ambient pressure and up to $\approx$430 K. Prominently, in the β-form, the spin on TDAE$^+$ cation, that was missing in the α-form, revives with the ESR g-shift in-between RbC_{60} polymer and TDAE$^+$ radical cation.

INTRODUCTION

A discovery of ferromagnetic transition at 16 K in TDAE-C_{60} [1] provoked a vast number of investigations to unveil its magnetic property and mechanism. In the early stage of research, as-grown powder form of TDAE-C_{60}, which is a mixture of ferromagnetic α- and paramagnetic α'-TDAE-C_{60}, was used, which gives rise to a variety of models, such as itinerant ferromagnetism [2], superparamagnetism [3], spin glass [4], and weak ferromagnetism, to interpret the curious experimental results. Later on, it was found that the physical properties of this material are extremely sensitive to the sample form and annealing procedure [5,6]. This fact could suggest the merohedral ordering of C_{60} balls plays an important role in appearance of the ferromagnetism. Actually, Kambe et al. reported an observation of X-ray superlattice peaks developed below 180 K and suggested a dimerization of unit cell size [7]. It is impressive that the structure of α- and α'-phases is very similar to

---

[1] e-mail: mizoguchi@phys.metro-u.ac.jp

each other, except for symmetry lowering of α form caused by C_{60}'s merohedral ordering at low temperatures [7,14,15]. In this report, we will propose a model based on antiferro-orbital ordering of Jahn-Teller distorted C_{60} balls, consistent with the present experimental results under pressure [8–12], and demonstrate the physical properties of the polymerized β-phase.

RESULTS AND DISCUSSION

Fig. 1 demonstrates the remarkable difference in the temperature dependence of ESR linewidth among the three phases, α, α' and β. This difference might be due to different exchange interactions between the electron spins, dominated by different merohedral ordering and/or bondings between the neighboring C_{60}'s. Fig. 2 shows a narrowing of ESR linewidth under pressure. This would come from an increase of exchange interaction with decreasing C_{60} distances.

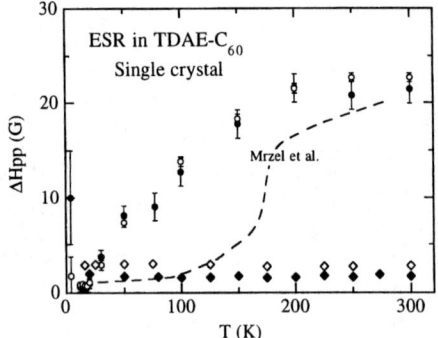

FIGURE 1. Temperature dependence of ESR peak-to-peak line width in the various crystal structures of TDAE-C_{60}. The open and closed circles: α-phase, the dashed curve: α'-phase [6], and the open and closed diamonds: β-phase [11,12].

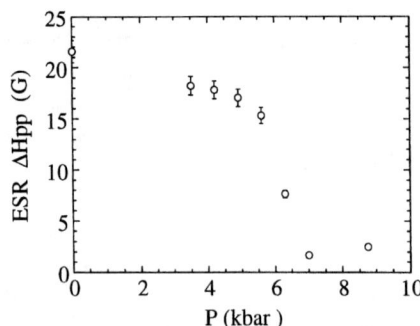

FIGURE 2. The pressure dependence of ESR linewidth in α-TDAE-C_{60} at rt.

TABLE 1. The distances between C_{60}'s or neighboring chains at 89 K for the new 1D polymer of β-TDAE-C_{60} [11,12].

		r_1 (Å)	r_2 (Å)	r_3 (Å)	N_S	g
TDAE-C_{60}	·New 1D polymer (β)	9.17± .07	13.69	15.70	2	2.0028
	·Ferromagnet (α)	9.99	13.00	15.86	1	2.0005
[2+2] polymer	o-Rb_1C_{60}	9.138	10.107	14.173	1	2.0012

Fig. 3 shows the pressure dependence of the Curie temperature measured under various static pressures generated with a CuBe clamped-type cell and Daphne 7373 oil [8,11,12]. The indicated pressures for $T \leq 100$ K are estimated using the reported data [13]. The external magnetic field was applied less than 30 Gauss to ensure negligible perturbation to T_C. Fig. 3 looks remarkably insensitive to the pressure in comparison with the reported data for a powdered sample [2]. T_C survives up to 7 kbar, approximately following a parabolic formula represented by the solid curve, which is a prediction by a model of antiferro-orbital ordering of Jahn-Teller distorted C_{60}'s with appropriate parameters [8]. This model is consistent with the recently determined structure [7,15]. Further, ESR evidence of Jahn-Teller distortion of C_{60} was reported in a model compound to observe Jahn-Teller splitting by reducing the exchange interaction between C_{60}'s [16]. X-ray analysis in a similar compound suggested a stabilization of static Jahn-Teller distortion by symmetry lowering induced by the merohedral ordering [17].

Table 1 shows physical properties of polymerized β-phase [11,12]. From the c-axis lattice constant, it is concluded that the β-phase is a [2+2] cycloadditive polymer phase. At ambient pressure, the polymer phase is confirmed to be stable up to $\approx$430 K for 30 min through the ESR linewidth at 300 K. Since the β-phase seems to have a triclinic structure, a single crystal converted to the β-phase is no longer single crystal because of multiple possibilities for a conversion to the β-phase from

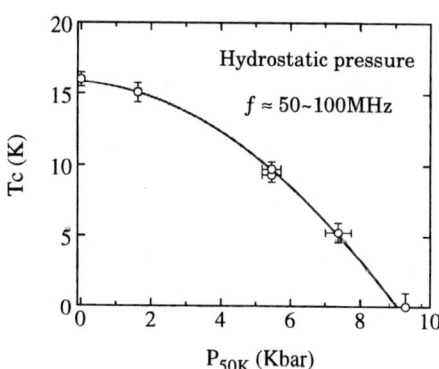

FIGURE 3. T_C versus P_{50K} in TDAE-C_{60}. The solid curve indicates the theoretical prediction with the data up to 7.4 kbar [8,11,12].

the α-phase. Thus, a powder diffraction study is in progress to study the crystal structure of the β-phase.

ACKNOWLEDGMENTS

K. M. greatefuly thanks to Prof. M. Yamashita and Dr. T. Ishii for utilization of X-ray apparatus. This work is supported by Grant-in-Aid for Scientific Research on the Priority Area "Fullerenes and Nanotubes" by the Ministry of Education, Science, Sports and Culture of Japan.

REFERENCES

1. Allemand, P.-M., Khemani, K.C., Koch, A., Wudl, F., Holczer, K., Donovan, S., Gruner, G., and Thompson, J.D., *Science* **253**, 301-303 (1991).
2. Sparn, G., Thompson, J.D., Allemand, P.-M., Li, Q., Wudl, F., Holczer, K., and Stephens, P.W., *Solid State Commun.* **82**, 779-782 (1992).
3. Dunsch, L., Eckert, D., Frohner, J., Bartl, A., and Muller, K.-H., *J. Appl. Phys.* **81**, 4611-4613 (1997).
4. Venturini, P., Mihailovic, D., Blinc, R., Cevc, P., Dolinsek, J., Abramic, D., Zalar, B., Oshio, H., Allemand, P.M., Hirsch, A., and Wudl, F., *Int. J. Mod. Phys. B* **6**, 3947-3951 (1992).
5. Blinc, R., Pokhodnia, K., Cevc, P., Arcon, D., Omerzu, A., Mihailovic, D., Venturini, P., Golic, L., Trontelj, Z., Luznik, J., Jeglicic, Z., and Pirnat, J., *Phys. Rev. Lett.* **76**, 523-526 (1996).
6. Mrzel, A., Cevc, P., Omerzu, A., and Mihailovic, D., *Phys. Rev. B* **53**, R2922-2925 (1996).
7. Kambe, T., Nogami, Y., and Oshima, K., *Phys. Rev. B* **61**, R862-865 (1999).
8. Kawamoto, T., Tokumoto, M., Sakamoto, H., and Mizoguchi, K., submitted to *J. Phys. Soc. Jpn*.
9. Kawamoto, T., and Suzuki, N., *Synth. Met.* **86**, 2387-2388 (1997).
10. Kawamoto, T., *Solid State Commun.* **101**, 231-235 (1997).
11. Mizoguchi, K., Machino, M., Sakamoto, H., Kawamoto, T., Tokumoto, M., Omerzu, A., and Mihailovic, D., to be published.
12. Mizoguchi, K., Machino, M., Sakamoto, H., Tokumoto, M., Kawamoto, T., Omerzu, A., and Mihailovic, D., to be published in *Synth. Metals*.
13. Murata, K., *Rev. Sci. Instrum.* **68**, 2490-2493 (1997).
14. Narymbetov, B., Kobayashi, H., Tokumoto, M., Omerzu, A., and Mihailovic, D., *Chem. Commun.* 1511-1512 (1999).
15. Narymbetov, B., Omerzu, A., Kabanov, V.V., Tokumoto, M., Kobayashi, H., and Mihailovic, D., *Nature* **407**, 883-885 (2000).
16. Bietsch, W., Bao, J., Ludecke, J., and Smaalen, S.V., *Chem. Phys. Lett.* **324**, 37-42 (2000).
17. Launois, P., Moret, R., Souza, N.-R.D., Azamar-Barrios, J.A., and Penicaud, A., *Eur. Phys. J. B* **15**, 445-450 (2000).

Synthesis and Structure of Alkaline Earth and Rare Earth Metal doped C_{70}

Taishi Takenobu, Yoshihiro Iwasa, Takayoshi Ito, and Tadaoki Mitani

Japan Advanced Institute of Science and Technology, Tatsunokuchi, Ishikawa 923-1292, Japan

Abstract. We have investigated the structure sequence of alkaline earth (A=Ba, Sr) and rare earth metal (R=Eu) doped C_{70} binary system. X-ray diffraction measurements revealed that there exist at least four stable phases at x=3, 4, 6, and 9 in A_xC_{70} and two stable phases at x=3, and 9 in R_xC_{70}. Among them, structural models are presented for Ba_4C_{70}, Sr_3C_{70}, and Eu_3C_{70}. Ba_4C_{70} takes an analogous structure to orthorhombic Ba_4C_{60}. Sr_3C_{70} and Eu_3C_{70} have monoclinic cell and their diffraction patterns are quite similar to that of Sm_3C_{70}, which involves a unique C_{70}-metal-C_{70} dimer structure. Preliminary results of Raman spectroscopy and magnetization measurement suggest the highly reduction state for A_9C_{70} and ferromagnetic interaction for Eu_xC_{70}.

INTRODUCTION

Fullerene C_{60} is known to form a vast variety of compounds by intercalation of alkali, alkaline earth, and rare earth metals. They display intriguing divers features including rather high T_C superconductivity as well as unique magnetic properties. Moreover, one of the unique properties of C_{60} fullerides is the wide reduction state from $(C_{60})^{0-}$ to $(C_{60})^{12-}$. Because of such a wide tunability of the valance states, C_{60} can be regarded as an "electronic sponge". In contrast to the rich properties of C_{60}, much less properties have been reported on the second abundant fullerene C_{70}. The studies on C_{70} so far have been limited in the alkali metal intercalation, and the reduction state of C_{70} is known up to $(C_{70})^{9-}$ [1]. To achieve higher reduction state intercalation of alkaline earth and rare earth metals should be promising. Another interesting aspect of using rare earth metals as intercalants is the magnetic properties. Here we pay atteion to Eu, since magnetic interaction and long range ordering is expected.

In this paper, we present the result of synthesis and preliminary structural study of C_{70} intercalation compounds of barium, strontium, and europium metals. The x-ray measurement revealed that the nominal composition of x=3, 4, 6, and 9 in Ba_xC_{70}, and composition of x=3 and 9 in Sr_xC_{70} and Eu_xC_{70} yield stable phases.

EXPERIMENTAL

Samples with the nominal concentration of M_xC_{70} (M=Ba, Sr, Eu) were prepared by a reaction of C_{70} powder and metals sealed in a quartz tube [2,3]. High-purity metals were broken into powders and mixed with C_{70} in a controlled atmosphere glove box. Powder mixtures were pressed to pellets and loaded into quartz tubes. Quartz tubes were then sealed under a vacuum of 10^{-6} torr. Heat treatments were carried out at temperatures 400-650°C, and for periods ranging from days to months. Highly uniform samples with less carbides were prepared by a two week annealing at 550°C with frequent intermittent grinding in A_xC_{70} (A=Ba, Sr) and by a month anneal at 450°C in Eu_xC_{70}. The samples were then sealed in 0.5 mm diameter capillaries for x-ray powder diffraction, Raman spectroscopy, and SQUID measurements. A synchrotron radiation x-ray powder diffraction experiment was carried out with imaging plates as detectors at the BL-1B, Photon Factory, KEK and BL02B2, SPring-8.

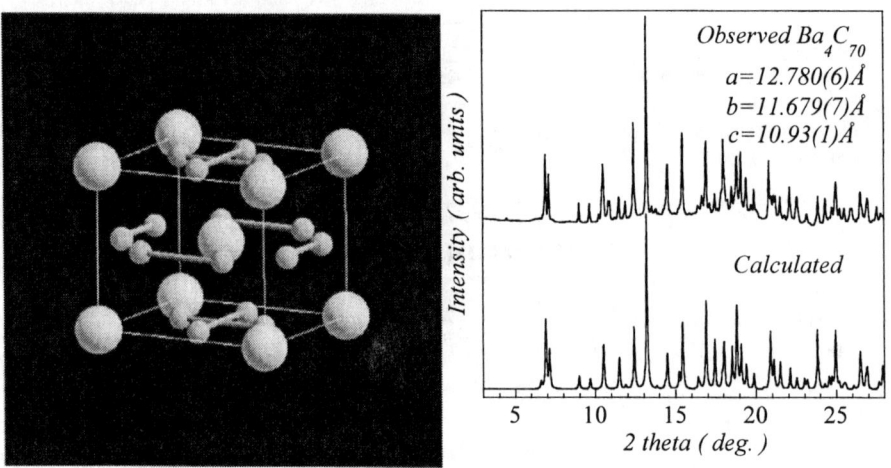

FIGURE 1. The structure model (left) and the observed and calculated diffraction patterns for the nominal Ba_4C_{70} (right) are shown. In left figure, big balls and small balls represent hypothetical spheres of C_{70} and Ba metals, respectively.

RESULT AND DISCUSSION

We observed four stable phases at x=3, 4, 6 and 9 in Ba_xC_{70} by the x-ray powder diffraction. The diffraction patterns of the other composition are explained by mixtures of these stable phases. Ba_3C_{70} and Ba_9C_{70} show identical patterns to those we reported previously [2,3]. Ba_4C_{70}, however, shows a completely different pattern. Most of the diffraction peaks of Ba_4C_{70} could be indexed as a orthorhombic cell with a lattice constant of a=12.780(6)Å, b=11.679(7)Å, and c=10.93(1)Å. This unit cell dimension and the intensity distribution are quite similar to those of Ba_4C_{60}, indicating that the structure of Ba_4C_{70} is a body-centered-orthorhombic structure. Figure 1 shows the observed and calculated diffraction pattern for the nominal Ba_4C_{70}. Intensity calculations were made using the Lazy-Pulverix software (K. Yvon, W. Jeitschko, E. Parthe, unpublished). In this calculation, we used the Ba_4C_{60} like orthorhombic cell (*Immm* space group), where the configuration of Ba-site is similar to that of Ba_4C_{60}, and C_{70} cage was hypothetically treated as a spherical shell (Fig. 1). The amazing similarity for the main peak position and the intensity distributions means that our structure model is reliable within the spherical shell model.

In Sr_xC_{70} and Eu_xC_{70} series, we found that x=3 and 9 yield stable phases. First, we discuss on the x=3 phases. Figure 2 shows the comparison of diffraction patterns of the new compounds (Sr_3C_{70} and Eu_3C_{70}) with those of the known compounds, Ba_3C_{70} and Sm_3C_{70} [2-4]. It is noted that only Ba_3C_{70} shows a totally different diffractogram. The patterns of Eu and Sm compounds are very similar, and the Sr_3C_{70} pattern is also similar to those of Eu_3C_{70} and Sm_3C_{70}. The structure of Ba_3C_{70} is a deformed *bcc*, while Sm_3C_{70} takes a monoclinic cell which is regarded as a deformed *fcc*. These results indicate that only Ba_3C_{70} takes a *bcc*-derived structure, while the other three form an *fcc*-derived structure. This situation is similar to the case of C_{60} compounds, where fullerene arrangement is controlled by the ionic radii of intercalated divalent species. (Ionic radii are 1.32Å, 1.31Å, 1.36Å, 1.49Å for Sr^{2+}, Eu^{2+}, Sm^{2+}, and Ba^{2+}, respectively) Because Eu^{2+} is an ^8S configuration (J=7/2, L=0, gJ=7 and μ_{eff}=7.815μ_B), a strong magnetization and a possible magnetic ordering are expected for Eu_3C_{70}. A ferromagnetic interactions between magnetic moments are observed by preliminary SQUID measurements for Eu_3C_{70} and Eu_9C_{70}. A magnetic studies of these samples are in progress.

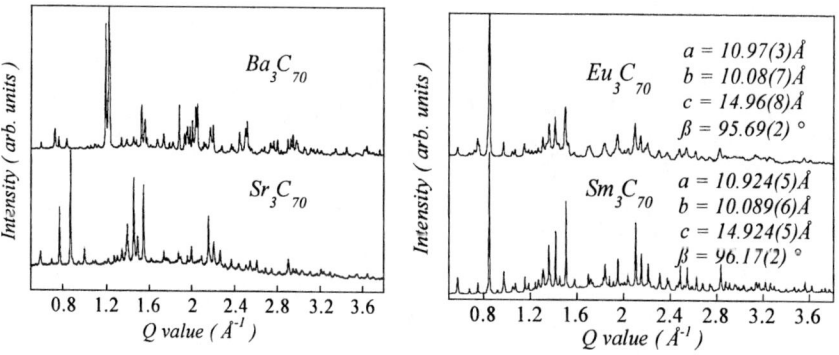

FIGURE 2. The diffraction patterns of A_3C_{70} (A=Ba, Sr : left) and R_3C_{70} (R=Eu, Sr : right) are compared.

The nominal $x=9$ samples are a metal-saturated phase since the systematic synthesis showed that all the samples of $x\geq 9$ yielded the same diffraction patterns. The diffraction patterns identically indicate a cubic symmetry and peak distributions are consistent with face-centering. Because the diffraction patterns of Ba_9C_{70}, Sr_9C_{70} and Eu_9C_{70} are very similar to K_9C_{70} [1] and $K_{8+x}C_{84}$ [5], it is very likely that M_9C_{70} (M=K, Ba, Sr, Eu) has a similar crystal structure. A common feature of both materials is a K_8 cube in the octahedral site of the fcc fullerene lattice. If we assume that metals are divalent in crystal, the charge state of C_{70} is estimated to be $(C_{70})^{18-}$. Preliminary measurements of Raman spectroscopy show the large red shift of carbon-carbon stretching mode in C_{70} molecule and suggest a charge transfer between metals and C_{70}. Such a high valence state is of significant interest to be investigated in the future.

In summary, we succeeded to synthesize Ba_xC_{60} with the nominal $x=3$, 4, 6, and 9, Sr_xC_{70} with $x=3$ and 9, and Eu_3C_{70} with $x=3$ and 9. Ba_4C_{70} had the orthorhombic cell, which is very close to that of Ba_4C_{60}. Sr_3C_{70} and Eu_3C_{70} are monoclinic cell and their diffraction patterns are quite similar to that of Sm_3C_{70}, which all common for metals investigated (Ba, Sr, Eu). Surprisingly, The structure of metal saturated phase $x=9$ are possibly identical for all metals.

ACKNOWLEDGMENTS

We thank E. Nishibori, M. Takata, M. Sakata, Y. Murakami, T. Muro, D. H. Chi, S. Moriyama and M. Miyake for experimental collaborations. This work has been supported by the JSPS "Future Program", RFTF96P00104, and also by the Grant-In-Aid for Scientific Research on the Priority Area "Fullerenes and Nanotubes" by the Ministry of Education, Sport, Science and Culture of Japan.

REFERENCES

1. Kobayashi, M. et. al., Phys. Rev. B **48**, 16877- (1993).
2. Takenobu, T. et. al., Mol. Cryst. and Liq. Cryst. **340**, 617-622 (2000).
3. Chen, X. H. et. al., J. Am. Chem. Soc. **122**, 5729-57325 (2000).
4. Dam, H. C. et. al., to be published.
5. Allen, K. M. et. al. , J. Am. Chem. Soc. **120**, 6681- (1998).

Structural and Physical Properties of Lithium Fullerides Li_xC_{70}

Hiroyasu Kumada, Mototada Kobayashi, Yuichi Akahama, and Haruki Kawamura

*Department of Material Science, Himeji Institute of Technology,
3-2-1 Koto, Kamigori, Ako, Hyogo 678-1297, Japan*

Abstract. We report the x-ray powder diffraction measurements for Li_xC_{70} with nominal composition x between 6 and 47. The profile showed on single phase with $x \geq 25$ and assigned to be an orthorhombic structure with a=17.11 Å, b=13.74 Å and c=13.86 Å. The magnetic susceptibility for $Li_{25}C_{70}$ was measured by ESR spectrometer. The temperature independent paramagnetic contribution for $Li_{25}C_{70}$ was fairly small and estimated to be 7.0×10^{-5} emu/mol. This shows the absence of Pauli paramagnetic.

INTRODUCTION

The heavy alkali metal intercalation in to fullerene C_{60} yields compounds with stoichiometry of A_xC_{60}(x=1,3,4,6. A=K,Rb,Cs). Among these compounds both superconductors (x=3) and magnetic materials(x=1) are found. Li has the smallest ionic radius in alkali metals. The x-ray diffraction of Li_xC_{60} has been reported and x for Li-saturated Li_xC_{60} was suggested to lie above x=24 [1]. The tetrahedral and octahedral voids of the pristine fcc-C_{70} crystal are larger than those of the C_{60} crystal, respectively. A saturated phase for heavy alkali metal fullerides A_xC_{70}(A=K,Rb,Cs) is found to be A_9C_{70}. Li_xC_{70} is expected to have larger composition than x=9. Consequently, Li_xC_{70} is expected to show novel physical properties depending on the doping level. Here we report syntheses and results of x-ray diffraction and ESR measurement for Li_xC_{70}.

EXPERIMENTAL

Li-fullerides Li_xC_{70} were synthesized by the mixture of C_{70} powder (99% up purity) and Li metal (99.9% purity) with nominal compositions x between 6 and 47. The mixture was introduced into a tantalum tube in an Ar glove box. Then it was sealed into a Pyrex glass tube after evacuating. It was heated in an electronic furnace at 623 ~ 693 K for 72 ~ 494 hours. Li_xC_{70} synthesis under high pressure at room temperature(RT) has also been tried. C_{70} powder and excess Li metal were loaded into DAC (diamond anvil cell) in an Ar glove box. X-ray powder diffraction measurements were carried out with Mo rotor x-ray generator and an imaging plate detector. The synchrotron radiation(SR) at KEK-PF(BL1B) and SPring-8 (BL02-B2 and BL04-B2) were also used. The magnetic susceptibility was measured by using Bluker ESP-300E ESR spectrometer at X band(IMS, Institute for Molecular Science). The specimen was sealed into a quartz tube together with helium gas for thermal exchange.

RESULTS AND DISCUSSION

The x-ray powder diffraction profiles for Li_xC_{70} (x=10, 20, 40) at RT are shown in Fig. 1. $Li_{10}C_{70}$ and $Li_{20}C_{70}$ were assigned to a fcc structure with a=14.67Å and a=14.50Å, respectively. $Li_{40}C_{70}$ was assigned to an orthorhombic structure with a=17.11 Å, b=13.74 Å and c=13.86 Å. Further, $Li_{25}C_{70}$ was assigned to an orthorhombic structure with a=17.11 Å, b=13.74 Å and c=13.95 Å. $Li_{35}C_{70}$ was assigned to an orthorhombic structure with a=17.06Å, b=13.75Å and c=13.85Å. It is concluded that structure of Li-saturated phase is an orthorhombic and the saturated composition ratio lie between x=20 and x=25.

The x-ray powder diffraction profiles for Li-saturated phase between 5K and 723K are shown in Fig. 2. There was no structural transition between 5K and 623K. The profile began to change around 673K and the profile at 723K can be assigned to a fcc structure with a=14.89 Å. This may show Li de-intercalation. Peaks from Li carbide(Li_2C_2) were observed above 623K. The carbide was often detected in a long time annealed sample.

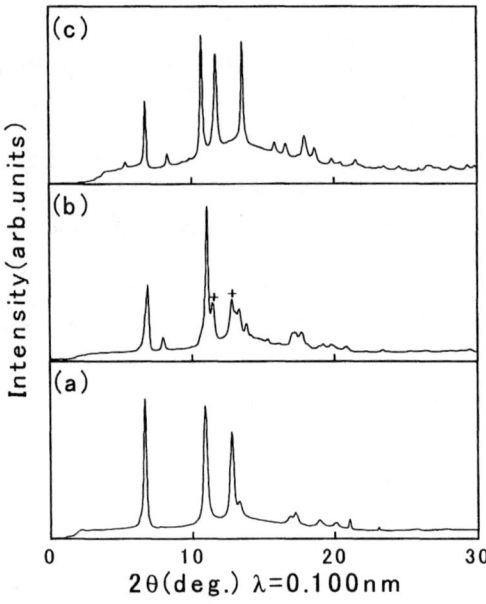

FIGURE 1. The x-ray powder diffraction profiles for Li_xC_{70} at RT (a)x=10, (b)x=20 and (c)x=40 using SR. + symbols indicate unidentified peaks.

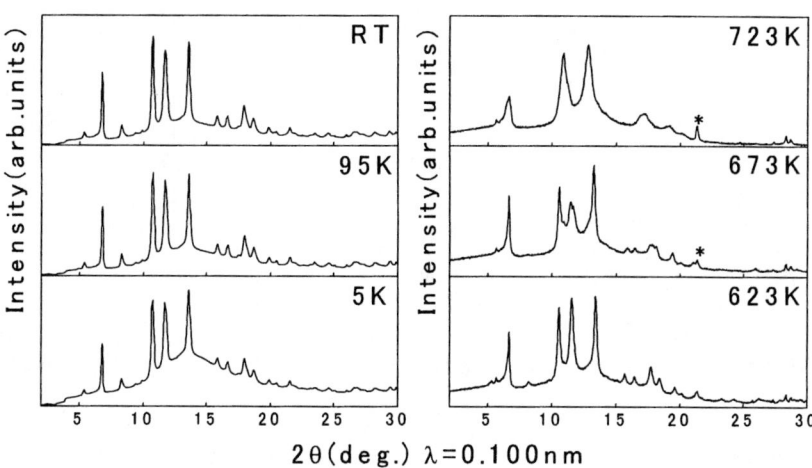

FIGURE 2. The x-ray powder diffraction profiles for Li-saturated phase between 10K and 723K. * describes Li carbide peaks.

We tried to synthesis Li_xC_{70} at RT under high pressure by using DAC in order to suppress Li carbide formation. The x-ray powder diffraction profiles for mixture of Li and C_{70} at RT are shown in Fig. 3. The profile at 0.17GPa can be assigned to an orthorhombic structure with a=17.01 Å, b=13.66 Å and c=13.89 Å. The profile at 3.93GPa can be assigned to an orthorhombic structure with a=16.77 Å, b=13.28 Å and c=13.61 Å. There was no structural transition between 0.17GPa and 3.93GPa. The profiles have not changed by heating up to 150°C. These orthorhombic structure is similar to that by annealing.

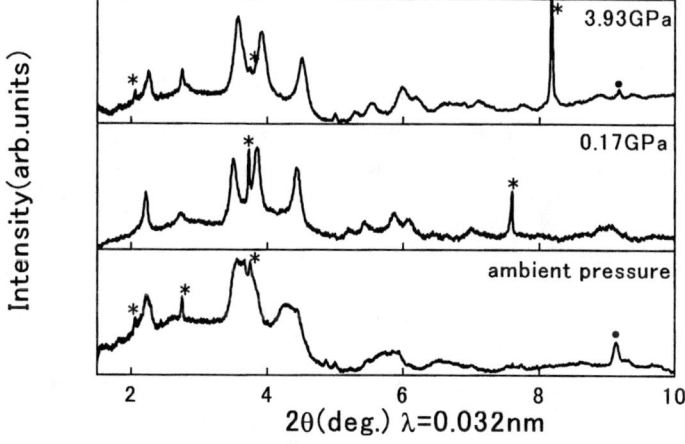

FIGURE 3. The x-ray powder diffraction profiles at RT for mixture of Li and C_{70} in DAC using SR radiation. * and ● describe peaks from diamond and gasket, respectively.

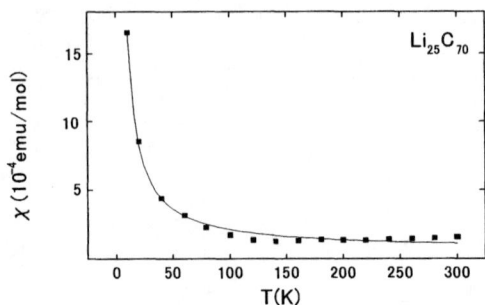

FIGURE 4. Temperature dependence of ESR spin susceptibility for $Li_{25}C_{70}$. Solid line shows fitting to the Curie-Weiss law.

The temperature dependence of the magnetic susceptibility χ for $Li_{25}C_{70}$ by using ESR spectrometer is shown in Fig. 4. The χ was obtained by integrating ESR intensity. The temperature-independent term χ_s was obtained by fitting the results to the formula

$$\chi = \frac{C}{T-\theta} + \chi_s \qquad (1)$$

where C is the Curie constant, θ is the Weiss temperature and χ_s is the temperature independent component of spin susceptibility. The spin concentrations calculated from C is 0.0311 spin per C_{70} molecule. The temperature independent paramagnetic contribution and the Weiss temperature for $Li_{25}C_{70}$ was estimated to be 7.0×10^{-5} emu/mol and -0.95K, respectively. The small χ_s suggests $Li_{25}C_{70}$ to be insulator. Li_xC_{70} with various x are now under investigations.

SUMMARY

Lithium fullerides Li_xC_{70} can be successfully synthesized by both annealing and applying high pressure at RT. Li-saturated phase has an orthorhombic structure and the saturated composition x lie between 20 and 25. This structure did not change between 5K and 623K. Li atoms are de-intercalated at 723K. ESR measurement for $Li_{25}C_{70}$ suggests insulating character of this Li-saturated phase.

ACKNOWLEDGMENTS

This work was supported by the Grant-in-Aid for Scientific Research on the Priority Area "Fullerene and Nanotubes" by the Ministry of Education, Science, and Culture of Japan.

REFERENCES

1. Cristofolini.L., Ricco.M., and Renzi.R.De., *Phys.Rev.* **B59**, 8343-8346 (1999).

The Structural and Magnetic Properties in TDAE-fullerene System

Kokichi Oshima*†, Yoshio Nogami*†, Takashi Kambe†,
Nobuaki Nagao† and Motoyasu Fujiwara*

*Graduate School of Natural Science and Technology,
† Department of Physics, Faculty of Science,
Okayama University, Okayama, 700-8530 Japan

Abstract. The structural and magnetic properties are discussed on the TDAE-fullerene system for C_{60} and C_{70}. The single crystal structural analysis reveals that the TDAE-C_{70} system contains solvent molecules (toluene), and the C_{70} molecules in the unit cell are dimerized. The room temperature structure is monoclinic $P2_1/n$, and crystal parameters are a=18.82(3)Å, b=16.38(2)Å, c=18.73(3)Å and $\beta = 117.58(5)°$. The same annealing procedure used in TDAE-C_{60} does not alter the magnetic behavior in TDAE-C_{70}. A preliminary report is given for the TDAE-C_{60} structural study in magnetic fields up to 7 T. The possible method that directly reveals the relation between magnetism and the molecular arrangement is proposed.

INTRODUCTION

The origin of the ferromagnetism in TDAE-C_{60} has been discussed long, but the explanation is controversial and still open to question. We have firstly experimentally shown the possible low temperature structural phase change related to the ferromagnetism in this compound [1]. The recent report on the low temperature structural analysis insists on the peculiar low temperature ordering of C_{60} at the lowest temperatures. And the authors gave a model with randomly occupied C_{60} molecules along the three-fold axis [2]. The ESR study at high pressures by the group including the same authors have proposed a different theoretical model [3], therefore the final mechanism explaining the ferromagnetism is still open to question. It seems that the difficulty in structural analysis is due to the rotating C_{60} molecules even at fairly low temperatures ($\sim$ 150 K). Therefore, it is probable that the low temperature structure is dependent on cooling processes. We still need a consistent model that can quantitatively explain the magnetic properties. We discuss in this paper on the structural and magnetic properties of the TDAE-fullerene system for C_{60} and C_{70}. Though the magnetic properties of TDAE-C_{70} have been discussed implicitly assuming a similar crystal structure as TDAE-C_{60} so far, but

TABLE 1. The room temperature crystal parameters of TDAE-C_{70} toluene.

Space group	a(Å)	b(Å)	c(Å)	β(deg)
Monoclinic $P2_1/n$	18.82(3)	16.38(2)	18.73(3)	117.58(2)

the actual study on TDAE-C_{70} system has been absent to our knowledge. Our room temperature X-ray analysis has shown that the TDAE-C_{70} system has a different crystal structure, and it contains solvent molecules (toluene). We also show in this paper a preliminary experimental structural study in magnetic fields for TDAE-C_{60}. The magnetic and structural measurements are usually performed in the different experimental run, and it seems that the low temperature structure is dependent on the cooling processes. Therefore, it is informative if we can obtain the low temperature structural data in the magnetic field.

Sample Preparation and Magnetic Properties

Single crystals have been prepared using the slow mixing of TDAE and the C_{60} or C_{70} resolved in the toluene solvent. In several weeks we obtain small single crystals (0.5x0.5x0.1 mm^3) with plate-like faces. Magnetic properties are studied using a SQUID magnetometer. The magnetic properties and their temperature dependencies are consistent with the previous reports [4]. The magnetic measurements on the one single crystal for TDAE-C_{70} are still on the way. The annealing effect on the magnetism in TDAE-C_{60} is well-known [1]. The annealing procedure used to study the ferromagnetism in TDAE-C_{60} made no effect in the magnetic properties of TDAE-C_{70}. As shown below, crystals contain toluene molecules in the structure. Therefore, we should try different sample preparation procedures to study the possibility to get the system without toluene. The crystal growth conditions are nearly the same in TDAE-C_{60} and TDAE-C_{70} cases. We have reported the systematic contraction of the unit cell parameters due to the annealing procedure in TDAE-C_{60} [1]. We think the possibility still to exist that TDAE-C_{60} system also includes solvent molecules in the structure. If the TDAE-C_{60} structure has a space where toluene molecules can easily fit in, and if the space is randomly occupied, we cannot identify the solvent by the X-ray analysis. We should reconsider the possibility if the solvent molecule can easily go out the structure by the annealing procedure. To confirm the possible effect due to solvent molecules, the experiment such as to check the effect on magnetism immersing the well-annealed sample in the solvent will be helpful.

Room Temperature Structure of the TDAE-C_{70} System

The crystal parameters for TDAE-C_{70} are shown in Table 1. The structure analysis has been performed using a TEXSAN (Rigaku co.) software on the data

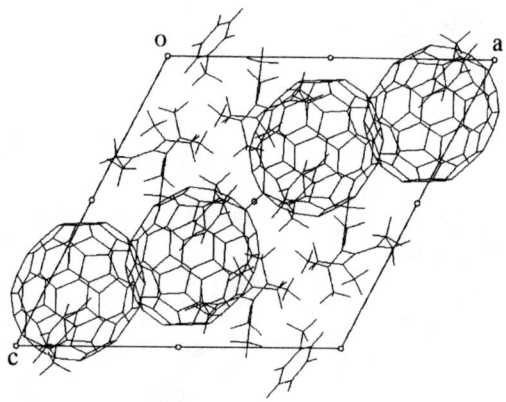

FIGURE 1. The projection of the structure along b axis in TDAE-C_{70} toluene.

obtained by IP (Imaging Plates). At present the R_1-factor for the analysis using the 4175 reflections is 0.11. The C_{70} molecules are not rotating in contrast to the case of TDAE-C_{60}. This may be due to the lower symmetry of the C_{70} molecule and the toluene molecule packing the vacant space in this system. The characteristic point is the dimerized C_{70} molecules in a unit cell. The shortest intradimer c-c distance is less than 1.7 Å and the shortest interdimer distance is more than 3.2 Å. The angles between C_{70} longer axes are 0° between intradimer molecules, and 23° between interdimer molecules. In spite of the dimerization of the C_{70} molecules TDAE-C_{70} does not become nonmagnetic at the lowest temperatures measured. Therefore, we should study the origin of the moment, and the lower temperature properties. There is also a possibility that TDAE molecules are magnetic.

The Structural Analysis under High Magnetic Fields

We present here a preliminary study on the structure in the high magnetic field. It has been reported that the magnetic field effect exists for the ordering of C_{60} molecules [5]. The relation of the superstructure in the well-annealed single crystals to the ferromagnetism [1], [2] can be unambiguously studied. Due to the superstructure, the room temperature C-centered cell transforms into the lower temperature primitive cell. If the structure is responsible for the ferromagnetism, the structure should be enhanced by the magnetic field. The Bragg reflection can be identified as shown in Fig. 2 using the new cryostat with a 1.5 K cooling insert and a 7 T persistent mode superconducting magnet. As can be seen in the figure, we see strong Debye rings due to the six Be windows around the crystal, therefore we need more tries to identify superstructure signals reducing the background noises. There also exist strong noises around the center due to the direct beam. The low temperature structural measurements in conjunction with the annealing in the magnetic field just above the temperature where the C_{60} molecules seize the rotating motion, we would get the information on the relation between the molecular structure and the magnetism for TDAE-fullerene systems.

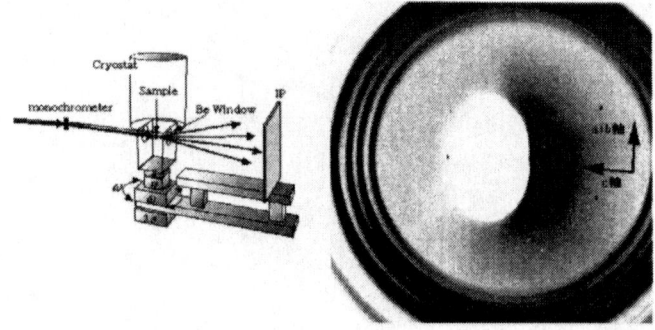

FIGURE 2. The X-ray system with the magnetic field and the Bragg reflection for TDAE-C_{60} in the 7 T field.

Conclusion

We have obtained for the first time the room temperature structure of the TDAE-C_{70}. The unit cell contains solvent molecule in contrast to our anticipation that the TDAE-C_{70} may have the nearly same structure as the TDAE-C_{60}. Therefore, we should seek the possibility again if we can obtain ferromagnetic TDAE-C_{70} system without toluene molecules. We have also shown the possibility to measure the crystal structure under the magnetic field using a newly developed X-ray apparatus. It is possible to obtain more definitive structural information to relate the structural ordering of fullerene molecules. We acknowledge the support by Mr. Shimomoto in constructing a new apparatus. We also thank for the V.B.L.(Venture Business Laboratories) of the Okayama University for the X-ray measurement.

REFERENCES

1. Kambe, T., Nogami, Y., and Oshima, K., *Phys. Rev.* **B 61**, R862 (2000).
2. Narymbetov, B., Omerzu, A.,Kabanov, V.V., Tokumoto, M.,Kobayashi, H., and Mihailovic, D., *Nature* **407**, 883 (2000).
3. Mizoguchi, K., Machino, M., Sakamoto, H.,Tokumoto, M.,Kawamoto, T.,Omerzu, A., and Mihailovic, M.,*Synth. Met.* in press (2001).
4. Tanaka, K.,Zakhidov, A.A.,Yoshizawa, K., Okahara, K., Yamabe, T., Yakushi, K., Kikuchi, K., Suzuki, S., Ikemoto, I., and Achiba, Y., *Phys. Rev.* **B 47**, 7554 (1993).
5. Mihailovic, D., Acron, D., Venturini, P., Blinc, R., Omerzu, A., and Cevc, P.,*Science* **268**, 400 (1995).

Synthesis and Structure of All-Carbon Bisfullerene C_{121}

Hidekazu Shimotani, Jian Wang, Nita Dragoe,* and Koichi Kitazawa*

*Department of Applied Chemistry, University of Tokyo and * Japan Science and Technology Corporation, JST-CREST, Tokyo 113-8656, Japan*

Abstract. We report the synthesis of the a bisfullerene C_{121}, the first all-carbon molecule to contain a homofullerene unit. This unsymmetrical bisfullerene was obtained as the major product of the solid-state thermolysis of a mixture of $C_{60}CBr_2$ and C_{60}. Theoretical calculations showed that the stability of the homofullerene isomer of C_{121} is higher than that of the more common isomer, the methanofullerene.

INTRODUCTION

Fullerene dimers and other bisfullerenes have attracted attention as models for fullerene polymers, *e.g.* C_{120},[1] or as intermediates for the formation of endohedral fullerenes.[2] We are interested in the synthesis and study of *all-carbon* derivatives of [60]fullerene, particularly in C_{121} and C_{122}.[3] Osterodt and Vogtle observed these species in mass spectroscopic studies of dibromomethano-[60]fullerene, $C_{60}CBr_2$, and proposed structures for them.[4] The preparation and isolation of C_{122} have also been reported by Strongin and co-workers, using an ingenuous reaction of fullerene with "carbon atom".[5] We describe here the synthesis, separation, and characterization of a C_{121} fullerene, **1b**, which is the first all-carbon fullerene to contain a homofullerene unit. Among the products of the thermolysis of dibromomethano-[60]fullerene we have also identified the high-symmetry C_{121} (D_{2d}, **1a**) and C_{122} (D_{2h}, **1d**) bisfullerenes whose structures are shown in Figure 1.

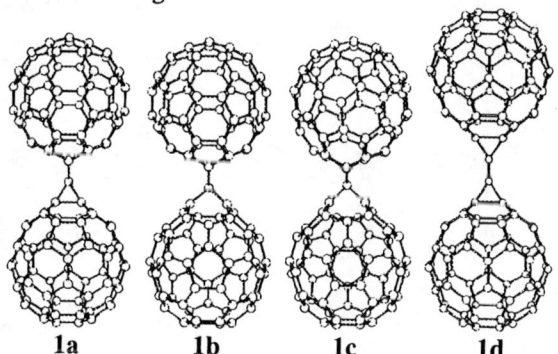

FIGURE 1. C_{121} [6,6]-[6,6] (**1a**), [6,6]-[6,5] (**1b**), and [6,5]-[6,5] (**1c**). C_{122} [6,6]-[6,6] (**1d**).

RESULTS AND DISCUSSION

Theoretical calculations[6]

Considering the possibilities of connecting a carbon atom to two [60]fullerene frameworks to produce C_{121}, three isomers can be envisioned (**1a**, **1b**, and **1c**, Figure 1): [6,6]-[6,6], [5,6]-[6,6], and [5,6]-[5,6], where [6,6] denotes a closed methanofullerene structure and [5,6] signifies an open, homofullerene structure. In order to compare the stabilities of the three isomers, we calculated the total energy (E_{tot}), and the relative values of standard enthalpy of formation ($\Delta(\Delta H_f^\circ(T))$) and standard Gibbs's free energy of formation ($\Delta(\Delta G_f^\circ(T))$) of **1a-c** (Figure 2(a)). The calculated $\Delta(\Delta G_f^\circ(T))$ show **1b** should be more stable than **1a** and **1c** by 2.52 kJ/mol and 21.78 kJ/mol at 298.15K, respectively. The differences increase at 750 K (the temperature of formation of C_{121}) to 7.89 kJ/mol and 24.73 kJ/mol.

The relative stability of **1a** and **1b** seems to be contrary to the general tendency that methanofullerenes are more stable than homofullerenes. The difference of E_{tot} calculated at B3LYP/6-31G(d) between methanofullerene form and homofullerene form of some experimentally accessible [60]fullerene derivatives are shown in Figure 2(b). We assumed that the difference of stability tendency between dimers and monomers could be explained by the strain energy (SE) around the bridging carbon atom of dimers, because formation of spiro structure releases additional SE.

We presumed that the SE of a spiro moiety of C_{12} approximately consists of the SE of two C_{60} methanobridges and the SE produced by the combination of two methanobridges (SE_{spiro}). The evaluations of SE_{spiro} were done using methanofullerene and homofullerene form of $C_{60}CH_2$ using B3LYP/6-31G(d) method.

The calculated SE_{spiro} of **1b** is smaller than that of **1a** by 24.343 kJ/mol, which is comparable to the difference of the E_{tot} of homofullerenes from that of methanofullerenes. Therefore, the cost of having a homofullerene moiety is compensated for by the SE_{spiro}.

SE_{spiro} of **1c** is smaller than **1a** by 33.792 kJ/mol, but this does not compensate for the presence of two homofullerene moieties (ca. 24-76 kJ/mol). That is the reason that **1c** is much less stable than both **1a** and **1b**.

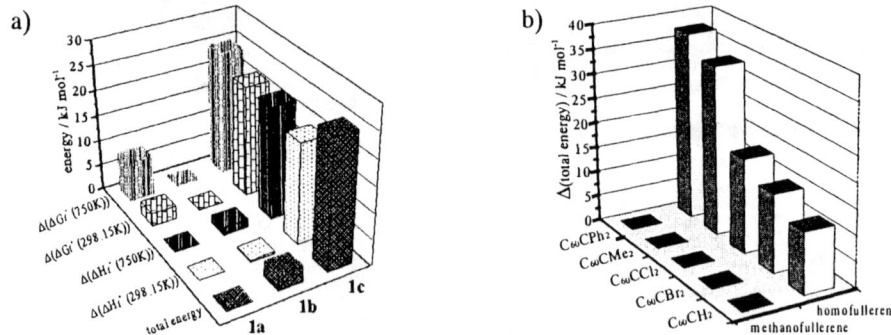

FIGURE 2. (a) The relative values of E_{tot}, $\Delta(\Delta H_f^\circ(298.15\ K))$, $\Delta(\Delta H_f^\circ(750\ K))$, $\Delta(\Delta G_f^\circ(298.15\ K))$, $\Delta(\Delta G_f^\circ(750\ K))$. (b) The relative values of E_{tot} of several experimentally accessible methanofullerenes and their homofullerene form.

Synthesis and characterization of C_{121}

For the synthesis of C_{121}, we used solid state thermolysis of a mixture of [60]fullerene and $C_{60}CBr_2$ up to 750 K. The separation of the mixture was made by using preparative GPC or HPLC by using recycling.

For the separation of the dimers, we used two preparative 5PBB HPLC columns (20×250 mm) with ODCB as eluent and recycling. Figure 3 shows the separation of the bisfullerenes by this method.

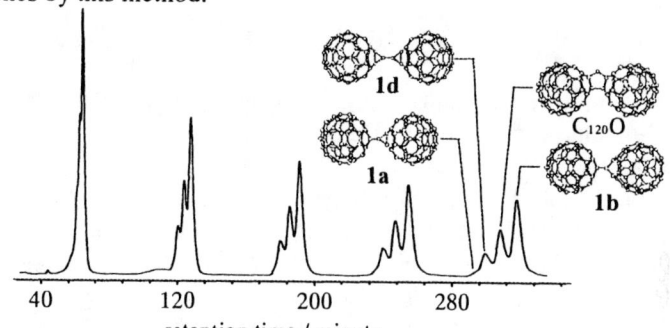

FIGURE 3. The HPLC of fraction II on Cosmosil Buckyprep column eluted with toluene.

The structure of **1b** was assigned based on ^{13}C-NMR, UV-VIS, MALDI-TOF MS and FT-IR experiments.[7]

Compelling evidence about the structures of **1b** and **1d** was obtained from ^{13}C-NMR experiments of samples prepared by mixing $C_{60}CBr_2$ in which the cyclopropane bridge carbon atoms were 99% ^{13}C-enriched (while the other 60 carbon atoms were not ^{13}C enriched) with 15% ^{13}C enriched [60]fullerene. The ^{13}C-NMR spectrum of **1d** showed the cyclopropylidene resonance as a singlet at 128.6 ppm while the ^{13}C-enriched **1b** showed for the resonance at 56.4 ppm two types of spin-spin couplings, broadened and of about 7-8% in intensity. The presence of these two types of spin-spin couplings, of about 13 and 23 Hz is indicative of both sp^3-sp^3 and sp^3-sp^2 couplings. This is the result of the scrambling of the ^{13}C enriched [60]fullerene cages between the two possibilities, *i.e.* closed and open structures.

This scrambling can arise from the presence of a symmetrical intermediate during the reaction process as follows:

-an interconversion of **1a** to **1b** which would be non-selective relative to the fullerene cages (enriched or not) thus yielding bisfullerenes in which both types of cages, methanofullerene or homofullerene, have ^{13}C-enriched carbon atoms

-an "inversion" of the **1b** structure, passing through **1a** as an intermediate.

The **1b** formed as a result of the nucleophilic attack of $C_{60}CBr^-$ should have only the homofullerene cage enriched with ^{13}C carbon atoms. Therefore, no sp^3-sp^3 couplings should be observed in the ^{13}C-NMR. Given the fact that we observed sp^3-sp^3 couplings for the bridge structure of **1b** and considering that the conversion of **1b** to **1a** and back to **1b** is unlikely, we suggest that the mechanism involves a carbene attack by C_{61} on C_{60} followed by isomerization of **1a** to **1b**.

CONCLUSIONS

We present experimental data consistent with the fact that a C_{121} isomer **1b** with one homofullerene cage is more stable than the corresponding symmetrical isomers, **1a** and **1c**, with two methanofullerene cages or two homofullerene cages. These findings present the unique possibility to study of 58- vs 60 p-electron chromophores and their intramolecular interactions.

EXPERIMENTAL

^{13}C-NMR spectra were obtained with a Jeol JNM270 (1H-270 MHz) spectrometer in CS_2, 1,2-dichlorobenzene-d4 (ODCB), or 1-chloronaphthalene solutions with Varian spectrometers (1H frequencies of 300 and 500 MHz).

HPLC experiments were performed with preparative (20×250) and analytical Cosmosil Buckyprep columns with toluene as eluent, or with a Cosmosil 5PBB column (10×250 and 20×250) with chlorobenzene or ODCB as eluents. The HPLC experiments were performed with a preparative JAI-LC 908 System, an analytical Jasco instrument or with a diode-array MD1515-Jasco HPLC equipped with both gradient and recycling options.

THEORETICAL METHODS

All calculations were made using the Gaussian 98W package. Frequency analysis was made at STO-3G, NMR-GIAO calculations were made at the HF/6-31(d) level, and geometry optimizations and total energy calculations were made using the Becke-style 3 Parameter Density Functional Theory, using the Lee-Yang and Parr correlation functional, with a 6-31G(d) basis set. In the calculations, vibrations and rotations were treated as harmonic vibrations and rigid rotors, respectively. Vibrations were scaled for calculations of thermal contribution to enthalpy and entropy.

REFERENCES

1. Komatsu, K.; Wang, G. -W.; Murata, Y.; Tanaka, T.; Fujiwara, K. *J. Org. Chem.* **63**, 9358 (1998).
2. Patchkovskii, S.; Thiel, W. *J. Am. Chem. Soc.* **120**, 556 (1998).
3. Dragoe, N.; Tanibayashi, S.; Nakahara, K.; Nakao, S.; Shimotani, H.; Xiao, L.; Kitazawa, K.; Achiba, Y.; Kikuchi, K; Nojima, K. *Chem. Commun.* 85 (1999).
4. Osterodt, J.; Vogtle, F. *Chem. Commun.* 547 (1996).
5. Fabre, T. S.; Treleaven, W. L.; McCarley, T. D.; Newton, C. L.; Landry, R. M.; Saraiva, M. C.; Strongin, R. M. *J. Org. Chem.* **63**, 3522 (1998).
6. Details of the calculation will be reported later. Submitted to *J. Phys. Chem.*.
7. Dragoe, N.; Shimotani, H; Wang, J.; Iwaya, M.; de Bettencourt-Dias, A.; Balch, A.L.; Kitazawa, K. *J. Am. Chem. Soc.* (2001). To be published.

Photoinduced Polymerization in Crystalline C_{60} via Multi-Photoexcitation: Lattice-Relaxation and Energy-Transfer of Excitons

Masato Suzuki

*Department of Physics, Graduate School of Science,
Osaka City University, Sumiyosi-ku, Osaka, 558-8585, Japan*

Abstract. We numerically investigate adiabatic natures of the lattice-relaxation and the energy-transfer of photogenerated excitons in crystalline C_{60} so as to clarify the mechanism of the photoinduced polymerization processes via multi-photoexcitation in this material. In our theory, we deal with the π-electrons together with the interatomic effective potentials. Using a cluster model, we calculate the adiabatic potential energy surfaces of Frenkel and charge-transfer (CT) excitons individually, relevant to the photoinduced dimerization processes occurring in a face centered cubic crystal of C_{60}. The potential surfaces of the Frenkel excitons lead to the conclusion that structural defects are expected to exist at low temperatures even in the single crystal as an intrinsic property of this crystal. From the analysis of the potential surfaces of the CT excitons, it is confirmed that the CT exciton relaxes down to its self-trapped state, wherein the adjacent two molecules get close together. We also investigate the energy-transfer of the photogenerated excitons in the crystalline C_{60} wherein the defects and/or the dimer have already been produced by the previous photoexcitations, so as to clarify how the defects or the dimer can cause nucleus of the successive structural changes that are induced by the subsequent multi-photons.

INTRODUCTION

As is well known, the C_{60} crystal undergoes the polymerization above T_c =260 K by the optical excitations with visible or UV light [1]. In this polymer phase, it was proposed that the adjacent molecules are covalently linked by a four-membered ring joining the two molecular cages through the photochemical [2+2] cycloaddition reaction. Thus, the intermolecular bond changes from the van der Waals type to the covalent one as a consequence of the optical excitation.

In order to see effects of the optical excitations on the polymerization process, it is useful to classify the excitons into three types; free, Frenkel, and charge-transfer (CT) excitons, according to the distribution patterns of the electron-hole pair in the molecular crystal. It is considered that the interaction between the CT exciton and the lattice phonons is strong compared with the case of Frenkel exciton, and this interaction brings about the large intermolecular lattice relaxations of the CT excitons. This means that the CT followed by the lattice relaxation will be the key mechanism of photodimerization.

In the present paper, we investigate the adiabatic natures of the lattice relaxation and the energy-transfer of the photogenerated excitons in the crystalline C_{60}, so as to clarify the mechanisms of the photoinduced polymerization in this crystal.

MODEL HAMILTONIAN AND CALCULATION METHODS

In order to clarify the mechanism of the photodimerization processes in crystalline C_{60}, we investigate a many-π-electron system described by the following model Hamiltonian ($\equiv H$), wherein the elastic energies between carbons are taken into account by effective potentials. H is given as ($\hbar \equiv 1$)

$$H = H_{\text{intra}} + H_{\text{inter}}, \quad (1)$$

$$H_{\text{intra}} = \sum_{l,i,\sigma} \varepsilon_{li}\, a^\dagger_{li,\sigma} a_{li,\sigma} + \sum_{l,i>j,\sigma} (-T_{llij}\, a^\dagger_{li,\sigma} a_{lj,\sigma} + \text{H.c})$$

$$+ U \sum_{l,i} n_{li,\alpha} n_{li,\beta} + \sum_{l,i>j,\sigma,\sigma'} V_{llij}\, n_{li,\sigma} n_{lj,\sigma'} + \sum_{l,i>j} \omega_{llij}, \quad (2)$$

$$H_{\text{inter}} = \sum_{l>l',i,j,\sigma} (-T_{ll'ij}\, a^\dagger_{li,\sigma} a_{l'j,\sigma} + \text{H.c}) + \sum_{l>l',i,j,\sigma,\sigma'} V_{ll'ij}\, n_{li,\sigma} n_{l'j,\sigma'} + \sum_{l>l',i,j} \omega_{ll'ij}, \quad (3)$$

$$n_{li,\sigma} \equiv a^\dagger_{li,\sigma} a_{li,\sigma}. \quad (4)$$

H_{intra} denotes the Hamiltonian for the isolated molecules, and H_{inter} is the intermolecular interaction. l indicates the lth C_{60} molecule and l' is its nearest neighbor ones. i or j represents the carbon in each C_{60} molecule. $a^\dagger_{li,\sigma}$ ($a_{li,\sigma}$) is the creation (annihilation) operator of a π-electron with spin $\sigma(=\alpha,\beta)$ at ith carbon site in lth molecule. The first term in Eq.(2) means the site-diagonal part of the electron-phonon interaction, and ε_{li} is its interaction parameter. It is assumed to be inversely proportional to the distance ($\equiv r_{ll'ij}$) between lith carbon and other $l'j$th ones. T_{llij} or $T_{ll'ij}$ denote the resonance transfer integral of π-electrons between carbon atoms. Its dependency on $r_{ll'ij}$ is assumed to be an exponential function. U and V_{llij} ($V_{ll'ij}$) denote the intrasite and intersite Coulombic repulsive energies, respectively. For $V_{ll'ij}$, we use the Ohono potential. ω_{llij} and $\omega_{ll'ij}$ are the elastic energies between carbon sites, and are defined by the Lennard-Jones potential together with the $1/r_{ll'ij}$-type potential.

In the calculation of the ground state, we use the mean-field theory for interelectron interactions. Within this mean-field Hamiltonian ($\equiv H_{\text{HF}}$), $n_{li,\sigma}$ and $a^\dagger_{li,\sigma} a_{l'j,\sigma}$ are replaced by their averages, $<n_{li,\sigma}>$ and $<a^\dagger_{li,\sigma} a_{l'j,\sigma}>$, which are unknown parameters to be determined later self-consistently. In order to take the electron-hole correlation into account in the excited state, we take the expectation values of $H - H_{\text{HF}}$ within the basis of the one-electron excited states obtained from H_{HF}.

Let us determine the unknown parameters included in the model Hamiltonian so as to reproduce the well-known experimental results for the bulk crystal. The parameter values determined are summarized in Table I of our previous paper [2].

ADIABATIC POTENTIAL ENERGY SURFACES FOR PHOTODIMERIZATION

In order to calculate the potential energy surfaces relevant to the photoinduced dimerization process, we use a cluster which consists of 54 C_{60} molecules arranged on the fcc lattice points. Using the cluster, we define a configuration space with respect to the lattice distortion that expresses the structural changes accompanied with the [2+2] cycloaddition reaction in the crystal. Then, we calculate the adiabatic potential energy surfaces of the lowest singlet exciton so as to determine the final structure of the dimer in the crystalline C_{60}. After this calculation, we finally can get the most stable dimerized structure as summarized in Table II of Ref. 2.

Next, let us determine the relaxation path of the exciton which starts from the Franck-Condon state in fcc and terminates at the dimerized structure determined above. To this end, we connect these two states in the multidimensional configuration space by an extremal path that minimizes the energy barrier between them. This path can be uniquely determined. The adiabatic potential energy curves of the Frenkel and the CT excitons relevant to the dimerization process have been depicted in Figure 10 of Ref.2.

The overall feature of the adiabatic potential energy curves calculated here are quite characteristic in the following sense. The potential curves of the lowest singlet and triplet excitons, that is the Frenkel excitons, change approximately parallel to that of the ground state through the structural changes from the fcc to the dimer. Moreover, these curves are quite uneven with several energy minimum points. Therefore, these metastable structures are expected to exist as the structural defects with a relatively long lifetime at low temperatures even in the single crystal as an intrinsic property in such molecule crystals. From the potential curves of CT excitons, we can see that the CT exciton have a tendency to attract the adjacent two molecules immediately after the photoexcitation, and it will relax down to its self-trapped state. Because the two molecules close to each other as a result of the self-trapping, it is concluded that the CT between adjacent two molecules is one of the trigger mechanisms for the photoinduced dimerization.

Let us briefly mention how the dimerization is completed in the crystalline C_{60}. The energies of the fcc and the dimerized structure in the crystal are almost equal, and the energy barrier between them has been computed to be about 2.5 eV. So, it is difficult to dimerize through the relaxation of the relatively low-energy single exciton. Therefore, the multiphotons are needed to achieve the dimerization. That is, the dimerization will be completed when the energy of such metastable states as the defects exceeds the energy barrier by using the stepwise multi-photoexcitation.

ENERGY TRANSFER IN CRYSTAL WITH DEFECTS AND DIMER

Let us consider the possible mechanisms of the energy-transfer of the excitons in the crystalline C_{60} to the defects and/or the dimer which have already been produced by the previous photoexcitations, so as to clarify how the defects or the dimer can cause nucleus of the successive structural changes that are induced by the subsequent photons.

In general, in such molecular crystals as C_{60}, the exciton energy is considered to be transferred through the Förster mechanism or the exciton-transfer. However, the lowest singlet exciton in C_{60} molecule is dipole forbidden to the ground state so that the Förster mechanism is less possibly the mechanism of energy-transfer even in the crystal. Contrary to this, the excitons possibly hop to the other molecules through the weak resonance transfer integral between molecules. Hence, we hereafter focus only on the exciton-transfer as the mechanism of the energy-transfer in this crystal.

On the basis of the Fermi's golden rule, let us qualitatively compare the transition probabilities of exciton from the bulk to the defects or the dimer with the hopping probability of exciton in the single crystal, that is the bulk to bulk transition. From the local density of states (LDOS) of the Frenkel and the CT excitons created in the bulk, the three typical defects, and the dimer, we here calculate the overlap intensities of the LDOS between bulk and bulk ($\equiv I_0$), bulk and defects ($\equiv I_1$, I_2, and I_3), and bulk and dimer ($\equiv I_4$) for the cases of the Frenkel and the CT excitons. These overlaps approximately express the probabilities of the resonance transfer of the excitons between them.

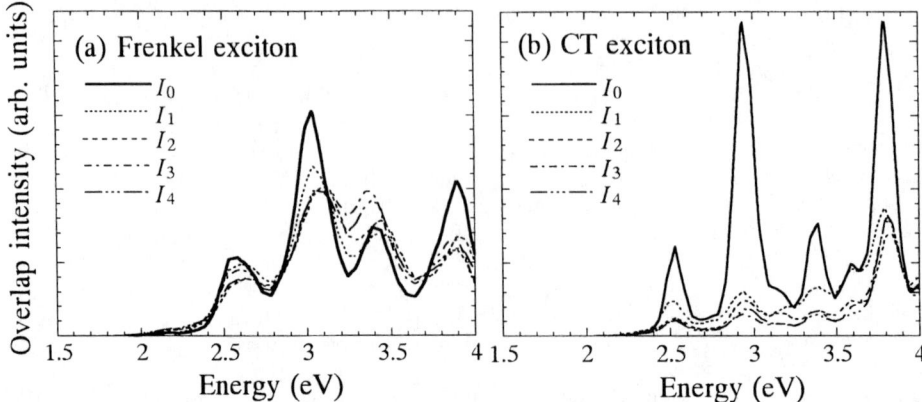

FIGURE 1. The overlap intensities of the local density of states between bulk and bulk ($\equiv I_0$), bulk and defects ($\equiv I_1$, I_2, and I_3), and bulk and dimer ($\equiv I_4$) for the cases of (a) Frenkel and (b) CT excitons.

The overlap intensities of LDOS are shown in Figs.1(a) and (b) as a function of exciton energy ; (a) is the case of the Frenkel exciton and (b) is the CT exciton. From these figures we can see the qualitative properties of the exciton-transfer as follows. It can be seen from Fig.1(a) that the overlap intensities I_1, I_2, I_3, and I_4 are higher than I_0 in the energy regions below 2.5 eV and around 3.1 to 3.7 eV. Therefore, in these energy regions, it is quite possible for the Frenkel excitons to transfer from the bulk to the defects or the dimer through the resonance interaction. Contrary to this, as shown in Fig.1(b) I_1, I_2, I_3, and I_4 has the intensities smaller than I_0. This means that the CT exciton in the bulk is less trapped resonantly in the defects or the dimer than the Frenkel exciton.

SPECULATION FOR PHOTOPOLYMERIZATION VIA MULTI-PHOTOEXCITATION

The followings are speculated from the above results. As seen in the previous sections, the CT exciton has a strong tendency to induce the local lattice distortion through the lattice relaxation of exciton, and then the defects or the dimer are newly created in the single crystal. On the other hand, the Frenkel exciton can not be expected to cause the lattice relaxation, and it has a strong tendency to transfer to the defects or the dimer. As the results of these processes, it is concluded that the exciton energies gather at the defects or the dimer, and the collected excess energy can induce the novel lattice relaxations around them. If this process is repeatedly occurred with the help of the multi-photoexcitations, finally the macroscopic structural changes will be induced. This is one of the microscopic mechanisms of the successive structural changes and the photoinduced structural phase transition due to the stepwise multi-photoexcitation in the crystalline C_{60}.

REFERENCES

1. Rao,A.M., Zhou,P.Z., Wang,K.-A., Hager,G.T., Holden,J.M., Wang,Y., Lee,W.-T., Bi,X.X., Eklund,P.C., Cornett,D.S., Duncan,M.A. and Amster,I.J., *Science* **259**, 955 (1993).
2. Suzuki,M., Iida,T. and Nasu,K., *Phys. Rev.* B **61**, 2188 (2000).

The Morphology of Vapor Grown C_{60} Crystals as an Ideal Example of the Gibbs-Wulff's Law

E. Schönherr[1], K. Matsumoto[2] and K. Murakami[2]

[1]*Max-Planck-Institut für Festkörperforschung*
Heisenbergstr. 1, D 70569 Stuttgart, Germany

[2]*Research Institute of Electronics, Shizuoka University,*
Johoku 3-5-1, Hamamatsu 432-8011, Japan

Abstract. C_{60} crystals grown from the vapor phase display only faces of a cubo-octahedron. The shape is derived from the Gibbs-Wulff's law by estimating the surface energies from the heat of evaporation and from considering the van der Waals interaction between first, second, third and fourth neighbored C_{60} molecules. In addition the critical size of a C_{60} nucleus formed on a fused silica wall is estimated from measured critical supercoolings.

INTRODUCTION

The fullerene C_{60} crystallizes by sublimation in the face centered cubic structure $Fm\bar{3}m$ with a lattice constant of a = 14.17 Å [1]. Up to now the observed crystal morphology contains only {100} and {111} faces [2-5].
Since the molecules are aligned along straight lines and the interactive forces are of pure van der Waals type, C_{60} crystal appears as an ideal model to test the Gibbs-Wulff's law.
For the present investigation the crystals were grown by the Pizzarello technique [4]. The Miller indices of the C_{60} crystals were determined either with a two-circle goniometer or with a scanning electron microscope [6]. The estimation of the critical supercooling for heterogeneous seed formation on fused silica is described in ref. [7]

RESULTS AND DISCUSSIONS

Equilibrium form

All investigated twin-free C_{60} crystals displayed only {100} and {111} faces. Nuclei formed in twinned positions on a (111) face led to additional {221} and {115} faces, which were original {100} and {111} faces respectively [4]. Since always both cubic and octahedral faces have been observed it is assumed that the equilibrium form is a cubo-octahedron.

Estimation of surface energies

It follows from the basic coordinates of the fcc structure that an inner molecule at the position (000) is surrounded by 12 first-nearest, 6 second-nearest, 24 third-nearest, 12 fourth-nearest neighbored molecules. In the case of solid C_{60}, the molecules are attracted by pure van der Waals forces [8]. When all neighbors up to the fourth are considered, the interaction energy of an inner molecule becomes

$$E_{int} = 12\varphi_1 + 6\varphi_2 + 24\varphi_3 + 12\varphi_4. \qquad (1)$$

where φ_n are the respective single van der Waals interaction energies. Assuming a Lennard-Jones law [8], the potentials φ_n approximately decrease with the 6th power of the distance $r_n = a\sqrt{n/2}$. In that case the potential ratio becomes $\varphi_n/\varphi_1 = n^{-3}$ and the single interaction energies follow from equ. (1) as $\varphi_n = (E_{int}/13.826)n^{-3}$.

Neglecting the weak force interaction of the C_{60} vapor molecules with the solid, the interaction energy assumes the heat of evaporation, i. e. $E_{int} = 173 \pm 1$ kJ/mol = $(2.87 \pm 0.02) \times 10^{-12}$ erg/molecule according to the vapor pressure measurements in refs. [9, 10].

As a consequence of pure van der Waals interaction, the energy of a crystal surface is given by the energy of missing interaction forces. The number N_b of the forces to the (000) molecule in the (100), (110) or (111) face and thus the number N_d of missing forces is found by representing the faces and planes in the Hess' normal form. The evaluated numbers of intermolecular forces and surface energies are listed in table 1. In the case of the (110) face, an additional free first neighbor interaction results from each molecule in the neighbored (220) plane. Thus the specific surface energies of the (100), (110) and (111) faces become

$$\sigma_{100} = 2/a^2(4\varphi_1 + \varphi_2 + 12\varphi_3 + 4\varphi_4) = 96.07 \text{ erg/cm}^2, \qquad (2a)$$
$$\sigma_{110} = 2/(\sqrt{2}\,a^2)(6\varphi_1 + 2\varphi_2 + 10\varphi_3 + 6\varphi_4) = 98.46 \text{ erg/cm}^2, \qquad (2b)$$
$$\sigma_{111} = 4/(\sqrt{3}\,a^2)(3\varphi_1 + 3\varphi_2 + 9\varphi_3 + 3\varphi_4) = 89.93 \text{ erg/cm}^2. \qquad (2c)$$

Since faces with high energy will not appear, the equilibrium shape will form either a cube, octahedron or a cubo-octahedron. In the last case the edge of the original octahedron may be a_o and a_c the length removed from a_o by creating a {100} face. The {111} and {100} faces have the distances $h_o = a_o/\sqrt{6}$ and $h_c = (a_o - a_c)/\sqrt{2}$ from a center point respectively. From the Gibbs-Wulff's law $\sigma_i/h_i = $ const. [11] the ratio

$$\frac{a_c}{a_o} = \beta = 1 - \frac{\sigma_{100}}{\sqrt{3}\sigma_{111}} \qquad (3)$$

is obtained. Only {111} faces will appear if a_c becomes zero, i. e. $\sigma_{111}=\sigma_{100}/\sqrt{3}$. In the case of an original cube with the edge b_c and the cut off length b_o resulting from a (111) face, the central point distances become $h_c=b_c/2$ and $h_o=(3b_c/2-b_o)/\sqrt{3}$. The Gibbs-Wulff's law leads then to the relation $b_o/b_c=(3\sigma_{100}-\sqrt{3}\sigma_{111})/(2\sigma_{100})$. Only {100} faces will appear if b_o becomes zero, i. e. $\sigma_{111}=3\sigma_{100}/\sqrt{3}$. Both faces will be present if $a_c>0$ and $b_o>0$, i. e. for $3/\sqrt{3}>\sigma_{111}/\sigma_{100}>1/\sqrt{3}$. The same relation (3) is derived from the condition of the minimum total energy of a cubo-octahedron in ref. [12]. Equations (2) and (3) lead to $a_c/a_o=0.383$, similar to the geometrical measurements of 0.3±0.1 on crystals.

According to the equations (2) the (111) face has the lowest surface energy. In the case of the Pizzarello technique this is confirmed by the observation that the initial contact plane of a C_{60} seed is a (111) face [4].

$\varphi/(10^{-15}\text{erg/molecule})$	A_0	A_1	A_2	A_3	N_b	N_d	N_s/a^2	$E_s/(\text{erg/cm}^2)$
$\varphi_1(100) = 208.2$	4	4	-	-	8	4	2	82.96
$\varphi_2(100) = 26.03$	4	-	1	-	5	1	2	2.59
$\varphi_3(100) = 7.712$	-	8	4	-	12	12	2	9.22
$\varphi_4(100) = 3.253$	4	-	4	-	8	4	2	1.30
$\varphi_1(110) = 208.2$	2	4	1	-	7	5	$2/\sqrt{2}$	73.33
$\varphi_1(220) = 208.2$						1	$2/\sqrt{2}$	14.67
$\varphi_2(110) = 26.03$	2	-	2	-	4	2	$2/\sqrt{2}$	3.67
$\varphi_3(110) = 7.712$	4	4	2	4	14	10	$2/\sqrt{2}$	5.43
$\varphi_4(110) = 3.253$	2	-	4	-	6	6	$2/\sqrt{2}$	1.37
$\varphi_1(111) = 208.2$	6	3	-	-	9	3	$4/\sqrt{3}$	71.84
$\varphi_2(111) = 26.03$	-	3	-	-	3	3	$4/\sqrt{3}$	8.98
$\varphi_3(111) = 7.712$	6	6	3	-	15	9	$4/\sqrt{3}$	7.98
$\varphi_4(111) = 3.253$	6	-	3	-	9	3	$4/\sqrt{3}$	1.12

Table 1. Number of interaction forces of a C_{60} molecule in the (100), (110) and (111) surface faces. A_0 is the number of intermolecular forces in the considered face, A_1, A_2 and A_3 that in the first, second and third neighbored plane, respectively. N_b and N_d are the numbers of interaction forces and missing forces per molecule. N_s is the number of molecules per a^2, where a is the lattice constant. E_s is the specific surface energy of the (hkℓ) face resulting from $N_sN_d\varphi_n(hk\ell)$, where φ_n is the energy per interaction force of the n-th nearest neighbors.

Estimation of critical sizes

When E_{ad} is the adhesive energy of the (111) face, the distance to the center point of that face changes from h_o to $h_{ad}=[(\sigma_{111}-E_{ad})/\sigma_{111}]a_o/\sqrt{6}$ according to Wulff's law. The (111) face increases from $F_{111}=\sqrt{3}(1-3\beta^2)a_o^2/4$ to $F_{ad}=F_{111}+3e\beta a_o+3f(1-2\beta)a_o+3f\sqrt{e^2-f^2}$ with $e=(h_o-h_{ad})\text{tg}(\alpha_1)$, $f=(h_o-h_{ad})\text{tg}(\alpha_2)$ where α_1 is the angle between [$\bar{1}2\bar{1}$] and [010] and α_2 that between [$1\bar{2}1$] and [$1\bar{1}1$]. The volume of the cubo-octahedron $V=\sqrt{2}(1-3\beta^3)a_o^3/3$ changes with $\Delta V=(h_o-h_{ad})(F_{111}+F_{ad}+\sqrt{F_{111}F_{ad}})/3$. The {100} face, i. e. $F_{100}=\beta^2 a_o^2$, and the {111} faces, which are adjoining to the (111) face decrease by $\Delta F_{100}=\beta a_o \sqrt{e^2+(h_o-h_{ad})^2}$ and $\Delta F_{111}=b[(1-2\beta)a_o+b\bullet tg(\alpha_3)/2]$ respectively, where b is given by $b=(h_o-h_{ad})/\cos(\alpha_2)$ and α_3 by the angle between [$\bar{1}12$] and [121].

The total Gibbs free energy of the heterogeneous seed with the contact (111) face results to

$$G_{tot}(a_o)=(-V+\Delta V)E_{cri}+(7F_{111}-3\Delta F_{111})\sigma_{111}+(6F_{100}-3\Delta F_{100})\sigma_{100}-F_{ad}(E_{ad}-\sigma_{111})$$

where $E_{cri}=(3.13\pm0.71)\times10^7$ erg/cm^3 is the critical free energy of heterogeneous seed formation on a fused silica wall [7]. The critical sizes are reached and the nucleus becomes stable when G_{tot} assumes a maximum. They are listed as a function of a prescribed adhesive energy for $E_{cri}=3.13\times10^7$ erg/cm^3 in Table 2.

$E_{ad}/10^{-5}$ erg/cm^2	$a_o^*/$ 10^{-6} cm	$h_o^*/$ 10^{-6} cm	$h_{ad}^*/$ 10^{-6} cm	$h_c^*/$ 10^{-6} cm	$V^*/$ 10^{-16} cm^3	$N^*/10^6$	$G_{max}/10^{-8}$ erg
80	1.50	6.12	0.68	6.54	7.62	1.11	1.20
40	1.43	5.82	3.23	6.21	9.53	1.39	1.49
0	1.41	5.76	-	6.15	10.98	1.60	1.71

Table 2. E_{ad} is the prescribed adhesive energy of the adhering (111) face. The asterisk * indicates terms, which represent critical parameters at the maximum total Gibbs free energy $G_{max}(E_{ad})$. a_o is the edge length of the original octahedron. h_o, h_{ad}, and h_c, are the central distances of the octahedral, adhering and cubic faces. V is the volume and N the number of C_{60} molecules of the cubo-octahedron.

REFERENCES

1. Dresselhaus, M. S., Dresselhaus, G., Eklund, P. C., *Science of Fullerenes and Nanotubes*, Academic Press, San Diego, 1995.
2. Verheijen, M. A., Meekes, H., Meijer, G., Raas, E., Bennema, P., *Chem. Phys. Letters* **191**, 339-344 (1992).
3. Verheijen, M. A.,van Enckevort, W. J. P., Meijer, G., *Chem. Phys. Letters* **216**, 72-80 (1993).
4. Matsumoto, K., Schönherr, E., Wojnowski, M., *J. Crystal Growth* **135**, 154-156 (1994).
5. Tachibana, M., Michiyama, M., Sakuma, H., Kikuchi, K., Achiba, Y., Kojima, K., *J.Crystal Growth* **166**, 883-887 (1996).
6. Schönherr, E., Winckler, E., *J. Crystal Growth* **36**, 353-354 (1976).
7. Schönherr, E., Matsumoto, K., Wojnowski, M., *J. Crystal Growth* **146**, 227-232 (1995).
8. Girifalco, L. A., *J. Phys. Chem.* **96**, 858-861 (1992).
9. Schönherr, E., Matsumoto, K., Freiberg, M., *Fullerene Sience and Technology* **7**, 455-466 (1999))
10. Schönherr, E., Matsumoto, K., Murakami, K.,,*Fullerenes 2000, Chemistry and Physics of Fullerenes and Carbon Nanomaterials, Proceedings of Electrochem Society* **10**, 89-99 (2000).
11. Wulff, G., *Zeit. Krist. Mineral.* **34**, 449-530 (1901).
12. Wells, A. F., *Phil. Mag.* 37, 605-630 (1946).

Synthesis and Crystal Structure of Cocrystallite with Silver Octaethylporphyrin and C_{70}

Tomohiko Ishii,* Ryo Kanehama, Naoko Aizawa, Masahiro Yamashita,*
Hiroyuki Matsuzaka, Hitoshi Miyasaka, Takeshi Kodama,
Kouichi Kikuchi, Isao Ikemoto, Yoshihiro Iwasa,[†] and Motoo Shiro[‡]

*Department of Chemistry, Tokyo Metropolitan University & PRESTO (JST),
1-1, Minamiohsawa, Hachioji, Tokyo 192-0397, Japan*
[†]*Department of Physical Materials Science, Japan Advanced Institute of Science and Technology,
1-1, Asahidai, Tatsunokuchi, Ishikawa 923-1292, Japan*
[‡]*Rigaku Co., 3-9-12, Matsubara-cho, Akishima, Tokyo 196-8666, Japan*

Abstract. A new cocrystallite, which contains C_{70} and silver complex of octaethylporphyrin (Ag^{II}(OEP)), is prepared and characterized. *Syn*-formed configuration of Ag^{II}(OEP) molecule in Ag^{II}(OEP)•C_{70}•$CHCl_3$ cocrystallite was observed, suggesting that there is the strong face-to-face π-π interaction between two adjacent silver porphyrin molecules.

INTRODUCTION

Fullerenes C_{60} and C_{70} are quite exceptional molecules because of their high symmetries (12 pentagons arrayed in a 'football'-like structure) and their large abundance and variety of solid state properties in intercalation compounds [1,2]. The fullerenes have attracted much attention, not only due to their unique molecular structures, but also due to their chemical properties. A variety of C_{60} and C_{70} cocrystallites have now been synthesized and amongst these, the 'host-guest' compounds of fullerenes and other molecules have been discovered [3,4].

It has been reported that the football-shaped fullerene molecules, C_{60} and C_{70}, are not appropriate for cocrystallization with planar molecules, since a curving of the planar molecule to match the concave structure would be required in order to fit to the ball-shaped C_{60} and C_{70} molecules [5,6]. In our previous work, the first examples of Pd and Cu complexes of *anti*-formed octaethylporphyrin cocrystallized with C_{60} were reported [7]. Syntheses, X-ray structural analyses and measurements of the magnetic properties of the cocrystallites of C_{60} with several kinds of metal octaethylporphyrins consisting of the *anti*-formed octaethylporphyrins have been carried out. From these results, it is supposed that the orientation of the eight terminal ethyl groups on the metal octaethylporphyrin is very important in order to keep the planar surfaced molecule in the cocrystallite despite the curved surfaced C_{60}. Recently, Ag [8] and Ni

---

* Authors to whom correspondence should be addressed: Telephone: +81-426-77-2549.
Fax: +81-426-77-2525. Email: Tomohiko Ishii mail@tishii.com. Masahiro Yamashita yamashit@comp.metro-u.ac.jp

[9] complexes of octaethylporphyrin cocrystallites with C_{60}, as examples of *anti*-formed metal complexes of octaethylporphyrins, were observed to form solids with a remarkably close contact between the curved π surface of the fullerene and the planar π surface of the porphyrin. Significant strong intermolecular interaction has also been observed in the cobalt complex of tetra(bis-*tert*-butylphenyl)porphyrin (TBP) cocrystallized with C_{60} [10]. However, no cocrystallites of any *anti*-formed metal porphyrin complexes with C_{70} have been synthesized up to now. It is well known that the lattice of a pristine metal porphyrin [11] is very flexible and that many kinds of constitutional isomers of cocrystallites with C_{70} are expected to be obtained.

One of our research targets is to obtain the *anti*-formed metal octaethylporphyrin complexes with C_{70}. As one of the candidates involving such a system, we focus on a silver octaethylporphyrin ($Ag^{II}(OEP)$). Here we report on the synthesis and the lattice structure of silver complexes of octaethylporphyrin cocrystallized with C_{70} in order to create a new *anti*-formed configuration of octaethylporphyrin complexes with C_{70} having the strong intermolecular interaction between the metal porphyrin and the C_{70} molecules.

EXPERIMENTAL, RESULTS AND DISCUSSION

The compound reported here was obtained in a form suitable for single-crystal X-ray diffraction by the diffusion of a solution of the fullerene C_{70} in benzene into a solution of the silver octaethylporphyrin in chloroform.

The Cocrystallite of Silver Octaethylporphyrin with C_{70}, $Ag^{II}(OEP) \cdot C_{70} \cdot CHCl_3$

We can obtain the new compound consisting of C_{70} with silver octaethylporphyrin. The centric unit cell of the $P\overline{1}$ space group in the compound consists of a C_{70} molecule, one silver octaethylporphyrin and a solvent molecules of chloroform, as shown in Fig. 1. Unfortunately, the C_{70} cage is disordered even if it is at 103 K in this compound. Within this unit cell the fullerene is positioned symmetrically between the two $Ag^{II}(OEP)$ units. The distance from the silver atom to this C-C bond is 3.01 Å. The Ag-C distances are Ag(1)•••C(40), 3.01(2) Å; Ag(1)•••C(41), 3.32(2) Å. These distances are short enough to interact between neighboring fullerene and porpnyrin.

The most important difference from the silver compounds with C_{60} is that the cocrystallite consists of the *syn*-formed octaethylporphyrin and C_{70}. These eight ethyl groups of the silver octaethylporphyrin portion lie on the same side of the porphyrin, toward the fullerene. This structural feature is very similar to those reported in $Co^{II}(OEP) \cdot C_{70} \cdot C_6H_6 \cdot CHCl_3$, $Ni^{II}(OEP) \cdot C_{70} \cdot C_6H_6 \cdot CHCl_3$ and $Cu^{II}(OEP) \cdot C_{70} \cdot C_6H_6 \cdot CHCl_3$ [12].

In addition to these fullerene/porphyrin interactions, there are significant porphyrin/porphyrin contacts with either pairwise or face-to-face contact. The arrangement is facilitated by the positioning all of the ethyl groups on the same side of the porphyrin from the adjacent porphyrin.

Within this unit cell, there is a molecule of silver octaethylporphyrin and a C_{70}, *i.e.* the composition of $Ag^{II}(OEP):C_{70}$ is 1:1 in this compound. Most of the reported cocrystallites of metal octaethylporphyrins with C_{60} whose composition is 2:1, *e.g.* $2Co^{II}(OEP) \cdot C_{60} \cdot CHCl_3$, $2Zn^{II}(OEP) \cdot C_{60} \cdot CHCl_3$ and $2Zn^{II}(OEP) \cdot C_{60} \cdot 2C_6H_6$, have the *syn*-formed configuration. In these compounds, a strong face-to-face interaction between two adjacent metal octaethylporphyrins exists. Such a strong face-to-face interaction between two adjacent porphyrin planes is also expected even if its composition of $Ag^{II}(OEP):C_{60}$ is 1:1.

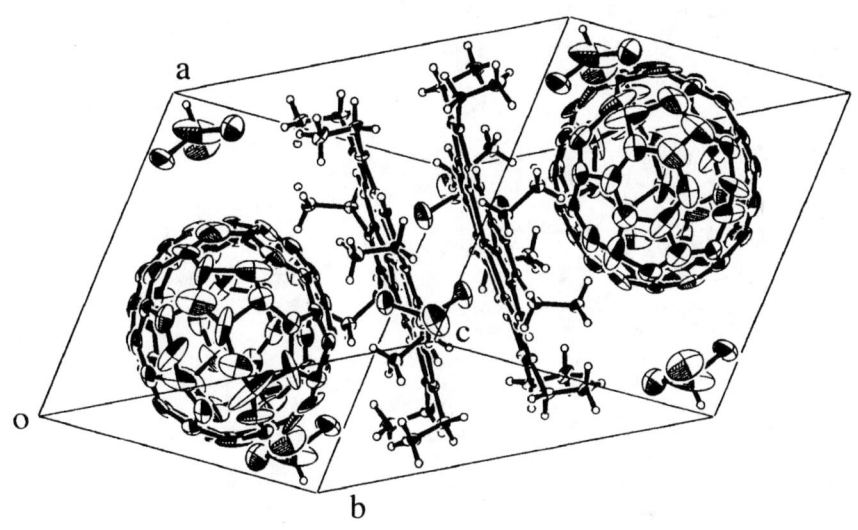

FIGURE 1. Molecular packing in $Ag^{II}(OEP) \cdot C_{70} \cdot CHCl_3$. P $\bar{1}$ (#2), $a = 14.4461(2)$, $b = 14.7561(4)$, $c = 18.9411(1)$ Å, $\alpha = 90.4099(4)$, $\beta = 87.4209(4)$, $\gamma = 87.4209(4)$ °, $V = 3541.5(1)$ Å^3, $Z = 2$.

It is well known that the lattice of a pristine metal octaethylporphyrin is very flexible. In some circumstances, the *anti*-formed configurations in the cases of $Cu^{II}(OEP)$ and $Ag^{II}(OEP)$ are more stable than the *syn*-formed configurations, which is predicted by the theoretical DV-Xα calculation [9]. After the prediction, only the cocrystallites containing of the *anti*-formed configuration of $Ag^{II}(OEP)$ can be obtained in $Ag^{II}(OEP) \cdot C_{60} \cdot C_6H_6$ [9]. However, the unexpected comfiguration of *syn*-formed $Ag^{II}(OEP)$ is observed in the cocrystallite of $Ag^{II}(OEP) \cdot C_{70} \cdot CHCl_3$. This result suggests that the face-to-face interaction between two adjacent $Ag^{II}(OEP)$ planes in the complex of $Ag^{II}(OEP) \cdot C_{70} \cdot CHCl_3$ is much stronger than that of $Ag^{II}(OEP) \cdot C_{60} \cdot C_6H_6$ cocrystallite. The molecular size of C_{70} is larger than that of C_{60}, which suggests that the chemical pressure of the $Ag^{II}(OEP)$ dimer in $Ag^{II}(OEP) \cdot C_{70} \cdot CHCl_3$ is expected to be larger than that in $Ag^{II}(OEP) \cdot C_{60} \cdot C_6H_6$. The C_{70} cage is disordered even at 103 K, implying that the chemical pressure would deepen hindered potentials, which makes C_{70} difficult to order. This intermolecular interaction is explained to be lower effectively due to the enlargement of the face-to-face interaction between porphyrin molecules in the complex of $Ag^{II}(OEP){\cdot}C_{70}{\cdot}CHCl_3$.

ACKNOWLEDGEMENTS

The authors are grateful to Profs. K. Sugiura, H. Imahori and Y. Sakata (Osaka Univ.) for their kind advice on the syntheses of metal octaethylporphyrins. This research was supported partly by a Grant-in-Aid for Scientific Research on Priority Area (Nos. 10149104 and 11165235), "Metal-Assembled Complexes" and "Fullerenes and Nanotubes Networks" from the Ministry of Education, Science and Culture, Japan.

REFERENCES

1. Hebard, A. F., Rosseinski, M. J., Haddon, R. C., Murphy, D. W., Glarum, S. H., Palstra, T. T. M., Ramirez, A. P., and Kortan, A. R., *Nature*, **350**, 600 (1991).
2. Allemand, P.-M., Khemani, K. C., Koch, A., Wudl, F., Holczer, K., Donovan, S., Gruner, G., and Thompson, J. D., *Science*, **253**, 301 (1991).
3. Reed, C., and Bolskar, R., *Chem. Rev.*, **100**, 1075 (2000).
4. Konarev, D., and Lyubovskaya, R., *Russ. Chem. Rev.*, **68**, 23 (1999).
5. Dyachenko, O. A., and Graja, A., *Fullerene Sci. Tech.*, **7(3)**, 317 (1999).
6. Hardie, M. J., and Raston, C. L., *Chem. Commun.*, 1153 (1999).
7. Ishii, T., Aizawa, N., Yamashita, M., Matsuzaka, H., Kodama, T., Kikuchi, K., Ikemoto, I., and Iwasa, Y., *J. Chem. Soc., Dalton Trans.*, **23**, 4407-4412 (2000).
8. Ishii, T., Aizawa, N., Yamashita, M., Matsuzaka, H., Ikemoto, I., Kikuchi, K., Kodama, K., and Iwasa, Y., *Synth. Met.*, (in press).
9. Ishii, T., Aizawa, N., Kanehama, R., Yamashita, M., Matsuzaka, H., Kodama, T., Kikuchi, K., and Ikemoto, I., *Inorg. Chim. Acta*, (in press).
10. Ishii, T., Kanehama, R., Aizawa, N., Yamashita, M., Matsuzaka, H., Miyasaka, H., Kodama, T., Kikuchi, K., and Ikemoto, I., *J. Am. Chem. Soc.*, (submitted).
11. Stolzenberg, A. M., Schussel, L. J., Summers, J. S., Foxman, B. M., and Petersen, J. L., *Inorg. Chem.*, **31**, 1678-1686 (1992). Pak, R., and Scheidt, W. R., *Acta Cryst.*, **C47**, 431-433 (1991). Cullen, D. L., and Meyer Jr., E. F., *Acta Cryst. Ser. B*, **32**, 2259 (1976). Senge, M. O., Forsyth, T. P., and Smith, K., *Z. Kristallogr.*, **211**, 176 (1996). Alexander, C. S., Rettig, S. J., and James, B. R., *Organomet.*, **13**, 2542 (1994). Hopf, F. R., O'Brien, T. P., Scheidt, W. R., and Whitten, D. G., *J. Am. Chem. Soc.*, **97**, 277 (1975).
12. Olmstead, M. M., Costa, D. A., Maitra, K. M., Noll, B. C., Phillips, L., Van Calcar, P. M., and Balch, A. L., *J. Am. Chem. Soc.*, **121**, 7090 (1999).

Photosensitized Oxygenation of Alkenes in the presence of bisazafullerene $(C_{59}N)_2$ and hydroazafullerene $C_{59}HN$

Nikos Tagmatarchis and Hisanori Shinohara[*]

Department of Chemistry, Nagoya University, Nagoya 464-8602, Japan

Abstract. Bisazafullerene $(C_{59}N)_2$ and hydroazafullerene $C_{59}HN$ photosensitize the reaction of alkenes with oxygen. 2-methyl 2-butene and α-terpinene undergo ene and Diels-Alder photooxygenation reactions, respectively, in the presence of minute amounts of azafullerenes to produce the corresponding peroxides.

One of the most spectacular discoveries in fullerenes research concerns with a simplest variant of C_{60} that can be envisaged by substituting one of the carbon atoms of the fullerene skeleton by a nitrogen atom. As a result of the different valencies of carbon and nitrogen, however, the resulting species is a radical. As such, it has been found to either rapidly dimerise yielding bisazafullerene $(C_{59}N)_2$ [1] or abstract a hydrogen atom forming the parent hydroazafullerene $C_{59}HN$ [2, 3]. The special molecular geometry of these materials and the presence of the electronegative nitrogen atom may be responsible for some unique photochemical and photophysical properties.

Not only the interesting mechanistic and synthetic aspects of singlet oxygen have been attracted considerable attention during the past years [4] but also its involvement to physiologically important oxidative damages and photodynamic damage of biological systems are among its numerous applications in medicinal chemistry. Therefore, we questioned whether they could act as photosensitizers, to promote the conversion of $^3O_2$ to $^1O_2$ upon UV irradiation [5]. *Ene* and *Diels-Alder* reactions are amongst the most widely used reactions of singlet oxygen with olefins. Hence, 2-methyl 2-butene and α- terpinene were subjected to heterofullerenes-sensitized photooxygenations.

A catalytic amount of the corresponding azafullerene (<10^{-3} % mol) dissolved in a benzene-d_6 solution was added to the alkene in an NMR tube. A stream of pure oxygen was passed through and the mixture was subsequently irradiated with a UV lamp equipped with a Kapton filter (cut-off wavelength <400nm) for 20 minutes. When hydroazafullerene $C_{59}HN$ was used as a sensitizer, the photooxygenation of 2-methyl 2-butene went to completion and afforded a 1:1 mixture of two allylic hydroperoxides **1** and **2** (Figure 1).

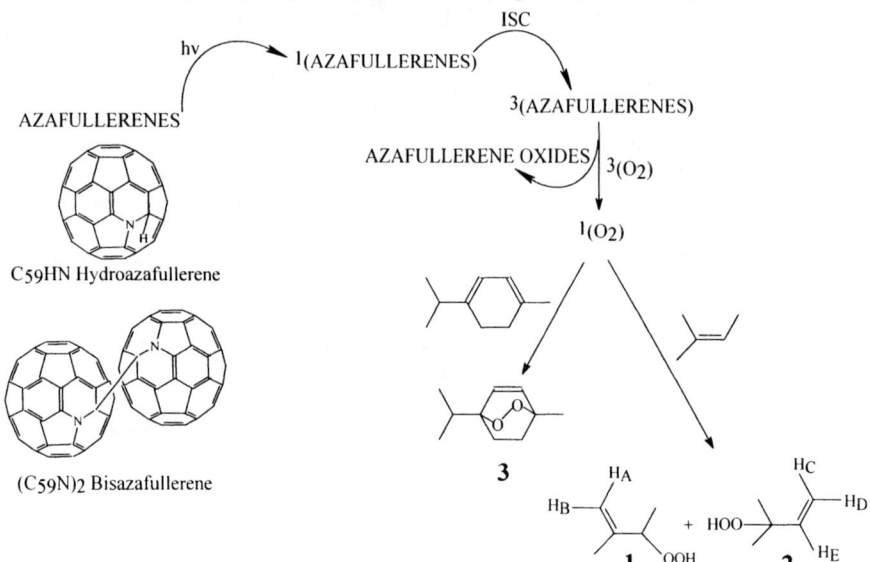

Figure 1. Schematic pathway for generation of singlet oxygen and reaction with the examined olefins

Similar results were obtained when bisazafullerene $(C_{59}N)_2$ was used as a photosensitizer. However, in the latter case, after irradiation the green reaction mixture faded its colour out and some insoluble materials appeared at the bottom of the tube. Under irradiation the weak interdimer bond of the bisazafullrene was disrupted and, in the presence of oxygen, finally oxidized. This was not surprising because we already knew that the C-C interdimer bond of $(C_{59}N)_2$ can easily be either photochemically or thermally cleaved. We have already found that hydroazafullerene is susceptible to oxygenation under ultraviolet irradiation [3]. In addition, when we performed the photooxygenations in the presence of a catalytic amount of 1,4-diazabicyclo[2.2.2]octane (DABCO), we hardly observed progress of the reaction and only after prolonged UV irradiation the formation of the corresponding hydroperoxides was identified in very low yields. This gives strong evidence that singlet oxygen is responsible for the photooxygenation reaction rather than some kind of azafullerenes – $^1(O_2)$ complex.

Following similar experimental procedures as described previously, we observed the formation of endoperoxide 3 (Figure 1) by using either the monomeric hydroazafullerene $C_{59}HN$ or the dimer bisazafullerene $(C_{59}N)_2$ as sensitizers for the photooxygenations of α- terpinene. However, Diels-Alder reaction between α-terpinene and oxygen did not proceed as smoothly as *ene* reaction of 2-methyl 2-butene.

The fact that both catalytic amounts of azafullerenes were needed and especially the presence of the fast completion of photooxygenation of 2-methyl 2-butene gives strong evidence that generation of singlet oxygen is high and similar to the amounts produced by the standard photooxygenation techniques especially with tetraphenylporphyrin and rosebengal. The turnover of the sensitizers is calculated to be more than 1000 for 2-methyl 2-butene and more than 100 for α-terpinene.

A representative pathway for the transformation of the examined alkenes to the corresponding peroxides **1, 2** and endoperoxide **3** by singlet oxygen. Hydroazafullerene $C_{59}HN$ or dimeric bisazafullerene $(C_{59}N)_2$ are electronically excited under UV irradiation to the singlet states where they undergo an intersystem crossing to the corresponding triplet states. The triplet states are then efficiently quenched by molecular oxygen and the so-produced singlet oxygen give *ene* and *Diels-Alder* reactions with the alkenes. Other related photochemical pathways for the electronically excited azafullerenes are presented in the energy level diagram in Figure 2.

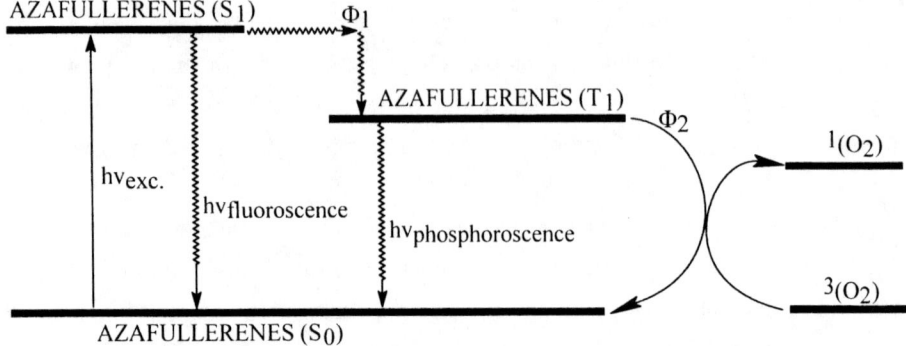

Figure 2. Energy level diagram for azafullerenes and their interaction with oxygen

The photosensitization of alkenes with singlet oxygen by hydroazafullerene $C_{59}HN$ is the first example of the use of this unique material in organic chemistry. Together with its dimeric derivative, $(C_{59}N)_2$, can be considered as useful and powerful photosensitizers. The unique spherical structures with the electronegative nitrogen atom incorporated into the conjugated fullerene skeleton differentiate them from conventional sensitizers which usually have flat-shaped conjugated networks. In addition, further functionalisation on their skeleton should lead to produce more soluble materials in a wide variety of solvents - including water - and thus may have important pharmaceutical applications. This would open a new interdisciplinary research in the development and construction of photosensitive material architectures.

ACKNOWLEDGEMENTS

N.T. thanks the Japan Society for the Promotion of Science (JSPS) for a Post Doctoral Fellowship for Foreigner Researchers. H.S. thanks JSPS for the Future Program on New Carbon Nano-materials.

REFERENCES

1. Hummelen, J. C.; Knight, B.; Pavlovich, J.; Gonzalez, R.; Wudl, F. *Science* **269**, 1554-1556 (1995).
2. Keshavarz-K, M.; Gonzalez, R.; Hicks, R. G.; Srdanov, G.; Srdanov, V. I.; Collins, T. G.; Hummelen, J. C.; Bellavia-Lund, C.; Pavlovich, J.; Wudl, F. *Nature* **383**, 147-150 (1996).
3. Tagmatarchis, N.; Shinohara, H.; Pichler, T.; Krause, M.; Kuzmany, H. *J. Chem. Soc. Perkin Trans. 2* 2361-2362 (2000).
4. Stratakis, M; Orfanopoulos, M. *Tetrahedron* 1595-1614 (2000).
5. Stephenson, L. M.; Grdina, M. J.; Orfanopoulos, M. *Acc. Chem. Res.* **13**, 419-425 (1980).

Transition of the heterofullerene $(C_{59}N)X$ to the monomeric phase of $C_{59}N$.

W. Plank[1], T. Pichler[1,2], S. Baes-Fischlmair[1], M. Krause[1], H. Kuzmany[1], N. Tagmatarchis[3], H. Shinohara[3]

[1] *Institut für Materialphysik der Universität Wien, Strudlhofg. 4, A-01090 Wien,*
[2] *Institut für Festkörper- und Werkstofforschung Dresden, Postfach 270016, D-01171 Dresden.*
[3] *Department of Chemistry, Nagoya University, Nagoya 464-8602, Japan*

Abstract.
The stability of the azafullerenes of the type $(C_{59}N)X$ was studied by using infrared, optical and Raman experiments. X represents either hydrogen or an other cage of $C_{59}N$. The solid phase of the dimer was shown to be stable up to 650 K followed by very slow degradation extending beyond 700 K. This is compared to our very latest results on the temperature stability of $C_{59}HN$ which was found to be stable only up to 540 K. A sudden change in the spectra at this temperature gives evidence for a transition to a new air stable phase which is claimed to be monomeric $C_{59}N$. When heated in vacuum to about 700 K this phase dimerizes into $(C_{59}N)_2$.

INTRODUCTION

This paper focuses on on ball doped fullerenes, namely the azafullerenes. Since the extra electron from the nitrogen should partly fill the empty t_{1u} derived band, the azafullerenes were expected to be molecular metals. But early experiments and quantum chemical calculations indicated that $C_{59}N$ is very unstable and dimerizes to $(C_{59}N)_2$ or reacts with hydrogen to $C_{59}HN$ [1]. We recently showed that the dimerized azafullerene is thermally stable up to 650 K [2]. At higher temperatures response from amorphous carbon appears in the spectra and indicates the partial breaking up of the cages but well expressed response of the dimer is still available up to 750 K [3]. We have performed similar experiments on $C_{59}HN$ and summarize the results together with a comparison with results on $(C_{59}N)_2$. This is followed by a characterisation of monomeric $C_{59}N$ by IR and Raman spectroscopy and some outcomes on its stability.

EXPERIMENTAL

The $(C_{59}N)_2$ and $C_{59}HN$ samples used in this study were prepared by a chemical route as described previously [4]. The material was either pressed into a pellet or drop coated on a gold coated silicon wafer and subsequently vacuum dried at 370 K. All temperature dependent Raman and IR studies were performed *in situ* in a vacuum better than 10^{-4} Pa. The samples were placed in purpose built sample chambers which allow heating to more than 800 K. The IR meassurements were performed with a Bruker 66 V spectrometer. A Dilor xy spectrometer with a liquid nitrogen cooled CCD was used for the detection of the Raman spectra.

RESULTS AND DISCUSSION

The substitution of a carbon atom with nitrogen decreases the symmetry of the cage. Symmetry analysis specifies C_{2h} for the dimer $(C_{59}N)_2$ and C_s for the monomers $C_{59}HN$ and $C_{59}N$. These differences in symmetry induce corresponding differences in IR and Raman spectra. The infrared spectrum of C_{60} is well understood, with only four IR active modes of symmetry F_{1u} at 527, 577, 1182 and 1428 cm^{-1}, respectively.

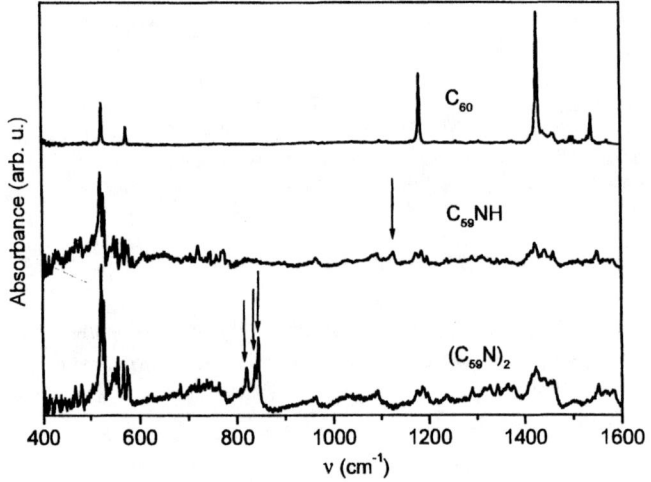

FIGURE 1. IR absorption spectra of C_{60} compared to the azafullerenes C_{60}, $C_{59}HN$ and $(C_{59}N)_2$.

Compared to C_{60} the spectra of the dimer and the monomers are by much richer since all 174 vibrational degrees of freedom from a C_{60} derived cage are Raman as well as IR allowed. This is clearly illustrated in Fig. 1 by the comparison of IR spectra of C_{60}, $C_{59}HN$ and $(C_{59}N)_2$. In the azafullerenes the IR active modes with F_{1u} symmetry are splitt into three components with strong IR symmetry. As a consequence of different molecular structures the mode positions are strongly dependent on the type of azafullerene. It was our motivation to use these differences for the identification of characteristic modes which allow the definitive identification of $C_{59}HN$ and $(C_{59}N)_2$. For $(C_{59}N)_2$ three additional absorption bands are observed at about 820-850 cm^{-1}. These are characteristic stretching modes of C-N and C-C bonds on the pentagonal rings. For C$_{59}$HN the bending mode of the C-H vibration at about 1125 cm^{-1} was used as a characteristic fingerprint of the compound. The stability of C$_{59}$HN was analyzed as a function of temperature by a series of *in situ* heating experiments for temperatures between room temperature and 700 K.

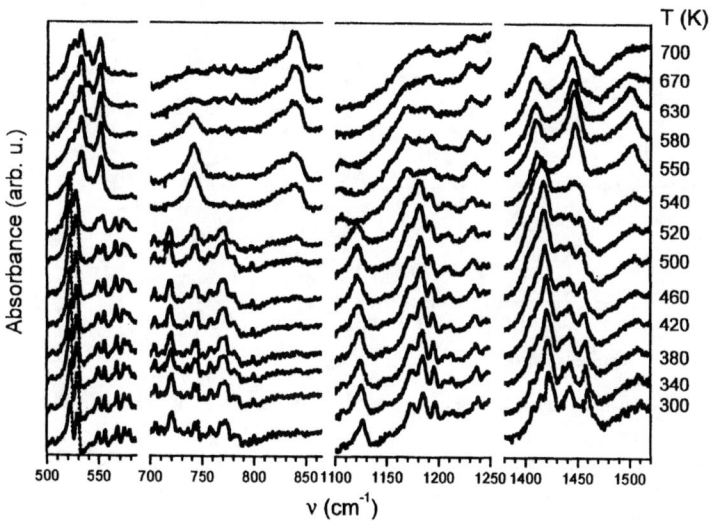

FIGURE 2. Temperature dependence of the IR response of C$_{59}$HN shown for characteristic spectral regions.

At each temperature the sample was equilibrated for one hour. The resulting IR absorption spectra are depicted in Fig. 2. Up to about 540 K the overall spectral shape remains the same indicating that C$_{59}$HN in the solid state and under vacuum conditions is stable up to this temperature. At 540 K the spectra change dramatically. The H$_g$(3,4) and the G$_g$(3,4) derived bands between 700 cm^{-1} and 800 cm^{-1} and the characteristical C-H band at 1125 cm^{-1} vanish and new bands appear at 745 cm^{-1}, 1452 cm^{-1} and 1505 cm^{-1}, as well as a broad structure at 820-850 cm^{-1}. At about 650 K the peaks at 745 cm^{-1} and 1505 cm^{-1} start diminishing while the broad peak at 840 cm^{-1} further increases. The observed changes in the IR absorption are interpreted as a consequence of an irreversible phase transformation to the monomeric azafullerene C$_{59}$N. Further changes in the spectra at temperatures close to 700 K can be understood as a transition to dimeric (C$_{59}$N)$_2$. Above 700 K the azafullerene starts to evaporate. The C$_{59}$N monomer generated at 540 K prooved to be stable when the temperature was reduced to 300 K exposed on air.

Fig. 3 depicts Raman spectra of C$_{59}$N together with spectra from C$_{59}$HN and (C$_{59}$N)$_2$. The splitting of the modes of C$_{59}$N is less dramatic as compared to the other azafullerenes. Characteristic peaks of the dimer like the mode at 625 cm^{-1}

miss in the case of the monomer and the radial breathing mode $A_g(1)$ at 490 cm^{-1} is strongly suppressed. The $A_g(2)$ pentagonal pinch mode is downshifted relative to C_{60} by 10 cm^{-1} to 1457.5 cm^{-1}.

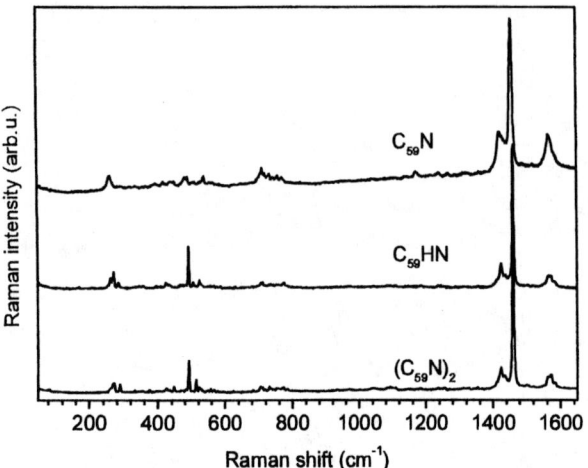

FIGURE 3. Overall Raman spectra of the three azafullerenes meassured with 514.5 nm excitation.

Additionally the $A_g(2)$ pinch mode of $C_{59}N$ is about twice as broad as for $C_{59}HN$ and $(C_{59}N)_2$. Such a broadening of the pinch mode has been observed for metallic fullerenes due to the presence of free charge carriers [6]. However, no strong absorption related to a charge carrier plasmon was found down to the mid IR region and we are not able to conclude from this experiments if $C_{59}N$ is a metal.

We thank the European Union for funding within the TMR Research Network 'FULPROP' (ERBFMRXCRT-970155). T. P. thanks the Austrian academy of sciences for an APART grant.

REFERENCES

1. M. Keshavarz-K, R. Gonzalez, R.G. Hicks, G. Srdanov, V.I. Srdanov, T.G. Collins, J.C. Hummelen, C. Bellavia-Lund, J. Pavlovich, F. Wudl, K. Holczer, Nature 383, 147 (1996).
2. H. Kuzmany, W. Plank, J. Winter, O. Dubay, N. Tagmatarchis, K. Prassides, Phys. Rev. B60, 1005 (1999).
3. M. Krause et al., submitted to J. Chem. Phys.
4. N. Tagmatarchis, H. Shinohara, T. Pichler, M. Krause, H. Kuzmany, J. Chem. Soc., Perkins Trans. 2, 2361 (2000).
5. W. Plank, T. Pichler, H. Kuzmany, O. Dubay, N. Tagnatarchis, K. Prassides, Eur. Phys. J. **17**, 33 (2000).
6. J. Winter, H. Kuzmany, Phys. Rev. B **52**, 7115 (1995).

Survey of Natural Fullerenes in Southwestern China

Eiji Ôsawa[*], Masaki Ozawa[*], Kazutoyo Chijiwa[†], Kouichi Hoyanagi[¶], Kazuyoshi Tanaka[#], and Michiko Kusunoki[§]

[*]*Department of Knowledge-based Information Engineering, Toyohashi University of Technology, Toyohashi 441-8580 Japan*
[†]*Department of Geology, Faculty of Science Education, Yamaguchi University, Yamaguchi 753-8513 Japan*
[¶]*Department of Geological Science, Shinshu University, Matsumoto 390-8621 Japan*
[#]*Department of Molecular Engineering, Kyoto University, Kyoto 606-8501 Japan*
[§]*Japan Fine Ceramics Center, FCT Central Research Department, Nagoya 456-8587 Japan*

Abstract. Unusually high concentration of C_{60}/C_{70} was discovered in a coal sample from China. Extensive analysis including TEM analysis suggests a surprising occurrence of the fullerenes in nature by shock heating, which we suggest to have been generated by coal gas explosion in the underground.

INTRODUCTION

Research in natural fullerenes is currently under some controversy since the danger of relying on the results of HPLC analysis alone has been clearly demonstrated [1]. Fullerenes claimed to exist in K-T boundary are now suspected in view the well-recognized sensitivity of C_{60} to UV light and air [2]. In this regard, recent reports on the unusually high contents of C_{60}/C_{70} (up to 0.1%) in coal from Yipinglang Mine, Lufeng, Yunnan, China [3,4] attracted our attention. Results of fullerene analysis at this level of concentration can be usually relied upon with high confidence, and the strongly reductive and anaerobic environments as well as the mild temperature and pressure during the coal formation process would provide ideal conditions to preserve fullerenes over the geological time span. On the other hand, the mild coalification conditions preclude the formation of fullerenes by any of the known mechanism [5].

We obtained the Yunnan coal sample (to be called *K1bE94* hereafter) from the discoverer, and soon confirmed as high as 30 ppm of C_{60}/C_{70} by HPLC analysis. The contents are considerably lower than reported, but still several orders of magnitude higher than the previously reported levels of other natural fullerenes (sub-ppm order) [6]. Direct LD-TOF-MS analysis of *K1bE94* coal supported the identity of fullerenes. At this point, it was clear that we encountered an interesting case of natural fullerene; hence we immediately set out to rationalize efficient fullerene formation in solid phase under mild conditions. In this presentation, we briefly outline how the origin of C_{60}/C_{70} Chinese coal was identified.

RESULTS

Combustion analysis of *K1bE94* did not indicate any unique feature, but only revealed it to be a typical bituminous coal with intermediate degree of carbonization: fixed carbon 68%, fuel ratio 3, heat of combustion 32 kJ/g with Si, Al, Fe and S as the major contaminant elements. More disappointing was highly local nature of the discovery. Namely the fullerene-containing coal sample was discovered only at one spot in a layer called *K1b*(*east*). Several dozens of coal samples taken from this and also from all other four strata in Yipinglang mine did not contain fullerenes. Nor about 100 other coal samples collected from various places of the world showed any significant amounts of fullerene upon HPLC analysis. Although the number of analysis is still relatively small compared to the great abundance of coal on the earth, we are inclined to assume that coal, as a rule, does not contain fullerenes.

To our surprise, however, TEM examination of *K1bE94* sample disclosed extraordinary microstructures in sharp contrast to the uncharacteristic amorphous structure of coal in general. The novel structures are identical with those of the smallest known carbon blacks, namely soot. Very small diameters (ca 10 nm) of the primary particles in the soot-like microstructure of *K1bE94* sample suggest that they have been exposed to about 2000 °C for milliseconds. It is difficult to think of any other cause than the shock heating to explain this mode of heating. Shock wave is known to occur in nature by thunder strike, meteorite impact, earthquake, tsunami, or volcanic eruption, and to generate a thin zone of high-temperature/high-pressure when passing through solid medium at supersonic velocity [7]. It is likely that the shock heating also responsible for the formation of C_{60}/C_{70}. Now our task is to identify the possible cause of shock wave that left the novel signatures in *K1bE94* coal.

We were first intrigued by local information on a crater-like geological formation near Yipinglang mine [8]. While direct fall of a sizable meteorite on coal field area is unlikely to leave combustible layers intact, posterior formation of coal from the shocked plants or other carbonaceous materials cannot be excluded. This faint possibility was examined by analyzing iridium, platinum and gold: meteorites are known to contain unusually high abundance of iridium relative to the other metals of platinum group [9]. Neutron activation analysis at the Atomic Pile JRR-3M of six coal samples collected near Yipinglang mine including *K1bE94* revealed no significant increase in iridium (precision: 0.4 ppb).

In the meantime two kinds of mineral crystals consisting of (K, Al, Si, O) and (Cu, Fe, S, O), respectively, have been detected and analyzed by electron diffraction analysis of *K1bE94* in TEM. However, the lattice constants and interlayer distances observed for these crystals do not match with any of those of the known coal minerals with the same or close elemental compositions such as allophane, metahaloysite, kaolinite, metakaolin, maskelynite and guildite. These features as well as some similarity of the observed compositions with mikasaite and melanterite suggest that these minerals are the shock-frozen and unknown polymorphs of aluminosilicate and sulfate minerals transformed under high-pressure high-temperature conditions of shock heating.

DISCUSSION

Metamorphoses of carbon phases by shock are well known, the best example being that of graphite into diamond by detonation [10-12]. If the heat is quickly dissipated while the pressure is still high, diamond does not equilibrate back to graphite. However, while the diamond formation is a thermodynamically driven change, phase transition will stop at the intermediate stage like fullerenes only when the shock was insufficient or suspended in the middle of metamorphosis. Interestingly enough, quite a few examples of shock transformation of some form of carbon into fullerenes have been reported. Wang and Cadman [13] carried out shock tube experiments by filling Ar gas containing 1% of oxygen in the low-pressure compartment, and filling He gas in the high-pressure compartment, and pulse-injecting benzene droplets in the former synchronously to a pulse current that bursts open a plastic diaphragm between the two compartments. Shock wave thus produced heated inside the tube up to 2400 K in 3.5 ms of combustion time to give soot, which contained 300 ppm of C_{60}. Novgorodova discovered novel tetragonal crystals of C_{60} in graphite taken from a deep-seated (25 km below the earth surface) xenolith, which is graphitized and phlogopitized spinelic peridotite, in a basaltoid pipe in Tajikistan [14]. She interprets the metamorphosis to have resulted from heat explosion during shear flow of graphite, a consequence of strong shock wave that passed through deep underground. Not only C_{60}/C_{70} but also multi-shell fullerenes, α-carbyne, p-diamond and other carbon forms have been observed by detonation of explosives in shock-generating devices in the presence of graphite, acetylene black, and silicon carbide [15].

Based on all the circumstantial evidence mentioned above, we propose that the most likely cause of shock heating that had occurred in *K1bE94* coal sample to form C_{60}/C_{70} and soot-like microstructure, is the *explosion of coal gas*. Coalfields of Southwestern China covering Provinces of Sichuan, Yunnan, and Guizhou lie above thick and gigantic basaltic layer and are rich in coal gas due to the thermal effects from below. In these areas, thunder strike or wildfire may have occasionally ignited coal gas at the exposed sites and led to underground explosion in concave places of coal strata. Such an explosion could potentially generate strong shock wave along the inside corridor of cave, thus causing intense but microscopic heat explosion through the shearing of coal surface.

The gas explosion hypothesis presented above fits well with the highly localized formation of soot-like microstructure and C_{60}/C_{70} observed for *K1bE94* coal as follows.
(1) These two products are shown to closely relate each other. The primary particles of soot is now believed to be highly defective onion-carbon where each shell of onion is defective giant fullerenes with C_{60} molecule as the smallest fullerene core [16-18].
(2) The shock metamorphosis of coal carbon into fullerenes by gas explosion will occur exclusively on the surface of coal layer. Hence, C_{60}/C_{70} is expected to evaporate or decompose quickly after excavation of coal under exposure to air

and light. Sharp decrease in the fullerene concentration was actually observed.
(3) 'Partial combustion' only at or near the surface of coal particle will not significantly affect the heat content of coal. The measured heat of combustion of *K1bE94* coal indeed did not show any change from the standard value of bituminous coal.

CONCLUSIONS

We suggest that the high abundance of natural fullerenes found in China is produced by shock heating from coal gas explosion. This route of fullerene formation has been sporadically observed in laboratories, but may be worthwhile to consider as a potential method of industrial production.

ACKNOWLEDGMENTS

We thank financial supports from Ministry of Education, Culture, Sports, Science and Technology through Grants-in-Aid for Scientific Research on Priority Areas (A, No. 11165222) and on Overseas Survey (B, No. 11691149). Professor Daniel Eylon, University of Dayton, first suggested me the methane gas explosion as a likely cause of underground fullerene formation. We are also indebted to Dr. Pao-Hsien Fang for the donation of coal samples and encouragements. Marubeni Corporation, China National Coal Industrial Department, Yunnan Coal Administration Bureau, and Yunnan Coal-field Geology Bureau, provided us with invaluable help during the survey tours carried out in November 1999 and October 2000. Professor K. Kondo and Dr. M. I. Novgorodova offered crucial instructions.

REFERENCES

(1) Taylor, R., and AbdulSada, A. K. *Fullerene Sci. Technol.*, **8**, 47-54 (2000). (2) Ôsawa, E. *Fullerene Sci. Technol.*, **7**, 637-652 (1999). (3) Fang, H.-P., Zhou, X., Tao, R., Wang, Q., Mu, C., and Wu, X. *Innov. Mat. Res.*, **1**, 129-134 (1996). (4) Fang, H.-P., and Wong, R. *Mat. Res. Innov.*, **1**, 130-132 (1997). (5) Kadish, K. M., and Ruoff, R. S. (eds.), *Fullerenes: Chemistry, Physics, and Technology*, John Wiley & Sons, Inc., New York, 2000. (6) Ôsawa, E., Slanina, Z., Ozawa, M., Zhao, X., and Saunders, M. "A Catalytic Mechanism of Fullerene Formation in Coal", in *Symposium on Recent Advances in the Chemistry and Physics of Fullerenes and Related Materials-1998*, edited by R. S. Ruoff, and K. M. Kadish, Electrochemical Society Proceedings, 99-12, Pennington, N. J., 1999, pp. 701-710. (7) Serre, D., *Hyperbolicity, Entropies, Shock Waves*, Cambridge University Press, Cambridge, 1999. (8) Private communication from Mr. Xiaoming Li, Yunnan Observatory, Chinese Academy of Sciences. (9) Schuraaytz, B. C. et al., Science, 271, 1573-1576 (1996). (10) Donnet, J.-B., Fousson, E., Samirant, M., Wang, T. K., Pontier-Johnson, M., and Eckhardt, A., *C. R. Acad. Sci. Paris, Chimie*, **3**, 359-364 (2000). (11) Hirai, H.; Kukino, S., and Kondo, K. *J. Appl. Phys.*, **78**, 3052-3059 (1995). (12) Lakoubovskii, K., Adraienssens, G. J., Meykens, K., Nesladek, M., Vul', A. Y., and Osipov, V. Y. *Diamond and Related Materials*, **8**, 1476-1479 (1999). (13) Wang, R., and Cadman, P. *Fullerene Sci. Technol.*, **3**, 553-563 (1995). (14) Novgorodova, M. I, *Geochem. Int.*, **37**, 896-904 (1999). (15) Yamada, K., Tanabe, Y., and Sawaoka, A. B. *Phil. Mag. A*, **80**, 1811-1828 (2000). (16) Kroto, H. W., and McKay, K., *Nature*, **331**, 328-331 (1988). (17) Iijima, S. *J. Phys. Chem.*, **91**, 3466-3467 (1987). (18) Ozawa, M.; Ôsawa, E., and Gotô, H. accompanying paper.

Effect of Chemical Treatment on the Structure of Ultradisperse Diamond and Onion-Like Carbon

A.E.Alexenskii [a], M.V.Baidakova [a], A.T.Dideikin [a], V.Yu.Osipov [a], E.Osawa [b], M.Ozawa [b], A.I. Shames [c], V.I.Siklitsky [a], A.Ya.Vul' [a,b]

[a] *Ioffe Physico-Technical Institute, Polytecnicheskaya 26, St.Petersburg, 194021Russia*
[b] *Toyohashi University of Technology, Tempaku-cho,Toyohashi,441-8580 Japan*
[c] *Department of Physics, Ben-Gurion University of the Negev, P.O.Box 653, 84 105 Be'er-Sheva, Israel*

Abstract. The effect of extraction conditions and a post-treatment in hydrogen atmosphere at high temperatures (HTT) on the surface structure and purity of ultradisperse diamond (UDD) have been studied by means of X-ray diffraction, small angle X-ray scattering, HRTEM and EPR.

INTRODUCTION

Ultradisperse diamond (UDD) is one of the few carbon products that can be produced in large amounts by the detonation method [1, 2]. It has been recently shown that UDD consists of crystalline diamond core of about 45 Å covered with a surface structure from sp^2-hybridized carbon atoms. The surface structure contains onion-like inner shell, which nearly encloses the core and piles of small graphene sheets [3, 4]. Here we studied the effect of extraction conditions and a post-treatment (HTT) in hydrogen atmosphere at high temperatures on the surface structure and purity of UDD by means of X-ray diffraction, small angle X-ray scattering, HRTEM and EPR.

EXPERIMENTAL

The study was done on UDD samples obtained from carbon contained in explosives, a mixture of TNT and hexogen (TNT/hexogene = 60/40), by the detonation technique. The pressure and temperature in the shock wave were within the region of thermodynamic stability of diamond ($P \geq 10$ GPa, $T \geq 3000$ K) [5]. The diamond phase was extracted by treating the carbon powder (detonation carbon) with nitric acid in an autoclave. The degree of the removal of the nondiamond phase was governed by the temperature of the treatment. After the reaction had come to an end and the material had been taken out of the autoclave, it was treated with distilled water until pH = 7 was reached. The UDD samples prepared in this way were identical in all parameters, except the amount of the amorphous phase coating the diamond core. The post-treatment was done by the annealing of UDD samples in a hydrogen flow at temperatures varied within the 720-1400 K for three hours.

X-ray diffraction measurements and small angle X-ray scattering were carried out on a RIGAKU diffractometer with copper radiation [3,4]. Room temperature (T = 297 K) EPR spectra were obtained using a Bruker EMX-220 digital X-band (9.4 GHz). Spectra were recorded at non-saturating microwave power of 200 µW and 100 kHz magnetic field modulation of 0.2 mT and 0.02 mT.

RESULTS and DISCUSSION

Fig. 1 shows HRTEM images of the detonation carbon (the starting sample of explosion produced carbon, before the treatment with nitric acid) and of the UDD powder after strong chemical cleaning. Fig.1 *a* displays the onion-like carbon characterized by partly destroyed graphitic sheets with the absence of diamond lattice observed. On the other hand, the UDD particles clearly reveal the diamond core covered with three shells of the onion-like carbon (Fig. 1b).

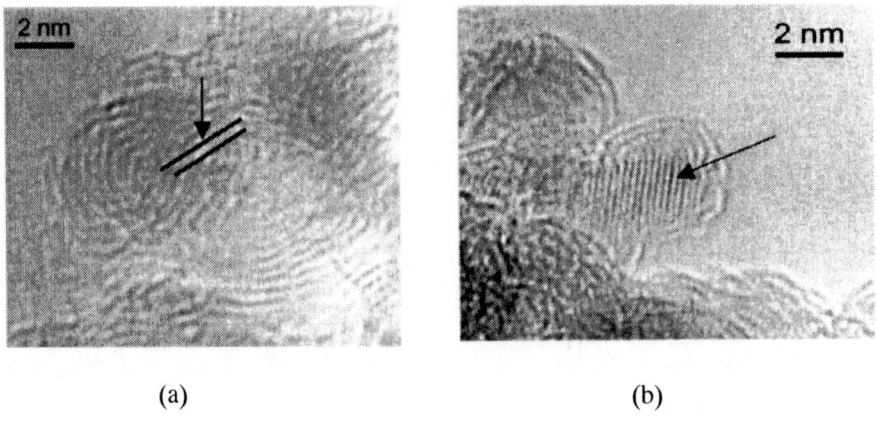

(a) (b)

FIGURE 1. HRTEM images illustrating the effect of chemical cleaning of the detonation carbon: (a) detonation carbon, (b) ultradisperse diamond particles after treatment of the detonation carbon with HNO_3 (70%, 250 ^{0}C). The distance between shells of the onion-like carbon shells marked is ca 0.35 nm (a). Diamond lattice of the UDD core is clearly seen and marked (b).

The experimental X-ray diffraction, scattering curves for both detonation carbon and UDD samples obtained at the same explosion conditions were presented earlier (see. Fig 1 in [3]). The broad symmetric diffraction maxima observed at the angles corresponding to (111) and (220) reflections from diamond-type lattice ($2\theta_{Br}$ = 43.9^0 $2\theta_{Br}$ = 75.3^0). At the same time the detonation carbon represents a broad diffraction maximum $2\theta_{Br} \approx 26^0$, which may be assigned to the reflection from the (0002) planes of a graphite-type lattice, which can be assigned to reflection from the (0002) planes of a graphite-type lattice. The half-width of this maximum is appropriate for scattering from spherical particles of sizes, which do not exceed 20 Å.

After chemical treatment the diffraction maximum corresponding to the graphite phase vanished and diffuse scattering (a halo) became evident at $2\theta_{Br} \approx 17^0$ ($q \approx 1.2$ Å^{-1}). We associated [3] the observed halo with scattering from graphitic plates in the onion-like structures. It is now supported by the TEM data.

Moreover the integrated-intensity ratio of the diamond (111) maximum ($2\theta_{Br} = 43.9^0$, $q \approx 3.05$ Å^{-1}), to the halo, depends on degree of chemical cleaning (temperature of a nitric acid) and gives a possibility to detect of the ratio of sp^3/sp^2 bonded carbon in UDD [6]. The method of such an estimation of sp^3/sp^2 bonded carbon in UDD showed that, after the annealing in hydrogen within the interval 450 – 720°C (in contrast with argon), the intensity of the halo that corresponds to the sp^2-bonded carbon decreases. It should be emphasized, this effect was observed only on the UDD prepared by the so-called "dry " detonation synthesis [7], when the sp^2 shell which covers the diamond core, has sufficient thickness.

The increase in the relative amount of sp^3-bonded carbon (in the diamond core of UDD) takes place only at T < 800 ^{0}C. But at T > 800 ^{0}C the content of sp^3 bonded carbon drops quickly. Such behaviour may be explained by the destruction of the graphite-like and onion-like carbon of the cluster shell and, probably, formation of the new diamond core layers at hydrogen annealing at 450 ^{0}C < T < 800 ^{0}C. The process of the destruction of the graphite-like shell upon annealing UDD samples at 420° C was confirmed by Raman spectroscopy data [6].

The EPR spectrum of the detonation carbon consists of two overlapping signals : an intensive broad line with g = 2.5 and peak-to-peak line width $\Delta H_{pp} \sim 170$ mT and a narrow line in the region of g ~ 2.0 (radical-type signal). The former clearly indicates the presence of certain amount of paramagnetic centers that do not relate to possible defects in graphite or diamond structure. Since the line is broad and structureless, the precise data on its origin is unavailable. However, such lines are typical for samples containing both paramagnetic and ferromagnetic impurities, which originate from such transition ions as Fe^{3+}, Cr^{3+}, Mn^{2+}, Co^{2+}, Ni^{2+} and Cu^{2+}. It is correlated to the X-ray diffraction data for detonation carbon [3]. The chemical treatment leads to the significant reduction of the broad line intensity. EPR spectra of UDD passed chemical treatment show significant reduction of magnetic impurities. Moreover, these EPR spectra produce the information on the efficacy of the chemical treatment procedure.

Narrow singlet EPR line with g = 2.0028 and $\Delta H_{pp} \sim 1$ mT may be referred to a broken, dangling C-C bond on the nano-diamond surface, yielding a localized unpaired electron. These carbon centered radicals are, presumably, detonation defects of sp^3-bonds

The actual quantity of these paramagnetic species was obtained by the numerical double integration of EPR signals. This quantity is minimal for the detonation carbon and corresponds to the concentration of paramagnetic centers of 10^{19} spins/g. All UDD samples undergone the chemical treatment show practically the same 10 times stronger EPR signal which corresponds to the radical concentration of 10^{20} spins/g.

In summary, we showed the effect of the chemical extraction and post-treatment in hydrogen atmosphere on the purity and structure of detonation carbon and ultradisperse diamond.

ACKNOWLEDGMENT

The work of the authors from Ioffe Institute was supported by the Russian State Research Program "Fullerenes and Atomic Cluster"

REFERENCES

1. N. R. Greiner, D. S. Phillips, and J. D. Johnson, F. Volk. *Nature*, **333**, 440 (1988).
2. A. I. Lyamkin, E. A. Petrov, A. P. Ershov, G. V. Sakovich, A. M. Staver, and V. M. Titov, Sov. Phys. Dokl. **33**, 705 (1988).
3. E. Aleksenskii, M. V. Baidakova, A. Ya. Vul', and V. I. Siklitsky. *Phys. Solid. State*, **41**, 669 (1999).
4. M.V.Baidakova, A.Ya.Vul', V.I.Siklitski. Chaos, **10,** 2153 (1999).
5. P.Bundy, Physica, **A 156,** 169 (1989)
6. A.E. Aleksenskii, M. V. Baidakova, A. Ya. Vul', A.T.Dideikin and V. I. Siklitsky, S.P.Vul'. Phys.Solid.State., **42,** 1575 (2000)
7. M.V.Baidakova, A.Ya.Vul', V.I.Siklitski, N.N.Faleev. Phys.Solid.State., **40,** 715 (1998)

Studies of Porphyrin-Fullerene Dyads with Oligoethylene Glycols Spacers in Solution

Reiko Ogura, Tatuo Toida, Katsunori Tsunoda, Hirofumi Yajima, and Tadahiro Ishii

Department of Applied Chemistry, Faculty of Science, Science University of Tokyo, Kagurazaka 1-3 Shinjuku-ku, Tokyo 162-8601, Japan

Abstract. Two novel Porphyrin-fullerene dyads (PFDs) were synthesized, in which flexible spacers of oligoethylene glycol chains were incorporated with different numbers of ethylene glycol moieties. And the emission and electrochemical properties were examined. From the estimations of the quantum yields and free energy, ΔG, for the electron transfer reaction in the excited state, our novel PFDs were suggested to have an efficient electron transfer capability required for compounds constituting artificial photosynthesis system.

INTRODUCTION

The fullerenes have the function of electron acceptors or even as electron accumulators. However, as the fullerenes have only a low absorbance at visible wavelengths, the molecular design of an efficient donor-acceptor dyad containing a fullerene moiety requires a donor capable of picking up light quanta in the visible region. Porphyrin-fullerene dyads (PFDs), which have the ability of photoinduced charge separation, have attracted a great deal of attention and expectation as the compounds constituting artificial photosynthesis systems [1]. Based on the objective to design PFDs with controllable electron transfer properties through conformational changes, we intended to synthesize two novel PFDs, in which flexible spacers of oligoethylene glycol chains were incorporated with different numbers of ethylene glycol moieties. In order to gain an insight into the molecular characteristics of the PFD_4 and PFD_6 dyads in the excited state, the emission and electrochemical properties were examined.

Experimental Section

Synthesis: The synthesis of PFD_4 was achieved according to the following procedures: (a) Bis(p-hydroxyphenyl)methano[60]fullerene (**1** in Scheme 1) was

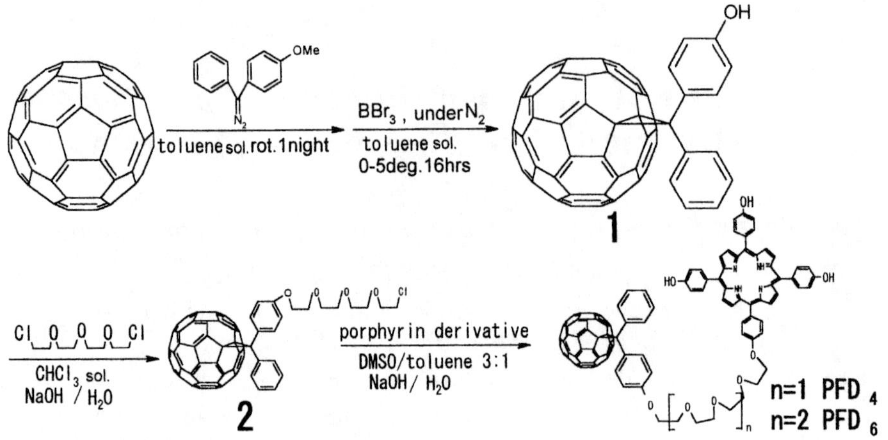

SCHEME 1. Synthesis of PFD$_4$ and PFD$_6$

obtained according to the method of Tezuka et al. [2]; (b) Subsequently, the primary hydroxyl groups in tetraethylene glycol were substituted for chlorine atoms and then made to react with **1**. Therefore, 1,10-dichloro-3,6,9-trioxaundecane (DCD) was obtained; (c) Compound **2** was synthesized by the reaction of **1** with DCD in the mixed solvent of toluene:DMSO = 1:3 with six hr of stirring, and then chromatographically isolated on silica gel; (d) Finally, **PFD$_4$** was synthesized by the reaction of **2** with 5,10,15,20-tetrakis (4-hydroxyphenyl)-21H,23H-porphine with three days of stirring at ambient temperature and then fractionated by chromatography on silica gel. **PFD$_6$** was synthesized using the same method with hexaethylene glycol instead of tetraethylene glycol.

Measurement: The absorption spectra were measured by a UV-2101PC (SHIMADZU) in the mixed solvent of toluene:DMSO = 1:3. Fluorescence spectra were measured using an F-4500 optical fluorescent photometer (Hitachi). Benzene was of spectrograde (KANTO)quality and was passed through a column of activated silica gel (Wako C-300) several times before the fluorescent measurement. The excited wavelength was fixed to the peak wavelength of the soret band for each sample. The electronic measurement was carried out by cyclic voltammetry (CV) using a potentiostat and a function generator HB-111 (HOKUTO DENKO Co.). All of the measurements were made using the solvent of toluene:acetonitril = 4:1, a supporting electrolyte of tetrabutylammonium perchlorate (TBAP), a reference Ag/Ag$^+$ electrode (BAS), and ferrocene as the standard [3].

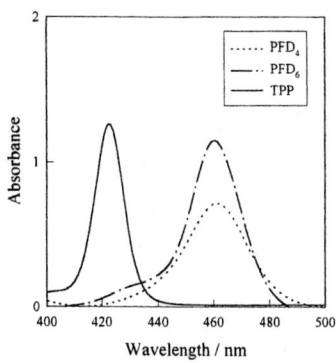

FIGURE 1. Absorption spectra of tetraphenylporphyrin(TPP) and fullerene-porphyrin dyads.

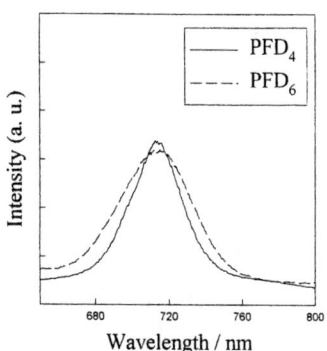

FIGURE 2. Fluorescence emission spectra of PFD_4 and PFD_6.

RESULTS AND DISCUSSION

Optical Properties: Figures 1 and 2 show the absorption and the fluorescence emission spectra of PFD_4 and PFD_6, respectively. Both PFDs exhibited soret bands at around 450nm. The fluorescence emission spectra were measured at the excitation wavelength of 450nm. The fluorescence of both dyads were strongly quenched compared with that of TPP. This may indicate that C_{60} is strongly involved in the quenching of the porphyrin in the excited singlet state. The fluorescence quantum yields (Φ) of PFD_4 and PFD_6, were determined to be 0.00079 and 0.00064, respectively, from Φ=0.13 for the TPP of the standard. It should be noted that the Φ value (0.00072) [4] for PFD with a rigid spacer coincided with that of PFD_4 or PFD_6. Therefore, although significant effects of the spacer length and flexibility on the emission properties in the steady state were not recognized, our developed PFDs were suggested to have an efficient electron transfer capability required for compounds constituting an artificial photosynthesis system.

Electrochemical Properties: In order to estimate the change in free energy, ΔG, for the electron transfer reaction in the excited state, the first reduction potential of the fullerene moiety and the first oxidation potential of the porphyrin moiety in both dyads were required. These were determined by cyclic voltammetric experiments with both dyads at ambient temperature, and these potentials are given in Table 1. ΔG

TABLE 1. Formal potentials (E / V) for compounds

Compounds	E^{Ox}_1	E^{Ox}_2	E^{Red}_3	E^{Red}_4	E^{Red}_5	E^{Red}_6
C_{60}			-0.719	-1.058	-1.683	
PFD_4	0.585	0.380	-0.680	-0.958	-1.599	-1.850
PFD_6	0.540	0.360	-0.615	-0.905	-1.549	-1.847

[kcal/mol] is given by Rehm-Weller equation (1):

$$\Delta G = 23.06 \, [(E^0(D^+/D) - E^0(A/A^-)) - w_p - \Delta G_{00} \quad (1)$$

where w_p is the work quantity of attraction, that is, the solvent effect in the polar solvent, and ΔG_{00} is used and the energy estimated from the (0,0) band energies of the reported values in TPP(1.9 eV) and C_{60}(2.0 eV) [5]. Therefore, the values of ΔG were calculated to be ΔG_{PFD4} = -11.71 [kcal/mol] and ΔG_{PFD6} = -13.83 [kcal/mol]. It is inferred that electron transfer from porphyrin to fullerene spontaneously occurs in the polar solvent and that the formation efficiency for the charge separation state of PFD_6 is larger than that of PFD_4. This result also indicates that the distance between the porphyrin ring and fullerene moiety in PFD_6 is smaller than in PFD_4, and the electrochemical properties of the dyads depend on the length of the spacers and the degree of freedom for conformations.

Our novel PFDs are expected to be able to control the electron transfer properties through the conformational change in the flexible spacers of the oligoethylene glycol chains induced by the inclusion of metal ions. To improve and rationalize the electron transfer and charge separation properties of our novel PDFs, further studies on the effects of the included species and solvents on the excited state properties are required. At present, dynamics studies on the electron transfer based on time-resolved optical measurements are currently in progress.

REFERENCES

1. Gust, D., Moore, T.A.,and Moore A.L., *Res.Chem.Intermed.*, **23**, 621-651(1997).
2. Tezuka, Y.,Kawasaki, N.,Yajima, H.,Ishii, T.,Oyama, T.,Takeuchi, K., Nakao, A., and Katayama, C., *Acta Cryst.*, **C52**,1008-1010(1996).
3. Dietel, E., Hirsch, A., Zhou, J., Rieker, A., *J.Chem.Soc.Perkin Trans.2*, 1357-1364(1998).
4. Liddell, P.A., Sumida, J.P., Macpherson, A.N., Noss, L., Seely, G.R., Clark, K.N., Moore, A.L., Moore, T.A. and Gust, D., *Photochem. Photobiolo.* **60,6**, 537-541(1994).
5. Kureishi, Y.,Tamiaki, H.,Shiraishi, H.,and Maruyama, K., *Biochem. and Bioenerg.*, 95-100(1999)

Magnetic-Field Induced Ferromagnetism in Bissilylated C_{60} by Pyrolysis

Yasukazu Kajihara*, Katsumi Tanigaki*, and Takeshi Akasaka[†]

Department of Material Science, Faculty of Science, Osaka City University
3-3-138 Sugimoto, Sumiyoshi-ku, Osaka 558-8585, Japan
[†] *Graduate School of Science and Technology, Niigata University*
8050 Igarashi, 2 nomachi, Niigata 950-2181, Japan

Abstract. Unique magnetic-field-induced ferromagnetism is observed in a solid of bissilylated C_{60} after heat treatment. The Curie temperature observed is higher than room temperature. The saturation moment is 9×10^{-3} μ_B and 0.045 μ_B per C_{60} molecule at 300 K and 1.9 K, respectively. Silyl substituents are confirmed to be cleaved during a heat process by means of ^{1}H-NMR and ^{13}C-NMR, and the electron spins left on C_{60} sites are responsible for the observed ferromagnetic properties.

INTRODUCTION

C_{60} has shown many physical and chemical properties. Physically, alkali, alkali-earth, or rare-earth metals doped C_{60}'s have been reported to be metals, superconductors, or ferromagnets depending on the dopants [1,2]. Chemically, modifications of organic substituents on C_{60} have been one of the important topics in organic chemistry [3]. It seems to be anticipated that unique electronic properties in solid state physics could arise if we use a chemically modified C_{60} with special functions. From many candidates so far, we have selected [t-BuPh$_2$Si]$_2$C$_{60}$ (Fig. 1) since the substituent group of t-BuPh$_2$Si can easily be cleaved under a low temperature-heating condition. This may give electron spins on C_{60} sites that can interact to be resulting in unique magnetic properties through C_{60} carbon networks.

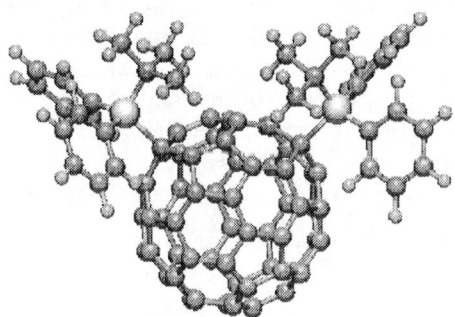

Fig. 1. [t-BuPh$_2$Si]$_2$C$_{60}$

The bissilylated C_{60} is reported to have a unique redox property and show almost one-electron transfer onto C_{60} from the silicon substituent, which is confirmed by cyclic voltammetry, because silicon shows high electron-releasing nature compared to that of carbon. We report here a ferromagnetic property induced by magnetic field in bissilylated C_{60} after heating at 873 K.

Experimental

The $[t\text{-BuPh}_2\text{Si}]_2 C_{60}$ (Fig. 1) was photo-chemically synthesized from 1,1,2,2-tetraphenyl-1,2-di-t-butyl-1,2-disilane and C_{60} in a toluene solution with a low-pressure mercury-arc lamp [4]. The pristine bissilylated C_{60} was confirmed to decompose leaving the silicon substituents as liquid by heating at 873 K for 20 minutes in the absence of oxygen. Phenyl and t-butyl protons were measured by ^{1}H-NMR spectroscopy to study the structure of the liquid detected after heating. The trace of sp^3 carbons of C_{60} was also studied by ^{13}C-NMR spectroscopy. Magnetic properties were studied with a Quantum Design MPMS7 apparatus under magnetic filed from 0 to 7 T. The chemical bonds created before and after heating was studied by IR spectroscopy in the state of powder solids using a KBr pellet method.

Results and Discussion

After heat treatment of the bissilylated C_{60} at 873 K, magnetization has been studied as a function of magnetic field (Fig. 2). After heating, the liquidized portion was first removed and the solid was isolated to be subjected to magnetic field measurements. As seen in Fig. 2, a steep increase in magnetization was observed at 1.9 K under low magnetic fields. The magnetization curve tends to saturate at higher magnetic fields, but it does not arrive at a plateau and instead increases continuously. This result strongly indicates that ferromagnetism can be induced by magnetic field in the isolated C_{60} solid. Surprisingly, such a large increase in magnetization can still be observed at room temperature. At room temperature, the magnetization saturates at about 4000 gauss as seen in the inset of Fig. 2. The value of the saturated magnetization measured at room temperature corresponds to 9×10^{-3} μ_B per C_{60} molecule. Since little saturation is seen in the M-B curve measured at 1.9 K, the full saturation magnetization moment was estimated from extrapolation of the curve, the value of which corresponds to 0.045 μ_B per C_{60} molecule.

Infrared measurements were carried out for the parent bissilylated C_{60} before and after heat treatment. As shown in Fig. 3, the spectrum clearly displays that the t-BuPh$_2$Si substituents are cleaved during the heating. It is also evident that the possibility of C_{60} polymerization is very little since the spectrum is almost the same as that of C_{60} and extra peaks, due to the break in the high I_h symmetry of C_{60}, do not appear.

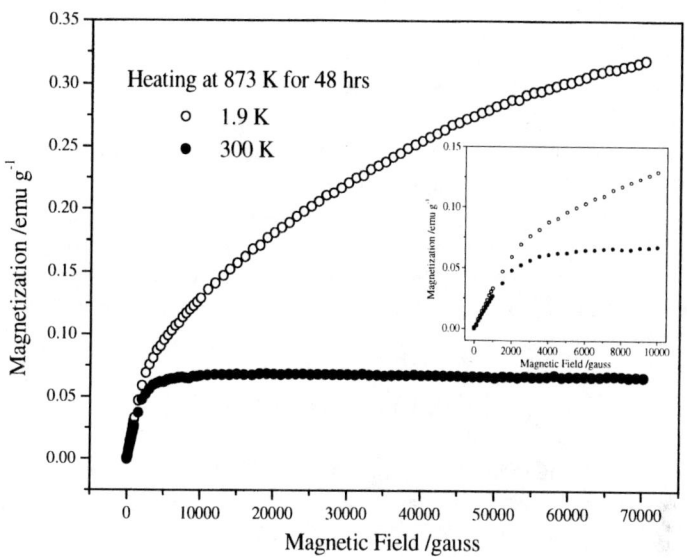

Fig. 2. Magnetic properties of the solid powder heated at 873 K for 48 hrs

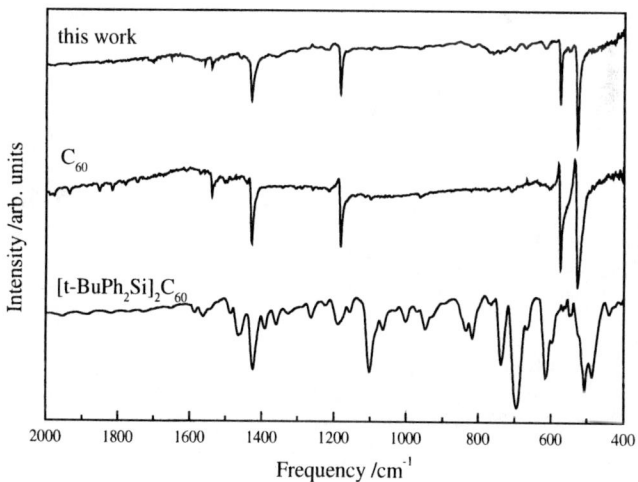

Fig. 3. IR spectrum

Conclusion

In conclusion, a unique ferromagnetic property induced by magnetic field was observed in the isolated solid in [t-BuPh$_2$Si]$_2$C$_{60}$ after heat treatment. The saturation moment is 9×10^{-3} μ$_B$ and 0.045 μ$_B$ per C$_{60}$ molecule at 300 K and 1.9 K, respectively. This value observed is quite high among other similar carbon materials so far reported [5]. IR spectrum of the solid showed that the unique ferromagnetism can not be ascribed to the polymer phase that may be created by heating. Instead, it is considered that the unique magnetic-induced properties observed from the heat-treated bissilylated C$_{60}$ may stem from the electron spins created in the crystal due to the cleavage of the substituents.

ACKNOWLEDGEMENTS

We thank Koichi. Komatsu for supplying the IR data of C$_{60}$ dimer made using a chemical reaction. Yutaka Maeda is acknowledged for technical assistance to synthesize [t-BuPh$_2$Si]$_2$C$_{60}$.

REFERENCES

1. Haddon, R. C. et al., *Nature* **350**, 320-322 (1991).
2. Allemand, P. M. et al., *Science* **253**, 301-303 (1991).
3. Haufler, R. E. et al., *J. Phys. Chem.* **94**, 8634 (1990).
4. Akasaka, T. et al., *J. Org. Chem.* **64**, 566-569 (1999).
5. Ito K., *Chem. Phys. Lett.* **1**, 235 (1967).

Characterization of Actinide Metallofullerenes

K. Akiyama [a], K. Sueki [a], Y-L. Zhao [a], H. Haba [b], K. Tsukada [b], T. Kodama [a], K. Kikuchi [a], T. Ohtsuki [c], Y. Nagame [b], H. Nakahara [a], and M. Katada [a]

[a] *Department of Chemistry, Tokyo Metropolitan University, Hachioji, Tokyo 192-0397, Japan*
[b] *Advanced Science Research Center, Japan Atomic Energy Institute (JAERI), Tokai, Ibaraki 319-1195, Japan*
[c] *Laboratory of Nuclear Science, Tohoku University, Mikamine, Taihaku, Sendai 980-8577, Japan*

Abstract. In this paper, the characterization of a number of actinoid metallofullerenes is reported. From the similarity of the HPLC elution behavior and identification by mass spectroscopy we observe that the electronic structures of U, Np, and Am metallofullerenes are extremely similar to that of $Ce@C_{82}$ and that the oxidation states of encapsulated Th and Pa have the value +3 or 4.

INTRODUCTION

The actinoids belong to the f block of the periodic table, the same block as the lanthanoid elements. In terms of inorganic chemistry, however, it is generally known that the most stable valence states of the actinoid elements are different from those of the lanthanoid elements. This variety in the valence states is caused by the close proximity of the energy level of the 5f orbital to that of the 6d orbital in the actinoid elements. It is therefore expected that the actinoid elements encapsulated in fullerenes show different behavior to the lanthanoid elements. From the viewpoint of inorganic chemistry and fullerene science, we have been interested in the investigation of the properties of actinoid metallofullerenes. The first synthesis of actinide fullerenes was reported by Guo et al. [1], who prepared uranium fullerenes and observed UC_{28} and UC_{60} in the primary soot. From the XPS study of the gross products, the authors concluded that the uranium atom encapsulated in the C_{2n} carbon cage was in the +4 valence state. However the observed data merely reflected the bulk properties of the material and not the property of the individual molecules. Recently, it was reported that the valence states of metal atoms are different depending on the surrounding carbon cage or on the number of encapsulated atoms. [2] It is important to understand the properties of isolated metallofullerene molecules. In this paper, the characterization of metallofullerenes for Th, Pa, U, Np, and Am is reported.

EXPERIMENTAL

^{234}Th was chemically separated from $^{238}UO_2(NO_3)_2 6H_2O$ by ion exchange. Radioactive tracers of ^{233}Pa, ^{237}U, ^{239}Np, and ^{240}Am were produced at the TANDEM accelerator of JAERI. These tracers and Ce were separated from fission products and other reaction products by anion and cation exchange methods. After the separation, the solutions including tracers were mixed. An ethanol solution, in which $La(NO_3)_3$ powder was dissolved as the carrier for reproducing the arc condition in which lanthanoid fullerenes are prepared, was added to the mixture. This solution was absorbed into a porous carbon rod and then was sintered at 800 °C. The soot containing the actinoid metallofullerenes was prepared by the arc-discharge method [3] using the carbon rod. The soot in the generator was recovered with CS_2 and filtered for removal of insoluble substances. The filtrates were evaporated to the point of dryness and then dissolved in toluene for HPLC injection. The samples were separated using 5PBB and Buckyprep HPLC columns, with toluene as an eluent. Fractions of the eluate were collected for every 2 minutes. The X and γ rays emitted from each sample were measured using a Low Energy Photon Spectrometer (LEPS) and a High Purity Ge semiconductor detector (HPGe). For the macroscopic production of U and Th metallofullerenes, a similar method was employed, except in this case uranium and thorium nitrates were used as the carriers.

RESULTS AND DISCUSSION

Figures a) and b) show the results of HPLC elutions, on a Buckyprep column and a 5PBB column, respectively, of the actinide fullerenes using radiotracers. In both figures, the elution behaviors of U, Np, and Am fullerenes were found to be extremely similar to that of Ce fullerenes as a contaminant from fission products with the largest elution peak at the retention time known for $M@C_{82}$ of trivalent lanthanoids. [4] The U species produced in macroscopic quantities at the elution peak was identified as $U@C_{82}$ by TOF/MS. The same retention times show that electronic states of the cages of these $M@C_{82}$ are similar to each other. [5] On the other hand, the retention times of metallofullerenes of Th and Pa were found to be late on a 5PBB column, and extremely

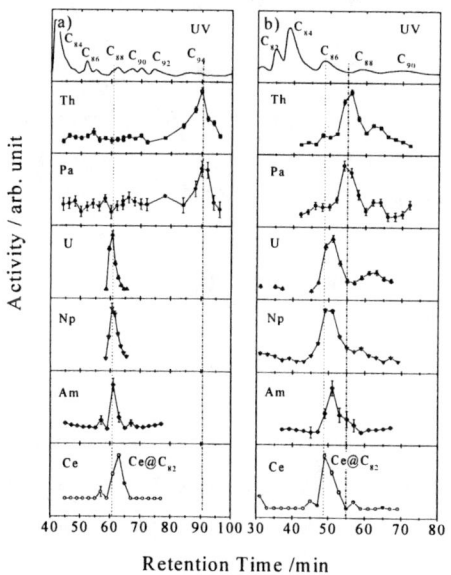

FIGURE. HPLC elution behavior of light actinide fullerenes: (a) The HPLC elution curve for the Buckyprep column elution with toluene at 3.2 ml/min. The major elution peaks of U, Np, and Am are observed at the retention time of around 60 min, which also corresponds to the retention time of $Ce@C_{82}$. (b) The HPLC elution curve for the 5PBB column elution by toluene at 6 ml/min. The major elution peaks of U, Np, and Am are observed at the retention time of around 50 min in agreement with the retention time for $Ce@C_{82}$.

late on a Buckyprep column in contrast to that for the above-stated group. As the result of identification by TOF/MS, this component of the observed elution peak for Th species has been found to be Th@C_{84}. The elution peak for Pa species is considered to be also Pa@C_{84} because of the similarity between the chromatographic behaviors on both columns. Stevenson et al. have reported the correlation between the retention time and the number of electrons on the fullerene cage for 5PBB column development. [6] According to their studies, the number of electrons on the cage of Th@C_{84} is same as that of C_{87} or C_{88} since this species is eluted between that of C_{86} and C_{88}. Thus, an extra three or four electrons are moved to the C_{84} cage from the encapsulated metal atom. In conclusion, it is expected that the oxidation state of encapsulated Th and Pa atoms inside C_{84} is +3 or +4.

ACKNOWLEDGEMENTS

This work was partly supported by the REIMEI Research Resources of the Japan Atomic Energy Research Institute and by Grant-in-Aid from the Ministry of Education, Culture, Sports, and Science of Japan. We thank to the crew of the JAERI TANDEM accelerator for providing the ^{6}Li ion beam used to produce the radiotracers.

REFERENCES

1. Guo, T., Diener, M. D., Chai, Y., Alford, M. J., Haufler, R. E., McClure, S. M., Ohno, T., Weaver, J. H., Scuseria, G. E., Smalley, R. E. *Science*, **257**, 1661 (1992).
2. Kikuchi, K., Akiyama, K., Sakaguchi, K., Kodama, T., Nishikawa, H., Ikemoto, I., Ishigaki, T., Achiba, Y., Sueki, K., Nakahara, H. *Chem. Phys. Lett.*, **319**, 472 (2000).
3. Sueki, K., Kikuchi, K., Akiyama, K., Sawa, T., Katada, M., Ambe, S., Ambe, F., Nakahara, H. *Chem. Phys. Lett.*, **300**, 140 (1999).
4. Ding, J., Yang, S. *Angew. Chem. Int. Ed. Engl.* **35**, 2234 (1996).
5. Akiyama, K., Zhao, Y-L., Sueki, K., Tsukada, K., Haba, H., Nagame, Y., Kodama, T., Suzuki, S., Ohtsuki, T., Sakaguchi, M., Kikuchi, K., Katada, M., Nakahara, H. *J. Am. Chem. Soc.*, **123**, 181 (2001)
6. Stevenson, S., Brubank, P., Harich, K., Sun, Z., Dorn, H.C., van Loosdrecht, P.H.M., deVries, M.S., Salem, J.R., Kiang, C.-H., Johnson, R.D., Bethune, D.S. *J. Phys. Chem. A*, **102**, 2833 (1998).

IV. FULLERENE POLYMERS

NMR Studies of Alkali-Doped C_{60} Polymers

Y. Maniwa[1], H. Ikejiri[1], H. Tou[1], S. Masubuchi[2], S. Kazama[2],
M. Yasukawa[3,4] and S. Yamanaka[4,5]

[1]*Department of Physics, Tokyo Metropolitan University, Minami-osawa, Hachi-oji, Tokyo 192-0397, Japan*
[2]*Department of Physics, Chuo Univ., Kasuga 1-13-27, Bunkyou-ku, Tokyo 112, Japan.*
[3]*Department of Materials-Science and Engineering, Kochi National College of Technology, 200-1, Monobe, Nankoku 783-8508, Japan.*
[4]*CREST, Japan Science and Technology Corporation (JST), Japan*
[5]*Department of Applied Chemistry, Faculty of Engineering, Hiroshima University, Hiroshima University, Higashi-Hiroshima 739-8527, Japan.*

Abstract. ^{23}Na- and ^{13}C-NMR of NaC_{60} heat-treated under high-pressure are presented, and the results are compared with those of Li_xC_{60} polymers. The ^{13}C-NMR spectra clearly showed the existence of sp^3 carbon atoms, evidencing for the transformation to the C_{60} polymer phase. The spin-lattice relaxation time, T_1, of ^{13}C-NMR suggested an electron localization effect, and the electronic density of states at the Fermi level at 4.2 K was estimated to be smaller than 0.6 states/eV/C_{60}/spin, which is smaller than 1/10 of that of a superconducting K_3C_{60}. It was also shown that the Na atoms are jumping among at least two different sites in the cavities of C_{60}-polymerized network above ~250K.

INTRODUCTION

Recently, alkali-doped C_{60} solid A_xC_{60}, where A = Li and Na, was heat-treated under high-pressure. The resultant materials were characterized at ambient pressure by X-ray diffraction (XRD), infrared (IR) spectra, solubility and electrical conductivity measurements [1, 2]. In the case of AC_{60} and A_4C_{60}, these measurements provided evidences for a formation of the 2D-C_{60} polymer phases, in which the molecules are linked through the [2+2] cycloaddition of carbon double bonds. The crystal structures have tetragonal (T) symmetry for A_4C_{60} and rhombohedral (3R) symmetry for AC_{60}, as illustrated in Fig. 1. These structures are the same as those for 2D-polymers

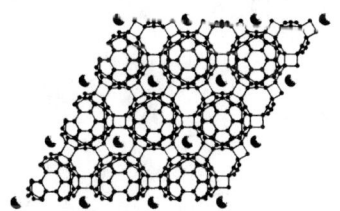 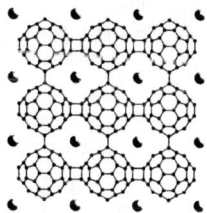

FIGURE 1. Proposed structures for AC_{60} (left) and A_4C_{60} (right) polymers. The ^{13}C NMR spectra confirmed these structures for LiC_{60}, NaC_{60} and Li_4C_{60}.

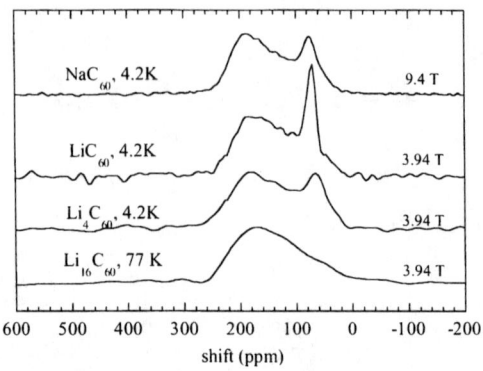

FIGURE 2. ^{13}C-NMR spectra of NaC$_{60}$, LiC$_{60}$, Li$_4$C$_{60}$ and Li$_{16}$C$_{60}$ at low temperature (4.2K or 77K). The applied magnetic field was 9.4T or 3.94T. The sharp signal around 70ppm indicates the existence of sp^3 carbons.

transformed from undoped solid C$_{60}$ under high-temperature and high-pressure [3, 4]. Only difference is that the alkali atoms are incorporated in the C$_{60}$ interstitial sites. In this paper, ^{23}Na- and ^{13}C-NMR of NaC$_{60}$ prepared by the high-pressure and high temperature treatment are presented and compared with those of C$_{60}$ polymers reported previously [5, 6].

EXPERIMENTAL RESULTS

Figure 2 shows examples of ^{13}C-NMR spectra in Li$_x$C$_{60}$ and NaC$_{60}$ at 4.2K. These materials were heat-treated at 573K for one hour under high-pressure, ~5GPa [1, 2]. The NMR spectra were taken by a conventional pulse Fourier-transform NMR technique. In order to check possible artificial distortions of the spectra, the NMR were carefully observed with many different pulse conditions. For example, we changed the pulse width, the frequency, the repetition time and the applied magnetic field (3.94T and 9.4T). From the ^{13}C powder spectra of the NaC$_{60}$ in Fig. 2, we can easily separate the sharp signal around 70ppm from the broad anisotropic signal ranging from 0 to 250ppm. It is found that this spectrum is very similar to those of 3R-LiC$_{60}$ and T-Li$_4$C$_{60}$ polymers [6], as shown, and also similar to those of the undoped C$_{60}$ polymers [5]. Therefore, the sharp signal can be assigned to the sp^3 carbon atoms and the broad signal to the sp^2 carbon, implying that the C$_{60}$ molecules are polymerized in the NaC$_{60}$ crystal.

Because the integrated intensity for each signal should be proportional to the number of carbon atoms, the proposed structures illustrated in Fig. 1 predict the signal intensity for the sp^3 carbon as 20% and 13.3% for the 3R and T structures, respectively. These intensities have been actually observed in LiC$_{60}$ and Li$_4$C$_{60}$ polymers, confirming the predicted structures [6]. On the other hand, the intensity in the NaC$_{60}$ seems to be much smaller than the predicted one for the 3R structure. This must be due to a very long spin-lattice relaxation time, T_1, for the sp^3 carbon as demonstrated in Fig. 3, where the ^{13}C-spectra taken at the several different repetition times, T_R, are shown. As the T_R increases, the sp^3 and sp^2 signals develop. But the sp^3 signal shows a much longer characteristic time than the sp^2 signal. Even at 21600sec (6 hour), the intensity still does not saturate. Therefore, the correct intensity ratio could not be reduced from the present spectra. The very long and very different values for the sp^3 and sp^2 carbons were also observed in the cases of the LiC_{60} and Li_4C_{60} polymers [6].

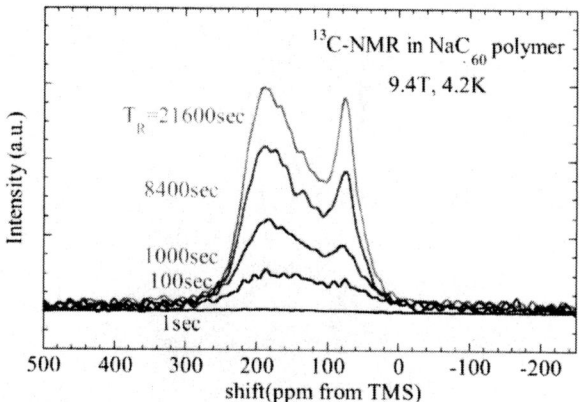

FIGURE 3. ^{13}C-NMR spectra in 3R-NaC_{60} polymer as a function of the repetition time T_R to observe the signal. The sp^3 signal around 70ppm develops with a longer characteristic time. The measurements were performed at 4.2K and an applied magnetic field of 9.4T. The T1 was obtained at ~200ppm for the sp^2 carbon and at ~70ppm for sp^3 carbon.

The peak intensity as a function of T_R, furthermore, gives us an estimate of the ^{13}C-T_1 as $T_1 > 2.5$h for the sp^2 carbon and $T_1 > 4$h for the sp^3 carbon. These values are quite longer than ~40sec at 4.2K in a metallic K_3C_{60}, for example. Assuming that hyperfine coupling constant between the ^{13}C nuclear spin at the sp^2 site and the conduction electron spin on the C_{60} molecules is the same as in the case of K_3C_{60}, we can obtain an upper limit for the density of states at the Fermi level, $N(E_F)$, using so-called Korringa relation, $1/T_1 \sim A^2 N(E_F)^2$, in the case of ordinal metals; $N(E_F) < \sim 0.6$ states/eV/C_{60}/spin. This value is only 7% of that of K_3C_{60} [7]. Combined with a fact that the T_1 of the sp^2 carbon was found to be strongly inhomogeneous, such small $N(E_F)$ suggests that the electronic state of the 3R-NaC_{60} polymer is semiconducting, and that the electrons transferd from the Na atoms to the C_{60} mopolymerized network are probably localized by a strong random potential induced by imperfect polymerization such as the broken bridging bonds.

Finally we discuss the ^{23}Na NMR spectra as a function of temperature (Fig. 4). The spectra were observed around 0ppm where the resonance of Na^+ ion in NaCl solution appears, suggesting that the Na atoms in the crystal are almost ionized. We also notice that a new peak around ~100ppm develops with increasing sample temperature above 250K. A very similar behavior was also observed in the 3R-LiC_{60} polymer. Therefore this may be a common feature of 3R-C_{60} polymers. One possible source for such behavior is jumping motion of the Na ions among the different sites in the crystal. The starting C_{60} solid has a face centered cubic structure, and there are two types of interstitial sites, O- and T-sites. Then, the C_{60} molecules are polymerized within the (111) plane to form the 3R-C_{60} polymer phase. The Na atoms mainly occupy the original O-sites, because the cavity of the O-site is larger than that of the T-site. However, with increasing temperature, some of the Na atoms may jump into the T-sites, resulting in the two ^{23}Na-NMR signals.

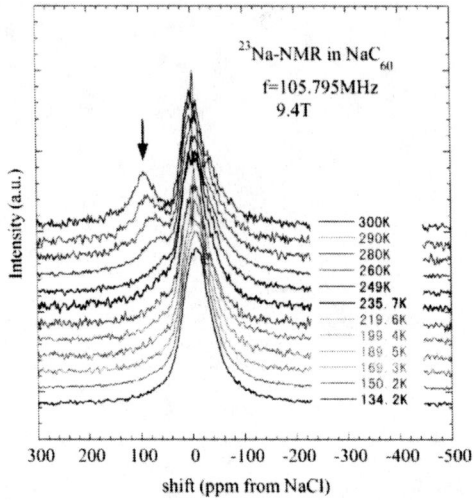

FIGURE 4. ^{23}Na-NMR spectra in 3R-NaC_{60} polymer as a function of temperature. With increasing temperature, a new signal appears around 100ppm indicated by an arrow. The measurements were performed at the applied magnetic field of 9.4T. The origin is the resonance position of 1M-NaCl solution. The positive shift implies the higher frequency shift at the constant field.

ACKNOWLEGMENTS

This work was supported in part by the Grant-in Aid for Scientific Research on the Priority Area "Fullerenes and Nanotubes" by the Ministry of Education, Science and Culture of Japan and CREST, Japan Science and Technology Corporation (JST), Japan.

REFERENCES

1. Yasukawa, M., and Yamanaka, S., *Fullerene Sci. Tech.*,**7,** 795 (1999); Yasukawa, M., and Yamanaka, S., *Mol. Cryst. Liq. Cryst.* **340**, 683 (2000).
2. Yasukawa, M., and Yamanaka, S., to be appeared in *Chem. Phys. Lett.*
3. Nunez-Regueiro, Marques, L., Hodeau, J.–H., Bethoux, O., Perroux, M., *Phys. Rev. Lett.*, **74.** 278 (1995).
4. Iwasa, Y., Arima, T., Fleming, R.M., Siegrist, T., Zho, O., Haddon, R.C., Rothberg, L.J., Lyons, K.B., Carter, Jr., H.L., Hebard, A.F., Tycko, R., Dabbagh, G., Krajewski, J.J., Thomas, G.A., Yagi, T., *Science* **264** , 1570 (1994).
5. Maniwa, Y., Sato, M., Kume, K., Kozlov, M.E., Tokumoto, M., *Carbon* **34**, 1287 (1996).
6. Maniwa, Y., Ikejiri, H., Tou, H., Yasukawa, M., and Yamanaka, S., to be appeared in *Synthetic Metals*.
7. Sato, N., Tou, H., Maniwa, Y., Kikuchi, K., Suzuki, S., Achiba, Y., Kosaka, M., Tanigaki, K..., *Phys. Rev.* **B58**, 12433 (1998).

Dimer Structure of Sm_3C_{70}

Hieu Chi Dam*, X. H. Chen†, T. Takenobu*, T. Itou*, Y. Iwasa*, T. Mitani*, E. Nishibori‡, M. Takata‡, and M. Sakata‡

*Japan Advanced Institute of Science and Technology, Tatsunokuchi,
Ishikawa 923-1292 Japan
†Structure Research Laboratory and Department of Physics, University of Science and Technology of China, Hefei, Anhui 230026, PRC
‡Department of Applied Physics, Nagoya University, Nagoya 464-8062, Japan.

Abstract. A single phase of rare earth doped C_{70} compounds, Sm_3C_{70}, has been first isolated. The Rietveld analysis of the synchrotron x-ray diffraction pattern of the nominal Sm_3C_{70} revealed that there exist extremely short Sm-C bonds. The short bonds are arranged so as to form a C_{70}-Sm-C_{70} dimer structure. The experiments of the high temperature x-ray diffraction show that this structure is maintained up to 850K, indicating that the bonds are extremely stable comparing to the conventional interfullerene bonds.

INTRODUCTION

Many attempts to dope a wide variety of atoms or molecules into fullerenes have been made since the discovery of superconductivity in K_3C_{60} by Hebard et al. It has been reported that alkali metals (K, Rb, Cs), alkali-earth metals (Ca, Sr, Ba), and rare-earth metals (Yb, Sm, Eu) can be intercalated into C_{60} solid. Intercalation of these metals into C_{60} solids yields various structural compounds with different physical properties. The basic structure of $R_{2.75}C_{60}$ (R= Yb, Sm) is face-centered cubic (fcc), but cation vacancy ordering in tetrahedral sites leads to a superstructure[1]. From the average nearest-neighbour R-C (R= Yb, Sm) distances, a strong, short-range, directional interaction between metal atoms and fullerene molecules was expected.

Relative to C_{60}, very little is known about the compounds of C_{70} so far, the next stable fullerene. The electronic structure has been reported for K-doped C_{70} and for Rb doped C_{70}. Electronic transport properties of K_xC_{70} thin films have been investigated. To our knowledge, no evidence of superconductivity has been reported, and no structural analysis of C_{70} fullerides has been reported either. In this paper, we report that intercalation of fullerene C_{70} with rare earth metals yields a novel fullerene dimer structure, in which two C_{70} molecules are bridged with intercalated samarium ions. The Rietveld analysis of the synchrotron x-ray diffraction pattern of the nominal Sm_3C_{70} revealed that there exist extremely short Sm-C bonds. The short bonds are coordinated so as to yields a C_{70}-Sm-C_{70} dimer structure. This structure maintained up to 850K, indicating that the bonds are extremely stable comparing to the conventional interfullerene bonds.

EXPERIMENT

The compound of Sm_3C_{70} has been synthesized by a solid state reaction by mixing a stoichiometric amount of Sm and C_{70} powders. The mixed powder was sealed in a quartz tube under high vacuum of 2×10^{-6} torr. Annealing was carried out at 450°C for 4 days and cooled down slowly to room temperature in a furnace. In a previous paper[2], we reported synthesis and structure of Sm_3C_{70}, but failed to determine the orientation of C_{70} molecule (in other words, coordinate of individual carbon atoms), due to the insufficient crystal quality. We improved the synthesis conditions and succeeded to obtain single phase materials free from any distortion and preferred orientations. The sample was sealed in a thin glass capillary of 0.3 mm in outer diameter. High resolution synchrotron x-ray powder diffraction data were collected on the BL02B2 at Super Photon Ring (Spring-8), Nishi-Harima, Japan, at room temperature. The incident x-ray was monochromatized at wavelength 0.8232 Å with a Si double crystal and collimated to 0.5 mm in diameter. An imaging plate was used for detection of diffraction rings, which was converted to the conventional spectrum by integrating the intensity along the Debye rings. The Rietveld refinement was carried out using the GSAS software (A. C. Larson and R. B. von Dreele, General Structural Analysis System, Los Alamos National Laboratory, 1985-1998), combined with a coordinate generator in Cerius2.

STRUCTURE ANALYSIS AND DISCUSSION

All the peaks of the x-ray diffraction pattern of the nominal Sm_3C_{70} can be indexed on a monoclinic cell with lattice parameter of a=14.86(1) Å, b=10.09(1) Å, c=10.92(1) Å, β=96.17(2)°. This unit cell is derived by a deformation of the fcc cell, where Sm ions occupy the tetrahedral and octahedral sites. In contrast to the ordinary occupation of the tetrahedral site, there are two Sm positions in the octahedral site, which are randomly occupied with the occupancy of 50%. In the previous analysis[2], we have determined the crystal structure within a spherical shell model, namely, the unit cell and the locations of Sm ions and the center of mass of C_{70}. Samples of improved quality allowed us more detailed analysis beyond this model. First, we found that the space group should be $P1$ rather than

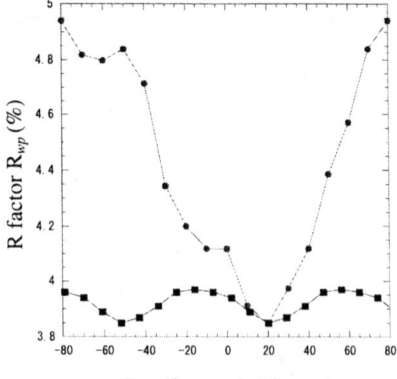

FIGURE 1. The dependence of R factors R_{wp} derived from $Sm_{2.78}C_{70}$ refinement depend on the rotation angles of C_{70}.

P 2. Though there was an inversion center on the basal plain in the previous model, the improved diffraction data clearly showed the lack of the inversion center, resulting in the symmetry reduction of space group to $P1$. The R-factor for the P 2 space group of $R_{wp}=8.23\%$ decreased to $R_{wp}=7.18\%$ for the $P1$ space group, by moving the tetrahedral Sm^{2+} from the symmetric positions. Second, the orientation of C_{70} molecules was determined within a rigid C_{70} model. We found the orientations of each C_{70} molecule, by calculating the R-factors as a function of rotation angles (Fig 1.) around the short (parallel to b axis, the cubic dots) and long axis iteratively (circles). For example, the rigid rotation around the five fold axis (long axis) of C_{70} molecule causes the variation of R_{wp} between 4.92% and 3.85%.

The final Rietveld refinement is shown in Fig. 2 for the 2θ range from $3°$ to $30°$, corresponding to a minimum d-spacing of 1.5 Å. The R-factors were $R_{wp}=3.85\%$, $R_I=7.62\%$. In the final refinement, Sm concentration was also optimized at $Sm_{2.78}C_{70}$. In this structure there exists two kinds of tetrahedral sites, one of which is approximately fully occupied and the other is half occupied. This site occupation and the refined chemical concentration are very similar to those in $Yb_{2.75}C_{60}$ (ref. 1), where the tetrahedral vacancies are ordered, forming a superlattice. In the present $Sm_{2.78}C_{70}$, absence of superlattice structure indicates that the tetrahedral vacancies are disordered.

Figure 3 displays the crystal structure of $Sm_{2.78}C_{70}$ viewed from <110> (<111>$_{fcc}$) direction. The orientational ordering of the C_{70} molecules causes a slight difference between b and c axes. The long axis of two C_{70} molecules in the unit cell are laid in the *ac* basal plane. The long axis of one C_{70} makes an angle of $20°$ with the c-axis, while the long axis of the other C_{70} is nearly parallel to the a-axis. The orientation of the former C_{70} may be responsible for the slightly longer c-axis than the b-axis, because the difference between the b and c axes are close to the difference (0.84 Å) between the long and short diameters of C_{70} molecule.

Another notable feature of $Sm_{2.78}C_{70}$ is the Sm-C distances. Since all the Sm ions are shifted from the center of the interstitial sites, Sm-C distances are rather small. The nearest neighbor Sm-C distance is 2.49 Å and 2.47 Å for the T- and O- sites,

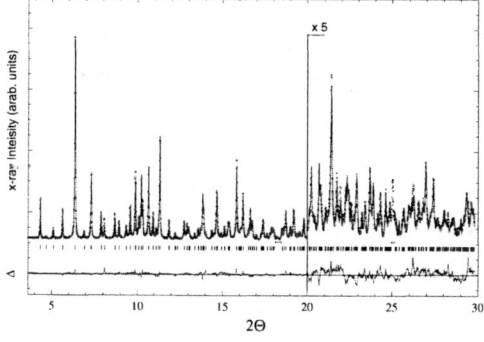

FIGURE 2. Best Rietveld refinement of $Sm_{2.78}C_{70}$ within a rigid C_{70} model. Top: the raw data (ticks) and the best-fit profile (solid lines), middle: peak positions of $Sm_{2.78}C_{70}$, and bottom; the difference between the experimental and calculated patterns.

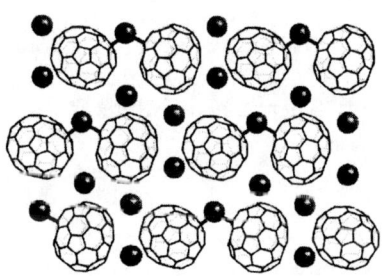

FIGURE 3. Crystal structure viewed from <111>$_{fcc}$ direction. Among four kinds of tetrahedral Sm ions, only one bridges C_{70} molecules.

respectively. These Sm-C distances are nearly identical to a typical Sm-C covalent bond in, for example, a carbide SmC_2. This suggests that an appropriate description of Sm-C bonds in $Sm_{2.78}C_{70}$ should be a covalent bonding, in sharp contrast to the conventional fullerene intercalation compounds. In particular, Sm ions shown by black spheres in Fig. 3 are significantly moved from the center of the T-site, being indicative of the covalent bonding. The short bonds are arranged so as to form a C_{70}-Sm-C_{70} dimer structure.

The experiments of the high temperature x-ray diffraction show that the diffraction pattern of Fig. 2 is qualitatively maintained up to 850K, followed by a drastic change of the diffraction pattern (Fig. 4). This result shows that the present dimer structure is amazingly stable, indicating that the Sm-C bond is much stronger than the direct interfullerene bonds. This new structural aspect is responsible to the bonding nature of rare earth metals, which was absent in alkali metals.

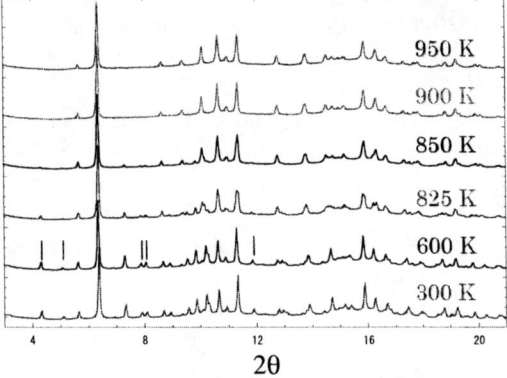

FIGURE 4. The temperature dependence of the diffraction pattern of $Sm_{2.78}C_{70}$.

Generally, rare earth fullerides provide an unexplored area of solid state science not only from the structural and but also from the magnetic or electronic point of view. The results presented here indicate that materials with new nanoscale network structures may be formed by combining fullerenes and rare earth metals. This type of structure should be of significant interest both for chemical and physical sciences, because the metal bridged fullerene polymers (dimers) are thermally stable, comparing to the conventional fullerene dimers/polymers.

ACKNOWLEDGMENTS

This work has been supported by the JSPS "Future Program" (RFTF96P00104), and by Priority Area Grant "Fullerenes and Nanotubes" from the Ministry of Education, Sport, Science, and Culture, Japan.

REFERENCES

1. E. Ozdas, E. et al., *Nature* **375**, 126 (1995).
2. X. H. Chen et al., *J. Am. Chem. Soc.*, **122**, 5729 (2000).

Photopolymerization of C_{60} crystal under high pressure

Masatoshi Sakai, Masao Ichida and Arao Nakamura

Department of Crystalline Materials Science, and Center for Integrated Research and Engineering, Nagoya University, Nagoya, 464-8603, Japan

Abstract. We have investigated a photopolymerization process in C_{60} crystals at room temperature under high pressure up to 1.1 GPa by means of Raman spectroscopy. Time variation of Raman intensity with laser irradiation time indicates the decrease of the monomer density and the generation of a polymerized phase with increasing irradiation time. The photopolymerization reaction coefficient analyzed by a rate equation model exhibits an exponential-type behavior with reduction of intermolecular distance, which indicates existence of a tunneling process in the reaction process.

INTRODUCTION

Polymerization of C_{60} molecules in solid phase is caused by high temperature treatment under high pressure or exposure to light. Low dimensional polymer phases with orthorhombic, tetragonal and rhombohedral structures have been synthesized by the pressure - temperature treatment method [1,2]. The laser irradiation on a C_{60} crystal at room temperature induces clusters of the polymerized phase in solid C_{60} [3].

The polymerization process is understood in terms of the cycloaddition reaction mechanism. The Schmidt rule for the "2+2" cycloaddition reaction requires parallel alignment of double bonds and an inter-carbon atom distance shorter than 4.2 Å. As the distance between the nearest carbon atom is 2.9 Å and molecules rotate freely in crystalline C_{60} at room temperature, the Schmidt rule is satisfied at ambient pressure. It is of interest to investigate dependence of photopolymerization rates on the intermolecular distance in solid phase. The crystalline C_{60} is a unique system for this aim because the π-electron distribution is quasi-spherical and highly symmetric.

In this paper, we have investigated pressure dependence of photopolymerization reaction coefficients at room temperature. We have carried out a time-gated Raman experiment under high pressure using a diamond anvil cell. The obtained dependence clearly shows an exponential-type behavior, which indicates the existence of a tunneling process on the potential energy surface.

EXPERIMENTS

C_{60} single crystals were prepared by a sublimation method using 99.98% purified C_{60} powder which was annealed in a quartz tube under dynamic vacuum at 470 K for

24 hours. The sample and ruby chip were loaded into a diamond-anvil cell filled with a pressure medium (silicone oil) to achieve hydrostatic pressure. The applied pressure was estimated from the emission peak energy of ruby chip. Raman scattering spectra and their time evolution were measured in the backscattering geometry using a double-grating monochromator equipped with a cooled CCD detector. A 514.5 nm line of Ar^+ laser was used as a light source for both photopolymerization and Raman scattering measurement.

RESULTS AND DISCUSSION

Figure 1 shows the time evolution of Raman spectra in the vicinity of Ag(2) mode in C_{60} crystal under laser irradiation at the power density of 93 mW/mm^2. In the Raman spectra measured at ambient pressure (Fig.1(a)), the Raman peak at 1469 cm^{-1} is ascribed to the pentagonal-pinch Ag(2) mode. The Raman intensity of Ag(2) mode decreases with increase of irradiation time, while a broad Raman band at 1459 cm^{-1} due to the Raman active mode of a polymerized phase grows up. As shown in Fig. 1 (b), the Ag(2) mode is observed at 1474 cm^{-1} when the hydrostatic pressure of 1.1 GPa is applied. Under high pressure, the Ag(2) mode peak shifts to the higher frequency side because of compression of C_{60} crystals [4]. The Raman intensity at 1474 cm^{-1} rapidly decreases with irradiation

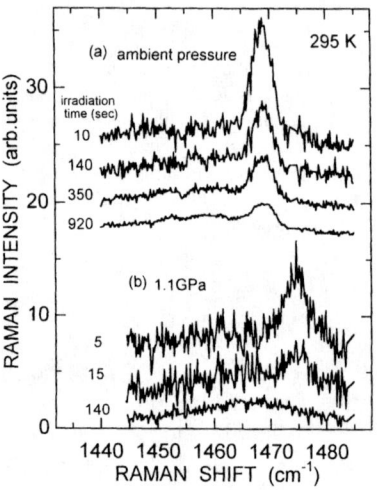

Fig. 1 Time evolution of Raman spectra in C_{60} crystal at ambient pressure (a) and 1.1 GPa (b).

time, and the peak disappears within 140 s. At 140 s the broad Raman band due to the polymerized phase dominates over the spectrum. This result suggests that the photopolymerization is more pronounced under high pressure.

Time variations of the Raman intensity of Ag(2) mode for different applied pressures are shown in Fig. 2. Comparing the behaviors measured at different pressures, we see that the Raman intensity decreases more rapidly at high pressure. The observed variation with irradiation time has been analyzed using a rate equation model taking into account both photo-generation and thermal dissociation of dimers and trimers. The detail of the model is described in our previous paper [5]. Rate equations of monomer (n_M), dimer (n_D) and trimer densities (n_T) are expressed by

$$\dot{n}_M = -k_{P1} \cdot n_M^2 + 2 \cdot k_{T1} \cdot n_D + k_{T2} \cdot n_T - k_{P2} \cdot n_M \cdot n_D$$
$$\dot{n}_D = \frac{1}{2} k_{P1} \cdot n_M^2 - k_{T1} \cdot n_D + k_{T2} \cdot n_T - k_{P2} \cdot n_M \cdot n_D$$
$$\dot{n}_T = k_{P2} \cdot n_M \cdot n_D - k_{T2} \cdot n_T$$

k_{P1} and k_{P2} are reaction rate constants of dimer and trimer generation, respectively. The reaction coefficients σ_1 and σ_2 are defined as $k_{P1}=\sigma_1 I$ and $k_{P2}=\sigma_2 I$, respectively. I is the power density of the laser light. k_{T1} and k_{T2} are thermal decomposition rates of dimer and trimer, respectively. Thermal decomposition rates are given by

$$k_{T1} = k_{T2} = \gamma \cdot \exp\left(-\frac{E_a}{k_B T}\right)$$

The following equation is further assumed:

$$N_0 = n_M + 2 \cdot n_D + 3 \cdot n_T$$

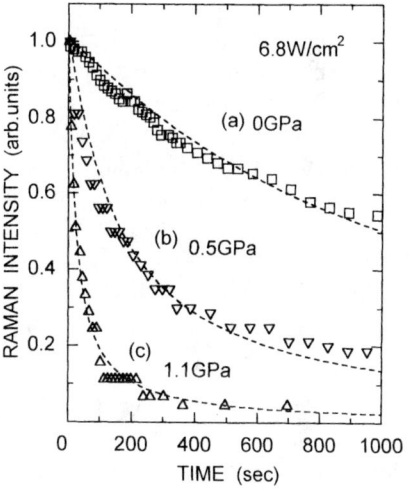

Fig. 2 Time variations of integrated intensity of Ag(2) mode for different pressures.

We calculated numerically the rate equations assuming that σ_1 and σ_2 have the same pressure dependence. The time variation of the Raman intensity was fitted to the numerical calculation changing σ_1 and σ_2 as adjustable parameters. Dashed lines in Fig. 2 show the fitted curves, and the obtained parameters are as follows: $\sigma_1 = 6.9 \times 10^{-37}$ J^{-1}m^5 and $\sigma_2 = 6.9 \times 10^{-36}$J^{-1}m^5 for ambient pressure, $\sigma_1 = 4.2 \times 10^{-36}$ J^{-1}m^5 and $\sigma_2 = 4.2 \times 10^{-35}$J^{-1}m^5 for 0.50 GPa and $\sigma_1 = 2.8 \times 10^{-35}$ J^{-1}m^5 and $\sigma_2 = 2.8 \times 10^{-34}$ J^{-1}m^5 for 1.10 GPa.

The pressure dependence of the reaction coefficient is shown in Fig.3. σ_1 exhibits an exponential-type behavior with increasing pressure. No significant change of σ_1 is noticed at the critical pressure of structural phase transition. The structural phase transition from a face centered cubic phase to a simple cubic phase takes place at about 0.2 GPa with increase of pressure at room temperature [6]. Dependence of σ_1 on the intermolecular distance between C_{60} molecules can be extracted from the pressure dependence of σ_1. We estimated an intermolecular distance (d) using the volume change with applied pressure reported by Pintschovius et al. [6] Thus obtained dependence of σ_1 on the intermolecular distance is shown in inset of Fig.3. The semi-logarithmic plot of σ_1 against d clearly shows an exponential dependence.

Here, we discuss the photopolymerization process exhibiting the exponential dependence of the reaction rate on the intermolecular distance. The theoretical model calculation revealed that there are multiple minima on the potential energy surface of the dimer generation via photoexcited states [7]. The experiment also showed existence of a potential barrier in the photopolymerization process. The effective barrier has been estimated as 0.24eV at ambient pressure [5]. Therefore, the exponential-type dependence on the intermolecular distance indicates that a tunneling process between the minima on the potential energy surface governs the reaction rate.

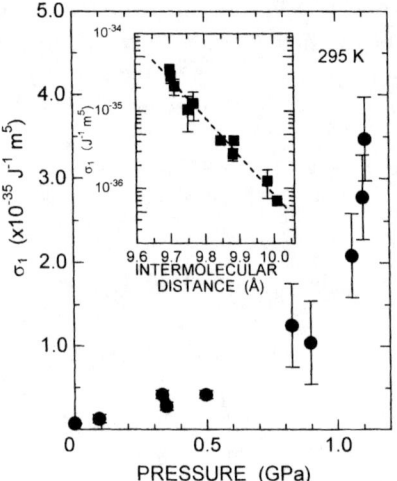

Fig. 3 Pressure dependence of reaction coefficient. Inset shows a semi-logarithmic plot of σ_1 against intermolecular distance

CONCLUSION

Variations of photopolymerization reaction rate with applied pressure in C_{60} crystals have been investigated by using time-gated Raman spectroscopy. The reaction coefficients of dimer and trimer generation were obtained from the experimental data and the rate equation analysis. The dependence of the reaction coefficient on the intermolecular distance exhibits an exponential-type behavior, which indicates the existence of the tunneling process on the adiabatic potential surface with multiple minima.

ACKNOWLEDGMENTS

We thank a Grant-in-Aid for Scientific Research from the Ministry of Education, Science, Sports and Culture, Japan. One of authors (M.S.) acknowledges the financial support by Research fellowships of Japan Society for Promotion of Science for Young Scientists and Toyota physical & Chemical Research Institute.

REFERENCES

1. Iwasa, Y., Arima, T., Fleming, R. M., Siegrist, T., Zhou, O., Haddon, R. C., Rothberg, L. J., Lyons, K. B., Carter Jr, H. L., Hebard, A. F., Tycko, R., Dabbagh, G., Krajewski, J. J., Thomas, G.A., Yagi, T., *Science* **264**, 1570-1572 (1994).
2. Davydov, V. A., Kashevarova, L.S., Rakhmanina, A. V., Senyavin, V. M., Ceolin, R., Szwarc, H., Allouchi, H., Agafonov, V., *Phys. Rev.* B **61**, 11936-11945 (2000).
3. Rao, A M., Zhou, Ping., Wang, Kai-An, Hager, G. T., Holden, J. M., Wang, Ying, Lee, W.-T., Bi, Xiang-Xin, Eklund, P.C., Cornett, D. S., Duncan, M. A., Amster, I. J., *Science* **259**, 955-957 (1993).
4. Snoke, D. W., Raptis, Y. S., Syassen, K., *Phys Rev.* B **45**, 14419-14422 (1992).
5. Sakai, M., Ichida, M., Nakamura, A. to be published in *Chem. Phys Lett.*
6. Pintschovius, L., Blaschko, O., Krexner, G., Pyka, N., *Phys. Rev.* B **59**, 11020-11025 (1999)
7. Suzuki, M., Iida, T., Nasu, K., *Phys. Rev.* B **61**, 2188-2198 (2000)

Out-of-plane and In-plane structures of the Cast Films of Long Alkyl Chain-Linked C_{60} via Phenyl Ring

M. Chikamatsu*, K. Kikuchi*, T. Kodama*, H. Nishikawa*, I. Ikemoto*, N. Yoshimoto¶, T. Hanada†, Y. Yoshida‡, N. Tanigaki‡ and K. Yase‡

*Department of Chemistry, Tokyo Metropolitan University, Hachiohji, Tokyo 192-0397, Japan
¶Faculty of Engineering, Iwate University, Ueda, Morioka 020-8551, Japan
†The Institute of Scientific and Industrial Research, Osaka University, Ibaragi, Osaka 567-0047, Japan
‡National Institute of Materials and Chemical Research, Tsukuba, Ibaragi 305-8565, Japan

Abstract. The films of dodecyl chain-linked C_{60} via *ortho*, *meta* and *para* positions on a phenyl group (**o-C12**, **m-C12** and **p-C12**) were prepared by casting the carbon disulfide solutions on the amorphous carbon coated silicon substrates. Out-of-plane and in-plane structures of these cast films were characterized by X-ray diffraction (XRD) and grazing incidence X-ray diffraction (GIXD) methods, respectively. As a result of XRD measurement, it was revealed that all the cast films take multilayer structures and the period of the layer structure of the **o-C12** film (32.5 Å) is much larger than that of the **m-C12** (23.2 Å) or the **p-C12** (23.3 Å) film. By GIXD measurement, it was found that C_{60} moieties form two-dimensional arrangement of the square lattice (a = 10.1 Å, γ = 90°) and pack densely in all the cast films.

INTRODUCTION

C_{60} solids not only exhibit interesting electrical and optical properties but are also promising candidates for a variety of applications ranging from transistor to diode, photoresist, optical limiter or photovoltaic device [1]. Recently, the films with the two-dimensional (2-D) C_{60} arrangement have attracted much attention [2]. So far we reported that a long alkyl chain-linked C_{60} via *ortho* position on a phenyl group takes a layer structure and forms the 2-D arrangement of C_{60} moieties in the cast film [3]. In order to investigate the layer structure for the long alkyl chain-linked C_{60} via a phenyl ring systematically, we prepared the cast films of dodecyl chain-linked C_{60} via *ortho* (**o-C12**), *meta* (**m-C12**) and *para* (**p-C12**) positions on a phenyl group (Fig. 1) and characterized out-of-plane and in-plane structures for these films.

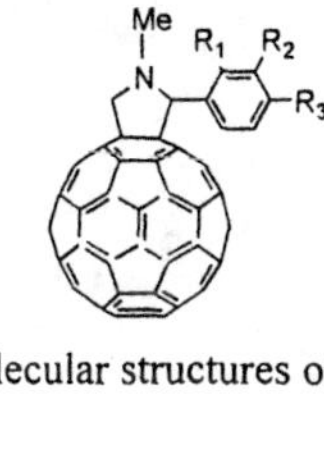

o-C12 : R_1=$(CH_2)_{11}CH_3$, R_2= R_3= H
m-C12 : R_2=$(CH_2)_{11}CH_3$, R_1= R_3= H
p-C12 : R_3=$(CH_2)_{11}CH_3$, R_1= R_2= H

FIGURE 1. Molecular structures of the dodecyl chain-linked C_{60} via a phenyl ring.

CP590, *Nanonetwork Materials*, edited by S. Saito et al.

EXPERIMENTS

Long alkyl chain-linked C_{60} via a phenyl ring (**o-C12**, **m-C12** and **p-C12**) were synthesized according to a similar route in the previous report [3, 4].

The films of **o-C12**, **m-C12** and **p-C12** were prepared by a cast method. The carbon disulfide solution of **o-C12**, **m-C12** or **p-C12** (0.15 ml) was dropped on the substrate using a pasteur pipette. Concentration of the solution was 4×10^{-3} mol/l. The substrate was amorphous carbon-coated silicon wafer (20 mm × 20 mm), which was made by the plasma polymerization equipment (NL-OP80NS, Nippon Laser & Electronics Lab.). Thickness of the amorphous carbon film was ~10 nm. Preparation of the uniform films of **o-C12**, **m-C12** and **p-C12** has been succeeded only on the amorphous carbon. These cast films were annealed at 120 °C for 12 hours.

For characterization of out-of-plane structure, X-ray diffraction (XRD) measurement of the cast films was carried out on a Rigaku Denki RU-300 using Cu K_α radiation (50 kV, 200 mA) with a curved graphite monochromator. The diffractions were measured from 1° to 30° in the 2θ-θ scan mode with 0.01°-step in 2θ and 0.6 second per step.

For characterization of in-plane structure, grazing incidence X-ray diffraction (GIXD) measurement of the cast films was carried out on a Rigaku Denki ATX-G using Cu K_α radiation (50 kV, 300 mA). The angle between the incident X-ray and the plane of the substrate was 0.2°. The diffractions were measured from 2° to 30° in the $2\theta\chi$-ϕ scan mode with 0.01°-step in $2\theta\chi$ and 0.6 second per step, where $2\theta\chi$ is the angle between the incident and the diffracted X-ray projected onto the plane of the substrate, and ϕ is the rotation angle of the sample around axis normal to the substrate.

RESULTS AND DISCUSSION

Fig. 2(a) shows XRD patterns of the **o-C12**, the **m-C12** and the **p-C12** films. In the **m-C12** and the **p-C12** films, 00l reflections were observed up to the seventh and no other reflection was observed within 30° in 2θ. These indicate that the **m-C12** and the **p-C12** films take the well-ordered layer structure, that is, the crystallites are preferentially oriented with the (001) plane parallel to the substrate. Peak positions of 00l reflections in the **m-C12** film nearly equal those in the **p-C12** film. The spacing of the (001) plane, d_{001}, of the **m-C12** and the **p-C12** films calculated by Bragg's equation is 23.2 and 23.3 Å, respectively.

On the contrary, in the **o-C12** film, 00l reflections were observed only up to the third and other reflection was also observed at $2\theta = 10.04°$. These indicate that the number of microcrystalline domains with the (001) plane parallel to the substrate is not enough and that some of the domains take different orientation. Compared with the **m-C12** and the **p-C12** films, peak positions of 00l reflections in the **o-C12** film shifted to small angle. As a result, the d_{001} of the **o-C12** film is 32.5 Å, which is much larger by 9.3 and 9.2 Å than those of the **m-C12** and the **p-C12** films, respectively.

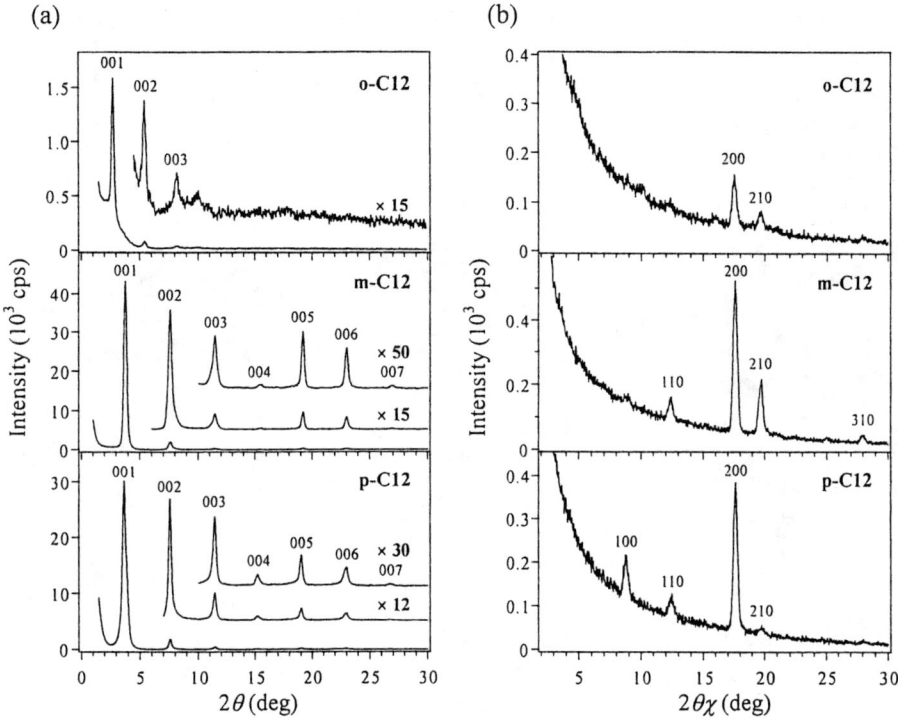

FIGURE 2. (a) XRD and (b) GIXD patterns of **o-C12** (top), **m-C12** (middle) and **p-C12** (bottom) cast films on amorphous carbon coated silicon substrates.

The large difference of the d_{001} is considered to depend on molecular conformation of each isomer. Then, minimum energy conformations of these molecules were examined by a MM2 calculation (Chem3D 3.5 program, Cambridge Soft Co.), and two-type minimum energy conformations were found for each isomer: a bent form and a stretched form. Molecular lengths of bent forms of **o-C12**, **m-C12** and **p-C12** are 21.8, 23.7 and 24.7 Å, respectively (space filling model). On the contrary, the lengths of stretched forms of **o-C12**, **m-C12** and **p-C12** are longer by 3.7, 5.1 and 5.1 Å than those of bent ones, respectively. From both the values of the d_{001} and the molecular length of obtained conformers, it is considered that **o-C12** takes the stretched form and that **m-C12** and **p-C12** take the bent forms in the multibilayer films (Fig. 3).

Fig. 2(b) shows GIXD patterns of the **o-C12**, the **m-C12** and the **p-C12** films. In all GIXD patterns, several $hk0$ reflections assigned the 2-D square lattice ($a - 10.1$ Å, $\gamma = 90°$) were observed. Since C_{60} diameter is 10.0 Å [5], it is considered that C_{60} moieties form the 2-D arrangement and pack densely in all the cast films (Fig. 4).

In conclusion, we succeeded in constructing two-type multibilayer structures of C_{60} derivatives by changing a linking position of a dodecyl chain to a phenyl group. **o-C12** takes the stretched form in a long-spacing (32.5 Å) multibilayer structure. On the other hand, **m-C12** and **p-C12** take the bent form in short-spacings (23.2 and 23.3 Å) multibilayer structures. In addition, it was found that C_{60} moieties form 2-D arrangement of the square lattice (a = 10.1 Å, γ = 90°) and pack densely in all the cast films.

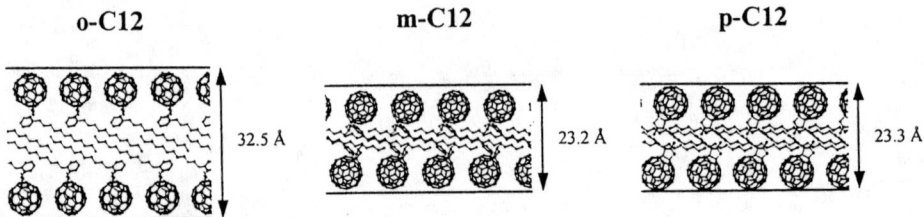

FIGURE 3. Illustration of one-period of multibilayer structures in the **o-C12** (left), the **m-C12** (center) and the **p-C12** (right) films.

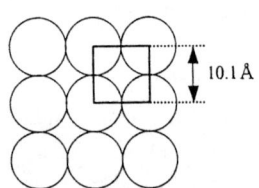

FIGURE 4. Illustration of in-plane arrangement of C_{60} moieties in the cast films.

ACKNOWLEDGMENTS

This work was partly supported by the Fund for Special Research Project at Tokyo Metropolitan University and by the Grants of the Ministry of Education, Science, Sports and Culture in Japan.

REFERENCES

1. Dresselhaus, M. S., Dresselhaus, G., and Eklund, P. C., *Science of Fullerenes and Carbon Nanotubes*, Academic Press, San Diego, 1996.
2. Mirkin, C. A., and Caldwell, W. B., *Tetrahedron* **52**, 5113-5130 (1996).
3. Chikamatsu, M., Hanada, T., Yoshida, Y., Tanigaki, N., Yase, K., Nishikawa, H., Kodama, T., Ikemoto, I., and Kikuchi, K., *Mol. Cryst. Liq. Cryst.* **316**, 157-160 (1998).
4. Maggini, M., Scorrano, G., and Prato, M., *J. Am. Chem. Soc.* **115**, 9798-9799 (1993).
5. Krätschmer, W., Lamb, L. D., Fostiropoulos, K., and Huffman, D. R., *Nature* **347**, 354-358 (1990).

V. ENDOHEDRALS

Structure of IPR-Violated Fullerene, $Sc_2@C_{66}$.

E. Nishibori*, M. Takata*, M. Sakata*, Chun-Ru Wang⁺,
M. Inakuma⁺ and H. Shinohara⁺

*Department of Applied Physics, Nagoya University, Nagoya 464-8603, Japan
⁺Department of Chemistry, Nagoya University, Nagoya 464-8603, Japan

Abstract. The endohedral structure of Isolated Pentagon Rule(IPR) violated scandium metallofullerene $Sc_2@C_{66}$ is determined by the maximum entropy method (MEM) combined with the Rietveld method using synchrotron radiation powder data. The result reveals for the exsistance of the fused pentagon first time. The fused pentagon locates close to the encapsulated Sc-dimmers. The Charge state of the encaged dimmer is also derermined $Sc_2^{2+}@C_{66}^{2-}$ form the MEM charge density. The Sc-Sc distance in Sc_2 cluster is 2.87(9) Å.

INTRODUCTION

Isolated-pentagon rule (IPR)[1,2] has been considered as the most important and essential rule in governing the geometry of fullerenes, stating that the most stable fullerenes are those in which all pentagons are surrounded by five hexagons. In fact, all the fullerenes produced, isolated and structurally characterized to date have been known to satisfy IPR[3-5]. There are no IPR fullerenes possible between C_{60} and C_{70}, and so the observation of any fullerenes in that range must violate the IPR. Recently, we have reported the first production, isolation of an IPR-violating scandium metallofullerene, $Sc_2@C_{66}$[6]. We succeeded in structure determination of IPR-violated fullerene, $Sc_2@C_{66}$, for the first time using synchrotron radiation powder data by the Maximum Entropy Method(MEM) combined with Rietveld analysis, MEM/Rietveld Method[5], and revealed that encapsulation of the Sc_2 dimer significantly stabilizes this otherwise extremely unstable C_{66} fullerene[6].

EXPERIMENT

The purity (99.8 %) of the material was confirmed by laser-desorption time-of-flight mass spectrometry. $Sc_2@C_{66}$ powder sample grown from toluene solvent was sealed in a silica glass capillary (0.3 mm inside diameter). X-ray powder pattern with good counting statistics was measured by the synchrotron radiation (SR) X-ray powder experiment with imaging plate (IP) as detectors at SPring-8 BL02B2(Fig.1). The exposure time on IP was 2 hours. The wavelength of incident X-rays was 0.75Å. The X-ray powder pattern of $Sc_2@C_{66}$ was obtained with a 0.02° step up to 20.3° in 2θ, which corresponds to 2.0 Å resolution in d-spacing. By pre-Rietveld analysis of the MEM/Rietveld Method, the $Sc_2@C_{66}$ crystal structure is determined as that of space group $Pmn2_1$(No.31); a=10.552(2)Å, b=14.198(2)Å, c=10.553(1)Å. The Result of the pre-Rietveld fitting is shown in Fig.2. The reliable factors of the pre-Rietveld fitting were R_{wp}=2.4%. and R_I=13.1%.

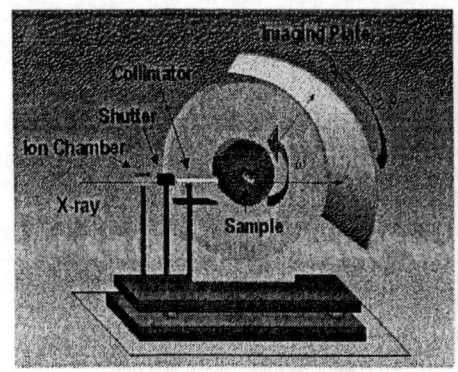

Capillary size : 0.1~0.4mm φ
Wavelength : 0.4~1.0Å
Exposure Time : 5min.~1 hour

2θ range : 0~80. degree
Temperature range: 20K~1000K

FIGURE .1 The large Debye-Scherrer Camera at BL02B2.

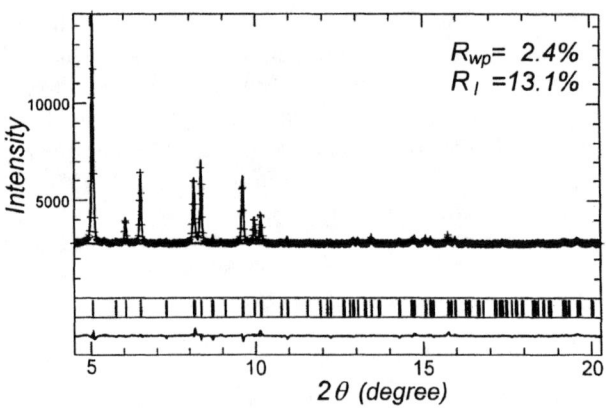

FIGURE 2. Fitting Result of pre-Rietveld Analysis of $Sc_2@C_{66}$ crystal.

Although there are a number of ways to violate IPR, the most straightforward way to do this is to generate the so-called "fused-pentagon" where pentagons are adjacent with each other. For 66-atom carbon cages with hexagonal and pentagonal faces, there are in total 4478 possible (non-IPR) structural isomers with $2 \times D_3$, $1 \times C_{3v}$, $18 \times C_{2v}$, $112 \times C_s$, $211 \times C_2$ and $4134 \times C_1$ symmetry[7]. Considering the observed 19-lines (5×2; 14×4) in the high resolution ^{13}C NMR spectrum of $Sc_2@C_{66}$, only 8 structural isomers of C_{66} with C_{2v} symmetry can satisfy this ^{13}C NMR pattern[6].

THE STRUCTURE OF $Sc_2@C_{66}$

The MEM 3-D electron density distribution of $Sc_2@C_{66}$ obtained by the MEM/Rietveld analysis[5,8] using the synchrotron X-ray diffraction is presented in Fig.3(a) together with an optimized geometry of $Sc_2@C_{66}$ based on the nonlocal density function B3LYP/Basis set [Sc(LanL2DZ); C(3-21G)] calculations (Fig.3(b)).

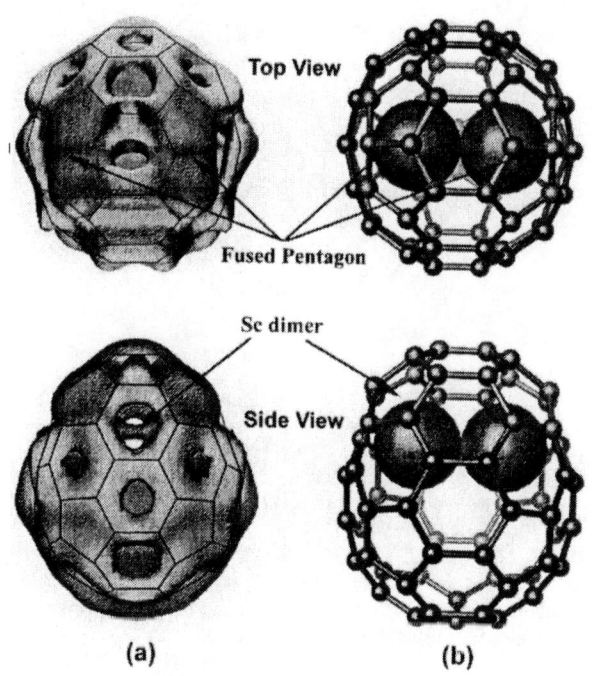

FIGURE 3. a) The X-ray structure of the IPR-violating $Sc_2@C_{66}$ fullerene with top view along the C_2 axis and side view. The equi-contour (1.4 e Å^{-3}) surface of the final MEM electron charge density. The Sc_2 dimer is colored in red. The two pairs of fused-pentagons are clearly seen. b)The calculated $Sc_2@C_{66}$ structures.

The final MEM charge densities, whose reliable factor is R_F=5.4%, clearly exhibit a pair of two-fold fused-pentagons on a C_{66}-C_{2v} cage that encapsulate a Sc_2 dimer; the most stable $Sc_2@C_{66}$ structure has the least number and degree of fused-pentagons out of the 4478 possible isomers.

$Sc_2@C_{66}$ shown in Fig. 3(b) contains two pairs of tow-fold fused pentagons to which the two Sc atoms are closely situated. The observed Sc-Sc distance is 2.87(9)Å, indicating the formation of a Sc_2 dimer in the C_{66} cage. The intrafullerene electron transfers in endohedral metallofullerenes have been known to play crucial roles in stabilizing the metallofullerenes[3-5,9,10]. The number of electrons in the area corresponding to Sc_2 dimer from the MEM charge density is 40.0(2) e, which is very close to that of $(Sc_2)^{2+}$ with 40 e. The *ab initio* calculation also indicates that the Sc_2 dimer donates two electrons to the C_{66} cage providing a formal electronic state of $(Sc_2)^{2+}@C_{66}^{2-}$. It is this charge-transfer interaction between the Sc_2 dimer and the fused pentagons that significantly decreases the strain energies caused by the pair of fused pentagons and thus stabilizes the fullerene cage. IPR is not necessarily a test for the stable geometry of endohedral metallofullerenes[6,9,11].

ACKNOWLEDGEMENTS

This work was supported by a Grant-In-Aid for Scientific Research from the Ministry of Education, Science, Sports and Culture. The synchrotron powder experiments were performed at the Spring-8 BL02B2 with the approval of the Japan Synchrotron Radiation Research Institute (JASRI).

REFERRENCES

1. Kroto, H., *Nature* **329**, 529-531 (1987).
2. Schmalz, T. G., Seitz, W. A., Klein, D. J. & Hite, G. E., *J.Am.Chem.Soc.* **110**, 1113-1127 (1988).
3. Shinohara, H., *et al.*, *Nature* **357**, 52-54(1992).
4. Yannoni, C. S., *et al.*, *Science* **256**, 1191-1192 (1992).
5. Takata, M., *et al.*, *Nature* **377**, 46-49 (1995); Takata, M., *et al.*,*Phys.Rev.Lett.* **83**, 2214-2217 (1999).
6. Wang, C-R., *et al.*, *Nature* **408**, 426-427(2000).
7. Fowler, P. W. & Manolopoulos, D.E. *An Atlas of Fullerenes* (Clarendon, Oxford, 1995) pp27 -42.
8. Takata, M., Nishibori, E. and Sakata, M., *Z.Kristallogr.* (2001) *in press*.
9. Kobayashi, K., Nagase, S., Yoshida, M. & Osawa, E., *J.Am.Chem.Soc.* **119**, 12693-12694 (1997).
10. Dorn, H. C., *et al.* in *Fullerenes: Recent Advances in the Chemistry and Physics of Fullerenes and Related Materials* (eds. K.M. Kadish & R.S.Ruoff) pp990-1002 (The Electrochemical Society, Pennington, 1998).
11. S, Stevenson. , *et al.*, *Nature* **408**, 427-428(2000).

Chemistry of Endohedral Metallofullerene Ions

T. Akasaka[1,2], T. Wakahara[1], S. Nagase[3], K. Kobayashi[3], M. Waelchli[4], K. Yamamoto[5], M. Kondo[1], S. Shirakura[1], Y. Maeda[1], T. Kato[2], M. Kako[6], Y. Nakadaira[6], X. Gao[7], E. Van Caemelbecke[7], and K. M. Kadish[7]

[1]*Graduate School of Science and Technology, Niigata University, Japan*
[2]*Institute for Molecular Science, Japan*
[3]*Department of Chemistry, Graduate School of Science, Tokyo Metropolitan University, Japan*
[4]*Bruker Japan, Japan*
[5]*Japan Nuclear Fuel Cycle Development Institute, Japan*
[6]*Department of Chemistry, The University of Electro-Communications, Japan*
[7]*Department of Chemistry, University of Houston, USA*

Abstract. The anion of La@C_{82} was electrochemically prepared and isolated. Anionic La@C_{82}(-) is very stable in water, even after exposure to air at room temperature. The high stability of La@C_{82}(-) is essentially due to its closed-shell electronic structure. As evidenced by the ESR analysis, La@C_{82}(-) is diamagnetic. These experimental findings are confirmed by density functional calculations. The cage structure of La@C_{82} was determined for the first time and shown to have C_{2v} symmetry based on the ^{13}C NMR measurements of the compound in its anionic form.

INTRODUCTION

Endohedral metallofullerenes have attracted special interest since their first proposal in 1985. Smalley and co-workers showed in 1991 that several lanthanum-containing fullerenes can be produced, and that extraction with toluene yields mostly La@C_{82}. Since then, La@C_{82} has been extensively investigated as the prototype of isolable metallofullerenes, but its structure is still unknown. Although M@C_{82} (M = group 3 or lanthanide metals) have been isolated and purified in macroscopic quantities, their instability in air has prevented detailed experimental characterization. Recently, we have accomplished the first preparation and isolation of the anion of La@C_{82} by an electrochemical method which was also used to generate the metallofullerene in its cationic form [1]. La@C_{82}(-) has an unusually high stability due to its diamagnetic nature originating from a closed-shell structure. This characteristic property has led to the first successful determination of the cage structure of La@C_{82} by ^{13}C NMR analysis, which is confirmed by density functional calculations [1].

RESULTS AND DISCUSSION

Bulk controlled potential electrolysis of La@C$_{82}$ was used to prepare the anion. No ESR signal was observed for La@C$_{82}$(-), unlike La@C$_{82}$, which indicates that La@C$_{82}$(-) is diamagnetic. The anion of the second isomer was also highly stable and diamagnetic. The cations of two La@C$_{82}$ isomers were prepared and isolated in the same way. To our knowledge, these are the first examples for the isolation of reduced or oxidized forms of M@C$_{82}$. As an extension, we have also prepared the anion and cation of two Pr@C$_{82}$ isomers that were recently purified [2].

La@C$_{82}$ shows broad absorption bands over the entire near-IR region down to 2300 nm because of its open-shell electronic structure, described formally as La^{3+}C$_{82}$. A significant color change from dark brown to dark green was observed during reduction of La@C$_{82}$. La@C$_{82}$(-) has an onset of a band around 1600 nm, a near-IR band at 930 nm, and a broad visible band at 580 nm. The color and absorption spectrum of La@C$_{82}$(-) did not change after 4 months in air, while those of La@C$_{82}$(+) was invariant for only several hours at room temperature under argon.

Because of the difficulty in preparing single crystals, NMR measurements are most useful for structural determination, but have not been utilized for La@C$_{82}$ due to its paramagnetic nature. However, the high stability and diamagnetic nature of La@C$_{82}$(-) allow NMR determination of the structure of La@C$_{82}$. The ^{139}La NMR spectrum of La@C$_{82}$(-) exhibits a single peak in d$_4$-ODCB at 300K with a linewidth of ~2600 Hz. The chemical shift at -470 ppm is close to that at -403 ppm observed for La$_2$@C$_{80}$. This may suggest that La has a similar formal charge in La@C$_{82}$(-) and La$_2$@C$_{80}$.

The C$_{82}$ fullerene has nine distinct isomers (C$_{3v}$ (a), C$_{3v}$ (b), C$_{2v}$, C$_2$ (a), C$_2$ (b), C$_2$ (c), C$_s$(a), C$_s$ (b), and C$_s$ (c)) that satisfy the isolated pentagon. Since the ^{13}C NMR study of C$_{82}$ shows only one isomer with C$_2$ symmetry is abundantly produced, it was once assumed that La was encapsulated inside the abundant isomer. Because of the three-electron transfer from La to C$_{82}$, however, it was recently predicted that encapsulation of La inside the C$_{2v}$, C$_{3v}$ (b), or C$_s$ (c) isomers is energetically much more favorable, which leads to C$_{2v}$, C$_{3v}$, and C$_s$ symmetry, respectively. These endohedral structures have 24 [17(4) + 7(2)], 17 [11(6) + 5(3) + 1(1)] and 44 [38(2) + 6(1)] nonequivalent carbons, respectively, where the values in parentheses is the relative intensity. The 125 MHz ^{13}C NMR spectrum of La@C$_{82}$(-) exhibits 17 distinct lines of near-equal intensity and 7 lines of half the intensity, verifying clearly that La@C$_{82}$ has C$_{2v}$ symmetry. This agrees with the fact that the C$_{2v}$ structure is energetically most stable.

The C$_{2v}$ structure of La@C$_{82}$ was optimized at the BLYP level and identified as an energy minimum from frequency calculations. La@C$_{82}$ is most energetically stabilized when La approaches the center of one hexagonal ring in C_{82} along the C_2 axis. This is consistent with the EXAFS study of La@C_{82} which shows that the number of nearest neighbor carbons is six. The same was calculated at the BLYP level for La@C_{82}(-) and La@C_{82}(+), confirming that La@C_{82} maintains C_{2v} symmetry even upon reduction and oxidation. The distances between La and the hexagonal carbons were calculated to be 2.638 and 2.646 Å for La@C_{82}. These differ little from those of 2.636 and 2.643 Å for La@C_{82}(-) and 2.640 and 2.649 Å for La@C_{82}(+), suggesting that the La position is only little changed upon either reduction and oxidation.

La@C_{82} has an open-shell structure. Reduction and oxidation take place on the carbon cage of La@C_{82}, leading to a closed-shell electronic structure of La@C_{82}(-) and La@C_{82}(+). Therefore, these ions have no radical character, as confirmed by the ESR study. This is consistent with the fact that La@C_{82}(-) is air-insensitive. The observation that La@C_{82}(+) is less stable in air than La@C_{82}(-) is probably due to the lower LUMO level of La@C_{82}(+).

In conclusion, La@C_{82} becomes diamagnetic and remarkably stable due to the closed shell structure obtained upon reduction. The ^{13}C NMR spectrum of the anion reveals that La@C_{82} has C_{2v} symmetry. ^{13}C NMR measurements of paramagnetic metallofullerenes in anionic forms may be widely applicable for structural determination. Isolation of an endohedral metallofullerene in its anionic form is an important stepping-stone on the way not only to the structural determination of neutral paramagnetic metallofullerenes but also for developing future applications.

ACKNOWLEDGMENT

This work was supported in part by a grant from the Asahi Glass Foundation and by a grant from the Ministry of Education, Science, Sports, and Culture of Japan. K. M. K. also acknowledges support of the Robert A. Welch Foundation (Grant E-680).

REFERENCES

1. T. Akasaka, et al., *J. Am. Chem. Soc.*, **122**, 9316 (2000).
2. T. Akasaka, et al., *Chem. Phys. Lett.*, **319**, 153 (2000).

Endohedral Metallofullerene

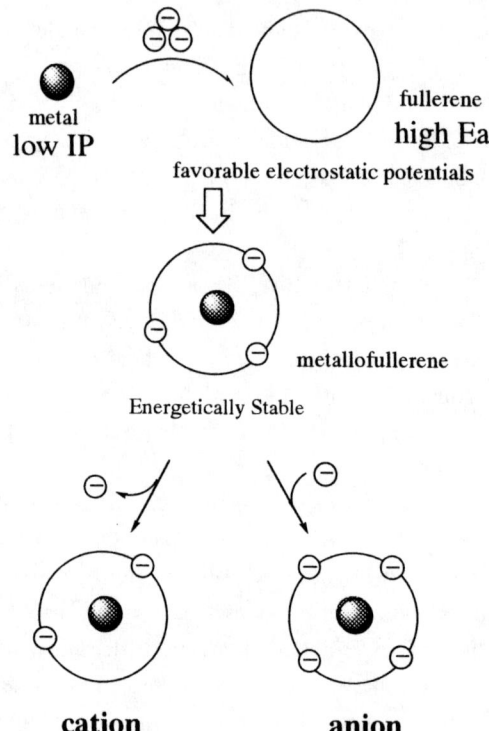

Anion and Cation Chemistry of Endohedral Metallofullerene

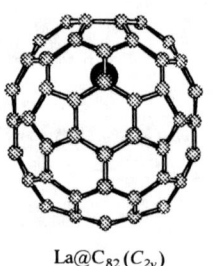

La@C$_{82}$ (C_{2v})

Spin Dynamics of Lanthanum Metallofullerenes

Shingo Okubo[1], Tatsuhisa Kato[1,2] *

[1] The Graduate University for Advanced Studies,
Okazaki, Myodaiji, 444-8585, Japan
[2] Institute for Molecular Science,
Okazaki, Myodaiji, 444-8585, Japan

Abstract. Enormous variety of electron spin resonance (ESR) spectra of La@C_ns was obtained in terms of g factor, hyperfine coupling constant, and line width. Line width of La@C_ns (n= 80, 82, and 84) at various temperature was analyzed by the theory on the basis of the hydrodynamic rotation in solution. The feature of the temperature dependence of the line width was almost interpreted by the hydrodynamics, and the electronic structure of La@C_n was deduced from the ESR parameters, which were obtained by the analysis. However in the cases of the isomer I of La@C_{80} and isomer II of La@C_{84} abnormally large line width was measured. The topological cage structure of La@C_{80} and La@C_{84} reflected on the specific spin dynamics.

INTRODUCTION

Since the first macroscopic production of endohedral metallofullerene, it has been of much interest to understand structure, chemical and physical properties of this new carbon based material [1]. In arc-generated soot, there are series of metallofullerenes (M_xC_n). In the case of extracts of soot containing Sc, Y, and La, ESR spectra showed existence of numbers of ESR active species [2-4]. For further investigation we need to obtain the series of pure materials. HPLC method has been widely used to obtain pure fullerene samples. [5] Full separation of topological isomers of each La@C_n component (n=76 to 90) was attempted by 2-stage HPLC separation with chlorobenzene eluent, and all species of La@C_n with even number n from 76 to 90 were detected. Among them La@C_{76}, La@C_{80}-I, II, La@C_{84}-I, and II were newly purified, La@C_{78}, La@C_{86}, and La@C_{88} were partially purified, and their ESR spectra were obtained. Among them a pair of purified isomers of La@C_{80}, La@C_{82}, and La@C_{84} were measured by ESR at various temperatures. The line width of ESR spectra in CS_2 solution was analyzed by the theory on the basis of hydrodynamics.

EXPERIMENTALS

HPLC separation was performed by the system of LC-908-C60 (Japan Analytical Industries). The first stage of separation was done using pentabromobenzyl column (COSMOSIL 5PBB, 20mm-i.d., 250mm-length, Nacarai Tesque Co.) with chlorobenzene eluent. To remove co-eluted hollow fullerenes and separate topological isomers for La@C_{76}, La@C_{84}, La@C_{80}-I, II, La@C_{84}-I, and II, second stage HPLC was carried out using 2-(pyrenyl) ethylsililated silica column (COSMOSIL 5PYE, 20mm-i.d., 250mm-length, Nacarai Tesque Co.) with chlorobenzene eluent.

The isolation of sample was confirmed by positive- and negative- LD-TOF mass spectrometry (KOMPACT MALDI IV, KRATOS) and purity of La@C_{76}, La@C_{80}-I, II, La@C_{84}-I, and II, were determined as more than 95%, that of La@C_{78} was about 60%, respectively. ESR spectra were recorded by ESR spectrometer (Bruker ESP300E and E500) at various temperatures. Samples were dissolved in CS_2, degassed by freeze-pump-thaw cycle, and sealed in thin wall quartz tube.

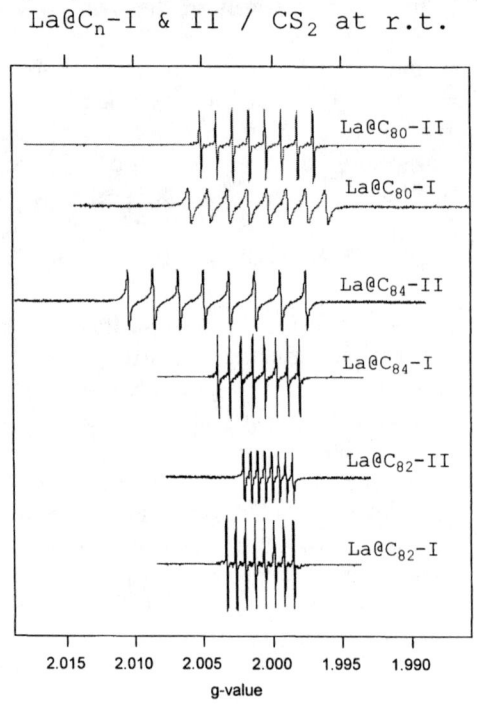

FIGURE 1.
ESR spectra of La@C_n isomers (n = 80, 82, and 84) at room temperature in CS_2 solution.

RESULTS AND DISCUSSIONS

Electron Spin Resonance (ESR) spectra of a pair of isomers of La@C_{80}, La@C_{82}, and La@C_{84} are shown in Figure 1. There are three important features of spectra measured at room temperature in CS_2 solution. At first isotropic 8 lines with equal intensity were obtained at room temperature, which was due to the hyperfine coupling (hfc) with La nucleus I=7/2. Second, hfc constant, which corresponds to the splitting between 8 lines, was very small. Thinking the big nuclear magneton of La, the splitting was extraordinarily small. Third the satellite structure, which was due to the hfc with ^{13}C nuclear on fullerene cage, were observed. 8 lines with equal intensities were characterized by the spectral averaging of hydrodynamics in solution. ESR parameters are in principle anisotropic in terms of the orientation of molecule. Though in frozen solution the spectra exhibited some anisotropy, at room temperature the anisotropy was averaged out by the hydrodynamic rotation. At the intermediate temperature insufficient averaging of the anisotropy gave specific spectral feature. 8 lines with equal intensities were changing with decreasing of temperature. At 170K the spectrum exhibited a strange fish bone pattern. [6] This pattern was due to the line broadening which depended on the quantum number of the La nuclear spin. This line broadening mechanism could be interpreted by the well defined hydrodynamic theory, and the anisotropy of the ESR parameters were deduced from the analysis of this spectral pattern. From the analysis the electronic structure of the isomer I of La@C_{82} was determined. 97.6% of spin density sited on the π orbital of outer fullerene cage, and 2.4% density remained on $5dz^2$ orbital of La nucleus. Tiny amount of density was on 6s orbital of La. As a result the electronic structure of La@C_{82} was described as La^{3+}@C_{82}^{3-}.

In almost all case of La fullerenes the line width in solution were characterized by the hydrodynamics, however, some exceptions were noticed. The line width of isomer I of La@C_{80} and isomer II of La@C_{84} were much bigger than the other at room temperature, as shown in Figure 1. From the analysis of the temperature dependence, these spectra exhibited the extra component of line width, which did not depend on the quantum number of La nuclear spin. Moreover these spectra exhibited very broad ^{13}C satellite structure. This meant that ^{13}C satellite lines were broaden by the extra and relative dynamics other than hydrodynamics of molecule.

The extra and relative dynamics would be the jumping among the Jahn-Teller (J-T) distorted structure of fullerene cage. Actually J-T effect could be expected in the case of fullerene cages of C_{80} and C_{84}. High symmetry of T_d, D_{6h}, and D_{3d} are possible for C_{84} cage, and I_h, D_{5d}, D_{5h}, and D_3 for C_{80}. For example, the T_d structure of C_{84} would be distorted to the lower symmetry of C_{2v}, and some jumping dynamics could happen among three distorted structures. The principal axis of the oval fullerene structure switched with this jumping, and the conjugation of the d-orbital of La metal would be modulated. As the result some extra component of line width came from the modulation.

In the case of the isomer II of La@C_{82}, the ESR spectrum also exhibited broad ^{13}C satellite structure, so the extra and relative dynamics other than hydrodynamics of molecule would be expected. The high symmetry of C_{3v} is possible for the C_{82} cage. However Akasaka et al. recently reported the cage symmetry of Cs for the isomer II of La@C_{82} by means of ^{13}C-NMR measurements of the anion of La@C_{82}-II. The consistent explanation is that the C_{3v} structure of La@C_{82}-II would be deformed to the lower symmetry Cs because of the J-T effect.

ACKNOWLEDGMENTS

The present research was financially supported by Grand-in-Aid of Scientific Research (No. 11165243) from the Japan Ministry of Education, Science, Sports and Culture. Thanks are due to the research Center for Molecular Materials, IMS, for assistance in obtaining the temperature dependent ESR spectra.

REFERENCES

(1) R. D. Johnson, M. S. deVries, J. Salem, D. S. Bethune, and C. S. Yannoni, *Nature*, **355**, 239(1992).
(2) M. Hoinkis, C. S. Yannoni, D. S. Bethune, J. R. Salem, R. D. Johnson, M. S. Crowder, and M. S. deVries, *Chem. Phys. Lett.* **198**, 461(1992).
(3) S. Suzuki, S. Kawata, H. Shirmaru, K. Yamauchi, K. Kikuchi, T. Kato, and Y. Achiba, *J. Phys. Chem.* **96**, 461(1992).
(4) S. Bandow, H. Kitagawa, T. Mitani, H. Inokuchi, Y. Saito, H. Yamaguchi, N. Hayashi, H. Sato, and H. Shinohara, *J. Phys. Chem.* **96**, 9609(1992).
(5) Yamamoto et al., the 20th Fullerene General Symposium Abstract, and references therein.
(6) T. Kato, S. Suzuki, K. Kikuchi, and Y. Achiba, *J Phys. Chem.*, **97**, 13425(1993)

Valence Change of Tm Atom in Metallofullerenes

Koichi Kikuchi*, Koichi Sakaguchi*, Ozawa Norio*, Takeshi Kodama*, Hiroyuki Nishikawa*, Isao Ikemoto*, Kenji Kohdate[§], Daiju Matsumura[§], Toshihiko Yokoyama[§], and Toshiaki Ohta[§]

*Department of Chemistry, Tokyo Metropolitan University, Hachioji, Tokyo 192-0397, Japan
[§]Department of Chemistry, University of Tokyo, Bunkyo-ku, Tokyo 113-0033, Japan

Abstract. The Tm LIII-edge XANES spectra were examined on $Tm@C_{82}$(I), $Tm@C_{84}$(III) and $Tm_2@C_{82}$(I). The absorption edge of $Tm_2@C_{82}$ is observed almost at the same position of Tm_2O_3, but those of $Tm@C_{82}$ and $Tm@C_{84}$ shifted about 7eV to the lower energy side. The 8eV energy shift was reported between the edges of TmS and TmTe, where Tm atom takes a trivalent and divalent state, respectively. Therefore Tm atom was confirmed to take the trivalent state in di-metallofullerenes, contrary to the divalent state in mono-metallofullerenes.

INTRODUCTION

Metallofullerenes have been attracting much attention in chemistry and physics since the first report of the production of $La@C_{82}$ [1]. In metallofullerenes, there is the electron transfer from metal atoms to fullerene cage, which is expected to produce the interesting phenomena. The systematic works of lanthanide-metallofullerenes have indicated that these elements can be divided into two groups from the HPLC elution behavior and the similarity of the absorption spectra [2-4]. The elements (La, Ce, Pr, Nd, Gd, Tb, Dy, Ho, Er and Lu) in one group take the trivalent state, since La is considered to take +3 state from various experiments [5-7]. On the other hand, the other lanthanide atoms (Sm, Eu, Tm and Yb) take the divalent state from the similarity of the absorption spectra between $M@C_{82}$ (M= Sm, Eu, Tm and Yb) and $Ca@C_{82}$ [2, 8, 9]. The XPS study has also confirmed the divalent character of Tm atom in $Tm@C_{82}$ [10].

Recently we found the good resemblance of UV-vis-NIR absorption spectra between the isomers of $Tm_2@C_{82}$ and $Er_2@C_{82}$, suggesting the trivalent state in $Tm_2@C_{82}$, contrary to the divalent state in $Tm@C_{82}$ [11]. In order to investigate the electronic state of Tm-metallofullerenes, their Tm LIII-edge XANES spectra were examined.

EXPERIMENTAL

The soot containing Tm-metallofullerenes was produced by direct-current (600 A) arc discharge of Tm/C composite rods under He flow (6 L/min) at 180 Torr. Both the empty fullerenes and metallofullerenes were extracted from the soot by refluxing for 8 hours with 1, 2, 4-trichrolobenzene. The isolation of Tm@C$_{82}$(I), Tm@C$_{84}$(III) and Tm$_2$@C$_{82}$(I) was carried out by the high-performance liquid chromatography (HPLC) method using two kinds of complimentary columns. Toluene or CS$_2$ was used as an eluent and a flow rate was 12 mL/min. In the initial separation step, a Buckyprep column (20mmϕ×250 mm, Nacalai Tesque) was employed to separate Tm@C$_{82}$(I) from Tm$_2$@C$_{82}$(I) and Tm@C$_{84}$(III). At second step, Tm@C$_{82}$(I) was separated from other two isomers by using two Buckyprep columns. At third and fourth steps two Buckyprep columns and a 5PBB column (20mmϕ×250 mm, Nacalai Tesque) were employed in each step to remove empty fullerenes. At second and third steps for Tm$_2$@C$_{82}$ and Tm@C$_{84}$ two Buckyprep columns and a 5PBB column were also employed in each step to remove empty fullerenes. At fourth step, two Buckyprep columns were again used to separate Tm$_2$@C$_{82}$ and Tm@C$_{84}$ from each other. The purity of an isolated species was checked by LD-TOF-MS.

The Tm LIII-edge XANES experiments were carried out using a Si(111) monochromator at Beam Line 12C of the Photon Factory in the National Laboratory for High Energy Physics. The XANES spectrum of Tm$_2$O$_3$, which is a reference as a trivalent state, was obtained by the transmittion method. Those of metallofullerenes were measured by fluorescence method with a Lytle ionization chamber. The metallofullerene samples were mounted into the hole (1mmϕ, depth 0.5mm) on the quartz plate.

RESULTS AND DISCUSSION

Figure 1 shows the Tm LIII-edge XANES spectra of Tm@C$_{82}$(I), Tm@C$_{84}$(III) and Tm$_2$@C$_{82}$(I) with that of Tm$_2$O$_3$. The inflection points E$_0$ of the Tm LIII-edge in the

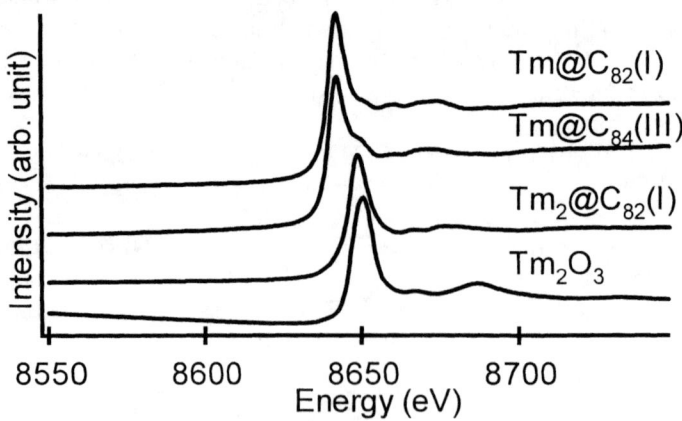

FIGURE 1. Tm LIII-edge XANES spectra of Tm@C$_{82}$(I), Tm@C$_{84}$(III) and Tm$_2$@C$_{82}$(I) with Tm$_2$O$_3$ as a reference.

spectra of Tm@C_{82}(I) and Tm@C_{84}(III) (8640 eV) are 7 eV lower than that of Tm_2O_3 (8647 eV). Such a Tm LIII-edge shift (8 eV) was reported between TmTe and TmS, where the oxidation states of Tm atom are +2 and +3, respectively [12]. Then it is concluded that Tm atom takes the divalent state in mono-metallofullerenes such as Tm@C_{82}(I) and Tm@C_{84}(III). This conclusion is consistent with the result of the former XPS study [10]. On the other hand, the inflection point in the XANES spectrum of Tm_2@C_{82}(I) (8647 eV) is almost the same as that of Tm_2O_3. This result confirms the trivalent state of Tm atoms in di-metallofullerenes such as Tm_2@C_{82}(I). It is also noteworthy that there is no peak splitting in its XANES spectrum. This suggests that two Tm atoms in Tm_2@C_{82}(I) have similar electronic structure.

There is the remaining problem; why does the valence state of Tm atom change due to the number of metal atoms inside? At present we speculate that this change of Tm atom may depend upon the ionic radius. For large ion such as La^{3+} (1.032Å), the production of M_2@C_{82} is not detected. The ionic radius of Tm^{3+} (0.880 Å) is smaller than that of Gd^{3+} (0.938 Å), which is the largest ion encapsulated into C_{82} cage as di-metallofullerenes, but the ionic radius of Tm^{2+} (1.03 Å) is bigger and almost the same as that of La^{3+}, which is not encapsulated into the C_{82} cage as di-metallofullerene. Therefore the divalent ion of Tm is too large, so the trivalent state may be necessary for Tm atom to be stably trapped into the C_{82} cage as di-metallofullerenes.

In conclusion, Tm atom was confirmed to take the trivalent state in di-metallofullerenes, contrary to the divalent state in mono-metallofullerenes from their XANES spectra. Such a valence change due to the number of the inside atoms is one of interesting properties of metallofullerenes.

ACKNOWLEDGMENTS

The authors thank Science Create Ltd. for the use of their soot generators. This work was partly supported by the Japanese Ministry of Education, Science, Sports, and Culture for the grants-in-aid and by Ishikawa Foundation for Carbon Science and Technology.

REFERENCES

1. Chai, Y., Guo, T., Jin, G., Haufler, R. E., Chibante, L. P. F., Fure, J., Wang, L., Alford, J. M., and Smalley, R. E., *J. Phys. Chem.* **95**, 7564 (1991).
2. Kikuchi, K., Sueki, K., Akiyama, K., Kodama, T., Nakahara, H., Ikemoto, I., and Akasaka, T., in *Recent Advances in the Chemistry and Physics of Fullerenes and Related Materials*; vol. 4, edited by K. M. Kadish and R. S. Ruoff, Electrochem. Soc. Inc., Pennington, NJ, 1997, p408.
3. Sueki, K., Akiyama, K., Yamauchi, T., Sato, W., Kikuchi, K., Suzuki, S., Katada, M., Achiba, Y., Nakahara, H., Akasaka, T. and Tomura, K., *Fullerene Science. Tech.* **5**, 1435 (1997).
4. Akiyama, K., Sueki, K., Kodama, K., Kikuchi, K., Ikemoto, I., Katada, M., and Nakahara, H., *J. Phys. Chem.* **A104**, 7224 (2000).
5. Johnson, R. D., de Vries, M. D., Salem, J. R., Bethune, D. S., and Yannoni, C. S., *Nature (London)* **355**, 239 (1992).
6. Hino, S., Takahashi, H., Iwasaki, K., Matsumoto, K., Miyazaki, T., Hasegawa, S., Kikuchi, K., and Achiba, Y., *Phys. Rev. Letters*, **71**, 4261 (1993).

7. Suzuki, T., Maruyama, Y., Kato, T., Kikuchi, K., and Achiba, Y., *J. Am. Chem. Soc.* **115**, 11006 (1993).
8. Xu, Z., Nakane, T. and Shinohara, H., *J. Am. Chem. Soc.* **118**, 11309 (1996).
9. Kirbach, U., and Dunsch, L., *Angew. Chem. Int. Ed. Engl.* **35**, 2380 (1996).
10. Pichler, T., Golden, M. S., Knupfer, M., Fink, J., Kirbach, U., Kuran, P., and Dunsch, L., *Phys. Rev. Letters* **79**, 3026 (1997).
11. Kikuchi, K., Akiyama, K., Sakaguchi, K., Kodama, T., Nishikawa, H., Ikemoto, I., Ishigaki, T.,. Achiba, Y, Sueki, K., and Nakahara. H., *Chem. Phys. Letters* **319**, 472 (2000).
12. Launois, H., Rawiso, M., Holland-Moritz, E., Pott, R., Wohlleben, D., *Phys. Rev. Letters* **44**, 1271 (1980).

Electronic Structure of Eu@C_{60}

Shugo Suzuki, Mizuho Kushida, Satoshi Amamiya, Susumu Okada, and Kenji Nakao

Institute of Materials Science, University of Tsukuba, Tsukuba 305-8573, Japan

Abstract. The electronic structure of Eu@C_{60} is investigated theoretically. The method used in the present study is the scalar relativistic full-potential linear-combination-of-atomic-orbitals method based on the density functional theory within the local density approximation. It is found that the reaction Eu+C_{60} → Eu@C_{60} is exothermic with the reaction energy of about −3 eV. Eu is placed at about 1.2 Å away from the center of the C_{60} cage. Two 6s electrons in the Eu atom are transfered to the C_{60} molecule as conduction electrons while seven 4f electrons remain in the Eu atom and thus the Eu atom in Eu@C_{60} exists as Eu^{2+}. The coupling between the conduction electrons and the 4f electrons is antiferromagnetic. The RKKY interaction between the Eu spins mediated by the conduction electrons through this antiferromagnetic coupling is very weak.

INTRODUCTION

The C_{60} compounds synthesized so far have shown interesting properties; superconductivity, ferromagnetism, antiferromagnetism, etc [1]. There has been, however, a missing area of the research of the compounds, i.e., the study on the endohedral C_{60} compounds. The lack is due to the difficulty of the synthesis of this kind of compounds. It is desirable to extend the investigation over the properties of the endohedral C_{60} compounds because other endohedral fullerene compounds have been found to show exotic behaviors.

Recently, the successful synthesis of Eu@C_{60} have been reported [2]. This enables one to study the properties of this compound in detail; the Eu L_{III}-edge X-ray absorption near edge structure (XANES) spectra have shown that the Eu atom in Eu@C_{60} exists as Eu^{2+} and furthermore the Eu L_{III}-edge X-ray absorption fine structure (XAFS) analysis has shown that the Eu atom is placed at 1.4 Å away from the center of the C_{60} cage.

Since the Eu atom has 4f electrons, the magnetic behavior expected for Eu@C_{60} should be of great interest. In particular, the nature of the magnetic coupling between the 4f electrons and the conduction electrons donated by the Eu atoms must be crucial to understand the magnetic properties of the material. Moreover, the resultant RKKY interaction between the Eu spins mediated by the conduction electrons through this magnetic coupling is also one of the important properties to be studied.

The purpose of the present study is to investigate the electronic structure of Eu@C_{60} theoretically. The method employed is the scalar relativistic full-potential linear-combination-of-atomic-orbitals (LCAO) method. In particular, we concentrate on the magnetic properties of Eu@C_{60}: the coupling between the $4f$ electrons and the conduction electrons and also the RKKY interaction between the Eu spins.

METHOD

The basis for the method of calculations is the density functional theory and the local-spin-density approximation is employed [3,4]. We have used the full-potential LCAO method that can take account of the scalar relativistic effects originated in the Darwin and mass-velocity terms [5]. We use a single basis function for a core orbital while we use double basis functions for a valence orbital. The atomic orbitals employed for the C atoms are the $1s$, $2s$, and $2p$ orbitals of a neutral C atom and the $2s$, $2p$ orbitals of a C^{2+} atom. Also, the atomic orbitals employed for the Eu atom are the $1s$, $2s$, $2p$, $3s$, $3p$, $3d$, $4s$, $4p$, $4d$, $4f$, $5s$, $5p$, $5d$, and $6s$ orbitals of a neutral Eu atom and the $6p$ orbitals of a Eu^{+} atom and the $5d$ and $6s$ orbitals of a Eu^{2+} atom and the $6p$ orbitals of a Eu^{3+} atom.

First, the geometry of the Eu@C_{60} molecule is optimized by using the conjugate gradient optimization scheme without any restriction on the geometry such as an impose of a symmetry on the geometry. Next, the electronic structure of the Eu@C_{60} crystal is calculated under the assumption that the molecules are assembled on a fcc lattice with the same lattice constant as the pristine C_{60} crystal. The magnetic coupling between the $4f$ electrons and the conduction electrons is then investigated in detail. Finally, the RKKY interaction is calculated by using a unit cell which includes two Eu@C_{60} molecules with parallel or antiparallel Eu spin configurations.

RESULTS AND DISCUSSION

First, the results of the calculations on the Eu@C_{60} molecule are given below. It is found as a result of the geometry optimization that the Eu atom is displaced at about 1.2 Å away from the center of the C_{60} cage. Accordingly, the molecule has a electric dipole moment; the calculated value is about 1.9 D. Two $6s$ electrons are donated to the C_{60} cage as valence electrons. On the other hand, seven $4f$ electrons remain on the Eu atom. This means that the Eu atom in the Eu@C_{60} molecule exists as Eu^{2+}. It is also found that the magnetic coupling between the valence electrons on the C_{60} cage and the $4f$ electrons in the Eu atom is antiferromagnetic; the spin direction of the valence electrons is opposite to that of the $4f$ electrons. The magnitude of this antiferromagnetic coupling is found to be about 0.1 eV.

Now we show the electronic structure of the Eu@C_{60} crystal calculated under the assumption that the Eu@C_{60} molecules are assembled in a fcc lattice with the same lattice constant of the pristine C_{60} crystal, their electric dipole moments being

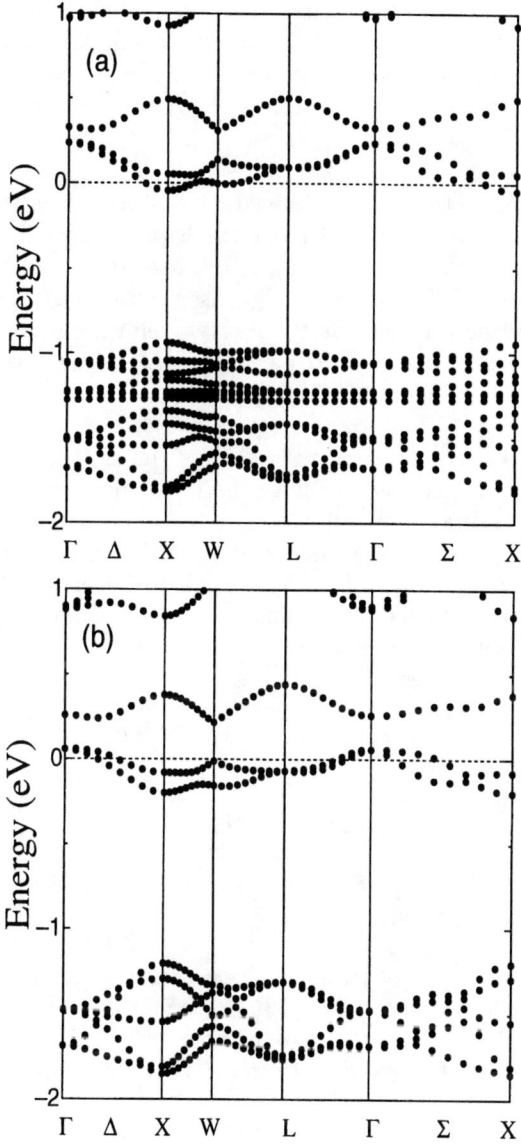

FIGURE 1. Band structure of Eu@C_{60} for (a) up-spin electrons and (b) down-spin electrons. The dotted line shows the Fermi level.

aligned in the ⟨111⟩ direction. Fig. 1(a) shows the band structure for the up-spin electrons and Fig. 1(b) shows that for the down-spin electrons. The dotted line in the figure indicates the Fermi level. The conduction bands are originated in the lowest unoccupied molecular orbitals (LUMO) of the C_{60} molecule. The five bands at the bottom in each figure are also originated in the highest occupied molecular orbitals (HOMO) of the C_{60} molecules. Furthermore, it should be noted that seven dispersion-less bands just above the HOMO-derived bands are originated in the $4f$ orbitals of Eu; these seven $4f$-derived bands are of the up-spin electrons while those of the down-spin electrons are at about 2 eV above the Fermi level and are not seen in the figure.

Several conclusions can be deduced from the above results. Firstly, two $6s$ electrons in Eu are donated into the LUMO-derived bands while seven $4f$ electrons in Eu still remain in Eu. This means that, also in the Eu@C_{60} crystal, the Eu atoms exist as Eu^{2+}. Secondly, the magnetic coupling between the conduction electrons and the $4f$ electrons is antiferromagnetic. That is, the majority-spin conduction electrons are down-spin electrons while the minority-spin conduction electrons are up-spin electrons. Although the solid-state effect reduces the magnitude of the spin polarization of conduction electrons slightly, the magnetic coupling is expected to be still strong as in the Eu@C_{60} molecule.

We further study one of the magnetic properties of the Eu@C_{60} crystal, i.e., the RKKY interaction. To this end, we calculate the total energies of two configurations of the Eu spins by using a unit cell consisting of two Eu@C_{60} molecules; one configuration is ferromagnetic one while the other is antiferromagnetic one in which we assume alternate stacking of up- and down-spin Eu@C_{60} layers along the ⟨111⟩ direction. As a result, we find that the ferromagnetic spin configuration is slightly stable more than the antiferromagnetic one; the energy difference is about 0.001 eV per Eu atom. However, this is beyond the accuracy of our calculations and accordingly we should regard the result as the indication that the RKKY interaction between the Eu spins is very weak.

REFERENCES

1. Ehrenreich, H., Spaepen, F.,(Eds.), *Solid State Physics*, Vol. 48, New York: Academic Press, 1994 and references therein.
2. Inoue, T., Kubozono, Y., Kashino, S., Takabayashi, Y., Fujitaka, K., Hida, M., Inoue, M., Kanbara, T., Emura, S., and Uruga, T., *Chem. Phys. Lett.* **316**, 381 (2000).
3. Hohenberg, P., and Kohn, W., *Phys. Rev.* **136**, B864 (1964).
4. Kohn, W., and Sham, L. J., *Phys. Rev.* **140**, A1133 (1965).
5. Suzuki, S., and Nakao, K., *J. Phys. Soc. Jpn.* **69**, 532 (2000).

Electronic Structures of the La@C_{82} Crystals by the Relativistic LCAO Method

S. Amamiya, S. Okada, S. Suzuki, and K. Nakao

Institute of Materials Science, University of Tsukuba, Tsukuba 305-8573, Japan

Abstract. We report on the electronic structure of La@C_{82} crystals by using the scalar relativistic full-potential linear-combination-of-atomic-orbitals method in the density functional theory. We find a metallic solution with half-filled energy band at the Fermi level. The band width around the Fermi level in La@C_{82} crystal is about 0.3 eV which is not affected by the arrangement of the C_{82} cage.

INTRODUCTION

Encapsulation of metal atoms inside fullerene cages, endohedral metallofullerene, are attracting great interest in the field of fullerene science. It might be expected to have novel properties which are not expected in hollow fullerenes, such as superconductivity, ferroelectrics, nonlinear optical response, and so on.. In particular, La@C_{82} has been widely studied as a typical monometallofullerene ever since the first success in extraction [1]. Electron spin resonance studies have shown that there is two dominant signals corresponding to two La@C_{82} isomers [2,3]. Furthermore, the recent ^{13}C NMR measurement of La@C_{82} anion reported that the main isomer of the extracted La@C_{82} has C_{2v} geometry [4]. The x-ray diffraction experiments on La@C_{82} powder, i.e., combination of the Rietveld analysis and the maximum entropy method, have also revealed that one of the extracted La@C_{82} has C_{2v} geometry [5]. In addition to the isolated La@C_{82} molecule, a solvent-free single crystal of La@C_{82} has been also synthesized by Watanuki et al [6]. They have revealed that the crystalline La@C_{82} has the face-centered cubic lattice structure ($a_0 = 15.78$Å) at room temperature. However, the details of the electronic structures are still unknown.

In the present study, we investigate the electronic structures of La@C_{82} crystals possessing C_{2v} cage symmetry by the scalar relativistic density-functional theory.

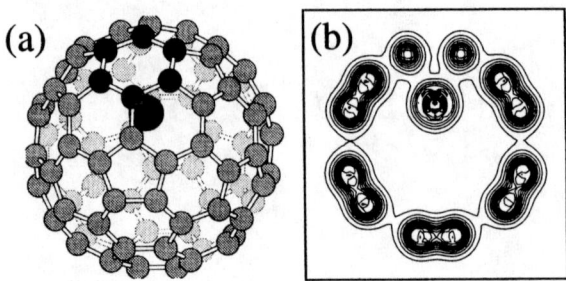

FIGURE 1. (a) The structure of the La@C_{82} molecule with C_{2v} cage symmetry. The C and La atoms are denoted by light and dark shaded circles, respectively. The pale shaded circles denote the nearest neighbor C atoms on the hexagon. (b) Contour plot of the total valence charge density of the La@C_{82} molecule. The contour lines are drawn from 0.0 eÅ^{-3} with 0.16 eÅ^{-3} intervals.

CALCULATION METHODS

To study the geometry and the electronic structure of La@C_{82}, we employ the scalar relativistic full-potential linear-combination-of-atomic-orbitals method which is based on the density functional theory with the local-density approximation [7]. We use not only the atomic orbital of neutral atoms but also those of charged atoms to consider the variational flexibility. The atomic orbitals used for C atoms are $1s$, $2s$, and $2p$ atomic orbitals of neutral C atoms and $2s$ and $2p$ atomic orbitals of C^{2+} atoms. Also, the atomic orbitals used for a La atom are $1s$, $2s$, $2p$, $3s$, $3p$, $3d$, $4s$, $4p$, $4d$, $4f$, $5s$, $5p$, $5d$, and $6s$ atomic orbitals of a neutral La atom, $5d$ and $6s$ atomic orbitals of a La^{2+} atom, and $6p$ atomic orbital of La^{+} and La^{3+} atoms. Exchange-correlation energy of interacting electrons is considered with functional form fitted to the Ceperley-Alder results [8]. We use 2064 and 4128 points per C and La atoms, respectively, to perform the three-dimensional numerical integration in the real space. Also, one k-point, i.e., Γ-point, is used for the integration in first Brillouin zone. The geometry of La@C_{82} molecule is optimized by the conjugate-gradient minimization scheme. In the optimized geometries, forces acting on the atoms are less than 0.5 eV/Å.

RESULTS AND DISCUSSION

We optimize the geometry of the La@C_{82} molecule without any restriction. We show the optimized structure and the total valence charge density of the La@C_{82} molecule in Fig. 1 (a) and (b), respectively. We find that the La atom is below the center of the hexagon and on the C_2-axis of the C_{82} cage. The distance between the La atom and the nearest neighboring C atom on the hexagon is 2.58 Å. This distance is too large to generate an inter-atomic covalent bond. In fact, by analyzing the charge distribution of the La@C_{82} molecule, the charge between the La and C atoms is considerably smaller than that on the C-C bonds (Fig. 1). Hence, the strong covalent bond between La and C is unlikely to take place and the La atom may freely move in the cage at room temperature. The calculated reaction energy for La@C_{82} is -8.4 eV. The reaction energy is defined by the difference between the total energy of La@C_{82} and the sum of the total energies of an isolated La atom and a C_{82} molecule. This result shows that the reaction La + $C_{82} \rightarrow$ La@C_{82} is exothermic. The electric dipole moment of the La@C_{82} molecule obtained is 2.2 D per molecule which is almost the same as that of H_2O (1.9 D). It is most likely that the electric dipole moment plays an important role to determine the arrangement of La@C_{82} molecules in the crystalline form. The La@C_{82} molecule is found to have the open shell electronic structure and the highest occupied state is singly occupied. The sum of spin density of the C_{82} cage is about 1.0 per molecule. The polarized electrons are mostly distributed on the carbon cages and extended over the whole network of the cage.

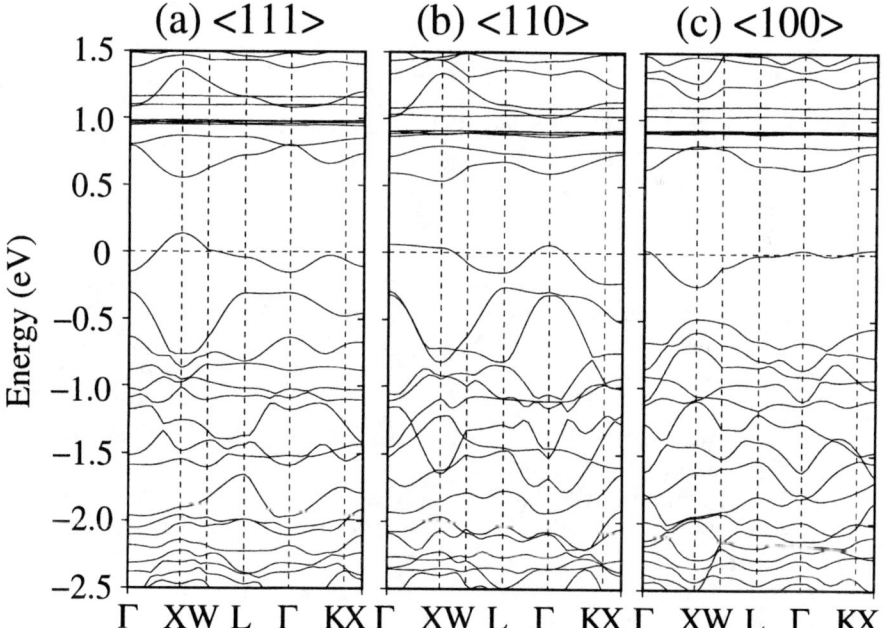

FIGURE 2. Electronic structures of La@C_{82} crystals. Energies measured from the Fermi level energy.

Figure 2 shows the electronic structures of La@C_{82} crystals. In this calculation, we assume that the La@C_{82} molecule are aligned in the fcc structure with experimentally observed lattice parameters [6]. Furthermore, we also assume that the electric dipole moments of the La@C_{82} molecules are aligned in one direction. We study, here, the three models of La@C_{82} crystals, in which the electric dipole moments are parallel to the (a) $\langle 111 \rangle$, (b) $\langle 110 \rangle$, and (c) $\langle 100 \rangle$ axes. The electronic structures of three models are qualitatively same. We find common features obtained as shown below. The electron spin is not polarized : We only find a metallic solution with a half-filled band at the Fermi level. The band width around the Fermi level is about 0.3 eV which is comparable to that of the fcc C_{60} [9]. The energy bands corresponding to the $4f$, $5d$, and $6s$ orbitals of the La atom are located at about 1 eV above the Fermi level. These energy bands exhibit extremely small dispersion in the whole Brillouin zone.

ACKNOWLEDGEMENTS

Part of this study is supported by the Grant-in-Aid for Scientific Research from the Ministry of Education, Science, Sports and Culture of Japan and the grant from Research and Development Applying Computational Science and Technology of Japan Science and Technology Corporation (ACT-JST).

REFERENCES

1. Chai, Y., Guo, T., Jin, C., Haufler, R. E., Chibante, L. P. F., Fure, J., Wang, L., Alford, J. M., and Smalley, R. E., *J. Phys. Chem.* **95**, 7564 (1991).
2. Suzuki, S., Kawata, S., Shiromaru, H., Yamauchi, K., Kikuchi, K., Kato, T., and Achiba, Y., *J. Phys. Chem.* **96**, 7159 (1992).
3. Bandow, S., Kitagawa, H., Mitani, T., Inokuchi, H., Saito, Y., Yamaguchi, H., Hayashi, N., Sato, H., and Shinohara, H., *J. Phys. Chem.* **96**, 9609 (1992).
4. T. Akasaka, T. Wakahara, S. Nagase, K. Kobayashi, M. Waelchli, K. Yamamoto, M. Kondo, S. Shirakura, S. Okubo, Y. Maeda, T. Kato, M. Kako, Y. Nakadaira, R. Nagahata, X. Gao, E. Van Caemelbecke, K. M. Kadish, J. Am. Chem. Soc. **122**, 9316 (2000).
5. Nishibori, E., Takata, M., Sakata, M., Tanaka, H., Hasegawa, M., and Shinohara, H., *Chem. Phys. Lett.* **330**, 497 (2000).
6. Watanuki, T., Fujiwara, A., Ishii, K., Shibata, T., Suematsu, H., Nakao, H., Fujii, Y., Kawada, H., Murakami, Y., Kikuchi, K., Achiba, Y., and Maniwa, Y., *Photon Factory Activity Report* **14**, 403 (1996).
7. Suzuki, S., and Nakao, K., *J. Phys. Soc. Jpn.* **69**, 532 (2000).
8. Ceperley, D. M., and Alder, B. J., *Phys. Rev. Lett.* **45**, 566 (1980).
9. Saito, S., and Oshiyama, A., *Phys. Rev. Lett.* **66**, 2637 (1991).

Low-energy Electron Energy Loss Spectroscopy of Monolayer and Thick La@C$_{82}$ Films Grown on MoS$_2$ Substrates

Keiji Ueno*, Yasunori Uchino*, Ken-ichi Iizumi*, Atsushi Koma*, Koichiro Saiki†, Yasuhira Inada¶, Kiyoe Nagai¶, Yoshihiro Iwasa¶, and Tadaoki Mitani¶

*Department of Chemistry, The University of Tokyo, 7-3-1, Hongo, Bunkyo-ku, Tokyo 113-0033, Japan
†Department of Complexity Science and Engineering, The University of Tokyo, 7-3-1, Hongo, Bunkyo-ku, Tokyo 113-0033, Japan
¶Japan Advanced Institute of Science and Technology, Tatsunokuchi, Ishikawa 923-1292, Japan

Abstract. Monolayer and multilayer La@C$_{82}$ films were epitaxially grown on MoS$_2$ substrates, and their growth features together with electronic states were investigated using atomic force microscopy and electron energy loss spectroscopy. It was found that the interaction between a La@C$_{82}$ molecule and a MoS$_2$ surface is stronger than that between La@C$_{82}$ molecules. The electronic state of the initial La@C$_{82}$ layer on the MoS$_2$ surface differs from that of the successively grown bulk-like layers due to excess electrons in the carbon cage of La@C$_{82}$.

INTRODUCTION

Electronic structure of organic molecules adsorbed on a substrate is usually drastically modified if covalent or ionic type interaction exists between them. In contrast, physical adsorption through so-called 'van der Waals interaction', causes far smaller influence on the electronic structure of organic molecules. However, the electronic structure of these physisorbed molecules may differ from that of molecules in its pure molecular crystal, because the substrate and grown materials have heterogeneous electronic structures. Electronic perturbation on adsorbed molecules could occur at the hetero-interface even through the van der Waals interaction, and a unique phenomenon may occur at the initial stage of the adsorption of organic molecules.

In this paper, we report on the electronic structure of monolayer and multilayer La@C$_{82}$ epitaxial films grown on MoS$_2$ substrates. MoS$_2$ has a two-dimensional layered structure with no active dangling bond on its cleaved surface. Therefore, La@C$_{82}$ molecules adsorb on the MoS$_2$ surface through the 'weak' interaction, then freely migrate, and form epitaxial layers with the same molecular interval as that of the bulk single crystal [1]. Successive La@C$_{82}$ layers also epitaxially grow on underlying layers. However, we have found from atomic force microscopy (AFM) observation that the growth mechanism of the initial La@C$_{82}$ layer on the MoS$_2$ substrate differs from that of successively grown layers. Furthermore, difference was also found in the electronic structures of the first and the successive La@C_{82} layers by the low-energy electron energy loss spectroscopy (LEELS) measurement. The existence of strong interaction between La@C_{82} molecules and the MoS$_2$ substrate has been suggested, as discussed below.

EXPERIMENTAL

La@C_{82} mixed with soot was produced by an arc discharge of La$_2$O$_3$-loaded carbon composite rods under a partial atmosphere (200 torr), followed by a two-step extraction via refluxing, firstly in orthodichlorobenzene and then in pyridine. The La@C_{82} was isolated by a high performance liquid chromatography with toluene as an eluent. La@C_{82} dissolved in CS$_2$ solution was transferred into an alumina crucible, and CS$_2$ was vaporized by heating at 70°C in atmosphere. It was carefully heated at 300°C for 24 hours under ultrahigh vacuum (UHV) in order to remove the residual solvent.

The MoS$_2$ substrate with a size of 5×10×0.1 mm^3 was cut from a natural molybdenite crystal. It was cleaved in atmosphere and thermally cleaned by heating at 400°C for 30 min under the UHV condition. Then La@C_{82} molecules were evaporated from the crucible heated at 430°C. During the growth of La@C_{82}, the substrate temperature was kept at 180°C.

LEEL spectra were measured with a double-pass cylindrical mirror-type electron energy analyzer (ULVAC-PHI 15-255G). Typical primary electron energy (E_p) was 20 eV. In the LEELS measurement, the full width at half maximum of the elastic peak was as narrow as 0.5 eV. AFM images of grown films were measured using SEIKO instrument SPI-3800 and SPA-300 system in atmosphere.

RESULTS AND DISCUSSION

Figure 1 indicates AFM images of La@C_{82} films grown on MoS$_2$ substrates. The

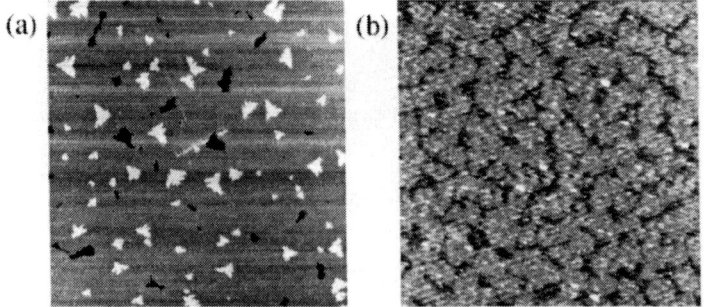

FIGURE 1. AFM images of La@C_{82} films grown on MoS$_2$ substrates (5 μm square). The nominal coverage of La@C_{82} is about 1.1 ML in (a), and 2.0 ML in (b). In (a), dark holes are uncovered MoS$_2$ surface surrounded by the initial La@C_{82} layer, and bright small islands are the second layer La@C_{82}. In (b), many gray domains are the second layer La@C_{82} grown on the initial La@C_{82} layer (darkest region), and some bright dots are the third layer La@C_{82}.

nominal coverage of La@C$_{82}$ is about 1.1 monolayer (ML) in (a), and about 2.0 ML in (b). As shown in Fig. 1(a), the initial La@C$_{82}$ layer almost completely covers the MoS$_2$ surface, and many small islands of the second La@C$_{82}$ layer grow on it. On the contrary, as shown in Fig. 1(b), these domains of the second layer do not completely cover the surface of the initial La@C$_{82}$ layer, and the growth of the third layer has already started. Therefore, the growth mode of the La@C$_{82}$ film on the MoS$_2$ substrate is the so-called Stranski-Krastanov type. In our previous studies of the epitaxial growth of C$_{60}$ on MoS$_2$ substrates [2], the initial C$_{60}$ layer did not completely cover the MoS$_2$ surface, and the growth of the second layer started at the substrate temperature of 170°C. Namely, the growth mode was the Volmer-Wever type. These results suggest that the interaction between a La@C$_{82}$ molecule and the MoS$_2$ surface is stronger than that between La@C$_{82}$ molecules, or that between a C$_{60}$ molecule and the MoS$_2$ surface. Then the interface energy between La@C$_{82}$ and MoS$_2$ is high enough to cause the Stranski-Krastanov mode growth.

Figure 2 shows LEEL spectra of 1 ML (a) and 2 ML (b) La@C$_{82}$ films grown on MoS$_2$ substrates. Probing depth of the LEELS measurement with $E_p = 20$ eV is less than 1 nm [3], so that features of these spectra originate mainly from the grown La@C$_{82}$ layers. As shown in the figure, energy loss peaks **B ~ F** appear at the same positions in both spectra. However, the spectral shape around 0 ~ 2 eV energy loss region is quite different. A distinct energy loss peak **A** appears in the spectrum of 1 ML La@C$_{82}$. In the spectrum of 2 ML La@C$_{82}$, however, the tail of an elastic peak broadens, and no clear peak appears at the position of peak **A**.

In the case of epitaxial C$_{60}$ films on MoS$_2$ substrates, LEELS of the initial layer did not show any difference from that of a thick C$_{60}$ film [4]. This indicates that the

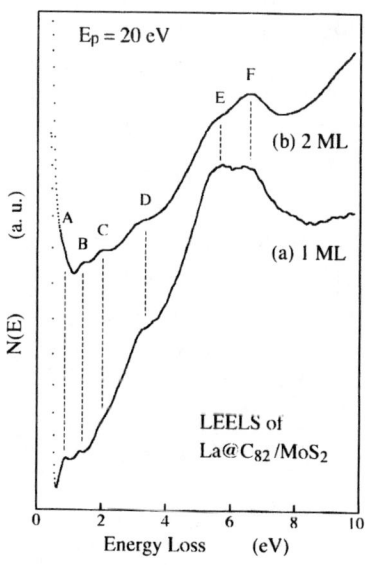

FIGURE 2. LEEL spectra of 1 ML (a) and 2 ML (b) La@C$_{82}$ films grown on MoS$_2$ substrates.

electronic perturbation from the MoS_2 surface is very small so that the initial C_{60} layer has the same electronic structure with a bulk crystal. Therefore, the result of the LEELS measurement also suggests that $La@C_{82}$ molecules in the initial layer suffer strong interaction from the MoS_2 surface.

These differences in LEEL spectra suggest two possibilities for its origin. One is that the electronic states which correspond to the energy loss process for peak **A** do not exist in the 2 ML $La@C_{82}$ film. The other is that some inelastic scattering processes with low energy loss (0.5 ~ 1 eV) diminish in the 1 ML $La@C_{82}$ film so that peak **A** hidden in the tail of the elastic peak becomes apparent. In our previous paper [1], we identified peak **A** in the 1 ML $La@C_{82}$ film as the transition from the HOMO-derived band to the SOMO-derived band of the $La@C_{82}$ film with the consideration of previous experimental [5] and theoretical [6, 7] works. This transition should exist also in the 2 ML $La@C_{82}$ film. Therefore the latter possibility is plausible. Although peak **A** really exists in the LEELS of 2 ML $La@C_{82}$, it appears only as a small hump on the broadened tail of the elastic peak.

As the origin of the change in the elastic peak width, we think as follows; the carbon cage of a $La@C_{82}$ molecule has three more electrons transferred from an encapsulated La atom than a C_{82} molecule, and they occupy HOMO and SOMO levels of $La@C_{82}$. These excess electrons around the Fermi level of $La@C_{82}$ interact with empty states of MoS_2, and the distinct perturbation on the electronic structure of the $La@C_{82}$ molecule is brought about. This seems to cause disappearance of the inelastic scattering with low energy loss and the narrowing of the elastic peak in the 1 ML $La@C_{82}$ film, which results in the appearance of the hidden peak **A**. However, this phenomenon does not occur in the successively grown $La@C_{82}$ layers or in the initial C_{60} layer adsorbed on the MoS_2 surface, because they suffer only small interaction from underlying materials. We suppose the interaction between the $La@C_{82}$ molecule and the MoS_2 surface has the character of the charge transfer but not of the covalent bond, because most energy loss features appear at same positions in both LEEL spectra (a) and (b).

REFERENCES

1. Iizumi, K., Ueno K., Uchino Y., Saiki, K., Koma, A., Inada, Y., Nagai, K., Iwasa, Y., and Mitani, T., *Phys. Rev. B* **62**, 8281-8285 (2000).
2. Sakurai, M., and Koma, A., *Trans. Mater. Res. Soc. Jpn.* **14B**, 1145-1148 (1994).
3. Koma, A., and Yoshimura, K., *Surf. Sci.* **174**, 556-560 (1986).
4. Iizumi, K., Ueno K., Saiki, K., and Koma A., *Appl. Surf. Sci.* **169-170**, 141-145 (2001).
5. Hino, S., Takahashi, H., Iwasaki, K., Matsumoto, K., Miyazaki, T., Hasegawa, S., Kikuchi, K., and Achiba, Y., *Phys. Rev. Lett.* **71**, 4261-4263 (1993).
6. Nagase, S., and Kobayashi, K., *Chem. Phys. Lett.* **214**, 57-62 (1993).
7. Nagase, S., Kobayashi, K. and Asaka, T., *Bull. Chem. Soc. Jpn.* **69**, 2131-2142 (1996).

Photophysical and Photochemical Properties of Higher Fullerenes and Endohedral Metallofullerenes

Osamu Ito[†], Mamoru Fujitsuka[†], Takeshi Akasaka[‡], and Kazunori Yamamoto[#]

[†]*Institute for Chemical Reaction Science, Tohoku University, Sendai 980-8577, Japan (CREST Japan)*
[‡]*Graduate School of Science and Technology, Niigata University, 960-2181, Japan*
[#]*Japan Nuclear Cycle Development Institute, Tokaimura, Ibaraki 319-1194, Japan*

Abstract. Photophysical and photochemical properties of higher fullerenes (C_{76}, C_{82} and C_{84}) and endohedral metallofullerenes ($La@C_{82}$ and $La_2@C_{80}$) have been examined by using a laser flash photolysis method. Transient absorption bands of the triplet excited states of higher fullerenes appeared in the near-IR region. Their intersystem crossing quantum yields were lower than 0.05, which were smaller than those of C_{60} and C_{70}. Photoinduced reduction and oxidation processes were also confirmed by transient absorption spectroscopy in the near-IR region. The reaction rate constants for the photoinduced processes were rationalized on the basis of the free energy change for the electron transfer processes. As for $La@C_{82}$ and $La_2@C_{80}$, transient absorption bands of $La@C_{82}$ showed two steps decay, which were attributed to the excited states in the different spin multiplicity.

INTRODUCTION

Photophysical and photochemical properties of fullerenes have been widely investigated. Nowadays, the excitation-relaxation processes of C_{60} and C_{70} have been well established. From the view point of photochemistry, C_{60} and C_{70} are good electron acceptors and many photoinduced reactions have been reported. On the other hand, studies on photophysical and photochemical properties of higher fullerenes and endohedral metallofullerenes are quite rare, since generation yield of these fullerenes are quite small. Similarly, excitation and relaxation processes of endohedral metallofullerenes have not been reported. In the present paper, we reported the results of our recent laser flash photolysis studies on higher fullerenes(C_{76}, C_{82} and C_{84}) and endohedral metallofullerenes

(La@C_{82} and La_2@C_{80}) (Fig. 1). Laser flash photolysis analyses disclosed quite different properties of these fullerenes from those of C_{60} and C_{70}.

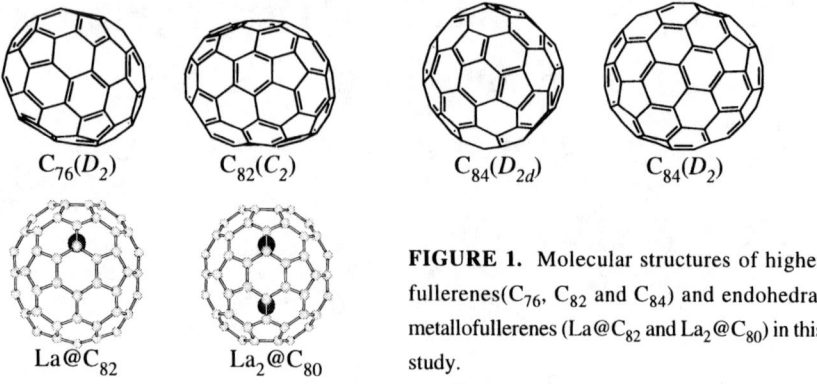

FIGURE 1. Molecular structures of higher fullerenes(C_{76}, C_{82} and C_{84}) and endohedral metallofullerenes (La@C_{82} and La_2@C_{80}) in this study.

EXPERIMENTAL

Transient absorption spectra of fullerenes were observed by nanosecond laser excitation at 532 or 355 nm (fwhm 6 ns). Probe light in the visible and near-IR regions were detected with Si-PIN and Ge-APD detectors, respectively, as described elsewhere [1]. Samples were deaerated by Ar-gas bubbling.

RESULTS AND DISCUSSION

Transient absorption spectra of C_{84} (D_{2d} and D_2 isomers) are shown in Fig. 2. These absorption bands can be attributed to T-T absorption bands, since these absorption bands

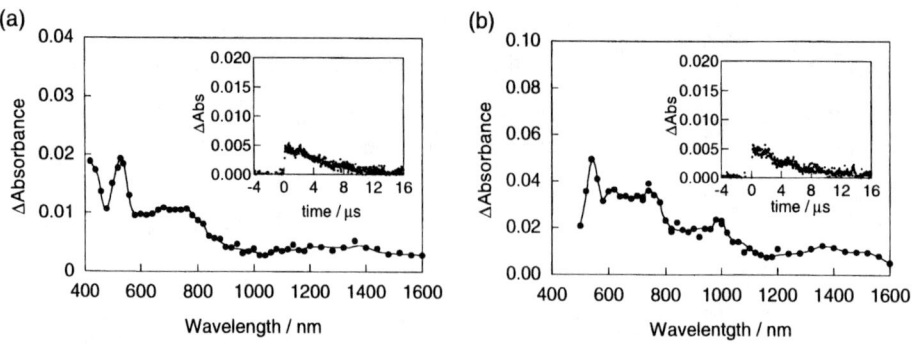

FIGURE 2. Transient absorption spectra of (a) $C_{84}(D_{2d})$ and (b) $C_{84}(D_2)$ in deaerated toluene at 100 ns by laser excitation at 355 nm. Insets: Absorption-time profiles at 760 nm.

TABLE 1. Photophysical Properteis of Higher Fullerenes.

	C_{60}	C_{70}	$C_{76}(D_2)$	$C_{82}(C_2)$	$C_{84}(D_{2d})$
Singlet Properties					
Energy / eV	1.7	1.8	1.3	~1.0	~1.1
τ_F / ns [a]	1.2	0.62	1.7		
Φ_F [b]	3.2×10^{-4}	8.5×10^{-4}	_[c]	_[c]	_[c]
Triplet Properties					
Energy / eV	1.5	1.5	1.0~1.1	0.9~1.0	0.9~1.0
λ_{TT} / nm [d]	700, 400	960, 680	840, 620, 540	750, <500	760, 530
ε_T / $M^{-1}cm^{-1}$ [e]	1.6×10^4	6.5×10^3	2.0×10^4	2.4×10^4	1.6×10^4
τ_{T0} / μs	143	11800	9.6	56	20
Φ_{isc} [f]	1.0	0.9	0.05	<0.01	<0.01

[a] Fluorescence lifetime. [b] Quantum yield of fluorescence. [c] No emission. [d] Peak positions of T-T absorption bands. [e] Extinction coefficient of T-T absorption band. [f] Quantum yield of the intersystem crossing process.

were quenched in the presence of triplet energy acceptor, such as β-carotene. It should be stressed that transient absorption spectra of Fig.2 show different spectral features, indicating triplet properties depend on their symmetry of higher fullerenes. Triplet absorption bands of other higher fullerenes were summarized in Table 1. The intrinsic triplet lifetimes (τ_{T0}) were obtained as in Table 1. Shorter triplet lifetimes than C_{60} and C_{70} seem to be one of the characteristics of higher fullerenes, while singlet lifetime of C_{76} is similar to that of C_{60}. Other singlet and triplet properties are also summarized in Table 1.

It is well known that C_{60} and C_{70} act as good electron acceptors in their excited states. Photoinduced electron transfer processes to higher fullerenes were also examined with some organic donors such as tetramethylphenylenediamine (TMPD). Fig. 3 (a) is transient absorption spectra of C_{82} and TMPD upon laser irradiation. Spectra show that radical anion ($C_{82}^{\bullet -}$, 860 nm) and radical cation (TMPD$^{\bullet +}$, 600nm) were generated with decay of

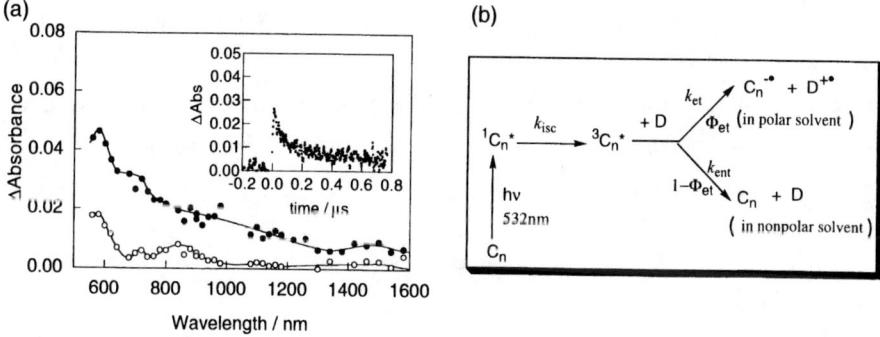

FIGURE 3. (a) Transient absorption spectra of C_{82} and TMPD in deaerated benzonitrile at 50 ns (filled circles) and 500 ns (open circles) by laser excitation at 532 nm. Inset: Absorption-time profile at 860 nm. (b) Scheme for photoinduced electron transfer processes.

triplet of C_{82}, indicating electron transfer via triplet excited state of C_{82} (Fig. 3 (b)). The reaction rate constants of electron transfer of higher fullerenes were smaller than those of C_{60} and C_{70} due to smaller free energy changes of the electron transfer processes.

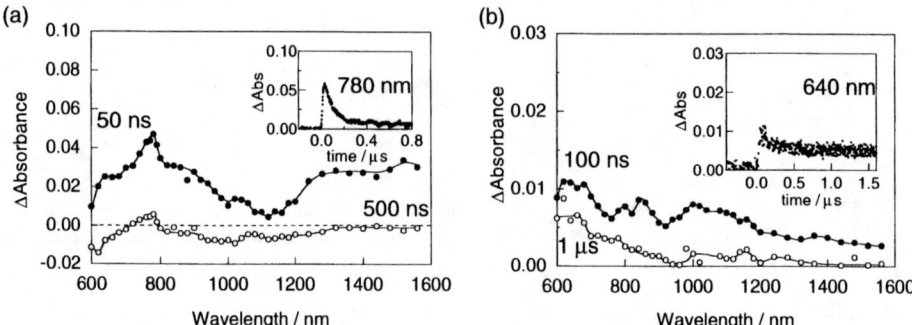

FIGURE 4. Transient absorption spectra of (a) La@C_{82} and (b) La$_2$@C_{80} in deaerated trichlorobenzene by laser excitation at 355 nm. Insets: Absorption-time profiles.

The transient absorption spectra of La@C_{82} were obtained as Fig. 4 (a). Immediately after the laser irradiation, the transient absorption spectrum showed a sharp band at 780 nm with broad bands around 1500 and 840 nm. As shown in the inset of Fig. 4 (a), transient absorption bands decayed according to two components. The fast and slow decay-components will be excited doublet and quartet states, respectively, since the ground state of La@C_{82} is doublet. Similarly, La$_2$@C_{80} shows two-step decay (Fig. 4 (b)). The fast and slow components will be the excited singlet and triplet states, respectively, since La$_2$@C_{80} is closed shell. It is interesting to note that both fast and slow components of La@C_{82} and La$_2$@C_{80} were accelerated in the presence of oxygen, indicating energy transfer to oxygen.

La@C_{82} forms stable radical anion. It was confirmed that photoexcited La@C_{82} donates electron to oxygen or octyl viologene to generate neutral La@C_{82}.

In conclusion, the properties of higher fullerenes and endohedral metallofullerenes are quite different from those of C_{60} and C_{70}. Many novel and interesting properties are expected for these rare fullerenes.

REFERENCE

1. Fujitsuka, M., Watanabe, A., Ito, O, Yamamoto, K. and Funasaka, H. *J. Phys. Chem. A*, **101**, 4840-4844 (1997).

VI. CLATHRATES

Silicon and Germanium Clathrates with Magnetic Elements

Katsumi Tanigaki[†,*1], Tetsuji Kawaguchi[†], Atsushi Nagai[†] and Masahiro Yasukawa[‡]

[†] *Material Science, Graduate School of Science, Osaka City University*
3-3-138, Sugimoto, Sumiyoshi, Osaka 558-8585, Japan
*PRESTO, JST
[‡]*Department of Applied Chemistry, Faculty of Engineering, Hiroshima University,*
Higashi-Hiroshima 739-8527, Japan

Abstract. We have made a new approach to design a novel magnetic material system on a basis of silicon and germanium nano-cluster crystals with polyhedral cage structure, where d- and f-block elements Mn and Ce are introduced endohedrally or exohedrally together with alkaline-earth metal Ba encapsulated inside the cage. In this system, the elements thus introduced act as the independent sources of magnetic- and conduction- electrons with having their strong correlations, giving rise to electron spin ordering in spite of the long-distant magnetic-atom position. Since these aspects can be realized with nano-scale control in clathrates, clathrates with magnetic elements will provide a new scientific stage for shedding a new light on magnetism.

INTRODUCTION

New aspects in magnetism have arisen recently with the advent of nano-materials. Nano-cage structures directed to magnetism have become a very important issue, as seen in endohedral-fullerenes [1] and in a series of rare-earth boron compounds [2,3]. Giant magneto-resistance in manganese copper-oxides has also gained intense interests from the viewpoint of the unique interactions between magnetic- and conduction- electrons [4,5]. Considering the importance of the nano scale control in magnetic interactions, it seems quite exciting to introduce magnetic elements to cluster crystals like fullerenes and clathrates which have polyhedral network structure of C, S, Ge and Sn [6-8]. We have made a new approach to design a novel magnetic material on a basis of silicon/germanium nano-cluster crystals with polyhedral cage structure, where d- and f-block magnetic elements as well as alkaline-earth metal Ba can be incorporated in the precise atomic sites of the crystals. A new family of magnetic materials are anticipated to be realized via

---
[1)] To whom correspondence should be addressed.

interactions between conduction- and magnetic- electrons through nano-structure polyhedral-networks [10]. In the present paper, we will show that the idea described above can be realized by demonstrating two new systems of $Ba_6Ce_2Au_4Si_{42}$ [11] and $Ba_8Mn_2Ge_{44}$ [12] as shown in Fig.1, where the interactions between magnetic d-/f- and conduction- electrons play an important role for spin-ordering. In these compounds, the long-distant d-/f-electron spins can interact with each other through nano-scale spacing of isotropic three-dimensionality, leading to the occurrence of a unique ferromagnetic spin-ordering.

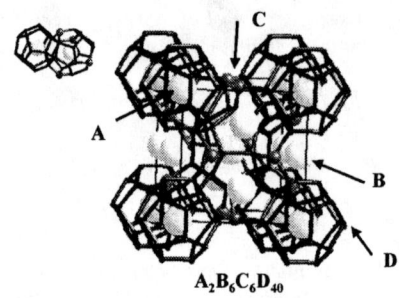

FIGURE 1. General crystal structure of type 1 clathrates. Here, the site D is Si or Ge, Ba is encapsulated in the crystal site A, d-block element Mn recides at the sites denoted by C and f-block element Ce is accommodated at the site B.

RESULTS AND DISCUSSION

Preliminary Rietveld refinement using a Cerius^2 program has been performed for $Ba_6Ce_2Au_4Si_{42}$ and $Ba_8Mn_2Ge_{44}$ and has confirmed the crystal structures that we proposed. The lattice parameters of $Ba_6Ce_2Au_4Si_{42}$ and $Ba_8Mn_2Ge_{44}$ measured at room temperature are 10.42 and 10.689 Å, respectively with a space group of Pm3n. It may important to point out that Ce can preferentially be accommodated inside the larger Si_{24} cage. This is because the Au atoms residing at the crystal sites C (see the structure of clathrates shown in Fig.1) can interact with Ce so that it can be encapsulated into a Si_{24} cage.

In magnetic measurements, when temperature was decreased from 16 K to 1.8 K under 20 G, a spontaneous magnetization was observed with a steep increase at 6.5 K and 10 K for $Ba_6Ce_2Au_4Si_{42}$ and $Ba_8Mn_2Ge_{44}$, respectively as seen in Fig.3. Hysteresis was also observed when the magnetic field was scanned in a loop of 3000 G at 1.8 K. These results show that a ferro-magnetic transition indeed occurs in these crystals. This is very much in contrast to the fact that $CeSi_2$, a well known Ce-Si compound consisting of planar three-coordinated silicon, is paramagnetic. It is surprising that a large coercive force (ca. 1400 G) as seen in the inset of Fig.3

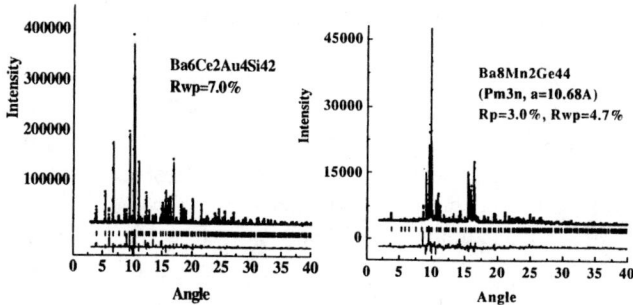

FIGURE 2. Xray diffraction spectra of $Ba_6Ce_2Au_4Si_{42}$ (left) and $Ba_8Mn_2Ge_{44}$ (right) measured by high energy radiation. Rietveld fittings are included in the figures

was observed in a dilute magnetic system (9 wt% of Ce) of $Ba_6Ce_2Au_4Si_{42}$, while it is is quite small in the case of $Ba_8Mn_2Ge_{44}$. This contrast may reflect the difference in either d- and f- electron systems or magnetic shielding due to the encapsulation environment.

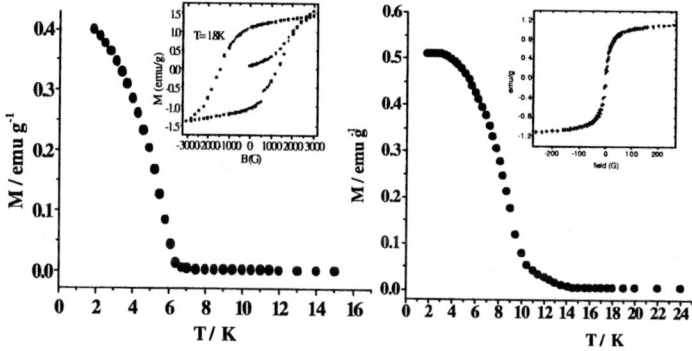

FIGURE 3. Magnetic moment measured by SQUID under a low magnetic-filed of 20 G for $Ba_6Ce_2Au_4Si_{42}$ (left) and $Ba_8Mn_2Ge_{44}$ (right).

Although it may be difficult to have a satisfactory understanding about the mechanism of the arising ferro-magnetic interactions as well as the other unusual magnetic behaviors presented here at the moment, one of the most important factors responsible for the occurrence of the magnetic orderings would be the interactions between itinerant d-/f- electrons on Mn/Ce and conduction electrons from Ba. Considering the large interval between Mn/Ce atoms (8.3 Å on the average), it is plausible to suppose here that the spin ordering of the d-/f- electrons can result from their RKKY type interactions via the conduction electrons spreading over the polyhedral-cluster network.

CONCLUSION

Magnetic d- and f- block elements were successfully introduced into the type-I Si and Ge clathrates. The Mn can be placed in to the position connecting Ge_{20} polyhedrons and Ce endohedrally inside the Si24 cage, with accurate control of their positions. A unique ferromagnetism was realized in these systems. We believe that this family will give one a good opportunity for understanding the fundamental aspects in magnetism as well as for applications to magnetic devices from the nano materials point of view.

Acknowledgements

First sincere thanks should be given to S. Yamanaka (CREST, Japan Science and Technology Corporation) who has given us continuous and helpful discussion. We are grateful to the staff members at SPring8 (beamline BL02B2) and KEK (PF-BL1B) for X-ray measurements. We also would like to thank K. Murata and H. Ishii for useful discussion. The project was supported by The Japan Society for the Promotion of Science (MIRAIKAITAKU-Project), the Grant-in-Aid for Scientific Research on the Priority Area "Fullerenes and Nanotubes" by the Ministry of Education, Science, and Culture of Japan.

REFERENCES

1. T. Ogawa, T. Sugai, and H. Shinohara, J. Am. Chem. Soc. **122**, 3538 (2000).
2. T. Susaki *et al.*, Phys. Rev. Lett. **82**, 992 (1999).
3. D. P. Young *et al.*, Nature **397**, 412 (1999).
4. S. Jin *et al.*, Science **264**, 413 (1994).
5. Y. Tokura *et al.*, J. Phys. Soc. Jpn. **63**, 3931 (1994).
6. J.S. Kasper, P. Hagenmuller, M. Pouchard, and C. Cros, Science **150**, 1713 (1965).
7. C. Cros, M. Pouchard, and P. Hagenmuller, J. Solid State Comm. **2**, 570 (1970).
8. H. Kawaji, H. Horie, S. Yamanaka, and M. Ishikawa, Phys. Rev. Lett. **74**, 1427 (1995).
9. R.F.W. Herrmann, K. Tanigaki, T. Kawaguchi, S. Kuroshima, and O. Zhou, Phys. Rev. B **60**, 13245 (1999).
10. K. Tanigaki, Proceeding in the Electrochemical Society, Motreal: Chemistry and Physics of Fullerenes and Carbon Nanomaterials, Proceeding Volume 10, p.184, 2000.
11. T. Kawaguchi, K. Tanigaki and M. Yasukawa, Phys. Rev. Lett. **85**, 3189 (2000).
12. T. Kawaguchi, K. Tanigaki and M. Yasukawa, Appl. Phys. Lett. **77**, 3438 (2000).

A new silicon clathrate compound: $I_8Si_{46-x}I_x$

Edouard Reny[+,*], Shoji Yamanaka[+,#], Christian Cros[♦] and M. Pouchard[♦]

[+] *Department of Applied Chemistry, Hiroshima University, 739-8527 Higashi-Hiroshima, Japan*
[#] *CREST, Japan Science and Technology Corporation (JST), Japan.*
[*] *Japanese Society for the Promotion of Science (JSPS), Japan*
[♦] *Institut de Chimie de la Matière Condensée de Bordeaux, 33608 Pessac cedex, France*

Abstract. The high pressure and high temperature technique is a powerful tool to synthesize new clathrate structures. In this paper, we report the synthesis of the first binary silicon clathrate doped with an electronegative element: $I_8Si_{46-x}I_x$ with $x = 1.5 \pm 0.5$. Some chemical and structural analyses of this compound are presented as well as a short discussion about its synthesis process.

INTRODUCTION

During the past fifteen years, the development of carbon chemistry started a rush for research on fullerene compounds and derivatives such as metallo-fullerenes, C_{60} polymers or carbon nanotubes. A fulleride cage can be defined as a closed polyhedra containing 12 pentagonal faces and a number of hexagonal faces different from 1 (*i.e.* the pentagon rule). Such polyhedra also exists for other group IV elements of the periodic table, such as silicon, germanium or tin, but in these latter cases, they are stacked together in a regular pattern by sharing common faces. The obtained crystals are called clathrates.

The type I silicon clathrate lattice is an arrangement of 2 pentagonal dodecahedra (Si_{20}) and 6 tetrakaidecahedra (Si_{24}) offering 8 sites per unit cell for guest alkali or alkaline earth atoms (M) as shown in figure 1. The resulting formula is therefore M_8Si_{46}. Note the existence of the type II silicon clathrate formed of an arrangement of 8 Si_{20} and 16 Si_{28} polyhedra per unit cell. In the centre of these cages, some atoms can also be trapped leading to the composition M_xSi_{136} where x defines a concentration domain between 0 and 24 [1].

Some previous experimental and theoretical investigations carried out on the 2 types of networks triggered great hopes in optoelectronic, electronic, thermoelectric [2] and mechanical properties [3], closely linked to their peculiar structures. Consequently, it is of primary importance to develop new ways of synthesis and to try to enlarge the clathrate family to other compositions.

Up to 1998, silicon clathrates were mostly synthesized by thermal decomposition of Zintl silicides. Recently, a new synthesis route has developed, high pressure high temperature (HPHT) synthesis. For instance, a silicon clathrate doped with only barium, Ba_8Si_{46}, was obtained for the first time under a pressure of 3 GPa and at 800 °C. This compound showed a superconducting transition (T_c) at 8 K (highest T_c ever found in silicon clathrates) [4]. The present study reports the first synthesis process for a silicon clathrate doped with an electronegative element.

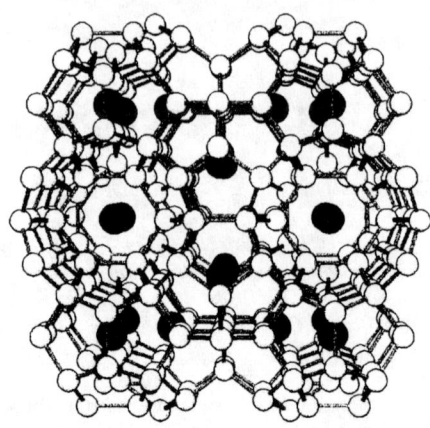

FIGURE 1. The type I clathrate structure hosting in the centre of the Si_{20} and Si_{24} building blocks alkali or alkaline earth guest atoms. The resulting formula is M_8Si_{46}.

SYNTHESIS

A mixture of iodine and silicon was prepared in an ideal stoichiometric ratio of I/Si = 8/46 and finely ground. The resulting powder was placed in an h-BN cell (8mm in inner diameter and 6mm in length) which was in turn placed in a carbon tube heater and in a pyrophyllite cube as a pressure media. A cubic multi-anvil press was used (Riken model CP10). The BN cell was heated electrically by the carbon heater while the temperature was monitored by a Pt/Pt-Rh thermocouple placed under the cell.

The pressure and temperature applied on the sample were respectively 5 GPa and 700 °C for 1 hour: The resulting product consisted of a mixture of 4 phases, Si and iodine doped silicon clathrate as major phases and air sensitive SiI_2 and SiI_4 as minor phases. The iodides could be washed away by ethanol. The yield of the remaining clathrate phase was determined to be 42 % in mass. Surprisingly however, an addition of only 1 % of iodine doped silicon clathrate to the starting mixture and application of the same HPHT conditions showed an increase in the yield of the clathrate formation up to 90%. This suggests that the addition of a seed is greatly effective for the growth of crystals. The new clathrate compound was found to be slightly more stable than diamond silicon in hot alkali solution. Therefore it was possible to remove most of the remaining silicon using some 0.2M sodium hydroxide solution [5].

Some more investigations have been carried out in an attempt to grow some large size crystals under pressure. The starting material consisted of a mixture of the previous starting material (I and Si in a ratio I/Si = 8/46) and bismuth tri-iodide in a weight ratio of 1/1. After a 5 GPa, 700 °C HPHT treatment, the final product consisted of iodine doped silicon clathrate as a major phase, BiI_3 and elemental Bi as minor phases, and traces of an air reactive unknown red coloured phase. However, no usable single crystal was found.

CHEMICAL ANALYSIS

In order to perform a chemical analysis, the purified clathrate sample was decomposed by a concentrated HF/HNO$_3$ solution in a sealed Polytetrafluoroethylene (PTFE) container at 700 °C. The silicon component was entirely dissolved and diluted with a saturated boric acid solution, while the iodine component had precipitated in the form of elemental crystals, which were in turn dissolved in potassium iodide. Inductively coupled plasma (ICP) analysis atomic emission spectroscopy was performed in order to determine the silicon content using a Perkin-Elmer Optima 3000 apparatus. To determine the iodine concentration by ICP, a separate solution was prepared, but this time the iodine crystals formed were dissolved in a sodium thyosulfate solution. The measurements showed a recovery ($C_{Si}\% + C_I\%$) of 97%. The iodine content was determined to be 48.6 ± 1.4 %. The corresponding formula deduced from these results is $I_{9.5\pm0.5}Si_{44.5\pm0.5}$. The sample was heated up to 800 °C for several hours under vacuum, and a weight loss uniquely attributed to iodine of 48.1 ± 1.0 % was observed. This value is in good agreement with the chemical analysis.

STRUCTURAL ANALYSIS

The powder diffraction pattern was collected on a MAC Science model M18XHF spectrometer using a graphite monochromated CuKα radiation and was collected at every 0.02° over the range 25-120° (2θ) by a step scan mode with a time of acquisition of 0.6 seconds per point. Voltage and intensity of the source have been fixed at 40kV and 200mA.

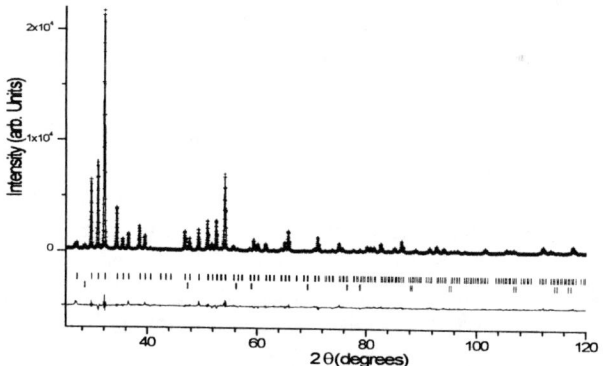

FIGURE 2. X-ray diffraction data for $I_8Si_{46-x}I_x$ with x=1.8. Discontinuous (+) and continuous line represent respectively the experimental data points and the calculated spectra. The upper tick marks indicate the calculated reflection positions for the clathrate phase and the lower tick marks the calculated reflection positions for some remaining traces of silicon. The lower continuous line represents the difference.

In a 25°-120° (2θ) angular range, 342 reflections were obtained and were indexed on the basis of a cubic unit cell with a = 10.4195(7) Å which is, to our knowledge, the highest value ever observed for a binary silicon clathrate. The space group $Pm\overline{3}n$ was assumed and various structural hypothesis, concerning the sites occupied by the extra x iodine atoms in $I_8Si_{46-x}I_x$ have been investigated. The only reasonable fit (S = 2.06, R_{wp} = 0.11) was obtained with a structural model, in which the iodine occupy all silicon cages (2a and 6d positions) and 11 % of the 16i silicon positions. The other hypothesis, where some iodine replace silicon in the 6c and 24k positions, led to very strong disagreement between the calculated and experimental data. The refined XRD plot can be seen in figure 2.

DISCUSSION

The fact that usual synthesis by a simple mixture of the elements does not produce a yield in silicon clathrate is believed to be a consequence of various concurrent reactions occurring under the applied conditions. Some trivial calculations show that the volume decrease for SiI_2, SiI_4 and $I_{9.5}Si_{44.5}$ are 15.5%, 13.2% and 14.4%, respectively, compared to the starting material components. Such similar values could partially explain the great difficulty to obtain a clathrate single phase. Once some SiI_2 or SiI_4 are formed, these quite stable compounds (under these conditions) create a lack in iodine preventing a regular growth of the clathrate phase. One solution could be to increase the ratio of iodine but this attempt resulted in favouring the SiI_4 formation.

The use of BiI_3 as a crystal growth environment in the synthesis of the iodine doped silicon clathrate revealed itself to be efficient for the reactivity of iodine: the small yield of the SiI_2 and SiI_4 formation showed that the iodine attack on silicon must have been weakened. On the other hand, the clathrate phase formation is strongly favoured by the presence of BiI_3: it is likely that the equilibrium constants of the decomposition of bismuth iodide and of the clathrate formation are related.

CONCLUSION

Since the discovery of the clathrate phase of silicon, various compositions have been investigated and revealed that depending on the doping element, the properties of the compound can change drastically. However, only alkali or alkaline-earth have been known to be binary silicon clathrates dopants. This work opens, with the successful synthesis of an iodine doped silicon clathrate by HPHT, a new field to be explored in silicon clathrate chemistry and shows that other groups of the periodic table should be considered as potential dopants for these peculiar structures.

REFERENCES

1. J. S. Kasper, P. Hagenmuller, M. Pouchard and C. Cros, *Science*, **3704**, 1713 (1965).
2. G. S. Nolas, J. L. Cohn, G. A. Slack and S. B. Schujman, *Appl. Phys. Lett.*, **73**, 178 (1998)
3. A. S. Miguel, P. Keghelian, X. Blase, P. Melinon, A. Perez, J. P. Itie, A. Polian, E. Reny, C. Cros and M. Pouchard, *Phys. Rev. Lett.*, **83**, 5290 (1999).
4. S. Yamanaka, E. Enishi, H. Fukuoka and M. Yasukawa, *Inorg. Chem.*, **39**, 56 (2000).
5. E. Reny, S. Yamanaka, C.Cros and M. Pouchard, *Chem. Commun.*, **24**, 2505 (2000).

NMR Studies of Silicon Clathrate Compounds

Hirokazu Sakamoto[a], Hideki Tou[a], Yutaka Maniwa[a], Hiroyoshi Ishii[a], Edouard Reny[b], and Shoji Yamanaka[b]

[a]Department of Physics, Tokyo Metropolitan University, Hachioji, Tokyo 192-0397, Japan
[b]Department of Applied Chemistry, Hiroshima University, Higashi-Hiroshima, Hiroshima 739-8527, Japan

Abstract. ^{29}Si and ^{137}Ba NMR measurements were applied to silicon clathrate compound $Ba_8Ag_xSi_{46-x}$ (x = 0 ~ 6) to reveal the electronic structure and the superconducting mechanism. The large Knight shifts in the superconducting samples indicated the large density of states at Si and Ba sites. In Ba_8Si_{46}, the temperature dependence of the Knight shift at Si site in the normal state was reproduced by an activation type plus constant between 50 K and 300 K, which suggests that the electrons at the Fermi level in the sharp band are thermally exited to the Si_{46} states. The opening of an energy gap was confirmed from the decrease of the spin-lattice relaxation time of ^{29}Si NMR with decreasing temperature. The temperature dependences of the Knight shift and the spin-spin relaxation rate are unconventional below the superconducting transition temperature.

INTRODUCTION

Silicon clathrate compounds have three dimensional network of Si atom connected through sp^3 covalent bonding. The crystalline structure consists of Si polyhedral cages, Si_{20} dodecahedron, Si_{24} tetrakaidecahedron and Si_{28} hexakaidecahedron, which can contain alkaline metals and barium etc. They show various properties such as metal-insulator transition [1, 2] and superconductivity according to the network structure or the kind of contained metals. The superconducting transition was discovered in Ba-codoped $Ba_6Na_2Si_{46}$ by Kawaji et al. [3], which is the first superconductor with sp^3 covalent bonding. Furthermore $Ba_8T_xSi_{46-x}$ (T = Au, Ag, Cu) was observed also to transform to superconducting state when x < 3 [4]. In this material Si atoms in the six-membered ring connecting Si 20 clusters are replaced by T atoms. The highest transition temperature of 8 K is obtained at x = 0. We have systematically studied in these materials to reveal the electronic states and the superconducting transition mechanism mainly with NMR measurements. In $Ba_6Na_2Si_{46}$ ^{29}Si and ^{137}Ba NMR spectra showed large Knight shift, implying that the hybridization between Si 3p and Ba 5d orbitals is important for superconductivity [5, 6] as predicted theoretically [7]. In addition it was indicated that $Ba_6Na_2Si_{46}$ is the weak coupling superconductor.

In this report ^{29}Si and ^{137}Ba NMR measurements were carried out in $Ba_8Ag_xSi_{46-x}$ (x = 0 ~ 6) to investigate the electronic structure of silicon clathrate compounds and the symmetry of the superconductivity. These lead to understanding of electronic structure of network systems.

EXPERIMENTAL

NMR measurements, ^{29}Si and ^{137}Ba spectra, spin-lattice and spin-spin relaxation time, were performed using a home-built Fourier transform spectrometer at the magnetic field of 3.9 T between 2.4 K and 300 K. The powder samples were synthesized as described in Ref. 4 and packed into Si free tubes instead of an ordinary NMR glass tube.

RESULTS AND DISCUSSION

Large Knight shifts of ^{29}Si and ^{137}Ba NMR were observed in superconducting $Ba_8Ag_xSi_{46-x}$ (x < 3) at room temperature, indicating the large density of states at Fermi energy $N(E_F)$ at each site and the strong hybridization of the Si_{46} states with the Ba 5d states [8]. This is consistent with the results of the band calculation [7]. Although there are three structures; S1', S2' and S3' corresponding to S1, S2 and S3 in Ref. 5, in ^{29}Si spectra, the ratio of the intensities of 3 Si sites is not equal to that of the number of the sites. This discrepancy may be related to the recent X-ray observation suggesting that nearly 10% of the Ba atoms in the Si_{20} cages are missing [9]. The temperature dependence of the shifts at S1' and S2' sites in Ba_8Si_{46} in the range of 50 K ~ 300 K is represented by an activation type $\delta_1+\delta_2 exp(-\Delta E/kT)$ with an activation energy $\Delta E \approx 20$ meV [8]. This temperature variation may suggest that the electrons in the sharp band in which E_F lies are thermally excited into the Si-like conduction band and the width of the sharp band is estimated to be 20 meV. The presence of the sharp band also agrees with the band calculation.

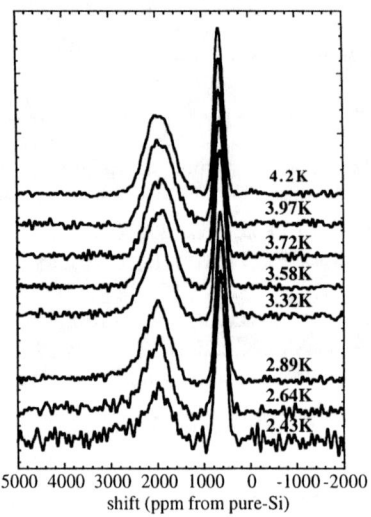

FIGURE 1. ^{29}Si NMR spectra in Ba_8Si_{46} at the magnetic field of 3.9 T at the temperature range of 2.43 K ~ 4.2 K. The shift is measured from the ^{29}Si resonance of pure Si. The superconducting transition temperature at 3.9T is about 3.5 K.

Figure 1 shows ^{29}Si NMR spectra in Ba_8Si_{46} measured at the magnetic field of 3.9 T. The origin of the shift is the ^{29}Si resonance of pure Si. The superconducting transition temperature is about 3.5 K under this magnetic field. As seen in Fig. 1 no decrease of the Knight shift was observed with decreasing temperature in the superconducting region, where the shift is expected to reduce in an ordinary superconductor. The finite Knight shift at 0 K may result from spin-orbit interaction [10], which is considered to be large, taking account of the essential role of 5d orbitals of the heavy Ba atoms in this material. If the spin-orbit scattering length is assumed to be nearly equal to the distance between Ba atoms, the shift is estimated to decrease by about 5 % at 0 K. In fact the scattering length should be longer, giving rise to the larger shift which may be observed. Another possible origin of no decrease of the shift could be that this material is a spin triplet superconductor [11]. Moreover the spin-spin relaxation rate of ^{29}Si NMR indicates also unusual temperature dependence, increasing rapidly just below the transition temperature (not shown in this report). Thus the experimental results suggest that Ba_8Si_{46} may not be an ordinary BCS type superconductor.

The temperature dependence and the existence of the coherence peak of $1/T_1$, where T_1 is a spin-lattice relaxation time, are important information of the symmetry of superconductivity. $1/T_1T$ of ^{29}Si NMR in Ba_8Si_{46} at 3.9 T as a function of temperature is shown in Fig. 2; closed squares for $\approx$ 2000 ppm and open squares for $\approx$ 820 ppm spectra in Fig. 1. Above the transition temperature indicated by an arrow $1/T_1T$'s are almost constant, which means the electronic states are metallic as expected. On the other hand, in the superconducting region the decrease of $1/T_1$ implies the opening of an energy gap, however, the temperature range of this measurement is not wide enough to clarify the symmetry. Moreover, although the coherence peak is not observed, it may be suppressed by the external magnetic field. Therefore the experiments in the lower temperature range and at the lower magnetic field are needed to conclude the symmetry of superconductivity.

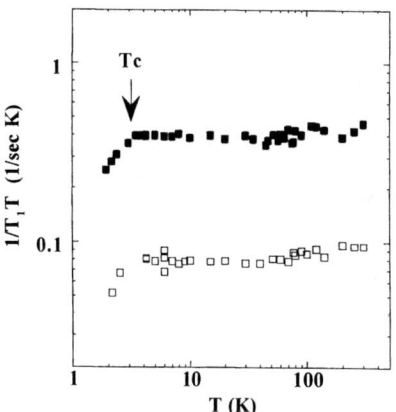

FIGURE 2. Temperature dependence of $1/T_1T$; T_1 is ^{29}Si NMR spin-lattice relaxaion time obtained at 3.9 T. Closed squares: $\approx$ 2000 ppm spectra in Fig. 1, Open squares: for $\approx$ 820 ppm spectra in Fig. 1.

CONCLUSION

In superconducting silicon clathrate compounds $Ba_8Ag_xSi_{46-x}$ ($x < 3$), the importance of the hybridization of the Si 3p states with the Ba 5d states was indicated from ^{29}Si and ^{137}Ba NMR experiments. Below the transition temperature, the Knight shift and the spin-spin relaxation time of ^{29}Si NMR in Ba_8Si_{46} with the highest transition temperature of 8 K show unusual temperature dependence, which suggests that this material may not be the ordinary BCS type superconductor. To reveal the symmetry of superconductivity, NMR at the lower temperature and the lower magnetic field is now undertaken.

ACKNOWLEDGEMENTS

This work is supported by Grant-in-Aid for Scientific Research on the Priority Area "Fullerenes and Nanotubes" by the Ministry of Education, Science, Sports and Culture of Japan.

REFERENCES

1. Cros, C., Pouchard, M., and Hagenmuller, P., *Bull. Soc. Chim. Fr.*, **2**, 379-386 (1971).
2. Mott, N. F., *J. Solid State Chem.*, **6**, 348-351(1973).
3. Kawaji, H., Horie, H., Yamanaka, S. and Ishikawa, M., *Phys. Rev. Lett.*, **74**,1427-1429(1995).
4. Enishi, E., *Proc. 15th Fullerene Symp.* 264(1998)
5. Shimizu, F., Maniwa, Y., Kume, K., Kawaji, H., Yamanaka, S., and Ishikawa, M., *Phys. Rev.*, B**54**, 13242-13246 (1996).
6. Maniwa, Y., Sakamoto, H., Tou, H., Aoki, Y., Sato, H., Shimizu, F., Kawaji, H., and Yamanaka, S., *Mol. Crys. Liq. Cryst*, **341**, 497-502(2000).
7. Saito, S., and Oshiyama, A., *Phys. Rev.* B**51** 2628- (1995).
8. Sakamoto, H., Tou, H., Ishii, H., Maniwa, Y., Reny, E. A., and Yamanaka, S., *Physica*, C**341-348**, 2135-2136 (2000).
9. Kitano, A., Yonemura, M., Moriguchi, K., Fukuoka, H., and Yamanaka, S., Sectional Meeting of Phys. Soc. Jpn. Sep. 2000.
10. MacLaughlin, D. E. "Magnetic Resonance in the Superconducting State" in , *Solid state Physics*. **31**, edited by H. Ehrenreich, T. Seitz, and D. Turnbull, Academic Press, New York, 1976, pp1-69.
11. Ishida, K., Mukuda, H., Kitaoka, Y., Asayama, K., Mao, Z. Q., Mori, Y., and Maeno, Y., *Nature*, **396**, 658-660(1998).

A Sign of Superconductivity in Li-doped α-rhombohedral Boron

Atsushi Oguri*, Kaoru Kimura*, Akihiko Fujiwara[†],
Masami Terauchi[¶] and Michiyoshi Tanaka[¶]

*Department of Advanced Materials Science, the University of Tokyo, 7-3-1 Hongo, Bunkyo-ku, Tokyo 113-0033, Japan
[†]Department of Physics, the University of Tokyo, 7-3-1 Hongo, Bunkyo-ku, Tokyo 113-0033, Japan
[¶]Research Institute for Scientific Measurements, Tohoku University, 2-1-1 Katahira, Aoba-ku, Sendai 980-8577, Japan

Abstract. The reports on a superconducting transition in metal-doped C_{60} and Si_{20} solids have suggested a possibility of superconductivity in metal-doped B_{12} solids which is constructed from clusters having the same symmetry as C_{60} and Si_{20}. Here we report a sign of the metal transition and superconductivity in Li-doped α-rhombohedral Boron, which consists of B_{12} clusters. The magnetic susceptibility dropped beginning at about 36 K.

INTRODUCTION

The crystalline structure of α-rhombohedral Boron (α-rh.B) is a distorted face-centered cubic lattice of icosahedral B_{12} clusters, being similar to the fcc structure of C_{60}. The B_{12} cluster has the same icosahedral symmetry as C_{60} and Si_{20} clusters. In C_{60} and Si_{20} solids, superconductivity was discovered [1,2]. These three cluster solids have some general and particular electronic properties summarized in Table 1.

It was reported that even a metal transition in Li-doped β-rh.B (Li_xB_{105}) doesn't occur, which consists of same B_{12} clusters [3]. However, existence of a metal transition and superconductivity in Li-doped α-rh.B (Li_xB_{12}) is theoretically predicted, based on the *ab initio* total-energy pseudopotential calculations in the local density approximation [4]. In the present paper, we report a sign of the metal transition and superconductivity in Li_xB_{12}.

EXPERIMENTAL PROCEDURES

α-rh.B was prepared by crystallizing amorphous boron [5]. Amorphous boron was placed in a tantalum tube and the tube was enclosed in a quartz tube at

TABLE 1. Comparison of the three cluster solids

		B_{12}	Si_{20}	C_{60}
group of elements		group III	group IV	
general		· constructed from basic clusters with icosahedral symmetry		
		· relatively large interstitial spacings between clusters in solid		
particular		covalent inter-cluster bonding		van der Waals type inter-cluster bonding
		unsaturated intra-cluster bonding orbital	saturated intra-cluster bonding orbital	
		peculiar form of bonding 3-center covalent bond [a]	simple form of bonding	
			$\sim sp^3$	$\sim sp^2$

[a] The highest electron density point is at a center of a triangle that three boron atoms make. [6]

14cmHg pressure of argon. The tube was heated up to the temperature of 1473 K and maintained for 10 hours.

The practical methods for Li-doping were essentially the same as described by Matsuda *et al.* (1995) [3]. α-rh.B and lithium metal bits were placed apart at each end of a tantalum tube. This treatment was carried out in helium atmosphere using a globe box to avoid oxidation of lithium. The tube was enclosed in an evacuated quartz tube. The quartz tube was heated up to the varied temperature between 1373 and 1493 K and kept at the temperature.

Electronic energy-loss spectroscopy (EELS) measurements were conducted to obtain an information of a metal transition. The lithium leakage during measurements was kept minimum by using a liquid nitrogen specimen holder and having the short recording time of 90 seconds.

The magnetic susceptibility was measured by a SQUID magnetometer. The sample was first cooled to 2 K in zero field and data were taken in a field on warming it (zero-field-cooled, ZFC). Then data were taken in the same field on cooling to 2 K (field-cooled, FC).

RESULTS AND DISCUSSION

As a consequence of the Rietveld refinement for X-ray powder diffraction data using monochromatized Cu $K\alpha$ radiation, the lithium occupancies of octahedral and tetrahedral sites are 15(2)% and 19(4)%, respectively ($R_{WP} = 13.83\%$), *i.e.* $x = 0.53$ in Li_xB_{12}. The lattice constants were refined to $a_{\text{hex}} = 4.920(2)$Å and $c_{\text{hex}} = 12.570(2)$Å. The changes of them by Li-doping are less than 0.1%.

Figure 1 shows results of EELS measurement on α-rh.B and Li_xB_{12}. The spectrum (A) for pure α-rh.B showed good agreement with the density of states of the conduction band obtained by using *ab initio* calculation [7]. (B) was measured after applying electron beam to Li_xB_{12} to remove lithium from it and is essentially

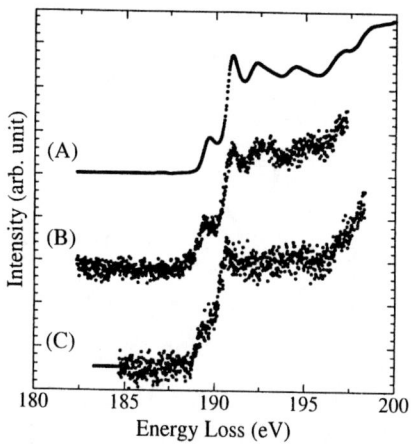
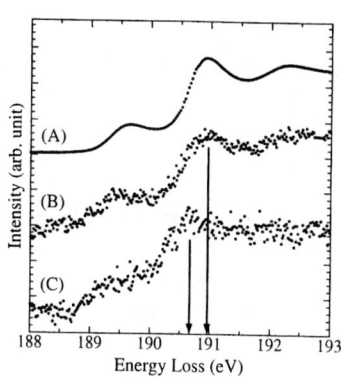

FIGURE 1. The boron K edge excitation spectrums measured by EELS. The energy resolution is 0.26 eV. The energy scale is expanded in the right figure. (A)α-rh.B (B_{12}) (B)a sample after applying electron beam to Li_xB_{12}(C)Li_xB_{12}

the same as (A). Compared with (B), the (C) spectrum for Li_xB_{12} shifts to a lower energy side by 0.3-0.4 eV and the spectrum around 189 eV becomes sharper due to Li-doping. The width of the sharp intensity increase was almost equal to the energy resolution of 0.26 eV. These indicate a chemical shift by an increase of electrons in conduction bands and appearance of a Fermi-edge by Li-doping. The observed chemical shift and Fermi-edge confirmed the metal transition. The appearance of Fermi-edge at the first peak in the rigid band for Li_xB_{12} ($0 < x \leq 1$) is also consistent with the calculated results [4].

The temperature dependence of the magnetic susceptibility measured by SQUID is shown in Fig. 2(a). The positive shift for Li_xB_{12} is considered to be originated from an unknown ferromagnetic impurity. The drop of the magnetic susceptibility of Li_xB_{12} only for ZFC at an onset T_c of about 36 K suggests the Meissner effect. The volume fraction of the superconducting phase is estimated to be ∼0.02%. The magnetization of Li_xB_{12} shown in Fig. 2(b) was measured in the varied applied magnetic field. The magnetization of the ferromagnetic impurity is canceled where it is supposed not to depend temperature. The shape of the magnetization curve at 20 K is well explained by a mixture of Pauli paramagnetic and superconducting phases, though that at 37.5 K shows only the former behavior.

CONCLUSION

The observed chemical shift and Fermi-edge measured by EELS indicate the metal transition in α-rh.B by Li-doping. The drop of the magnetic susceptibility with T_c around 36 K and the shape of the magnetization curve suggest the existence of a superconducting phase.

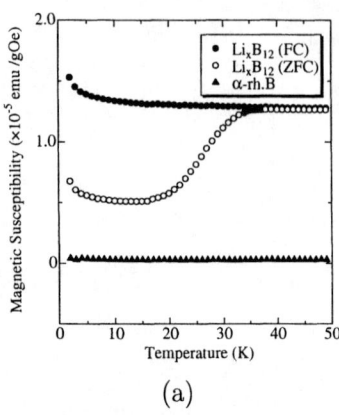
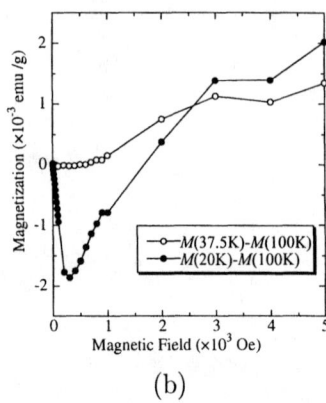

FIGURE 2. (a)Temperature dependence of Magnetic susceptibility of α-rh.B (B_{12}) for zero-field-cooled (ZFC) and Li_xB_{12} for ZFC and field-cooled (FC). (b)Magnetization vs. applied magnetic field for sample 1. The data of magnetization at 100 K were subtracted from those at each temperature to cancel the temperature independent portion of magnetization.

ACKNOWLEDGMENTS

The Authors thank Mr. S. Satoh of Cryogenic Center in the University of Tokyo for SQUID measurement in the early stage of this work and Mr. A. Ogino for experimental assistance. We appreciate Dr. T. Nakayama and Prof. H. Suematsu for valuable discussions. This work was partially supported by Grant-in-Aid for Scientific Research on Priority Area from Ministry of Education, Science, and Culture of Japan.

REFERENCES

1. Hebard A. F. , Rosseinsky M. J. , Haddon R. C. , Murphy D. W. , Glarum S. H. ,Palstra T. T. M. , Ramirez A. P. and Kortan A. R. , *Nature*, **350**, 600 (1991).
2. Kawaji H. , Horie H. , Yamanaka S. and Ishikawa M. , *Phys. Rev. Lett.* **74**, 1427 (1995).
3. Matsuda H. , Nakayama T. , Kimura K. , Murakami Y. , Suematsu H. , Kobayashi M. and Higashi I. , *Phys. Rev. B* **52**, 6102 (1995).
4. Gunji S. and Kamimura H. , *Phys. Rev. B* **54**, 13665 (1996).
5. Ugai J. A. and Soloviev N. E. , "Methods of Preparation of α-Rhombohedral Boron," in *Boron and Refractory Borides*, edited by Matkovich V. I. , Springer-Verlag, 1977, pp.227.
6. Fujimori M. , Nakata T. , Nakayama T. , Nishibori E. , Kimura K. , Takata M. and Sakata M. , *Phys. Rev. Lett.* **82**, 4452 (1999).
7. Terauchi M. , Kawamura Y. , Tanaka M. , Takeda M. and Kimura K. , *J. Solid State Chem.* **133**, 156 (1997).

VII. OTHER MOLECULAR MATERIALS

Laser Induced Dissociation of Linear C_6 and Reorientation of Trapping Sites in Solid Neon

Tomonari Wakabayashi[1], Aik-Loong Ong and Wolfgang Krätschmer

Max Planck Institut für Kernphysik, Postfach 103980, Heidelberg D-69029, Germany

Abstract. Carbon clusters are formed from carbon vapor of resistively heated graphite rods. The clusters C_n are trapped in a matrix of solid Ne at 7 K. High resolution infrared (IR) absorption spectra of the matrix sample show a distinct fine structure for each vibrational band of clusters C_n, suggesting different types of trapping sites. In the dissociation experiment, the matrix sample is irradiated by an ultraviolet (UV) laser tuned to the wavelength of an electronic transition of a linear cluster. The IR and UV absorption spectra are recorded before and after the irradiation. We found that the intensity of the IR absorption line of linear C_6 at 1958.7 cm^{-1} decreases when the matrix sample is irradiated at a wavelength of 235 nm, near the maximum of a strong UV band. As a result of this exposure, also the UV band diminishes. The intensity decrease of the IR and UV band is correlated and indicates decomposition of linear C_6. We also observed changes in the fine structure of the IR absorptions, which indicate reorientation of trapping sites upon the electronic excitation.

INTRODUCTION

Small carbon clusters C_n (n=2-30) are possible precursors of fullerenes and nanotubes. The determination of the relative abundance of clusters upon resistive heating or laser ablation is crucial to understand the formation mechanism of larger clusters. However, the spectroscopic information is still limited for these smaller molecules [1]. So far, the assignment of absorption features of neutral clusters C_n has been performed by matrix isolation spectroscopy [2-6] and by high resolution spectroscopy in the gas phase, particularly in the IR [1,7]. Kurtz and Huffman noted an intensity correlation between an UV absorption band at 247 nm and an IR absorption line at 1952 cm^{-1} in Ar matrices [2]. The experiment was followed by the observation of rotationally resolved spectra at 1959.8585 cm^{-1} in the gas phase to conclude that the carrier of the above IR line is linear C_6 [7]. Very recently, Maier and co-workers assigned an UV band peaking at 235 nm in Ne matrices to linear C_6 using a mass selection technique and *ab initio* quantum chemical calculations [3].

In the present work we found that in Ne matrices photodissociation of linear C_6 occurs efficiently by irradiation with UV laser pulses tuned at 235 nm (5.3 eV). The UV/IR correlation obtained by the UV exposure confirms the assignment of the UV band. Further we investigated the changes in the fine structure of the IR absorptions.

---

[1] A guest researcher from Division of Chemistry, Kyoto University, Kyoto 606-8502, Japan.

EXPERIMENT

The experimental setup is described elsewhere [8]. Briefly, carbon clusters from a resistively heated graphite rod were co-deposited with excess neon on a rhodium-coated sapphire mirror cooled at 7 K. The UV and IR absorption spectra were recorded by a reflection geometry. The spectral resolution was 2 nm for the UV (EG&G Model 1236 OMA 0.5 Meter Spectrograph with a deuterium lamp as a light source) and 0.1 cm^{-1} for the IR (Bruker IFS 113v with a globar light source and a lN$_2$ cooled MCT detector). The frequency-doubled output from a tunable dye laser (Lambda Physik FL3001 with a FL32 SHG crystal and Coumarin 47 which is pumped by an excimer laser SMG201 operated at 308 nm) was directed to the matrix sample for the excitation of clusters. The UV and IR absorption were recorded after the irradiation and compared with those before the irradiation.

RESULTS

Figure 1 shows UV absorption spectra before (dotted line) and after (solid line) irradiation at 235.0 nm (~0.5 mJ/pulse for 2 hours). Linear clusters C_n (n = 6-9, 11, and 13) show strong absorption bands in this spectral range [3,4], whereas a band at 317 nm presumably belongs to cyclic C_{10} [3]. The band at 235 nm in Fig. 1 is reduced by about 20 % in intensity, while the others remain almost unchanged. This indicates selective decomposition of linear C_6. Figure 2 shows the IR absorption spectra (a) before and (b) after the irradiation. Both spectra correspond to the UV spectra in Fig. 1. The assignments in Fig. 2, all for linear clusters C_n, are taken from the literature [5,6]. Linear C_6 has two infrared active stretching modes, namely v_4 at 1958.7 cm^{-1} and v_5 at 1199.3 cm^{-1} in Ne matrices [5]. The coincidence of the decrease of the v_4 intensity in Fig. 2 confirms the assignment of the 235 nm UV band to linear C_6.

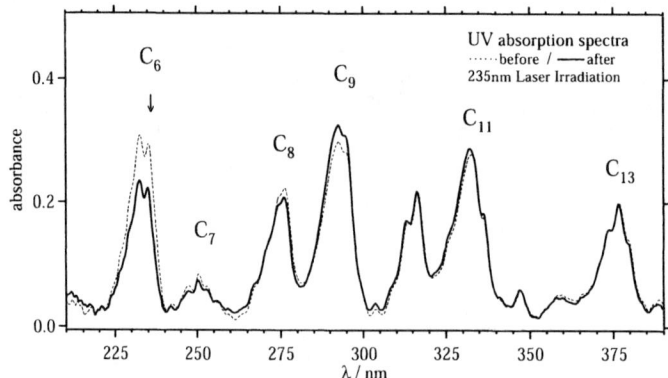

FIGURE 1. UV absorption spectra of carbon clusters C_n in solid Ne before (dotted line) and after (solid line) 235 nm laser irradiation. The arrow indicates the $^3\Sigma_u^- - ^3\Sigma_g^-$ transition of linear C_6.

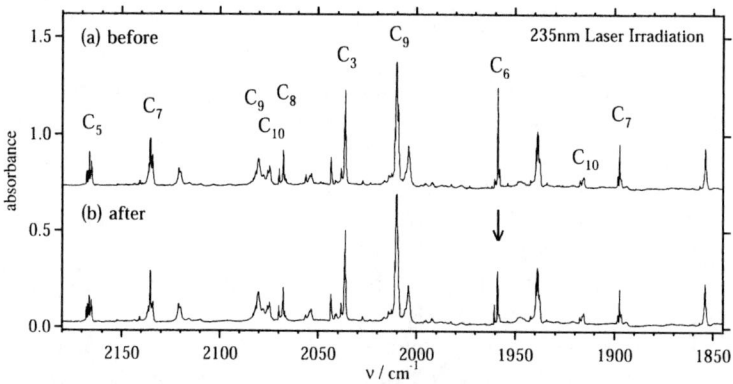

FIGURE 2. IR absorption spectra of carbon clusters C_n in solid Ne (a) before and (b) after 235 nm laser irradiation. The arrow indicates the peak of the ν_4 stretching mode of linear C_6. The spectra correspond to those of Fig. 1.

Figure 3(a) shows an expanded portion of Fig. 2 focusing on the ν_4 stretching mode of linear C_6. The upper trace in Fig. 3(a) is the spectrum before the 235 nm irradiation and the lower after the irradiation. The spectrum consists of a main peak at 1958.7 cm^{-1} having a shoulder at 1958.9 cm^{-1} and of satellite peaks at higher (1960.5 cm^{-1}) and lower frequencies (1957.9 and 1957.7 cm^{-1}). All these peaks belong to linear C_6. Upon irradiation, the main peak and the satellites at lower frequencies decreased, but the satellite at a higher frequency increased in intensity (see arrows in Fig. 3(a)). The shoulder peak in the main band remained unchanged. Figure 3(b) shows spectra of the ν_5 stretching mode of linear C_6 before (upper) and after (lower) irradiation. Similar structures are seen in both Figs. 3(a) and (b), although the separation of a ν_5 satellite peak at 1200.4 cm^{-1} from the main peak at 1199.3 cm^{-1} is smaller than that of ν_4. The spectral changes for the ν_5 mode in Fig. 3(b) are similar to those for the ν_4 in Fig. 3(a).

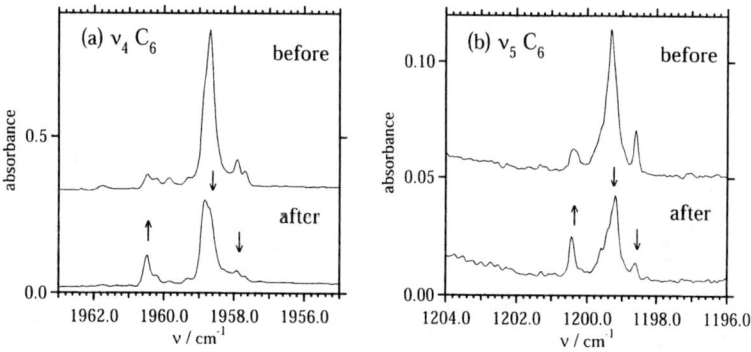

FIGURE 3. IR absorption spectra of (a) ν_4 and (b) ν_5 stretching modes of linear C_6 in solid Ne before and after 235 nm laser irradiation. The downward (upward) arrows indicate peaks whose intensity decreases (increases) upon irradiation.

DISCUSSIONS

After irradiation, in both the ν_4 and ν_5 line, the height of the main peaks decreased by about 50 %. This fact might contradict to the observation in the UV spectra in Fig. 1 where the intensity reduction was about 20 %. However, if we add all contributions from the main peak, the satellites, and the shoulders, the total intensity decrease is roughly 20 % consistent with the observation in the UV spectra. A possible explanation is that a part (~20 %) of linear C_6 clusters is decomposed and another fraction (~30 %) is not, but, for the latter, rearrangements of surrounding neon atoms occur to cause frequency shifts in the IR.

It is natural to consider that the change of molecular geometry upon the electronic excitation induces the reorientation of trapping sites to make the vacancy larger. Trapped in such environment, the molecules may vibrate differently from those in a tight trapping site. As a consequence, the vibrational frequency may shift to the blue.

Here we try to estimate the size of trapping sites. Using a typical van der Waals (vdw) diameter of sp^2 carbon, ~3.5 Å, and a C-C bond length, ~1.3 Å, the molecular volume of linear C_6 is roughly ~84 Å^3. Similarly, for a linear array of n neon atoms, the volume can be calculated as ~66 Å^3 for n=3 and ~90 Å^3 for n=4, provided a vdw diameter of 3.15 Å for Ne. Therefore, a linear C_6 molecule can accommodate in a vacancy of at least four neon atoms. In this case, the length of the vacancy ~12.6 Å is sufficiently large for linear C_6 (~10 Å). However, other vacancies are also possible, e.g. vacancies of a rhombic shape. These sites require more than four neon atoms to be removed, because the length of the vacancy should exceed that of linear C_6.

At the moment we cannot specify which type of vacancy corresponds to the main peak or the satellite at a higher frequency in the IR. However, the aforementioned conjecture is worth while to be investigated further experimentally and theoretically, because this may provide a key for understanding the IR-absorption fine structures and molecular motions in the matrix. In addition, as is demonstrated in the present work for the case of linear C_6, the selective excitation of a specific UV absorption band can be extended also to other carbon clusters.

REFERENCES

1. Van Orden, A., and Saykally, R. J., *Chem. Rev.* **98**, 2313-2357 (1998).
2. Kurtz, J., and Huffman, D. R., *J. Chem. Phys.* **92**, 30-35 (1990).
3. Grutter, M., Wyss, M., Riaplov, E., Maier, J. P., Peyerimhoff, S. D., and Hanrath, M., *J. Chem. Phys.* **111**, 7397-7401 (1999).
4. Forney, D., Freivogel, P., Grutter, M., and Maier, J. P., *J. Chem. Phys.* **104**, 4954-4960 (1996).
5. Smith, A. M., Agreiter, J., Härtle, M., Engel, C., and Bondybey, V. E., *Chem. Phys.* **189**, 315-334 (1994).
6. Freivogel, P., Grutter, M., Forney, D., and Maier, J. P., *Chem. Phys.* **216**, 401-406 (1997).
7. Hwang, H. J., Van Orden, A., Tanaka, K., Kuo, E., Heath, J. R,. and Saykally, R. J., *J. Mol. Spectrosc.* **79**, 769-776 (1993).
8. Cermak, I., Förderer, M., Cermakova, I. Kalhofer, S., Stopka-Ebeler, H., Monninger, G., and Krätschmer, W., *J. Chem. Phys.* **108**, 10129-10142 (1998).

Spiral Carbon Nanoparticles

Masaki Ozawa, Eiji Ôsawa*, and Hitoshi Gotô

Department of Knowledge-based Information Engineering, Faculty of Engineering, Toyohashi University of Technology, Tempaku-cho, Toyohashi 441-8580 Japan

Abstract. Carbon nanoparticles in Archimedean spiral form are observed under TEM during the formation of onion-like carbon nanoparticles from various carbon materials by irradiation of intense electron beam, and their structural model is presented. The spiral particles are considered as intermediates in the formation process of onions. Revision of logarithmic spiral model for the primary soot particle by Kroto-McKay is suggested.

INTRODUCTION

Carbon nanoparticles appear in two distinct forms: nested polyhedron often with large void in the center, and onion or Russian doll that is a concentric assembly of quasi-spherical shells [1]. In both forms inter-shell distances are invariably close to that of graphite, 0.335 nm. In spite of their potential importance as multi-shell fullerenes, these particles were never isolated pure until recently and all available information came from TEM images. There was a strong belief that onion-like carbons were metastable species, which emerged after repeated observation of well-developed onions decomposing extensively upon prolonged irradiation of intense electron beam within TEM [2]. Tomita [3] broke the myth by isolating macroscopic amounts of small onions by heating ultra-disperse diamond and confirmed their stability in air; hence we duly expect resurge of interests in this class of fullerenes.

While searching for more readily available starting material for the preparation of onion-like carbons [4], we observed a novel spiral form of carbon nanoparticles that appear as intermediate responsible for the development of onion-like structure. This paper briefly describes the observations and discusses its significance.

RESULTS

Figure 1a shows a spiral carbon nanoparticle that appeared after brief irradiation of intense electron beam upon commercial carbon black of ultra-fine grade: there is an ill-shaped but closed structure in the center, probably C_{60} or other single-shell fullerene, but major portion of dark line can be traced to a single and continuous spiral up to the second outermost shell. Schematic tracing of spiral is shown in Fig 1b. What is clearly seen here is the difficulty in discerning spiral from

---

* Author to whom correspondence should be addressed to OsawaEiji@aol.com

onion form. In fact we realize a number of published TEM images of onion-like carbons are actually spirals. To our knowledge, Ru [5], Kuznetsov [6], and Donnet [7] recognized the spiral form and reproduced their images in their publications, but none of them commented on the significance of their finding. We surmise that the conscious oversight is due to difficulty in conjecturing 3D atomistic structure of spiral particle from sp^2-hybridized carbon atoms.

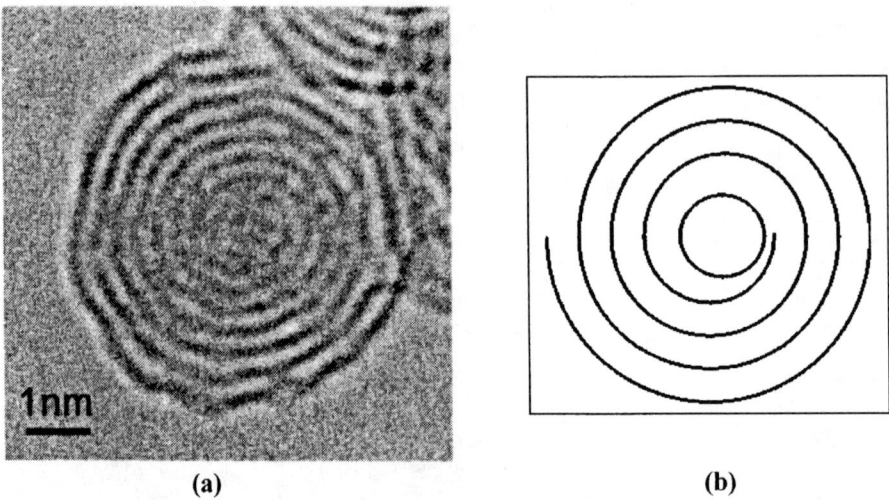

Fig. 1. (a) Image of a spiral carbon nanoparticle formed after brief irradiation of commercial carbon black of ultra-fine grade, Tokablack #8500 with average diameter of primary particle 14 nm (manufactured by Tokai Carbon Company), with electron beam having an estimated current density of 150 A/cm^2 at an accelerating voltage of 200 KeV in a transmission electron microscope JEOL JEM-2010. (b) Schematic trace of dark lines.

We noticed that spirals changed into onions and *vice versa* with considerable ease upon continued irradiation of electron beam under the same conditions as mentioned in Fig 1a. It is clear that valence isomerization pathways are available for the reversible interconversion between the two distinct geometric forms. This observation helped us to construct a realistic chemical model for the new form of carbon nanoparticle, with the following assumptions: (1) these two forms are of comparable stabilities, (2) the spiral-onion interconversion is thermally allowed, inter-shell reaction with relatively low energy barrier, (3) the reaction does not involve too much nuclear movement, and (4) spiral structure contains as small number of dangling bonds as possible. We derived our first, albeit still tentative, model by applying Diels-Alder reactions across neighboring shells. The smallest two-turn model, C_{300} was built from $C_{60}@C_{240}$, then geometry-optimized with a semi-empirical MO method (MOPAC/AM1), under RHF scheme in order to avoid insurmountable computational-technical difficulties. We are still far from evaluating reliable energies, but the resulting 3D spiral cage geometry does not seem too strained even at this stage. We realize that our spiral model could have a large number of geometrical isomers, and that they can be discerned as spiral only when looked down along one of the three Cartesian axes. In other words, spirals may look like onions most of the time in TEM.

We have so far examined carbon blacks, fullerene blacks, and ultra-disperse diamonds as the target of electron bombardment, and always found spirals. It appears that spiral carbon nanoparticles occur frequently, probably always, during the formation of onion-like carbon nanoparticles.

DISCUSSION

If we recall the fact that scroll models have long been proposed for the structure of carbon nanotubes [8], the frequent observation of spiral carbon nanoparticles is not be really surprising. We may also note that spiral forms appear frequently in nature, from molecules (helicene, protein, DNA), crystals (diamond), chemical reaction (Belousov-Zhabotinskii system), spider's nest, plants (phyllotaxis in sunflower and pine cone), shells (nautilus), living organs (cochlea, finger prints), to outer space (galaxy). Spirals seen in living organisms are generally related with the growth phenomenon, hence the logarithmic type dominates, in which the distance between the neighboring arcs increases exponentially but the shapes of arcs remain similar within any radial angle, reflecting the self-reproduction mechanism of life. In our case as well, spiral form offers a conceptually attractive model for the intermediates in the growth process of onion-like carbon nanoparticles. Probably for this reason, Kroto and McKay depicted a logarithmic spiral to illustrate growing primary particles of soot in their 1988 paper [9]. However, we actually observed in our work only Archimedean spirals, in which the distances between the neighboring curves are all equal. This implies dominant importance of van der Waals attraction between neighboring layers of spiral in nanoparticles. We suggests a slight modification to the Kroto-McKay model of soot particle, namely Archimedean rather than logarithmic spiral should be adopted. Our model does not contradict with Donnet's previous model of complex spiral for soot particle [7]. We have actually observed double spirals in our TEM experiments.

Discovery of Archimedean spiral polymorph solves another enigma peculiar to multi-shell fullerenes: since our old work [10] it is now well-recognized that the molecular formula $C_{60}@C_{240}@C_{540}@\cdots C_{60n^2}@$ is the unique solution for any multi-shell fullerene whatever shape it takes, but the question is how and when the number of atoms is counted? For example, as Ugarte observed, when a polyhedral carbon nanoparticle was irradiated with intense electron beam, the inside of polyhedron first became amorphous, then quasi-spherical shells grow from outside to inside [11]. This means that the first shell must contain exactly $60n^2$ carbon atoms. One might ask how does the first shell know the total number of shells (n) in the onion before the inside shells are formed? What actually happens is as follows: first an defective onion is formed, often with incomplete shells rapidly repeating opening and closing motions, but then the whole onion changes into a spiral, which finally produces a correct onion. We have actually observed this process. Thus, the spiral intermediate framework plays the role of a self-template to bring onion into the most stable configuration. Further mechanistic details of the template action will be discussed elsewhere.

CONCLUSIONS

We find spiral carbon nanoparticles to form more often than previously thought during the formation of quasi-spherical onion-like carbon nanoparticles from various sources including carbon black and polyhedral carbon nanoparticles upon electron beam irradiation. It is possible to construct a reasonable 3D molecular model for spiral carbon nanoparticle containing only a few dangling bonds from pentagonal and hexagonal network of sp^2-hybridized carbon atoms. Spiral particles appear to have some relevance as the intermediates in the formation of onion-like carbon nanoparticles.

ACKNOWLEDGMENTS

We thank Ministry of Education, Culture, Sports, Science and Technology for financial support through a Grant-in-Aid for Scientific Research on Priority Areas (A, No. 11165222). We are indebted to Professor D. Tománek for helpful discussion and suggestions.

REFERENCES

1. Lozovik, Y. E., and Popov, A. M. "Formation of fullerene, onions and other nanometer size carbon clusters," in *Physics of Clusters*, edited by V. D. Lakhno and G. N. Chuev, World Scientific Publ. Co., Singapore, 1998, pp. 1-55.
2. Banhart, F. *Rep. Progr. Phys.*, **62**, 1181-1221 (1999).
3. Tomita, S., Fujii, M., Hayashi, S., and Yamamoto, K. *Diamond & Rel. Mater.*, **9**, 856-860 (2000).
4. Ôsawa, E., and Ozawa, M. in '*Proceedings of the Tenth Symposium on Beam Engineering of Advaced Material Syntheses Including Bio-Medical Materials and Treatments (BEAMS 1999),*' November 24-25, **1999**, Kyoto, p. 137-140.
5. Ru, Q., Okamoto, M., Kondo, Y., and Takayanagi, K. *Chem. Phys. Lett.*, **259**, 425-431 (1996).
6. Kuznetsov, V. L., Chuvilin, A. L., Butenko, Y. V., Mal'kov, I. Y., and Titov, V. M. *Chem. Phys. Lett.*, **222**, 343-348 (1994).
7. Donnet, J.-B. *Rubber Chem. Technol.*, **71**, 323-341 (1998).
8. Maniwa, Y. *et al.*, This Proceeding.
9. Kroto, H. W., and McKay, K. *Nature*, **331**, 328 (1988).
10. Yoshida, M.; Ôsawa, E. *Fullerene Sci. Technol.* **1**, 55-74 (1993).
11. Ugarte, D. *Carbon*, **33**, 989-993 (1995).

Electronic Structures of Carbyne Model Compounds

S. Hino[1,2], Y. Okada[1], K. Iwasaki[1,2], M. Kijima[3], H. Shirakawa[3]

[1] Graduate School for Science and Technology, Chiba University, Chiba 263-8522 Japan
[2] Faculty of Engineering, Chiba University, Chiba 263-8522 Japan
[3] Institute of Material Science, Tsukuba University Tsukuba 305-8573 Japan

Abstract. Photoelectron spectra of compounds having carbyne skeletal structures have been measured. The photoelectron spectra of the compounds having a cumulene type carbon bonds (=C=C=C=C=) reveals that the highest occupied molecular orbital (HOMO) level becomes shallower in accordance with the increase of the carbon chain length. The spectra of two compounds having $-C \equiv C-C \equiv C-$ bond were reproduced well by MO calculation, which allows us to estimate the electronic structure of much longer carbon chain. The calculation indicates the same carbon chain length dependence in polyyne type compounds as are observed in cumulene type compounds.

INTRODUCTION

Fullerenes and nanotubes constitute nano-scale carbon networks. Apart from them, one dimensional carbon chain, so-called carbyne, could be another nano-scale carbon network[1]. Carbyne can be classified into two categories: one is polyyne type having $-C \equiv C-C \equiv C-$ structure and the other is cumulene type having =C=C=C=C= structure. Carbyne is believed to be unstable so that it cannot be isolated with macroscopic quantity. Because of this nature, carbyne has not been investigated intensively, but it is considered as an ultimate form of carbon nanotubes so that its application such as field emitter, molecular conductive wire and fibers with mechanical strength are expected.

A large difference is expected in the electronic structures of two carbyne chains, but its experimental investigation is difficult because of its instability. We have tried to elucidate the electronic structure of carbyne itself by measuring ultraviolet photoelectron spectra (UPS) of carbyne model compounds. We present UPS of polyyne type carbyne model compounds and discuss their electronic structure with an aid of molecular orbital (MO) calculation.

EXPERIMENTAL

Carbyne model compounds having cumulene structure measured in this work are tetraphenylallene (TPA), tetraphenylbutatriene (TPBT), bis(2-2'-biphenylene)-butatriene (BBBT) and tetraphenylhexapentatriene (TPHP), and compounds having polyyne structure are bis(trimethylsilyl)butadiyne (BTSB) and diphenylbutadiyne (DPB). Their molecular structures are shown in Fig. 1.

Photoelectron spectra were measured with a Kratos XSAM-800 photoelectron spectrometer. Excitation light source was HeI resonance line. Most specimens for the measurement were thin films prepared by vacuum deposition onto a gold disk. A TPHP film was prepared by application of TPHP powder on the disk, since it was not stable when heated to sublime.

MO calculation was carried out using Fijitsu WinMOPAC version 2 program module. Ionization potentials of cumulene type compounds were obtained with optimized molecular geometry using PM3 Hamiltonian. Geometry of polyyne type compounds was optimized with MNDO Hamiltonian then their ionization potentials were calculated with PM3 Hamiltonian.

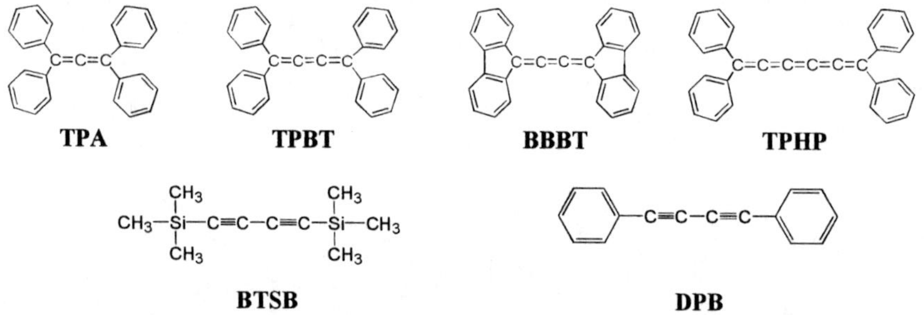

FIGURE 1. Molecular structures of carbyne model compounds. tetraphenylallene (TPA), tetraphenylbutatriene (TPBT), bis(2-2'-biphenylene)-butatriene (BBBT), tetraphenylhexapentatriene (TPHP), bis(trimethylsilyl)butadiyne (BTSB) and diphenylbutadiyne (DPB).

RESULT AND DISCUSSION

Photoelectron Spectra of Cumulene Type Compounds

Photoelectron spectra of cumulene type model compounds together with that of benzene[3] are shown in Fig. 2. Four structures labeled A – D are observed in the spectra. While structures B – D of these compounds are observed at almost same energy region with the same intensity, structure A shifts change its position. The longer the chain length extends, the lower the peak of structure A shifts.

Except for BBBT, because of steric hindrance phenyl groups and cumulene chain are not coplanar[2]. This molecular arrangement reduces resonance interaction between the benzene ring and the cumulene chain so that their photoelectron spectra might be divided into two parts. Structure B – D seems to derive from electrons on the benzene ring, since the spectrum of benzene resembles the spectra of these structures. Therefore, it could be concluded that structure A is derived from the cumulene chain. Extension of the chain length reduce the ionization potential energy and probably induces chain instability. This could be one of the reasons why carbyne could not be isolated with macroscopic quantity.

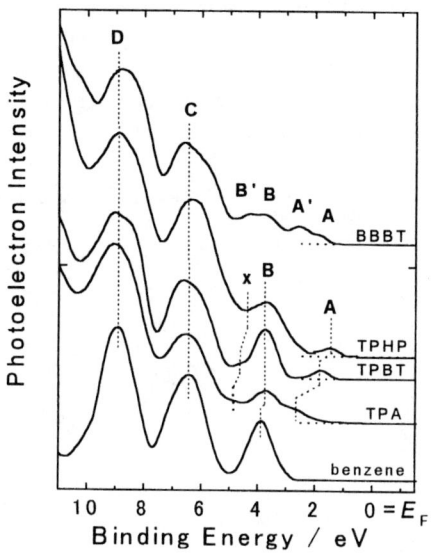

FIGURE 2. Photoelectron spectra of TPA, TPBT, TPHP and BBBT are shown. A spectrum of solid benzene is also show for comparison.

MO Calculation of Cumulene type Compounds

Calculation reveals that the HOMO level that forms structure A is actually on the cumulene chain. There are another two MO's attributed to the cumulene chain of 4 carbon atoms and they locate at the energy region where the HOMO of benzene lies. They are observed as a small hump located at around 4.7 eV.

Although there is a dependence of the even and odd number of carbon atoms, principally the HOMO shifts towards lower binding energy side when chain length increases. This shift does not saturate even the number of carbon atoms exceeds more than 30.

Photoelectron Spectra of Polyyne Type Compounds

Figure 3 shows photoelectron spectra of BTSB and tetramethylsilane (TMS) $Si(CH_3)_4$[4]. While only two structures labeled X and Y are observed in the spectrum of $Si(CH_3)_4$, there are four structures labeled A – D in the spectrum of BTSB. Spectroscopic analysis of the $Si(CH_3)_4$ spectrum indicated that structure X is attributed to electrons constituting the bonds between Si and C and structure Y to methyl groups[4]. Structure D might be due to methyl groups, since binding energy of structures D and Y coincides. Usually binding energy of π-electrons is small so that structures B and C might be σ-electron origin. Bonds between Si and C can be classified into two types; Si – C ≡ and Si – CH$_3$ bonds. The former is one third of the latter and the intensity ratio of structure B to C is 1 : 3. Therefore, structure B can be ascribed to the former bond and structure C to the latter. This deduction is verified by MO calculation. Thus, structure A can be attributed to the polyyne chain.

Measurement of photoelectron spectra of much longer polyyne chain has not been done although it is under the way. MO calculation suggests that the HOMO level does not become shallow when the number of carbon atoms exceeds more than 10.

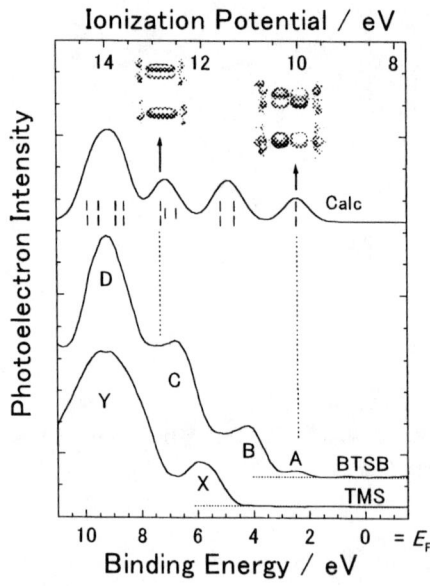

FIGURE 3. Photoelectron spectra of BTSB and tetramethylsilane (TMS) are shown. Calculated ionization potentials (bar) and a Gaussian broadened spectrum is shown together with electron density of the orbitals derived from polyyne structure.

ACKNOWLEDGMENTS

This work is supported by a Grant-in-Aid for Scientific Research on the Priority Area "Fullerenes and Nanotubes" (No. 11165208) from the Ministry of Education, Science, Sports and Culture of Japan.

REFERENCES

1. Whittker, A. G., *Science* **200**, 763-764 (1978).
2. Berkovitch-Yellin, Z. and Leiserowitz, L., *Acta Cryst.* **B33**, 3657-3669 (1977).
3. Demuth J. E., and Eastman, D.E., *Phys. Rev. Lett.* **32**, 1123-1127 (1974).
4. Seki, K., and Inokuchi, H., *Bull. Chem. Soc. Jpn.* **56**, 2212-2219 (1983).

Magnetic Properties of K-absorbing zeolite LTA

H. Kira[*], H. Tou[*], Y. Maniwa[*], and Y. Murakami[†]

[*]*Department of Physics, Tokyo Metropolitan University, Minami-osawa, Hachi-oji, Tokyo,192-0397, Japan*

[†]*Photon Factory, National Lab. for High Energy Physics, Tsukuba, Ibaraki, 305-0801, Japan.*

Abstract. The ^{29}Si-NMR measurements on several samples of potassium loaded zeolite LTA (K_x/K_{12}-$Al_{12}Si_{12}O_{48}$), where x ($1 \leq x \leq 7$) is loaded potassium atoms per α-cage, were carried out. It was found that NMR line-shape at 4.2K significantly depends on the x. These spectra were divided into different 3 types; diamagnetic phase for $x \leq 2$, magnetic ordered phase for $3 \leq x \leq 5$, and paramagnetic phase for $x \geq 6$. The magnetic phase diagram for K_x/K_{12}-LTA was determined by NMR.

INTRODUCTION

Zeolite LTA, (M_{12} $Al_{12}Si_{12}O_{48}(H_2O)_n$), where M denotes the kind of alkali-metal, is synthetic hydrated aluminosilicates. The zeolite LTA has a framework structure consisting of SiO_4 and AlO_4 tetrahedra sharing the oxygen atoms and forming two kinds of three-dimensional cages called α and β-cages. The K-loaded zeolite A (K_x/K_{12}-LTA), where x is the number of additional potassium atoms per α-cage, was found to show a ferromagnetic (FM) like transition around 7K by Nozue et al. in 1992 [1]. The 4s-electrons of the guest K-atoms are delocalized in the α-cage and are believed to be responsible for the observed magnetic properties.

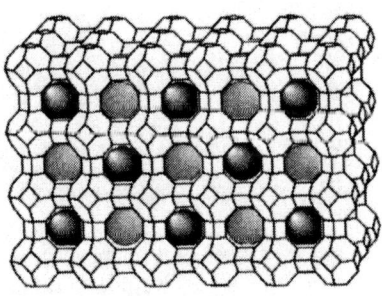

FIGURE 1. Schematic representation of K-absorbed zeolite LTA. There are two kinds of α-cage with different K density, as shown by black and white circles.

The DC magnetic susceptibility measurement under high magnetic fields suggested the presence of antiferromagnetic interaction among the localized magnetic moments. The X-ray diffraction (XRD) study [1] showed the super-lattice structure (Fig.1) that develops with increasing x, and well correlated with the FM magnetic moment [1,2]. The ^{29}Si-NMR measurements confirmed that a clear magnetic transition occurs in the bulk of sample [3].

Based on these observations, the magnetic state for $3 \leq x \leq 5$ may be considered to be ferrimagnetism or Spin-canting magnetism. However, it seems that the magnetic phase diagram for K_x/K_{12}-LTA has not yet been established and a possibility of magnetic impurity for the observed transition has not yet been definitely ruled out. In this study, we employed NMR method to determine the magnetic phase diagram.

NMR Measurements

First, we describe the ^{29}Si-NMR spectra of K_x/K_{12}-LTA (x=1 to 7). The pulse-Fourier transform NMR measurements were carried out at the magnetic field of 3.93 T, using a conventional spin-echo sequence, $\pi/2$-π-echo. All measurements were done on the powder samples. The sample quality was checked by X-ray diffraction measurements using synchrotron radiation at Spring-8 and PF facilities in Japan. The DC magnetic susceptibility measurement by SQUID magnetometer clearly showed a FM-like transition around 6K in the sample of x=3 to 5. No sign for the FM order was observed in the other samples down to 2K. However, this does not imply that there is no antiferromagnetic (AF) state. It is sometimes difficult to distinguish the AF state from the paramagnetic state, particularly in poor quality samples. Therefore we performed NMR measurement which sensitively probe the local magnetic field.

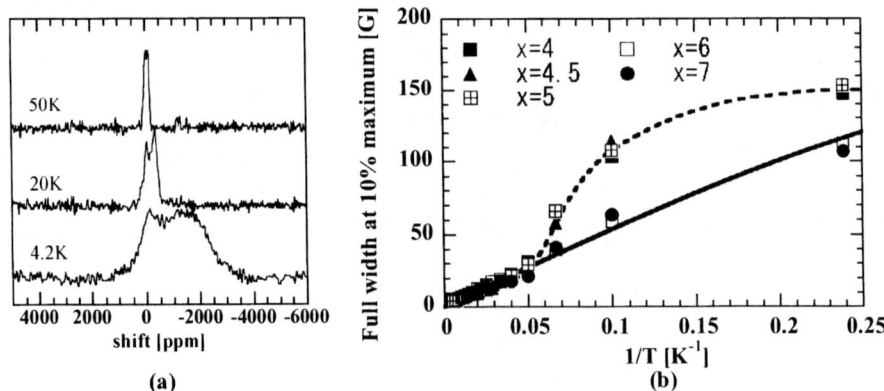

FIGURE 2. (a) ^{29}Si-NMR spectra in K_5/K_{12}-LTA at 4.2K, 20K and 50K. (b) Full width at the 10% maximum of the ^{29}Si-NMR spectra in K_x/K_{12}-LTA as a function of temperature.

As an example of temperature dependence, the ^{29}Si-NMR spectra of the K_5/K_{12}-LTA are shown in Fig.2a. The origin of the shift was defined by a resonance position where ^{29}Si-NMR appears in semiconducting diamond silicon. These spectra are regarded as the sum of the sharp signal around 0 ppm and the broad signal with negative shift. The signal around 0ppm hardly shifts from RT to 4.2 K, so these are probably assigned to diamagnetic phases such as insufficient reacted-LTA and deteriorated K_x/K_{12}-LTA. The other broad signal shifts to the lower frequency (to the right hand side) as the temperature decreases.

Fig.2b shows the temperature dependence of the width of the broad signal. Here, the width was defined as the full width at the 10% maximum. Then, it is found that the width can be nearly described with a Brillouin function in high temperatures, as shown in Fig. 2(b) by a solid line. However, in the K_4/K_{12}-LTA, $K_{4.5}/K_{12}$-LTA and K_5/K_{12}-LTA, it suddenly deviates from the Brillouin function below ~17K. This is one of the strong evidence for the transition to a magnetic order. These results are consistent with previous paper [3] for x=4.5, 6. Thus, we confirmed from ^{29}Si-NMR measurements that the samples absorbing more than 6 K-atoms per α-cage are paramagnetic and there is clear FM like transitions in those absorbing 4 to 5 K-atoms per α-cage.

^{29}Si-NMR spectra at 4.2K in K_x/K_{12}-LTA with x=1~7 are shown in Fig.3. The line-shape of the broad signal in K_6/K_{12}-LTA and K_7/K_{12}-LTA are different from that of K_5/K_{12}-LTA. However, it should be noted that they have the same line-shape at 20K in the paramagnetic state. Therefore, in the samples x=3-5, the line-shape changes from the asymmetrical shape like that of K_6/K_{12}-LTA material to the symmetrical shape shown by the K_5/K_{12}-LTA material at 4.2K. Decreasing the K contents from x=5 to 3 yields a decrease in the broad ordered signal, though the width seems to be unchanged. In K_1/K_{12}-LTA and K_2/K_{12}-LTA, the sharp signal dominates the observed NMR intensity. Therefore, these materials should be diamagnetic.

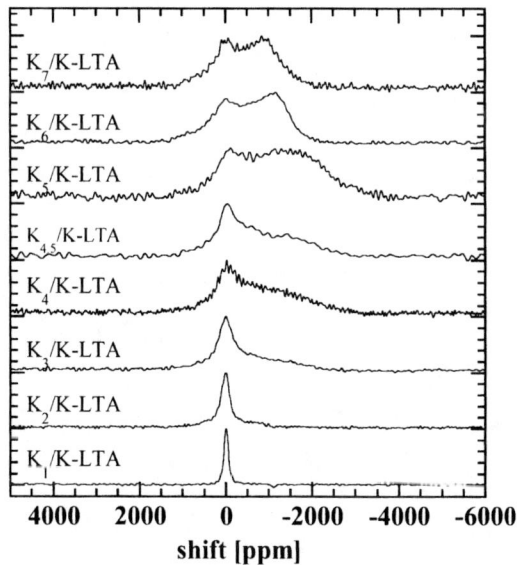

FIGURE 3. ^{29}Si-NMR spectra at 4.2K in K_x/K_{12}-LTA with x=1~7. In K_1/K_{12}-LTA and K_2/K_{12}-LTA, diamagnetic phase seems to be dominant. In K_3/K_{12}-LTA, magnetic phase is 60% of the spectrum.

The Magnetic phase diagram

From the above NMR experiments, we are led to the phase diagram shown in Fig. 4. Here, the results of X-ray and neutron diffraction measurements on the super-lattice reflection assigned to mass density wave (MDW) [2] are also summarized. The neutron diffraction measurement confirmed that the superstructure maintains down to 1.8K for ferromagnetic samples. On the other hand, the high temperature XRD measurements confirmed that it can exist up to 600K.

In conclusion, it was confirmed that the low concentric materials ($x<3$) are intrinsically diamagnetic in the present works. Therefore, there must be some mechanism for the formation of spin singlet states for $x<3$. This would be a future problem to be solved.

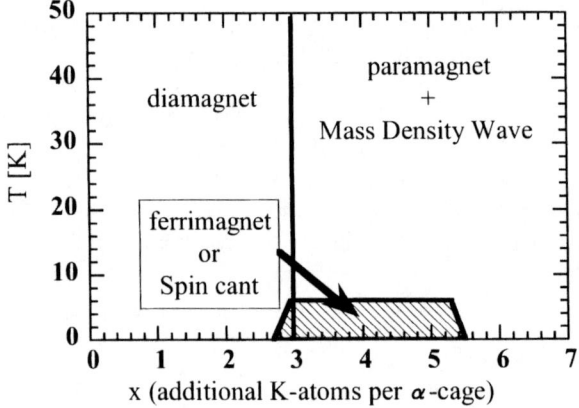

FIGURE 4. The magnetic phase diagram for K_x/K_{12}-LTA determined by NMR.

ACKNOWLEDGMENTS

This work was supported in part by the Grant-in Aid for Scientific Research by the Ministry of Education, Science and Culture of Japan.

REFERENCES

1. Y. Maniwa. H. Kira, F. Shimizu, Y. Murakami, J. Phys. Soc. Jpn. **68**, (1999) 2902.
2. Y. Nozue, T. Kodaira and T. Goto, Phys. Rev. Lett. **68**, (1992) 3789.
3. H. Kira, H. Tou, Y. Maniwa, Y. Murakami, to be published.

Mechanism of Magnetism in Stacked Nanographite: Theoretical Study

Kikuo Harigaya*[†], Naoki Kawatsu[†], and Toshiaki Enoki[†]

*Electrotechnical Laboratory, Umezono 1-1-4, Tsukuba 305-8568, Japan
[†]Tokyo Institute of Technology, Ookayama 2-12-1, Meguro-ku 152-8551, Japan

Abstract. Antiferromagnetism in stacked nanographite is investigated with using the Hubbard-type model. The A-B stacking is favorable for the hexagonal nanographite with zigzag edges, in order that magnetism appears. Next, we find that the open shell electronic structures can be origins of the decreasing magnetic moment with the decrease of the inter-graphene distance, as experiments on adsorption of molecules suggest.

INTRODUCTION

Nanographite systems, where graphene sheets of the orders of the nanometer size are stacked, show novel magnetic properties, such as, spin-glass like behaviors [1], and the change of ESR line widths while gas adsorptions [2]. Recently, it has been found [3] that magnetic moments decrease with the decrease of the interlayer distance while water molecules are attached physically.

In this paper, we consider the stacking effects in order to investigate mechanisms of antiferromagnetism using the Hubbard-type model with the interlayer hopping t_1 and the onsite repulsion U. We will show that the A-B stacking is favorable for the hexagonal nanographite with zigzag edges, in order that magnetism appears. Next, we show that the open shell electronic structures, coming from functional units and/or geometrical effects, can be origins of the decreasing magnetic moment with adsorption of molecules. Details will be reported elsewhere [4].

CLOSED SHELL ELECTRON SYSTEMS

First, we report the total magnetic moment per layer for the A-B stacked hexagonal nanographite shown in Fig. 1 (a). The first and second layers are displayed by

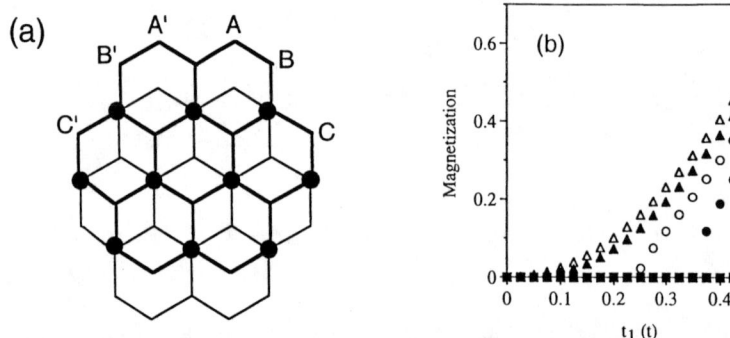

FIGURE 1. (a) A-B stacked hexagonal nanographite with zigzag edges. (b) The absolute magnitude of the total magnetic moment per layer as a function of t_1. The onsite interaction is varied within $1.8t$ (closed squares) $\leq U \leq 2.3t$ (open triangles). The interval of U between the series of the plots is $\Delta U = 0.1t$.

the thick and thin lines, respectively. In each layer, the nearest neighbor hopping t is considered. Each layer has closed shell electron systems when the layers do not interact mutually, because the number of electrons is equal to the number of sites. The interlayer hopping t_1 is assigned at the sites with closed circles. The model is solved with the unrestricted Hartree-Fock approximation, and antiferromagnetic solutions are obtained. Figure 1 (b) shows the absolute value of the total magnetic moment per layer as functions of t_1 and U. As increasing U, the magnitude of the magnetization increases. The magnetic moment is zero at the smaller t_1 region for $1.9t$ (open squares), $2.0t$ (closed circles), and $2.1t$ (open circles). The magnetic moment is zero only at $t_1 = 0$ for $U = 2.2t$ (closed triangles) and $2.3t$ (open triangles). We can understand the parabolic curves as a change due to the Heisenberg coupling proportional to t_1^2/U.

We have also calculated for the simple A-A stacking. We have not found any finite magnetization in this case. This is a remarkable difference between the A-A and A-B stackings, and is a new finding of this paper. The A-B stacking should exist in nanographite systems, because the exotic magnetisms have been observed in recent experiments [1-3]. The decrease of the interlayer distance while attachment of water molecules makes t_1 larger. However, it is known that the magnetism decreases while the attachment of molecules [3]. The calculation for the closed electron systems cannot explain the experiments even qualitatively.

OPEN SHELL ELECTRON SYSTEMS

Here, we consider the Hubbard-type model for systems which have open shell electronic structures when a nanographene layer is isolated. One case is the effects of additional charges coming from functional side groups. The next case is the roles of the standing magnetic moments due to the geometrical origin.

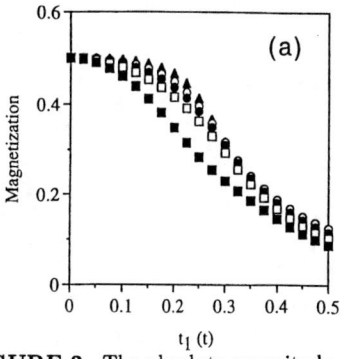 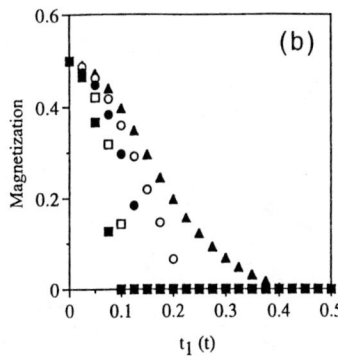

FIGURE 2. The absolute magnitude of the total magnetic moment per layer as a function of t_1 for the system with a site potential $E_s = -2t$, (a) at the site A and (b) at the site B. The site positions are displayed in Fig. 1 (a). In (a), the onsite interaction is varied within $0.6t$ (closed squares) $\leq U \leq 1.8t$ (closed triangles). The interval of U between the series of the plots is $\Delta U = 0.3t$. In (b), it is varied within $1.0t$ (closed squares) $\leq U \leq 2.0t$ (closed triangles). The interval of U between the series of the plots is $\Delta U = 0.25t$.

The active functional groups are simulated with introducing a site potential E_s [5] at edge sites. When $E_s > 0$, the site potential means the electron attractive groups. When $E_s < 0$, the electron donative groups are simulated because of the increase of the electron number at the site potentials. Here, we take $E_s = -2t$, and one additional electron per layer is taken account. Figure 2 displays the absolute values of total magnetic moment per layer. In Fig. 2 (a), the site potentials locate at the site A in the first layer [Fig. 1 (a)], and at the symmetry equivalent site in the second layer. In Fig. 2 (b), the site potential exists at the site B. The total magnetization is a decreasing function in both figures. The decrease is faster in Fig. 2 (b) than in Fig. 2 (a). The site B is neighboring to the site with the interaction t_1, and thus the localized character of the magnetic moment can be affected easily in this case. The decease of magnetization by the magnitude $30 - 40\%$ with the water molecule attachment [3] may correspond to the case of Fig. 2 (b).

Next, we look at the magnetism of stacked "triangulenes". The "triangulene" has the geometry displayed in Fig. 3 (a), and there are nine hexagonal rings [6]. The Lieb's theorem [7] says that the total spin S_{tot} of the repulsive Hubbard model of the A-B bipartite lattice is $S_{tot} = \frac{1}{2}|N_A - N_B|$, where N_A and N_B are the numbers of A and B sites. We find $S_{tot} = 1$ for the single triangulene. Figure 3 (b) displays the absolute magnitude of the total magnetic moment per layer for the A-B stacking with the vertical shift [Fig. 3 (a)]. The total magnetic moment is a decreasing function with respect to t_1. As we discuss in detail [4], there appear strong local magnetic moments at the zigzag edge sites, and they give rise dominant contributions to the magnetism of each layer. In the triangulene case, most of the edge sites are neighboring to the sites with the interaction t_1 in Fig. 3 (a). The interactions of the edge sites with the neighboring layers are strong, and the itinerant characters of electrons become larger as increasing t_1. Therefore, the

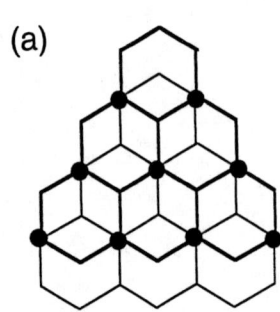 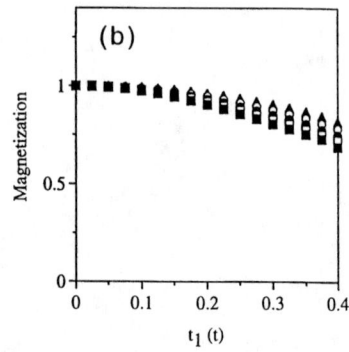

FIGURE 3. (a) A-B stacked triangulene with vertical shift. (b) The absolute magnitude of the total magnetic moment per layer as a function of t_1. The onsite interaction is varied within $0.4t$ (closed squares) $\leq U \leq 2.0t$ (closed triangles). The interval of U between the series of the plots is $\Delta U = 0.4t$.

magnetic moment is a decreasing function in Fig. 3 (b).

The present two calculations agree with the experiments, qualitatively. We can explain the decrease of magnetism in the process of adsorption of molecules [3]. Thus, the open shell electronic structures due to the active side groups and/or the geometrical origin are candidates which could explain the exotic magnetisms.

SUMMARY

Antiferromagnetism in stacked nanographite has been investigated with the Hubbard-type model. The A-B stacking is favorable for the hexagonal nanographite with zigzag edges, in order that magnetism appears. Next, we have found that the open shell electronic structures can be origins of the decreasing magnetic moment with adsorption of molecules.

REFERENCES

1. Y. Shibayama et al., Phys. Rev. Lett. **84**, 1744 (2000).
2. N. Kobayashi et al., J. Chem. Phys. **109**, 1983 (1998).
3. N. Kawatsu et al., Meeting Abstracts of the Physical Society of Japan **55** Issue 1, 717 (2000).
4. K. Harigaya, J. Phys.: Condens. Matter **13**, in press (2001); cond-mat/0010043; cond-mat/0012349.
5. K. Harigaya, A. Terai, Y. Wada, and K. Fesser, Phys. Rev. B **43**, 4141 (1991).
6. G. Allinson, R. J. Bushby, and J. L. Paillaud, J. Am. Chem. Soc. **115**, 2062 (1993).
7. E. H. Lieb, Phys. Rev. Lett. **62**, 1201 (1989); ibid. **62**, 1927 (1989).

PARTICIPANT LIST

Yohji Achiba
: Department of Chemistry, Tokyo Metropolitan University
1-1 Minami-Ohsawa, Hachi-oji, Tokyo 192-0397, Japan
achiba-yohji@c.metro-u.ac.jp

Hiroki Ago
: National Institute of Materials and Chemical Research
Tsukuba, Ibaraki 305-8565, Japan
ago@nimc.go.jp

Markus Ahlskog
: Low Temperature Laboratory, Helsinki University of Technology
P.O.B 2200 FIN-02015 Espoo, Finland
markus@boojum.hut.fi

Yoshio Akai
: Department of Physics, Tokyo Institute of Technology
2-12-1 Ookayama, Meguro-ku, Tokyo 152-8551, Japan
akai@stat.phys.titech.ac.jp

Takeshi Akasaka
: Graduate School of Science and Technology, Niigata University
Igarashi, Niigata 950-2181, Japan
akasaka@gs.niigata-u.ac.jp

Kazuhiko Akiyama
: Department of Chemistry, Tokyo Metropolitan University
1-1 Minami-Ohsawa, Hachi-oji, Tokyo 192-0397, Japan
aki@comp.metro-u.ac.jp

Voul Alexandre
: Department of Knowledge-based Information Engineering, Toyohashi University of Technology
1-1 Hibari-ga-oka, Tempaku-cho, Toyohashi, Aichi 441-8580, Japan
vul@cochem2.tutkie.tut.ac.jp

Satoshi Amamiya
: Institute of Materials Science, University of Tsukuba
1-1-1 Tennoudai, Tsukuba 305-8573, Japan
ama@ims.tsukuba.ac.jp

Kay Hyeok An
: Semiconductor Physics Research Center, Jeonbuk National University
664-14 1ga, Duckjin-Dong, Duckjin-Gu, Jeonju, Jeonbuk, Jeonju 561-756, Korea
khan@sprc2.chonbuk.ac.kr

Tsuneya Ando
: Institute for Solid State Physics, University of Tokyo
5-1-5 Kashiwanoha, Kashiwa, Chiba 277-8581, Japan
ando@issp.u-tokyo.ac.jp

Yoshinori Ando
: Department of Materials Science and Engineering, Meijo University
1-501 Shiogamaguchi, Tenpaku-ku, Nagoya 468-8502, Japan
yando@meijo-u.ac.jp

Nobuyuki Aoki
: Material Technology, Chiba University
1-33 Yayoi-cho, Inage-ku, Chiba 263-8522, Japan
n-aoki@xtal.tf.chiba-u.ac.jp

Yoshinobu Aoyagi
: Department of Information Processing, Tokyo Institute of Technology

4259 Nagatsuda, Midori-ku, Yokohama 226-8503, Japan
aoyagi@postman.riken.go.jp

Sivaram Arepalli
G. B. Tech./ Lockheed Martin
P. O. Box 58561, Mail Stop C61 Houston, TX 77258, U.S.A
sivaram.arepalli1@jsc.nasa.gov

Yuji Awano
Fujitsu Laboratories Ltd.
10-1 Morinosato-Wakamiya, Atsugi, Kanagawa 243-0197, Japan
awano@flab.fujitsu.co.jp

Shunji Bandow
Japan Science and Technology Corporation, Department of Materials Science and Engineering, Meijo University
1-501 Shiogamaguchi, Tenpaku-ku, Nagoya 468-8502, Japan
bandow@meijo-u.ac.jp

B. Batlogg
ETH Zurich
Zurich, CH-8093, Switzerland
batlogg@solid.phys.ethz.ch

Mark Baxendale
Condensed Matter & Materials Physics, Department of Physics & Astronomy, University College London
Gower St., London WC1E 6BT, UK
m.baxendale@ucl.ac.uk

Kee Joo Chang
Department of Physics, Korea Advanced Institute of Science and Technology
Taejon 305-701, Korea
kchang@mail.kaist.ac.kr

Dam Hieu Chi
Japan Advanced Institute of Science and Technology
1-8-5-105 Asahidai, Tatsunokuchi, Ishikawa 923-1292, Japan
dam@jaist.ac.jp

Tadahiko Chida
Institute of Materials Science, University of Tsukuba
1-1-1 Tennoudai, Tsukuba, Ibaraki 305-8573, Japan
tchida@ims.tsukuba.ac.jp

Masayuki Chikamatsu
Graduate School of Science, Tokyo Metropolitan University
1-1 Minami-Ohsawa, Hachi-ohji, Tokyo 192-0397, Japan
mchika@comp.metro-u.ac.jp

Jaeuk Chu
Microelectronics Laboratory, Samsung Advanced Institute of Technology
Suwon 440-600, Korea
jkchu@sait.samsung.co.kr

Marvin L. Cohen
Department of Physics, University of California
Berkeley, CA 94720, USA
cohen@jungle.berkeley.edu

Anne Debarre
Laboratoire Aime Cotton, CNRS
Batiment 505, 91405 Orsay, France
anne.debarre@lac.u-psud.fr

Lixin Dong

Department of Microsystem Engineering, Nagoya University
Furo-cho, Chikusa-ku, Nagoya 464-8603, Japan
dong@robo.mein.nagoya-u.ac.jp

Nita Dragoe
Department of Applied Chemistry, University of Tokyo
7-3-1 Hongo, Bunkyo-ku, Tokyo 113-8656, Japan

Tomotaka Ezaki
Ise Electric Corporation
1-13-8 Toranomon, Minato-ku, Tokyo 105-0001, Japan

Amir Farajian
National Institute of Materials and Chemical Research
1-1 Higashi, Tsukuba, Ibaraki 305-8565, Japan
keivan@imr.edu.

J. Fink
Institut for Solid State Research, IFW Dresden
P.O. Box 270016, D-01171 Dresden, Germany
J.Fink@ifw-dresden.de

L. Forro
Swiss Federal Institute for Technology, Physics Department/IGA
Lauzanne, CH-1015, Switzerland
forro@igahpse.epfl.ch

Ryosuke Fujii
Department of Chemistry, Tokyo Metropolitan University
1-1 Minami-Ohsawa, Hachi-oji, Tokyo 192-0397, Japan
ryosuke@comp.metro-u.ac.jp

Satoshi Fujiki
Department of Chemistry, Okayama University
3-1-1 Tsushima-naka, Okayama 700-8530, Japan
fuziki@cc.okayama-u.ac.jp

Akihiko Fujiwara
Department of Physics, University of Tokyo
7-3-1 Hongo, Bunkyo-ku, Tokyo 113-0033, Japan
fujiwara@phys.s.u-tokyo.ac.jp

Joshua Fujiwara
Wako Research Center, Honda R&D Co. Ltd.
1-4-1 Chuo, Wako-shi 351-0193, Japan
josf@f.rd.honda.co.jp

Ryuji Fujiwara
Department of Physics, Tokyo Metropolitan University
1-1 Minami-Ohsawa, Hachi-oji, Tokyo 192-0397, Japan
fujir@comp.metro-u.ac.jp

Dmitri Golberg
National Institute for Research in Inorganic Materials
1-1 Namiki, Tsukuba, Ibaraki 305-8568, Japan
golberg@nirim.go.jp

Hajime Goto
Fundamental Technology Division, HONDA Technical Research Institute
1-4-1 Chuo, Wako-shi, Saitama 350-0198, Japan

Noriaki Hamada
Department of Physics, Science University of Tokyo
2641 Yamazaki, Noda, Chiba 278-8510, Japan
hamada@ph.noda.sut.ac.jp

In Taek Han

Display Laboratory, Samsung Advanced Institute of Technology
P.O. Box 111, Suwon 440-600, Korea
ithan@sait.samsung.co.kr

Jong Hun Han
Nanotechnology Center, Iljin Nanotech Co. Ltd.
KayangTechnoTown 1487, Kayang-Dong, Kangseo-Ku, Seoul 157-203, Korea
jhhan@iljin.co.kr

Toshifumi Hara
Department of Material Science, Himeji Institute of Technology
3-2-1 Kouto, Kamigohri, Hyogo 678-1297, Japan
t-hara@sci.himeji-tech.ac.jp

Kikuo Harigaya
Electrotechnical Laboratory
1-1-4 Umezono, Tsukuba, Ibaraki 305-8568, Japan
harigaya@etl.go.jp

Abdou Hassanien
Electrotechnical Laboratory
1-1-4 Umezono, Tsukuba, Ibaraki 305-8568, Japan
hasanien@etl.go.jp

Koichi Hata
Department of Electrical & Electronic Engineering, Mie University
Tsu, Mie 514-8507, Japan
hata@is.elec.mie-u.ac.jp

Tobias Hertel
Department of Physical Chemistry, Fritz-Haber-Institut der Max-Planck-Gesellschaft
Faradayweg 4-6, D-14195 Berlin, Germany
hertel@fhi-berlin.mpg.de

Shojun Hino
Department of Information and Image Sciences, Chiba University
Inage-ku, Chiba 263-8522, Japan
hino@image.tp.chiba-u.ac.jp

Kaori Hirahara
Japan Science and Technology Corporation, Meijo University
1-501 Shiogamaguchi, Tenpaku-ku, Nagoya 468-8502, Japan
kaori_h@meijo-u.ac.jp

Tatsuki Hiraoka
Department of Material Science, Himeji Institute of Technology
3-2-1 Kouto, Kamigohri, Hyogo 678-1297, Japan
sh970069@cmta.himeji-tech.ac.jp

Jin Hirosawa
Institute of Materials Science, University of Tsukuba
1-1-1 Tennoudai, Tsukuba, Ibaraki 305-8573, Japan
jin@ims.tsukuba.ac.jp

Houjin Huang
Frontier Science Laboratories, Sony Corporation
2-1-1 Shin-sakuragaoka, Hodogaya-ku, Yokohama 240-0036, Japan
Houjin.Huang@jp.sony.com

Masao Ichida
CIRSE, Nagoya University
Chikusa-ku, Furo-cho, Nagoya 464-8603, Japan
ichida@cirse.nagoya-u.ac.jp

Kenji Ichimura

Graduate School of Natural Science and Technology, Kumamoto University
2-39-1 Kurokami, Kumamoto 860-8555, Japan
ichimura@gpo.kumamoto-u.ac.jp

Koichi Ichimura
Division of Physics, Hokkaido University
Kita 10 Nishi 8 Kita-ku, Sapporo 060-0810, Japan
ichimura@sci.hokudai.ac.jp

Sumio Iijima
Department of Material Science and Engineering, Meijo University
Shiogamaguchi, Tempaku-ku, Nagoya 468-8502, Japan
iijimas@meijo-u.ac.jp

Hideo Ikejiri
Department of Physics, Tokyo Metropolitan University
1-1 Minami-Ohsawa, Hachi-oji, Tokyo 192-0397, Japan
vanhclen@comp.metro-u.ac.jp

Takehiko Ishiguro
Physics Department, Kyoto University
Kitashirakawa, Kyoto 606-8502, Japan
tishi@scphys.kyoto-u.ac.jp

Kenji Ishii
Department of Physics, University of Tokyo
7-3-1 Hongo, Bunkyo-ku, Tokyo 113-0033, Japan
ishii@suematsu.phys.s.u-tokyo.ac.jp

Tomohiko Ishii
Department of Chemistry, Tokyo Metropolitan University
1-1 Minami-Ohsawa, Hachi-oji, Tokyo 192-0397, Japan
mail@tishii.com

Kouichi Ito
Hiroshima Prefectural Institute of Industrial Science and Technology
3-10-32 Kagamiyama, Higashi-Hiroshima 739-0046, Japan
ito@sankaken.gr.jp

Osamu Ito
ICRS, Tohoku University
Katahira, Sendai 980-8577, Japan
ito@icrs.tohoku.ac.jp

Takayoshi Ito
Japan Advanced Institute of Science and Technology
1-8-5-105 Asahidai, Tatsunokuchi, Ishikawa 923-1292, Japan

Yoshihiro Iwasa
Japan Advanced Institute of Science and Technology
1-8-5-105 Asahidai, Tatsunokuchi, Ishikawa 923-1292, Japan
iwasa@jaist.ac.jp

Kentaro Iwasaki
Department of Image Science, Chiba University
1-33 Yayoi-cho, Inage-ku, Chiba 263-8522, Japan
iwasaki@image.tp.chiba-u.ac.jp

Goo-Hwan Jeong
Department of Electronic Engineering, Tohoku University
05 Aoba, Aramaki, Aoba-ku, Sendai, Japan
hatak17@ec.ecei.tohoku.ac.jp

Hee Jin Jeong
Semiconductor Physics Research Center, Jeonbuk National University
664-14 1ga, Duckjin-Dong, Duckjin-Gu, Jeonju, Jeonbuk, Jeonju 561-756, Korea

khan@sprc2.chonbuk.ac.kr

Kwang Seok Jeong
   Microelectronics Laboratory, Samsung Advanced Institute of Technology
   Suwon, Korea
   gsjeong@sait.samsung.co.kr

Balachandran Jeyadevan
   Department of Geoscience and Technology, Tohoku University
   Katahira, Aoba, Sendai 980-8577, Japan
   jeya@ni4.earth.tohoku.ac.jp

Jifei Jia
   Hiroshima Prefectural Institute of Industrial Science and Technology
   3-10-32 Kagamiyama, Higashi-Hiroshima 739-0046, Japan
   jiajifei@sankaken.gr.jp

Ado Jorio
   Massachusetts Institute of Technology
   13-3017, 77 Massachusetts Av., Cambridge, MA 02139, USA
   adojorio@mgm.mit.edu

Yasukazu Kajihara
   Department of Material Science, Osaka City University
   3-3-138 Sugimoto, Sumiyoshi-ku, Osaka 558-8585, Japan
   kajihara@sci.osaka-cu.ac.jp

Kenjiro Kanamitsu
   Department of Physics, Tokyo Institute of Technology
   2-12-1 Ookayama, Meguro-ku, Tokyo 152-8551, Japan
   kenjiro@stat.phys.titech.ac.jp

Akinobu Kanda
   Institute of Physics, University of Tsukuba
   1-1-1 Tennodai, Tsukuba, Ibaraki 305-8571, Japan
   kanda@lt.px.tsukuba.ac.jp

Yohko Kasuya
   ICORp-JST
   C/O NEC 34 Miyukigaoka, Tsukuba, Ibaraki 305-8501, Japan

Hiromichi Kataura
   Department of Physics, Tokyo Metropolitan University
   1-1 Minami-Ohsawa, Hachi-ohji, Tokyo 192-0397, Japan
   kataura@phys.metro-u.ac.jp

Takashi Kato
   Institute for Fundamental Chemistry
   Takano-nishi, Hiraki-cho, Kyoto 606-8103, Japan
   kato@ifc.or.jp

Tatsuhisa Kato
   Institute for Molecular Science
   Myodaiji, Okazaki 444-8585, Japan
   kato@ims.ac.jp

Shinji Kawasaki
   Faculty of Textile Science & Technology, Shinshu University
   3-15-1 Tokida, Ueda 386-8567, Japan
   skawasa@giptc.shinshu-u.ac.jp

A. Kazaoui
   Nanostructured Photoconversion Materials Group, National Institute of Materials and Chemical Research
   1-1 Higashi, Tsukuba, Ibaraki 305-8565, Japan
   kazaoui@nimc.go,jp

Koichi Kikuchi
: Department of Chemistry, Tokyo Metropolitan University
: 1-1 Minami-Ohsawa, Hachi-oji, Tokyo 192-0397, Japan
: kikuchi-koichi@c.metro-u.ac.jp

Dong Ho Kim
: Department of Physics, Yeungnam University
: 214-1 Dae-dong, Kyungsan 712-749, Korea
: dhkim@yu.ac.kr

Won Seok Kim
: Semiconductor Physics Research Center, Jeonbuk National University
: 664-14 1ga, Duckjin-Dong, Duckjin-Gu, Jeonju, Jeonbuk, Jeonju 561-756, Korea
: khan@sprc2.chonbuk.ac.kr

Kaoru Kimura
: Department of Advanced Materials Science, University of Tokyo
: 7-3-1 Hongo, Bunkyo-ku, Tokyo 152-8551, Japan
: bkimura@phys.mm.t.u-tokyo.ac.jp

Hiroshi Kira
: Department of Physics, Tokyo Metropolitan University
: 1-1 Minami-Ohsawa, Hachi-oji, Tokyo 192-0397, Japan
: Kira@comp.metro-u.ac.jp

Keiichi Kitazume
: Department of Physics, Tokyo Metropolitan University
: 1-1 Minami-Ohsawa, Hachi-oji, Tokyo 192-0397, Japan
: kI286ta@comp.metro-u.ac.jp

Takushi Kizuka
: Research Center for Advanced Waste & Emission Management, Nagoya University
: Furo-cho, Chikusa-ku, Nagoya 464-8603, Japan
: j46110a@nucc.cc.nagoya-u.ac.jp

Mototada Kobayashi
: Department of Material Science, Himeji Institute of Technology
: 3-2-1 Kouto, Kamigohri, Hyogo 678-1297, Japan
: kobayashi@sci.himeji-tech.ac.jp

Shigenori Kobayashi
: Department of Physics, Tokyo Metropolitan University
: 1-1 Minami-Ohsawa, Hachi-oji, Tokyo 192-0397, Japan
: phys@hk.airnet.ne.jp

Mathieu Kociak
: Laboratoire de Physique des Solides
: Batiment 510, Universite Paris-Sud, Orsay 91405, France
: kociak@lps.u-psud.fr

Masamichi Kohno
: Faculty of Engineering, University of Tokyo
: 2-11-16 Yayoi, Bunkyo-ku, Tokyo 113-8656, Japan
: kohno@photon.t.u-tokyo.ac.jp

Nobuaki Kojima
: Toyota Technological Institute
: 2-12-1 Hisakata, Tempaku, Nagoya 468-8511, Japan
: nkojima@toyota-ti.ac.jp

Takeshi Komada
: Department of Chemistry, Tokyo Metropolitan University
: 1-1 Minami-Ohsawa, Hachi-oji, Tokyo 192-0397, Japan
: kodama-takeshi@c.metro-u.ac.jp

Fumitaka Kosha
: Department of Materials Chemistry, Hosei University
3-7-2 Kajinocho, Koganei, Tokyo 184-8584, Japan
i00r2108@k.hosei.ac.jp

Akira Koshio
: c/o NEC Corporation
34 Miyukigaoka, Tsukuba, Ibaraki 305-8501, Japan
koshio@nlp.jst.go.jp

Wolfgang Kraetschmer
: Max-Planck Institute for Nuclear Physics
PO Box 103980, Heidelberg 69029, Germany
kraetschmer@mpi-hd.mpg.de

Yoshihiro Kubozono
: Department of Chemistry, Okayama University
3-1-1 Tsushima-naka, Okayama 700-8530, Japan
kubozono@cc.okayama-u.ac.jp

Hiroyasu Kumada
: Department of Material Science, Himeji Institute of Technology
3-2-1 Kouto, Kamigohri, Hyogo 678-1297, Japan
h.kumada@sci.himeji-tech.ac.jp

Vijay Kumar
: Institute for Materials Research, Tohoku University
2-1-1 Katahira, Aoba-ku, Sendai 980-8577, Japan
kumar@imr.edu

Shogo Kuno
: Department of Electrical & Electronic Engineering, Mie University
1515 Kamihama-cho, Tsu, Mie 514-8507, Japan

Hiroyuki Kurachi
: Ise Electric Corporation
728-23 Tumura-cho, Ise, 516-1103, Japan
hkurachi@itron-ise.co.jp

H. Kuzmany
: Institut fuer Materialphysik, Universitaet Wien
Wien Struddlhofgasse, A-1090 Wien 4, Austlia
kuzman@ap.univie.ac.at

Toru Kuzumaki
: Department of Materials Science, University of Tokyo
7-3-1 Hongo, Bunkyo-ku, Tokyo 113-8656, Japan
kuzumaki@micro.mm.t.u-tokyo.ac.jp

Cheol Jin Lee
: School of Electrical Engineering, Kunsan National University
San 68, Miryong-dong, Kunsan 573-701, Korea
cjlee@ks.kunsan.ac.kr

Young Hee Lee
: Department of Physics, Jeonbuk National University
Jeongju 561-756, Korea
leeyy@sprc2.chonbuk.ac.kr

Yun Hi Lee
: Korea Institute of Science and Technology
Hawolgok-dong 39-1, Sungpuk-gu, Soul 136-791, Korea
lyh@kist.re.kr

Seong Chu Lim
: Semiconductor Physics Research Center, Jeonbuk National University

664-14 1ga, Duckjin-Dong, Duckjin-Gu, Jeonju, Jeonbuk, Jeonju 561-756, Korea
choiyc@sprc2.chonbuk.ac.kr

Steven G. Louie
Department of Physics, University of California
Berkeley, CA 94720, USA
sglouie@uclink.berkeley.edu

David Luzzi
Department of Materials Science, University of Pennsylvania
3231 Walnut St., Philadelphia, PA 19104-6272, USA
luzzi@lrsm.upenn.edu

Masayoshi Machino
Department of Physics, Tokyo Metropolitan University
1-1 Minami-Ohsawa, Hachi-oji, Tokyo 192-0397, Japan
machino@comp.metro-u.ac.jp

Atsutaka Maeda
Department of Basic Science, University of Tokyo
3-8-1, Komaba, Meguro-ku, Tokyo 153-8902, Japan
maeda@maildbs.c.u-tokyo.ac.jp

Yutaka Maniwa
Department of Physics, Tokyo Metropolitan University
1-1 Minami-Ohsawa, Hachi-oji, Tokyo 192-0397, Japan
maniwa@phys.metro-u.ac.jp

Shigeo Maruyama
School of Engineering, University of Tokyo
7-3-1 Hongo, Bunkyo-ku, Tokyo 113-8656, Japan
maruyama@photon.t.u-tokyo.ac.jp

Yusei Maruyama
Department of Materials Chemistry, Hosei University
Kajinocho, Koganei, Tokyo 184-8584, Japan
maruyama@k.hosei.ac.jp

Terauchi Masami
Research Institute for Scientific Measurements, Tohoku University
2-1-1 Katahira, Aoba-ku, Sendai 980-8577, Japan
terauchi@rism.tohoku.ac.jp

Shin-ichi Masubuchi
Department of Physics, Chuo University
Kasuga, Bunkyo-ku, Tokyo 112-8551, Japan
masubuch@phys.chuo-u.ac.jp

Takeo Matsui
Kansai Research Institute Inc.
17 Chudori, Minamimachi, Shimogyo-ku, Kyoto 600-8813, Japan
matsui@kyoto.kansai-ri.co.jp

Takanori Matsumoto
Department of Physics, Tokyo Institute of Technology
2-12-1 Ookayama, Meguro-ku, Tokyo 152-8551, Japan
matumoto@stat.phys.titech.ac.jp

Takashi Minakata
Chemistry & Chemical Process Laboratory, Asahi Chemical Industry Co Ltd.
1-3-1 Yakoh, Kawasaki-ku, Kawasaki 210-0863, Japan
minakata.tb@om.asahi-kasei.co.jp

Nobutsugu Minami
National Institute of Materials and Chemical Research
1-1 Higashi, Tsukuba, Ibaraki 305-8565, Japan

minami@nimc.go.jp

Takashi Miyake
   Department of Physics, Tokyo Institute of Technology
   2-12-1 Ookayama, Meguro-ku, Tokyo 152-8551, Japan
   miyake@stat.phys.titech.ac.jp

Kenji Mizoguchi
   Department of Physics, Tokyo Metropolitan University
   1-1 Minami-Ohsawa, Hachi-oji, Tokyo 192-0397, Japan
   mizoguchi@phys.metro-u.ac.jp

Jeong-Mi Moon
   Semiconductor Physics Research Center, Jeonbuk National University
   664-14 1ga, Duckjin-Dong, Duckjin-Gu, Jeonju, Jeonbuk, Jeonju 561-756, Korea
   choiyc@sprc2.chonbuk.ac.kr

Takao Mori
   National Institute for Research in Inorganic Materials
   1-1 Namiki, Tsukuba, Ibaraki 305-0044, Japan
   moritk@nirim.go.jp

Satoru Motohashi
   Department of Materials Chemistry, Hosei University
   3-7-2 Kajinocho, Koganei, Tokyo 184-8584, Japan
   i9903227@k.hosei.ac.jp

Indrajit Mukhopadhyay
   Faculty of Textile Science and Technology, Shinshu University
   Ueda 386-8567, Japan
   Indra@pmac103.shinshu-u.ac.jp

Takahiko Muneyoshi
   Material Design Center Displays, Central Research Lab., Hitachi Ltd.
   1-280 Higashi-koigakubo, Kokubunji, Tokyo 185-8601, Japan
   mune@crl.hitachi.co.jp

Hideoki Murakami
   Faculty of Education, Tokyo Gakugei University
   Nukii Kitamachi, Koganei, Tokyo 184-8501, Japan
   murakami@u-gakugei.ac.jp

Kenji Murakami
   Research Institute of Electronics, Shizuoka University
   3-5-1 Johoku, Hamamatsu 432-801, Japan
   rskmura@ipc.shizumoa.ac.jp

Nobukata Nagasawa
   Department of Physics, University of Tokyo
   7-3-1 Hongo, Bunkyo-ku, Tokyo 113-0033, Japan
   nagasawa@exciton.phys.s.u-tokyo.ac.jp

Arao Nakamura
   Department of Applied Physics, Nagoya University
   Furo-cho, Chikusa-ku, Nagoya 464-8603, Japan
   nakamura@nuap.nagoya-u.ac.jp

Takeshi Nakanishi
   Department of Applied Physics, Delft University of Technology
   Lorentzweg 1, 2628 CJ Delft, The Netherlands
   T.Nakanishi@tn.tudelft.nl

Kenji Nakao
   Institute of Material Science, University of Tsukuba
   1-1-1 Tennodai, Tsukuba, Ibaraki 305-8573, Japan
   nakao@ims.tsukuba.ac.jp

K.L. Narayanan
: Semiconductor Laboratory, Toyota Technological Institute
2-12-1 Hisakata, Tempaku-ku, Nagoya 468-8511, Japan
klnbabe@hotmail.com

Mant Ton Nguyen
: Japan Advanced Institute of Science and Technology
1-1 Asahidai, Tatsunokuchi, Ishikawa 923-1292, Japan
mnguyen@jaist.ac.jp

Fumiyuki Nihey
: NEC Fundamental Research Laboratories
34 Miyukigaoka, Tsukuba, Ibaraki 305-8501, Japan
nihey@frl.cl.nec.co.jp

Eiji Nishibori
: Department of Applied Physics, Nagoya University
1 Furo-cho, Chikusa, Nagoya 464-8603, Japan
eiji@mcr.nuap.nagoya-u.ac.jp

Ryoichi Nishida
: Research & Development Department, Osaka Gas Co. Ltd.
6-19-9 Torishima, Konohana-ku, Osaka 554-0051, Japan
nishida@osakagas.co.jp

Yuichiro Nishina
: Ishinomaki-Senshu University
1 Shin-mito, Minamisakai, Ishinomaki 986-0031, Japan

Yoko Noda
: Mitsubishi Corporation
2-6-3 Marunouchi, Chiyoda-ku, Tokyo 100-8086, Japan
yoko.noda@j.mitsubishicorp.com

Kazushige Nomura
: Division of Physics, Hokkaido University
Kita-ku, Sapporo 060-0810, Japan
knmr@phys.sci.hokudai.ac.jp

Yuichi Ochiai
: Department Materials Technology, Chiba University
1-33 Yayoi-cho, Inage-ku, Chiba 263-8522, Japan
ochiai@xtal.tf.chiba-u.ac.jp

Hironori Ogata
: Institute for Molecular Science
Myodaiji, Okazaki 444-8585, Japan
hogata@ims.ac.jp

Takayuki Ogawa
: Department of Geoscience and Technology, Tohoku University
Aoba, Aramaki, Aobaku, Sendai 980-8579, Japan
ogawa@ni4.earth.tohoku.ac.jp

Reiko Ogura
: Department of Applied Chemistry, Science University of Tokyo
1-3 Kagurazaka, Shinjuku-ku, Tokyo 162-0825, Japan
j1300618@ed.kagu.sut.ac.jp

Atsushi Oguri
: Department of Advanced Material Science, University of Tokyo
7-3-1 Hongo, Bunkyo-ku, Tokyo 113-8656, Japan
oguri@phys.mm.t.u-tokyo.ac.jp

Yutaka Ohmura
: Takiron Co. Ltd

1455 Kariya, Mitsu-cho, Hyogo 671-1393, Japan
ohmura@takiron.ac.jp

Kaoru Ohno
Department of Physics, Yokohama National University
79-5 Tokiwadai, Hodogaya-ku, Yokohama 240-8501, Japan
ohno@ynu.ac.jp

T. Ohtsuki
Laboratory of Nuclear Science, Tohoku University
1-2-1 Mikamine, Taihaku-ku, Sendai 982-0826, Japan
ohtsuki@LNS.tohoku.ac.jp

Susumu Okada
Institute of Material Science, University of Tsukuba
1-1-1 Tennodai, Tsukuba 305-8573, Japan
sokada@moose.ims.tsukuba.ac.jp

Toshiya Okazaki
Department of Chemistry, Nagoya University
Furo-cho, Chikusa-ku, Nagoya 464-8602, Japan
okazaki@nano.chem.nagoya-u.ac.jp

Toshihiko Osaki
National Industrial Research Institute of Nagoya
Hirate-cho, Kita-ku, Nagoya 462-8530, Japan
tosaki@nirin.go.jp

Eiji Osawa
Department of Knowledge-based Information Engineering, Toyohashi University of Technology
1-1 Hibari-ga-oka, Tempaku-cho, Toyohashi, Aichi 441-8580, Japan
user862134750@aol.com/osawa@cochem2.tutkie.tut.ac.jp

Kokichi Oshima
Department of Physics, Okayama University
3-1-1 Ttsushima-naka, Okayama 700-8530, Japan
oshima@science.okayama-u.ac.jp

Norio Ozawa
Department of Chemistry, Tokyo Metropolitan University
1-1 Minami-Ohsawa, Hachi-oji, Tokyo 192-0397, Japan
ozawa@comp.metro-u.ac.jp

Lujin Pan
Faculty of Engineering, Osaka-Furitu University
1-1 Gakuen, Sakai, Osaka 599-853, Japan
pan@dd.pe.osakafu-u.ac.jp

Jeung Hoon Park
Nanotechnology Center, ILJIN Nanotech Co. Ltd.
Kayang Techno Town 1487, Kayang-Dong, Kangseo-Ku, Seoul 157-810, Korea
cnt-park@iljin.co.kr

Young Soo Park
Semiconductor Physics Research Center, Jeonbuk National University
664-14 1ga, Duckjin-Dong, Duckjin-Gu, Jeonju, Jeonbuk, Jeonju 561-756, Korea
khan@sprc2.chonbuk.ac.kr

Wolfgang Plank
Universitaet Wien
Strudlhofgasse 4, Wien 1090, Austria
PLANCK@AP.UNIVIE.AC.AT

K. Prassides
School of Chemistry, Physics and Environmental Science, University of Sussex

Falmer, Brighton, BN1 9QJ, UK
K.Prassdes@susx.ac.uk

Lu-Chang Qin
JST-ICORP Nanotubulite Project, c/o NEC Corporation
34 Miyukigaoka, Tsukuba, Ibaraki 305-8501, Japan
qin@frl.cl.nec.co.jp

Edouard Reny
Department of Applied Chemistry, Hiroshima University
1-1-1 Kagamiyama, Higashi-Hiroshima 739-8527, Japan
RENY@ipc.hiroshima-u.ac.jp

Stephan Roche
Departement de Recherche Fondamentale sur la Matiere Condensee Commissariat, a l'Energie Atomique (DRFMC/SPSMS-CEA)
17 avenue des Martyrs 38042, Grenoble, France
sroche@cea.fr

Riichiro Saito
University of Electro-Communications
1-5-1 Chofugaoka, Chofu, Tokyo 182-8585, Japan
rsaito@ee.uec.ac.jp

Susumu Saito
Department of Physics, Tokyo Institute of Technology
2-12-1 Ookayama Meguro-ku, Tokyo 152-8551, Japan
saito@stat.phys.titech.ac.jp

Yahachi Saito
Department of Electrical & Electronic Engineering, Mie University
1515 Kamihama-cho, Tsu, 514-8507, Japan
saito@elec.mie-u.ac.jp

Koichi Sakaguchi
Department of Chemistry, Tokyo Metropolitan University
1-1 Minami-Ohsawa, Hachi-oji, Tokyo 192-0397, Japan
guchi@comp.metro-u.ac.jp

Masatoshi Sakai
Center for Integrated Research in Science and Engineering, Nagoya University
Furo-cho, Chikusa-ku, Nagoya 464-8603, Japan
sakai@nano.nuap.nagoya-u.ac.jp

Hirokazu Sakamoto
Department of Physics, Tokyo Metropolitan University
1-1 Minami-Ohsawa, Hachi-oji, Tokyo 192-0397, Japan
sakamoto@phys.metro-u.ac.jp

Masahito Sano
Chemotransfiguration Project, Japan Science and Technology Corporation
2432 Aikawa-cho, Kurume 839-0861, Japan
mass@jst.ktarn.or.jp

Hideki Sato
Department of Electrical & Electronic Engineering, Mie University
1515 Kamihama-cho, Tsu, Mie 514-8507, Japan
sato@elec.mie-u.ac.jp

S. Sawada
Department of Physics, Kwansei Gakuin University
Nishinomiya 662-0886, Japan
s-sawada@kwansei.ac.jp

Erich Schoenherr
Research Institute of Electronics, Shizuoka University

    3-5-1 Johoku, Hamamatsu 432-8011, Japan
    shoen@rie.shizumoa.ac.jp

Rahul Sen
    Department of Chemistry, Tokyo Metropolitan University
    1-1 Minami-Ohsawa, Hachi-oji, Tokyo 192-0397, Japan
    rahul@comp.metro-u.ac.jp

Chang Woo Seo
    Nanotechnology Center, ILJIN Nanotech Co. Ltd.
    Kayang Techno Town 1487, Kayang-Dong, Kangseo-Ku, Seoul 157-810, Korea
    sai0424@iljin.co.kr

Hiroyuki Shiba
    Department of Physics, Tokyo Institute of Technology
    2-12-1 Ookayama, Meguro-ku, Tokyo 152-8551, Japan
    shiba@stat.phys.titech.ac.jp

Nobutaka Shimizu
    ULVAC PHI Inc }
    Enzo 370, Chigasaki-shi 253-0084, Japan
    nshimizu@phi.com

Hidekazu Shimotani
    Department of Applied Chemistry, University of Tokyo
    7-3-1 Hongo, Bunkyo-ku, Tokyo 113-8656, Japan
    tt07216@mail.ecc.u-tokyo.ac.jp

Hisanori Shinohara
    Department of Chemistry, Nagoya University
    1 Furo-cho, Chikusa-ku, Nagoya 464-8603, Japan
    nori@nano.chem.nagoya-u.ac.jp

Ferenc Simon
    Budapest University of Technology and Economics
    PO Box 91, Budapest, 1521, Hungary
    simon@esr1.fkf.bme.hu

Odile Stephan
    Laboratoire de Physique des Solides
    Batiment 510, Universite Paris-Sud, Orsay 91405, France
    stephan@lps.u-psud.fr

Keisuke Sueki
    Department of Chemistry, Tokyo Metropolitan University
    1-1 Minami-Ohsawa, Hachi-oji, Tokyo 192-0397, Japan
    sueki-keisuke@c.metro-u.ac.jp

Hiroyoshi Suematsu
    Department of Physics, University of Tokyo
    7-3-1 Hongo, Bunkyo-ku, Tokyo 113-0033, Japan
    suematsu@phys.s.u-tokyo.ac.jp

Tomohiro Suetsuna
    Faculty of Engineering, University of Tokyo
    7-3-1 Hongo, Bunkyo-ku, Tokyo 113-0033, Japan
    kk06118@mail.ecc.u-tokyo.ac.jp

Kenji Suzuki
    Department of Materials Chemistry, Hosei University
    3-7-2 Kajinocho, Koganei, Tokyo 184-8584, Japan
    sodium2000@geocities.co.jp

Masato Suzuki
    Department of Physics, Osaka City University
    Sumiyosi-ku, Osaka 558-8585, Japan

suzuki@sci.osaka-cu.ac.jp

Satoru Suzuki
: NTT Basic Research Laboratories
3-1 Morinosato-Wakamiya, Atsugi, Kanagawa 243-0198, Japan
ssuzuki@will.brl.ntt.co.jp

Shinzo Suzuki
: Department of Chemistry, Tokyo Metropolitan University
1-1 Minami-Ohsawa, Hachi-oji, Tokyo 192-0397, Japan
suzuki-shinzo@c.metro-u.ac.jp

Syugo Suzuki
: Institute of Materials Science, University of Tsukuba
1-1-1 Tennoudai, Tsukuba, Ibaraki 305-8573, Japan
shugo@ims.tsukuba.ac.jp

Hidekatsu Suzuura
: Institute for Solid State Physics, University of Tokyo
5-1-5 Kashiwanoha, Kashiwa, Chiba 277-8581, Japan
suzuura@issp.u-tokyo.ac.jp

Masaru Tachibana
: Faculty of Science, Yokohama City University
22-2 Seto, Kanazawa-ku, Yokohama 236-0027, Japan
tachiban@yokohama-cu.ac.jp

Nikos Tagmatarchis
: Chemistry Department, Nagoya University
Furo-cho, Chikusa-ku, Nagoya 464-8602, Japan
nikos@nano.chem.nagoya-u.ac.jp

Takenobu Taishi
: Department of Physical Material Science, Japan Advanced Institute of Science and Technology
1-1 Asahidai, Tatsunokuchi, Ishikawa 923-1292, Japan
takenobu@jaist.ac.jp

Yasuhiro Takabayashi
: Department of Chemistry, Okayama University
3-1-1 Tsushima-naka, Okayama 700-8530, Japan
kourin@cc.okayama-u.ac.jp

Masaki Takata
: Department of Applied Physics, Nagoya University
Furo-cho, Chikusa-ku, Nagoya 464-8603, Japan
takata@nuap.nagoya-u.ac.jp

Junji Takeuchi
: Molecular Materials Science, Osaka City University
3-3-138 Sugimoto, Sumiyoshi-ku, Osaka 558-8585, Japan
takej@sci.osaka-cu.ac.jp

Hirofumi Takikawa
: Department of Electrical and Electronic Engineering, Toyohashi University of Technology
Toyohashi, Aichi 441-8580, Japan
takikawa@eee.tut.ac.jp

Ryo Tamura
: Department of Physics, University of Tokyo
7-3-1 Hongo, Bunkyo-ku, Tokyo 113-0033, Japan
ryo@phys.s.u-tokyo.ac.jp

Z.K. Tang
: Department of Physics, Hong Kong University of Science and Technology

Clear Water Bay, Kowloon, Hong Kong
phzktang@ust.hk

Katsumi Tanigaki
Department of Material Science, Osaka City University
3-3-138 Sugimoto, Sumiyoshi-ku, Osaka 558-8585, Japan
tanigaki@sci.osaka-cu.ac.jp

Paul Tchenio
Laboratoire Aime Cotton, CNRS
Universite Paris XI, Orsay 91405, France
paul.tchenio@lac.u-psud.fr

Hiroshi Tokumoto
Joint Research Center for Atom Technology
1-1-4 Higashi, Tsukuba, Ibaraki 305-8562, Japan
htokumot@jrcat.or.jp

Madoka Tokumoto
Electrotechnical Laboratory
1-1-4 Umezono, Tsukuba, Ibaraki 305-8568, Japan
madoka@etl.go.jp

Hideki Tou
Department of Physics, Tokyo Metropolitan University
1-1 Minami-Ohsawa, Hachi-oji, Tokyo 192-0397, Japan
tou@phys.metro-u.ac.jp

Kazuhito Tsukagoshi
Semiconductor Laboratory, The Institute of Physical and Chemical Research (RIKEN)
2-1 Hirosawa, Wako-shi, Saitama 350-0198, Japan
tsuka@postman.riken.go.jp

Keiji Ueno
Department of Chemistry, University of Tokyo
7-3-1 Hongo, Bunkyo-ku, Tokyo 113-0033, Japan
kei@chem.s.u-tokyo.ac.jp

Koichiro Umemoto
Department of Physics, Tokyo Institute of Technology
2-12-1 Ookayama, Meguro-ku, Tokyo 152-8551, Japan
umemoto@stat.phys.titech.ac.jp

Seiji Uryu
Institute for Solid State Physics, University of Tokyo
5-1-5 Kashiwanoha Kashiwa, Chiba 277-8581, Japan
uryu@issp.u-tokyo.ac.jp

Tomonari Wakabayashi
Max-Planck-Institut fuer Kernphysik
Postfach 103980, D-69029 Heidelberg, Germany
Tomonari.wakabayashi@mpi-hd.mpg.de

Peng Wang
Hiroshima Prefectural Institute of Industrial Science and Technology
3-10-32 Kagamiyama, Higashi-Hiroshima 739-0046, Japan
johnwang@sankaken.gr.jp

Eiichirou Watanabe
Semiconductor Laboratory, The Institute of Physical and Chemical Research (RIKEN)
2-1 Hirosawa, Wako-shi, Saitama 351-0198, Japan
eiichiro@postman.riken.go.jp

Hiroki Watanabe

DENSO Corporation
500-1 Nanzan, Komenogicho, Nisshin-shi 470-0111, Japan
hwatana@rlab.denso.co.jp

Hiroyuki Watanabe
Chiba University
2-584 Kemigawa-cho, Hanamigawa-ku, Chiba 262-0023, Japan
watanabe@lepton.chiba-u.ac.jp

Kazuyuki Watanabe
Department of Physics, Science University of Tokyo
1-3 Kagurazaka, Shinjuku-ku, Tokyo 162-8601, Japan
kazuyuki@rs.kagu.sut.ac.jp

Tatsuya Yaguchi
Institute for Solid State Physics, University of Tokyo
5-1-5 Kashiwanoha, Kashiwa, Chiba 277-8581, Japan
tatsuya@issp.u-tokyo.ac.jp

Tokio Yamabe
Institute for Fundamental Chemistry
Takano-nishi, Hiraki-cho, Kyoto 606-8103, Japan

Nobuhide Yoneya
The Institute of Physical and Chemical Research (RIKEN)
2-1 Hirosawa, Wako-shi, Saitama 351-0198, Japan
nyoneya@postman.riken.go.jp

Masako Yudasaka
c/o NEC Corporation
34 Miyukigaoka, Tsukuba, Ibaraki 305-8501, Japan
yudasaka@frl.cl.nec.co.jp

Alex Zettle
Department of Physics, University of California
Berkeley, CA 94720, USA
azett@socrates.berkeley.edu

Minfang Zhang
International Cooperative Research Project of Japan Science Technology
Corporation, c/o NEC Corporation
34 Miyukigaoka, Tsukuba, Ibaraki 305-8501, Japan
minfang@nlp.jst.go.jp

Yisong Zheng
Institute for Solid State Physics, University of Tokyo
5-1-5 Kashiwanoha Kashiwa, Chiba 277-8581, Japan
zheng@issp.u-tokyo.ac.jp

Otto Zhou
Department of Physics and Astronomy, University of North Carolina
CB 3255 Phillipps Hall, Chapel Hill, NC27599, USA
zhou@physics.unc.edu

AUTHOR INDEX

A

Achiba, Y., 15, 27, 51, 81, 121, 165, 177, 189, 197, 233
Ahlskog, M., 141
Aizawa, N., 409
Akahama, Y., 325, 389
Akasaka, T., 433, 465, 489
Akiyama, K., 313, 437
Alexenskii, A. E., 425
Amamiya, S., 477, 481
Ambe, F., 313
Ambe, S., 313
An, K. H., 39, 209, 217, 221, 241
Ando, T., 149, 261, 269
Ando, Y., 7, 31
Aoki, N., 273
Aoyagi, Y., 265
Arai, F., 71
Arepalli, S., 11

B

Bae, D. J., 39, 209, 221, 241
Baes-Fischlmair, S., 417
Baidakova, M. V., 425
Bando, Y., 67
Bower, C., 193

C

Chan, C. T., 125
Chang, K. J., 169, 245
Chen, X. H., 447
Chen, Z., 253
Chida, T., 337
Chijiwa, K., 421
Chikamatsu, M., 455
Choi, C. H., 75
Choi, Y. C., 35, 39, 221, 241
Cohen, M. L., 297
Cros, C., 499
Cumings, J., 107

D

Dam, H. C., 447
Dideikin, A. T., 425
Dong, L., 71
Dragoe, N., 397
Dresselhaus, G., 129, 185
Dresselhaus, M. S., 129, 185

E

Enoki, T., 529
Enomoto, R., 273
Enomoto, S., 313
Esfarjani, K., 237, 253, 257

F

Farajian, A. A., 253, 257
Fasel, R., 181
Files, B., 11
Fink, J., 87
Fleming, L., 95
Frauenheim, T., 117
Fudo, H., 177
Fujiki, S., 345, 357
Fujitsuka, M., 489
Fujiwara, A., 189, 317, 507
Fujiwara, M., 393
Fukuda, T., 71
Fülöp, F., 329

G

Gao, B., 95
Gao, X., 465
Garaj, S., 329
Gogia, B., 361
Golberg, D., 67
Golden, M. S., 87
Goto, H., 517
Grueneis, A., 81

H

Haba, H., 437
Hadjiev, V., 11
Hafner, J. H., 129, 185
Hakonen, P., 141
Hamada, N., 201
Han, J. H., 59
Hanada, T., 455
Hara, T., 325
Harada, A., 19
Harigaya, K., 529
Hashi, Y., 237
Hata, K., 113
Hatakeyama, R., 3
Hertel, T., 181
Heun, S., 193
Hibi, Y., 31
Hino, S., 521
Hirahara, K., 7, 31, 165
Hirosawa, J., 341
Holmes, W., 11
Horiike, Y., 281
Hoyanagi, K., 421
Hsu, W. K., 249
Hulman, M., 81
Hunter, M., 129, 185

I

Ichida, M., 121, 451
Ichihashi, T., 137
Ichimura, K., 197, 225, 229
Iijima, S., 7, 31, 137, 165, 205, 237
Iizumi, K., 485
Ikejiri, H., 443
Ikemoto, I., 409, 455, 473
Imaeda, K., 225, 229
Inada, Y., 485
Inakuma, M., 461
Inokuchi, H., 225, 229
Ishibashi, K., 273
Ishida, H., 3
Ishii, H., 503
Ishii, K., 317, 345, 357
Ishii, T., 237, 409, 429
Ito, O., 489
Ito, S., 31
Ito, T., 177, 385

Itou, T., 447
Iwasa, Y., 177, 353, 369, 373, 385, 409, 447, 485
Iwasaki, K., 521

J

Jang, Y. T., 75
Jánossy, A., 329
Jeong, H. J., 35, 39, 221, 241
Jeong, S. Y., 35
Jeyadevan, B., 3, 47
Jin, C. W., 225
Jorio, A., 129, 185
Ju, B. K., 75

K

Kabanov, V. V., 321
Kadish, K. M., 465
Kajihara, Y., 433
Kako, M., 465
Kambe, T., 393
Kanda, A., 265
Kanehama, R., 409
Kashino, S., 345, 357
Kasuya, A., 3, 47, 157, 161
Katada, M., 313, 437
Kataura, H., 15, 27, 51, 81, 87, 121, 165, 177, 189, 197, 233
Kato, T., 465, 469
Kawaguchi, T., 495
Kawamoto, T., 381
Kawamura, H., 325, 389
Kawana, M., 133
Kawasaki, S., 23, 249
Kawatsu, N., 529
Kawazoe, Y., 63, 237, 253, 277
Kazama, S., 233, 443
Kazaoui, S., 15
Kijima, M., 521
Kikuchi, K., 165, 313, 409, 437, 455, 473
Kim, D. H., 75
Kim, D-H., 169
Kim, E. K., 75
Kim, K. S., 35, 209, 221
Kim, S. T., 75

Kim, W. S., 217, 221, 241
Kimura, K., 133, 507
Kira, H., 525
Kitano, H., 353
Kitazawa, K., 397
Kitazume, K., 377
Kiyokura, T., 193
Kizuka, T., 281, 285
Kleinhammes, A., 95
Knupfer, M., 87
Kobayashi, H., 321
Kobayashi, K., 465
Kobayashi, M., 325, 357, 389
Kobayashi, S., 309
Kodama, T., 165, 409, 437, 455, 473
Kohdate, K., 473
Kojima, N., 349
Koma, A., 485
Komiyama, S., 23
Kondo, M., 465
Korecq, L., 329
Kosaka, M., 309, 377
Kosha, F., 157
Kramberger, C., 81
Krätschmer, W., 51, 165, 291, 513
Krause, M., 417
Kriebel, J., 181
Kubozono, Y., 317, 345, 357
Kudryashov, I., 213
Kumada, H., 389
Kumagai, T., 157
Kumar, V., 277
Kurashima, K., 67
Kurata, C., 313
Kushida, M., 477
Kusunoki, M., 421
Kuzmany, H., 81, 417
Kuzumaki, T., 281, 285
Kyotani, T., 249

Li, G. D., 125
Li, Z. M., 125
Lieber, C. M., 129, 185
Lim, S. C., 39, 209, 221, 241
Liu, H. J., 125
Liu, X., 87
Louie, S. G., 101
Lyu, S. C., 55

M

Machino, M., 381
Maeda, A., 353
Maeda, T., 161
Maeda, Y., 465
Maniwa, Y., 189, 197, 233, 369, 377, 443, 503, 525
Maruyama, Y., 63, 157, 365
Masubuchi, H., 233
Masubuchi, S., 233, 443
Masumoto, K., 63
Matsuda, N., 15
Matsumoto, K., 405
Matsumura, D., 473
Matsuo, R., 353
Matsuoka, Y., 189
Matsushita, A., 205
Matsuzaka, H., 409
Mayou, D., 145
McClure, T., 129, 185
Mihailovic, D., 321, 381
Mikami, M., 257
Minami, N., 15
Mitani, T., 177, 353, 369, 373, 385, 447, 485
Miwa, K., 353
Miyake, M., 373
Miyamoto, Y., 173
Miyano, K., 189
Miyano, R., 31
Miyasaka, H., 409
Mizoguchi, K., 309, 381
Mizuno, S., 121
Moon, J. M., 209, 241
Moos, G., 181
Motohashi, S., 365
Mukhopadhyay, I., 249
Muneyoshi, T., 43
Murakami, K., 405

L

Lee, C. J., 55, 59
Lee, C. W., 75
Lee, K.-H., 59
Lee, S. M., 117
Lee, T. J., 55
Lee, Y. H., 35, 39, 75, 117, 209, 217, 221, 241

Murakami, Y., 525
Muro, T., 373
Muroga, N., 369

N

Nagai, A., 495
Nagai, K., 485
Nagame, Y., 437
Nagao, N., 393
Nagasawa, N., 213
Nagase, S., 465
Nagayama, M., 157
Nakadaira, Y., 465
Nakahara, H., 313, 437
Nakamura, A., 121, 451
Nakanishi, T., 149
Nakao, K., 333, 337, 341, 477, 481
Nakayama, Y., 19
Narymbetov, B., 321
Nath, K. G., 193
Nihey, F., 137
Nikolaev, P., 11
Nishibori, E., 447, 461
Nishikawa, H., 455, 473
Nishina, Y., 47, 157, 161
Nogami, Y., 393
Nomura, K., 197

O

Ochiai, Y., 273
Ogata, H., 345, 365
Ogawa, N., 189
Ogawa, T., 47
Ogino, A., 133
Ogino, T., 193
Ogura, R., 429
Oguri, A., 507
Ohmori, S., 23
Ohno, K., 63
Ohshita, Y., 349
Ohta, T., 473
Ohtsuki, T., 63, 437
Okada, S., 173, 201, 333, 477, 481
Okada, Y., 521
Okai, M., 43
Okino, F., 23, 249

Okubo, S., 469
Omerzu, A., 321, 381
Ong, A. L., 513
Oogama, K., 313
Ootuka, Y., 265
Osawa, E., 421, 425, 517
Osawa, M., 197
Oshima, K., 393
Oshiyama, A., 173
Osipov, V. Y., 425
Ozawa, M., 421, 425, 517
Ozawa, N., 473

P

Paalanen, M., 141
Pan, L., 19
Park, C. J., 245
Park, Y. S., 35, 209, 217, 221, 241
Peisert, H., 87
Pekker, S., 329
Peterlik, H., 81
Pichler, T., 81, 87, 417
Plank, W., 81, 417
Pouchard, M., 499

Q

Qin, L. C., 205

R

Reny, E., 499, 503
Roche, S., 145
Rockenbauer, A., 329
Roschier, L., 141
Rubio, A., 145

S

Saiki, K., 485
Saito, R., 129, 185
Saito, S., 173, 201, 305
Saito, Y., 113
Sakaguchi, K., 473
Sakai, M., 451

Sakakibara, T., 31
Sakamoto, H., 309, 381, 503
Sakata, M., 447, 461
Sasaki, S., 43
Sasaki, T., 205
Sato, N., 3
Sato, T., 67
Sato, Y., 3, 47, 161
Schönherr, E., 405
Seifert, G., 117
Sen, R., 27, 51
Shames, A. I., 425
Shiga, K., 63
Shikano, K., 63
Shimoda, H., 95, 369
Shimotani, H., 397
Shin, J. K., 75
Shin, Y. M., 35, 39, 221
Shinoda, K., 47
Shinohara, H., 413, 417, 461
Shirakawa, H., 521
Shirakura, S., 465
Shiro, M., 409
Siklitsky, V. I., 425
Sim, H. S., 245
Sim, H-S., 169
Simon, F., 329
Sluiter, M., 277
Son, K. H., 55
Sueki, K., 313, 437
Suematsu, H., 189, 317, 345, 357
Suzuki, K., 133, 157, 365
Suzuki, M., 401
Suzuki, S., 27, 51, 165, 189, 193, 197, 233, 333, 337, 341, 477, 481
Suzuura, H., 269

T

Tagmatarchis, N., 413, 417
Takabayashi, Y., 345, 357
Takagi, S., 365
Takagi, T., 3
Takahashi, H., 157
Takakura, A., 113
Takano, Y., 19
Takata, M., 447, 461
Takenobu, T., 353, 369, 373, 385, 447
Takeuchi, J., 361

Takikawa, H., 31
Tamura, R., 153
Tanaka, K., 421
Tanaka, M., 133, 507
Tang, J., 205
Tang, X. P., 95
Tang, Z. K., 125, 213
Tanigaki, K., 309, 361, 377, 433, 495
Tanigaki, N., 455
Tao, Y., 31
Tarkiainen, R., 141
Terauchi, M., 133, 507
Tohji, K., 3, 47, 157, 161
Toida, T., 429
Tokumoto, M., 321, 381
Tomita, A., 249
Tou, H., 369, 377, 443, 503, 525
Touhara, H., 23, 249
Triozon, F., 145
Tsuda, S., 213
Tsukada, K., 437
Tsukada, M., 153
Tsukagoshi, K., 265
Tsunoda, K., 429

U

Uchino, Y., 485
Ueno, K., 3, 485
Umemoto, K., 305

V

Van Caemelbecke, E., 465
Vul', A. Y., 425

W

Waelchli, M., 465
Wakabayashi, T., 513
Wakahara, T., 465
Walters, D. M., 87
Wang, C. R., 461
Wang, J., 397
Wang, N., 125
Watanabe, K., 365
Watanabe, Y., 193

Wu, Y., 95

Y

Yaguchi, T., 43, 261
Yajima, H., 429
Yamaguchi, H., 51
Yamaguchi, M., 349
Yamaji, M., 201
Yamamoto, K., 465, 489
Yamanaka, S., 443, 499, 503
Yamashita, M., 409
Yao, A., 23
Yase, K., 455
Yasukawa, M., 443, 495
Yatsu, Y., 177

Yokoyama, T., 473
Yoo, J. E., 55, 59
Yoo, S. C., 59
Yoshida, Y., 455
Yoshimoto, N., 455
Yudasaka, M., 137, 205

Z

Zettl, A., 107
Zhang, M., 19
Zhao, X., 7
Zhao, Y. L., 313, 437
Zhou, O., 95, 193
Zhu, W., 193